Flood Handbook

Flood Handbook

Analysis and Modeling

Edited by
Saeid Eslamian and Faezeh Eslamian

CRC Press
Taylor & Francis Group
Boca Raton London New York

CRC Press is an imprint of the
Taylor & Francis Group, an **informa** business

First edition published 2022
by CRC Press
6000 Broken Sound Parkway NW, Suite 300, Boca Raton, FL 33487-2742

and by CRC Press
2 Park Square, Milton Park, Abingdon, Oxon, OX14 4RN

ISBN: 9781138614765 (hbk)
ISBN: 9781032201870 (pbk)
ISBN: 9780429463938 (ebk)

DOI: 10.1201/9780429463938

Typeset in Times
by Deanta Global Publishing Services, Chennai, India

Dedication

To David George Doran (1992) my late PhD Co-Supervisor from UNSW, Australia having more than 30 technical papers on Flood Data Measurement

Contents

PART VII   Flood Regionalization

PART VIII   Flood Soft Computing

Editors

Saeid Eslamian is a full professor of environmental hydrology and water resources engineering in the Department of Water Engineering at Isfahan University of Technology, Iran, where he has been since 1995. His research focuses mainly on statistical and environmental hydrology in a changing climate. In recent years, he has worked on modeling natural hazards, including floods, severe storms, wind, drought, pollution, water reuses, sustainable development and resiliency, etc. Formerly, he was a visiting professor at Princeton University, New Jersey, and the University of ETH Zurich, Switzerland. On the research side, he started a research partnership in 2014 with McGill University, Canada. He has contributed to more than 1K publications in journals, books, and technical reports. He is the founder and chief editor of the *International Journal of Hydrology Science and Technology* (IJHST). Eslamian is now associate editor of six important publications: *Journal of Hydrology* (Elsevier), *Ecohydrology and Hydrobiology* (Elsevier), *Water Reuse* (IWA), *Arabian Journal of Geosciences* (Springer), *International Journal of Climate Change Strategies and Management* (Emerald), and *Journal of the Saudi Society of Agricultural Sciences* (Elsevier). Professor Eslamian is the author of approximately 65 books and Special Issues and 300 book chapters.

Professor Eslamian's professional experience includes membership on editorial boards, and he is a reviewer of approximately 100 *Web of Science* (ISI) journals, including the ASCE *Journal of Hydrologic Engineering*, ASCE *Journal of Water Resources Planning and Management*, ASCE *Journal of Irrigation and Drainage Engineering*, *Advances in Water Resources*, *Groundwater*, *Hydrological Processes*, *Hydrological Sciences Journal*, *Global Planetary Changes*, *Water Resources Management*, *Water Science and Technology*, *Eco-Hydrology*, *Journal of American Water Resources Association*, *American Water Works Association Journal*, etc. UNESCO has also nominated him for a special issue of the *Ecohydrology and Hydrobiology Journal* in 2015.

Professor Eslamian was selected as an outstanding reviewer for the *Journal of Hydrologic Engineering* in 2009 and received the EWRI/ASCE Visiting International Fellowship in Rhode Island (2010). He was also awarded outstanding prizes from the Iranian Hydraulics Association in 2005 and the Iranian Petroleum and Oil Industry in 2011. Professor Eslamian has been chosen as a distinguished researcher of Isfahan University of Technology (IUT) and Isfahan Province in 2012 and 2014, respectively. In 2016, he was also a candidate for national distinguished researcher in Iran.

He has also been the referee of many international organizations and universities. Some examples include the US Civilian Research and Development Foundation (USCRDF), the Swiss Network for International Studies, the Majesty Research Trust Fund of Sultan Qaboos University of Oman, the Royal Jordanian Geography Center College, and the Research Department of Swinburne University of Technology of Australia. He is also a member of the following associations: American Society of Civil Engineers (ASCE), International Association of Hydrologic Science (IAHS), World Conservation Union (IUCN), GC Network for Drylands Research and Development (NDRD), International Association for Urban Climate (IAUC), International Society for Agricultural Meteorology (ISAM), Association of Water and Environment Modeling (AWEM), International Hydrological Association (STAHS), and UK Drought National Center (UKDNC).

Professor Eslamian finished Hakimsanaei High School in Isfahan in 1979. After the Islamic Revolution, he was admitted to IUT for a BS in water engineering and graduated in 1986. After graduation, he was offered a scholarship for a master's degree program at Tarbiat Modares University, Tehran. He finished his studies in hydrology and water resources engineering in 1989. In 1991, he was awarded a scholarship for a PhD in civil engineering at the University of New South Wales, Australia. His supervisor was Professor David H. Pilgrim, who encouraged him to work on "Regional Flood Frequency Analysis Using a New Region of Influence Approach." He earned a PhD in 1995 and returned to his home country and IUT. In 2001, he was promoted to associate professor and in 2014 to full professor. For the past 26 years, he has been nominated for different positions at IUT, including university president consultant, faculty deputy of education, and head of department. Eslamian is now director of the Center of Excellence in Risk Management and Natural Hazards (RiMaNaH).

Professor Eslamian has made three scientific visits to the United States, Switzerland, and Canada in 2006, 2008, and 2015, respectively. In the first, he was offered the position of visiting professor by Princeton University and worked jointly with Professor Eric F. Wood at the School of Engineering and Applied Sciences for one year. The outcome was a contribution in hydrological and agricultural drought interaction knowledge by developing multivariate L-moments between soil moisture and low flows for northeastern US streams.

Recently, Professor Eslamian has published the 11 handbooks with Taylor & Francis (CRC Press): the three-volume *Handbook of Engineering Hydrology* in 2014, *Urban Water Reuse Handbook* in 2016, *Underground Aqueducts Handbook* (2017), the three-volume *Handbook of Drought and Water Scarcity* (2017), *Constructed Wetlands: Hydraulic Design* (2019), *Handbook of Irrigation System Selection for Semi-Arid Regions* (2020), and *Urban and Industrial Water Conservation Methods* (2020).

An Evaluation of Groundwater Storage Potentials in a Semiarid Climate and *Advances in Hydrogeochemistry Research* by Nova Science Publishers are also his book publications in 2019 in 2020 respectively. The two-volume *Handbook of Water Harvesting and Conservation* (Wiley) and *Handbook of Disaster Risk Reduction and Resilience* (New Frameworks for Building Resilience to Disasters) are early 2021 book publications of Professor Eslamian.

He has been appointed as a World Top 2% Researcher by Stanford University, California, in for 2019 and 2020.

 **Faezeh Eslamian** holds a PhD in Bioresource Engineering from McGill University. Her research focuses on the development of a novel lime-based product to mitigate phosphorus loss from agricultural fields. Faezeh completed her bachelor's and master's degrees in civil and environmental engineering from Isfahan University of Technology, Iran, where she evaluated natural and low-cost absorbents for the removal of pollutants such as textile dyes and heavy metals. Furthermore, she has conducted research on worldwide water quality standards and wastewater reuse guidelines. Faezeh is an experienced multidisciplinary researcher with an interest in soil and water quality, environmental remediation, water reuse, and drought management.

Contributors

Mohd Shahrizal Ab Razak
Department of Civil Engineering, Faculty of
Engineering
Universiti Putra Malaysia
Selangor, Malaysia

Iftekhar Ahmed
Department of Civil and Environmental
Engineering
Prairie View A&M University
Prairie View, TX

Hasan Albo-Salih
Department of Civil, Environmental and
Sustainable Engineering
Arizona State University
Tempe, AZ

Hydar Lafta Ali
Centre for Restoration of the Iraqi Marshlands
and Wetlands
Ministry of Water Resources
Baghdad, Iraq

Haireti Alifu
Department of Civil Engineering
Shibaura Institute of Technology
Tokyo, Japan

Mina Arianpour
Department of GIS-RS, Yazd Branch
Islamic Azad University
Yazd, Iran

Mohammadtaghi Avand
Department of Watershed Management
Engineering
Tarbiat Modares University
Noor, Iran

Marta Barszczewska
Institute of Meteorology and Water
Management
National Research Institute
Warsaw, Poland

Brij Bhushan
Department. of Applied Research
Mewar University
Chittorgarh, India

Rim Chérif
Institut Supérieur des Sciences et
Technologies de l'Environnement
Université de Carthage
Borj Cédria, Tunisia

Mehran Dadashzadeh
Department of Water Engineering
University of Tabriz
Tabriz, Iran

Mariana Madruga de Brito
Department of Urban and Environmental
Sociology
UFZ-Helmholtz Centre for Environmental
Research
Leipzig, Germany

Saeid Eslamian
Department of Water Engineering
Isfahan University of Technology
Isfahan, Iran
and
Center of Excellence in Risk Management and
Natural Hazards
Isfahan University of Technology
Isfahan, Iran

Md Golam Rabbani Fahad
Department of Civil, Construction,
and Environmental
Engineering
University of Alabama at Birmingham
Birmingham, AL

Jihui Fan
Institute of Mountain Hazard and
Environments
China Academy of Science
Chengdu, China

Majid Galoie
Civil Engineering Department
Buein Zahra Technical University
Buin Zahra, Iran
and
Institute of Mountain Hazard and
    Environments (CAS)
Chengdu, China

Emna Gargouri-Ellouze
Laboratoire de Modélisation en Hydraulique et
    Environnement
Université de Tunis El Manar
Tunis, Tunisia

Mustafa Goodarzi
Agricultural Engineering Research Department
Markazi Agricultural and Natural Resources
    Research and Education Center, AREEO
Arak, Iran

Colin H. Green
Flood Hazard Research Centre
Middlesex University
London, UK

Aziz Hassan
Civil Engineering Department
Technische Universität Berlin
Berlin, Germany

Reinhard Hinkelmann
Chair of Water Resources Management and
    Modeling of Hydrosystems
Technische Universität Berlin
Berlin, Germany

Ali Akbar Jamali
Department of GIS-RS and Watershed
    Management
Islamic Azad University
Maybod, Iran

Saeid Janizadeh
Department of Watershed Management
    Engineering
Tarbiat Modares University
Noor, Iran

Daniel Jato-Espino
GREENIUS Research Group
Universidad Internacional de Valencia
Valencia, Spain

Fazlul Karim
Managing Water Ecosystems Group
CSIRO Land and Water Flagship
Canberra, Australia

Maryam Karimi
Department of Civil, Construction, and
    Environmental Engineering
University of Alabama at Birmingham
Birmingham, AL

Zaved K. Khan
Murray-Darling Basin Authority,
Australian Government
Canberra, Australia

Akinola Adesuji Komolafe
Department of Remote Sensing and GIS
Federal University of Technology
Akure, Nigeria

Zhong Liu
NASA Goddard Earth Sciences Data and
    Information Services Center
Goddard Space Flight Center
Greenbelt, MD
and
Center for Spatial Information Science and
    Systems
George Mason University
Fairfax, VA

Hämmerling Mateusz
Department of Hydraulic and Sanitary
    Engineering
Poznań University of Life Sciences
Poznań, Poland

Larry W. Mays
School of Sustainable Engineering and the
    Built Environment
Arizona State University
Tempe, AZ

Darshan Mehta
Civil Engineering Department
Government Engineering College
Surat, India

Venkatesh Merwade
Lyles School of Civil Engineering
Purdue University
West Lafayette, IN

D. Meyer
NASA Goddard Earth Sciences Data and
    Information Services Center
Goddard Space Flight Center
Greenbelt, MD

Thamer Ahamed Mohammed
Civil Engineering Department
University of Baghdad
Baghdad, Iraq

Artemis Motamedi
Civil Engineering Department
Buein Zahra Technical University
Buin Zahra, Iran
and
Institute of Mountain Hazard and
    Environments (CAS)
Chengdu, China

Rouzbeh Nazari
Department of Civil, Construction, and
    Environmental Engineering
University of Alabama at Birmingham
Tempe, AZ

Vahid Nourani
Center of Excellence in Hydroinformatics
University of Tabriz
Tabriz, Iran

Dana Ostrenga
NASA Goddard Earth Sciences Data and
    Information Services Center
Goddard Space Flight Center
Greenbelt, MD
and
Adnet Systems Inc.
Greenbelt, MD

Ilhan Özgen-Xian
Lawrence Berkeley National
    Laboratory
NASA
Berkeley, CA

Jayantilal N. Patel
Civil Engineering Department
Sardar Vallabhbhai National Institute of
    Technology
Surat, India

Pravin Patil
Department of Soil and Water
    Engineering
Maharana Pratap University of Agriculture and
    Technology
Udaipur, India

Saied Pirasteh
Department of Geography and
    Environmental
    Management
University of Waterloo
Waterloo, ON

Keyur Prajapati
Civil Engineering Department
Shree Swami Atmanand
    Saraswati Institute of
    Technology
Surat, India

Ayesha S. Rahman
School of Computing, Engineering, and
    Mathematics
University of Western Sydney
Penrith, Australia

Balqis Mohamed Rehan
Department of Civil Engineering
Universiti Putra Malaysia
Selangor, Malaysia

Siddharth Saksena
Department of Civil and Environmental
    Engineering
Virginia Tech
Blacksburg, VA

Andrey Savtchenko
NASA Goddard Earth Sciences Data and
    Information Services Center
Goddard Space Flight Center
Greenbelt, MD
and
Adnet Systems Inc.
Greenbelt, MD

Thomas J. Scanlon
mts-cfd.com
Stewarton, UK

Abhishek Sharma
Wisdom Tree Arts
New Delhi, India

Priyanka Sharma
Soil and Water Engineering Department
Maharana Pratap University of
    Agriculture and
    Technology
Udaipur, India

Yasuyuki Shimizu
Hydraulic Research Laboratory
Hokkaido University
Sapporo, Japan

Franz Simons
Bundesanstalt für Wasserbau
Karlsruhe, Germany

Zaborowski Stanisław
Poznań University of Life Sciences
Poznań, Poland

Gokmen Tayfur
Department of Civil Engineering
Izmir Institute of Technology
Izmir, Turkey

William Teng
NASA Goddard Earth Sciences Data and
    Information Services Center
Goddard Space Flight Center
Greenbelt, MD
and
Adnet Systems Inc.
Greenbelt, MD

Akshay R. Thorvat
Department of Civil and Environmental
    Engineering
Kolhapur Institute of Technology's College of
    Engineering (Autonomous)
Kolhapur, India

Kałuża Tomasz
Department of Hydraulic and Sanitary
    Engineering
Poznań University of Life Sciences
Poznań, Poland

Franziska Tügel
Chair of Water Resources Management and
    Modeling of Hydrosystems
Technische Universität Berlin
Berlin, Germany

Mehdi Vafakhah
Department of Watershed Management
    Engineering
Tarbiat Modares University
Noor, Iran

Bruce Vollmer
NASA Goddard Earth Sciences Data and
    Information Services Center
Goddard Space Flight Center
Greenbelt, MD

Jean-François Vuillaume
JAMSTEC Yokohama Earth Simulator
Yokohama, Japan

Sahita Waikhom
Civil Engineering Department
Government Engineering College
Surat, India

Conrad Wasko
Department of Infrastructure Engineering
University of Melbourne
Melbourne, Australia

Jennifer Wei
NASA Goddard Earth Sciences Data and
    Information Services Center
Goddard Space Flight Center
Greenbelt, MD

S. M. Yadav
Civil Engineering Department
Sardar Vallabhbhai National Institute of
 Technology
Surat, India

Badronnisa Yusuf
Department of Civil Engineering
Universiti Putra Malaysia
Selangor, Malaysia

Part I

An Introduction to Flooding and Humans

1 Floods and People

Colin H. Green and Saeid Eslamian

CONTENTS

1.1 INTRODUCTION

Floods impact people and their livelihoods and so also the economy. People interpret floods, seek understanding of the causal chains involved that result in flooding, and thus how to intervene effectively. People interact with each other through language and other symbolic systems with the purpose of deciding how to cope with, adapt to, or reduce the risk of a flood occurring. We co-act in various ways to intervene to improve coping or adaptation or to reduce the risk of flooding. To decide, we have invented complex social structures and different forms of government arrangements, and to act, we have invented a wide variety of organizations: mutuals, firms, and governmental agencies in some structural arrangement. Management of flood risk is thus ultimately about individuals and social relationships, the forms and success of flood risk management depending upon the governance arrangements, the organizations created, and the skills of the individuals involved (Nelson and Winter, 1982).

Indeed, without people, there would be no floods. There would still be the inherent hydrological variability but there would be no one there to label one extreme as a "flood," no one there to be adversely affected this extreme or the other extreme of "drought," no one there deciding and acting to modify either the environment or their behavior to reduce the impacts of these extremes, and no one there to try to influence the others. It may seem pedantic to say that floods are a human construct but we look at floods through a telescope whose focus is defined by cognitive, cultural, sociological, and ideological boundaries. Pragmatically, recognizing those boundaries and their biases may suggest better means of adapting to and coping with the floods. Otherwise we are trapped within those boundaries and it is difficult to think outside of those limits. For example, the label of "flood defense" implicitly implies that we are at war with nature. This tended to exclude approaches involving adapting to or coping with floods.

DOI: 10.1201/9780429463938-2

Whilst a hydrological definition of a "flood" is "an overflow of a large amount of water beyond its normal limits, especially over what is normally dry land" (EU Flood Directive EC, 2007), this is not terribly useful, not least because it does not imply any reason for concern. Nor does it define a baseline as either the average annual low water level or the average annual flood. From the perspective of people, a "flood" is simply water in the wrong place in the wrong amounts at the wrong time. This distinguishes between flood irrigation and desirable floods, such as the annual flood which nourished Egyptian ancient agriculture, and the situation in which the presence of water creates adverse consequences to people and our activities. In the inherent intra- and inter-year variability of rainfall and hence river flows, a flood defines a discontinuity in the consequences either between the desired and the undesired or between the inconvenient and the harmful. Societies that rely upon flood recession irrigation often distinguish between the desirable annual flood and the undesirable extreme flood.

Secondly, flooding expresses our relationships with each other, as does water management more generally. Some flooding is the result of the rain falling on the land in question and saturating the soil, a particular problem in agriculture since it is the saturated soil in the root zone that damages the plants (Shaxson and Barber, 2003). But much of flooding is with other people's water, the runoff from their land being conveyed by rivers or sewers to the area flooded. This is especially true of urban areas whose paved areas and roofs convert a large fraction of rainfall into runoff, often resulting in surface water flooding, surcharging sewers, or overflowing watercourses. There is, not surprisingly, often a conflict as to who is responsible for preventing or alleviating this flooding. Do those whose actions create the flooding have an absolute right to use their land purely in their own interest or have they a responsibility to those downhill or downstream? In early 20th-century England, taxing upland farmers to partly fund flood alleviation for downstream land users was strongly resisted by those upland farmers (Royal Commission on Land Drainage, 1927). Similarly, the Chicago school of geographers' approach to flood risk management placed the burden of flooding primarily upon those who are flooded (White, 1945).

Given the problems with defining floods, it is necessary also define what is meant here by "people." Individuals have their own capacities, personalities, and interests but live and work in informally or formally organized groups in a cultural setting which creates roles for individuals. For flood risk management, of particular concern at the individual level of response is the proportion of the population who have one or more disabilities. In the UK, the proportion is nearly 20% of the population, including about 10% each of the population with a mobility problem or with lifting or carrying (Office for Disability Issues, 2014). The number of people with a chronic health problem such as heart disease is about 25% of the population but multi-morbidity is common (Long Term Health Conditions Team, 2016). Chronic health problems are the most common amongst those over 60 (58%) and amongst those with low incomes where the severity is also greater. Associations of vulnerability with both the elderly and low incomes may then simply be picking up the effects of poor prior health. Personality, and the associated attitudes and beliefs, have also been found to be significant in coping with stressful events (Chen et al., 2017).

Individuals live and work in groups, starting with households and in the progressively more complex forms of communities and societies, and in more formally constructed forms of mutuals (Wagret, 1967), firms (Hart, 2011), civil society (Savage and Pratt, 2013), and government hierarchies (Finer, 1970). We interact with each other through symbolic systems, most obviously language, and seek to influence each other. Most notably, we commonly both make decisions and to co-act in groups, starting with households (Roberts, 1991) but then through larger groups including firms and governments. Those larger groups are structured, they are organized and involve internal interaction. We act in groups, whether this be in the form of all but the smallest firm, the mutuals which have played such a central role in the history of water management (Wagret, 1967), or the formal structures of government.

If individuals seek to influence the decisions of the groups of which they are part, those groups also influence the individual; individuals are given roles (Goffman, 1959) and functional responsibilities by those groups, and may act as a representative of that group. Hence, an individual may simultaneously have the role of a parent and a spouse, of being an engineer, and acting as a representative of a government agency on an international catchment management body. Gender is the most obvious example of a socially constructed role (Marecek et al., 2004).

At the broadest level, culture is the set of widely shared beliefs and thus interpretations of the world, which translate into behaviors that are common to members of that group (Spencer-Oatey, 2012). Part of that set of beliefs are shared values, those moral principles which are argued on religious or other grounds ought to govern behavior (Marecek et al., 2004). Thus, Islam emphasizes duties to the other people and other species as all are aspects of the deity's creation (Deen, 1990). The Preamble to the Constitution of France 1946 stated that there will be social solidarity in the face of natural disasters; this necessarily influences the approach to flood risk management in France, as can the emphasis on social harmony in the other countries (Wang, 2008). Conversely, the way in which flooding is addressed simultaneously illustrates the dominant pattern of social relationships within that society and our relationship with nature.

As is logical, this cultural pattern-making is reflected in social norms about interpersonal relations and behavior. These social norms exist as informal rules as well as those formalized into law. Two key points here are, firstly, that social norms are sometimes powerful influences on even trivial behavior (Melnyk et al., 2010). Secondly, interventions may either act with existing social norms, exploiting these norms to change behavior, or contradict them. Resistance to actions that contradict those norms can be strong, as shown in the numerous experiments on social dilemmas theory (Liebrand, 1992) in which individuals take action at a cost to themselves to penalize those who violate social norms. But choices of technology either reflect existing social norms or create new social relationships at variance with those social norms. Formal rules take the form of laws, the forms and nature of which then vary widely across cultures, notably in regard to water management (Caponera, 1992). In particular, what are often termed "property rights" are now recognized as rules about the interpersonal relationships articulated through access to things (Worthington, 2003).

1.2 COGNITIONS AND CULTURAL CONSTRUCTIONS

Cognition is the framework through which we seek to understand and interpret the world; to find or impose order upon it, to find patterns. Our cognitive capabilities are limited; Miller, in his classic paper (Miller, 1956), observed that the experimental evidence was that the average person in the average situation could deal with only seven variables at once. Kahneman and Tversky (Kahneman, 2011) found a number of other cognitive factors which frame the way in which the world is interpreted. Simultaneously, behaviors are often determined by habits (Duhigg, 2012), which themselves are a form of cognition.

Cognition is not therefore neutral but an interpretation where that interpretation is itself also restricted by our cultural setting (DiMaggio, 1997). The so-called Whorf-Sapir hypothesis that differences in language determine differences in what can be thought, cognition, is now seen as an over-simplification (Kay and Kempton, 1984). But equally, different languages do imply or reflect a different interpretative structure on the world (Lakoff, 1987). At the same time, it is argued that different cultures do have different ways of thinking and consequently of acting (Hall, 1959). Equally, words in one language may mean quite different things in different cultures. Nannette Ripmeester (www.expatica.com/uk/insider-views/Doing-business-The-Dutch-with-the-British_102530.html) provided an entertaining translation of what someone whose first language is English would mean by a word or phrase versus what a Dutch listener might understand them to mean.

Moreover, the different disciplines, using Geertz (1993) definition of culture as "a system of inherited *conceptions* expressed in symbolic forms by means of which men communicate, perpetuate, and develop their knowledge about and attitudes toward life," (epmphasis added) are themselves different cultures. For example, the expectation of engineers is that they design and build physical objects. The outcome of asking an engineer, lawyer, ecologist, land use planner, or sociologist to determine the best means of reducing the risk of flooding in a given area should be expected to be different. That the different disciplines use, if not different languages, at least the different dialects makes interdisciplinary working more difficult. For example, in working with an ecologist on a project to assess the benefits of coastal protection (Penning-Rowsell et al., 1992), as an economist, my definition of the problem was that of valuing specific sites for their ecological value. It took me some time to understand her framing of the problem was that the value of any individual site was only determined by its contribution to the ecosystem as a system, the whole determining the contributing value of the part.

We are left with a potentially complex set of connections between language, culture, and the inherent structure of cognitive functions. In turn, how we interpret the world determines how we seek to intervene to reduce the risk of flooding; we seek to determine the causes of flooding and hence to decide the best means of reducing the risk of flooding.

Seeking to determine causalities and other relationships might be argued to be the definition of rationality. Flood victims almost inevitably, therefore, seek to determine the cause of the flood that they experienced. They may seek physical explanations; flooding in Bangladesh has in the past been blamed erroneously upon deforestation in Nepal. Or blame may be cast upon institutional failings; that those organizations with responsibility for dealing with one or another aspect of flood risk management failed to perform their duties properly for one or another reason. In early civilizations including Babylon, floods were interpreted as divine punishments and such labels as "flood defense" interpret nature as inherently malevolent.

The difference between local knowledge and local interpretation should be emphasized. Local populations can have long-run knowledge of the hydrological and morphological behavior of their watercourses. However, their interpretation of the causes of specific events needs to be judged in terms of the evidence and alternative explanations, the use of what Toulmin (1958) defined as "argument." A sufficient interpretation is then one which addresses the system interactions which are expressed by the event, here a flood; a consistent interpretation is one which matches the available evidence and is not contradicted by the other evidence or interpretations of that evidence. Given that floods are the product of runoff from a catchment as the result of a variable, rainfall, a sufficient explanation needs to be one that addresses the dynamics of the catchment and not simply one extreme of streamflow or rainfall.

How then can we then think about floods in a way which is consistent and sufficient? Somewhat speculatively, it may be argued that some principles are:

- We should always think in terms of how to intervene in systems, starting with catchments but also considering interactions with the other systems, those of human behavior and ecosystems. Systems are characterized by dynamic behavior so considering flood risk management outside of the overall variability of rainfall and streamflows is misleading. Systems and their interactions often result in non-linear relationships, including discontinuities (Sawyer, 2005). Assuming linear and additive responses by the other variables in response to a change in another variable is a dangerous assumption. Specifically, it is sensible to look for inflection points and discontinuities in responses.
- Since catchments are systems that have spatial as well as time-varying characteristics, a key principle is not to simply move floods about. Shifting a flood onto someone else was in the past sometimes done deliberately (Harrison and Mooney, 1993) but 19th-century flood risk management interventions often had this effect, incidentally, as was the case with the Rhine.
- An advantage in thinking about floods from a complex systems perspective is that there can be the possibility of finding synergistic forms of intervention, ones that address several different problems with one intervention. Reducing agricultural runoff may be a byproduct of treating the more critical problem of reducing soil erosion (Montgomery, 2007) whilst the adoption of green roofs or some other forms of sustainable drainage systems (SuDS) can reduce the heat island effect in cities, reduce air pollution, and lower energy consumption (Banting et al., 2006). Doing more with less, sustainable development, requires that we make every euro do the work of two or three through such synergistic means.

- Conversely, the effects of one intervention on different systems may also be antagonistic; permanent storage of water changes the downstream flow regime most obviously in the case of reservoirs. Some forms of wetlands (Mitsch and Gosselink, 2000) will also change the downstream flow regime.
- A key series of questions in considering any form of intervention is: how will it fail, under what conditions, and with what consequences (Green, 2003). The potential for failure includes the consequences of errors in the predictions of the key variables governing its performance including hydrological, engineering, and human failure. Any system whose success depends upon humans always performing every task without error will fail sooner or later, the only question is how often (Smith, 2005). Hence, it has been argued that flood risk management interventions should be designed from a consideration of failure (Green et al., 1993); in particular, that failure should be such as to take place with a reasonable warning lead time, slowly and in a relatively benign way. The principle should be that failure should not result in a more severe flood than if the intervention had not been made in the first place. Hence, rather than adopt any "design standard of protection," the logic is to focus upon what will happen when the intervention is overwhelmed or otherwise fails: to look at "above design standard" benefits. Different forms of intervention have widely different consequences when they are overwhelmed (Green et al., 2000). Thus, two traditional forms of flood risk management in Japan were discontinuous dykes which allowed progressive controlled flooding of the areas behind them and dykes with an intended overtopping point (Iwasada et al., 1999). A wider traditional approach to designing for failure was the strategy adopted in Zeeland: a dyke ring with the local population mobilized to maintain the dyke, settlements built on artificial mounds, "terps," and boats kept on the terps for transport during floods (van de Ven, 2004). A modern version of such a "layered strategy" might consist of flood warnings, flood storage, flood proofing, and a support mechanism prepared for the recovery period.

1.3 DISCOURSE

Communication has multiple purposes including passing on information but also for building and maintaining social relationships. Communication is also used to influence others, to change the way that they interpret the world or ourselves and so change their behavior; as a form of power, the use of discourse (Hajer, 1995). This includes debate and negotiation. This functionality was recognized by the Greeks and Romans as the discipline of rhetoric (Aristotle, 1991); it is now more generally to be recognized in the marketing both of products and political positions (Nelson, 2004). More generally, all communication has intentionality, and "information" is a somewhat misleading term as it implies both a commonality of interpretation and prior understanding. For example, the purpose of communicating about floods is to change what those at risk do in a way that will help them adapt and to cope more effectively with the risk and actuality of floods (MacGillivray and Livesey, 2015).

Two questions are thus: firstly, what is the most effective way of communicating (Kasperson, 2014)? Secondly, if language is a tool for influencing other people, what are the ethically permissible limits to using language in this way (Habermas, 1995)? Underpinning these two questions is the fundamental question of how a language can work to communicate (Wittgenstein, 1958) if the two parties start with different understandings, including of the meanings of individual words themselves.

1.4 SOCIAL RELATIONSHIPS

Social relationships are articulated through the causes of flooding, in terms of added runoff and exported floods; through responses to flooding in terms of the different forms of support to those who experience flooding; and actions to reduce the risk of flooding, including cost-sharing. These relationships are very clearly articulated in the ways in which decisions are made and the way in which interventions are undertaken.

1.5   DECISION-MAKING

It has been argued (Green, 2003) that there are two necessary preconditions for the existence of a choice and hence the existence of a decision. These preconditions are conflict and uncertainty so that if the conflict can be resolved and confidence be obtained that one course of action should be preferred to all of the others, then the choice has been made. When more than one individual is involved in a choice, then a choice potentially involves resolving one or more forms of interpersonal conflict, those between the values and interests. Different forms of conflict resolution have been proposed (Fisher et al., 1991) but a key requirement is that of procedural justice (Lind and Tyler, 1988). Even small children expect that household decision-making will be "fair" (Butler et al., 2005).

The concept of fairness underpins the commitment to stakeholder engagement (Green and Penning-Rowsell, 2010); that those who will be affected by the course of action potentially adopted should have a say in the decision that is made. Difficult questions in stakeholder engagement are then who is a stakeholder, should each have an equal say in the process, and how to include the interests of the general taxpayer who frequently have to pay a significant share of the costs of the option adopted?

Uncertainty, doubt as to which option should be adopted, is the other precondition for the existence of a choice. It has to be reduced, where information is conventionally defined as that which reduces uncertainty (Cherry, 1966). What is information thus depends upon the recipient and the potential array of actions open to them.

This is to specify the abstract nature of choice, the human side of the problem is who should make the choice, how the choice is to be made, and what choice is made given that the gains and losses will be distributed unequally. This is the problem of governance (OECD, 2015) where the form of governance has to match three facets: the nature of the problem, the form of the technological intervention eventually adopted, and social relationships, as expressed in social norms. A complication is the nature of the problem and the technological options available change over time.

Governance is simultaneously the combination of rules and power (Green and Eslamian, 2014). If the rules create the structure, the process is one of power in its different forms. What powers may be used, for what purposes by whom is governed by the rules. This process is now increasingly centered upon stakeholder engagement, but one more widely requiring legitimacy/and accountability (Lindberg, 2009). Self-evidently, humans design the structure, and the process of governance is performed by humans, the process being heavily influenced by competing discourses. If governance is the structure, and government one set of organizations in that structure, the process is that of politics (Crick, 1964). All have to be considered, all have to be appropriate.

1.6   ACTION

We can either modify the environment or change ourselves; change the way in which we use catchments and flood plains. Modifying the environment largely involves organizations that have the physical resources and financing to undertake the planned works. Changing the way of behaviors requires organizations first to discover which powers are most effective in changing the behaviors of the target group and then have those powers. Giving and using powers to change other people's behavior obviously raises questions of justice – whether it is right and proper to have and use those specific powers. For example, we can askwhether it is right to institute mandatory evacuation at gunpoint when a flood is forecast.

1.7   RELIABILITY

An important issue in the consideration of interventions and governance is how likely they are to perform successfully. This requires understanding how likely and under what circumstances failure is likely to occur. Human failure either during the design, implementation, or operation of the intervention is the proximate or ultimate cause of system failure. This failure may either be individual or one of organization design and performance where organizational design can have a strong influence on the individual error rates, both random and particularly systemic.

1.8 HUMAN ERROR RATES

Multiple factors affect the rates at which we make random errors but even in simple tasks, error rates are of the order of one per 10,000 tasks. Entering a ten-digit number in a calculator has an error rate of 1:20 whilst an untrained person is likely to fail to act correctly in an emergency within one minute up to nine out of ten times (Smith, 2005).

It is necessary to look particularly carefully at human error rates when considering non-structural options. Whereas concrete is very robust, unless there was a human failure in design or construction, the reliability of a land-use planning system depends upon the performance of the humans who administer it. In addition, such a system is dependent both upon adequate resourcing, where operational costs are always the first thing to be cut in a budget crisis, and low levels of corruption. A contributor to the potential failure of contingent floodproofing is the risk of the lack of anyone there available to install it when needed.

1.9 ABUSES OF POWER

Boulding (1989) defined power as "the ability to change the future," but it also includes the ability to resist change. It has been argued (Green and Eslamian, 2014) that power and rules sit in a yin-yang relationship, the role of rules being to limit and shape the uses of power positively so that the allocation of power is appropriate to the three facets of governance. For the social relationship facet, the appropriate use of power can be defined as being in accordance with social norms or higher principles of morality, notably where these have been formalized into law. Abuse then is the use of power outside of these norms or principles, an illicit gain being made by the holder of power in exchange for misusing that power: it is a relationship. "Corruption" then occurs when the holder of a public office exploits firms or individuals or others for personal gain (Transparency International, nd). Firms exploiting others for excess profits through monopoly or cartel power is labeled "rent seeking" (Tullock, 1967). Individuals within an organization may then misuse their power over others in a range of ways including sexual abuse, bullying, and fraud.

The consequences of the abuse of power are, firstly, to prevent the development of a country at the rate that would otherwise be possible. Secondly, and probably more importantly, by violating moral norms, it reduces public trust and undermines the legitimacy of a nation's government, where both "trust" (Mollering, 2001) and "legitimacy" (Johnson et al., 2006) are difficult to define in relationship terms but also key ones.

Flood risk management is an area that provides scope for abusive practices including providing flood alleviation measures in one area rather than another, in the granting of design or construction contracts, in approvals of proposed works, or as to the satisfactory completion of specific works. Construction is an area particularly prone to such abuse, WEF (2016) reporting estimates of 10–30% of all of the construction activity as involving corrupt practices.

Probably the key question about power abuse in the form of corruption is: why is there not more of it? Many countries had long periods of such abuse after which the incidence apparently fell (Knight, 2016). Understanding how such changes occurred can provide insights into how abuse can be reduced.

1.10 ORGANIZATIONAL FAILURE

Whilst governance is a key to effective flood risk management (Green and Eslamian, 2014), the framework of power and rules so established has to be populated by organizations appropriately structured, and those organizations staffed with people having the appropriate skills (Eggers and Kaplan, 2013). Hence, an apparently appropriate governance arrangement may fail either because the organizations are poorly structured or the members of those organizations lack the appropriate skills (Nelson and Winter, 1982). Conversely, it would appear that apparently poor governance arrangements can still produce quite successful flood risk management when the organizations are satisfactory and their members have good skills (Green, 2017). Given the multiple organizations generally involved, the ability to communicate and negotiate with the other organizations, as well as with members of the same organization, is a necessary capability (Fisher et al., 1991).

The nature of the problem means that flood risk management organizations must be prepared for infrequent events and here failure will generally have the most serious consequences. Hence, the problem is to design organizations that will perform well in routine conditions, and not fail in extreme events. Here, there are a number of studies of organizational failure (Perrow, 1984; Reason, 1997; Turner, 1978) in addition to an enormous literature on organizational design itself.

1.11   VULNERABILITY

Floods reduce people's well-being (Stiglitz et al., 2009) directly by disrupting their lives or livelihoods; the flow of well-being falls. Floods also reduce their well-being by damaging or destroying their homes and contents, this stock loss reducing the flow loss of well-being which normally is consequent upon their home. Such stock losses, commonly termed "direct damages," are significant because resources have to be diverted from other purposes to repair or replace those damages. Current consumption, borrowing, or, where they exist, savings have to be diverted in this way. Future consumption and well-being have to be sacrificed in order to recover current well-being to its level prior to the flood. Thus, "direct damages" are not a loss in themselves but are a proxy measure of the loss of well-being. Thirdly, direct damages are significant because the dwelling has a mediating effect between the flood and people, either reducing or amplifying the effects of the flood on well-being. For example, the traditional adaptations to a risk of flooding in the form of building on stilts, or floating homes, reduce the effect of a flood; timber balloon frame homes or those built of adobe amplify them. Floods in consequence both impact individuals and social relationships; increases in domestic violence are reported along with substance abuse (Morris-Oswald and Simonovic, 1997). Evacuation disrupts and potentially destroys friendship and kinship ties (Milne, 1977).

Whilst there is a danger of confounding a stock loss, damage to a dwelling, with the flow loss, the trajectory of well-being over time, it has been found that impacts to well-being such as health, disruption, and stress are commonly reported by flood victims as being more severe than the damage to their home and contents (Green and Penning-Rowsell, 1989).

The utility of the term "vulnerability" then lies in the degree to which it enables us to differentiate between those who are most affected by which characteristics of a stressor in which circumstances. It is useful to the extent to which it enables us to target the appropriate forms of intervention to the most vulnerable groups. Unfortunately, the term has become so vaguely defined as to be almost useless in analysis (Green and Penning-Rowsell, 2007) and is often taken either to include resilience or resilience is taken to include vulnerability. Rather than confound the two, a potential differentiation is that vulnerability is the extent of the initial change in well-being by the target population as a result of some given stress. Resilience is the degree to which the target population recovers and the time taken for it to recover to its prior trajectory of well-being. That resilience may be promoted by external support. In each case, it may be as useful to determine which people are least vulnerable or most resilient as to identify who is most vulnerable or least resilient.

Figure 1.1 is the commonly used illustration of how a shock or stressor may result in a change in some dependent variable. The initial shock depresses the level of that dependent variable; the level of that dependent variable then over time hopefully progressively returns to the prior trajectory of the dependent variable. Vulnerability then contributes to the difficulties of recovery but resilience

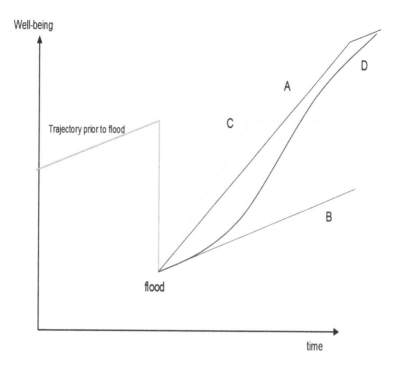

FIGURE 1.1   Time well-being relation.

is determined by other factors as well. In turn, it may be argued that the appropriate definition of vulnerability is one that is defined back from the appropriate means of enabling recovery.

Since the definition of vulnerability is useful to the extent that it indicates the forms of intervention that are most likely to be effective in reducing the high levels of vulnerability, it is also a form of discourse (Green and Penning-Rowsell, 2007). To be useful, the starting point in considering vulnerability is whether vulnerability is a characteristic or state of the population in question, or inherently defines the relationship between that population and a particular stressor. States, such as poverty or acute poor health, could be changed and so reduce vulnerability: for example, the poor could either be given financial support in the event of a flood or their incomes could be raised generally. Characteristics of a particular population cannot be changed in the short term: for instance, the age of neither the very young nor very old could be changed if either age group were found to be particularly vulnerable. Interventions would be limited to removing such vulnerable people from the stressor, or by targeting assistance at such vulnerable groups so as to promote resilience rather than to reducing vulnerability. Thirdly, vulnerability may be either defined as existing in relation to all stressors or vulnerability may be specific only to some stressors but not to others. An example of such a relationship vulnerability is hemophilia.

An index of vulnerability would be useful once we are clear about what we intend by a definition of vulnerability. A number of attempts have been made to derive such an index but deriving such an index presents a number of further problems. Firstly, the appropriate contributing variables have to be identified. Secondly, the appropriate functional relationship between each variable and vulnerability has to be determined. Whilst a linear, additive function tends to be the default model adopted, the appropriate form has instead to be determined. For example, the standard model of the effects of stress (Lazarus, 1966) is cumulative with a discontinuity. The variables may interact in different ways instead of simply being substitutes as the additive model requires. As well as multiplicative, for example, interactions, there may be other mediating or moderating interactions (Kim et al., 2001). To take a simpler instance, the risk to life from a flood when not in a building is a complex function of the velocity and depth of flow (Abt et al., 1989).

For example, a very simple model of household vulnerability is:

$$\text{Vulnerability} = \text{stressor} * \text{built form} * \text{household characteristics}$$

The number of potential variables and their possible interactions means that deriving the appropriate functional equation will depend upon having a very large sample survey from which to derive a statistical relationship. Actually having a vulnerability index that is useful requires that it can be applied using readily available data, such as census data. In turn, that means that the largely socio-economic and demographic factors must comprise either all of the variables in the vulnerability equation or be reasonable proxies for the determining variables. Consequently, it has to exclude other possible determinants such as personality (Chen et al., 2017), prior health status (Yzermans et al., 2005), prior life events, and concurrent stressors (Horowitz et al., 1979). If some determinants are excluded, then the results of using the vulnerability index will be biased in unknown ways.

A key question that has not so far been addressed here is: what is the dependent variable? What is it that we deem to be vulnerable to some stressor? If there is more than one potential dependent variable then vulnerability will be relative (unless exactly the same variables in exactly the same combination determine the vulnerability of every dependent variable). Further, a follow-up question needs to be whether that dependent variable is a stock or flow and hence whether vulnerability as defined results in a change in the state of the stock or the level of the flow. A third question is: what is the target population for which we seek to define vulnerability? Is it a person, household, community, society, firm, economy, or something else? An individual firm, for example, may be vulnerable without the local, regional, or national economy being vulnerable to the shock in question.

1.12   INTERVENTIONS

We can either seek to modify the environment or change our behavior, including the ways in which we use land. Changing the environment or behavior can be attempted either upstream of the location at risk or at the point at risk. The timing of interventions can be before a flood occurs, during the flood, or after the flood. In each case, there are a variety of possible forms of intervention and so a very large number of individual interventions that can be attempted, and an even larger number of potential strategies, combinations of interventions. There is an increasing emphasis on adopting a layered approach, of combining intervention strategies so that the failure of one intervention is counterbalanced by another form of intervention (Green et al., 2000). The traditional approach in Zeeland described above is a classic example of a layered approach. The ways in which the environment can be modified are described elsewhere in this handbook but these involve different forms of power than seeking to change behavior. But non-structural interventions are not the only ones that involve changing the behavior of people; environmental modification does so too but not necessarily that of the people at risk of flooding. Introducing means to reduce runoff or store and release flood peaks both affect the way in which upstream land users can make use of their land as each requires both space and place. All forms of intervention therefore involve a decision to intervene and a means of changing behavior, most drastically by moving people off their land or out of their home. Simply charging upstream inhabitants for the quantity of runoff they generate, or towards the cost of providing flood alleviation measures downstream, equally result in changes in their behavior since they now have less money for the other things.

Seeking to change people's behavior requires understanding how power works (Green and Eslamian, 2014) and which form of power will be most effective in the circumstances to induce the desired change in behavior. To change other people's behavior requires first recognizing why they adopt the behaviors they do, the nature of the barriers to their making a change, and providing the appropriate and sufficient incentives to enable them to make the change. Whilst the nature of power and its reciprocal, rules, have been discussed elsewhere, the key question is: which of the many forms of power (Green, 2003) will be most effective in overcoming the barriers to change?

For example, Toronto determined that the least cost method of reducing the risk of flooding of basements and to avoid the cost of new capital works was to disconnect roofs from sewers. It initially adopted a policy of voluntary disconnection, roofs instead being discharged to rain gardens or other sinks, with subsidies being provided to cover the very low costs to individual householders of disconnection. The takeup rate proved to be too slow and the voluntary program was replaced by a mandatory program which achieved very rapid results (Toronto Water, 2007).

It also raises the question of: is this use and form of power just? Hence, the issue of governance is central: who makes the choice as to whether and how to intervene, the form of intervention adopted, and how that intervention is to be implemented.

1.13   BEFORE A FLOOD

To a degree, floods are created by people, firstly by choosing to settle by rivers and secondly by activities that increase the probability and magnitude of extreme flows. Traditionally, particularly in temperate climates, many flood plains offered competitive advantages over other areas as areas for arable farming and hence for preferential settlement. They often offered a combination of flat land, alluvial soils, water supply, and the cheapest form of transport by boat. There was and remains a global scarcity of fertile soils, there being a world average of 0.19 ha/p (https://data.worldbank.org/indicator/AG.LND.ARBL.HA.PCx).

Floods themselves are a product of climate*landform*land use: only the latter is generally open to intentional human modification, by changing the proportion of precipitation that becomes runoff. But the effect of land use is largely restricted to small catchments and relatively high-frequency events (Calder, 2000). However, the expansion of impermeable areas in urban areas does create both flood problems within those areas and exports floods downstream locally. More significant than changes in land use is the human-induced climate change.

1.14   DURING A FLOOD

The scope for actions to reduce the effects of a predicted flood depends upon the warning lead time, the precision and accuracy of the warning. The appropriateness and success of the action undertaken depends in turn upon prior knowledge of the possible appropriate actions and the content of the warning. Hence, flood warnings are the last step in a chain of information communication.

There is a fundamental difference between a flood forecast and a flood warning; a flood forecast is a prediction of flooding, a flood warning provides the necessary information for those at risk to take appropriate action. The intention of flood warnings is to make it more likely that those at risk will avoid dangerous actions (e.g., traveling through floodwaters on foot or in a vehicle) and more likely that they will adopt the appropriate behavior. Adopting the appropriate behavior requires knowing what is the most appropriate behavior given the circumstances and having the abilities and resources to undertake that behavior.

The first key component of a flood warning is providing sufficient lead time to execute the appropriate action. Unfortunately, there is an inherent trade-off between warning lead time and the reliability of the warning (Krzyaztofowicz et al., 1994). Secondly, it needs to be sufficiently specific to target those actually at risk. Thirdly, as part of a flood awareness program, it needs to set out the appropriate behavior in the circumstances of the recipient. What is the appropriate behavior depends on the nature of the flooding and the circumstances of those at risk. As in all of the communication, it has to start with an understanding of recipients including their expectations, needs, and capabilities. Not all those at risk will be able to carry out any specific behavior; those who are disabled or in poor health are unlikely to be able to self-evacuate or undertake contingent flood-proofing. Thus, in the early days of flood warnings in the UK, warnings were justified on the rationale that flood losses would be reduced by giving recipients time to carry furniture and fittings above the expected flood level, rather than either attempt to flood-proof the property or to evacuate.

Now, when a risk to life is not anticipated, the advice is to move portable valuable items and those things which will help recovery above the flood level (www.abi.org.uk/products-and-issues/topics-and-issues/flooding/preparing-for-a-flood/).

Since a flood warning to be effective is the outcome of a chain of probabilities (CNS, 1991), the overall success rate of any flood warning system will fall significantly below 100% when measured as the proportion of those at risk who took the appropriate action.

1.15   RESILIENCE

The critical issue is the extent and speed of recovery after a flood. Hence, what factors promote recovery, what are the determinants of resilience? Resilience is not directly observable, all that can be observed is the trajectory of recovery. Some effects of a flood cannot be undone, death is the self-evident example. The loss of memorabilia, physical memories in the form of photographs and heirlooms, is also irreversible. Flood victims have spoken of their whole life being destroyed by the loss of memorabilia. The psychological impacts of extreme stressors may also persist for very long periods and the health effects of floods have been found to persist for a substantial length of time (Tunstall et al., 2006).

As a label, resilience has also been subject to the same meaning inflation as vulnerability, the meaning being inflated unless exactly the same variables determine each and every aspect of the meaning assigned to resilience. Hence, it is unsafe to include in resilience such other desirables as adaptation and learning. As a label, "resilience" risks assuming that the same characteristics apply to human subjects to a shock such as a flood as determine physical resilience, such as that of a rubber ball. Or, that human resilience is akin to the definition of ecological resilience (Holling, 1973). The obvious question on resilience is: what provides the capacities that enable a resilient response? The four obvious candidates are:

- Excess capacity over everyday needs.
- Personality and cultural factors.
- Reserves that can be mobilized.
- External aid.

In each case, other opportunities have to be sacrificed in order to divert the resources to recover from the flood. It is these sacrificed opportunities that determine the magnitude of the economic costs in the addition to the loss of well-being until it recovers to its prior trajectory. The use of excess capacity depends upon its existence and obviously, households and household members vary in the degree to which there is any excess capacity. The possible areas in which there may be excess capacity are: time and related human energy demands, and disposable income. Cleaning and repair take time and energy. The general finding is that women under normal circumstances have less time left for leisure and recreational activities than men (unstats.un.org/unsd/gender/timeuse/index.html). Female-headed households should therefore be expected to have less available capacity for recovery activities than pair-based households and diverting time to recovery may displace income-generating activities. Women also bear a disproportionate share of the burden of recovery (Moench et al., 2003). The elderly and those with prior poor health or disabilities can also be expected to have less surplus energy to undertake the physical recovery activities. The stress of the flood itself may also result in reduced energy as a result of depression and anxiety (Figure 1.2).

The use of surplus income to recover obviously depends on there being any and the amount of surplus income. For example, in the UK, the ratio of discretionary expenditure from the highest income decile household to the lowest decile is 6:1 (Office for National Statistics, 2018). Thus, for a given loss, it would take six times as long for a household in the lowest income decile to replace that loss as one in the highest income decile. This is to assume that the need to eat out would not increase as it might well do so when their home has been flooded, and neither would the other needs.

Personality (Ripley, 2009; Xiang et al., 2016) and cultural factors have been found to be significant both in terms of relative vulnerability and resilience.

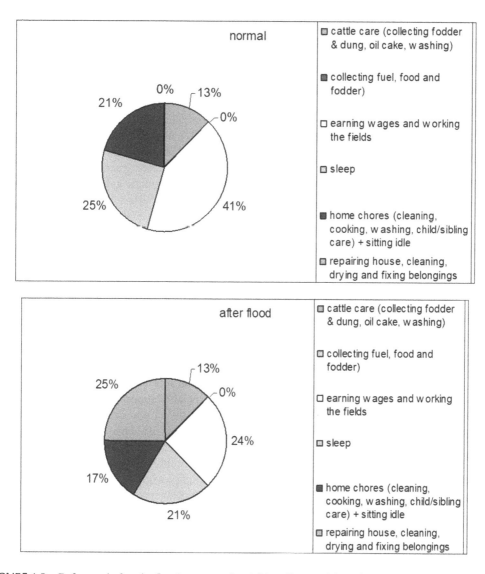

FIGURE 1.2 Before and after the flood: percent of activities. (Source: Moench et al., 2003.)

Reserves are most clearly seen in money. Here, reserves may be in the form of savings or the capacity to borrow. In Japan, significant savings are held as a precaution against the occurrence of a natural disaster (Sawada, nd). More generally, both the extent and form of savings, and hence the degree to which they can be rapidly mobilized, varies between countries (Ynesta, 2008). Similarly, the extent of the savings available for mobilization is likely to be dependent upon household income; in the UK, the lowest income decile households have negative savings (Crossley, 2016). The alternative to using savings is to borrow from family and friends as reported in Bangladesh (Jahan, 2000). Borrowing obviously depends upon the capacity to borrow either from informal sources as in Bangladesh or formal sources. In the latter case, the capacity to borrow depends on the security and extent of income so those in informal work with insecure incomes may be limited to moneylenders demanding extortionate rates. Both drawing down savings and borrowing involve sacrificing future consumption for current consumption. An implication is that in terms of assessing potential resilience, it may be appropriate to report the direct damages for households in terms of losses as a proportion of either discretionary income or of savings.

External aid can come in multiple forms including social support however this is defined (Murphy, 1988), including emotional and psychological support, provision of temporary accommodation, labor, and most obviously financial support. That financial support can either be direct in the form of payments or indirect through tax relief or low-interest loans or subsidized flood insurance.

Disasters demonstrate social relationships in action and particularly those between the population and their government. In China, traditionally, Emperors ruled under the Mandate of Heaven, a disaster being a sign that this mandate had been withdrawn. If a flood appears to imply the abandonment by the government of that population to their fate, both that population and the wider population will draw their own conclusions about whether their governments are acting in a way consistent with social norms. More generally, disasters are the time when the public judges their governments; hence, it is critical that governments are seen to care; visits to major disasters by senior members of government being mandatory. Secondly, that they act to aid recovery and, in doing so, demonstrate their competence. If they fail in a major test, the conclusion is likely to be drawn that they also fail in their routine tasks.

1.16 SUMMARY AND CONCLUSIONS

Understanding the causes and mechanisms of flooding is only part of the problem; another side of the problem is understanding how floods affect people and social relationships. The applied physical sciences are not enough on their own.

More particularly, decisions and actions are undertaken by groups of people interacting. Hence, a key question is how to interact more effectively, what are the appropriate forms of social relationships, and the skills required to build and maintain those social relationships. Communication is as much an issue of relationships as it is about content. Hence we have to develop better governance arrangements, better organizations, and teach the appropriate skills. This question is made more difficult first by the fragmenting of knowledge into different disciplines, and hence the need for interdisciplinary working (Committee on Facilitating Interdisciplinary Research, 2005) if sustainable flood risk management is to be delivered. Multidisciplinary working was argued above to be a variety of cross-cultural working and involve a series of competencies (Deardorff, 2009). Secondly, we seek to make the change in the face of external change, and the changes we make produce further changes, including climate change. Hence, there is a requirement to change how we decide and what we do: to learn, adapt, and innovate. What we knew and the skills we had are always under the threat of obsolescence.

Thirdly, who decides, how decisions are made, and what decisions are made all involve social relationships including such difficult issues as legitimacy (Johnstone et al., 2006), trust (Möllering, 2001), and justice (Pettit, 1980). Rather than the traditional approach of decisions being made by scientific bureaucracies on the basis of technical knowledge, addressing these issues is now one of stakeholder engagement in which scientific knowledge is democratized: transdisciplinary science (Bracken et al., 2015). This requires multiple skills so interpersonal skills are as important as technical skills. In short, FRM without recognition of the central role of people is a purely self-indulgent exercise.

REFERENCES

Abt, S.R., Wittler, R.J., Taylor, A., and Love, D.J. 1989. Human stability in a high flood hazard zone, *Water Resources Bulletin*, 25(4): 881–890.

Aristotle, G.A. 1991. *On Rhetoric: A Theory of Civic Discourse*, Oxford University Press, Oxford, UK.

Banting, K., Johnston, R., and Kymlicka, W. 2006. Do multiculturalism policies erode the welfare state? An empirical analysis. In: Banting, K., Kymlicka, W. (eds) *Multiculturalism and the Welfare State: Recognition and Redistribution in Contemporary Democracies*, Oxford University Press, New York, UK, pp. 49–91.

Boulding, K.E. 1989. *Three Faces of Power*, Sage Publications, Newbury Park, CA, USA.

Bracken, C., Rajagopalan, B., Alexander, M., and Gangopadhyay, S. 2015. Spatial variability of seasonal extreme precipitation in the western United States, *Journal of Geophysical Research: Atmospheres*, 120: 4522–4533.

Butler, D., Kokkalidou, A., and Makropoulos, C.K. 2005. Supporting the siting of new urban developments for integrated urban water resource management. In: Hlavinek P., Kukharchyk, T. (eds) *Integrated Urban Water Resources Management, NATO Scientific Series*, Springer, Berlin, Germany, pp. 19–34.

Calder, I.R. 2000. Land use impacts on water resources, background paper 1. In: FAO Electronic Workshop on Land-Water Link Ages in Rural Watersheds, 18 September, Italy.

Caponera, D.A. 1992. *Principles of Water Law and Administration, National and International*, Balkema, Rotterdam, UK.

Chen, B., Yang, Q., Zhou, S., Li, J.S., and Chen, G.Q. 2017. Urban economy's carbon flow through external trade: Spatial-temporal evolution for Macao, *Energy Policy*, 110: 69–78.

Cherry, C. 1966. *On Human Communication*, MIT Press, Cambridge, MA, USA.

CNS Scientific and Engineering Services. 1991. *Benefitcost Analysis of Hydrometric Data—River Flow Gauging*, Department of the Environment, Foundation for Water Research, Marlow, Canada.

Committee on Facilitating Interdisciplinary Research. 2005. *Facilitating Interdisciplinary Research*, National Academies Press, Washington, DC, USA.

Crick, B. 1964. *In Defence of Politics*, Pelican, Harmondsworth, UK.

Crossley, L. 2016. Review of the information, http://lcrossley.com/blog/2016/09/26/the-information/.

Deardorff, D.K. 2009. Implementing intercultural competence assessment. In: D.K. Deardorff (eds) *The SAGE Handbook of Intercultural Competence*, Sage, Thousand Oaks, CA, USA, pp. 477–491.

Deen, M.Y.I. 1990. Islamic environmental ethics, law, and society. In: Engel, J.R., Engel, J.B. (eds) *Ethics of Environment and Development*, Belhaven, London, UK.

DiMaggio, P. 1997. Culture and cognition, *Annual Review of Sociology*, 23: 263–287.

Duhigg, C. 2012. The power of habit: Why we do what we do in life and business, https://www.sciencenews.org/article/power-habit-why-we-do-what-we-do-life-and-business-charles-duhigg.

Eggers, J.P., and Kaplan, S. 2013. Cognition and capabilities: A multi-level perspective, *Academy of Management Annals*, 7(1): 295–340, https://doi.org/10.5465/19416520.2013.769318.

EU Flood Directive EC. 2007. *Flood Risk*, https://ec.europa.eu/environment/water/flood_risk/implem.htm.

Finer, S.E. 1970. Almond's concept of "the political system", https://openlibrary.org.

Fisher, A.G., Murray, E.A., and Bundy, A.C. 1991. *Sensory Integration Theory and Practice*, F.A. Davis, Philadelphia, PA, USA.

Geertz, C.J. 1993. Ethnic conflict: Three alternative terms, *Common Knowledge*, 2(3): 54–65.

Goffman, E. 1959. *The Presentation of Everyday Life*, Doubleday, New York, UK.

Green, C.H. 2003. *Handbook of Water Economics: Principles and Practice*, John Wiley & Sons, UK.

Green, C.H. 2017. Competent authorities for the flood risk management plan – reflections on flood and spatial planning in England, *Journal of Flood Risk Management*, 10: 195–204.

Green, C., and Eslamian, S. 2014. Water governance. In: Eslamian, S. (ed) *Handbook of Engineering Hydrology*, Ch. 24, Vol. 3: Environmental Hydrology and Water Management, Taylor and Francis, CRC Group, Cleveland, OH, USA, pp. 461–483.

Green, C.H., and Penning-Rowsell, E.C. 1989. Flooding and the quantification of 'Intangibles', *Water and Environment Journal*, 3(1): 27–30.

Green, C.H., and Penning-Rowsell, E.C. 2007. Socio-economic drivers, cities and science. In: Thorne, C.R., Evans, E.P., Penning- Rowsell, E. (eds) *Future Flooding and Coastal Erosion Risks*, Thomas Telford, London, UK, pp. 116–131.

Green, C., and Penning-Rowsell, E.C. 2010. Stakeholder engagement in flood risk management. In: Pender G., Faulkner, H. (eds) *Flood Risk Science and Management*, Wiley-Blackwell, New Jersey, USA, pp. 372–385.

Green, C.H., Parker, D.J., and Penning-Rowsell, E.C. 1993. Designing for failure. In: Merriman, P.A., Browitt, C.W.A. (eds) *Natural Disasters: Protecting Vulnerable Communities*, Thomas Telford, London, UK.

Green, C.H., Parker, D.J., and Tunstall, S.M. 2000. Assessment of flood control and management options. In: *WCD Thematic Review IV. 4 Prepared as an Input to the World Commission on Dams*, Cape Town, South Africa.

Habermas, J. 1995. Reconciliation through the public use of reason: Remarks on John Rawls's political liberalism, *The Journal of Philosophy*, 92: 109–131.

Hajer, M. 1995. *The Politics of Environmental Discourse: Ecological Modernization and the Policy Process*, Clarendon Press, Oxford, UK.

Hall, E.T. 1959. *The Silent Language*. Doubleday and Company, New York, USA.

Harrison, R.W., and Mooney, Jr., J.F. 1993. *Flood Control and Water Management in the Yazoo-Mississippi Delta, Social Research Report Series, 93–95*, Social Science Research Center, Mississippi State University, Mississippi State, MS, USA, p. 15.

Hart, R.E. 2011. An inverse relationship between aggregate northern hemisphere tropical cyclone activity and subsequent winter climate, *Geophysical Research Letter*, 38(1), https://doi.org/10.1029/2010GL045612.

Holling, C.S. 1973. Resilience and stability of ecological systems, *Annual Review of Ecology and Systematics*, 4(1): 1, https://doi.org/10.1146/annurev.es.04.110173.000245.

Horowitz, M.J., Wilner, N., and Alvarez, W. 1979. Impact of event scale: A measure of subjective distress, *Psychosomatic Medicine*, 41: 209–218.

Iwasada, M., Sasaki, K., and Murakami, M. 1999. The damage of riverstructure and natural bank protection in Kochi flood disaster in 1998. In: *Shikoku-based Affiliate of Japan Society of Civil Engineers*, Takamatsu, Shikoku, Japan, pp. 128–129.

Jahan, S. 2000. Economic growth and human development: Issues revisited. In: Keynote Speech at a Conference Hosted by the St. Mary's University, Institute for International Development, Halifax, Canada.

Johnson, C., Dowd, T.J., and Ridgeway, C.L. 2006. Legitimacy as a social process, *Annual Review of Sociology*, 32: 53–78.

Johnson, M.R., van Vuuren, C.J., Visser, J.N.J., Cole, D.I., de Wickens, H.V., Christie, A.D.M., Roberts, D.L., and Brandl, G. 2006. Sedimentary rocks of the Karoo Supergroup. In: Johnson, M.R., Anhaeuser, C.R., Thomas, R.J. (eds) *The Geology of South Africa*, Geological Society of South Africa and Council for Geosciences, Johannesburg, Pretoria, South Africa, pp. 461–499.

Kahneman, D. 2011. *Thinking, Fast and Slow*, 1st edition, Farrar, Straus and Giroux, New York.

Kasperson, R.E. 2014. Four questions for risk communication. *Journal of Risk Research*, 17(10): 1233–1239.

Kay, P., and Kempton, W. 1984. What is the sapir-whorf hypothesis? *American Anthropologist*, 86(1): 65–79.

Kim, G., Burnett, W.C., Dulaiova, H., Swarzenski, P.W., and Moore, W.S. 2001. Measurement of Ra-224 and Ra-226 activities in natural waters using a radon-in-air monitor, *Environmental Science & Technology*, 35: 4680–4683.

Knight, C. 2016. New Zealand's rivers. In: *An Environmental History*, Canterbury University Press, Christchurch, New Zealand.

Krzysztofowicz, R., Kelly, K.S., and Long, D. 1994. Reliability of flood warning systems, *Journal of Water Resources Planning and Management*, 120(6): 906–926.

Lakoff, G. 1987. *Women, Fire, and Dangerous Things: What Categories Reveal About the Mind*, The University of Chicago Press, Chicago, IL, USA.

Lazarus, R.S. 1966. *Psychological Stress and the Coping Process*, McGraw-Hill, New York, USA.

Liebrand, W.B.G. 1992. How to improve our understanding of group decision making with the help of artificial intelligence, *Acta Psychologica*, 80(1–3): 279–295.

Lind, E.A., and Tyler, T.R. (eds) 1988. *The Social Psychology of Procedural Justice*, Springer, New York, USA.

Lindberg, S. 2009. Olemisen Rytmi. In: Hirvonen, A., Linberg, S. (eds) *Mika Mimesis? Philippe Lacoue-Labarthen Filosofinen Teatteri*, Tutkijaliitto, Helsinki, Finland, pp. 19–37, https://docplayer.net /20884558-Improving-access-to-psychological-therapies-london-recruitment-2016-information-for -applicants.html.

Long Term Conditions Team. 2012. *Long Term Conditions Compendium of Information*, 3rd edition, Department of Health, London, UK.

MacGillivray, A., and Livesey, H. 2015. *Delivering Benefits through Evidence. Public Dialogues on Flood Risk Communication*. Evaluation Report. Environment Agency, Horizon House, Canada.

Marecek, J., Crawford, M., and Popp, D. 2004. On the construction of gender, sex, and sexualities' In: Eagly, A.H., Beall, A.E., Sternberg, R.J. (eds) *The Psychology of Gender*, Guilford Press, New York, NY, US, pp. 192–216.

Melnyk, S.A., Davis, E.W., Spekman, R.E., and Sandor, J. 2010. Outcome-driven supply chains, *MIT Sloan Management Review*, 51(2): 33.

Miller, G.A. 1956. The magical number seven plus or minus two: Some limits on our capacity for processing information, *Psychological Review*, 63(2): 81–97, https://doi.org/10.1037/h0043158.

Milne, G. 1977. Cyclone tracy: I Some consequences on the evacuation for adult victims, *Australian Psychologist*, 12: 39–54.

Mitsch, W.J., and Gosselink, J.G. 2000. *Wetlands*, 3rd editon, John Wiley, New York, USA.

Moench, M., Dixit, A., Janakarajan, S., Rathore, M.S., and Mudrakartha, S. 2003. *The Fluid Mosaic: Water Governance in the Context of Variability, Uncertainty and Change*, Institute for Social and Environmental Transition, Boulder, CO, USA.

Mollering, G. 2001. The nature of trust: From Georg Simmel to a theory of expectation, interpretation and suspension, *Sociology*, 35(2): 403–420.

Montgomery, D.R. 2007. Soil erosion and agricultural sustainability, *Proceedings of the National Academy of Sciences*, 104(33): 13268–13272.

Morris-Oswald, T., and Simonovic, S.P. 1997. *Assessment of the Social Impact of Flooding for Use in Flood Management in the Red River Basin*, Report prepared for the IJC red River Basin Task Force, Winnipeg, Canada.

Murphy, K.J. 1988. Aquatic weed problems and their management: A review. I. The worldwide scale of the aquatic weed problem, *Crop Protection*, 7: 232–248.

Nelson, A. 2004. *African Population Database*, UNEP GRID, Sioux Falls, USA.

Nelson, R.R. 1982. *Winter SG. An Evolutionary Theory of Economic Change*, Belknap, Cambridge, UK.

OECD. 2015. *Disaster Risk Financing: A Global Survey of Practices and Challenges*, OECD Publishing, Paris, France, http://doi.org/10.1787/9789264234246-en.

Office for Disability Issues. 2014. Guidance: Accessible communication formats. Available from www.gov.uk /government/publications/inclusive-communication/accessible-communication-formats.

Office for National Statistics. 2018. 2011 census: Usual resident population by five-year age group and sex, local authorities in the United Kingdom. Retrieved from https://www.ons.gov.uk/peoplepopulationand community/populationandmigration/populationestimates/datasets/2011censuspopulationandhousehol destimatesfortheunitedkingdom.

Penning-Rowsell, E.C., Green, C.H., Thompson, P.M., Coker, A.M., Tunstall, S.M., Richards, C., and Parker, D.J. 1992. *The Economics of Coastal Management: A Manual of Assessment Techniques*, Belhaven, London, UK.

Perrow, C. 1984. *Normal Accidents: Living With High-Risk Technologies*, Basic Books, New York, UK.

Pettit, R. 1980. *Judging Justice: An Introduction to Contemporary Political Philosophy*, Routledge, London, UK.

Reason, J. 1997. *Managing the Risks of Organizational Accidents*, Ashgate, Aldershot, UK.

Ripley, B. 2009. *Tree: Classification and Regression Trees. R Package Version 1.0-27.* [www document] URL http://CRAN.R-project.org/package=tree (Site visited on 14 August 2010).

Roberts, A.R. 1991. Conceptualizing crisis theory and the crisis intervention model. In: Roberts, A.R. (ed) *Contemporary Perspectives on Crisis Intervention and Prevention*, Prentice-Hall, Englewood Cliffs, NJ, USA, pp. 3–17.

Royal Commission on Land Drainage. 1927. *Report of the Royal Commission on Land Drainage in England and Wales*, HM Stationery Office, London, UK.

Savage, O., and Pratt, B. 2013. The history of UK civil society. International NGO Training and Research Centre, Briefing Paper 38, Online. Available at: https://www.intrac.org/wpcms/wp-content/uploads /2016/09/Briefing-Paper-38-The-history-of-UK-civilsociety.pdf [Accessed 14 April 2018].

Sawyer, S. 2005. *Resampling Data: Using a Statistical Jackknife*, Washington University, Washington, DC, USA.

Shaxson, F., and Barber R. 2003. *Optimizing Soil Moisture for Plant Production: The Significance of Soil Porosity*, FAO Soils Bulletin, 79, UN-FAO, Rome, Italy.

Smith, D.E. 2005. Evidence for secular Sea surface level changes in the Holocene raised shorelines of Scotland, UK, *Journal of Coastal Researchm*, 42: 26–42.

Spencer-Oatey, H. (ed).2012. *What is Culture? A Compilation of Quotations, GlobalPAD Core Concepts*, https://warwick.ac.uk/fac/cross_fac/globalpeople2

Stiglitz, J.E., Sen, A., and Fitoussi, J.-P. 2009. *Report by the Commission on the Measurement of Economic Performance and Social Progress*, Commission on the Measurement of Economic Performance and Social Progress, Paris. www.stiglitz-sen-fitoussi.fr.

Toronto Water, General Manager. 2007. *Wet Weather Flow Master Plan Implementation Report 2006*, Toronto Water, Toronto, Canada.

Toulmin, S.E. 1958. *The Uses of Argument*, Cambridge University Press, Cambridge, UK.

Tullock, G. 1967. The general irrelevance of the general impossibility theorem, *Quarterly Journal of Economics*, 81: 256–270.

Tunstall, S.M., Tapsell, S.M., Green, C.H., Floyd, P., and George, C. 2006. The health effects of flooding: Social research results from England and Wales, *Water Health*, 4(3): 365–380.

Turner, B.A. 1978. *Man-Made Disasters*, Wykeham Publications Ltd., London, UK.

Van de Ven, G.P. 2004. *Man Made Lowlands, History of Water Management and Land Reclamation in the Netherlands*, 4th edition, Matrijs, Utrecht, The Netherlands.

Wagret, P. 1967. *Polderlands*, Methuen, London, UK.

Wang, X.L. 2008. Penalized maximal F test for detecting undocumented mean shift without trend change, *Journal of Atmospheric and Oceanic Technology*, 25(3): 368–384.

WEF (World Economic Forum). 2016. *The Global Risk Report 2016*, WEF, Geneva, Switzerland.

White, G.F. 1945. *Human Adjustments to Floods, Department of Geography Research*, Paper No. 20 29, The University of Chicago, Chicago, USA.

Wittgenstein, L. 1958. *Philosophical Investigations*, Basil Blackwell, Oxford, UK.

Worthington, S. 2003. *Equity*, Oxford University Press, Oxford, UK.

Xiang, H., Zhang, Y., and Richardson, J.S. 2016. Importance of riparian zone: Effects of resource availability at land-water interface, *Riparian Ecology and Conservation*, 3(1): 1–17.

Ynesta, I. 2008. *Households' Wealth Composition Across OECD Countries and Financial Risks Borne by Households*, Financial Market Trends, ISSN 1995–2864,© OECD.

Yzermans, C.J., Donker, G.A., Kerssens, J.J., Dirkzwager, A.J.E., Soeteman, R.J.H., and Ten Veen, P.M.H. 2005. Health problems of victims before and after disaster: A longitudinal study in general practice, *International Journal of Epidemiology*, 34: 820–826.

Part II

Food Observation and Modeling Uncertainty

2 Importance of Hydrological and Meteorological Measurements and Observations in the Implementation of the Paris Agreement and the Katowice Climate Package

Marta Barszczewska

CONTENTS

2.1 INTRODUCTION

During the night from December 14–15, 2018, the difficult but extremely important 24th Conference of the Parties to the United Nations Framework Convention on Climate Change (COP 24) was ended. The aim of the UN climate summit in Katowice was to develop a rulebook, i.e., a "road map" for the implementation of the Paris Agreement agreed in 2015, which aimed at limiting the global temperature rise.

At COP 21 in Paris, on December 12, 2015, the UNFCCC Parties reached a groundbreaking agreement on combating climate change and accelerating and intensifying the actions and investments needed for a sustainable, low-carbon future. The Paris Agreement is based on a convention and for the first time, unites all countries on a common cause to make ambitious efforts to combat climate change and to adapt to its effects while supporting developing countries to achieve this goal. It can therefore be said that it sets a new course for global climate efforts.

DOI: 10.1201/9780429463938-4

The main objective of the Paris Agreement is to strengthen the global response to the threat of climate change by maintaining global temperature rise in this century at much less than 2° C above pre-industrial levels and continuing the efforts to limit the temperature rise even to 1.5° C. In addition, the agreement aims to increase the capacity of countries to cope with the impact of climate change and to ensure that financial flows are consistent with low greenhouse gas emissions and resilience to climate change. To achieve these goals, appropriate mobilization and funding, a new technological framework, and strengthened capacity building should be established, thus supporting the activities of developing and most vulnerable countries, in accordance with their own national goals. The agreement also provides for an improved transparency framework for the actions taken and provided support.

The Paris Agreement commits all Parties to make all efforts through "national determined contributions" (NDCs) and to strengthen these efforts in the coming years. This includes requirements that all Parties regularly submit about their emissions and about the efforts to implement them. It was also assumed that every five years a global stocktake of the situation would be made in order to assess the collective progress in achieving the goal of the agreement and in informing about further individual actions of the Parties.

The Paris Agreement entered into force on November 4, 2016, 30 days after the so-called "double threshold" (ratification of 55 countries, which account for at least 55% of global emissions). From that time, 184 countries from 197 Parties to the Convention ratified and continued to ratify the agreement (as of December 18, 2018).

For the Paris Agreement to become fully operational, in Paris a work program to develop principles, procedures, and guidelines for a wide range of issues was launched. Since 2016, the Parties have cooperated with each other in subsidiary bodies (APA, SBSTA, and SBI) and in various bodies. The Conference of the Parties serving as the meeting of the Parties to the Paris Agreement (CMA) met for the first time in connection with the COP 22 in Marrakech (in November 2016) and adopted the first two decisions. The work program, in accordance with the assumption, was completed during COP 24 in Katowice.

The Paris Agreement, adopted by Decision 1/CP.21, addresses the key areas needed to combat climate change. Some of the key aspects of the decision are outlined below:

- Long-term temperature target (art. 2): The Paris Agreement aims at strengthening the global response to climate change, confirming the objective of limiting the global temperature rise to much less than 2° C while aiming to limit growth to 1.5° C.
- Global peaking and "climate neutrality" (art. 4): To achieve this goal in terms of temperature, the Parties strive to achieve a global increase in climate-neutral gases (greenhouse gas emissions – GHG) as soon as possible, recognizing that the time for developing-country Parties to achieve a balance between anthropogenic emissions and GHG will be much longer and will be achieved in the second half of the century.
- Mitigation (art. 4): The Paris Agreement establishes binding commitments of all Parties in the preparation, transmission, and maintenance of the national determined contribution (NDC) and the implementation of national measures to achieve them. It also provides that the Parties will provide information necessary for clarity and transparency through their national data centers every five years. To establish a solid foundation for achieving the higher ambitions, each successive NDC will represent progress beyond the previous ones and reflect the highest possible ambition. Developed countries should continue to play a leading role, pursuing absolute reduction targets throughout the economy, while developing countries should continue to increase their efforts to mitigate climate change and are encouraged to pursue the overall economic goals over time in the light of different national circumstances.
- Sinks and reservoirs (art. 5): The Paris Agreement also encourages the Parties to maintain and improve, respectively, sinks and greenhouse gas storage tanks referred to in article 4 par. 1 d) of the Convention, including the increase of forest area.

- Voluntary cooperation/market and non-market approaches (art. 6): The Paris Agreement recognizes the possibility of voluntary cooperation between the Parties that allows for more ambitious goals and establishes principles – including transparency, environmental integrity, and cooperation, which includes an international transfer of knowledge on mitigation. In addition, it establishes a mechanism that contributes to the mitigation of greenhouse gas emissions, promotes sustainable development, and sets the framework for a non-market approach to sustainable development.
- Adaptation (art. 7): The Paris Agreement sets the global goal for adaptation – increasing adaptability, strengthening resilience, and reducing vulnerability to climate change in the context of a temperature target. Adaptation to significantly strengthen national adaptation efforts, including through international cooperation and support, recognizes that adaptation is a global challenge. All parties should engage in adaptation, including through the formulation and implementation of National Adaptation Plans (NAPs), and submit and periodically update the adaptation communication, describing their priorities, needs, plans, and actions. The adaptation efforts of developing countries should be recognized.
- Loss and damage (art. 8): The Paris Agreement recognizes the importance of preventing, minimizing, and eliminating loss and damage related to the adverse effects of climate change, including extreme weather events and slow events, and the role of sustainable development in reducing the risk of loss and damage. The Parties are to increase their understanding, action, and support, including through the Warsaw International Mechanism (adopted during COP 19), on the basis of cooperation and facilitation in relation to loss and damage related to the adverse effects of climate change.
- Finance, technology, and capacity-building support (art. 9, 10, and 11): The Paris Agreement confirms the commitments of developed countries to support the efforts of developing country Parties to build clean, climate-resilient futures contracts while encouraging for the first time voluntary contributions from other Parties. Providing resources should also aim to achieve a balance between adaptation and mitigation. In addition to reporting on funds already provided, developed countries are committed every two years to provide indicative information about future support, including projected levels of public finance. The Agreement also provides that the Financial Mechanism of the Convention, including the Green Climate Fund (GCF), will be used for the Agreement. International cooperation on the development of climate-friendly technologies and transfer and capacity building in developing countries has also been strengthened – a technological framework has been established under the agreement, and capacity-building activities will be strengthened, inter alia, through strengthened support for capacity building in developing countries.
- Education, training, social awareness in the field of climate change, social education and public access to information (art. 12) are also to be strengthened under the agreement.
- Transparency (art. 13), implementation, and compliance (art. 15): The Paris Agreement is based on a solid transparency and accounting system to ensure the clarity of the activities and support of the Parties, taking into account the flexibility of the various possibilities of the Parties. In addition to reporting information on mitigation, adaptation, and support, the agreement requires that the information submitted by each Party be subject to international technical expertise. The agreement also includes a mechanism that will facilitate the implementation and promotion of compliance in a non-discriminatory and non-punishable way that will be updated annually through the CMA reports.
- Global stocktake (art. 14): "Global inventory," which is to be held in 2023, and then every five years, will assess the collective progress towards achieving the goal of the agreement in a comprehensive and easy-to-understand way. It will be based on the best available science and its long-term global goal. Its result will inform the Parties on the update and improvement of their activities and on supporting and strengthening international cooperation in the field of climate-related activities.

One of the fundamental elements of the Paris Agreement are activities undertaken in the area of adaptation, and their implementation at the national level would not be possible without meteorological and hydrological monitoring carried out by the Institute of Meteorology and Water Management – National Research Institute, under the Water Law Act, as a national hydrological meteorological survey.

The lack of a common position of experts emphasizes the importance and necessity of conducting regular analyses of the state of the environment, based on a unified methodology. Conducting measurements and observations by the national hydrological and meteorological survey covers all physical processes taking place in the atmosphere and hydrosphere, which are important in planning activities that slow down the processes of climate change, as well as in planning activities aimed at adapting to existing changes, and the methodology used for meteorological and hydrological analyses is reflected in the international systems that create an integrated system of data transmission, processing, and its collection.

Measurement data, in accordance with the scope applicable to the type and the nature of station specified in the Ordinance of the Minister of the Environment of November 6, 2008, on standard procedures for collecting and processing information by the national hydrological and meteorological survey and national hydrogeological survey, originate from a ground-based observation and measurement system, international data exchange, remote sensing, radar and satellite systems, as well as the results of hydrological and meteorological models that enable their mutual verification.

Coherent systems and methods for conducting measurements and observations, as well as independent analyses conducted in many countries, allow creating the scenarios and models of climate change and determining trends, and finally reveal the need to immediately implement measures actioning the recommendation of the Paris Agreement. As a result of the work of scientists in the field of climate change, on the basis of long-term measurement and observation sequences, from Member countries of the World Meteorological Organization (WMO), a special report on the impact of climate change on 1.5° C was developed.

2.2   IPCC REPORT 2018 (SR15)

The World Meteorological Organization (WMO) established by the Parties of the United Nations Framework Convention on Climate Change (UNFCCC) in 1950 is the leading international organization for the climate observation of the Global Climate Observing System (GCOS). The scope of the WMO's activities covers issues in the area of managing the development and improvement of global meteorology, climatology, and operational hydrology. In 1963, the WMO established the World Weather Watch (WWW), acting as a global system for measuring, exchanging, and processing meteorological data, which form the basis for the development of forecasts and warnings against dangerous phenomena occurring in the atmosphere and hydrosphere. Under the guidance of the WMO, as well as the WWW, ranges are set for dates of observations being conducted, as well as research and scientific programs being conducted, i.e., the World Climate Program (WCP), the World Climate Research Program (WCRP), the International Panel of Climate Change (IPCC), or the International Decade of Natural Disaster Reduction (IDNDR).

The task of the IPCC is to systematically carry out meteorological observations and measurements in selected and unchanged-for-many-decades measurement points enabling climate monitoring from the ground and early detection of its contemporary changes. Measurements also concern the lower atmosphere layers: troposphere, stratosphere, and sea-level changes. Observations clearly confirm climate change by increasing the global average temperature of air and oceans, melting the snow cover and glaciers, increasing the level of the sea and oceans. On the basis of the data of the national meteorological services and surveys and other institutions, it was found that:

- The average global temperature during 12 of the last 13 years (1995–2007) was included in the series of 13 warmest years since the beginning of instrumental measurements. The average global temperature has increased by 0.74° C since the beginning of the 20th century.

- The average level of the world's ocean increased by 1.8 mm per year in the period 1961–2003 and by 3.1 mm per year from 1993.
- In the last 25 years, the Arctic Ocean has lost 17% of the ice cover. Arctic sea ice will reduce its surface in the next 30–40 years.
- The carbon dioxide content has increased by 36% since the beginning of the industrial age. The concentration of other greenhouse gases such as methane has also increased.

At the request of the UNFCCC, the Intergovernmental Panel on Climate Change (IPCC) developed and adopted in October 2018 a special report on the impact of climate change of 1.5° C (SR15), which states that global warming is likely to increase to 1.5° C above the level from the pre-industrial period between 2030 and 2052 if no action is undertaken and the insulation will continue to grow at the currently observed rate. SR15 summarizes, on the one hand, the impact of 1.5° C warming (equivalent to 2.7° F) on the functioning of the planet, and on the other hand, the necessary steps to limit global warming.

Even assuming full implementation of the conditional and unconditional national determined contribution made by the Parties in the Paris Agreement, net emissions would increase in comparison to 2010, which would lead to a warming of around 3° C to 2100 and a further increase in later years. Limiting the insulation to below or close to 1.5° C would require a reduction of net emissions by about 45% by 2030 and a net-zero by 2050 (i.e., maintaining the total cumulative emissions within the budget of carbon dioxide emissions). Even to limit global warming to below 2° C, CO_2 emissions should fall by 25% by 2030 and by as much as 100% by 2075.

The paths (i.e., scenarios and budgets of mitigation options) that would allow for such a reduction by 2050 allow for the production of about 8% of global electricity from gas and 0–2% based on coal (compensated by carbon capture and storage). It is predicted that on these paths renewable energy sources will provide 70–85% of electricity in 2050, and the share of nuclear energy should show an increasing trend. It is also assumed that the other actions will be taken at the same time: for example, emissions of gases other than CO_2 (such as methane, black coal, nitrous oxide) should be similarly reduced. In order to determine the necessary emission reduction, scenarios were assumed that the demand for energy will remain unchanged, reduced up to 30%, or offset by an unprecedented method of carbon dioxide removal, which will only be developed while developing and implementing new policies and research to improve agricultural efficiency and industry.

2.3   THE INFLUENCE OF TEMPERATURE INCREASE ON INDIVIDUAL ELEMENTS OF THE ENVIRONMENT

2.3.1   THE EFFECT OF WARMING 1.5° C OR 2° C

According to the report, with global warming of 1.5° C, there will be an increased risk for health, livelihoods, food security, water supply, human security and economic growth. Impact vectors include reduced yields and food quality. The effect of temperature increase on livestock was also noted through changes in feed quality, disease spread and availability of water resources. In addition, the risk of certain diseases transmitted by temperature vectors, such as malaria and dengue, is predicted to increase.

Limiting global warming to 1.5° C compared to 2° C can reduce the number of people exposed to climate risks and vulnerable to poverty by as much as several hundred million by 2050. Climate risk as global warming increases depends on geographic location, level of development and vulnerability, and the speed and extent of climate change mitigation and adaptation actions to climate change. For example, urban heat islands reinforce the impact of heatwaves especially felt in cities. In SR15 it was found that subtropical countries of both hemispheres have the greatest impact on economic growth.

2.3.2 Weather, Sea Level, and Ice

In many regions and seasons, warming is higher than the global annual average, e.g., 2–3 times higher in the Arctic. Warming is generally higher on land than on the ocean and correlates with extreme temperatures (which are to warm up to twice as much on land than the average global surface temperature) as well as extreme rainfall (both heavy rains and droughts). The estimated risk levels mostly increased compared to the previous IPCC report from 2008.

The global average sea level is projected to increase (compared to 1986–2005) by 0.26 to 0.77 m in the perspective of up to 2100 for global warming by 1.5° C and an additional around 0.1 m for 2° C. The difference of 0.1 m may correspond to ten million or more people exposed to the associated risk. Sea level rise will continue after 2100, even if global warming is reduced to 1.5° C. Irreversible instability of ecosystems in Antarctica and the Greenland ice cover will be observed, which results in an increase in sea level with a scale impossible to accurately estimate. The Arctic ice-free summer is predicted once a century. Limiting global warming to 1.5° C instead of 2° C will prevent the thawing of permafrost in the range from 1.5 to 2.5 million km^2.

2.3.3 Ecosystems

The decrease in the annual global marine fisheries catches in 1.5 or 3 million tons in the case of global warming of 1.5° C or 2° C is projected by a single global model of fisheries quoted in the report. It is predicted that coral reefs will drop by another 70–90% with an increase of 1.5° C, and at 2° C even by more than 99%. Of the 105,000 known species, 18% of insects, 16% of plants, and 8% of vertebrates, according to estimates, will lose more than half of their specific geographical coverage for global warming by 2° C.

It is estimated that 4% or 13% of the land surface of the earth switches ecosystems transformation from one type to another, respectively on a temperature increase of 1.5° C or 2° C. Tundra at high latitude and boreal forests are particularly vulnerable to degradation and loss caused by climate change, and woody shrubs are already entering the tundra and will contribute to the continuation of further warming of these areas.

Positive evaluation of the system for monitoring climate change and its effectiveness will be possible only in the case when it will be possible to plan the appropriate adaptation, eliminating or reducing the risk by all of the countries. To carry out such an assessment it is necessary to have historical, current, and forecasted hydrological and meteorological information about changing environmental conditions, which is provided by the national survey, which is also closely cooperating with the sector responsible for crisis management in the country.

2.4 NATIONAL HYDROLOGICAL AND METEOROLOGICAL SURVEY OF THE INSTITUTE OF METEOROLOGY AND WATER MANAGEMENT – NATIONAL RESEARCH INSTITUTE

2.4.1 Historical Outline of 100 Years of Hydrological and Meteorological Survey

Polish hydrological measurements and observations date back to 1717 and cover the Odra River in Wrocław. The first regular observations, the records of which have been preserved, come from 1760 and concern the Vistula in Torun, and subsequently Warsaw and Tczew, which have been conducted continuously since 1799.

At the beginning of the 19th century, regular observations of the water level in the Odra basin began on the Warta, Notec, Drawa, Nysa Luzycka, or Nysa Klodzka, and later in the Vistula basin on the Dunajec, Poprad, Sana, Narew, Bug, Brda, and Bzura. In the rivers entering the Baltic Sea, regular observations of the water level began the latest, from 1901, on Reda, Slupia, Leba, and Lyna. The first flow measurements in Poland were made in 1833 on the Odra river in Wrocław, then in 1836 on the Vistula river near Swiecie. The systematic flow measurements made in the upper Vistula basin come from 1885, in particular on Small Vistula, upper Vistula (from Przemsza to Cracow), as well as on the following villages: Sola, Skawa, Raba, Dunajec, Wisloka, and San with Wislok.

The Polish meteorological measurement and observation network, called the Florentine network, was established in 1654 and included 11 stations in Paris and Warsaw. Further meteorological stations were built in 1779, also in Warsaw, and in 1792 in Cracow. In Cracow, in 1883, the first systematic measurements of insolation were started, and then these measurements were carried out in Wroclaw, Warsaw, Kolobrzeg, and at the top of Sniezka Mountain. The oldest series of atmospheric pressure and air temperature measurements in Poland, which covers the years 1710–1721, is the David von Grebner series of measurements from Wroclaw.

Formally, in July 1885, the Warsaw Meteorological Network was established, which formed the ground for establishing in January 1919, on the basis of the Regulation of the Council of Ministers the National Meteorological Institute. Two years later, the tasks and goals of the hydrological survey were also defined.

In 1932, the Council of Ministers adopted a resolution on the merger of the Hydrographic Institute with the National Meteorological Institute into one national state organism conducting hydrological and meteorological surveys, but due to the scale of destruction in the measurement and observation network infrastructure during World War II its entry into force was significantly delayed.

In 1946, the Council of Ministers adopted another important resolution for the history of the national hydrological and meteorological survey, under which the National Hydrological and Meteorological Institute was established, which in December 1959 was transformed into the Institute of Water Management.

The undertaken modernization activities and the introduction of new technologies of measurement meant that in those days the Institute of Water Management was one of the most modern hydrological and meteorological surveys that had a system of quickly warning the community about dangerous hydrological and meteorological phenomena.

The undertaken activities along with the subsequent definition of the scope of the work allowed Poland to entrust the role of the national hydrological and meteorological survey under the Water Law Act to the Institute of Meteorology and Water Management, established by Resolution No. 338/72 of the Council of Ministers of December 30, 1972, on the connection of National Hydrological and Meteorological Institute with the Institute of Water Management.

Conducted monitoring of the atmosphere and hydrosphere, along with the simultaneous use of the warning system against dangerous phenomena, i.e., hurricanes, floods, storms, hailstorms, downpours, temperature changes, drought, or atmospheric pollution, turned out to be a valuable tool during the flood in 1997, but also it allowed defining the further investment needs in the area of a hydrological and meteorological network.

With the favor of the government and the financial support of the World Bank, the national hydrological and meteorological survey was designed and significantly modernized. As a result of these works, the basic measurement and observation network was modernized, including 61 hydrological and meteorological stations, eight meteorological radars, a satellite data reception station, and nine light detection sensors. In addition, modernization was carried out at 795 water gauge stations and 978 rainfall stations (including 150 stations with daily registration and data transmission).

According to the Ordinance of the Council of Ministers of September 13, 2010, the Institute of Meteorology and Water Management in Warsaw was granted the status of the National Research Institute.

2.4.2 The Current State of the IMGW-PIB Measurement and Observation Network

Currently, the Institute of Meteorology and Water Management – National Research Institute has an observation and measurement network, forecasting offices, a telecommunications system, and a system for collecting and disseminating data. The national hydrological and meteorological survey carried out at IMGW-PIB continuously provides government authorities, society, and the national economy with up-to-date information within three main systems:

1) A measurement and observation system consisting of a terrestrial hydrological-meteorological measurement and observation network, which consists of approximately 1,750 measurement stations throughout the country:
   - Hydrological stations network;
   - Meteorological stations network;
   - Aerological measurement stations network;
   - Actinometric measurement stations network;
   - Limnological stations network;
   - Evaporometer stations network;
   - Phenological observations network;
   - POLRAD weather radar system;
   - PERUN lighting detection system;
   - A system for receiving and processing data from meteorological satellites.
2) A data transfer system used primarily to collect data from the measurement and observation network and to alert governmental administration units about dangerous phenomena. It is also used to exchange meteorological information within the international communication network GTS (Global Telecommunication System) as well as to exchange information and products between IMGW-PIB units. The data transmission system consists of international and national leased lines, satellite connections, and mobile telephone and radio lines.
3) A data processing, forecasting, and warning system consisting of central and regional meteorological and hydrological forecasting and protection centers; operational and historical databases system; numerical, statistical, and conceptual systems of meteorological, hydrological, and prognostic models; system of data dissemination, forecasts, and warnings for central and provincial decision-making bodies and other users.

It is worth noting that IMGW-PIB, through regular measurements carried out in accordance with the guidelines and recommendations of WMO throughout Poland based on the network of hydrological and meteorological measurement stations, contributes to the development of both methods of warning against dangerous phenomena, but also allows for the development of science in the area of unprecedented or unrecognized processes and phenomena.

Implementation of the Paris Agreement and the Katowice Climate Package in Poland would not be possible without proper mobilization and allocation of key resources in the decision-making process. In the case of extreme phenomena, the occurrence of which in recent years has significantly increased, the hydrological and meteorological protection system plays the key role in interaction with the crisis management system. Experiences with meteorological and hydrological hazards that hit Poland in the 21st century have shown the need to strengthen activities both in the area of adaptation to them as well as the possibilities of their detection for the effective use of information by entities responsible for the security of society.

Measurement data collected, processed, and made available by IMGW-PIB significantly improve the ability of early identification and assessment of potential risks through the usage of modern tools for the detection of extreme weather phenomena.

Correct and reliably performed tasks of the national hydrological and meteorological survey allow providing the public administration bodies, as well as the whole society, with up-to-date data on the state of the atmosphere and hydrosphere, forecasts, and warnings in both everyday and threat situations. These activities, in order to ensure consistency and correctness of interpretation, as well as for planning activities implementing the Paris Agreement by recipients and users of forecasting systems around the world, are carried out in cooperation with hydrological and meteorological surveys and services within the World Meteorological Organization.

It should be emphasized that strengthening the protection and developing new solutions is a continuous process for IMGW-PIB, implemented, among others, by increasing the number of measurement and observation stations under the national hydrological and meteorological survey, improving the quality and availability of data from existing stations, and developing tools for detecting the extreme weather phenomena (Cox et al., 2014).

2.5 CLIMATE CHANGE FORECAST FOR POLAND

The data collected and processed by the Institute are the basis both for determining the level of achievement of the Paris Agreement, but also for monitoring, forecasting, and planning to take necessary action in the territory of the country.

The results of forecasts, models, and analyses show that by 2030 climate change will have both positive and negative impacts on the environment, society, and economy.

A positive effect of the forecasted increase of average air temperature will be the extension of the growing season and the possibility of growing new plant species, shortening the heating season, and extending the tourist season. On the other hand, however, negative effects on the natural environment due to the extension of the growing season should be noted as the frost season will appear in less favorable stages of plant growth.

However, the negative consequences of climate change, which will affect biodiversity through the increase in the share of alien species, including invasive species, will predominate.

Serious adverse climatic changes are associated with adverse changes in hydrological conditions. The annual rainfall sums do not change significantly, but their nature will be more uneven – the extension of no-rainfall periods and rapid torrential rains.

Trends indicate a decrease in groundwater level, which will be reflected in biodiversity and natural resources, and in particular will affect the operational work of water reservoirs and will threaten the wetland areas. Changes will also be observed in the winter season by reducing the time of retention and thickness of the snow cover, intensification of evaporation, and, consequently, changes in the water balance of the country.

At the same time, the effect of the forecasted climate changes will be increasing the frequency of extreme phenomena: heavy rains that threaten flooding or landslides, but also water shortages caused by drought. The negative effects of climate change are also increased processes of overgrowth of water plants and the eutrophication of inland, transitional, and coastal waters, increasing the threat to human life and health, increasing the demand for electricity in the summer, and reducing the cooling capacity of power plants.

Undoubtedly a challenge for sustainable development in Poland and in the world is to adapt to the changing climate by improving the resilience of individual sectors of the economy. Climate change will have a huge impact primarily on the water management of the country, and taking into account the fact that Poland's water resources are well below average – even in Europe – and the relatively low efficiency of their use, it is predicted that some regions will have to face the problem of water supply.

The adoption of adaptation actions, and the same implementation of the Paris Agreement, will require all sectors of the economy. It is therefore necessary to take adaptation actions to extreme weather events, such as strong winds, rainstorms and storms, floods, landslides, snow and ice, and extreme temperatures.

There is also a requirement to observe a forecasted increase in the intensity and frequency of sea storms as well as an increase in wave heights in the Baltic Sea region, and thus increased coastal erosion, floods from the sea, and an increase in the salinity of groundwater in lower areas. Climate change will also be caused by mild winters – reduction of the ice cover, which is a natural protection against storm surge, and reduction of the edge's resistance to shore blurring.

The expected global temperature increase will cause the migration of new, unprecedented species, with the simultaneous withdrawal of native species, unsuitable for high temperatures and drought periods. The impact of climate change on the species of trees and their condition will also be significant.

2.6 SUMMARY AND CONCLUSIONS

Climate change should be seen as a risk and an element compulsorily taken into account when creating regulatory mechanisms and investment plans against the real threat of weather anomalies and long-term changes that may limit the availability of food, drinking water, and energy, thus increasing competition for resources and migration of people from risk areas such as the island countries.

Conducting constant, reliable, unified monitoring by hydrological and meteorological survey and services is one of the basic tools that provide the data and information on the current and forecasted situation, and thus inform the planning activities that minimize the negative effects of climate change and lead to the full implementation of the Paris Agreement.

REFERENCES

Cox, J. P., Shaeri Karimi, S., and Eslamian, S., 2014, Optimum Hydrometric Site Selection, in *Handbook of Engineering Hydrology, Ch. 22, Vol. 1: Fundamentals and Applications*, edited by Eslamian, S., Taylor and Francis, CRC Group, Florida, USA, 471–483.

IPCC, 2007, The Fourth Assessment Report of the IPCC, available at https://www.ipcc.ch/assessment-report /ar4/ (January 21st, 2019).

IPCC, 2018, Impacts of 1.5°C of Global Warming on Natural and Human Systems, Special Report on Global Warming of 1.5°C, available at https://www.ipcc.ch/site/assets/uploads/sites/2/2018/12/SR15_Chapter3 _Low_Res.pdf (January 21st, 2019).

Kozłowska-Szczęsna, T., Limanówka, D., Niedźwiedź, T., Ustrnul, Z., and Paczos, S., 1993, *Thermal Characteristic of Poland*, 18, Polish Academy of Sciences Institute of Geography and Spatial Organization, Poland, 73 pp.

Kurowska-Łazarz, R., Szulc, W., Woźniak, B., Piotrowska, M., and Drożdżyńska, J., 2015, *Vademecum Meteorological Measurement and Observations*, Institute of Meteorology and Water Management, National Research Institute, Poland, 69 pp.

Ministry of the Environment, 2008, *Standard Procedures for Collecting and Processing Information, Ordinance of the Minister, November 6th, 2008*, National Hydrological and Meteorological Survey and National Hydrogeological Survey, Poland.

Szumiejko, F., and Wdowikowski, M., 2016, IMGW-PIB Monitor as a Source of Information on Dangerous Meteorological and Hydrological Phenomena for Crisis Management Needs, *Obronność–Zeszyty Naukowe Wydziału Zarządzania, Dowodzenia Akademii Sztuki Wojennej*, 2(18), 14.

Szumiejko, F., Wdowikowski, M., Hański, A., Kańska, A., Mielke, M., Aneszko, J., Jankowska, I., and Wydrych, M., 2015, *Vademecum–Hydrological Observations and Maesurements*, Institute of Meteorology and Water Management, Poland.

The Council of Ministers, 1972, *Resolution on Connection of National Hydrological and Meteorological Institute with the Institute of Water Management*, 338/72, December 30th.

The Council of Ministers, 2010, *On Granting the Status of the National Research Institute*, The Institute of Meteorology and Water Management, September 13th, Warsaw, Poland.

UNFCCC, 2016, Decision 1/CP.21 Adoption of the Paris Agreement, *FCCC/CP/2015/10/Add.1 Report of the Conference of the Parties on its Twenty-First Session*, Paris, France, 30 November–13 December 2015, pp. 21–36.

3 Flood Observation Uncertainty

Jean-François Vuillaume and Haireti Alifu

CONTENTS

3.1 INTRODUCTION

Floods can be observed by a variety of tools: ground-based platforms, aircraft and satellite remote sensing, and even with cameras on stationary high-altitude balloons. Moreover, there are varied monitoring targets such as discharge, water flooded extent, water depth, and flood dynamics. However, each method has uncertainty. Also, floods present several categories such as flash flood (associated with heavy rainfall), coastal flood (associated with cyclone storm surge), groundwater flooding (associated with the rising of groundwater), and more classic river flooding due to over-banking and floodplain systems (Opolot, 2013). The time and length scales of flood widely depend on the river size, orientation/shape, slope, geology, vegetation (seasonal period), and land use. Flood duration exhibits a time scale of a day up to several months and a change in water space scale of 10–100 m. However, the current observation system is not able to capture such variability globally due to the limited coverage of the path, swath, and revisiting time of satellites (Bates et al., 2014).

Rivers are complex systems often associated with a river-floodplain system such as the Grande in Bolivia illustrated in Figure 3.1. It consists of a main channel with multiple meanders that is either active or abandoned. The river system also consists of a levee, an active sand deposit, and floodplain wetland. This geometry illustrates a dynamic view of the river-flood system.

Floods propagate along rivers and are characterized by discharge, water height, duration, and floodplain flooding. Flood observations require either a local or large global scale that can be observed from space/airborne in order to represent the river system at different scales. Large river basin floods will require a large extent of observation, such as microwave or optical remote sensing. However, large river airborne monitoring is still necessary to obtain continuous observation at high resolution that can report flood dynamics. The monitoring of flood extent using remote sensing techniques has great advantages (cost-effective tool, multi-temporal observation of large area, able to monitor the flood in remote areas) compared to a conventional hydrological monitoring system that has limited use in flood forecasting, mapping, and emergency response (i.e., ground observations). Therefore, similar to the Global Precipitation Monitoring mission (Hou et al., 2014), global water

DOI: 10.1201/9780429463938-5

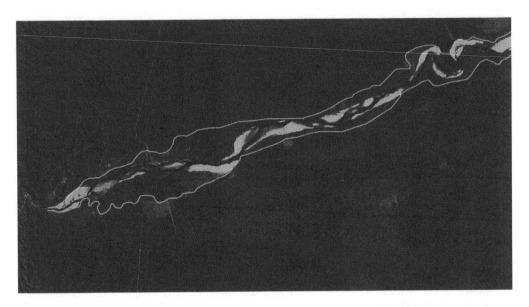

FIGURE 3.1 Tri-Decadal Global LANDSAT Orthorectified Pan-Sharpened ETM Mosaic (1999–2003) of the Grande (Amazon basin) in Bolivia. The floodplain boundary is shown as a line. (Source: US Geological Survey and downloaded from their Global Visualization Viewer [GLOVIS].)

mapping such as the Constellation of Small Satellites for Mediterranean basin Observation (Cosmo-Sky, Pulvirenti, et al., 2011) or Satellite Pour l'Observation de la Terre (SPOT) constellations (Yésou et al., 2003) become more and more needed. In addition, a smooth satellite data merging system for fast processing (Irwin et al., 2017; Chen et al., 2018) is also very useful.

The capability to monitor floods is strongly linked to the level of development of a country due to the cost of the equipment and its maintenance. Several flood national agencies from developing countries face institutional, technical, and financial challenges for flood observation (Ekeu-wei and Blackburn, 2018) despite the fact that several worldwide observation data are available free of charge. Moreover, it requires support from global multi-agency networks and the United Nations for flood monitoring and prevention activity.

Remote sensing provides considerable support in data collection, in remote, inaccessible, and sparse locations in particular when provided as an open-access supply. For example, the JRC-ECMWF (Joint Research Center of the EU European Centre for Medium-Range Weather Forecasts) with Global Flood Awareness System (GLOFAS, www.globalfloods.eu/) or the Global Flood Initiative (GFI) provides emergency observations support that can benefit every country.

However, the lack of ground measurement reduces the potential of quantitative estimation of discharge that can be critical for modeling and projection. Indeed, discharge observation stations provide calibration for flood depth and flood extension mapping. Moreover, it is crucial for spared data regions (i.e., the Congo river basin, the second largest in the world; Alsdorf et al., 2016). Table 3.1 summarizes the broad sensors used to monitor floods (Chen et al., 2018). It covers the detection and monitoring of the maximum flood extent mapping, flood risk assessment, damage assessment, and flood dynamics analysis.

Optical sensors such as LANDSAT 5/7/8, Moderate-Resolution Imaging Spectroradiometer (MODIS), SPOT, or IKONOS permit a clear distinction between water bodies and other land types due to the different reflectivity of water. The early attempt (70s) started with the launch of the Earth Resources Technology Satellite, ERTS (Rango and Salamonson, 1974). Importantly, optical observations can only be acquired during the daytime and are strongly affected by clouds which often occurred during the heavy weather associate flood event (i.e., cyclone). Moreover, high temporal resolution flood observation can provide inundation dynamics information (Martinis and Rieke, 2015).

TABLE 3.1

Temporal/Spatial Resolution of Sensors Used for Different Purposes in Flood Monitoring (Chen et al., 2018)

Type	Satellite/Sensor	Temporal/ Spatial Resolution	Use
Space-borne	RADARSAT 1/2 (C-band SAR sensor)	24 days 3–100 m	Map of the flood monitor extent (Zhou et al., 2000; Bonn and Dixon, 2005)
	TERRA/AQUA (MODIS)	Daily 250–1,000 m	MODIS-based flood detection, mapping, and measurements (Brakenridge and Anderson, 2006)
	TERRASAR-X (X-band SAR sensor)	11 days 0.25–40 m	Flood detection (Martinis et al., 2009; Mason et al., 2010)
	HJ 1A/1B (CCD sensors)	4–31 days 30–300 m	Waterbody mapping (Lu et al., 2011)
	LANDSAT series (5/7/8)	16 days 30–120 m	Delineate the maximum flood extent (Wang, 2004; UNITAR/UNOSAT, 2015) and flood risk analysis (Demirkesen et al., 2007)
	EO-1 (ALI, Hyperion, LAC)	16 days 10–30 m	Flood detection and monitoring (Ip et al., 2006; Rudorff et al., 2009)
	QUICKBIRD (optical sensor)	1–20 days 0.6–2.88 m	Flood mapping and risk assessment (Shamaoma et al., 2006; Hadjimitsis, 2007)
	IKONOS-2 (optical sens or)	1–14 days 0.82–3.2 m	Assist flood risk and flood damage assessment (Sande et al., 2003)
	SPOT 5 (optical sensor)	26 days 2.5–20 m	Flood extent mapping (Yésou et al., 2003)
Airborne	Airborne radar (microwave sensors)	Anytime 0.26 m to several meters	Flood mapping and analysis (Chandran et al., 2006; Pope et al., 1992)
	Unmanned aerial vehicle (optical sensor)	Anytime 0.05–0.2 m	Application in flood disasters (Li et al., 2010)
	Hot-air balloon (optical sensor /video camera)	Anytime	Low altitude observations of floods (Hariyanto et al., 2009)
Ground	Water level gauge		Standard instrument of hydrologic observation station
	Tachometer		Standard instrument of hydrologic observation station
	Rain and stream gauge		Standard instrument of hydrologic observation station

Source: Chen et al., 2018.

Due to cloudy weather conditions during flood events, active radars are practical tools that are not affected by weather conditions. However, it is necessary that the satellite is passing at the location of interest at the time of the flood. Compared to the optical sensors, Synthetic Aperture Radar (SAR) has the advantage that it can acquire data under any weather condition (day, night, or cloudy).

Optical sensors are unable to detect inundation areas covered by vegetation because most of the light was reflected by vegetation. However, SAR can penetrate the vegetation cover and detect flooded water surfaces. Nevertheless, SAR presents some limitations because of mixed signals (partially saturated soil or mud-flows, tree interaction) caused by the signal reflected by floods. The accuracy depends on the operator's subjective interpretation of the backscatter threshold that represents flooded areas, but canopy-covered, partially saturated, or mud-flooded areas or a region with strong winds can distort the signal backscattering (Matgen et al., 2007). Therefore, it can result in an underestimation of flooding extent. However, it is permitted to observe the water retreat during long periods with regular updates.

Major issues arise to observe peak floods and to map flood extent. It is needed to support emergency, damage assessment, or future flood prevention planning. Inundation area detection in urban and vegetated areas is limited by resolution and penetration capacity. Moreover, uncertainty in the mapping ranges from null (completely uncertain because not seen), subject to interpretation (depending on the skill of the interpreter and ground observation), and high accuracy (range of the spatial resolution or the sensor). Considering the frequency of revisiting time, it is most likely that only a few visible pictures will be usable: all Earth observation satellite systems have time constraints for the acquiring of mapping and uplinking the acquisition to the satellite. Optical satellites for monitoring of flooding have the shortest revisit time of one day (Kumar and Reshmidevi, 2013) while in contrast, the shortest revisit time for SAR is six days (Prigent et al., 2016) and < 12 hours for the COSMO-SkyMed constellation (Grandoni et al., 2014).

Table 3.2 gives a general overview of available remote sensing methods with their resolution, swath, temporal sampling, and limitation.

The timing of observation acquisition is critical for the uncertainty associated with floods. The classical use of the post-flood wrack mark (debris disposed of by the flood) cannot provide a fully

TABLE 3.2

Optical, SAR, Passive Microwave, UAS/UAV, and Field Mission Associated Techniques to Map Water Surfaces and Their Main Characteristics

Technique	Spatial Resolution (m)	Swath (km)	Temporal Sampling	Contamination	Remarks
VIS/NIR/SWIR	Down to 1 m	11.3–2,330	Down to daily if cloud-free	Clouds, vegetation	Open water only Regional to global scale
Active microwave SAR	1–500	10–100	Up to 6 days with sentinel 1A/B	Vegetation (to some extent)	Regional applications mostly so far
Passive microwave	~10,000	Wide	Daily	Vegetation (to some extent)	Detection of open water as well as wetlands
Aircraft-VIS/LIDAR	0.1–5	2	Real-time, minutes	Vegetation (to some extent)	Expensive survey
UAS UAV	0.04–0.05 ~0.1	0.1–1	0.8–1.2s	Clouds, vegetation	Batteries life time per survey (30 min) (Opolot 2013; Pakoksung and Takagi, 2016; Tamminga, 2016)
Field mission	< 0.01	0.01		Vegetation (to some extent)	

Source: modified from Prigent et al., 2016.

accurate measurement because the mark can be deposited during post flooded retrieval wave (Bates et al., 2014). The accuracy of the flood extension is based on the spatial-temporal resolution of the sensor and the flood mapping methods. It can range from field survey (very high spatial-temporal resolution and accuracy with limited area), aircraft, and space-borne. For example, a case study of the Poyang Lake (China) by Wu and Liu (2015) based on the combination of DEM (5 m resolution) with MODIS reported a water level fluctuation root mean square error (RMSE) from 0.79 m to 1.09 m, 10% of the maximum variability of the lake (Wu and Liu, 2015).

Depth of flooding is another important observation parameter that can be acquired indirectly with altimetry radar such as ENVISAT, JASON, or the future SWOT (Durand et al., 2010) with centimeter vertical accuracy. It required digital elevation methods (DEM) and altimetry measurement. The DEM map is superposed with the flood extent map introduced previously (Pinel et al., 2015). However, the uncertainty in the DEM constrains the final uncertainty of the flood mapping (Jung and Jasinski, 2015). Also, altimetry measurements are used to calibrate flood modeling that uses DEM as input.

Some recent flood observation technology development focuses on increasing the spatial-temporal resolution of observation, such as river with a smaller width of 100 m and below (Surface Water Ocean Topography Mission) planned for 2020 (Biancamaria et al., 2016), the use of unmanned aircraft systems (UAS) and unmanned aircraft vehicles (UAV) that can fly under cloud level with continuous observation at very high temporal resolution and urban flood observation that required very high resolution (~10 cm). On the other side, some research investigated the use of a "citizen contributions to science" approach (picture of flooded area, local camera record of flow) that can constitute crucial ground data for validation in complement to quantitative discharge stations (Assumpção et al., 2018).

Finally, remote sensing methods for flood observations are subject to uncertainty, which is crucial to understand. Indeed, detection methods affect the accuracy of the flood extension, the water height, and the flood dynamics which condition prevention and rescue plans. Furthermore, post-processing methods can be applied to improving the flooding area using past flooding extension data or by combining the result derived from observation and numerical simulation. The result from Stephens et al. (2012) indicates that, in the absence of validation data, integrating of LiDAR-DEM with ERS-2 SAR image (from 2006) provided considerable uncertainty, with elevation errors showing significant spatial correlation. Reducing uncertainty can be achieved by the combination of multi-sensors and modeling to provide better flood hazard maps (Giustarini et al., 2015). However, it required combining heterogeneous geospatial data (Chen et al., 2015) that required new approaches.

We had to make a selective sorting of the important information relevant to flood observation uncertainty as well as the articles that we considered relevant. For this reason, we have carried out bibliographical research as thoroughly as possible. This cannot be considered complete due to the large amount of information available on the subject. Despite this, not all quality articles on the subject have probably been identified. The aim of this chapter is to gather dispersed information and references on flood observation uncertainty.

3.2 CRUCIAL CONSIDERATION ON FLOOD UNCERTAINTY

Uncertainty related to flood observation varies considering the purpose of the observations, i.e., emergency situation, damage estimation or observation, etc. Flood observation may face emergencies and require fast processing at the cost of accuracy or constraints of data availability, algorithm efficiency, and satellite information selection.

However, it may also benefit the use of expensive aircraft surveys/UAV/UAS to get higher accuracy of flood observation. Moreover, the acquisition time may be crucial to support modeling flood teams in calibration data for further flood forecast modeling (Grimaldi et al., 2016; Schumann et al., 2009). Based on the previous discussion, timing of the flood and timing of observation, surface extension, sensor resolution, cloud contamination, and vegetation cover have also to be considered when evaluating the uncertainty of the observations.

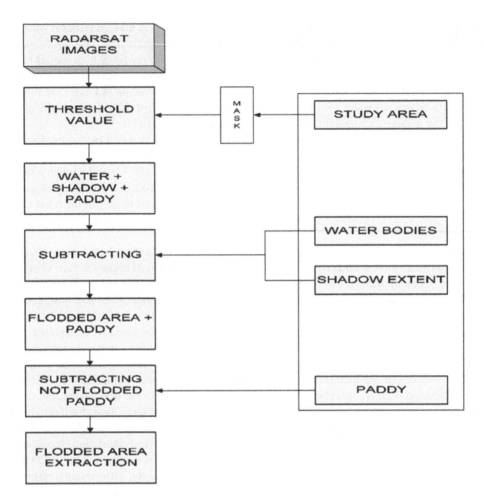

FIGURE 3.2 Flowchart for flooded area extraction from RADARSAT during-flood image. (Source: Lawal et al., 2014.)

Flood observation uncertainties are present along with the flow of flood mapping extent and depth estimation acquisition and processing (Figures 3.2 and 3.3) such as the raw remote sensing data for the flood extension, the DEM for the stage uncertainty, or the interpretation of the water body. For instance, SAR is affected by speckle noise due to the extremely rough surface. Errors from geo-referencing required projection on a flat plain with a Geographical Information System (GIS), and ortho-rectification corrects systematic sensor errors, as well as platform geometry errors when sensors not pointing at the nadir location and unavoidable uncertainty may occur.

Overall, uncertainties are present in SAR backscatter measurements, ground perturbation (wind, tree, building masking water, terrain geometry), pre-processing (geo-referencing, ortho-rectification, speckle removal from SAR), spatial averaging, spatial resolution, temporal averaging, retrieval of flood extent from thresholds, image classification, retrieval water levels, classified images, and DEM for classification between flooded and non-flooded areas.

3.3 DISCHARGE OBSERVATIONS AND MEASUREMENTS

A classical flood observation-gauging system usually consists of a height measurement from an elevation reference system. It is one of the early attempts to estimate flood magnitude. Figure 3.4 illustrates a statue located in Paris under a bridge that became a "citizen" qualitative measurement

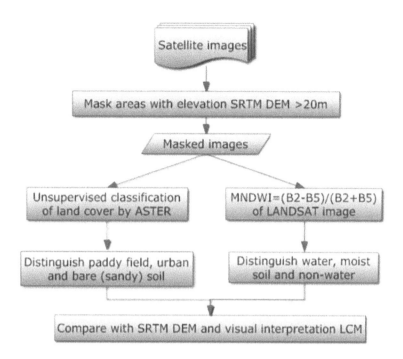

FIGURE 3.3 Flow chart of the process of DEM and ASTER/LANDSAT masking. (Source: Ho et al., 2010.)

of flood since 1856. Figure 3.5 illustrates manual quantitative water level retrieval during flooding in Bangladesh. A more advanced system can use an automatic recording system (Doppler profiler). However, the availability of data remained a challenge, and the Global Monitoring Discharge on the Ground (GRDC) is trying to close this gap (Fekete et al., 2012). The GRDC constitutes an important dataset for flood observations.

A cross-section profile of discharge permits the construction of typical rating curves from height measurement and therefore derived discharge measurement. Uncertainty in the estimation of the

FIGURE 3.4 June 2016 Paris inundation indicator, the "Zuave." (Source: Wikimedia commons [https://commons.wikimedia.org/wiki/].)

FIGURE 3.5 Team member Manual George takes a water level reading. (Source: photo courtesy of Dr. Khan; https://sites.nationalacademies.org/pga/PEER/PEERscience/PGA_146755.)

water height as well as the profile cross-section occurs, resulting in uncertainty in the river flow estimation. Discharge is often difficult to acquire during flooding due to adverse meteorological conditions and the risk to destroy the measurement station. Furthermore, accurate stage height observation is more critical for smaller rivers (with less discharge) than large rivers. However, the knowledge of the rating curve can be used when combined with satellite remote sensing (altimetry measurement) to deduce discharge from space. "Uncertainty analysis indicates that existing satellite-based sensors can measure water-surface width, water-surface elevation, and potentially the surface velocity of rivers with accuracies sufficient to provide estimates of discharge with average uncertainty of less than 20%" (Bjerklie et al., 2003). Figure 3.6 illustrates the uncertainty obtained when discharge is extrapolated to high flow (flood event) from past records leading to underestimation or overestimation of discharge.

Large-scale particle image velocimetry (LSPIV) is becoming more common in river research. For example, a case of river flash flood registered a discharge of 300 m³ s⁻¹ concluding that the LSPIV velocities throughout the river cross-section were found to be in good agreement (± 10%) with concurrent measurements by Doppler profiler (Le Coz et al., 2010). For higher discharges ranging above from 300 up to 2,500 m³ s−1, LSPIV discharges show acceptable agreement

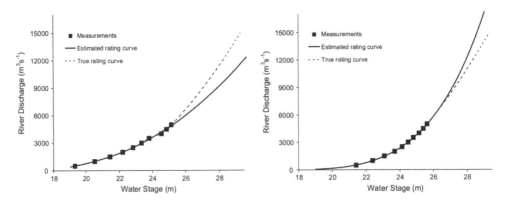

FIGURE 3.6 Example of how errors induced by the extrapolation of the rating curve often lead to either a systematic underestimation (left panel) or overestimation (right panel). (Source: Di Baldassarre et al., 2012.)

(< 20%) with the rating curve. However, detrimental image conditions or flow unsteadiness during the image sampling period can lead to larger deviations ranging from 30–80%. Similar systems can be used with LSPIV cameras mounted to a UAV, results from Lewis et al. (2018) indicate that the velocity difference between fixed and UAS cameras are less than 10%. Moreover, differences between LSPIV-derived and reference discharge values are generally less than 20%. From another perspective, public monitoring of flow indicates that online video (i.e., YouTube, DailyMotion) past flood records from citizens showed the difference in time-averaged longitudinal velocity was less than 5% compared to the original camera full resolution (Le Boursicaud et al., 2016). This accuracy suggests that flood flow records based on crowd-sourced citizens can provide useful information on flood discharge events despite their "random" changing location.

3.4 FLOOD EXTENT REMOTE SENSING

The remote sensing technique has been a crucial technology for flood extent monitoring since the 70s, despite its discontinuity and uncertainty. Figure 3.7 illustrates a recent case of flood mapping in Bangladesh using both SENTINEL-1 and LANDSAT-8 optical sensors for flood extent mapping for pr- and post-flood events of July 27, 2015 (provided by UNITAR/UNOSAT).

While the previous case in Bangladesh illustrates an example of space-borne flood extent mapping, Table 3.3 describes some of the most important sensors used for flood monitoring including visible, passive, and active microwave sensors.

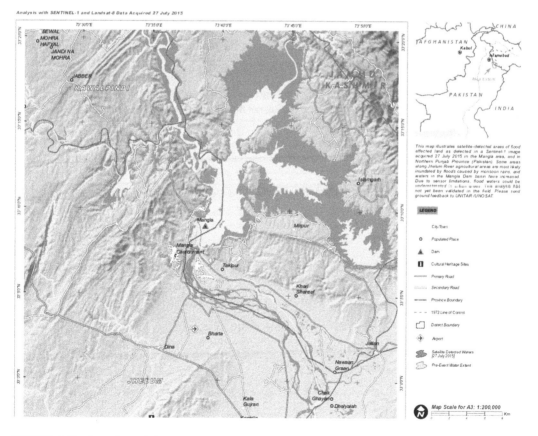

FIGURE 3.7 Overview of floodwaters in Magla area, and northern Punjab (Pakistan) analyzed with SENTINEL-a and Landsat-8 data acquired July 27, 2015. (Source: UNITAR/UNOSAT, 2015.)

TABLE 3.3
Some of the Important Satellites and Sensors Used for Flood Monitoring with Their Special Correlation and Frequency of Covering

Sensor	Satellite/ Aircraft	Special Resolution (m)	Revisit interval (day)	Coverage Swath (km)	Cloud Penetration	References
LANDSAT TM	Landsat 4/5/7/8	30	16	185	Thin clouds	UNITAR/UNOSAT, 2015
IRS LISS-3	IRS 1C/1D	23.5	24	142	Thin clouds	Pandiammal et al., 2015
SPOT 6/7	SPOT	1.5–6	1	60	Thin clouds	Blasco et al., 1992
AVHRR	NOAA	~1,100	1	2,900	Thin clouds	Lacava et al., 2018
MODIS	Terra	250	1	2,330	Thin clouds	Lacava et al., 2018
	IKONOS	4			Thin clouds	Sande et al., 2003
Airborne scanner	Aircraft	5	–		Thin clouds	Yun, 2016
Panchromatic aerial photos/ camera ADS40	Aircraft	15 cm–1 m			Thin clouds	Klemas, 2015

Source: modified from Kumar and Reshmidevi, 2013.

3.4.1 Visible Observations

The availability of multi-temporal satellite data allows the monitoring of floods over large areas at low cost. The identification and mapping of water surfaces have direct use of optical remote sensing because of the spectral reflectance property difference between water and other land types. Figure 3.8 indicates the reflectance difference between water bodies with vegetation and soil. It is the main feature of visible/near-infrared mapping of the water body: water appears as black for wavelength > 0.8 m.

Optical retrieval is the most straightforward remote sensing method, but only permitted during daylight without clouds. Water absorbs most of the near- and mid-infrared (NIR) and middle infrared

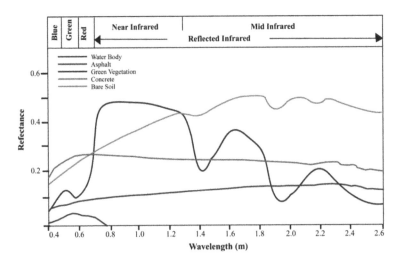

FIGURE 3.8 Spectral signatures for dry bare soil, green vegetation, clear water, asphalt, and concrete. (Source: Rahman and Di, 2017 adapted from Smith, 2001 and Herold et al., 2003.)

TABLE 3.4

Satellite Mission Featuring Optical/NIR Sensor

Satellite Data	Agency	Years of Operation	Spatial Resolution (m)	Coverage Swath (km)	Revisit Interval (days)	Cost/km² (US $)
WV 2	DIGITAL	2009–present	1.84	16.4	1.1	$17.5 (min 25 km²) archive
WV 3	GLOBE	2014–present	1.24			
WV 4	GLOBE	2016–present	1.24			$27.5 (min 100 km²) new
SPOT 5	AIRBUS	2002–present	2.5–5	60	5–26	$1.20 (min 500 km²) archive
SPOT 6	defence and	2012–present	1.5–6			
SPOT 7	space	2014–present	1.5–6			$1.75 (min 100 km²) new
IKONOS-2	GeoEye Company of Virginia NASA	2000–2015	1–4	11.3	3	$10 (min 25 km²) archive
Landsat 5	NASA	1984–2013	15–30	185	16	Free
Landsat 7		1999–present	15–30			
Landsat 8		2013–present	15–30			
MODIS Terra	NASA	1999–present	250–1,000	2,330*10	0.5–1	Free
MODIS Aqua		2002–present				
IRS LISS-3	IRS 1C/1D	2003–present	23.5	142	1	Free

Source: modified from Grimaldi et al., 2016 and Kumar and Reshmidevi, 2013.

wavelengths. Therefore, multi-spectral visible to infrared imaging is very useful for flood mapping. Table 3.4 summarizes the satellite missions featuring optical sensors such as WV2, WV3, SPOT 5, 6, 7, Landsat, and MODIS with operation period, ground resolution, revisit interval, and cost.

SPOT, IKONOS, and World-View provide high spatial resolution for flood monitoring (~1 m). Unfortunately, commercial optical satellite images are expensive and limited by persistent cloud covers. For example, World-View (WV-2) study over the Nørreå river valley in Denmark on a 3 km river section for 12 flood extent maps derived indicated an overall accuracy ranging from 77% to 95% accuracy (Malinowski et al., 2015).

The 30 m meter resolution LANDSAT ETM+ can be used to simply map flooded areas. However, its low temporal frequency of 16 days makes it difficult to map flooded areas. Therefore, optical observation presents large uncertainty for the estimation of the maximum flood extension. Moreover, it requires extended post-treatment to extrapolate flooding areas at the time of interest based on the latest observation data. Nevertheless, it presents a large spatial coverage suitable for the large flood system.

A large number of previous studies focused on more regional flood monitoring using various types of remote sensing images. For instance, MODIS (250 m) is seen as the primary input for the Dartmouth Flood Observatory (http://floodobservatory.colorado.edu/ (Adhikari et al., 2010). Two MODIS sensors are currently operational aboard the Terra and Aqua satellites (Doxaran et al., 2009). As an example, a study reported the detection of a flood event in December 2013 in Basilicata and Puglia regions (Southern Italy) based on various optical satellite data (Suomi National

Polar-Orbiting Partnership [SNPP] Visible Infrared Imaging [VIIRS], MODIS, and Landsat ETM+) and compared the results with each other, i.e., MODIS and Landsat onDecember 5, 2013 (Lacava et al., 2018). As a result, this study indicated that integration of VIIRS and MODIS (250–375 m spatial resolution) allowed for the discrimination of a flood area extent up to 80 km², lower than that potentially detectable by using a high spatial resolution sensor like the ETM+ (about 24% greater). Moreover, the use of airborne observation (equipped with a high-resolution camera) can cover the shortage in resolution from space-borne remote sensing.

Aircraft equipped with state-of-the-art high-resolution digital cameras, like the ADS40 (Klemas, 2015)m can provide very high spatial resolution of flood extension (from 0.15 m up to 1 m). Moreover, aerial reconnaissance is effective for determining the spatial extent of coastal and river flooding but requires specific missions and that is cost-expensive compared to the large amount of free available multi-mission satellite images.

3.4.2 PASSIVE MICROWAVE IMAGERY

The passive microwave imagery has a similar concept to thermal remote sensing; a passive microwave sensor detects the naturally emitted microwave energy within its field of view. Thus, passive microwave sensors can distinguish water from land because all objects such as land/water emit different magnitudes of microwave energy. The passive microwave imagery presents a large special coverage 20–100 km due to its large angular beam and therefore has to be limited to large catchments because of low resolution. Nevertheless, data has a higher temporal resolution (daily) and is free of charge. The Global Disaster Alerting Coordination System (GDACS) flood detection website provides a daily flood detection map and uses passive microwave remote sensing (i.e., AMSR-E and TRMM sensors): www.gdacs.org/flooddetection/ (DeGroeve, 2010). Table 3.5 illustrated two examples of passive microwave missions provided by NASA/JAXA.

3.4.3 ACTIVE MICROWAVE IMAGERY

The Synthetic Aperture Radar (SAR) is the main active microwave imagery reliable for river width < 1 km (Schwatke et al., 2015). The SAR-derived observations are the most efficient methods for both extension and depth of flood. Data is cloud-free, but is not directly retrieved and requires post-treatment compared to an optical one. SAR-derived observations are subject to interpretation error because of human/automatic detection based on the classification method/ground knowledge. Table 3.6 shows detailed information on SAR satellite imaging.

Figure 3.9 illustrates the radar backscatter mechanism for short (e.g., C and X) and long (e.g., P and L) wavelengths for various surface cover under non-flood and flooded conditions.

Finally, Figure 3.10 illustrates the spectral domain of typical satellite-based SAR (TerraSAR-X, COSMO-SkyMed, RADARSAT, SENTINEL, ALOS-2) including their band type (X, C, S, and

---

TABLE 3.5
Passive Microwave Examples

Mission/Satellite	Agency	Years of Operation	Ground Resolution (km)	Revisit (days)	Cost/km2 (US $)
TRMM-TMI	NASA/JAXA	1997–2014	5 × 7 up to 72 × 43	1 – 2	Free
AMSR-E on board of AQUA	NASA	2002	6 × 4 up to 74 × 43	1 – 2	Free

Source: Grimaldi et al., 2016.

TABLE 3.6
Satellite Missions Featuring SAR

Satellite Mission	Agency	Years of Operation	Spatial Resolution (m)	Revisit Interval (Days)	Polarization	Wavelength	Cost per Scene (US $)
ERS-1	ESA	1991–2000	25	35	VV (single)	C	Free
ERS-2	ESA	1995–2011	25	35	VV (single)	C	Free
JERS 1	JAXA	1992–1998	18	44	HH (single)	L	Free
RADARSAT-1	CSA	1995–2013	8–100	24	HH (single)	C	1,155 –3,465
RADARSAT-2	CSA	2007–present	3–100	24	Full	C	2,770–6,000
ENVISAT-ASAR	ESA	2002–2012	30–1000	35	Single or dual	C	Free
ALOS-PALSAR	JAXA	2006–2012	10–100	46	Single or dual	L	44/free
ALOS-PALSAR 2	JAXA	2014–present	3–100	14	Single or dual	L	1,335–4,450 archive 2,670–5,785 new
COSMO-SkyMed	ASI	2007–present	15–100	4–16	Single or dual	X	720–5,310
TerraSAR-X	DLR	2007–present	1–16	11	Full	X	1,216–5,970
TANDEM-X	DLR	2010–present	3	11	Full	X	1,216– 5,970
KOMPSAT-5	KARI	2013–present	1–20	28	Full	X	800–3,300
Sentinel 1A Sentinel 1B	ESA	2014–present 2016–present	5–100	6–12	Full	C	Free
SWOT*	NASA-CNES- CSA ASC UK SPACE AGENCY	2020–		21	Full	X	Free*

*Not launched yet

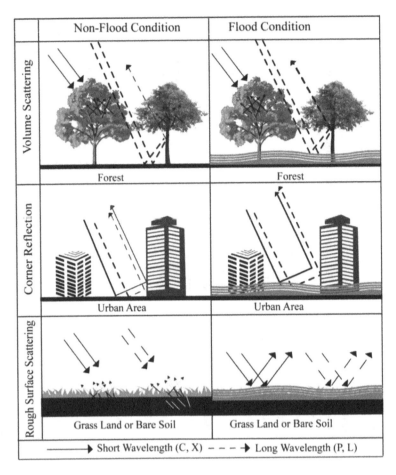

FIGURE 3.9 Schematic illustration of radar backscatter mechanism for short (e.g., C, X) and long (e.g., P, L) wavelengths for various surface covers under non-flood and flooded conditions. (Source: Rahman and Di, 2017.)

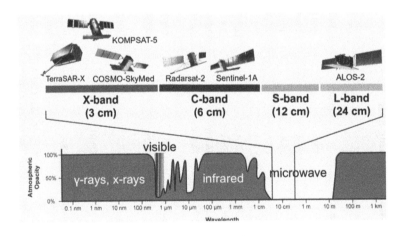

FIGURE 3.10 Atmospheric windows and current SAR missions. (Yun et al., 2016.)

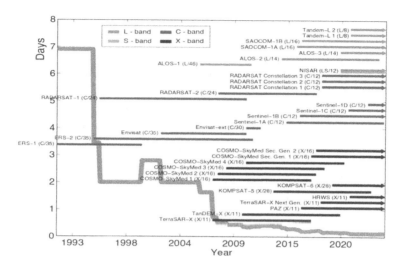

FIGURE 3.11 Evolution of the SAR Data Acquisition Latency evolution for 1991–2022. Grayscale band illustrates the related missions available. (Yun, 2016.)

L). Moreover, Figure 3.11 highlights the strong decrease of satellite data acquisition latency from ALOS L-band (6–7 days), RADARSAT C-band (5–6 days), Sentinel C-band (4–5 days), COSMO-Sky X-band (2–3 days), and TerraSAR/TanDEM/KOMPSAT X-band (1–2 days) in the recent years.

However, flood mapping using the SAR technique involves inaccuracies from flood under canopy, soil moisture, and wind that are reflected in the choice of backscatter threshold coefficient. Overall, the classification and conversion of backscatter coefficient into the type of area is subject to uncertainty and careful calibration when available from direct observation or optical remote sensing observation. Figures 3.12–3.13 illustrate the uncertainty associated with threshold backscatter values. It allows the differentiation between flooded and potentially flooded using T, the temperature of the threshold value based on the backscatter coefficient (Figure 3.12). Figure 3.13 illustrates the potential distribution and temporal change of backscatter coefficients during flooding.

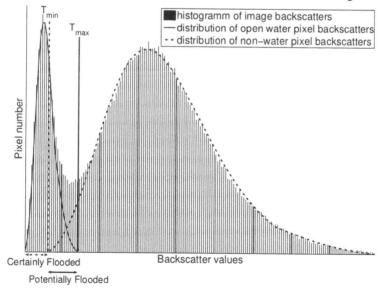

FIGURE 3.12 Illustration of overlapping of the water bodies and other land-use type distributions on the histogram of a flood SAR image and proposed threshold values used for flood extent mapping. (Source: Hostache et al., 2009.)

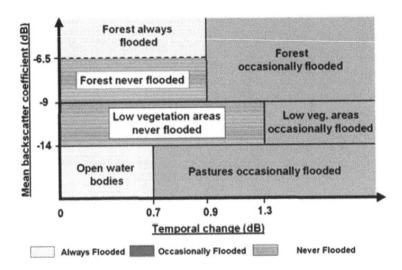

FIGURE 3.13   Summary of thresholds used by the multi-temporal classifier scheme for flood dynamics and vegetation mapping, on absolute change and absolute values. (Source: Martinez and Le Toan, 2007.)

All SAR presents great potential in detecting flooding beneath vegetation even with X-band:

However, the use of fixed empirical threshold values for automating the detection of partially submerged vegetation of various types is nearly impossible due to the dependency of the SAR signal return over partially flooded vegetation on various system and environmental parameters.

(Martinis and Rieke, 2015)

Thus, the most critical step is the selection of pre-flood data, especially for classes of high backscatter variability such as crop-land (Du et al., 2018). The results obtained on five watersheds in the United States, the 30-m L-band retrievals showed a spatial accuracy with a commission error of 31.46% and an omission error of 30.20% when compared to independent water maps derived from Landsat-8 imagery. Moreover, multi-satellite combination methods (i.e., combined satellite information from SMAP brightness temperature, AMSR2, and Landsat) can improve the observation of the global surface water inundation dynamics (Tong et al., 2018). The Soil Moisture Active Passive (SMAP) surface water data provides one to three daily global repeat observations.

There are several directions of research focused on reducing uncertainty produced by the SAR in terms of time due to the high revisit time and the mismatch between flood and observation as well as flooded vegetation SAR treatment to improve the flooded extent mapping under a canopy that are affected by backscattering from vegetation while being flooded area. First, it should be noted that obviously the backscatter coefficient varies during flooding. Secondly, a range of backscatter coefficients can represent similar conditions as illustrated by Figures 3.11 and 3.12. Finally, the processing of multiple SAR images to obtain continuous data at different time and space scales constitutes a challenge and an alternative/complement to satellite constellations such as CosmoSky (four satellites operated within a constellation). It required the development of a processing platform and pre-processing data for integrating them into a useful product. Early attempts at solving this issue can be found in the literature, using SAR images with a frequency of up to 12 hours compared to the usual six days (Grandoni et al., 2014).

3.5   FLOOD STAGE LEVEL REMOTE SENSING

As discussed in the previous section, river systems are composed of two main parts: the river and the floodplain. While rivers can be observed on a regular basis, floodplain flooding occurs only at one sporadic time. The flood level can be obtained by indirect methods using radar altimeters, light detection and ranging (LiDAR), and SAR remote sensing method. The space-borne radar altimeter is able to measure the water surface altimetry at high vertical accuracy but usually with low horizontal coverage (ENVISAT, JASON). In addition to the surface flood elevation, flood depth estimation requires the knowledge of the elevation of the terrain before flooding to deduct the flood level. Therefore, the accuracy of DEMs is critical data as it permits the computation of flooded stages when combined with SAR images.

Shuttle Radar Topography Mission (SRTM) is the most main DEM data used for hydrological studies. It provides data from 56S to 60N at 1–3 arc-sec resolution (~30–90 m) and an average vertical accuracy of 6 m (Farr et al., 2007). Error in the accuracy is mainly caused by speckle noise, stipple noise, vegetation, and buildings. Nevertheless, the accuracy of the floodplain currently provided by the SRTM can be larger than the flood signal (Bates et al., 2014) and require post-processing treatment for correction. Moreover, channel depths are tied closely to the local amplitude of the passing main river flood wave. A study on the floodplain channel morphology and networks of the middle Amazon River pointed out that 96% of the floodplain channels are not wide enough to be represented well, or at all, in the 90 m resolution of SRTM (Trigg et al., 2012). Table 3.7 provides a list of missions dedicated to ground surface elevation estimation that exhibits vertical error from 5 m (AW3D30) up to 26–30m for GMTED2010 with large spatial resolution from 30 m (SRTM, ASTER GEM, and AW3D30) to 1,000 m (ACE 2 GDEM and GTOPO 30).

Radar altimeter presents a long record of the observation mission such as TOPEX/Poseidon (1992–2005), ERS-2 (1995–2011), ENVISAT (2002–2012), Jason-1 (2001–2013), 2 (2008), and 3 (January 2016), Sentinel-3A-B-C (2016, 2017, and planned 2024), and SWOT (planned 2022) with a revisit time of 10–35 days. The missions present a global coverage with high accuracy; Jason 3 and Sentinel 3 achieved a 0.03 m vertical uncertainty but with a revisit time of 10–35 days. The vertical extension of space-borne radar altimetry shall be largely improved with the future SWOT mission. Table 3.8 lists the main characteristics of the past and current altimetry mission.

Finally, Table 3.9 provides a list of study examples that estimate the uncertainty of flood extent. The RMSE obtained vary from 0.1 to 2 m depending on the methods used and the validation data. The table highlights the importance of validation data and the low error obtained with LiDAR DEM.

TABLE 3.7
Open Source Digital Elevation Models Properties and Case Studies

DEM	Spatial Resolution (m)	Vertical Error (m)	Case Study	Reference
SRTM	30 and 90 m	±16	Damoda River, India	Sanyal et al, 2013Rodriguez et al., 2006
ASTER GDEM	30	±25	Lake Tana, Ethiopia	Tarekegn et al., 2010Tachikawa et al., 2011
ACE 2 GDEM	1,000	>10	Balkan Peninsula, Croatia	Varga and Bašić, 2015
GTOPO 30	1,000	9–30	Balkan Peninsula, Croatia	Varga and Bašić, 2015
AW3D30	30	±5	Sindh and Balochsitan, Pakisatan	Tadono et al., 2014Jilani et al., 2017
GMTED2010	250	26–30	Shikoku, Japan	Danielson and Gesch, 2011Pakoksung and Takagi 2016

Source: from Ekeu-wei and Blackburn, 2018.

TABLE 3.8
Altimetry Mission Main Characteristics

S/N	Mission	Ground Footprint (m)	Revisit Time (days)	Operation Timeline	Accuracy (m)	References
1	TOPEX/ Poseidon	~600	9.9	1993–2003	0.35	Frappart et al., 2006
2	ERS-1	~5,000	35	1991–2000	N/A	Da Silva et al., 2010
3	ERS-2	~400	35	1995–2003	0.55	Frappart et al., 2006
4	ENVISAT	~400	35	2002–2012	0.28	Frappart et al., 2006
5	Jason-1	~300	10	2002–2009	1.07	Jarihani et al., 2015
6	ICE Sat/GLAS	~70	–	2003–2009	0.10	Urban and Schutz, 2008
7	Cryosat-2	~300	369	2010*	<SRTM (30 m)	Schneider et al., 2016
8	Jason-2	~300	10	2008*	0.28	Jarihani et al., 2015
9	SARAL/Altika	~173	35	2013*	0.11	Schwatke, et al., 2015
10	Sentinel 3 SRAL	~300	27	2016*	0.03	Fu et al., 2009
11	Jason-3	~300	10	2016*	0.03	ESA, 2018
12	SWOT	~10–70	21	2020*	0.10	Fu et al., 2009

Source: adapted from Ekeu-wei and Blackburn, 2018 and O'Loughlin et al., 2016.
Notes: S/N=sequential number; current=*, future=+,SRTM=shuttle topography mission.

Airborne remote sensing is a very important and complementary tool for space-borne remote sensing. It can acquire data at higher resolution, with vertical accuracy of ~5 cm root mean square error (RMSE) and with continuous time and space. However, DEM acquired by airborne is expensive, and therefore available at limited locations.

High-resolution optical cameras and LiDAR became the privileged sensor of aircraft. LiDAR as a topographic data source can yield information at a resolution better than 2.0 m and a precision that is better than 0.15 m (Yu and Lane, 2006). In the case of the airborne observation of the Iowa 2008 flood, the RMSE of the LiDAR-derived floodwater surface profile to high-water marks was 30 cm, the consistency between the two flooded areas derived from LiDAR and SPOT imagery was 72% (81% if suspicious isolated ponds in the SPOT-derived extent were removed), and LiDAR-derived flood extent had a horizontal resolution of ~3 m (Chen et al., 2017).

It should be noted that DEMs present absolute bias, stripe noise, speckle noise, and tree height bias that are not always removed in a hydrological study. To this purpose, the Multi-Error-Removed Improved-Terrain (MERIT) DEM was developed by removing multiple error components from the existing space-borne DEMs (such as SRTM3 v2.1 and AW3D-30m v1) was completed. It improved the floodplain elevation accuracy (Yamazaki et al., 2017). Moreover, for small rivers, the surrounding topography can have a major impact on the echo shape returned to the altimeter and create inaccuracy in the data.

TABLE 3.9

Reported Water Level Retrieval Performance Based on Remotely Sensed Flood Extent with DEM Fusion and Their Accuracies for Rural Areas

Method	Error (RMSE)	Validation Data	Source
Landsat TM-derived flood extent superimposed on topographic contours for volume estimation	± 21%	Field data	Gupta and Banerji, 1985 Wang et al., 2002
ERS SAR flood extent overlain on topographic contours	0.5 m–2 m	Field data	Oberstadler et al., 1997
ERS SAR flood extent overlain on topographic contours	Up to 2 m	Model outputs	Brakenridge et al., 1998
Inter-tidal area shorelines from multiple ERS images superimposed onto simulated shoreline heights	Mean error of 0.2–0.3 m	Field data	Mason et al., 2001
Flooded vegetation maps from combined airborne L- and C-SAR integrated with LiDAR vegetation height map	Around 0.1 m	Field data	Horritt et al., 2003
Integration of high-resolution elevation data with event wrack lines	< 0.2 m	Model outputs	Lane et al., 2003
Fusion of RADARSAT-1 SAR flood edges with LiDAR	Correlation coefficient of 0.9	TELEMAC-2D model outputs; mean between maximum and minimum estimation	Mason et al., 2003
Complex fusion of flood aerial photography and field-based water stages from various floodplain structures	0.4–0.7 m	Field data	Raclot, 2006
Fusion of ENVISAT ASAR flood edges with LiDAR and interpolation modeling	< 0.2 m	Field data	Matgen et al., 2007
Fusion of ENVISAT ASAR flood edges with LiDAR and regression modeling	< 0.35 m/0.7 m/1.07 m	1D model outputs	Schumann et al., 2008
Fusion of ENVISAT ASAR flood edges with LiDAR/topographic contours/ SRTM and regression modeling	0.3 m; 0.5 m uncertainty	Mean between maximum and minimum estimation; aerial photography	Hostache et al., 2009
Fusion of hydraulically sensitive flood zones from ENVISAT ASAR imagery and LiDAR	Mean error of up to 0.5 m	One field gauge	Mason et al., 2009
Fusion of TerraSAR-X flood edge with LiDAR (rural area)	Error of 1.17 m (note that relative changes in levels from TerraSAR-X and aerial photography were mapped with an accuracy of 0.35 m compared to gauge data)	One field gauge	Zwenzner and Voigt, 2009
Fusion of uncertain ENVISAT ASAR WSM flood edges with SRTM heights	Error in median estimate of 0.8 m	LiDAR-derived water levels	Schumann et al., 2010
LIDAR	0.5 and 0.4 cm	Gauge altimetry data on the side of the lake	Zlinszky et al., 2017

Source: modified from Schumann et al., 2015.

The accuracy of the ICESat laser altimetry was tested in the river confluence of Tapajos and Amazon rivers, Brazil (Urban and Schutz, 2008). Results indicate a root mean square error of river elevations is 3 cm under clear conditions, 8 to 15 cm under partly-cloudy skies, and 25 cm under heavy clouds.

3.6 RESULTS AND DISCUSSIONS

Flood observation enclosed discharge measurement, space- and airborne as well as more recent methods such as the one based on UAV/UAS. Discharge measurement presents large uncertainty primarily when a flood occurs. SAR measurement uncertainty occurs due to the radiance measurement and the post-processing of the primary signal. Then, uncertainty appears during the flood mapping when optical sensors are used under the cloud cover condition. Flood under canopy, mudflood, saturated soil, and wind backscatter contamination present large uncertainty. However, extent map validation can obtain good performance with globally 10–20% of error. Altimetry remote sensing permits a very high measurement of water level over the river (0.03 m for Jason-3). Moreover, river depth required the use of a DEM that presents high inaccuracy (SRTM, ± 16m). Therefore, flood depth mapping products of the merging of the flood extent mapping and DEM inherit from their respective uncertainties. SAR short revisiting time can be reduced by the use of constellations satellites.

Considering flood observation revisit intervals, space-borne presents strong limitations. Airborne can monitor flooding in a continuous way but at a high monetary cost which can partially be covered by cheaper UAV/UAS systems. That gap can be partially solved by using a numerical model that is calibrated based on the previous observation data. However, numerical models add uncertainties from physical limitations, bias, and input data uncertainty. On other hand, real-time monitoring requires real-time data that requires optimization. Therefore, it relies on strategy to merge data information and retrieval algorithm speed as well as parallelization and computation methods. Attempts to optimize remote sensing data for real-time operation can be found in Hu et al. (2018), where an example for the Jinsha River basin is provided to facilitate efficient and effective observational planning for flood satellite sensors. The following discussion covers the recent use of UAV/UAS, urban flood dynamics, and geomorphology change during flood events.

3.6.1 UAV/UAS/BALLOON

UAV provides some complementary methods to space-borne remote sensing. The use of UAV is adaptable to a large extent of spatial scales, ranging from small irrigation channels up to large rivers with breadths > 100 m. Yun (2016) proposes a summary of aerial platforms for flood monitoring under various weather conditions. It introduces UAV types such as the Helikite able to fly under all-weather operations. However, limitation exists for UAV like the small coverage area of the river system and a battery system that has a short time (~30 min). Balloon platforms can be used at fixed locations to provide high resolution of flood areas based on the optical sensor (Hariyanto et al., 2009). The advantages of the low cost of operation, continuous temporal monitoring, as well as simplicity of data retrieval can also benefit to reduce the uncertainty of flood observation. A study on mapping the flood in Yuyao, China indicated that optical sensors mounted on UAV can extract the inundated areas precisely with a 94% overall accuracy (Feng et al., 2015).

3.6.2 URBAN FLOOD

Urban flood observation has specific challenges such as (1) flood plains that do not play a direct role, (2) the need for high-resolution data (less than 5 m), and (3) shadow building removal that require specific methodology due to the low angle of the SAR method. One of the main challenging issues is to distinguish the shadow of high-rise buildings from water bodies and another is to use a LiDAR dataset depending on its availability.

A primary observation is the discharge in urban areas that can be acquired by high-resolution cameras such as the flow observation achieved by velocimetry (LSPIV). LSPIV is an optical method that can take advantage of a bidding frame to set up an observation system (Tauro et al., 2016). However, deploying instruments in the flow are often infeasible due to high velocity and abundant sediment transport.

Giustarini et al. (2013) investigated the Severn River flood (UK) in July 2007 based on the very high-resolution SAR sensor onboard TerraSAR-X as well as airborne photography. It highlights the advantages and limitations of the proposed method. In fact, the percentage of under-detected flood pixels rises from 15.6% to 16.2%. In a second study, a specific case of a one-in-150-years flood near Tewkesbury, United Kingdom, in 2007 indicates that 76% of the urban water pixels were correctly detected that are visible to TerraSAR-X (Mason et al., 2010). It is also associated with a false positive rate of 25%. If all the urban water pixels were considered, including those in shadow and layover regions, these figures fell to 58% and 19%.

For SAR methods, a pre-processing method is a key step to improve uncertainty and reduction in particular for shadow-affected flooded areas. For instance, a SAR/LiDAR (in the single-image case) double scattering approach proved to be equally successful at detecting urban flooding and when using change detection between flooded and un-flooded images. This approach overcomes the need for a high-resolution DEM and a SAR simulator for determining shadow regions that are not visible to the satellite.

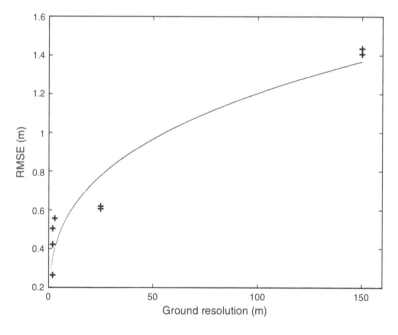

FIGURE 3.14 Relationship between spatial resolution and water level accuracy for urban areas of the type analyzed here (based on the results of this study). Temporal frequency, i.e., satellite revisit times, is inversely linked to spatial resolution and so the value of each satellite sensor depends really on the user's requirements, the actual availability of topographic data (whether from LiDAR, InSAR, or SRTM), and the choice of spatial as well as temporal scales. For instance, if near-real-time monitoring over larger spatial scales is the priority, then ASAR-WSM is the preferred option at the cost of data accuracy. If, however, the user is interested in greater accuracies and much smaller scales and is satisfied with less frequent revisit times, then TerraSAR-X, with its 11-day repeat cycle, would be a more obvious choice (Sources: Schumann et al., 2009; Schumann et al., 2011.)

TABLE 3.10

Channel Change Detection Analysis in Response to Floods: Remote Sensing Data, Methods, and Main Aspects to Be Taken into Account for Detecting Changes

Change Detection	Data	Method	Feasibility		
			Size of Detected Feature/Data Resolution	Change Magnitude	Errors
Channel width	Aerial and satellite imagery	Multi-temporal and spatial analysis	+++	+++	+
Channel pattern	Aerial and satellite imagery	Multi-temporal and spatial analysis	+++	++	+
Channel shifting	Aerial and satellite imagery	Multi-temporal and spatial analysis	+++	+++	+++
Bank retreat	Aerial and satellite imagery	Multi-temporal and spatial analysis	+++	+++	+++
Submerged geomorphic feature	Airborne/space-borne hyperspectral imagery	Imagery classification comparison: field survey	++	++	+
Topography	Airborne/space-borne hyperspectral imagery	Bathymetry mapping	+	+++	+++
Erosion/ deposition	LiDAR or photogrammetry derived DEMs/DTMs	DEM of Difference (DoD)	++	+++	+++

Source: Righini and Surian, 2018.

Notes: Feasibility of detecting geomorphic changes considering three aspects: the relationships (i.e. ratio) between the size of the detected feature and spatial resolution of imagery, the magnitude of geomorphic change, and errors due to georectification, digitalization, etc. The importance of each aspect is assessed as follows: + negligible or little importance, ++ moderate importance, +++ high or very important.

On the other hand, it requires a reference image with the same imaging characteristics as the flood image. Yang et al. (2017) proposed an automated urban water extraction method with an omission error of 9.26%. Compared with Normalized Difference Water Index (NDWI), the urban water extraction method (UWEM) significantly improved accuracy by lessening commission and omission errors by 73% stable performances than NDWIs in a range of thresholds near zero. It reduces the over- and underestimation issues which are often accompanied by the indicators such as the NDWI.

Light detection and ranging (LiDAR) is the most adopted method to obtain high-quality terrain information (Chen et al., 2012). As an example, at present the UK Environment Agency LiDAR data cover 62% of England and Wales with horizontal spatial resolution ranging from 0.25 m to 2 m and a vertical accuracy between 5 cm and 15 cm. The method successfully detected double scattering due to flooding in the single-image case with 100% classification accuracy (albeit using a small sample set) and un-flooded curves with 91% classification accuracy (Mason et al., 2014). Chen et al. (2017) investigated the agreement between wetted areas derived from the LIDAR (centimeter accuracy) and SPOT visible image for the 2008 Iowa city flood. The results gave 72% of agreement, and it increased to 81% if isolated ponds were removed from the SPOT-derived flood extent. Schumann et al. (2009) investigated the logarithmic relation between DEM horizontal resolution and the RMSE of water level. The result illustrated by Figure 3.14 was obtained by eight space-borne SAR and aerial photographic images of flooding of the UK town of Tewkesbury acquired over an eight-day period in summer 2007.

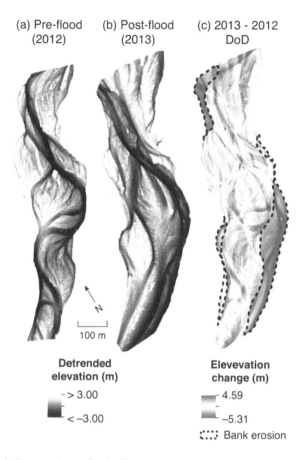

FIGURE 3.15   Detrended pre- and post-flood DEMs and DEM differencing. Detrended elevations reflect the removal of valley slope and normalization relative to mean reach elevation; positive values are above mean elevation, negative values are below mean elevation. (Source: Tamminga, 2016.)

3.6.3   FLOOD GEOMORPHOLOGY CHANNEL CHANGE

An important point is the change of the geomorphology of channels, levees that produces channel shifts, bank retreat, and erosion because it strongly impacts the DEM acquired previously. Table 3.10 summarizes the main change and the qualitative error estimation of geomorphological change detection occurring after floods. Airborne and satellite imagery are needed and can provide updates in river and floodplain geometry that can be used to update modeling infrastructure and investigate the future areas at risk. The airborne systems are the main sources of geomorphological change observations.

Figures 3.15–3.16 illustrate the river/floodplain geomorphological changes after flood acquired with UAV LiDAR and an airborne optical camera. Both figures highlight changes in the floodplain extension and changes in the river position with the floodplain as well as sand deposit. Therefore, geomorphological river processes provide a dynamic view of the river-floodplain system while continually creating uncertainty in DEMs acquired previously.

3.7   SUMMARY AND CONCLUSIONS

Flood observation uncertainties are crucial data to appreciate the accuracy of flood discharge, extent, and depth observations. In this chapter, we have illustrated the emergence of remote sensing as an observation method which, however, still relies on in situ measurements for its calibration. Despite the strong improvement of observations, limitation remains to be able to capture continuous flood

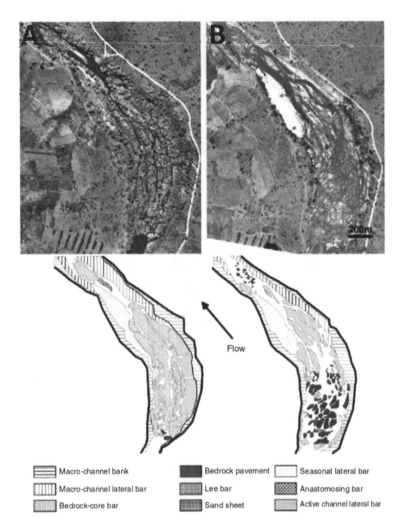

FIGURE 3.16 Example of cohesive mixed anastomosed channel type on the Sabie River, showing (a) preflood and (b) change in morphologic unit composition following the extreme flood of 2000 on aerial photographs (1:10,000 scale). (Source: Heritage et al., 2004.)

observation mainly due to the return period of the satellite. Recent use of UAV/UAS and balloons have provided high temporal and spatial resolution making possible the continuous observation of flood geomorphology channel changes as well as urban flooding that presents specific challenges. Overall, optical and satellite aperture radar usually achieves an accuracy of flood extent of over 80% when compared to ground observation. Flood depth RMSE reached < 0.5m when LIDAR data are used for digital elevation models combined with flood extent. The development of UAV/UAS/balloon and public monitoring is promising local solutions for high resolution. Urban flood observation remains a challenge due to building shadows. Moreover, geomorphology channel change occurs during flood events, which affects the knowledge of the DEM.

REFERENCES

Adhikari, P., Hong, Y., Douglas, K.R., Kirschbaum, D., Gourley, J.J., Adler, R.F., and Brakenridge, G.R. 2010. A digitized global flood inventory (1998–2008): Compilation and preliminary results, *Natural Hazards*, 55: 405–422.

Alsdorf, D., Beighley, E., Laraque, A., Lee, H., Tshimanga, R., O'Loughlin, F., Mahé, G., Dinga, B., Moukandi, G., and Spencer, R.G.M. 2016. Opportunities for hydrologic research in the Congo Basin, *Reviews of Geophysics*, 54: 378–409, https://doi.org/10.1002/2016RG000517.

Assumpção, T.H., Popescu, I., Jonoski, A., and Solomatine, D.P. 2018. Citizen observations contributing to flood modelling: Opportunities and challenges, *Hydrology and Earth System Sciences*, 22: 1473–1489, https://doi.org/10.5194/hess-22-1473-2018.

Bates, P.D., Pappenberger, F., and Romanowicz, R.J. 2014. Uncertainty in flood inundation modelling. In: *Applied Uncertainty Analysis for Flood Risk Management*, Imperial College Press, London, UK, pp. 232–269, https://doi.org/10.1142/9781848162716_0010.

Biancamaria, S., Lettenmaier, D.P., and Pavelsky, T.M. 2016. The SWOT mission and its capabilities for land hydrology. In: *Remote Sensing and Water Resources*, https://doi.org/10.1007/978-3-319-32449-4_6.

Bjerklie, D.M., Dingman, L.S., Vorosmarty, C.J., Bolster, C.H., and Congalton, R.G. 2003. Evaluating the potential for measuring river discharge from space, *Journal of Hydrology*, 278: 17–38, https://doi.org/10.1016/S0022-1694(03)00129-X.

Blasco, F., Bellan, M.F., and Chaudhury, M.U. 1992. Estimating the extent of floods in Bangladesh using SPOT data, *Remote Sensing of Environment*, 39(3): 167–178, ISSN 0034-4257, https://doi.org/10.1016/0034-4257(92)90083-V.

Bonn, F., and Dixon, R. 2005. Monitoring flood extent and forecasting excess runoff risk with RADARSAT-1 data, *Nature Hazards*, 35(3): 377–393.

Brakenridge, R., and Anderson, R. 2006. Modis-based flood detection, mapping and measurement: The potential for operational hydrologial applications. In: Marsalek, J., Stancalie, G., Balint, G. (eds) *Transboundary Floods: Reducing Risks through Flood Management. Nato Science Series: IV: Earth and Environmental Sciences*, vol. 72, Springer, Dordrecht, The Netherlands.

Brakenridge, G.R., Tracy, B.T., and Knox, J.C. 1998. Orbital SAR remote sensing of a river flood wave, *International Journal of Remote Sensing*, 19(7): 1439–1445.

Chandran, R.V., Ramakrishnan, D., Chowdary, V.M., Jeyaram, A., and Jha, A.M. 2006. Flood mapping and analysis using air-borne synthetic aperture radar: A case study of July 2004 flood in Baghmati river basin, Bihar, *Current Science*, 90(2): 249–256.

Chen, A.S., Evans, B., Djordjević, S., and Savić, D.A. 2012. A coarse-grid approach to representing building blockage effects in 2D urban flood modelling, *Journal of Hydrology*, 426–427: 1–16, ISSN 0022-1694, https://doi.org/10.1016/j.jhydrol.2012.01.007.

Chen, B., Krajewski W.F., Goska R., and Young, N. 2017. Using LiDAR surveys to document floods: A case study of the 2008 Iowa flood, *Journal of Hydrology*, 553: 338–349, ISSN 0022-1694, https://doi.org/10.1016/j.jhydrol.2017.08.009.

Chen, N., Zhou, L., and Chen, Z.A. 2015. Sharable and efficient metadata model for heterogeneous earth observation data retrieval in multi-scale flood mapping, *Remote Sensing*, 7:9610–9631, https://doi.org/10.3390/rs70809610.

Chen, Z., Chen, N., Du, W., and Gong, J. 2018. An active monitoring method for flood events, *Computers & Geosciences*, 116: 42–52, ISSN 0098-3004, https://doi.org/10.1016/j.cageo.2018.04.009.

Danielson, J.J., and Gesch, D.B. 2011. *Global Multi-Resolution Terrain Elevation Data 2010 (GMTED2010)*, United States Geological Survey, Reston, VA, USA.

Da Silva, J.S., Calmant, S., Seyler, F., Rotunno Filho, O.C., Cochonneau, G., and Mansur, W.J. 2010. Water levels in the Amazon basin derived from the ERS 2 and ENVISAT radar altimetry missions, *Remote Sensing of Environment*, 114: 2160–2181.

DeGroeve, T. 2010. Flood monitoring and mapping using passive microwave remote sensing in Namibia, *Geomatics, Natural Hazards and Risk*, 1(1): 19–35, https://doi.org/10.1080/19475701003648085.

Demirkesen, A.C., Evrendilek, F., Berberoglu, S., and Kilic, S. 2007. Coastal flood risk analysis using Landsat-7 ETMþ imagery and SRTM DEM: A case study of izmir, Turkey, *Environmental Monitoring and Assessment*, 131(1–3): 293–300.

Di Baldassarre, G., Laio, F., and Montanari, A. 2012. Effect of observation errors on the uncertainty of design floods, *Physics and Chemistry of the Earth, Parts A/B/C*, 42–44: 85–90, ISSN 1474-7065, https://doi.org/10.1016/j.pce.2011.05.001.

Doxaran, D., Froidefond, J.-M., Castaing, P., and Babin, M. 2009. Dynamics of the turbidity maximum zone in a macrotidal estuary (the Gironde, France): Observations from field and MODIS satellite data, *Estuarine, Coastal and Shelf Science*, 81: 321–332.

Du, J., Kimball, J.S., Galantowicz, J., Kim, S., Chan, S.K., Reichle, R., Jones, L.A., and Watts, J.D. 2018. Assessing global surface water inundation dynamics using combined satellite information from SMAP,

AMSR2 and Landsat, *Remote Sensing of Environment*, 213: 1–17, ISSN 0034-4257, https://doi.org/10 .1016/j.rse.2018.04.054.

Durand, M., Rodriguez, E., Alsdorf, D.E., and Trigg, M. 2010. Estimating river depth from remote sensing swath interferometry measurements of river height, slope and width, *IEEE Journal of Selected Topics in Applied Earth Observations and Remote Sensing*, 3(1): 20–31, https://doi.org/10.1109/JSTARS.2009 .2033453.

Ekeu-wei, I., and Blackburn, G. 2018. Applications of open-access remotely sensed data for flood modelling and mapping in developing regions, *Hydrology*, 5(3): 39, https://doi.org/10.3390/hydrology5030039.

ESA (European Space Agency). 2018. Altimetry instrument payload. Available online: https://sentinel.esa.int/ web/missions/sentinel-3/instrument-payload/altimetry (accessed on 20 August 2018).

Farr, T.G., Rosen, P.A., Caro, E., Crippen, R., Duren, R., Hensley, S., Kobrick, M., Paller, M., Rodriguez, E., Roth, L., and Seal, D. 2007. The shuttle radar topography mission, *Reviews of Geophysics*, 45: RG2004, https://doi.org/10.1029/2005RG000183.

Fekete, B.M., Looser, U., Pietroniro, A., and Robarts, R.D. 2012. Rationale for monitoring discharge on the ground, *Journal of Hydrometeorology*, 13: 1977–1986, https://doi.org/10.1175/JHM-D-11-0126.1.

Feng, Q., Gong, J., Liu, J., and Li, Y. 2015. Flood mapping based on multiple end member spectral mixture analysis and random forest classifier? The case of Yuyao, China. *Remote Sensing*, 7: 12539–12562, https://doi.org/10.3390/rs70912539.

Fu, L.-L., Alsdorf, D., Rodriguez, E., Morrow, R., Mognard, N., Lambin, J., Vaze, P., and Lafon, T. 2009. The SWOT (Surface Water and Ocean Topography) mission: Spaceborne radar interferometry for oceano-graphic and hydrological applications. In: *OCEANOBS'09 Conference*, 16 pp. 21–25 September 2009, Venice Convention Centre, Venice-Lido, Italy.

Frappart, F., Calmant, S., Cauhopé, M., Seyler, F., and Cazenave, A. 2006. Preliminary results of ENVISAT RA- 2- derived water levels validation over the Amazon basin, *Remote Sensing of Environment*, 100: 252–264.

Giustarini, L., Hostache, R., Matgen, P., and Schumann, G.J.P. 2013. A change detection approach to flood mapping in urban areas using TerraSAR-X, *IEEE Transactions on Geoscience and Remote Sensing*, 51(4): 2417–2430, https://doi.org/10.1109/TGRS.2012.2210901.

Giustarini, L., Chini, M., Hostache, R., Pappenberger, F., and Matgen, P. 2015. Flood hazard mapping combin-ing hydrodynamic modeling and multi annual remote sensing data, *Remote Sensing*, 7: 14200–14226, https://doi.org/10.3390/rs71014200.

Grandoni, D. Battagliere, M.L., Daraio, M.G., Sacco, P., Coletta, A., Di Federico, A., and Mastracci, F. 2014. Space-based technology for emergency management: The COSMO-SkyMed constellation contribution, *Procedia Technology*, 16: 858–866, ISSN 2212-0173, https://doi.org/10.1016/j.protcy.2014.10.036.

Grimaldi, S., Li, Y., Pauwels, V.R.N., and Walker, J.P. 2016. Remote sensing-derived water extent and level to constrain hydraulic flood forecasting models: Opportunities and challenges, *Surveys in Geophysics*, 37(5): 977–1034, https://doi.org/10.1007/s10712-016-9378-y.

Gupta, R.P., and Banerji, S. 1985. Monitoring of reservoir volume using LANDSAT data, *Journal of Hydrology*, 77: 159–170.

Hadjimitsis, D.G. 2007. The use of satellite remote sensing and GIS for assisting flood riskassessment: A case study of the Agriokalamin Catchment area in Paphos-Cyprus, *Proceedings of SPIE*, 6742: 67420Z.

Hariyanto, H., Santoso, H., and Widiawan, A.K. 2009. Emergency broadband access network using low alti-tude platform. In: Proceedings of the International Conference on Instrumentation, Communications, Information Technology, and Biomedical Engineering, Bandung, Indonesia, 23–25 November, p. 6.

Heritage, G.L., Large, A.R.G., Moon, B.P., and Jewitt, G. 2004. Channel hydraulics and geomorphic effects of an extreme flood event on the Sabie River, South Africa, *Catena*, 58: 151–181.

Herold, M., Gardner M., Noronha, V., and Roberts, D. 2003. Spectrometry and hyperspectral remote sensing of urban road infrastructure, *Online Journal of Space Communication*, 3: 1–29.

Ho, T.K., Umitsu, M., and Yamaguchi, Y. 2010. Flood hazard mapping by satellite images and SRTM DEM in the Vu Gia–Thu Bon Alluvial Plain, Central Vietnam, *International Archives of the Photogrammetry, Remote Sensing and Spatial Information Science*, 38(8): 275–280.

Horritt, M.S., Mason, D.C., Cobby, D.M., Davenport, I.J., and Bates, P.D. 2003. Waterline mapping in flooded vegetation from airborne SAR imagery, *Remote Sensing of Environment*, 85(3): 271–281.

Hostache, R., Matgen, P., Schumann, G., Puech, C., Hoffmann, L., and Pfister, L. 2009. Water level estimation and reduction of hydraulic model calibration uncertainties using satellite SAR images of floods, *IEEE Transactions on Geoscience and Remote Sensing*, 47(2): 431–441.

Hou, A.Y., Kakar, R.K., Neeck, S., Azarbarzin, A.A., Kummerow, C.D., Kojima, M., Oki, R., Nakamura, K., and Iguchi, T. 2014. The global precipitation measurement mission, *Bulletin of the American Meteorological Society*, 95: 701–722, https://doi.org/10.1175/BAMS-D-13-00164.1.

Hu, C., Li, J., Lin, X., Chen, N., and Yang, C. 2018. An observation capability semantic-associated approach to the selection of remote sensing satellite sensors: A case study of flood observations in the Jinsha River Basin, *Sensors*, 18: 1649.

Ip, F., Dohm, J.M., Baker, V.R., Doggett, T., Davies, A.G., Casta, O.R., Chien, S., Cichy, B., Greeley, R., Sherwood, R., Tran, D., and Rabideau, G. 2006. Flood detection and monitoring with the autonomous sciencecraft experiment onboard EO-1, *Remote Sensing of Environment*, 101(4): 463–481.

Irwin, K., Beaulne, D., Braun A., and Fotopoulos, G. 2017. Fusion of SAR, optical imagery and airborne LiDAR for surface water detection, *Remote Sensing*, 9(9): 890, https://doi.org/10.3390/rs9090890.

Jarihani, A.A., Callow, J.N., McVicar, T.R., Van Niel, T.G., and Larsen, J.R. 2015. Satellite- derived Digital Elevation Model (DEM) selection, preparation and correction for hydrodynamic modelling in large, low- gradient and data- sparse catchments, *Journal of Hydrology*, 524: 489–506.

Jilani, R., Munir, S., and Siddiqui, P. 2017. Application of ALOS data in flood monitoring in Pakistan. In: Proceedings of the 1st PI Symposium of ALOS Data Nodes, Kyoto, Japan, 10 July 2017.

Jung, H.C., and Jasinski, M.F. 2015. Sensitivity of a floodplain hydrodynamic model to satellite-based DEM scale and accuracy: Case study—The Atchafalaya Basin, *Remote Sensing*, 7: 7938–7958, https://doi.org /10.3390/rs70607938.

Klemas, V. 2015. Remote sensing of floods and flood-prone areas: An overview, *Journal of Coastal Research*, 31(4): 1005–1013.

Kumar, D.N., and Reshmidevi, T.V. 2013. Remote sensing applications in water resources, *Journal of the Indian Institute of Science*, 93(2): 163–188, http://journal.library.iisc.ernet.in/index.php/iisc/article/view /1088/2316

Lacava, T., Ciancia, E., Faruolo, M., Pergola, N., Satriano, V., Tramutoli, V. 2018. Analyzing the December 2013 Metaponto Plain (Southern Italy) flood event by integrating optical sensors satellite data, *Hydrology*, 5: 43.

Lane, S.N., James, T.D., Pritchard, H., and Saunders, M. 2003. Photogrammetric and laser altimetric reconstruction of water levels for extreme flood event analysis, *Photogrammetric Record*, 18(104): 293–307.

Lawal, D.U., Matori, A.N., Yusf, K.W., Hashim, A.M., and Balogun, A.L. 2014. Analysis of the flood extent extraction model and the natural flood influencing factors: A GIS-based and remote sensing analysis. *IOP Conference Series: Earth and Environmental Science*, 18(1), IOP Publishing: 012059.

Le Boursicaud, R., Pénard, L., Hauet, A., Thollet, F., and LeCoz, J. 2016. Gauging extreme floods on YouTube: Application of LSPIV to home movies for the post-event determination of stream discharges, *Hydrological Processes*, 30: 90–105, https://doi.org/10.1002/hyp.10532.

Le Coz, J., Hauet, A., Pierrefeu, G., Dramais, G., and Camenen, B. 2010. Performance of image-based velocimetry (LSPIV) applied to flash-flood discharge measurements in Mediterranean rivers, *Journal of Hydrology*, 394(1–2): 42–52, ISSN 0022-1694, https://doi.org/10.1016/j.jhydrol.2010.05.049.

Lewis, H.W., Sanchez, J.M.C., Graham, J., Saulter, A., Bornemann, J., Arnold, A., Fallmann, J., Harris, C., Pearson, D., Ramsdale, S., de la Torre, A.M., Bricheno, L., Blyth, E., Bell, V.A., Davies, H., Marthews, T.R., O'Neill, C., Rumbold, H., O'Dea, E., Brereton, A., Guihou, K., Hines, A., Butenschon, M., Dadson, S.J., Palmer, T., Holt, J., Reynard, N., Best, M., Edwards, J., and Siddorn, J. 2018. The UKC2 regional coupled environmental prediction system. *Geoscientific Model Development*, 11: 1–42.

Li, C., Li, S., Wang, H., and Lei, T. 2010. The research on unmanned aerial vehicle remote sensing and its applications. In: Proceedings of the 2nd International Conference on Advanced Computer Control, Shenyang, Liaoning, China, 27–29 March, pp. 644–647.

Lewis, H.W., Sanchez, J.M.C., Graham, J., Saulter, A., Bornemann, J., Arnold, A., Fallmann, J., Harris, C., Pearson, D., Ramsdale, S., de la Torre, A.M., Bricheno, L., Blyth, E., Bell, V.A., Davies, H., Marthews, T.R., O'Neill, C., Rumbold, H., O'Dea, E., Brereton, A., Guihou, K., Hines, A., Butenschon, M., Dadson, S.J., Palmer, T., Holt, J., Reynard, N., Best, M., Edwards, J., and Siddorn, J. 2018. The UKC2 regional coupled environmental prediction system, *Geoscientific Model Development*, 11: 1–42.

Lu, S., Wu, B., Yan, N., and Wang, H. 2011. Water body mapping method with HJ-1A/B satellite imagery, *International Journal of Applied Earth Observation and Geoinformation*, 13(3): 428–434.

Malinowski, R., Groom, G., Schwanghart, W., and Heckrath, G. 2015. Detection and delineation of localized flooding from WorldView-2 multispectral data, *Remote Sensing*, 7: 14853–14875, https://doi.org/10 .3390/rs71114853.

Martinez, J.M., and Le Toan, T. 2007. Mapping of flood dynamics and spatial distribution of vegetation in the Amazon floodplain using multitemporal SAR data, *Remote Sensing of Environment*, 108(3): 209–223, ISSN 0034-4257, https://doi.org/10.1016/j.rse.2006.11.012.

Martinis, S., and Rieke, C. 2015. Backscatter analysis using multi-temporal and multi-frequency SAR data in the context of flood mapping at River Saale, Germany, *Remote Sensing*, 7: 7732–7752, https://doi.org/10 .3390/rs70607732.

Martinis, S., Twele, A., and Voigt, S. 2009. Towards operational near real-time flood detection using a split-based automatic thresholding procedure on high resolution TerraSAR-X data, *Natural Hazards and Earth System Sciences*, 9(2): 303–314.

Mason, D.C., Bates, P.D., and Dall'Amico, J.T. 2009. Calibration of uncertain flood inundation models using remotely sensed water levels, *Journal of Hydrology*, 368: 224–236.

Mason, D.C., Cobby, D.M., Horritt, M.S., and Bates, P.D. 2003. Floodplain friction parameterization in two-dimensional river flood models using vegetation heights derived from airborne scanning laser altimetry, *Hydrological Processes*, 17: 1711–1732.

Mason, D.C., Davenport, I.J., Flather, R.A., Gurney, C., Robinson, G.J., and Smith, J.A. 2001. A sensitivity analysis of the waterline method of constructing a digital elevation model for intertidal areas in ERS SAR scene of Eastern England, *Estuarine, Coastal and Shelf Science*, 53: 759–778.

Mason, D.C., Giustarini, L., Garcia-Pintado, J., and Cloke, H.L. 2014. Detection of flooded urban areas in high resolution synthetic aperture radar images using double scattering, *International Journal of Applied Earth Observation and Geoinformation*, 28: 150–159, ISSN 0303-2434, https://doi.org/10.1016/j.jag.2013.12.002.

Mason, D.C., Speck, R., Devereux, B., Schumann, G.J.-P., Neal, J.C., and Bates, P.D. 2010. Flood detection in urban areas using TerraSAR-X, *IEEE Transactions on Geoscience and Remote Sensing*, 48(2): 882–894, https://doi.org/10.1109/TGRS.2009.2029236.

Matgen, P., Schumann, G., Henry, J.B., Hoffmann, L., and Pfister, L. 2007. Integration of SAR-derived inundation areas, high precision topographic data and a river flow model toward real-time flood management, *International Journal of Applied Earth Observation and Geoinformation*, 9(3): 247–263.

Oberstadler, R., Hönsch, H., and Huth, D. 1997. Assessment of the mapping capabilities of ERS-1 SAR data for flood mapping: A case study in Germany, *Hydrological Processes*, 10: 1415–1425.

O'Loughlin, F.E., Neal, J., Yamazaki, D., and Bates, P.D. 2016. ICESat-derived inland water surface spot heights, *Water Resources Research*, 52: 3276–3284.

Opolot, E. 2013. Application of remote sensing and geographical information systems in flood management: A review, *Research Journal of Applied Sciences, Engineering and Technology*, 6: 1884–1894.

Pakoksung, K., and Takagi, M. 2016. Digital elevation models on accuracy validation and bias correction in vertical, *Modeling Earth Systems and Environment*, 2: 1–13.

Pandiammal, C., Senthil, J., and Anand, P.H. 2015. Tsunami flood damages assessment in Cuddalore district using remote sensing technology, *Advances in Applied Science Research*, 6(8): 96–100.

Pinel, S., Bonnet, M.P., Santos da Silva, J., Moreira, D., Calmant, S., Satgé, F., and Seyler, F. 2015. Correction of interferometric and vegetation biases in the SRTMGL1 spaceborne DEM with hydrological conditioning towards improved hydrodynamics modeling in the Amazon Basin, *Remote Sensing*, 7: 16108–16130, https://doi.org/10.3390/rs71215822.

Pope, K.O., Sheffner, E.J., Linthicum, K.J., Bailey, C.L., Logan, T.M., Kasischke, E.S., Birney, K., Njogu, A.R., and Roberts, C.R. 1992. Identification of central Kenyan Rift Valley fever virus vector habitats with Landsat TM and evaluation of their flooding status with airborne imaging radar, *Remote Sensing of Environment*, 40(3): 185–196.

Prigent, C., Lettenmaier, D.P., Aires, F., and Papa, F. 2016. Toward a high-resolution monitoring of continental surface water extent and dynamics, at global scale: From GIEMS (Global Inundation Extent from Multi-Satellites) to SWOT (Surface Water Ocean Topography), *Surveys in Geophysics*, 37: 339, https://doi.org/10.1007/s10712-015-9339-x.

Pulvirenti, L., Chini, M., Pierdicca, N., Guerriero, L., and Ferrazzoli, P. 2011. Flood monitoring using multi-temporal COSMO-SkyMed data: Image segmentation and signature interpretation, *Remote Sensing of Environment*, 115(4): 990–1002, ISSN 0034-4257, https://doi.org/10.1016/j.rse.2010.12.002.

Raclot, D. 2006. Remote sensing of water levels on floodplains: A spatial approach guided by hydraulic functioning, *International Journal of Remote Sensing*, 27(12): 2553–2574.

Rahman, M.S., and Di, L. 2017. The state of the art of spaceborne remote sensing in flood management, *Natural Hazards*, 85: 1223–1248, https://doi.org/10.1007/s11069-016-2601-9.

Rango, A., and Salamonson, V.V. 1974. Regional flood mapping from space, *Water Resources Bulletin*, 10: 473–489.

Righini, M., and Surian, N. 2018. Remote sensing as a tool for analysing channel dynamics and geomorphic effects of floods. In: Refice, A., D'Addabbo, A., Capolongo, D. (eds) *Flood Monitoring through Remote Sensing. Springer Remote Sensing/Photogrammetry*, Springer, Cham, Germany.

Rodriguez, E., Morris, C.S., and Belz, J.E. 2006. A global assessment of the SRTM performance, *Photogrammetric Engineering and Remote Sensing*, 72: 249–260.

Rudorff, C.M., Galvao, L.S., Novo, E.M.L.M. 2009. Reflectance of floodplain waterbodies using EO-1 Hyperion data from high and receding flood periods of the Amazon River, *International Journal of Remote Sensing*, 30(10): 2713–2720.

Sande, C.J.V.D., Jong, S.M.D., and Roo, A.P.J.D. 2003. A segmentation and classification approach of IKONOS-2 imagery for land cover mapping to assist flood risk and flood damage assessment, *International Journal of Applied Earth Observation and Geoinformation*, 4(3): 217–229.

Sanyal, J., Carbonneau, P., and Densmore, A. 2013. Hydraulic routing of extreme floods in a large ungauged river and the estimation of associated uncertainties: A case study of the Damodar River, India, *Nature Hazards*, 66: 1153–1177.

Schneider, R., Godiksen, P.N., Villadsen, H., Madsen, H., and Bauer-Gottwein, P. 2016. Application of CryoSat- 2 altimetry data for river analysis and modelling, *Hydrology and Earth System Sciences*, 19: 1–19.

Schumann, G.J.-P., Bates, P.D., Horritt, M.S., Matgen, P., and Pappenberger, F. 2009. Progress in integration of remote sensing derived flood extent and stage data and hydraulic models, *Reviews of Geophysics*, 47: 20.

Schumann, G.J.-P., Bates, P.D., Neal J.C., and Konstantinos M.A 2015. Chapter 2 - Measuring and mapping flood processes. In: Shroder, J.F., Paron, P., G. Di Baldassarre (eds) *Hydro-Meteorological Hazards, Risks and Disasters*, Elsevier, pp. 35–64, ISBN 9780123948465, https://doi.org/10.1016/B978-0-12-394846-5.00002-3.

Schumann, G.J.-P., Matgen, P., Cutler, M.E.J., Black, A., Hoffmann, L., and Pfister, L. 2008. Comparison of remotely sensed water stages from LiDAR, topographic contours and SRTM, *ISPRS Journal of Photogrammetry and Remote Sensing*, 63: 283–296.

Schumann, G.J.-P., Neal, J.C., and Bates, P.D. 2010. Global floodplain inundation simulation with low resolution space-borne data. In: *Hydrology and Remote Sensing Symposium 2010 September 27–30*, IAHS, IAHS, Jacksonhole, WY, USA, pp. 139–147.

Schumann, G.J.-P., Neal, J.C., Mason, D.C., and Bates, P.D. 2011. The accuracy of sequential aerial photography and SAR data for observing urban flood dynamics, a case study of the UK summer 2007 floods, *Remote Sensing of Environment*, 115(10): 2536–2546, ISSN 0034-4257, https://doi.org/10.1016/j.rse.2011.04.039.

Schwatke, C., Dettmering, D., Börgens, E., and Bosch, W. 2015. Potential of SARAL/AltiKa for Inland water applications, *Marine Geodesy*, 38: 626–643.

Shamaoma, H., Kerle, N., and Alkema, D. 2006. Extraction of flood-modelling related base-data from multi-source remote sensing imagery. In: Proceeding ISPRS Mid-term Symposium, Enschede, Netherlands, 8–11 May, p. 7.

Stephens, E.M., Bates, P.D., Freer, J.E., and Mason, D.C. 2012. The impact of uncertainty in satellite data on the assessment of flood inundation models, *Journal of Hydrology*, 414–415: 162–173, ISSN 0022-1694, https://doi.org/10.1016/j.jhydrol.2011.10.040.

Smith, R.B. 2001. Introduction to remote sensing of the environment, WWW.Microimages.Com.

Tadono, T., Ishida, H., Oda, F., Naito, S., Minakawa, K., and Iwamoto, H. 2014. Precise global DEM generation by ALOS PRISM, *ISPRS Annals of the Photogrammetry, Remote Sensing and Spatial Information Sciences*, 2(4): 71–76.

Tamminga, A. 2016. *UAV-Based Remote Sensing of Fluvial Hydrogeomorphology and Aquatic Habitat Dynamics (T)*, University of British Columbia. Retrieved from https://open.library.ubc.ca/collections/ubctheses/24/items/1.0315349.Attribution-NonCommercial-NoDerivatives 4.0International.

Tarekegn, T.H., Haile, A.T., Rientjes, T., Reggiani, P., and Alkema, D. 2010. Assessment of an ASTER- generated DEM for 2D hydrodynamic flood modeling, *International Journal of Applied Earth Observation and Geoinformation*, 12: 457–465.

Tachikawa, T., Kaku, M.; Iwasaki, A., Gesch, D.B., Oimoen, M.J., Zhang, Z., Danielson, J.J., Krieger, T., Curtis, B., and Haase, J. 2011. *ASTER Global Digital Elevation Model Version 2-Summary of Validation Results*, NASA, Washington, DC, USA.

Tauro, F., Olivieri, G., Petroselli, A., Porfir M., and Grimaldi, S. 2016. Flow monitoring with a camera: A case study on a flood event in the Tiber River, *Environmental Monitoring and Assessment*, 188: 118, https://doi.org/10.1007/s10661-015-5082-5.

Tong, X., Luo, X., Liu, S., Xie, H., Chao, W., Liu, S., Liu, S., Makhinov, A.N., Makhinova, A.F., and Jiang, Y. 2018. An approach for flood monitoring by the combined use of Landsat 8 optical imagery and COSMO-SkyMed radar imagery, *ISPRS Journal of Photogrammetry and Remote Sensing*, 136: 144–153.

Trigg, M.A., Bates, P.D., Wilson, M.D., Schumann, G.J.-P., and Baugh, C.A. 2012. Floodplain channel morphology and networks of the middle Amazon River. *Water Resources Research*, 48(10): W10504, https://doi.org/10.1029/2012WR011888.

UNITAR/UNOSAT. 2015. *Overview of Flood Waters in Mangla Area. And Northern Punjab (Pakistan)*, https://unosat-maps.web.cern.ch/unosat-maps/PK/FL20150723PAK/UNOSAT_A3_Multan_200k _portrait_20150727.pdf. Retrieved the 20 September 2018.

Urban, T.J., and Schutz, B.E. 2008. Neuenschwander, A.L. A survey of ICESat coastal altimetry applications: Continental Coast, Open Ocean Island, and Inland River, *TAO: Terrestrial, Atmospheric and Oceanic Sciences*, 19: 1–19.

Varga, M., and Bašić, T. 2015. Accuracy validation and comparison of global digital elevation models over Croatia, *International Journal of Remote Sensing*, 36: 170–189.

Wang, Y. 2004. Using Landsat 7 TM data acquired days after a flood event to delineate the maximum flood extent on a coastal floodplain, *International Journal of Remote Sensing*, 25(5): 959–974.

Wang, Y., Colby, J.D., and Mulcahy, K.A. 2002. An efficient method for mapping flood extent in a coastal floodplain using Landsat TM and DEM data, *International Journal of Remote Sensing*, 23(18): 3681–3696.

Wu, G., and Liu, Y. 2015. Combining multispectral imagery with in situ topographic data reveals complex water level variation in China's largest freshwater lake, *Remote Sensing*, 7: 13466–13484, https://doi.org /10.3390/rs71013466.

Yamazaki, D., Ikeshima, D., Tawatari, R., Yamaguchi, T., O'Loughlin, F., Neal, J.C., Sampson, C.C., Kanae, S., and Bates P.D. 2017. A high accuracy map of global terrain elevations, *Geophysical Research Letters*, 44(11): 5844–5853, https://doi.org/10.1002/2017GL072874.

Yang, F., Guo, J., Tan, H., and Wang, J. 2017. Automated extraction of urban water bodies from ZY-3 multi-spectral imagery, *Water*, 9: 144.

Yésou, H., Clandillon, S., Allenback, B., Bestault, C., and DeFraipont. 2003. A constellation of advantages with SPOT SWIR and VHR SPOT 5 data for flood extent mapping during the September 2002 Gard event (France). In: IGARSS, Toulouse, July 21–25 2003, p. 3.

Yu, D., and Lane, S.N. 2006. Urban fluvial flood modelling using a two-dimensional diffusion-wave treatment, part 1: Mesh resolution effects, *Hydrological Processes*, 20: 1541–1565, https://doi.org/10.1002/ hyp.5935.

Yun, S.H. 2016. *Synthetic Aperture Radar for Rapid Flood Extent Mapping*, ARIA Team, Jet Propulsion Laboratory, California Institute of Technology, NASA, https://arset.gsfc.nasa.gov/sites/default/files/ disasters/Advanced2016/Flood-L2-Week3-SAR.pdf.

Zhou, G., Luo, J., Yang, C., Li, B., and Wang, S. 2000. Flood monitoring using multi-temporal AVHRR and RADARSAT imagery, *PE&RS, Photogrammetric Engineering & Remote Sensing*, 66(5): 633–638.

Zlinszky, A., Boergens, E., Glira, P., and Pfeifer, P. 2017. Airborne laser scanning for calibration and validation of inshore satellite altimetry: A proof of concept, *Remote Sensing of Environment*, 197: 35–42, ISSN 0034-4257, https://doi.org/10.1016/j.rse.2017.04.027.

Zwenzner, H., and Voigt, S. 2009. Improved estimation of flood parameters by combining space based SAR data with very high resolution digital elevation data, *Hydrology and Earth System Sciences*, 13: 567–576.

4 Flood Modeling and Forecasting Uncertainty

Jean-François Vuillaume and Akinola Adesuji Komolafe

CONTENTS

4.1 INTRODUCTION

Uncertainty analysis in flood modeling and forecast is very important in order to ensure near-accurate prediction of future flood risks for effective and adequate response in flood emergencies and disaster risk reduction. Uncertainty occurs from either random or systematic error originating from model uncertainty and data input uncertainty. Therefore, various uncertainties from simulation models, observations, and forcing data make it challenging to obtain accurate flood forecasts for the required lead-time for small rivers. Random statistical errors can be removed by statistical average, while model systematic uncertainty occurs due to simplification of models or systematic data bias errors. However, uncertainty analysis is indispensable for complex environmental systems. To address the uncertainty from different sources needs to be identified and quantified or ranked

DOI: 10.1201/9780429463938-6

qualitatively and then prioritized. Then, uncertainties should be properly communicated (Beven et al., 2015; Dottori et al., 2013).

In flood modeling and forecast, the target is to provide the map of risk zones caused by possible river channel changes, monitoring of floods in real time, and for civil security. Furthermore, as remote sensing data and other spatial data can be integrated to improve the forecast such as the water extend and/or depth, modeling and forecast must be combined, integrated, and synchronized all with the necessary parameters to predict the areas and depths of flooding. However, real- and non-real-time referring to flood modeling and flood forecast reflect different objectives with different approaches to uncertainty. For instance, compromise has to be found between the quality of the initial data, duration of the forecast, and its accuracy of forecasting in a timely constrains environment when used for real-time prediction. Real-time constraint allows for no verification for future products and supposes a well-calibrated system on previous cases. Overall, flood modeling improvement is used to improve flood forecast as one of its main components. Moreover, uncertainty analysis is useful as it allows more extensive analysis of models and data; it is a fundamental issue particularly in flood modeling and forecast (Pappenberger and Beven, 2006). The increased attention to uncertainty estimation recognizes the limitation and uncertainty in data use in modeling. It makes the use of models more complex but provides the potential to increase the accuracy of our conclusions when drawing conclusions on flood modeling in the present and future forecast (Juston et al., 2013).

Flood modeling can be used to test the catchment response to scenarios such as discharge increase, land-use change, or change on river-floodplain roughness. Paradoxically, the failed model helps us to learn more than the "good" model. We can both learn about the model structure representation of the hydrological flow as well as the influence of the uncertainty in the observation by a probabilistic response outcome. The main finding of a non-fails model is that it potentially fits the set of parameters chosen. However, it has been shown by several studies that optimization is not unique making the model more robust to epistemic error. Therefore, it sounds more "right" when validation used observation uncertainties. Moreover, uncertainty drives for additional data because we can consider the feasibility of reducing uncertainty, suggesting that uncertainty estimation will positively drive model improvement.

Flood forecast requires flood modeling as a module of its flow chart since flood forecast implies rainfall forecast and land-surface model and often post-processing in the case of ensemble and data assimilation. Rainfall forecast relies on a Numerical Weather Prediction (NWP) model that targets different scales (global, regional) and lead-times (now-cast, extended range, seasonal) that present large uncertainties in the calibration. Moreover, it may benefit from additional processing such as dynamic downscaling, parameterization, or assimilation, which can strongly constrain the rainfall output both in spatial location, temporal occurrence, and rainfall quantity. Therefore, uncertainty can be large and can occur from input data, boundary conditions, and model limitations as well as random error.

Furthermore, the rainfall/runoff or the land surface model (LSM) that inputs rainfall to output run-off presents uncertainties in the exact location of surface type controlling the infiltration, the thickness of soil, and their properties well as the energy balance such as the different canopy properties (intersection, evapotranspiration). Similarly, a simplified conceptual rainfall-runoff such as a tank model is based on assumptions about soil properties and is constrained by calibration on past events. Therefore, remote sensing has provided valuable data to improve the static parameter associated with the land surface model. Nevertheless, scaling and comparison with point scale measurement remains a challenge. For instance, in the case of LSM, only a few measurements of the flux tower of radiation that can be used to calibrate land surface model exist (Larsen et al., 2016).

Fully dynamically coupled models are generally not used because initial conditions are considered as the main driver for short forecasts. Moreover, the influence of soil moisture as the initial condition is still a debated topic. Experiments at the Valescure experimental catchment (France) on the use of soil moisture satellite measurement using SM2RAIN revealed the improvement of the flood discharge simulation up to 15% in mean and 34% in median Nash Sutcliffe (Massari et al., 2014). The measurement itself is subject to uncertainty and is generally achieved through satellite. In addition, static parameters such as roughness coefficient and channel topography are still not available while topography is available but often not released as public data. The Manning coefficient presents strong uncertainty and derives from local point laboratory measurement. As discussed previously, the flood modeling presents large uncertainty as well that derived from discharge uncertainty itself influenced by the runoff and rainfall uncertainty.

Finally, flood forecast uncertainty/probability faces challenges in communication, as required improvement of communication for proper visualization of uncertainty requires training. Ensemble forecast requires post-treatment and specific metrics to evaluate its performance (i.e., Continuous Ranked Probability Score and sharpness). Demerit et al. (2010) showed that operational flood forecasters understand the skill, operational limitations, and informational value of products in a variety of different and sometimes contradictory ways. Pappenberger et al. (2013) suggested that it requires discharge, lead-time date fully written, warning alert and/or return period, observations, uncertainty representation, worst/best scenario, meta-information (location, provider, contact), and the risk measure in term of cost and population affected.

4.2 UNCERTAINTY

4.2.1 Identifying Sources of Uncertainty

There are many sources of uncertainty in flood inundation modeling and in flood forecast but they fundamentally differ from each other: natural and epistemic uncertainty. Natural uncertainty stems from variability of the underlying stochastic processes whereas epistemic uncertainty results from incomplete knowledge about the process under study. Most of the natural uncertainties can be treated formally by statistical probabilistic methods, although "Epistemic uncertainties, including subjective uncertainty, lack-of-knowledge uncertainty, and errors, imply that the nature of the uncertainty may not be consistent in time or space, and may therefore be difficult to treat formally by probabilistic methods" (Freer et al., 2011). Uncertainty can be associated with different sources such as epistemic model structure (Apel et al., 2004) or uniform formulae such as Manning's or Chezy's coefficients (Pappenberger et al., 2005). Model parameters can also be sources of uncertainty such as channel conveyance coefficients (Romanowicz and Beven, 2003) as well as model inputs like rainfall-runoff (Loveridge and Rahman, 2014) and inflow (Savage et al., 2014). Moreover, numerical schemes can be sources of uncertainty (Pappenberger et al., 2005). Finally, validation data of the flood model can be a source of uncertainty with satellite data (Stephens et al., 2012; Werner et al., 2005). Figure 4.1 illustrates the location of the sources of uncertainty in flood inundation mapping.

In Butts et al. (2004), further model structure uncertainty details were highlighted such as process descriptions, coupling of the processes, numerical discretization, representations of the spatial variability-zones, grids, sub-catchments, etc., element scale and sub-grid process representations including distribution functions, different degrees of lumping, effective parametrization, etc., interpretations and classifications of soil type, geology land use cover, vegetation, etc.

Tables 4.1 and 4.2 summarize the meaning and the sources of uncertainty present in models. They distinguish between aleatory and epistemic error. The aleatory type of uncertainty is related to the timing of natural events and the inability of the model to capture natural variability and categories as stochastic or aleatory (Willis, 2014).

Uncertainty can be addressed with a sensitivity analysis (Saltelli et al., 2000) that investigates the main uncertainty of the model based on its impact on the results (discharge, water level, flood, extension). A classical method consists of a perturbed parameter at a time with a small amount or at some fixed point; a global method in which all variables or parameters are varied simultaneously over their entire feasible space, and interactions between them are assessed. Sensitivity studies have mainly focused on the uncertainties sourced from model parameters and inputs. Although friction

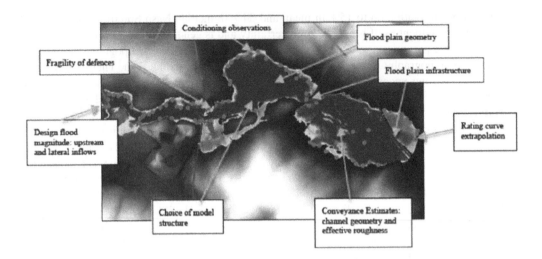

FIGURE 4.1   Sources of uncertainty in flood inundation mapping. (Source: Beven et al., 2015.)

parameters are usually considered the most influential and are therefore given first consideration in

---

TABLE 4.1
A Classification of Different Types of Uncertainty

Type of Uncertainty	Description
Aleatory	Uncertainty with stationary statistical characteristics.
Epistemic (system dynamics)	Uncertainty arising from a lack of knowledge about how to represent the catchment system in terms of both model structure and parameters.
Epistemic (forcing and response data)	Uncertainty arising from lack of knowledge about the forcing data or the response data with which model outputs can be evaluated.
Epistemic (disinformation)	Uncertainties in either system representation or forcing data that are known to be inconsistent or wrong.
Semantic/linguistic	Uncertainty about what statements or quantities in the relevant domain actually mean.
Ontological	Uncertainty associated with different belief systems.

Notes: see complete list in Beven, 2016.

---

model calibration. Hall et al. (2005) and Pappenberger et al. (2005, 2008) use a regional study that suggests different methods can lead to completely different rankings.

Therefore, it is impossible to draw clear conclusions about the relative sensitivity of different factors outside of the fact that parameters present dynamic coupling with each other and are difficult to estimate individually. Shin et al. (2016) showed, for a wide class of conceptual rainfall-runoff models, that climate input period, quantity of predictive interestm and objective function affect parameter sensitivity rankings. However, the major sensitivities can be identified. It is the lesser ones that may change with circumstance, making the use of a sensitivity analysis and precise specification of the quantity of interest, we would argue, still important to understand the relative and crucial sources of uncertainty. Hydrological models present uncertainties such as rainfall forecasts. A vast number of regional and national hydro-meteorological centers have flood forecasting and early warning systems in place based on weather predictions (see Alfieri et al., 2014 for a recent review of European systems).

TABLE 4.2

Sources of Uncertainty in the Model System Separated into Aleatory and Epistemic Uncertainty

Modules	Variability	Incomplete Knowledge (Aleatory Uncertainty) (Epistemic Uncertainty)
Extreme value statistics	• **Annual maximum discharge**	• Measurement errors • Plotting Positions formulae • Selection of data and partial series • *Selection of distribution function of annual maximum discharge* • *Sampling uncertainty of annual maximum discharge* • Parameter estimation for distribution function
Routing	• Changes in river channel over time	• Parameter estimations • Error in model selection
Stage-discharge relation	• Hysteresis during a flood wave • Changes in river channel over time	• Measurement error • *Parameter estimation* • Error in model selection
Levee failure	• Spatial variation of levee geometry • Substrate distribution	• *Measurement errors of levee geometry* • Variability estimations of levee parameters (geometry, substrate, breach width, turf) • *Dimension of levee breaches* • Turf quality of levee cover
Tributaries	• *Correlation main river - tributaries*	• Extent of correlation, measurement error
Damage estimation	• Building use and value • Spatio-temporal course of inundation	• Method of assessing values of buildings and contents • Error in damage model selection in the polder • Parameter estimation (e.g., stage-damage curves)

Notes: sources printed in plain boldface are considered in the Monte-Carlo Framework for the risk and uncertainty analysis. Sources in italic are considered in scenario calculations (Apel et al., 2004).

4.2.2 QUANTIFICATION OF UNCERTAINTY OF HYDROLOGICAL MODELS

Quantification of hydrological model output (mainly discharge) for floods has attracted attention in recent years. There are available methods such as GLUE and the formal Bayesian method, two of the most popular (Jin et al., 2010). They derive from what is traditionally called sensitivity analysis but require more extensive computation. To some extent, both methods provide results with similarity when a higher threshold value (> 0.8) is used. A summary of the Generalized Likelihood Uncertainty Estimation (GLUE) can be found in the paper by Beven and Binley (2014). The main idea consists of investigating the likelihood of some output model based on knowledge of the range of input parameters with the goal of finding optimal parameters. In the early example used for GLUE, four parameters were used to calibrate the model such as saturated hydraulic conductivity (K_s), saturated moisture content (θ_s), initial soil moisture potential (φ_{in}), and overland flow roughness coefficient (f). Then, a 500,000 realization of the model using TOPMODEL was performed to estimate the discharge output obtained with the range of inputs used indicating a 9% and 95% limits envelope.

An example of GLUE performed on the Abolabbas catchment (284 km^2) in Iran using a mixed conceptual and physical-based rainfall-runoff model (AFFDEF) is illustrated in Figure 4.2. The output of GLUE for four flood cases is illustrated in Figure 4.3 and highlights the uncertainty associated with flood discharge.

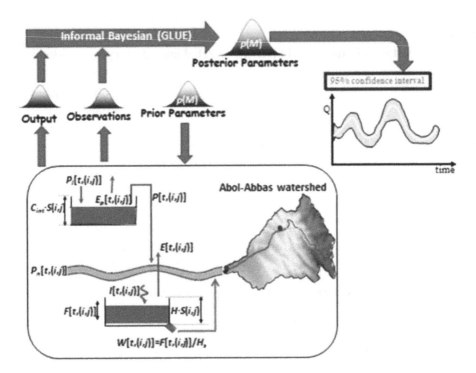

FIGURE 4.2 Schematic representation of GLUE algorithm linked to AFFDEF. A portion of the AFFDEF. (Source: model is adopted from Moretti and Montanari, 2007; Pourreza-Bilondi and Samadi, 2016.)

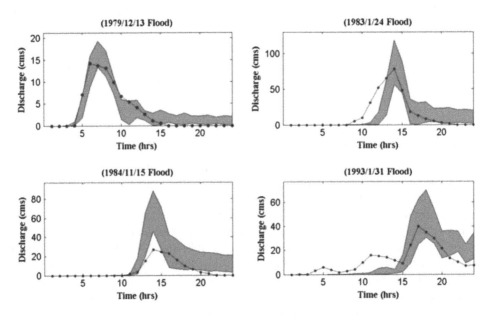

FIGURE 4.3 95% posterior simulation uncertainty ranges with the GLUE algorithm for four events. The dark dots represent the observed data; the shaded area shows the predictive uncertainty. (Source: Pourreza-Bilondi and Samadi, 2016.)

4.3 UNCERTAINTY IN FLOOD MODELING

The purpose of flood modeling is to transform discharge output from a land surface model (LSM) estimating run-off or flood frequency analysis in distributed flood water depth, water extent, and flow velocity. Flood models are essential to fully understand the limitations of the predictions and inundation models and estimation of flood risk. They often use numerical models for solving momentum and mass conservation equations such as Saint-Venant and rely on the digital elevation model (DEM) for the constraints of the floodplain and channel cross-section and a rough estimation of roughness. Moreover, channel cross-sections are often either unavailable particularly in developed countries or placed under confidentiality issues. Therefore, in the global model channel depth and width are estimated using empirical power laws while assuming the geometry of the channel (i.e., rectangular).

The equation remains valid when (1) the horizontal length scale is larger than the vertical one, and (2) the vertical pressure gradient is hydrostatics (Bates et al., 2014). Basically, its core is a hydrodynamics model that solves conservation and momentum equations in 1D, 1D–2D, 2D, and 3D dimensions. Hydrodynamic (HD) modeling tools are widely used for floodplain inundation modeling to high accuracy, but they are resource-intensive, making them impractical to use for large catchments (Ticehurst et al., 2015). It solves the classical equation of Saint-Venant conservation of mass and momentum. 2D can suit floodplain 2D requirements and they are the most widely used. Nevertheless, model results are difficult to evaluate due to the scarcity of observed data and computational efficiency. Modeling can lead to false confidence given by high-resolution outputs, as accuracy is not necessarily increased by higher precision (Dottori et al., 2013).

However, uncertainty in both input data and model parameter is large and difficult to separate as it interacts non-linearly making uncertainty analysis complex. Nevertheless, uncertainty estimation allows the delivery of confidence in the model output, provides space for further improvement, and opens the door to probability estimation that has become more common and more useful in recent years. Input data for flood modeling comes from discharge gauge stations or remotely sensed derived discharge that presents large uncertainty because it is difficult to estimate, in particular during the flooding period. Among the parameters, the roughness estimation remains one of the major uncertainties.

Table 4.3 summarizes the results of a study that used flood modeling to model the water extent or the water high of floods. It highlights the variety of domain sizes and validation methods to estimate the uncertainty associated with the model. Flood extent is usually well captured at ~80% except for the storm surge cases and the Mekong.

Reduction of uncertainty and uncertainty estimation are two approaches associated with modeling. Reduction of uncertainty is generally achieved through a calibration-validation approach. Uncertainty estimation is generally understood as the difference between a modeled and an observed flood. But, uncertainty can also be expressed by multi-model assessment of the same flood. For example, the multi-inundation model ensemble drew interest in the development of global model flood models (GFMs) that are important for international disaster risk management. Results from multi models in Africa with recent large scale flood events (Lokoja, Nigeria; Idah, Nigeria; and Chemba, Mozambique) indicate a critical success index of individual models across the three regions ranging from 0.45 to 0.7, and the percentage of flood captured ranges from 52% to 97% (Bernhofen et al., 2018). Firstly, an ensemble model performs similarly to the best individual and aggregated model. Secondly, the multi-model ensemble permits us to estimate the uncertainty associated with floods. Trigg et al. (2016) in a study over Africa using six global models conclude that there is around 30–40% agreement on flood extent and significant differences in hazard magnitude and spatial pattern between models, notably in deltas, arid/semiarid zones, and wetlands. Figure 4.4 illustrates the agreement between models. Furthermore, the multi-model approach can be developed for regional floods for uncertainty estimation.

TABLE 4.3

Flood Modelling Examples

Location	Model	Size	Calibration	Validation	Score on Variable	References
Oti River basin, West Africa	LISFLOOD-FP	140 km	Discharge NSE = 0.87	Discharge NSE = 0.94	Flood extent 64%	Komi et al., 2017
Damodar River, India	LISFLOOD-FP	110 km			0.77	Sanyal et al., 2013
Central Amazon river basin section, Brazil	LISFLOOD-FP	240 × 125 km	Gauged data observation	JERS-1 SAR flood extent Altimetry radar satellite	72% high water 0.99 m water stage 23% low water 3.17 m RMSE water stage error	Wilson et al., 2007
Bay of Bengal storm surge	LISFLOOD-FP	561.6 km × 396.9 km		In situ measurements and remotely sensed observations of water level	41–43%	Lewis et al., 2013
Ob River, Siberia	LISFLOOD-FP	900 km	0.75–0.94			Biancamaria et al., 2009
Veliky Ustyug at the Northern Dvina River (Europe); (2) Mezhdurechensk at the Tom River (Siberia); (3) Blagoveschensk at the Amur River (Far East)	STREAM_2D	~10 × 10 km	Observed discharge	Observed discharge	88–93%	Belikov et al., 2015
Lower Paraná River (Argentina)	CTSS8-FLUSED	208 km	Water level 0.8–0.9 NSE NSE 0.38 (low fir)		Water level 0.8–0.9 NSE	Garcia et al., 2015
Mundeni Aru River Basin (Sri-Lanka)	LISFLOOD-FP	70 km			80% surface compared to SAR	Amarnath et al., 2015
River Po, Italy		98 km		Observation	90% flood extent	Mukolwe et al., 2016
Mekong River Basin	Rainfall-Runoff Inundation (RRI)	4,350 km	Discharge station/ Landsat7	Discharge station/ Landsat7 NSE = 0.86	Flood extent 67.50–68.27%	Try et al., 2018
Lakhimpur district, Assam State, India	CCHE2D	600 km²	SAR rs	SAR rs	85%	Singh et al., 2017

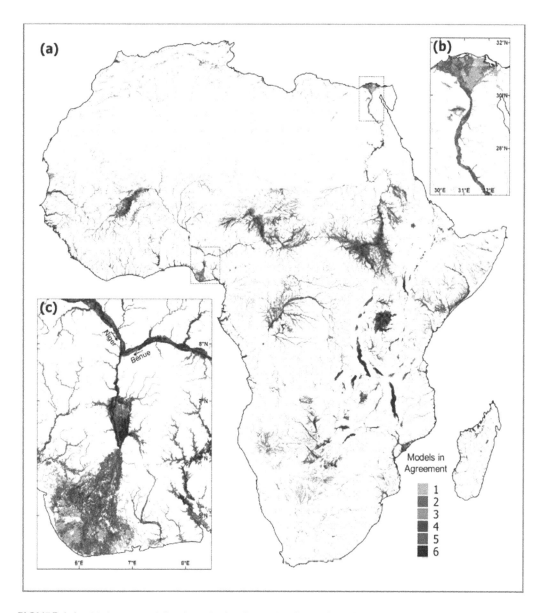

FIGURE 4.4 (a) Aggregated flood results for six models for a 1-in-100 year return period fluvial flood hazard for the African continent. Greyscale indicates how many models predict flooding. (b) Detail for the lower Nile. (c) Detail for the lower Niger, showing areas of strong agreement (narrow confined floodplains at the confluence of Benue and Niger Rivers) and areas of disagreement in the Niger coastal delta. (Source: Trigg et al., 2016.)

4.3.1 RAINFALL UNCERTAINTY

Rainfall input consists of stations extrapolated to a grid with the Thiesen method or grid-based remote sensing data that comes from a satellite. Problems occur depending on the distance between stations and the climate situation of the model. For instance, a tropical area exposed to a convective rainfall system will present high uncertainty in addition to being located in a poorly gauged area. Results indicated that gridded remote sensing-based input presents high uncertainty. Furthermore, large differences exist depending on location in the rainfall grid performance as illustrated by Sun et al. (2018). Yoshimoto and Amarnath (2017) tested the use of satellite rainfall estimates (SREs) as input in flood model for the Mundeni Aru River Basin in eastern Sri Lanka. The results found that SREs give satisfactory discharge results even if the volumes of precipitation of the SREs tended to be smaller than those of the gauged data. An approach to account for rainfall uncertainty is the use of the Monte Carlo approach (Felder and Weingartner, 2016). It allows for the generation of various spatio-temporal distributions of extreme precipitation. Then, rainfall can be tested on a rainfall-runoff model and then a hydrograph can be generated. In return, the best matching rainfall can be used as the optimal rainfall distribution.

4.3.2 LSM and Rainfall-Runoff

The land surface model requires input data that are not always completely available, and model structures are only a crude description of the underlying natural processes. Therefore, model parameters need calibration in the LSM model. Different model concepts (inter-flow, direct runoff), or rather the processes represented, such as infiltration, soil water movement, etc. are more or less dominating different sections of the runoff spectrum. Soil properties show high heterogeneity at different spatial scales and their correct characterization remains a crucial challenge (Baroni et al., 2017). Most models do not account for such transient characteristics inherent to the hydrograph (Cullmann et al., 2009). This falls together with uncertain input data (e.g., rainfall intensity on different scales and the rainfalls' spatial distribution, especially if rainfall is a predicted parameter.

Therefore, the uncertainty that appears on the forecast affects the accuracy of flood prediction, added to that which is possessed in the rainfall-runoff model. An example of discharge uncertainty using a distributed tank model instead of LSM by Kardhana and Mano (2009) estimates total uncertainty by quantifying mean error and standard deviation on the precipitation and discharge forecast. However, the result has shown that the precipitation forecast is more uncertain than discharges. Regarding lead time, it was observed that uncertainty is significantly increased after 12 hours and draws a common characteristic between both models.

4.3.3 Discharge Boundary Condition

Discharge input is one of the most fundamental observation inputs for flood models. The uncertainty of the discharge is affecting the uncertainty in the flood modeling. Bermúdez et al. (2018) studied the July 2007 floods of the River Severn in the UK using input from a gauge discharge station and reported that flood extent estimation was only achieved with discharge uncertainty estimating boundary condition uncertainty and local rainfall contributions. As discussed in Chapter 3 ("Flood Observation Uncertainty") (Vuillaume and Alifu, 2022), discharge derived from remote sensing is widely used and presents large uncertainty when discharge estimation is extrapolated to the flood condition. Using discharge estimated from ground measurement or derived with remote sensing water stage as a boundary condition for the model is linked to uncertainty that affects the flood model and can be used to estimate the uncertainty of the final flood map.

4.3.4 Roughness Coefficient

Among the parameters, roughness coefficient estimation remains one of the most uncertain. Coefficients are often derived from pebble count field surveys and empirical formulas and suffer from scaling when input in models. Furthermore, roughness has most likely strong variability inside a channel and in a floodplain which presents seasonal variability due to vegetation. Pappenberger et al. (2005) used the GLUE approach to investigate roughness uncertainty. Papaioannou et al. (2017) used the roughness coefficient uncertainty in hydraulic modeling for probabilistic flood inundation mapping at ungauged streams with the HEC-RAS model. Some attempts to derive vegetation roughness from space-borne observation have been conducted by Straatsma and Baptist (2008) in a river section of the River Waal, a distributary of the River Rhine in the Netherlands. The modeling results showed improvement of water levels.

4.3.5 Topographic Data Sources and Spatial Resolution

One of the most important data in flood modeling is topographic data such as the digital elevation model. Elevation determines the topographical and geomorphological characteristics of a watershed, which are very important factors that determine the flow direction and accumulation of water on the surface. However, the sources and the spatial resolution of the elevation data determine the accuracy and precision of the final output of flood simulation models. DEMs are generally derived from various sources such as remote sensing techniques (the Advanced Spaceborne Thermal Emission and Reflection Radiometer [ASTER], the Shuttle Radar Topography Mission [SRTM], and the Light Detection and Ranging (LiDAR) and ground surveys [e.g., topographic contour maps]). All these data are characterized by various uncertainties in flood modeling arising from different spatial resolutions used (Md Ali et al., 2015). Impacts of spatial resolution on the elevation models have been previously investigated (Hervoeut and Van Haren, 1996; Hardy et al., 1999; Horritt and Bates, 1999). An investigation by Bates and De Roo (2000) showed that high-resolution spatial resolution (100 m) resulted in better model performance in predicting water level than the coarse resolution of 1,000 m. The accuracy of flood simulation at spatial resolution is dependent on the scale of mapping and the types of flood models used. Spatial resolution has a great effect on the flood model's predictive ability. Generally, spatial resolution will affect (1) bulk flow characteristic, (2) inundation extent, (3) and calibration parameters and internal results of flood models (Hardy et al., 1999). In most river flood modeling, low DEMs are often used as inputs to models, however, in a complex urban system where small urban features (e.g., such as roads, buildings, embankments), which often affect flow are difficult to represent in the model for simulation, it is highly important to make use of high-resolution data.

The use of high resolution is highly complicated when modeling at regional or global scale. While it may be ideal, the computing time and resources often prevent the use of high resolution, rather most researchers often adopt the coarse data for simulation. However, possibilities of modeling at large to global scales using high resolution have been proposed by some researchers (Mateo et al., 2017). Models such as CaMa-Flood and GLOFRIS (Winsemius et al., 2013) were proposed to simulate a 1×1km grid. The LISFLOOD-FP model was used by Sampson et al. (2013) to simulate global flood hazards using high resolution (~90m). Despite the endeavor to simulate regional floods at high resolutions, some level of uncertainty is still visible (Mateo et al., 2017). It is therefore important in flood modeling to understand the level of uncertainties expected when simulating at low or high resolution; this will assist modelers in ensuring near-accurate prediction for water flow.

4.3.6 Flood Model Outputs Error

Output error arises from the differences between the simulated floodplain map and the observed. Although the uncertainty associated with the flood model outputs comes from input data errors and model calibration parameters, it is very essential to quantify the level of uncertainty in both the simulated and the observed data. According to Bates et al. (2014), in flood inundation problems, uncertainty or errors associated with modeling can be dealt with by analyzing the output error. In general, the objective of flood modeling is to generate a floodplain map in terms of flood characteristics such as water depths, velocity, and duration, which can be integrated with exposures and vulnerability models to develop a potential flood risk map for flood disaster risk reduction plan and policy decision. As such, the flood inundation map must be close to being accurate, hence the need for the quantification of the uncertainty level. In order to model the errors of the output result, conditioning model prediction is often performed by comparing the simulated with the observed with the assumption that the errors are additive and independent both in time and space. With these assumptions, a statistical analysis such as data assimilation is applied (Bates et al., 2014).

Most of the errors in the output of the flood model are input errors and are basically associated with the hydrological data supplied. Hydrograph shape selected as boundary condition, according to Alemseged and Rientjes (2007), affects the simulation results. In principle, the change in water volume in the model domain will cause corresponding differences in flow characteristics such as water depths and velocity. The velocity estimation, however, has been associated with lots of modeling complexities (Bates et al., 2014). Discharge hydrographs are derived in two ways: (1) through hourly or daily river gauge observations or water level measurements; (2) by simulation using rain-runoff or hydrological models. Uncertainty in the initial method arises from the deficiency of measuring gauges, incomplete data records (especially in developing countries), and ungauged rivers. For missing or incomplete data, most times hydrologists adopt interpolation or extrapolation methods to derive the discharges, which may not give an accurate and expected hydrograph shape, thereby resulting in over- or under-estimation of flood model outputs. Uncertainty in hydrological modeling according to Baldassarre and Montanari (2009) as listed by many hydrologists are (1) uncertainty in observation data such as rainfall and river levels, (2) calibration parameter uncertainty, and (3) structural uncertainty of the model originating from imperfect data. In order to ensure accuracy in flood model output, therefore, it is very important to estimate the uncertainties in the observed and simulate discharge data, which serve as inputs to the flood model.

Validation of the flood model outputs is very important. An event-based flood model validation is done using earth observation data such as LiDAR, radar, and some optical images (Komolafe et al., 2018). Images of the same flooding events are acquired and processed to obtain surface water distribution; this is overland on the simulated and compared. Other sources of flood model validation are through the previous newspapers and administration of questionnaires within the floodplain. These validation sources do, however, contain some elements of errors and uncertainty, which can lead to very wrong conclusions on the accuracy of the simulated flood hazards. For instance, the analysis of remote sensing data such as the derivation of surface moistures and water distribution is dependent on the bands/channels used and the thresholds of pixels adopted. Inaccurate mapping of the surface water distribution from the imagery could lead to lots of wrong assumptions about the efficiency of the flood model and the accuracy of the results output. Effective analysis and understanding of the level of uncertainty of the flood hazards map validation source can enhance near-accurate prediction of floods (Bates et al., 2014).

4.3.7 Flood Modeling Approaches

As stated earlier, flood inundation mapping employs the use of various hydrodynamics models in different dimensions 1D, 2D, and 3D. Hydrodynamic models are mathematical models that define fluid motion for solving water flow computationally by applying physical laws (Teng et al., 2017). They are the most used models for flood simulation as they are able to link river models with hydrological models and they provide distributed representations of flood flow which are very significant in the preparation of flood risk maps for planning and decision making (Teng et al., 2017). The choice of these models, which are dependent on many factors such as the data availability, catchment scale and size, river geometry, and study objectives, however, plays an important role in the accurate prediction of water flow. A one-dimensional model represents floodplain flow as one dimension along the river while 2D models represent floodplain flow in a two-dimensional field using continuity and dynamic wave momentum (motion equation):

$$\frac{\partial h}{\partial t} + h\frac{\partial V}{\partial x} = I \tag{4.1}$$

$$S_f = S_o - \frac{\partial h}{\partial x} - \frac{V}{g}\frac{\partial V}{\partial x} - \frac{1}{g}\frac{\partial V}{\partial t} \tag{4.2}$$

Where "h" is the flow depth and "V" is the depth-average velocity in one of the eight flow directions x. I, Sf, and So denote the excessive rainfall intensity and friction slope, which are determined by the Manning equation and the bed slope pressure gradient, respectively (Flo-2D, 2009; Komolafe et al. 2018). Three-dimensional modeling represents the floodplain flow in three dimensions. It has evolved in recent years as an approach to solving various complexity in floodplain flow dynamics such as modeling of vertical turbulence, vortices, and dam breaches. Although there have been advances in hydrodynamic modeling, the choice of models used can influence the expected output of the flood flow analysis. These models have both advantages and disadvantages and as such, there will be a need for models integration for accurate prediction of floods.

4.3.8 SPECIAL CASES

4.3.8.1 Urban Flood

This type of flood usually requires LiDAR-based DEM with a good understanding of the urban area affected by the flood. Results indicate high-accuracy performance such as in an urban basin in South Florida using the HEC-HMS model in a small urban basin in West Palm Beach, Florida, in the USA (Stella and Anagnostou, 2018). Another case study focused on the Kuala Lumpur 2003 urban flood also using LiDAR indicates that top-view LiDAR data cannot detect whether passages for flood waters are hidden underneath vegetated areas or beneath overarching structures such as roads, railroads, and bridges. Furthermore, hidden structures can play a major role in urban flood and will conduct a mismatch in both flood water depth and flood propagation pattern (Meesuk et al., 2015).

4.3.8.2 Arid Flood (i.e., Wadi)

An example provided by Saber and Habib (2015) on flash floods modeling for wadi systems illustrates the fundamental limit on the capacity of the rainfall-runoff model to reproduce the observed flow due to transmission loss in arid regions. Few field studies have investigated the transmission loses (Crerar et al., 1988; Hughes and Sami, 1992) even though Sorman and Abdulrazzak (1993) recognized it as one of the fundamentals in such an area and reported that Wadi Tabalah, southwest Saudi Arabia basin, experienced 75% of bed infiltration to the water table. Another example provided by Maref and Seddini (2018) on the Wadi Mekerra basin catchment situated in the northwest of Algeria shows it is characterized by a semiarid climate, convective thunderstorms, and ephemeral flow. Results indicate that the model underestimates peak runoff due to convective rainfall that may be very localized in space. Even achieving "good" Nash-Sutcliffe (NSE) during calibration and validation, the analysis depends strongly on the potential maximum retention parameter which is related to the land use type and varies significantly between seasons according to the vegetation cover dynamics, rainfall intensity, and drought.

4.3.8.3 Flood Modeling in the Humid Tropical

Humid tropics have unique characteristics in terms of water cycle and typically present strong shallow throughflow during storms. The humid tropical region presents much warmer and uniform temperature, large inter-annual and intra-seasonal variability, as well as strong spatial variability due to the predominance of the distributed convective system as the main rainfall provider in the tropics (Wohl et al., 2012). In addition, the process of run-off in this region comprises excessive infiltration, saturation, unsaturated subsurface flow, and saturated subsurface (Paul et al., 2002). In this region, especially in West Africa, due to the excessive rainfall during raining season, runoff is greatly increased because most catchments are wetted up and groundwater level rises to the surface. In this case, the soil moisture content is exceeded. In modeling the runoff, therefore, accurate

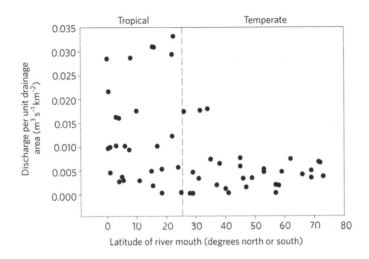

FIGURE 4.5 Average discharge per unit drainage area as a function of latitude at the river mouth, showing differences between tropical and temperate latitudes. The range of discharges from tropical latitudes is much greater and includes some of the largest riverine fluxes of water on the Earth. (Source: data on drainage area, latitude, and discharge taken from Herschy, 1998; from Wohl et al., 2012.)

characterization of the soil formation of the sub-surface strata is very essential (Paul et al., 2002). Tropical hydrology is less studied and understood than its counterparts such as mid-latitude or polar, resulting in high uncertainty on hydrological models. It results in a larger discharge spread compared to the temperate region as illustrated in Figure 4.5. It indicates that the region results in larger rainfall uncertainty that leads the larger discharge input uncertainty to flood modeling. Several rain-runoff models have been applied in many regions of the world, but very few have been applied to humid regions. Even the published results of model applications are characterized by uncertainty due to the complexity of the runoff processes and the availability of data. In developing countries, most hydrological data acquired by river basin authorities are less detailed and mostly inadequate. Rain and discharge data are mostly collected weekly or daily, which is insufficient for modeling sub-daily rain storms. Furthermore, uncertainty in flood modeling in the region is associated with inadequate information about the subsurface data such as groundwater flow, groundwater level, soil conductivity, and infiltration data. Other issues are the several ungauged river basins, which have been wrongly modeled with a high level of uncertainties and misleading results.

4.3.8.4   Ice Jam Flood and Glacial Lake Outburst Flood (GLOF)

Ice jam floods are associated with cold regions (i.e., Canada) on rivers and streams that are prone to ice-related flooding, but can occur during winter over mid-latitude regions such as Romania (Sabău et al., 2018). Rokaya et al. (2018) indicated that 60% of the rivers in the northern hemisphere experience significant seasonal effects of river ice. Ice jams form during breakup events and offer a challenging modeling challenge (Beltaos and Kääb, 2013). Ice jam events are highly variable and difficult to evaluate, especially with respect to the likelihood of site-specific re-occurrence and statistical frequency. Furthermore, field data are often unavailable and ice-related floodplain delineation is less developed. In addition, no standard for flood hazard delineation exists (Kovachis et al., 2017). However, an attempt to model ice jams was proposed by Beltaos et al. (2012) using the Hydrologic Engineering Center's River Analysis System (HEC-RAS) model, which can simulate ice jams. A calibrated model for operational application along the international Saint John River from Dickey, Maine, USA, to Grand Falls, New Brunswick, generates good results, but model parameters change from site to site. Large uncertainty remains in modeling and forecast of ice jams that highlight the limitation of the current modeling of such physical phenomena. Figure 4.6 illustrates

FIGURE 4.6 Upstream-looking views of toe (upper) and head (lower) of ice jam on May 3, 2014; note loose accumulation of ice floes near the head. (Source: courtesy of J. Straka and O. Gray of Parks Canada; Deltaus, 2018.)

the upstream-looking views of the toe (upper) and head (lower) of an ice jam that occurred on May 3, 2014, in Canada.

An example is an event that occurred in 2012 when a landslide from a moraine slope triggered a multi-lake outburst flood in the Artizon and Santa Cruz Valleys, Cordillera Blanca, Peru (Mergili et al., 2018). This phenomenon highlights the high sensitivity of process chain modeling to key inputs: release volume, basal friction, or the entrainment coefficient. The processes involved the input parameters and parameter results in high uncertainty as illustrated by Figure 4.7 which indicates the contour of the ice, therefore directly impacting the discharge below and therefore the resulting flooded area. It underlines the requirement of uncertain simulation results for decision-making as a result of the high uncertainty in flood modeling.

4.3.8.5 Ungauged Basin

Ungauged basins have strongly benefited attention from the Prediction in Ungauged Basin (PUB), IAHS initiative operating throughout the decade of 2003–2013. This was meant to improve the scientific knowledge and estimation of hydrological characteristics of the ungauged basins because of the poor behavior of the hydrology (Tim van et al., 2015). As a result, many data acquisition techniques and hydrological modeling approaches have been developed. Generally, methods commonly adopted in un-gauged sites to estimate flood frequency distribution are usually based on statistical or regression analyses performed on a limited number of hydrological parameters, treated as self-standing non-interacting pieces of information (Boni et al., 2007). Although many river basins worldwide are sufficiently gauged, ungauged basins are often associated with developing countries located in the arid/tropical area. This region suffers from a basically ineffective implementation of hydrological policies, lack of willpower and resources. Problems associated with ungauged basins are basically a lack of data for calibration and validation. According to Boni et al. (2007), lack of validation influences the validation process of hydrological results in an ungauged basin and also prevent the improvement of the model.

4.4 UNCERTAINTY IN FLOOD FORECASTING

Uncertainties of flood forecast inherit from the uncertainties present in the flood modeling system described before, as well as the uncertainty present in the land surface model (LSM) that computes the run-off over the water that is infiltrated, evaporated, and/or stored in the soil, and the numerical weather prediction (NWP) system. Furthermore, there is uncertainty with the lead time and the flood's nature (i.e., river flood and flash flood).

Due to large uncertainty in flood forecast derived from uncertainty in rainfall forecast, flood modeling, and post-processing, several cases exist when a flood was not predictable due to high

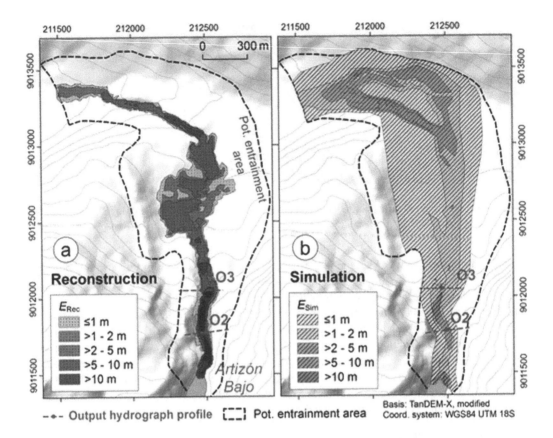

FIGURE 4.7 Reconstructed and simulated entrained heights in the potential entrainment area (Figure 4.6). (a) Reconstructed entrained height (E Rec); (b) simulated entrained height (E Sim). (Source: Mergili et al., 2018.)

uncertainty. Moreover, uncertainty will vary with the lead time considering the assumption that uncertainty will grow with the lead time following the behavior of rainfall forecast uncertainty. Therefore, multiple uncertainties exist in the flood forecast, presenting risks involving flood control decisions.

The ensemble flood forecast aims to give a range of possible discharges based directly on the range of possible weather conditions and quantitative rainfall prediction (QPF) obtained. Ensemble forecasting is a form of Monte Carlos analysis using random change on the initial condition of the model. The ensemble accounts for two sources of uncertainties, imperfect initial conditions which hold strong control over the forecast due to the chaotic nature of atmospheric equation evolution and the uncertainty in the model formation and the approximation of mathematics method to solve partial equations.

Uncertainty in streamflow forecast and therefore flood is represented using ensemble. While it does not represent the full uncertainty of the forecast it has a practical advantage to incorporate different sources of uncertainties in the forecast process. A second method that is growing in use is the assimilation of real-time stream gauge discharge to the hydrodynamics model to reduce the uncertainty of the model by merging the model and observations at a specific location.

Flood forecast targets the determination of both timing and intensity of flood. Uncertainty in flood forecast was poorly addressed before the implementation of ensemble forecast that uses multi-realization based on initial numerical weather prediction (NWP) ensemble, physical parameterization ensemble, or multi-model ensemble. The issue of ensemble provides (1) the uncertainty associated with the flood forecast and (2) the capability to predict events that can't be detected by deterministic forecast (Ferraris et al., 2002), particularity in the case of flash flood illustrated for the Mediterranean region.

Flood forecast can focus on single or multiple-point locations and estimate the river discharge or either provide a full 2D extent map that will provide the exact location of water with its flood depth, velocity, and discharge. Discharge forecast already constitutes one of the major pieces of forecast information for floods. Compared to past flood discharge levels, it consists of basic information to provide flood forecast information. Furthermore, the use of ensemble forecasts will give information on the uncertainties associated with the flood forecast.

The idea behind ensemble simulations relies on both the quantification of the uncertainty in hydrological forecasts and the implementation of ensemble-based data assimilation schemes for uncertainty reduction in non-linear systems. For instance, many studies with ensemble Kalman filter applications have considered the perturbation of forcing, states, and/or parameters by adding a Gaussian random number to deterministic values (Mendoza et al., 2012). Figure 4.8 illustrates the European Flood Awareness Systems Ensemble Prediction System with its components such as the meteorological forecasts, the hydrological model, and the post-processing and web interface.

Various studies of flood forecast on diverse basins exist; they consist of a deterministic/ensemble base rainfall forecast based on the numerical weather prediction model, feeding a land surface model that outputs data used in a river hydrodynamic model combined with a floodplain inundation model. All models present uncertainty and one approach to quantifying them is to use an ensemble rather than a deterministic rainfall forecast. While ensemble forecasts are usually produced at coarse resolution and therefore lower accuracy than the ensemble, ensemble provides greater information of the probability distribution of the rainfall and therefore the probability of streamflow discharge and inundation extent. Flood forecast uncertainty presents the characteristics of a cascade uncertainty model. Flood forecast consists basically of a rainfall forecast as output used as input in a land surface model that computes runoff and then discharge and floodplain inundation extent and depth. Primary uncertainty comes from the rainfall forecast as well as rainfall observation since most of the flood modeling used gridded rainfall observations that combine stations and satellite remote sensing products. Figure 4.9 illustrates the different uncertainties that are inputted in the forecast such as the rainfall, the parametrization of the model, and initial condition calibration that impact the output. Each part of the system presents uncertainties such as the forecast precipitation errors, the structural errors, and the output errors.

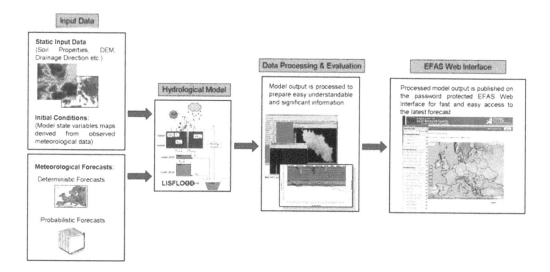

FIGURE 4.8 Ensemble prediction systems – key to longer flood warning times. (Source: EFAS, 2018.)

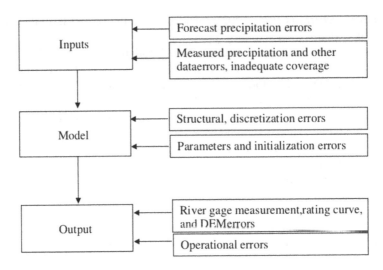

FIGURE 4.9   Error framework for rainfall-runoff models used in flood forecasting. (Source: Jain et al., 2018.)

4.4.1   Initial Condition/Real-Time Assimilation: Soil Moisture

Soil moisture uncertainty estimation is particularly useful for flood forecasting as it has a strong impact on flood occurrence. An analysis of the real-time forecasting of the Upper Danube (Nester et al., 2016) reported a model performance of 88% of the snow cover in the basin on more than 80% of the days.

4.4.2   Rainfall Forecast Uncertainty

Rainfall forecasting relies mainly on NWP which can be used at different lead times and therefore is associated with different sources of uncertainty conditioned by the primary driver of the model (initial condition or boundary condition). We can distinguish between:

- Short-term or nowcast (up to nine hours) that mainly rely on radar observation and targeting flash flood-forecasting.
- Short range to extended range of one to ten days that are first conditioned by initial condition and later with boundary conditions, therefore strongly affected by uncertainty in initial data such as pressure, temperature, soil moisture, and water vapor, for example.
- Sub-seasonal and seasonal forecast, strongly dependent on the boundary condition predicted by the global circulation model.

Furthermore, the NWP presents various performances depending on the latitude of the forecasted domain reflecting the lower skill of prediction at the equator. Both the lack of observations from several developing countries located at the equator and the high frequency of convective-based rainfall systems explained that situation. Uncertainty of flood forecast is also linked with the type of weather system that occurred. In fact, regarding heavy precipitation, it is well-known that accurate forecasts associated with deep moist convection are challenging due to uncertainties from NWP, and high sensitivity to misrepresentation of the initial atmospheric state (Lombardi et al., 2018). A well-known process to decrease the uncertainty of the NWP is to optimize the calibration of the system to the region as illustrated by the Weather Research and Forecasting's (WRF) various optimizations: for example, Italy (Avolio and Federico, 2018), Cyprus (Tymvios et al., 2018), Peru (Moya-Álvarez et al., 2018), India (Mohan et al., 2018), West Africa (Gbode et al., 2018). or based on the weather cluster circulation (Vuillaume and Hearth, 2018). The case of flash flood forecast presents high dependence on rainfall uncertainty. Edouard et al. (2018) indicates that flash flood's major source of uncertainty comes from rainfall forcing and convective-scale meteorological ensemble prediction systems. Moreover, model parameter and initial soil moisture also present strong uncertainty that limits the predictability of models.

4.4.3 THE HYDROLOGICAL/LAND SURFACE MODEL UNCERTAINTIES

The hydrological model or the land surface model is one of the key elements of the system at present. The impact of soil moisture uncertainty from the LSM is well known but still poorly addressed in flood forecast modeling due to the complexity of observation measurements that are remotely sensed and derived. One of the main difficulties that flood forecasters are faced with is the evaluation of how errors and uncertainties in forecasted precipitation propagate into stream flow forecast. In theory, these errors should be combined with the effects of different initial soil moisture conditions that will have a significant impact on the results of a flood forecast. Moreover, small catchments are strongly influenced by forecast rainfall, soil moisture, and ensemble size because of the short response of the river basin to variation.

4.4.4 ENSEMBLE SYSTEM

Ensemble-based theory takes advantage of stochastic simulation conditioned by the chaotic nature of weather forecasting to prove an estimation of the uncertainties of the rainfall and therefore the river discharge. The current system usually does not use a multi-hydrology model to also account for the uncertainty in the hydrodynamic model. Flood forecast uncertainty is usually highlighted with probability discharge on a river usually derived from the rainfall uncertainties from the numerical weather prediction model. The ensemble rainfall forecast usually reflects the uncertainty in the initial condition of the weather forecast or a multi-physics ensemble, or a multi-model that uses several parameterizations of the cumulus, microphysics, planetary boundary layer, and radiative transfer for the current system (Diomede et al., 2008). The ensemble is supposed to capture the prediction of discharge in the range of variability of the ensemble providing uncertainty in discharge that can be input in a flood model as described before. Therefore, flood forecast uncertainty usually focuses on the uncertainty in the input of the flood modeling part. Flood forecast is provided at a different scale that matches both natural phenomena such as flash flood associated with a strong convective activity that lasts a few hours and longer-term flood forecasts that are targeting floods associated with snow melt and soil moisture, i.e., physical phenomena that are seasonal. As expected, flood forecast uncertainty increases with the lead time for each type of forecast: flash flood (Creutin et al., 2013), extended forecast lead time (ten days; Alfieri et al., 2014), and seasonal streamflow forecast, therefore flood forecast.

Ensemble is currently deployed in several centers to estimate the probability of discharge in several rivers. The Ensemble Hydrological Prediction System is implemented at several centers (Wetterhall et al., 2013). For example, the European Forecast Awareness System (EFAS; Thielen et al., 2009) is a multi-model ensemble approach accounting for uncertainty from several weather forecasts (two deterministic high-resolution and two ensembles). To capture some of the uncertainty in the weather predictions, EFAS has been designed to operate with several numerical weather prediction (NWP) systems and then use the LISFLOOD hydrological model. Alifieri et al. (2014) indicated that skillful predictions are found in medium to large rivers over the whole ten-day range. Figures 4.10 and 4.11 illustrate the uncertainties associated with discharge estimation at the output of the ensemble providing an estimation of the uncertainty of discharge.

The performance of ensemble forecasts is usually estimated with Figures 4.12 and 4.13. As expected from such a global approach at a European scale, there are significant differences in skill for different rivers.

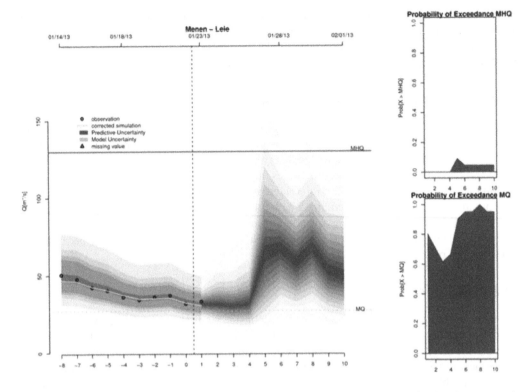

FIGURE 4.10 Post-processed ten-day discharge forecast at Menen on the River Leie showing summaries of the post-processed forecast along with the probability of exceeding the daily mean (MQ) and mean annual maxima (MHQ) observed discharges. (Source: Smith et al., 2016.)

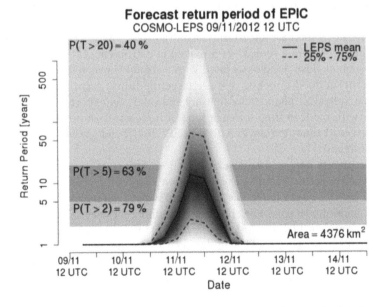

FIGURE 4.11 Return period plot of probabilistic EPIC forecast for 9/11/2012 12 UTC. Reporting point for the Piave river at the outlet, NE Italy. (Source: Smith et al., 2016.)

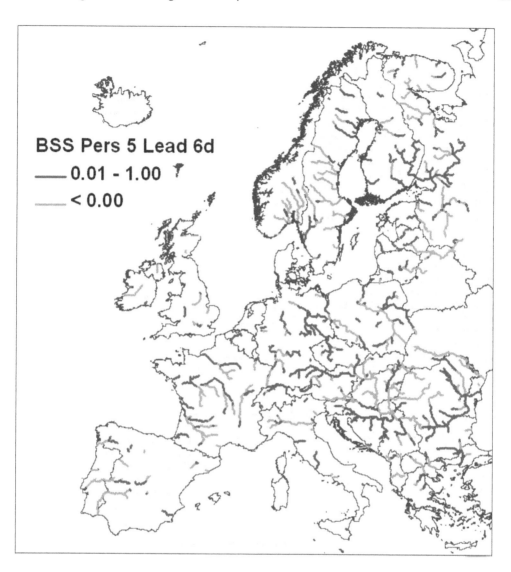

FIGURE 4.12 Areal distribution of BSS median values for six days of lead-time and persistence (> 5 EPS). (Source: Wetterhall et al., 2013.)

4.4.5 FLASH FLOOD

Flash flood is described as a rapid and extreme event often associated with a convective system that developed during a strong vertical gradient of temperature or an atmospheric flux of water vapor. Therefore, uncertainty in terms of lead time and rainfall quantity is high due to the limited amount of information available prior to the flood. Alfieri et al. (2014) investigated rainstorm events and flash floods in Europe denoting a probability of detection up to 90%, corresponding to 45 events correctly predicted, with an average lead time of 32 h. Further study by Alfieri et al. (2015) summarized the development of flash flood prediction. Flash flood systems are usually based on rainfall observations derived from rain gauge networks and weather radars, rather than on forecasts. The authors highlight the recent development of flash flood forecast using ensemble to manage the high uncertainty of location, intensity, and duration of flash floods. A study from Huang et al. (2015) on the full 2D hydrodynamic modeling of rainfall-induced flash floods in Lengkou catchment in Shanxi Province (China) indicates good performance of hydrological modeling but, with a stage

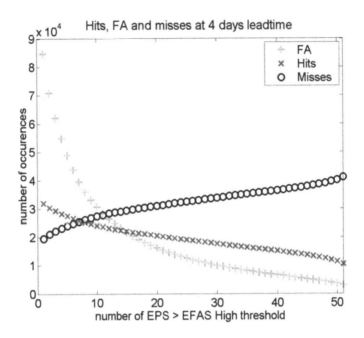

FIGURE 4.13   Absolute numbers reporting the three contingency TABLE fields "hits" (h [x]), "false alerts" (f [+]), and "misses" (m [o]) for at least 5 EPS in the previous forecasts at lead time four days. (Source: Wetterhall et al., 2013.)

and discharge hydrographs more sensitive to the Manning roughness and initial water content in the catchment than to infiltration.

4.5   HYDROLOGICAL DATA ASSIMILATION

Hydrological data assimilation that has been developed in recent years has pointed out that assimilation of streamflow observations can improve the hydrologic routing model prediction (Mazzoleni et al., 2018). Results from Neal et al. (2009) using data assimilation of from river discharge derived from synthetic aperture radar (SAR) indicates that discharge estimation from space assimilating water level measurements led to a 79% reduction in ensemble discharge uncertainty over the coupled hydrological hydrodynamic model alone. Multi-parametric variational data assimilation was applied to the uppermost area of the River Main in Germany. The multi-parametric assimilation shows an improvement of up to 23% for CRPS performance and approximately 20% in Brier Skill Scores with respect to the deterministic approach. It also improves the skill of the forecast in terms of a rank histogram and produces a narrower ensemble spread (Alvarado-Montero et al., 2017). There is a strong potential for data assimilation to handle the various sources of uncertainty in hydrological models. Sequential DA's objective is to update recursively the hydrological model states that can also be used for parameter correction (roughness, channel depth) in a modeling optimization framework and to update variables (discharge, water depth).If the ensemble DA assumption supposes observation error as Gaussian, then DA can't handle property the flood event. Therefore, advanced methods such as particle filtering that do not assume applicability to a non-Gaussian state model are used (Noh et al., 2013). An example of particle filter sequential assimilation for the 887 km² at the Katsura River catchment in Japan with 24 hours forecast water and energy transfer process (WEP model) indicates strong improvement of peak discharge estimation and substantial uncertainty reduction that degrades with forecast lead time.

Hydrological data assimilation usually refers to modeling of real-time assimilation of discharge or water depth measure either from gauge stations or remote sensing-based such as a rating curve.

The data assimilation permits to combine the uncertainty of the model and the uncertainty of the observation in an optimal manner and even take advantage of localization property (ensemble Kalman filter transform property) of observations to spread observation near the observations. Table 4.4 summarized some recent studies that used data assimilation mainly from remote sensing to improve water level (using altimetry satellite, synthetic aperture radar [SAR]), Manning coefficient (SAR), water extent (SAR), and flood extent (optical sensor like MODIS). The gain in terms of water level is focused on the downstream part of the river and qualitative improvement is usually observed with water level uncertainty, flood extent, or Manning coefficient.

Finally, a recent study (Mazzoleni et al., 2017) on data assimilation using crowd-sourced data reported the growing inclusion of citizens in participatory processes related to water resources management due to the spread of low-cost sensors. It supported that quality of data influence more the model than the irregularity of measurement. An experiment on Bacchiglione catchment (Italy) in May 2013 using a semi-distributed hydrological model demonstrates the usefulness of integrating crowd-sourced observations and concludes that the assimilation of crowd-sourced data can ensure high model performance for high lead-time values depending on their location (upstream), that crowd-sourced observations can be biased and inaccurate, and that citizen motivation as a "community of friends" has the best effect in the model performance. Yu et al. (2016) investigated the validating city-scale surface water flood modeling using crowd-sourced data through August 12, 2011, in the city of Shanghai, China. The results suggest that the model is able to capture the broad patterns of inundated areas at the city scale.

4.6   SUMMARY AND CONCLUSIONS

Flood modeling presents large uncertainties originating from epistemic (process, boundary, and parameter) and random error. The flood or inundation modeling can be one of the main modules of a flood forecasting system. Both types of modeling consist of sequential modeling from rainfall, land-surface model hydrology, and hydro-climatology-dynamics/inundation modeling. Therefore, uncertainty associated with one module affects the whole chain and the final product such as flood extent and flood depth.

Different approaches have been used to reduce and quantify uncertainty in both flood and forecasting. The starting point of the modeling is the calibration-validation approach that consists of reducing uncertainty by fitting parameters that minimize the error. However, studies have shown that this minimization is highly dependent on the calibration/validation data used. Moreover, the solutions of the optimization parameters can be equivalent while being very different despite the performance of modeling being high. Nevertheless, a likelihood of combination can be drawn when the problem is approached with the stochastic framework.

On the other hand, flood forecast is limited by the chaotic nature of weather forecast that is associated with the strong uncertainty in the initial condition and the equations used in weather forecasting. Regarding uncertainty, its quantification is revealed to be necessary. The ensemble method consists of estimating the uncertainty as a probability distribution of flood occurrence with the assumption that the ensemble member can express the full distribution of rainfall forecast. In addition, the uncertainty in ensemble can as well be reduced using some ensemble statistics post-processing methods that are able to reduce bias error (Gneiting et al., 2005; Pagano et al., 2013). A parallel approach to reduce uncertainty is data assimilation (i.e., ensemble Kalman transform filter EKTF, 4D variational), which can be either real-time (using stream flow observation) or be a tool to reduce uncertainty (i.e., Manning coefficient; Lai et al., 2014). However, improvement can be either very good or very limited since it depends on both the uncertainty of the model and the observation used for the assimilation.

Flood modeling has recognized uncertainty related to rainfall input, discharge boundary, digital elevation model (DEM) accuracy, and static parameters such as Manning coefficient and channel geometry. In addition, the hydrological model being strongly developed in the mid-latitude, the

TABLE 4.4

Summary of Relevant Studies where Satellite-Derived Data Was Assimilated within a Hydraulic and/or Hydrological Model

Study	Satellite Sensor/ Aquisition Frequency	Model	DA Method	Study Area and No. of In Situ River Gauges Used	Study Period	Objective/ Approach	Key Findings	Uncertainty Reduction
Guistarini et al., 2011	ERS-2 SAR and ENVISAT ASAR (WL)	HEC-RAS	Particle filter	19 km reach of the Alzette River (Luxembourg)	January 2003	Integration of water level data into a one-dimensional (1D) hydraulic model	The updating of hydraulic models through the proposed scheme	Water level at station. Overstimation of upstream station. Standard deviation reduction at all stastion, Mean error decrease −0.07 to 0.02 m and −0.14 to −0.06m. It increase increase 0.17 to 0.34 and 0.22 to 0.37 for 2 upstream stations
Hostache et al., 2010	SAR (RADARSAT-1) (WL)	Shallow water equations (2D-SWEs)	Variational data assimilation (4D-var)	28 km reach of teh Mosel River (France/ Germany) (3 gauges)	February 28, 1997	Assimilation of satellite-derived water levels in a 2D shallow water model	DA enhances model calibration, optimal to identify Manning friction coefficients in the river channel	± 40 cm average vertical uncertainty
Mason et al., 2012	SAR (TerraSAR-X)	Hydraulic model (not specified)	Not specified	Lower Severn and Avon rivers (UK) (2 gauges)	July 19 to August 1, 2007	Development of a methodology to employ waterline assimilation to correct the model state	Waterline levels from SAR images may be assimilated. The levels extracted from a SAR image of flooding agreed with nearby gauge readings	Good agreement with gauged level
Matgen et al., 2010	SAR (WL)	Coupled hydrologi-hydraulic (H-H)	Particle filter	19 km reach of the Alzette River (Luxembourg)		Development of a new concept for sequential assimilation of SAR-derived water stages into coupled H-H models	Significant uncertainty reduction of water level and discharge at the time step of assimilation	After stage error reduction for 48 h, 24 h, and 12 h assimilation window to 0.25–0.39 m, 0.39–0.19 m, and 0.39–0.18 m respectively for station located upstream and middle-stream of the channel

(Continued)

TABLE 4.4 (CONTINUED)
Summary of Relevant Studies where Satellite-Derived Data Was Assimilated within a Hydraulic and/or Hydrological Model

Study	Satellite Sensor/ Aquisition Frequency	Model	DA Method	Study Area and No. of In Situ River Gauges Used	Study Period	Objective/ Approach	Key Findings	Uncertainty Reduction
Lai et al., 2014	MODIS flood extent (250 m)/ daily	2D flood model (not specified)	4D-var	Huaihe River, flood detention area (180 m²) (1 gauge)	June 29 to July 15, 2007	Direct assimilation of the flood-extent data into a 2D flood model. A 4D-var method incorporated with a new cost function is introduced	Promising way of data assimilation for flood inundation modeling by using directly flood extent suitable for improving flood modeling in the floodplains or similar areas	Manning coefficient optimization
Zhang et al., 2013	GFDS (AMSR-E) + TRMM rainfall + WL/ daily	HyMOD	Ensemble Square Root Filter (EnSRF)	Cubango River basin (1 gauge)	2003–2005	Investigate the utility of satellite data estimates in improving flood prediction	Shows opportunities in integrating satellite data in improving flood forecasting by careful fusion of remote sensing and in situ observations	Significantly improved skill and detectability of floods as well as reduced false alarm rates. The Bias is improved from to, RMSE reduces from 68.33% to 29.50% Nash-Sutcliffe Coefficient of Efficiency (NSCE) goes up from 0.61 to 0.91

(Continued)

TABLE 4.4 (CONTINUED)
Summary of Relevant Studies where Satellite-Derived Data Was Assimilated within a Hydraulic and/or Hydrological Model

Study	Satellite Sensor/ Aquisition Frequency	Model	DA Method	Study Area and No. of In Situ River Gauges Used	Study Period	Objective/ Approach	Key Findings	Uncertainty Reduction
Revilla-Romero et al., 2016	GFDS (AMSR-E + TRMM) flood extent/daily	LISFLOOD	EnKF	African (6 gauges) and South America continent (95 gauges)	2003	Test the impact of satellite-derived daily surface water extent in continental hydrological modeling	Assess the potential of assimilation of GFDS data into large scale hydrological model. Largest were obtained by the locations with poorest skill scores on the deterministic runs. However, it might not be beneficial for all locations	NSE score improved on 61 out of 101 stations obtaining significant improvements in both the timing and volume of the flow peaks

Source: Adapted from Revilla-Romer et al., 2016.
Note: studies are listed in alphabetical order by author.

current model presents limitations for ice jam floods, arid/wadi floods, urban and tropical areas. Soil moisture processes in the arid and tropical areas are not fully understood. Urban areas required very high resolution because of building and infrastructure geometry. Moreover, flash floods, for instance, rely strongly on initial conditions such as soil moisture which are not well captured by current observation and require careful initial condition setting.

As a final consideration, forecasters must always exercise caution when interpreting the flood forecast system results due to its large uncertainty. Various factors such as the weather type characteristics of the event and the soil moisture initial condition can influence the uncertainty of the flood forecast. However, the influence of the ensemble size on the results cannot be (totally) neglected, but a reasonable range can reduce its impacts on the final forecast (Silvestro and Rebora, 2014).

REFERENCES

Alemseged, T.H., and Rientjes, T.H.M. 2007. Uncertainty issues in hydrodynamic flood modeling. In: Proceedings of the 5th International symposium on Spatial Data Quality SDQ, Modelling Qualities in Space and Time, ITC, Enschede, The Netherlands, 13–15 June, 2007. Enschede: ITC. 6 p.

Alfieri, L., Berenguer, M., Knechtl, V., Liechti, K., Sempere-Torres, D., and Zappa, M. 2015. Flash flood forecasting based on rainfall thresholds. In: Duan, Q., Pappenberger, F., Thielen, J., Wood, A., Cloke, H., Schaake, J. (eds) *Handbook of Hydrometeorological Ensemble Forecasting*, Springer, Berlin, Heidelberg, Germany.

Alfieri, L., Pappenberger, F., Wetterhall, F., Haiden, T., Richardson, D., and Salamon, P. 2014. Evaluation of ensemble streamflow predictions in Europe, *Journal of Hydrology*, 517: 913–922, ISSN 0022-1694, https://doi.org/10.1016/j.jhydrol.2014.06.035.

Alvarado-Montero, R., Schwanenberg, D., Krahe, P., Helmke, P., and Klein, B. 2017. Multi-parametric variational data assimilation for hydrological forecasting, *Advances in Water Resources*, 110: 182–192, ISSN 0309-1708, https://doi.org/10.1016/j.advwatres.2017.09.026.

Amarnath, G., Umer, Y.M., Alahacoon, N., and Inada, Y. 2015. Modelling the flood-risk extent using LISFLOOD-FP in a complex watershed: Case study of Mundeni Aru River Basin, Sri Lanka, *Proc. IAHS*, 370: 131–138, https://doi.org/10.5194/piahs-370-131-2015.

Apel, H., Thieken, A.H., Merz, B., and Blöschl, G. 2004. Flood risk assessment and associated uncertainty, *Natural Hazards and Earth System Sciences*, 4: 295–308, https://doi.org/10.5194/nhess-4-295-2004.

Avolio, E., and Federico, S. 2018. WRF simulations for a heavy rainfall event in southern Italy: Verification and sensitivity tests, *Atmospheric Research*, 209: 14–35, ISSN 0169-8095, https://doi.org/10.1016/j.atmosres.2018.03.009.

Baldassarre, G.D., and Montanari, A. 2009. Uncertainty in river discharge observations: A quantitative analysis, *Hydrology and Earth System Sciences*, 13. 913–921.

Baroni, G., Zink, M., Kumar, R., Samaniego, L., and Attinger, S. 2017. Effects of uncertainty in soil properties on simulated hydrological states and fluxes at different spatio-temporal scales, *Hydrology and Earth System Sciences*, 21: 2301–2320, https://doi.org/10.5194/hess-21-2301-2017.

Bates, P.D., and De Roo, A.P.J. 2000. A simple raster based model for flood inundation simulation, *Journal of Hydrology*, 236: 54–77.

Bates, P.D., Pappenberger, F., and Romanowicz, R.J. 2014. Uncertainty in flood inundation modelling. In: Beven, K.J. and Hall, J. (eds) *Applied Uncertainty Analysis for Flood Risk Management*, Imperial College Press, London, pp. 232–269. ISBN 978-1-84816-270-9.

Belikov, V.V., Krylenko, I.N., Alabyan, A.M., Sazonov, A.A., and Glotko, A.V. 2015. Two-dimensional hydrodynamic flood modelling for populated valley areas of Russian rivers, *Proceedings of IAHS*, 370: 69–74, https://doi.org/10.5194/piahs-370-69-2015.

Beltaos, S. 2018. The 2014 ice–jam flood of the Peace-Athabasca Delta: Insights from numerical modelling, *Cold Regions Science and Technology*, 155: 367–380, ISSN 0165-232X, https://doi.org/10.1016/j.coldregions.2018.08.009.

Beltaos, S., and Kääb, A. 2013. Estimating river discharge during ice breakup from near-simultaneous satellite imagery, *Cold Regions Science and Technology*, 98: 35–46, https://doi.org/10.1016/j.coldregions.2013.10.010.

Beltaos, S., Tang, P., and Rowsell, R. 2012. Ice jam modelling and field data collection for flood forecasting in the Saint John River Canada, *Hydrological Processes*, 26: 2535–2545, https://doi.org/10.1002/hyp.9293.

Bermúdez, M., Ntegeka, V., Wolfs, V., and Willems, P. 2018. Development and comparison of two fast surrogate models for urban pluvial flood simulations, *Water Resources Management*, 32(8): 2801–2815.

Bernhofen, M., Whyman, C., Trigg, M.A., Sleigh, P.A., Smith, A.M., Sampson, C.C., Yamazaki, D., Ward, P.J., Rudari, R., Pappenberger, F., Dottori, F., Salamon, P., and Winsemius, H.C. 2018. A first collective validation of global fluvial flood models for major floods in Nigeria and Mozambique, *Environmental Research Letters*, 13(10): 104007.

Beven, K.J. 2016. Facets of uncertainty: Epistemic uncertainty, non-stationarity, likelihood, hypothesis testing, and communication, *Hydrological Sciences Journal*, 61(9): 1652–1665, https://doi.org/10.1080/02626667.2015.1031761.

Beven, K.J., and Binley, A. 2014. GLUE: 20 years on, *Hydrological Processes*, 28: 5897–5918, https://doi.org/10.1002/hyp.10082.

Beven, K.J., Lamb, R., Leedal, D., and Hunter, N. 2015. Communicating uncertainty in flood inundation mapping: A case study, *International Journal of River Basin Management*, 13(3): 285–295, https://doi.org/10.1080/15715124.2014.917318.

Biancamaria, S., Bates, P.D., Boone, A., and Mognard, N.M. 2009. Large-scale coupled hydrologic and hydraulic modelling of the Ob river in Siberia, *Journal of Hydrology*, 379(1–2): 136–150, ISSN 0022-1694, https://doi.org/10.1016/j.jhydrol.2009.09.054.

Boni, G., Ferraris, L., Giannoni, F., Roth, G., and Rudari, R. 2007. Flood probability analysis for un-gauged watersheds by means of a simple distributed hydrologic model, *Advances in Water Resources*, 30(10): 2135–2144, ISSN 0309-1708, https://doi.org/10.1016/j.advwatres.2006.08.009.

Butts, M.B., Payne, J.T., Kristensen, M., and Madsen, H. 2004. An evaluation of the impact of model structure on hydrological modelling uncertainty for streamflow simulation, *Journal of Hydrology*, 298(1–4): 242–266, ISSN 0022-1694, https://doi.org/10.1016/j.jhydrol.2004.03.042.

Crerar, S., Fry, R.G., Slater, P.M., van Langenhove, G., and Wheeler, D. 1988. An unexpected factor affecting recharge from ephemeral river flows in SWA/Namibia. In: Simmers, I., Dordrecht, D. (eds) *Estimation of Natural Groundwater Recharge*, Reidel, Dordrecht, The Netherlands, pp. 11–26.

Creutin, J.D., Borga, M., Gruntfest, E., Lutoff, C., Zoccatelli, D., and Ruin, I. 2013. A space and time framework for analyzing human anticipation of flash floods, *Journal of Hydrology*, 482: 14–24, ISSN 0022-1694, https://doi.org/10.1016/j.jhydrol.2012.11.009.

Cullmann, J., Krausse, T., and Philipp, A. 2009. Communicating flood forecast uncertainty under operational circumstances, *Journal of Flood Risk Management*, 2: 306–314, https://doi.org/10.1111/j.1753-318X.2009.01048.x.

Demeritt, D., Nobert, S., Cloke, H., and Pappenberger, F. 2010. Challenges in communicating and using ensembles in operational flood forecasting, *Meteorological Applications*, 17: 209–222, https://doi.org/10.1002/met.194.

Diomede, T., Davolio, S., Marsigli, C., Miglietta, M.M., Moscatello, A., Papetti, P., Paccagnella, T., Buzzi, A., and Malguzzi, P. 2008. Discharge prediction based on multi-model precipitation forecasts, *Meteorology and Atmospheric Physics*, 101: 245, https://doi.org/10.1007/s00703-007-0285-0.

Dottori, F., Di Baldassarre, G., and Todini, E. 2013. Detailed data is welcome, but with a pinch of salt: Accuracy, precision, and uncertainty in flood inundation modeling, *Water Resources Research*, 49: 6079–6085, https://doi.org/10.1002/wrcr.20406.

Edouard, S., Vincendon, B., and Ducrocq, V. 2018. Ensemble-based flash-flood modelling: Taking into account hydrodynamic parameters and initial soil moisture uncertainties, *Journal of Hydrology*, 560: 480–494, ISSN 0022-1694, https://doi.org/10.1016/j.jhydrol.2017.04.048.

EFAS. 2018. EFAS concepts and tools, https://www.efas.eu/about-efas.html. Consulted the 7th September 2018.

Felder, G., and Weingartner, R. 2016. An approach for the determination of precipitation input for worst-case flood modelling, *Hydrological Sciences Journal*, 61(14): 2600–2609, https://doi.org/10.1080/02626667.2016.1151980.

Ferraris, L., Rudari, R., and Siccardi, F. 2002. The uncertainty in the prediction of flash floods in the Northern Mediterranean Environment, *Journal of Hydrometeorology*, 3: 714–727, https://doi.org/10.1175/1525-7541(2002)003<0714:TUITPO>2.0.CO;2.

Flo-2D. 2009. Flo-2D Reference manual. In: *FLO-2D Software*, I., ed.: Nutrioso.

Freer, J., Beven, K.J., Neal, J., Schumann, G., Hall, J., and Bates, P. 2011. Flood risk and uncertainty. In: *Risk and Uncertainty Assessment for Natural Hazards* (Vol. 9781107006195, pp. 190–233). Cambridge University Press, https://doi.org/10.1017/CBO9781139047562.008.

Garcia, M.L., Basile, P.A., Riccardi, G.A., and Rodriguez, J.F. 2015. Modelling extraordinary floods and sedimentological processes in a large channel-floodplain system of the Lower Paraná River (Argentina), *International Journal of Sediment Research*, 30(2): 150–159, ISSN 1001-6279, https://doi.org/10.1016/j.ijsrc.2015.03.007.

Gbode, I.E., Dudhia, J., Ogunjobi, K.O., and Ajayi, V.O. 2018. Sensitivity of different physics schemes in the WRF model during a West African monsoon regime, *Theoretical and Applied Climatology*, https://doi.org/10.1007/s00704-018-2538-x.

Giustarini, L., Matgen, P., Hostache, R., Montanari, M., Plaza, D., Pauwels, V.R.N., De Lannoy, G.J.M., De Keyser, R., Pfister, L., Hoffmann, L., and Savenije, H.H.G. 2011. Assimilating SAR-derived water level data into a hydraulic model: A case study, *Hydrology and Earth System Sciences*, 15: 2349–2365, http://dx.doi.org/10.5194/hess-15-2349-2011.

Gneiting, T., Raftery, A.E., Westveld, A.H., and Goldman, T. 2005. Calibrated probabilistic forecasting using ensemble model output statistics and minimum CRPS estimation, *Monthly Weather Review*, 133: 1098–1118, https://doi.org/10.1175/MWR2904.1.

Hall, J.W., Tarantolo, S., Bates, P., and Horrittm, M. 2005. Distributed sensitivity analysis of flood inundation model calibration, *ASCE Journal of Hydraulic Engineering*, 131(2): 117–126.

Hardy, R.J., Bates, P.D., and Anderson, M.G. 1999. The importance of spatial resolution in hydraulic models for floodplain environments, *Journal of Hydrology*, 216: 124–136.

Herschy, R.W. 1998. *Encyclopedia of Hydrology and Water Resources* (eds Herschy, R.W. and Fairbridge, R.W.), pp. 571–584.

Hervoeut, J.-M., and Van Haren, L. 1996. Recent advances in numerical methods for fluid flow. In: Anderson, M.G., Walling, D.E., Bates, P.D. (eds), *Floodplain Processes*, John Wiley and Sons, Chichester, UK.

Horritt, M.S., and Bates, P.D. 1999. Effects of spatial resolution on a raster based model of flood flow, *Journal of Hydrology*, 253: 239–249.

Hostache, R., Lai, X., Monnier, J., and Puech, C. 2010. Assimilation of spatially distributed water levels into a shallow-water flood model. Part II: Use of a remote sensing image of Mosel River, *J. Hydrol.*, 390: 257–268, http://doi.org/10.1016/j.jhydrol.2010.07.003.

Huang, W., Cao, Z., Qi, W., Pender, G., and Zhao, K. 2015. Full 2D hydrodynamic modelling of rainfall-induced flash floods, *Journal of Mountain Science*, 12: 1203, https://doi.org/10.1007/s11629-015-3466-1.

Hughes, D.A., and Sami, K.1992. Transmission losses to alluvium and associated moisture dynamics in a semi-arid ephemeral channel system in southern Africa, *Hydrological Processes*, 6: 45–53.

Jain, S.K., Mani, P., Jain, S.K., Prakash, P., Singh, V.P., Tullos, D., Kumar, S., Agarwal, S.P., and Dimri, A.P. 2018. A brief review of flood forecasting techniques and their applications, *International Journal of River Basin Management*, 16(3): 329–344, https://doi.org/10.1080/15715124.2017.1411920.

Jin, X., Xu, C., Zhang, Q., and Singh, V.P. 2010. Parameter and modeling uncertainty simulated by GLUE and a formal Bayesian method for a conceptual hydrological model, *Journal of Hydrology*, 383(3–4): 147–155, ISSN 0022-1694, https://doi.org/10.1016/j.jhydrol.2009.12.028.

Juston, J.M., Kauffeldt, A., Montano, B.Q., Seibert, J., Beven, K.J., and Westerberg, I.K. 2013. Smiling in the rain: Seven reasons to be positive about uncertainty in hydrological modelling, *Hydrol Processes*, 27: 1117–1122.

Kardhana, H., and Mano, A. 2009 Uncertainty on a short-term flood forecast with rainfall-runoff model. In: *Advances in Water Resources and Hydraulic Engineering*, Springer, Berlin, Heidelberg, Germany.

Komi, K., Neal, J., Trigg, M.A., and Diekkrüger, B. 2017. Modelling of flood hazard extent in data sparse areas: A case study of the Oti River basin, West Africa, *Journal of Hydrology: Regional Studies*, 10: 122–132, https://doi.org/10.1016/j.ejrh.2017.03.001.

Komolafe, A.A, Srikantha H., and Avtar R., 2018. Development of generalized loss functions for rapid estimation of flood damages: A case study in Kelani river basin, *Applied Geomatic*, 10(1): 13–30.

Kovachis, N., Burrell, B.C., Huokuna, M., Beltaos, S., Turcotte, B., and Jasek, M. 2017. Ice-jam flood delineation: Challenges and research needs, *Canadian Water Resources Journal/Revue canadienne des ressources hydriques*, 42(3): 258–268.

Lai, X., Liang, Q., Yesou, H., and Daillet, S. 2014. Variational assimilation of remotely sensed flood extents using a 2-D flood model, *Hydrology and Earth System Sciences*, 18: 4325–4339, http://doi.org/10.5194/hess-18-4325-2014.

Larsen, M.A.D., Refsgaard, J.C., Jensen, K.H., Butts, M.B., Stisen, S., and Mollerup, M. 2016. Calibration of a distributed hydrology and land surface model using energy flux measurements, *Agricultural and Forest Meteorology*, 217: 74–88, ISSN 0168-1923, https://doi.org/10.1016/j.agrformet.2015.11.012.

Lewis, M., Bates, P., Horsburgh, K.F, Neal, J., and Schumann, G. 2013. A storm surge inundation model of the northern Bay of Bengal using publicly available data, *Quarterly Journal of the Royal Meteorological Society*, 139: 358–369, https://doi.org/10.1002/qj.2040.

Lombardi, G., Ceppi, A., Ravazzani, G., Davolio, S., and Mancini, M. 2018. From deterministic to probabilistic forecasts: The 'Shift-Target' approach in the Milan Urban Area (Northern Italy). *Geosciences*, 8: 181, https://doi.org/10.3390/geosciences8050181.

Loveridge, M., and Rahman, A. 2014. Quantifying uncertainty in rainfall–runoff models due to design losses using Monte Carlo simulation: A case study in New South Wales, Australia, *Stochastic Environmental Research and Risk Assessment*, 28: 2149, https://doi.org/10.1007/s00477-014-0862-y.

Maref, N., and Seddini, A. 2018. Modeling of flood generation in semi-arid catchment using a spatially distributed model: Case of study Wadi Mekerra catchment (Northwest Algeria), *Arabiab Journal of Geosciences*, 11: 116, https://doi.org/10.1007/s12517-018-3461-2.

Mason, D.C., Schumann, G.J.-P., Neal, J.C., Garcia-Pintado, J., and Bates, P.D. 2012. Automatic near real-time selection of flood water levels from high resolution synthetic aperture radar images for assimilation into hydraulic models: A case study, *Remote Sensing of Environment*, 124: 705–716, http://dx.doi.org/10.1016/j.rse.2012.06.017.

Massari, C., Brocca, L., Moramarco, T., Tramblay, Y., and Lescot, J-F.D. 2014. Potential of soil moisture observations in flood modelling: Estimating initial conditions and correcting rainfall, *Advances in Water Resources*, 74: 44–53, ISSN 0309-1708, https://doi.org/10.1016/j.advwatres.2014.08.004.

Mateo, C.M.R., Yamazaki, D., Kim, H., Champathong, A., Vaze, J., and Oki, T. 2017. Impacts of spatial resolution and representation of flow connectivity on large-scale simulation of floods, *Hydrology and Earth System Sciences*, 21: 5143–5163.

Matgen, P., Montanari, M., Hostache, R., Pfister, L., Hoffmann, L., Plaza, D., Pauwels, V.R.N., De Lannoy, G.J.M., De Keyser, R., and Savenije, H.H.G. 2010. Towards the sequential assimilation of SAR-derived water stages into hydraulic models using the particle filter: Proof of concept, *Hydrology and Earth System Sciences*, 14: 1773–1785, http://doi.org/10.5194/hess-14-1773-2010.

Mazzoleni, M., Verlaan, M., Alfonso, L., Monego, M., Norbiato, D., Ferri, M., and Solomatine, D.P. 2017. Can assimilation of crowdsourced data in hydrological modelling improve flood prediction? *Hydrology and Earth System Sciences*, 21: 839–861, https://doi.org/10.5194/hess-21-839-2017.

Mazzoleni, M., Cortes Arevalo, V.J., Wehn, U., Alfonso, L., Norbiato, D., Monego, M., Ferri, M., and Solomatine, D.P. 2018. Exploring the influence of citizen involvement on the assimilation of crowdsourced observations: A modelling study based on the 2013 flood event in the Bacchiglione catchment (Italy), *Hydrology and Earth System Sciences*, 22: 391–416, https://doi.org/10.5194/hess-22-391-2018.

Md Ali, A., Solomatine, D.P., and Di Baldassarre, G. 2015. Assessing the impact of different sources of topographic data on 1-D hydraulic modelling of flood, *Hydrology and Earth System Sciences*, 19: 631–643.

Meesuk, V.M., Vojinovic, Z.V., Mynett, A.E., and Abdullah, A.F. 2015. Urban flood modelling combining top-view LiDAR data with ground-view SfM observations, *Advances in Water Resources*, 75: 105–117, ISSN 0309-1708, https://doi.org/10.1016/j.advwatres.2014.11.008.

Mendoza, P.A., McPhee, J., and Vargas, X. 2012. Uncertainty in flood forecasting: A distributed modeling approach in a sparse data catchment, *Water Resources Research*, 48: W09532, https://doi.org/10.1029/2011WR011089.

Mergili, M., Emmer, A., Juřicová, A., Cochachin, A., Fischer, J.-T., Huggel, C., and Pudasaini, S.P. 2018. How well can we simulate complex hydro-geomorphic process chains? The 2012 multi-lake outburst flood in the Santa Cruz Valley (Cordillera Blanca, Perú), *Earth Surface Processes and Landforms*, 43: 1373–1389, https://doi.org/10.1002/esp.4318.

Mohan, P.R., Srinivas, C.V., Yesubabu, V., Baskaran, R., and Venkatraman, B. 2018. Simulation of a heavy rainfall event over Chennai in Southeast India using WRF: Sensitivity to microphysics parameterization, *Atmospheric Research*, 210: 83–99, ISSN 0169-8095, https://doi.org/10.1016/j.atmosres.2018.04.005.

Moretti, G., and Montanari, A. 2007. AFFDEF: A spatially distributed grid based rainfall–runoff model for continuous time simulations of river discharge, *Environmental Modelling & Software*, 22(6): 823–836.

Moya-Álvarez, A.S., Martínez-Castro, D., Flores, J.L., and Yamina S. 2018. Sensitivity Study on the Influence of Parameterization Schemes in WRF_ARW Model on Short- and Medium-Range Precipitation Forecasts in the Central Andes of Peru, *Advances in Meteorology*, 2018: 16, https://doi.org/10.1155/2018/1381092.

Mukolwe, M.M., Yan, K., Di Baldassarre, G., and Solomatine, D.P. 2016. Testing new sources of topographic data for flood propagation modelling under structural, parameter and observation uncertainty, *Hydrological Sciences Journal*, 61(9): 1707–1715, https://doi.org/10.1080/02626667.2015.1019507.

Neal, J., Schumann, G., Bates, P., Buytaert, W., Matgen, P., and Pappenberger, F. 2009. A data assimilation approach to discharge estimation from space, *Hydrological Processes*, 23: 3641–3649, https://doi.org/10.1002/hyp.7518.

Nester, T., Komma, J., and Blöschl, G. 2016. Real time flood forecasting in the Upper Danube basin, *Journal of Hydrology and Hydromechanics*, 64(4): 404–414, https://doi.org/10.1515/johh-2016-0033.

Noh, S.J., Tachikawa, Y., Shiiba, M., and Kim, S. 2013. Sequential data assimilation for streamflow forecasting using a distributed hydrologic model: Particle filtering and ensemble Kalman filtering, floods: From risk to opportunity, *IAHS Publication*, 357: 341–349.

Pagano, T.C., Shrestha, D.L., Wang, Q.J., Robertson, D., and Hapuarachchi, P. 2013. Ensemble dressing for hydrological applications, *Hydrological Processes*, 27: 106–116, https://doi.org/10.1002/hyp.9313.

Papaioannou, G., Vasiliades, L., Loukas, A., and Aronica, G.T. 2017. Probabilistic flood inundation mapping at ungauged streams due to roughness coefficient uncertainty in hydraulic modelling, *Advances in Geosciences*, 44: 23–34, https://doi.org/10.5194/adgeo-44-23-2017.

Pappenberger, F., and Beven, K.J. 2006. Ignorance is bliss: Or seven reasons not to use uncertainty analysis, *Water Resources Research*, 42: W05302, https://doi.org/10.1029/2005WR004820.

Pappenberger, F., Beven, K., Horritt, M., and Blazkova, S. 2005. Uncertainty in the calibration of effective roughness parameters in HEC-RAS using inundation and downstream level observations, *Journal of Hydrology*, 302(1–4): 46–69, ISSN 0022-1694, https://doi.org/10.1016/j.jhydrol.2004.06.036.

Pappenberger, F., Beven, K.J., Ratto, M., and Matgen, P. 2008. Multi-method global sensitivity analysis of flood inundation models, *Advances in Water Resources*, 31(1): 1–14, ISSN 0309-1708, https://doi.org/10.1016/j.advwatres.2007.04.009.

Pappenberger, F., Stephens, E., Thielen, J., Salamon, P., Demeritt, D., Andel, S.J., Wetterhall, F., and Alfieri, L. 2013. Visualizing probabilistic flood forecast information: Expert preferences and perceptions of best practice in uncertainty communication, *Hydrological Processes*, 27: 132–146, https://doi.org/10.1002/hyp.9253.

Paul, C., Anne, G., Keith, B., and Jan, F. 2002. Rainfall-runoff modelling of a humid tropical catchment: The TOPMODEL approach, *Hydrological Processes*, 16: 231–253, https://doi.org/10.1002/hyp.341.

Pourreza-Bilondi, M., and Samadi, S.Z. 2016. Quantifying the uncertainty of semiarid flash floods using generalized likelihood uncertainty estimation, *Arabian Journal of Geosciences*, 9: 622, https://doi.org/10.1007/s12517-016-2650-0.

Revilla-Romero, B., Wanders, N., Burek, P., Salamon, P., de Roo, A. 2016. Integrating remotely sensed surface water extent into continental scale hydrology, *Journal of Hydrology*, 543: 659–670, ISSN 0022-1694, https://doi.org/10.1016/j.jhydrol.2016.10.041.

Rokaya, P., Budhathoki, S., and Lindenschmidt, K.E. 2018. Ice-jam flood research: A scoping review, *Natural Hazards*, https://doi.org/10.1007/s11069-018-3455-0.

Romanowicz, R., and Beven, K.J. 2003. Estimation of flood inundation probabilities as conditioned on event inundation maps, *Water Resources Research*, 39(3): 1073.

Sabău, D., Şerban, G., Kocsis, I., Stroi, P., and Stroi, R. 2018. Winter Phenomena (Ice Jam) on Rivers from the Romanian Upper Tisa Watershed in 2006–2017 Winter Season. In: Zelenakova, M. (eds) *Water Management and the Environment: Case Studies. WINEC 2017. Water Science and Technology Library*, vol. 86, Springer, Cham, UK.

Saber, M., and Habib, E. 2015. Flash floods modelling for wadi system: Challenges and trends. In: Melesse, A., Abtew, W. (eds) *Landscape Dynamics, Soils and Hydrological Processes in Varied Climates. Springer Geography*, Springer, Cham, UK.

Saltelli, A., Chan, K., and Scott, E.M. (eds) 2000. *Sensitivity Analysis*, Wiley, New York, USA, p. 475.

Sampson, C.C., Bates, P.D., Neal, J.C., and Horritt, M.S. 2013. An automated routing methodology to enable direct rainfall in high resolution shallow watermodels, *Hydrological Processes*, 27: 467–476, https://doi.org/10.1002/hyp.9515.

Sanyal, J., Carbonneau, P., and Densmore, A.L. 2013. Hydraulic routing of extreme floods in a large ungauged river and the estimation of associated uncertainties: A case study of the Damodar River, India, *Natural Hazards*, 66(2): 1153–1177, http://doi.org/10.1007/s11069-012-0540-7.

Savage, J., Bates, P., Freer, J., Neal, J., Aronica, G. 2014. Vulnerability, uncertainty, and risk: Quantification, mitigation, and management. CDRM 9. Second International Conference on Vulnerability and Risk Analysis and Management (ICVRAM) and the Sixth International Symposium on Uncertainty, Modeling, and Analysis (ISUMA) (eds, Beer, M., Au, S.-K., and Hall, J.W.

Shin, M. -J., Eum, H. -I., Kim, C. -S., and Jung, I. -W. 2016. Alteration of hydrologic indicators for Korean catchments under CMIP5 climate projections, *Hydrological Processes*, 30: 4517–4542, http://doi.org/10.1002/hyp.10948.

Silvestro, F., and Rebora, N. 2014. Impact of precipitation forecast uncertainties and initial soil moisture conditions on a probabilistic flood forecasting chain, *Journal of Hydrology*, 519: 1052–1067, ISSN 0022-1694, https://doi.org/10.1016/j.jhydrol.2014.07.042.

Singh, Y.K., Dutta, U., Prabhu, T.S.M., Prabu, I., Mhatre, J., Khare, M., Srivastava, S., and Dutta, S. 2017. Flood response system—A case study, *Hydrology*, 4: 30.

Smith, P., Pappenberger, F., Wetterhall, F., Thielen, J. Krzeminski, B., Salamon, P., Muraro, D., Kalas, M., and Baugh, C. 2016. On the operational implementation of the European Flood Awareness System (EFAS). ECMWF Technical Memorandum, number 778, April 2016.

Sorman, A.U., and Abdulrazzak, M.J. 1993. Infiltration—Recharge through Wadi beds in arid regions, *Hydrological Sciences Journal*, 38(3): 173–186.

Stella, J., and Anagnostou, E. 2018. Modeling the flood response for a sub-tropical urban basin in south Florida, *Tecnología y Ciencias del Agua*, 9(3): 128–142, https://doi.org/10.24850/j-tyca-2018-03-05.

Stephens, E.M., Bates, P.D., Freer, J.E., and Mason, D.C. 2012. The impact of uncertainty in satellite data on the assessment of flood inundation models, *Journal of Hydrology*, 414–415: 162–173, ISSN 0022-1694, https://doi.org/10.1016/j.jhydrol.2011.10.040.

Straatsma, M.W., and Baptist, M.J. 2008. Floodplain roughness parameterization using airborne laser scanning and spectral remote sensing, *Remote Sensing of Environment*, 112(3): 1062–1080, ISSN 0034-4257, https://doi.org/10.1016/j.rse.2007.07.012.

Sun, Q., Miao, C., Duan, Q., Ashouri, H., Sorooshian, S., and Hsu, K.-L. 2018. A review of global precipitation data sets: Data sources, estimation, and inter-comparisons, *Reviews of Geophysics*, 56: 79–107, https://doi.org/10.1002/2017RG000574.

Teng, J., Jakeman, A.J., Vaze, J., Croke, B.F.W., Dutta, D., and Kim, S. 2017. Flood inundation modelling: A review of methods, recent advances and uncertainty analysis, *Environmental Modelling and Software*, 90: 201–216.

Thielen, J., Bartholmes, J., Ramos, M.-H., and de Roo, A. 2009. The European flood alert system—Part 1: Concept and development, *Hydrology and Earth System Sciences*, 13: 125–140, http://doi.org/10.5194/hess-13-125-2009.

Ticehurst, C., Dutta, D., Karim, F., Petheram, C., and Guerschman, J.P. 2015. Improving the accuracy of daily MODIS OWL flood inundation mapping using hydrodynamic modelling, *Natural Hazards*, 78: 803, https://doi.org/10.1007/s11069-015-1743-5.

Tim van, E., Gert, M., Dirk, E., Marijn, P., and Hubert, S. 2015. Predicting the ungauged basin: Model validation and realism assessment, *Frontiers in Earth Science*, https://doi.org/10.3389/feart.2015.00062.

Trigg, M.A., et al. 2016. The credibility challenge for global fluvial flood risk analysis, *Environmental Research Letters*, 11(9).

Try, S., Lee, G., Yu, W., Oeurng, C., and Jang, C. 2018. Large-scale flood-inundation modeling in the Mekong River Basin, *Journal of Hydrologic Engineering*, 23(7): 05018011.

Tymvios, F., Charalambous, D., Michaelides, S., and Lelieveld, J. 2018. Intercomparison of boundary layer parameterizations for summer conditions in the eastern Mediterranean island of Cyprus using the WRF-ARW model, *Atmospheric Research*, 208: 45–59, ISSN 0169-8095, https://doi.org/10.1016/j.atmosres.2017.09.011.

Vuillaume, J.-F., and Alifu, H. 2022. *Flood Observation Uncertainty, 3-Volume Flood Handbook, Vol. 2: Flood Analysis and Modeling*, edited by Eslamian, S., Eslamian, F., Taylor and Francis, USA.

Vuillaume, J.-F., and Hearth, S. 2018. Dynamic downscaling based on weather types classification: An application to extreme rainfall in south-east Japan, *Journal of Flood Risk Management*, 11: e12340, https://doi.org/10.1111/jfr3.12340.

Werner, M., Blazkova, S., and Petr, J. 2005. Spatially distributed observations in constraining inundation modelling uncertainties, *Hydrological Processes*, 19: 3081–3096, https://doi.org/10.1002/hyp.5833.

Wetterhall, F., Pappenberger, F., Alfieri, L., Cloke, H.L., Thielen-del Pozo, J., Balabanova, S., Daňhelka, J., Vogelbacher, A., Salamon, P., Carrasco, I., Cabrera-Tordera, A.J., Corzo-Toscano, M., Garcia-Padilla, M., Garcia-Sanchez, R.J., Ardilouze, C., Jurela, S., Terek, B., Csik, A., Casey, J., Stankūnavičius, G., Ceres, V., Sprokkereef, E., Stam, J., Anghel, E., Vladikovic, D., Alionte Eklund, C., Hjerdt, N., Djerv, H., Holmberg, F., Nilsson, J., Nyström, K., Sušnik, M., Hazlinger, M., and Holubecka, M. 2013. HESS Opinions "Forecaster priorities for improving probabilistic flood forecasts", *Hydrology and Earth System Sciences*, 17: 4389–4399, https://doi.org/10.5194/hess-17-4389-2013.

Willis, T.D.M. 2014. Systematic analysis of uncertainty in flood inundation modelling, PhD Thesis, The University of Leeds, UK.

Wilson, M., Bates, P., Alsdorf, D., Forsberg, B., Horritt, M., Melack, J., Frappart, F., and Famiglietti, J. 2007. Modelling large-scale inundation of Amazonian seasonally flooded wetlands, *Geophysic Research Letter*, 34: L15404, http://dx.doi.org/10.1029/2007GL030156.

Winsemius, H.C., Van Beek, L.P.H., Jongman, B., Ward, P.J., and Bouwman, A. 2013. A framework for global flood risk assessments, *Hydrology and Earth System Sciences*, 17: 1871–1892, https://doi.org/10.5194/hess-17-1871-2013.

Wohl, E., Barros, A., Brunsell, N., Chappell, N.A., Coe, M., Giambelluca, T., … Ogden, F. 2012. The hydrology of the humid tropics, *Nature Climate Change*, 2(9): 655–662, https://doi.org/10.1038/nclimate1556.

Yoshimoto, S., and Amarnath, G. 2017. Applications of satellite-based rainfall estimates in flood inundation modeling—A case study in Mundeni Aru River Basin, Sri Lanka, *Remote Sens.* 9: 998.

Yu, D., Yin, J., and Liu, M. 2016. Validating city-scale surface water flood modelling using crowd-sourced data, *Environmental Research LettersOpen Access*, 11(12): 124011.

Zhang, Y., Hong, Y., Wang, X., Gourley, J.J., Gao, J., Vergara, H.J., and Yong, B. 2013. Assimilation of passive microwave streamflow signals for improving flood forecasting: A first study in Cubango River Basin, Africa, *IEEE Journal of Selected Topics in Applied Earth Observations and Remote Sensing*, 6: 2375–2390, http://doi.org/10.1109/JSTARS.2013.2251321.

Part III

Flood Modeling and Forecasting

5 Empirical Modeling in Flood Design Estimation

Akshay R. Thorvat, Jayantilal N. Patel, and Saeid Eslamian

CONTENTS

5.1 INTRODUCTION

Although the unit hydrograph(UH) theory was applied in runoff computation for nearly four decades, a uniform interpretation of its definition does not seem to exist. Sherman (1932), the originator of the theory, defines it as the hydrograph of surface runoff on a given basin due to an effective rain falling for a unit time. The UH is a traditional means of representing the linear system response at the watershed outlet to rainfall over the watershed (Maidment et al., 1996). The UH theory assumes that a catchment acts on an input of effective precipitation in a linear and time-invariant manner to produce an output of direct storm runoff (Dooge, 1959 cited in Bruen and Dooge, 1992). The shape of the UH reflects the runoff characteristics of the drainage basin (Wesley et al.,1987). An observed hydrograph at a gauged site may be a single peak simple hydrograph as a result of an isolated storm or a complex hydrograph produced by a series of storms of varying intensity. These single peak hydrographs and complex hydrographs are treated differently in order to derive UH from them.

There are many more sites in the same region where the hydrologic data are not available but the design flood estimates are needed (Thorvat and Mujumdar, 2011;Mujumdar et al., 2014). Simulation and prediction of storm runoff for the gauged or ungauged catchments is a prime concern of

DOI: 10.1201/9780429463938-8

hydrologists (Singh, 2005). Among the numerous techniques developed for the determination of runoff from limited data, the synthetic unit hydrograph (SUH) is of great significance for determining the runoff volume with respect to time (Bhunya et al., 2009). SUH is a tool to derive UH for the other gauging stations in the same catchment or for the other similar catchments for which runoff data are not available (Singh, 2000). SUH is used to arrive at the UH for ungauged catchments where rainfall and runoff data are not available or inadequate (e.g., at other points on the stream in the same catchment or for other catchments that have hydrological and meteorological conditions similar to that for which it has been calibrated) (Singh, 2000). The term "synthetic" in SUH denotes that the UH is derived from watershed characteristics rather than rainfall-runoff data. The linear approach used in the UH derivation is later expanded to the SUH method for flood predictions by correlating UH parameters to topographic parameters such as drainage area, length of the waterway, length to the centroid of the watershed, and slope of the watershed (Guo, 2006). The present study is an attempt in these directions.

The fundamental idea presumes that the practical connections among hydrological and geomorphological parameters whenever conversed in dimensionless terms can be utilized to make certain coherent derivations about the complex process of runoff generation. This will help in setting up a simple and rapid procedure to express theoretical equations in the form of dimensionless parameters and assist its utilization for the improvement of a strategy for professionals for flood design at ungauged sites using significant and easily available parameters.

5.2 HYDROLOGICAL APPROACHES IN THE WATERSHED SYSTEMS

The role of basin geomorphology in controlling the hydrological response of a river basin is known for a long time. Earlier works have provided an understanding of basin geomorphology–hydrology relationships through empirical relations. Many investigators have endeavored to relate the UH parameters to the catchment geomorphology and thus obtain the SUH. Popular methods of SUH either use a few points on a UH to manually fit a curve or use a dimensionless UH to get a smooth shape of SUH. A greater degree of subjectivity and labor is involved in fitting a smooth curve manually over a few points to get a SUH and at the same time to adjust the area under the SUH to unity (Singh, 2000). Among the available approximate methods for the derivation of SUH, the method of fitting a smooth curve manually through a few salient points of the UH is generally practiced (Singh, 1988; Bhunyaet al., 2005). A SUH is often constructed using peaking parameters. Bernard (1935) accomplished the transformation of rainfall into runoff through a distribution graph as a function of catchment characteristics. Snyder (1938) utilized the empirical equations to estimate salient points of the hydrograph, such as peak discharge, time to peak, base period, and UH widths at 50% and 75% of peak discharge. Commons (1942) developed a dimensionless hydrograph from a number of flood hydrographs from Texas, USA.

Mitchell (1948) studied 58 SUHs from Illinois and developed the summation curves which were grouped and plotted according to the basin areas with a probable error of 39%in the magnitude of the crest and 37.5%in timing. Taylor and Schwarz (1952) derived SUH for areas varying from 20 to 1,600 sq miles considering the average slope of the main channel and watershed characteristics. The US Soil Conservation Service (SCS, 1957) method gives the SUH shape from an average dimensionless hydrograph, thus avoiding manual fitting. The SCS's dimensionless UH is assumed to be invariant regardless of the shape, size, and location of the catchments, which may not be justified (Singh, 2000). Espeyet al. (1977) and Espey and Altman (1978) proposed the empirical equations for the time base of the UH and widths at 50% and 75% of the UH peak to define the UH (Singh, 2005). Knowledge of the variables, functions, and time series that define the hydrologic behavior of a river basin makes possible a number of projects and studies related to water availability (Pinheiro and Naghettini, 2013).

The various alternative approaches for the SUH derivation reviewed by Singh et al. (2014) include, traditional (or empirical), conceptual, probabilistic, and geomorphological. According to this review, the traditional SUH models are based on different empirical equations and have certain region-specific constants/coefficients varying over a wide range. The conceptual models are based on the continuity equation and linear storage discharge relationship. The probabilistic or probability distribution function-based models use a parametric approach and employ the density functions for SUH derivation. The geomorphological class uses basin geomorphology to develop UH for flood hydrograph modeling for ungauged basins.

Most of the empirical formulae were developed on the basis of limited experience or study made with liberal assumptions. To evolve a method to predict the SUH parameters with reasonable accuracy is a task engaging the attention of hydrologists, engineers, and researchers all over the world. It would be of immense value if formulae of general validity could be evolved in which only significant parameters of a limited number appear characterizing the basin system (CWC, 1973).

5.3   SYNTHETIC UNIT HYDROGRAPH (SUH) AS A TOOL IN CASES OF LIMITED DATA

Keeping this objective, the present study aims to develop various physiographic characteristics quantitatively and collection of requisite concurrent rainfall-runoff data of upper Kumbhi, Dhamani, Kadavi, and Hiranyakeshi basins in Maharashtra, India,for the derivation of UH by single event method. An attempt is made for systematic analysis to derive average representative UHs (ARUHs) with minimum root mean square error (E_{RMS}) and validated by the leave-one-out cross-validation (1CV) method. In order to develop the widespread empirical equations to determine peak discharge (Q_p) (m³/s), time to peak (T_p) (h), base period (T_B) (h), width at 50% and 75% of peak discharge (W_{50}) (h)and (W_{75}) (h) respectively of a SUH, a novel approach of using the Buckingham pi theorem of dimensional analysis is used. The significant parameters used in the analysis are length of the main channel (L) (km); length of the main channel between outlet and centroid of the watershed (L_c) (km); drainage area (A) (km²); slope of the main channel (S_0) (m/km); and standard durationof UH (t_R) (h), as these are the significant features which are instrumental in governing the infiltration, soil moisture content, vegetative growth and the rate at which water is supplied to the main stream as it proceeds to the outlet. These features also influence the period of rise (Gray, 1964). To demonstrate the goodness of this approach, the 1CV procedure is applied. The procedure is repeated as many times as the number of catchments to obtain the variability of errors in different catchments. The comparison of Q_p, T_p, and T_B expressed as percentage error and shape of the hydrograph with corresponding ARUHs is made in each case.

Hence, the Dimensional Analysis-Based Empirical Model (DAEMS) which predicts the SUH parameters in better agreement as compared to the other models considering the differences in Q_p, T_p, and T_B, is compared with the other synthetic methods. For the empirical synthesis of UH, four methods are tested, viz. Snyder's method, Mitchell's equation, Commons' method, and the Central Water Commission (CWC) method. Finally, the performance of the dimensional analysis-based approach is validated by comparing the results with the results obtained by other synthetic methods on the basis of minimum percentage error in peak discharge prediction and shape of the hydrograph. It is possible to impart a conceptual meaning to use this approach to determine key parameters of SUH along with the shape of the hydrograph to select the proper empirical method.

5.4   STUDY AREA

The study area considered is an upper catchment of Kumbhi, Dhamani, Kadavi, and Hiranyakeshi basins of Kolhapur district in Maharashtra state, India. The catchment of the upper Kumbhi basin is situated at the continental divide of the Sahyadri hill range on the leeward side. The river runs about 56 km from its origin up to the confluence. The terrain of the area under consideration is hilly and thickly covered by forest. The length (L) of the river considered for the study is about 27.49 km and the catchment area (A) up to the gauging site at Mandukali is about 121.62 km². It lies between latitude 16°33' north to 16°40' north and longitude 73°52' east to 74°00' east in Gaganbawada tahsil.

The average annual rainfall of the basin is 6,230 mm. Southwest monsoon from mid-June to mid-October accounts for the major rainfall of the present basin. July and August are the two rainiest months. The temperature in this region varies between a mean maximum of 31°C and a mean minimum of 18°C. The mean relative humidity is high during the monsoon season and comparatively low during the non-monsoon season. The Dhamani basin is comprised of horizontally disposed dark grey massive and amygdaloidal basalts of the Deccan trap formation. These basalts at higher levels have been lateralized. The length (L) of the river considered for the study is about 33.53 km and the catchment area (A) is about 156 km^2. It lies between latitude 16°25' north to 16°37' north and longitude 73°59' east to 74°10' east. The region receives an average annual rainfall of about 2,080 mm. The temperature in this region varies between a mean maximum of 38°C and a mean minimum of 15°C.

The Kadavi River is comprised of the catchment area (A) of 430 km^2 and flows eastwards to join the Warana River on the right bank flowing southeastwards. The Warana River continues to flow eastwards and join the Krishna River on the right bank near Sangli. The river rises in the dense mixed forest in the Kolhapur district of Maharashtra State and drains the eastern slopes of Western Ghats. The length (L) of the river considered in the present study is about 62.61 km. The basin lies between latitude 16°50' north to 17°10' north to longitude 73°45' east to 74°50' east. The climate of the basin is characterized by hot summer from March to May with a rainy season from June to October. Also, the area has some rains in the post-monsoon season. The average annual rainfall in this region is about 3,820 mm. The mean maximum temperature varies from 26.5°C to 37.4°C with a mean minimum temperature variation of 14.6°C to 22.3°C. The Hiranyakeshi basin of south Maharashtra is comprised of the catchment area (A) of 244.56 km^2 and lies between latitude 16°00' north to 16°18' north and longitude 74°00' east to 74°10' east in Ajara tahsil of Kolhapur district. The Sahyadri range lies in the west part of the basin and the slope of the region is decreasing from west to east. The Hiranyakeshi River flows from southwest to northeast direction and meets the Ghataprabha River in Karnataka before it meets Krishna. The length (L) of the river considered for the study is about 51.82 km. The mean maximum and mean minimum temperatures in the region are 34.5° C and 13.1° C respectively. The climate is moderate subtropical with an average annual rainfall of 3,140 mm.

The ArcGIS software is used in the present study. The boundaries of the catchments and all of the streams have been mapped at a scale of 1:50,000 from Survey of India topo sheets 47-G/16, 47-H/13, 14, 15, 48-I/1, and 47-L/1, 3, 4. The location map of upper Kumbhi, Dhamani, Kadavi, and Hiranyakeshi basins is shown in Figure 5.1. For the basins under study, the catchment characteristics are presented in Table 5.1. As the hourly rainfall data is required for the development of UH, records from self-recording tipping bucket-type rain gauges are collected from Hydrology Project Circle, Nashik in Maharashtra, India. River gauging stations situated at Mandukali, Patryachiwadi, Sarud, and Ajara for the upper Kumbhi, Dhamani, Kadavi, and Hiranyakeshi respectively record hourly stages. The data are of stages at one-hour intervals by Automatic Water Level Recorder (AWLR) and hourly discharge is collected. The hourly rainfall and hourly discharge data so collected areanalyzed and brought in the proper form and presented. In the present study, the data pertaining to 28 isolated storms for the upper Kumbhi basin, 18 isolated storms for the Dhamani basin, and tenisolated storms each for the Kadavi and Hiranyakeshi basins from June 1, 1999, to October 31, 2013, is selected for the derivation of UHs by single event method. Then the study is intended to derive the average representative UH for the selected basins as a whole.

5.5 MATERIALS AND METHODS

The climate of the Kolhapur region is mild and temperate. The climate in this region is wet tropical. The western part of the region is consistently cooler than the eastern part. In the hilly west, with enormous forest regions and substantial precipitation, the climate is consistently humid during the rainy season and cool in summer. The hill traps alter the hot winds prevalent in April, May, and

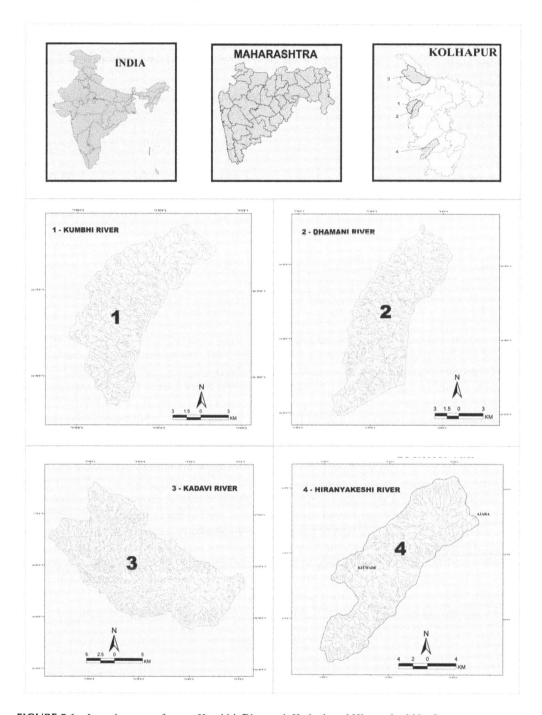

FIGURE 5.1   Location map of upper Kumbhi, Dhamani, Kadavi, and Hiranyakeshi basins.

June, and maintain a pleasant climate. The climatic year is divided into three seasons, specifically (1) a moderately warm wet season from June to October, (2) a cool dry season from November to February, and (3) a hot dry season from March to June. The mean maximum temperature in the region is 30.9°C and the mean minimum temperature is 19.0°C. The rainy season lasts from June to October and the climate is warm during these months. Rainfall in this season changes broadly from 500 mm in the northeast to 5000 mm in the southwest areaof the region.

TABLE 5.1

Catchment Characteristics of the Upper Kumbhi, Dhamani, Kadavi, and Hiranyakeshi Basins

Catchment Characteristics	Symbol	SI Unit	KumbhiBasin	DhamaniBasin	KadaviBasin	HiranyakeshiBasin
Drainage area of the basin	A	km^2	121.62	156.00	430.00	244.56
Length of main channel	L	km	27.49	33.53	62.61	51.82
Length between outlet and centroid of the basin	L_c	km	12.42	16.46	31.76	19.03
Slope of main channel	S_o	m/km	0.227	0.218	0.206	0.172
Mean basin slope	S_b	–	1:208	1:103	1:109	1:360

The present study aims to develop various physiographic characteristics quantitatively and derivation of UH by a single event method for the selected basins. The records of hourly rainfall from two self-recording rain gauge stations at Mandukali (MN) and Revechiwadi (RV) for the upper Kumbhi basin and at Ajara (AJ) and Kitawade (KT) for the Hiranyakeshi basin are collected. The Theissen's weighted areas for self-recording rain gauge stations at Mandukali, Revechiwadi, Ajara, and Kitawade are about 12.1 km^2, 109.52 km^2, 118.95 km^2, and 125.61 km^2 respectively. The average annual precipitation values at Madukali, Revechiwadi, Ajara, and Kitawade are 4,195.30 mm, 5,486.40 mm, 3,687.50 mm, and 2,609 mm respectively. For the Dhamani and Kadavi, the hourly storm data is collected from single self-recording rain gauge stations at Patryachiwadi and Sarud respectively. The average annual precipitation values at Patryachiwadi and Sarud are 2,080 mm and 3,820 mm respectively. For the chosen storms, the hourly discharges at gauging stations Mandukali, Patryachiwadi, Sarud, and Ajara corresponding to the recorded stages are noted. The criterion used for the UH derivation is that the flow at gauging stations should be at low stages at all times during the experimentation. Therefore, the river flows are chosen in the month of low flows, i.e., June, September, and October. It is necessary to separate the observed hydrograph into its component parts namely, the direct runoff and baseflow. A trial and error procedure is adopted to locate the starting point of the rising limb of direct runoff hydrograph (DRH) to be the same as the start of the effective rainfall. To mark the endpoint of direct runoff, an empirical equation is used.

$$N = 0.827A^{0.2} \tag{5.1}$$

Where, N (days) is the time measured from the time of peak discharge and A (km^2) is the drainage area (Subramanya, 2008). Thus, e.g., for the Dhamani basin with a drainage area of 156 km^2,

$$N = 0.827(156)^{0.2} = 2.27 \, \text{days}$$

i.e., approximately 54 h. This seems to be too long for the small basins. Thus the endpoint is fixed by observing the recession limb of the hydrograph where it flattens almost toa straight line and the separation line joints it asymptotically. For the selected isolated storms, the corresponding runoff hydrograph is plotted and the base flow is separated to determine the DRH ordinates. The ordinates of DRH are divided by the depth of effective rainfall (ER) to get the UH ordinates.

Some other key criterions considered for the derivation of UH are that the ER is assumed to be uniformly distributed over the entire drainage basin and within its duration. The hydrologic losses are uniform. The time distribution of surface runoff from the given storm period is independent of runoff from the antecedent storm period. The hydrologic system is assumed to be linear and time-invariant. According to these criteria, for the chosen isolated storms, having ER duration equal to the duration of UH to be derived, the single event method is used for the derivation of UH for the basin under study.

5.5.1 CRITERIA FOR SELECTING HYDROLOGIC DATA FOR DEVELOPMENT OF UH

Bernard (1935), Brater (1939), and Barnes (1959) have suggested several criteria to be followed in selecting the hydrologic data suitable for the UH development. These were summarized to formulate the basis of the following list of standards.

1. The rain must have fallen within the selected time unit and must not have extended beyond the period of rise of the hydrograph.
2. The storm must have been well distributed over the watershed, all stations showing an appreciable amount.
3. The storm period must have occupied a place of comparative isolation in the record.
4. The runoff following a storm must have been uninterrupted by the effects of low temperatures and unaccompanied by melting snow or ice.
5. The stage graphs or hydrographs must have a sharp, defined, rising limb culminating to a single peak and followed by an uninterrupted recession.
6. All stage graphs or hydrographs for the same watershed must show approximately the same period of rise.

5.5.2 STATISTICAL VALIDATION METHOD

The isolated storms having ER duration equal to the duration of UH to be derived, a single event method is used for the derivation of UH for the basins under study. The observed storms are considered to be statistically independent. As per the Central Water Commission (1983), the normal variation of about $\pm 20\%$ in T_p and $\pm 30\%$ in Q_p is allowed in case of finalizing the average representative UH. The values of Q_p and T_p of UHs falling out of prescribed ranges are excluded and the average of the remaining values of Q_p, T_p, and T_B is worked out. It is generally difficult to theoretically verify the statistical independence among different storm events. However, it can be intuitively justified if the selected rainfall-runoff events are not produced by the same storm system (Zhao et al., 1995). Validation is an important task in the process of developing a model. By validation, one evaluates how well the derived model can predict future events. Therefore it is very important to investigate the accuracy of prediction of DRH by using the derived UH applied to the available ER data. The cross-validation method is considered a general method which is a statistical method of validation. Zhaoet al.(1995) examined the predictability of the UH derived for watersheds in Taiwan by applying some statistical validation methods out of which, leave-one-out cross-validation (1CV) method is used in the present study. The essential steps involved in this method are described below:

1. If "R" is the total number of storm events considered in the original data set, then it can be written as $S = (S_1, S_2, S_3, ..., S_R)$, where S is the storm event.
2. One of the storm events is left out from the above original data set.
3. An average UH is determined considering remaining $(R-1)$ storm events.
4. Now, the effective rainfall data of the excluded storm is applied to the average UH derived in the previous step to predict the DRH of that storm.

5. The predicted DRH is compared with the actual observed DRH by calculating root mean square error (E_{RMS}).

5.5.3 ERROR FUNCTIONS IN PREDICTING DIRECT RUNOFF HYDROGRAPH (DRH) FOR THE SELECTED STORM EVENT

To find how efficiently the average representative UH predicts the DRH, the following error functions are calculated for the selected storm event. Let Q (m³/s) and $\bar{Q}$ (m³/s) be corresponding ordinates of actual DRH and predicted DRH; N is the number of DRH ordinates; Q_m (m³/s) is the mean discharge; E (%) is the efficiency in the DRH prediction; E_{RMS} (m³/s) is the root mean square error; E_p (%) is the percentage error in prediction of direct runoff peak discharge; and E_T (%) is the percentage error in prediction of direct runoff time to peak:

$$Q_m = \Sigma Q / N \tag{5.2}$$

$$E = \left[\left(Q - Q_m\right)^2 - \left(Q - \bar{Q}\right)^2 \middle/ \left(Q - Q_m\right)^2\right] \times 100 \tag{5.3}$$

$$E_{RMS} = \left(\sum_{n=1}^{N}\left(Q - \bar{Q}\right)^2 \middle/ N\right)^{1/2} \tag{5.4}$$

$$E_p = \left[\left(Q_p - \bar{Q}_p\right) \middle/ Q_p\right] \times 100 \tag{5.5}$$

$$E_T = \left[\left(T_p - \bar{T}_p\right) \middle/ T_p\right] \times 100 \tag{5.6}$$

In the present study, the data pertaining to 28 isolated storms for the upper Kumbhi basin, 18 isolated storms for the Dhamani basin, and tenisolated storms each for the Kadavi and the Hiranyakeshi basins are selected for the derivation of UHs by single event method. Then the study is intended to derive average representative UH for the selected basins as a whole. To investigate the accuracy of the prediction of DRH by using derived average UH applied to the available effective storm data; the 1CV method is used as it considers all of the storm events in calibration as well as in the validation process (Zhaoet al., 1995).

The average representative UHs using the 1CV method for the upper Kumbhi, Dhamani, Kadavi, and Hiranyakeshi basins are named ARUH-S1, ARUH-S2, ARUH-S3, and ARUH-S4 respectively. The study further deals with the development of widespread empirical equations to determine Q_p (m³/s), T_p (h), T_B (h), W_{50} (h), and W_{75} (h) of a SUH through a novel approach using the Buckingham pi theorem of dimensional analysis technique for the basins under study.

5.6 EMPIRICAL SYNTHESIS OF UH

The term "empirical" is adopted to infer that the graph is developed from empirical data and to avoid the possibility of misinterpretation conveyed by the words mean or average (Gray, 1961). In order to construct UH for such areas, empirical equations of regional validity which relate the salient hydrograph characteristics to the basin characteristics are available. UHs derived from such relationships are known as SUHs. A number of methods for developing SUHs are reported in the literature. It should, however, be remembered that these methods based on empirical correlations are applicable only to the specific regions for which they were developed and should not be considered as general relationships for use for all regions. Figure 5.2 shows the elements of SUH.

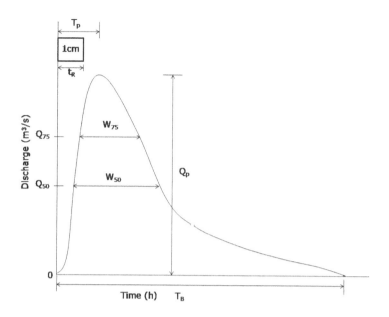

FIGURE 5.2 Elements of synthetic unit hydrograph.

5.6.1 SNYDER'S METHOD

Snyder (1938) developed a method that involves the construction of a SUH for a drainage basin with inadequate data. This UH is derived from measured drainage basin characteristics and is based solely upon such measurements. Snyder (1938) was perhaps the first to have established a set of formulae relating the physical geometry of the watershed to three basic parameters of the UH. These formulae were based on a study of 20 watersheds located in the Appalachian Highlands of the eastern United States, which varied in watershed size from 10 to 10,000 mi^2 (25.6 to 25,600 km^2) (Singh, 1994). The basic parameter that Snyder defined is (t_p) (h), the time of lag to peak. According to the Snyder method:

$$t_p = C_t \left(L \cdot L_c \right)^n \tag{5.7}$$

where L (km) is the basin length measured along the watercourse from the basin divide to the gauging station, L_c (km) is the distance along the main watercourse from the gauging station to a point opposite to the watershed centroid, and C_t is a regional constant representing watershed slope and storage effects. The values of C_t in Snyder's study ranged from 1.35 to 1.65. However, studies by many other investigators have shown that C_t depends upon the region under study and wide variations of the C_t values ranging from 0.3 to 6.0 have been reported (Chow et al., 1988; Subramanya, 2008; Das, 2010; Ponce, 1989). The catchment under study is a valley drainage area for which the values of constants C_t and n are given by Linsley et al. (1988) as 0.5 and 0.3 respectively.

The unit duration (t_R) (h) of the effective rainfall is given by:

$$t_R = t_p/5.5 \tag{5.8}$$

The peak discharge (Q_p) (m^3/s) is given by:

$$Q_p = 640 C_p A/t_p \tag{5.9}$$

The values of coefficient C_p range from 0.56 to 0.69 for Snyder's study areas and are considered as an indication of the retention and storage capacity of the watershed (Singh, 1994). In the present study, C_p is assumed to be average of the range 0.56 and 0.69, i.e., 0.625 for the valley drainage area. The base period (T_B) (d) is given by:

$$T_B = 5\left[\left(t_p/11\right)+t_p\right] \tag{5.10}$$

As an aid to sketching a reasonable UH, the US Army Corps of Engineers (1940) developed a relation between Q_p and width of UH at 50% (W_{50}) and 75% (W_{75}) of peak discharge as given below:

$$W_{50} = 2.14\left(q_p\right)^{-1.08} \tag{5.11}$$

$$W_{75} = 1.22\left(q_p\right)^{-1.08} \tag{5.12}$$

where q_p (m³/s/km²) is the unit peak rate of discharge of UH. However, in the present study, an attempt is also made to evaluate Snyder's coefficients C_t and C_p for the study area. These coefficients are evaluated by using the actual observed basin lag and peak discharge given by the average UH by the 1CV method.

5.6.2 MITCHELL'S EQUATION

Mitchell (1948) developed a method for constructing SUH based on the basin lag and the summation curve by studying UHs for 58 watersheds in Illinois with areas ranging from 10 to 3,090 mi². He proposed the following equations:

$$t = 2.80t_p^{0.81} \tag{5.13}$$

$$t = 1.05A^{0.6} \tag{5.14}$$

where t (h) is the time lag between the centroid of effective rainfall hyetograph and the centroid of DRH and A (mi²) is the drainage area. He has given UH ordinates in terms of distribution percentage which can be converted into the actual ordinates of UH by multiplying it by q_s.

$$q_s = A\ n_d/0.03719 \tag{5.15}$$

Where q_s represents the volume of surface runoff of 1 inch over the drainage area A (mi²) and n_d is the number of intervals per day. Mitchell (1948) developed the curves which were grouped and plotted according to three classifications: under 175 mi², 175 to 750 mi², and over 750 mi². The duration to be used for an actual (not synthetic) UH depends on the definition desired. If it is desired to adjust a SUH to a revised peak value of time or discharge, the result can be conveniently obtained by use of the summation curve. In the present study as the drainage area falls under the first classification, hence synthetic summation curve 80 with 0.10 lag is used to find the ordinates of UH.

5.6.3 COMMONS' DIMENSIONLESS HYDROGRAPH

From the literature, it is observed that various dimensionless hydrographs have been developed from different areas in different ways but still display a remarkable similarity in form and magnitude. This indicates that the dimensionless hydrographs have potential in applying these methods to basins outside the area in which they are developed. Thus by this procedure, a UH is reduced to a standard curve by plotting dimensionless coordinates or dimensionless sets of parameters. The dimensionless form eliminates the effect of basin size and much of the effect of basin shape. The purpose of using this method is for the development of a mean dimensionless hydrograph from recorded runoff hydrographs for individual basins under study and to compare its performance with the dimensional analysis-based approach.

Commons (1942) developed a dimensionless hydrograph from a study of major flood hydrographs in Texas. He found that for hydrographs with a single peak, the rate of rise was approximately 1.5 times as rapid as the rate of fall and the rise from approximately 33% to 95% of peak flow was nearly a straight line. The base period of the Commons' dimensionless hydrograph is divided into 100 units. The area under the curve being equal to 1,196.5 square units, the peak discharge is divided into 60 units (Commons, 1942; Craiget al.,1969; Hoffmeister and Weisman, 1977; Clevelandet al., 2006). The parameters derived in this method are unit time (T_u), volume (V) of 1 inch of runoff, unit discharge (Q_u), peak discharge (Q_p), and base period (T_B). He suggested the following relationships:

$$T_u = T_p / 14 \tag{5.16}$$

$$V = 23,23,200A \tag{5.17}$$

$$Q_u = V/\left(1196.5 T_u\, 3600\right) \tag{5.18}$$

$$Q_p = 60 Q_u \tag{5.19}$$

$$T_B = 100 T_u \tag{5.20}$$

To apply Commons' dimensionless hydrograph, some parameters of the desired UH must be known. Parameters thatcan be estimated in the field without setting up extensive recording equipment are the time to peak and base length for any storm. These estimates gathered from very few storm events could be used with the dimensionless graph to form a SUH for the selected basin.

5.6.4 CENTRAL WATER COMMISSION (CWC) METHOD

The Krishna and Pennar Subzone 3(h) is one of the 26 hydrometeorologically homogeneous sub-zones into which India has been divided, for developing the regional methodology for assessing the design flood of small and medium catchments. The flood estimation report of Krishna and Pennar Subzone 3(h) (Design Office Report No., K/6/1982) was published in the year 1982 by the Central Water Commission (CWC, 1983). The method used in the present study is a revision of the earlier method reported. The Central Water Commission, in association with the Ministry of Railways, the India Meteorological Department, and the Ministry of Transport, has revised the methodology for estimation of floods for small and medium catchments on a regional basis in the year 2000 (CWC, 2000) as:

$$t_p = 0.325 \left(L L_c / \sqrt{S_o}\right)^{0.447} \tag{5.21}$$

$$q_p = 0.996 \left(t_p\right)^{-0.497} \tag{5.22}$$

$$W_{50} = 2.389 \left(q_p\right)^{-1.065} \tag{5.23}$$

$$W_{75} = 1.415 \left(q_p\right)^{-1.067} \tag{5.24}$$

$$W_{R50} = 0.753 \left(q_p\right)^{-1.229} \tag{5.25}$$

$$W_{R75} = 0.558 \left(q_p\right)^{-1.088} \tag{5.26}$$

$$T_B = 7.392 (t_p)^{0.524} \qquad \qquad (5.27)$$

$$T_p = t_p + t_R/2 \qquad \qquad (5.28)$$

$$Q_p = q_p \times A \qquad \qquad (5.29)$$

5.6.5 DIMENSIONAL ANALYSIS TECHNIQUE

The approach or the technique used in this study is referred to as the Buckingham pi theorem method. In Buckingham's technique only the dimensions or the properties of the problem at hand are analyzed (Genick, 2013). The Buckingham pi theorem states that, if there are n variables (dependent and independent variables) in a dimensionally homogeneous equation and if these variables contain m fundamental dimensions (such as M, L, T, etc.) then the variables are arranged into $(n-m)$ dimensionless terms. These dimensionless terms are called "π" (pi) terms (Rajput, 2009).

The basic method of dimensional analysis is a rather informal, unstructured approach for determining dimensional groups. It depends on being able to construct a functional equation that contains all of the relevant variables, for which there is knowledge of the dimensions. The proper dimensionless groups are then identified by the thoughtful elimination of dimensions. A physical equation is a relationship between two or more physical quantities. In dimensional analysis, an equation expressing a physical relationship between quantities must be dimensionally homogeneous. A dimensionally homogeneous equation is one in which every independent, additive term in the equation has the same dimensions. This means that any one term can be solved as a function of all the others. But, it should be remembered that a dimensionally homogeneous equation is independent of the units of measurement being used (Dym, 2004). The dimension defines a physical quantity and stipulates the procedures required to transform that physical quantity into a physical magnitude. On the other hand, a standard unit provides a scale with which to obtain the fraction of or multiple of the standard unit that comprises the physical magnitude (Worstell, 2014). Dimensional analysis investigates the physical quantities; thus, it uses the intrinsic properties and not physical magnitudes (Pankhurst, 1964).

In design flood predictions, when resources such as technology, time, or funds do not permit a more rigorous method of hydrograph determination, there is a need for a reasonable alternative for the determination of a peak discharge and time to peak. Gray (1964) studied the interrelationships between certain geomorphic properties such as drainage area size, length and slope of the main stream, general land slope, and channel geometry on the time distribution of surface runoff of a watershed and demonstrated the use of the factors for synthesizing unit graphs for ungauged areas. Thus, linking quantitative geomorphology with basin hydrological characteristics can provide a simple way to assess the hydrological behavior of different basins, particularly ungauged ones. The physical characteristics of the drainage basin include the drainage area, basin shape, ground slope, and centroid (Singh et al., 2014).

The drainage area and stream lengths are measured from suitable topographical maps. Area is probably the most used parameter to represent the catchment characteristics of a catchment. It determines the potential runoff volume, provided the storm covers the whole area. Length is defined in more than one way: (1) the greatest straight-line distance between any two points on the perimeter, (2) the greatest distance between the outlet and any point on the perimeter, (3) the length of the main channel from its source to the outlet (L), or (4) the length of the main channel between the outlet and centroid of the watershed (L_c). It is assumed that the travel time is proportional to the channel length from the point under consideration to the catchment outlet (Singh, 2005). The main branch of the stream is taken to be that with the bigger contributing drainage area (Hoffmeister and Weisman, 1977). The watershed shape may influence the hydrograph shape, especially for small watersheds.

L_c is a measure of shape. As the upper portion of a basin has very little influence on flow prior to the peak occurrence (Black, 1972), the measurement of shape is, therefore, most important for the lower part of the basin, and this is recognized by using L_c in the analysis.The slope of the main channel (S_o) has a profound effect on the velocity of overland flow. It is also an important parameter in many watershed simulation models (Subramanya, 2008). These parameters have physical significance. Along with these physical characteristics, among climatic factors, the standard duration of UH (t_R) of given intensity also has a direct proportional effect on the volume of runoff. This effect is reflected in the rising limb and peak flow of UH.

For ungauged basins, the empirical synthesis of UH is usually accomplished using expressions thatrelate hydrological characteristics with easilymeasured catchment characteristics. Hence, the study is carried out to use the Buckingham pi theorem to develop the functional relationships between the key parameters T_p, Q_p, T_B, W_{50}, and W_{75} of SUH, and selected catchment characteristics viz. L (km), L_c (km), A (km²), and S_o (m/km) are measured from the topographic map, and t_R (h) obtained from comprehensive storm data of the respective basin.

- Time to peak (T_p):

$$f\left(T_p, t_R, L, L_c, A, S_o\right) = 0 \tag{5.30}$$

Here, C is a constant.

No. of variables = n = 6.

Dimensions of these variables are given as,

$$T_p = [T]; t_R = [T]; A = [L^2]; L = [L]; L_c = [L] \text{ and } S_o = [M^0 L^0 T^0]$$

Hence, here, no. of dimensions = m = 2.

Therefore, no. of π terms = $(n-m)$ = 6 −2 = 4.

$$f\left(\pi_1, \pi_2, \pi_3, \pi_4\right) = 0 \tag{5.31}$$

After selecting the length of the main channel (L)and standard duration of UH (t_R) as repeating variables from geometric and kinematic groups respectively we get,

$$\pi_1 = L^a t_R{}^b T_p \tag{5.32}$$

$$\pi_2 = L^a t_R{}^b L_c \tag{5.33}$$

$$\pi_3 = L^a t_R{}^b A \tag{5.34}$$

$$\pi_4 = L^a t_R{}^b S_o \tag{5.35}$$

Now, representing equation (5.32):

$$\pi_1 = L^a t_R{}^b T_p$$

In dimensional form equation (5.32) can be stated as follows:

$$[M^0 L^0 T^0] = [L]^a [T]^b [T]^1$$

Here, comparing the indices of the left-hand side and the right-hand side:

$$\text{For } L, \ 0 = a$$

$$\text{For } T, \ 0 = b + 1$$

Therefore, a = 0 and b = –1. Putting values of a and b in equation (5.32) we get:

$$\pi_1 = t_R^{-1} T_p$$

$$\pi_1 = \left(T_p / t_R\right) \tag{5.36}$$

Similarly, putting dimensions in equations (5.32), (5.33), (5.34), and (5.35) and then comparing the exponents of LHS and RHS we get:

$$\pi_2 = \left(L_c / L\right) \tag{5.37}$$

$$\pi_3 = \left(A / L^2\right) \tag{5.38}$$

$$\pi_4 = \left(S_o\right) \tag{5.39}$$

Therefore, putting equations (5.36), (5.37), (5.38), and (5.39) in equation (5.31) we get:

$$f\left[\left(T_p / t_R\right), \left(L_c / L\right), \left(A / L^2\right), \left(S_o\right)\right] = 0 \tag{5.40}$$

This can be written as:

$$\pi_1 = \psi\left(\pi_2, \pi_3, \pi_4\right) \tag{5.41}$$

To find the exact nature of (ψ) a power product relationship of the dimensionless groups is used as follows:

$$\pi_1 = \beta_1 \left(\pi_2^{\beta_2} \pi_3^{\beta_3} \pi_4^{\beta_4}\right) \tag{5.42}$$

Equation (5.42) represents a non-linear equation. If relationships are non-linear, there are two alternatives: (1) transform the data to make the relationships linear, or (2) use an alternative statistical model. By logarithmic transformation, the above non-linear equation is written as a linear equation and the coefficients are estimated in the same manner as for linear equation. A non-linear relation is reasonably well approximated by a linear equation within a particular range of the variables. Therefore, to solve equation (5.42), a logarithmic transformation is used as follows:

$$\log(\pi_1) = \log(\beta_1) + \beta_2 \log(\pi_2) + \beta_3 \log(\pi_3) + \beta_4 \log(\pi_4) \tag{5.43}$$

$$\log(T_p / t_R) = \log(\beta_1) + \beta_2 \log(L_c / L) + \beta_3 \log(A / L^2) + \beta_4 \log(S_o) \tag{5.44}$$

Here π_1 is treated as the dependent variable and the salient points of average representative UH (ARUH-S1) derived through the 1CV method are taken on LHS and catchment characteristics and slope of the upper Kumbhi basin are taken on RHS of the equation (5.44). By keeping LHS fixed, the values of constants ($\beta_1, \beta_2, \beta_3, \beta_4$) are found out by performing the trial-and-error procedure of equation (5.44) (Narvekar et al., 2013a, 2013b;Thorvat and Patel, 2015, 2016;Patel and Thorvat, 2016). Hence after the iterations, the following is the final form of the dimensionally homogeneous equation to find the time to peak (T_p) (h) as:

$$T_p = \frac{AL_c t_R}{L^3 S_o^{2.686}} = [T]$$ (5.45)

Similarly, equations for peak discharge (Q_p), base period (T_B), widths at 50% (W_{50}) and 75% (W_{75}) of peak discharge are derived by using "π" terms obtained in each case in a similar manner to that of the time to peak (T_p). The dimensionally homogeneous equation is independent of the system of units chosen to measure length and time (Dym, 2004). Following are the final form of dimensionally homogeneous equations obtained after the dimensional analysis.

- Peak discharge (Q_p):

$$f(Q_p, A, S_o, T_p) = 0$$ (5.46)

  By solving we get:

$$Q_p = \frac{A^{1.5} S_o^{1.408}}{T_p} = [L^3 T^{-1}]$$ (5.47)

- Base period (T_B):

$$f(T_B, T_p) = 0$$ (5.48)

  By solving we get:

$$T_B = 10.247 T_p \left[\frac{\left(\frac{Q_p T_p}{A^{1.5}}\right)^{0.102} \left(\frac{L_c}{L}\right)^{0.415}}{S_o^{0.460}} \right] = [T]$$ (5.49)

- Width at 50% peak discharge (W_{50}):

$$f(W_{50}, Q_p, A) = 0$$ (5.50)

  By solving we get:

$$W_{50} = 0.211 \left(\frac{A^{1.5}}{Q_p}\right) \left(\frac{L_c}{L S_o^{0.536}}\right) = [T]$$ (5.51)

- Width at 75% peak discharge (W_{75}):

$$f(W_{75}, Q_p, A) = 0$$ (5.52)

  By solving we get:

$$W_{75} = 0.120 \left(\frac{A^{1.5}}{Q_p}\right) \left(\frac{L_c}{L S_o^{0.536}}\right) = [T]$$ (5.53)

Similarly, the dimensional analysis approach is used to derive empirical equations to determine T_p, Q_p, T_B, W_{50}, and W_{75} of SUH for the other three basins under study. The salient points of average representative UHs viz. ARUH-S2, ARUH-S3, and ARUH-S4 derived by 1CV method for the Dhamani, Kadavi, and Hiranyakeshi basins respectively are taken on LHS of the equations. The catchment characteristics and slope of the respective basins are taken on the RHS of the equations in each case. A trial and error procedure is applied to compute β_1, β_2, β_3, and β_4 by keeping the LHS of the equations fixed. Further, in order to demonstrate the goodness of the Buckingham pi theorem method, the 1CV procedure is applied to all of the catchments under study. The procedure is repeated as many times as the number of catchments to obtain the variability of errors in the different catchments. For each model the comparison of peak discharge values and time to peak expressed as percentage error and shape of the hydrograph with corresponding average representative UHs obtained from the 1CV method in each case is made.

Hence, the Dimensional Analysis-Based Empirical Model (DAEMS) which predicts the SUH parameters in better agreement is compared with the other synthetic methods viz. Snyder's method, Mitchell's equation, Commons' method, and the CWC method on the basis of minimum percentage error in the peak discharge prediction and shape of the hydrograph.

5.7   RESULTS AND DISCUSSIONS

The present study is intended to give computationally easy and rapid methodology of utilizing a dimensional analysis-based approach to set up an empirical model to conclude SUH parameters for the ungauged basins under study. To exhibit the goodness of this methodology, the 1CV method is applied and the procedure is repeated to validate and exhibit the satisfactory performance of this approach. The comparison of peak discharge, time to peak, and base period expressed as percentage error and shape of the hydrograph with corresponding average representative UHs is made in each case. Finally, the results are compared with the results obtained by other methods as well as the observed UHs.

5.7.1   DEVELOPMENT OF AVERAGE REPRESENTATIVE UHs (ARUHs)

In the present study, the UHs derived for the upper Kumbhi, Dhamani, Kadavi, and Hiranyakeshi basins from actual storm data practically satisfy the assumptions made in the derivation of UH. As hourly rainfall and hourly discharge data are available, it is possible to choose the UH duration as one hour which is suitable for the basins under study and are analyzed to derive UH by the single event method. From the results, it is observed that there is a considerable variation in Q_p and T_p of UH computed for each isolated storm event. When the available data are limited, a method such as the 1CV method has significant appeal because it is based on a resampling procedure. By resampling, one can generate a simulated population (Zhao et al., 1995). Hence an attempt is made to develop an average representative UH with minimum root mean square error, which is useful to provide a tool for the prediction of design flood.

For the upper Kumbhi basin, it is observed that the UHs derived for 28 isolated storms events are not identical. Therefore, the average UH is evaluated to determine the average representative UH for this basin to predict DRH for this basin with reasonable accuracy. To facilitate the validation by the 1CV method, 28 average UHs are determined by leaving out one UH every time from 28 observed UHs and averaging the remaining 27 UHs. Average UH is determined by averaging Q_p, T_p, T_B, W_{50}, and W_{75} of peak discharge of individual UH. The loss rate study based on a∅ index approach using the actual data of flood hydrographs provided general guidance in arriving at the design loss rate. The $\emptyset$ index approach to runoff estimation is a simple tool and it represents the combined effects of interception, depression storage, and infiltration. Effective rainfall (ER) is the portion of rainfall that contributes to direct runoff. The ER is obtained by applying the loss rate. All of the UHs are validated by applying the ER of the left out storm to the corresponding average UH ordinates. Thus the ER of storm event (1) is applied to average UH_1, the ER of storm event (2) is applied to average UH_2, and so on for all the 28 storm events to compute DRHs in each case. Thus 28 computed DRHs are obtained and are compared with the corresponding observed DRHs in the validation part.

It is observed that for the upper Kumbhi basin, average UH derived from storm event (4) predicts DRH in better agreement with minimum root mean square error (E_{RMS}) as 2.637 m³/s and maximum efficiency (E) as 95.275%, named as ARUH-S1. Similarly, for the Dhamani basin 18 average UHs, for the Kadavi basin ten average UHs, and for the Hiranyakeshi basin ten average UHs are determined by the 1CV method. Then corresponding DRHs are computed and compared with the corresponding observed DRHs in the validation test for each basin. From the results, it is observed that for the Dhamani basin DRH prediction using the average UH from storm event (13) gives minimum E_{RMS} as 4.830 m³/s and maximum E as 87.371%, named as ARUH-S2. For the Kadavi basin DRH prediction using average UH from storm event (2) gives minimum E_{RMS} as 1.475 m³/s and maximum E as 97.365%, named as ARUH-S3. For the Hiranyakeshi basin average UH from storm event (1) predicts DRH in better agreement with minimum E_{RMS} as 4.882 m³/s and maximum E as 91.242%, named as ARUH-S4. The computed DRH parameters using ARUH-S1, ARUH-S2, ARUH-S3, and ARUH-S4, and corresponding observed DRH parameters and error functions in predicting DRHs for basins under study using the 1CV method are summarized in Table 5.2.

5.7.2 DIMENSIONAL ANALYSIS-BASED EMPIRICAL MODELS (DAEMS)

The size of the four basins used in the study varied from 121.62 km² to 430 km². In the present study, the Buckingham pi theorem of dimensional analysis technique has been employed in an effort to develop the complete functional relation between the many variables which describe the

---

TABLE 5.2

Computed and Observed DRH Parameters and Error Functions in Predicting DRHs for Different Basins Using 1CV Method

Validation Results	Symbol (SI Unit)	Kumbhi Basin	Dhamani Basin	Kadavi Basin	Hiranyakeshi Basin
Mean	Q_m (m³/s)	7.386	10.065	8.272	14.522
Computed direct runoff peak discharge	Q_{drp}' (m³/s)	43.754	57.109	29.726	71.829
Computed direct runoff time to peak	T_{drp}' (h)	3.905	2.856	4.952	3.905
Observed direct runoff peak discharge	Q_{drp} (m³/s)	51.324	59.301	28.354	69.112
Observed direct runoff time to peak	T_{drp} (h)	4.000	3.000	5.000	4.000
Efficiency of direct runoff prediction	E (%)	95.275	87.371	97.365	91.242
Root mean square error	E_{RMS} (m³/s)	2.637	4.830	1.475	4.882
Percentage error in direct runoff peak discharge	E_P (%)	14.749	3.696	−4.840	−3.931
Percentage error in direct runoff time to peak	E_T (%)	2.375	4.800	0.960	2.375

rainfall-runoff system for the study area. It is possible to formulate a dimensional analysis technique for hydrograph analysis but, the only thing is that various parameters involved in the problem should be known along with their interrelationships. The Dimensional Analysis-Based Empirical Models (DAEM) using the Buckingham pi theorem method to determine the SUH parameters are established for the upper Kumbhi, Dhamani, Kadavi, and Hiranyakeshi basins and named as, DAEMS-1, DAEMS-2, DAEMS-3, and DAEMS-4 respectively. Table 5.3 shows the summary of Dimensional Analysis-Based Empirical Models for basins under study. The usefulness of the derived SUH in the present study is highly dependent on the reliability of the UHs derived from the actual storm data. This study is undertaken to test the methods by which a representative basin could be used to develop a SUH for an ungauged basin within its hydrological region.

In order to calibrate and validate the performance of these empirical models, a 1CV method is employed to all of the catchments under study. The procedure is repeated as many times as the number of catchments to demonstrate the goodness of the method and to obtain the variability of errors in the various catchments. Each model is applied to the other basins under study and a comparison of peak discharge values and time to peak expressed as percentage error with corresponding average representative UHs (ARUHs) obtained from the 1CV method in each case is made. As per Hoffmeister and Weisman (1977), the base period and the shape of the UH can have a large effect on the design storm hydrograph derived by applying the UH to the design rainfall excess and the peak discharge prediction with an error of about ± 25% is assumed to be reasonable. After comparison, the results of SUH parameters and error functions using DAEMS-1, DAEMS-2, DAEMS-3, and DAEMS-4 and respective ARUHs are presented in Tables 5.4, 5.5, 5.6, and 5.7 respectively.

5.7.3 CONCLUDING REMARKS FOR DIMENSIONAL ANALYSIS-BASED EMPIRICAL MODELS

An identical UH would follow from any flood event if all of the conditions underlying the method as well as approximate calculations would reflect sufficiently the flood runoff and if the observed values would be used for the calculations without any considerable errors. It is obvious that these requirements cannot be performed strictly. However, the special literature accessible to us till now did not indicate which derivations have generally been ascertained and found to be justifiable.

There are three types of drainage basin similarities: (1) geometric similarity in terms of basin area, shape, main channel slope, and topography; (2) hydrologic similarity in terms of hydrologic processes such as rainfall, infiltration, runoff, and channel storage; and (3) geologic similarity in terms of the properties relating to the groundwater flow, soil erosion, porous media, sediment characteristics, and sediment transport. Hydrologic relations can be transferred to similar basins if basin similarity requirements are met. In the present study, one of the limitations may be that the watersheds did not possess exact geometric similarity. The Kadavi basin has a contributing catchment (430 km²) largerthan that of the other basins viz. upper Kumbhi (121.62 km²), Dhamani (156 km²), and Hiranyakeshi (244.56 km²) basins, which are considered as small/midsize basins having areas less than 250 km². For the Kadavi basin, the agreement of the individual graphs is very good and difficulty arose in developing the empirical graph. Thus, the large fluctuations in UHs derived for the Kadavi basin using DAEMS-1, 2, and 4 may be due to the large drainage area, the shape of the basin, the width of the watershed, and other influencing topographical, hydrological, meteorological, and geological factors.

For large catchments, rainfall is likely to vary spatially, either as a general storm of concentric isohyetal distribution covering the entire catchment with moderate rainfall or as a highly intensive local storm covering only a portion of the catchment. In the type of hydrological work used in this study, there are many possible sources of errors at any stage: (1) the accuracy of river flow measurements and the ratings applied to them since the water level-discharge relation will often be uncertain in the case of high water levels for lack of discharge measurements; (2) the accuracy of rainfall measurements in deriving an average excess rainfall over a basin, particularly during storm rainfall in connection with high wind velocities; (3) the application of the UH theory to these storm events;

TABLE 5.3

Summary of Dimensional Analysis-Based Empirical Models Using Buckingham Pi Theorem Method for Basins under Study

River Basin (Empirical Model)	T_p (h)	Q_p (m³/s)	T_B (h)	W_{50} (h)	W_{75} (h)
Kumbhi basin (DAEMS-1)	$\dfrac{AL_c t_R}{L^3 S_o^{2.686}}$	$\dfrac{A^{1.5} S_o^{1.408}}{T_p}$	$10.247 T_p \left[\dfrac{\left(\dfrac{Q_p T_p}{A^{1.5}}\right)^{0.102}\left(\dfrac{L_c}{L}\right)^{0.415}}{S_o^{0.460}}\right]$	$0.211\left(\dfrac{A^{1.5}}{Q_p}\right)\left(\dfrac{L_c}{LS_o^{0.536}}\right)$	$0.120\left(\dfrac{A^{1.5}}{Q_p}\right)\left(\dfrac{L_c}{LS_o^{0.536}}\right)$
Dhamani basin (DAEMS-2)	$\dfrac{AL_c t_R}{L^3 S_o^{2.453}}$	$\dfrac{A^{1.5} S_o^{1.399}}{T_p}$	$13.833 T_p \left[\dfrac{\left(\dfrac{Q_p T_p}{A^{1.5}}\right)^{0.102}\left(\dfrac{L_c}{L}\right)^{0.415}}{S_o^{0.480}}\right]$	$0.156\left(\dfrac{A^{1.5}}{Q_p}\right)\left(\dfrac{L_c}{LS_o^{0.587}}\right)$	$0.089\left(\dfrac{A^{1.5}}{Q_p}\right)\left(\dfrac{L_c}{LS_o^{0.587}}\right)$
Kadavi basin (DAEMS-3)	$\dfrac{AL_c t_R}{L^3 S_o^{2.828}}$	$\dfrac{A^{1.5} S_o^{1.516}}{T_p}$	$7.3067 T_p \left[\dfrac{\left(\dfrac{Q_p T_p}{A^{1.5}}\right)^{0.102}\left(\dfrac{L_c}{L}\right)^{0.415}}{S_o^{0.456}}\right]$	$0.103\left(\dfrac{A^{1.5}}{Q_p}\right)\left(\dfrac{L_c}{LS_o^{0.515}}\right)$	$0.059\left(\dfrac{A^{1.5}}{Q_p}\right)\left(\dfrac{L_c}{LS_o^{0.515}}\right)$
Hiranyakeshi basin (DAEMS-4)	$\dfrac{AL_c t_R}{L^3 S_o^{2.707}}$	$\dfrac{A^{1.5} S_o^{1.447}}{T_p}$	$8.478 T_p \left[\dfrac{\left(\dfrac{Q_p T_p}{A^{1.5}}\right)^{0.102}\left(\dfrac{L_c}{L}\right)^{0.415}}{S_o^{0.476}}\right]$	$0.162\left(\dfrac{A^{1.5}}{Q_p}\right)\left(\dfrac{L_c}{LS_o^{0.523}}\right)$	$0.092\left(\dfrac{A^{1.5}}{Q_p}\right)\left(\dfrac{L_c}{LS_o^{0.523}}\right)$

TABLE 5.4

Comparison of SUH Parameters and Error Functions Using DAEMS-1 and ARUHs for Basins under Study

UH Parameters	T_p (h)	Q_p (m³/s)	T_B (h)	W_{50} (h)	W_{75} (h)
Kumbhi: UH by DAEMS-1	3.90	42.61	46	6.64	3.78
Dhamani: UH by DAEMS-1	4.08	55.99	50	8.16	4.64
Kadavi: UH by DAEMS-1	3.88	248.68	49	8.95	5.09
Hiranyakeshi: UH by DAEMS-1	3.78	84.80	45	8.98	5.11
Kumbhi: ARUH-S1	4.00	42.56	46	6.65	3.79
Dhamani: ARUH-S2	3.00	51.24	47	7.12	4.06
Kadavi: ARUH-S3	5.00	102.15	41	10.11	5.76
Hiranyasheshi: ARUH-S4	4.00	78.16	39	7.34	4.18
Error functions (%)	T_p Error (%)	Q_p Error (%)	T_B Error (%)	W_{50} Error (%)	W_{75} Error (%)
Kumbhi: error in UH by DAEMS-1	2	0	0	0	0
Dhamani: error in UH by DAEMS-1	−36	−9	−7	−15	−14
Kadavi: error in UH by DAEMS-1	22	−143	−21	11	12
Hiranyakeshi: error in UH by DAEMS-1	5	−8	−14	−22	−22

TABLE 5.5

Comparison of SUH Parameters and Error Functions Using DAEMS-2 and ARUHs for Basins under Study

UH Parameters	T_p (h)	Q_p (m³/s)	T_B (h)	W_{50} (h)	W_{75} (h)
Kumbhi: UH by DAEMS-2	2.76	39.10	43	5.77	3.29
Dhamani: UH by DAEMS-2	2.86	51.25	47	7.12	4.06
Kadavi: UH by DAEMS-2	2.68	226.90	45	7.86	4.49
Hiranyakeshi: UH by DAEMS-2	2.51	76.57	39	8.04	4.59
Kumbhi: ARUH-S1	4.00	42.56	46	6.65	3.79
Dhamani: ARUH-S2	3.00	51.24	47	7.12	4.06
Kadavi: ARUH-S3	5.00	102.15	41	10.11	5.76
Hiranyasheshi: ARUH-S4	4.00	78.16	39	7.34	4.18
Error functions (%)	T_p Error (%)	Q_p Error (%)	T_B Error (%)	W_{50} Error (%)	W_{75} Error (%)
Kumbhi: error in UH by DAEMS-2	31	8	6	13	13
Dhamani: error in UH by DAEMS-2	5	0	0	0	0
Kadavi: error in UH by DAEMS-2	46	−122	−11	22	22
Hiranyakeshi: error in UH by DAEMS-2	37	2	0	−10	−10

(4) the variance within a hydrological region from the representative basin; and (5) the application of the SUH methods derived from the study of hydrographs for overseas basins and conditions. The fact cannot be changed by the correction made on the basis of the observed volume of runoff as this refers only to the total rainfall and does not eliminate an erroneous estimation of the effective rainfall within the temporal rainfall distribution. Also, the methodic error inherent in the linear method should be considered. When observing the hydraulic law of the flow process, one will arrive at a non-linear UH theory, however, its application will complicate the UH method considerably.

TABLE 5.6

Comparison of SUH Parameters and Error Functions Using DAEMS-3 and ARUHs for Basins under Study

UH Parameters	T_p (h)	Q_p (m³/s)	T_B (h)	W_{50} (h)	W_{75} (h)
Kumbhi: UH by DAEMS-3	4.82	18.85	38	7.11	4.07
Dhamani: UH by DAEMS-3	5.06	24.22	42	8.91	5.11
Kadavi: UH by DAEMS-3	4.85	104.30	41	10.08	5.77
Hiranyakeshi: UH by DAEMS-3	4.86	32.20	38	11.12	6.37
Kumbhi: ARUH-S1	4.00	42.56	46	6.65	3.79
Dhamani: ARUH-S2	3.00	51.24	47	7.12	4.06
Kadavi: ARUH-S3	5.00	102.15	41	10.11	5.76
Hiranyasheshi: ARUH-S4	4.00	78.16	39	7.34	4.18
Error functions (%)	T_p Error (%)	Q_p Error (%)	T_B Error (%)	W_{50} Error (%)	W_{75} Error (%)
Kumbhi: error in UH by DAEMS-3	−20	56	18	−7	−7
Dhamani: error in UH by DAEMS-3	−69	53	11	−25	−25
Kadavi: error in UH by DAEMS-3	3	−2	0	0	0
Hiranyakeshi: error in UH by DAEMS-3	−21	59	3	−52	−52

TABLE 5.7

Comparison of SUH Parameters and Error Functions Using DAEMS-4 and ARUHs for Basins under Study

UH Parameters	T_p (h)	Q_p (m³/s)	T_B (h)	W_{50} (h)	W_{75} (h)
Kumbhi: UH by DAEMS-4	4.03	39.56	40	5.39	3.06
Dhamani: UH by DAEMS-4	4.21	51.88	44	6.63	3.76
Kadavi: UH by DAEMS-4	4.01	229.79	43	7.29	4.14
Hiranyakeshi: UH by DAEMS-4	3.93	77.65	39	7.36	4.18
Kumbhi: ARUH-S1	4.00	42.56	46	6.65	3.79
Dhamani: ARUH-S2	3.00	51.24	47	7.12	4.06
Kadavi: ARUH-S3	5.00	102.15	41	10.11	5.76
Hiranyasheshi: ARUH-S4	4.00	78.16	39	7.34	4.18
Error functions (%)	T_p Error (%)	Q_p Error (%)	T_B Error (%)	W_{50} Error (%)	W_{75} Error (%)
Kumbhi: error in UH by DAEMS-4	−1	7	13	19	19
Dhamani: error in UH by DAEMS-4	−40	−1	6	7	7
Kadavi: error in UH by DAEMS-4	20	−125	−6	28	28
Hiranyakeshi: error in UH by DAEMS-4	2	1	0	0	0

The geology underlying a drainage basin affects the runoff characteristics of a drainage basin. In-depth knowledge of geology is important to thoroughly understand the hydrology of the drainage basin. Many drainage basins supply groundwater flow to streams either on a continuous basis or on an intermittent basis. Thus, groundwater hydrology is closely related to surface water hydrology.

Hence, the aforementioned factors are required to be taken into account while applying DAEMS-1, 2, and 4 to the Kadavi basin for the determination of the SUH parameters. Also, the aforementioned factors are the probable reasons that the DAEMS-3 for the Kadavi basin did not consistently over-predict or under-predict a hydrograph peak or give an accurate description or UH shape for upper Kumbhi, Dhamani, and Hiranyakeshi basins. The DAEMS-4 for the Hiranyakeshi basin proved reasonably accurate for upper Kumbhi and Dhamani basins producing SUHs within the range of the individual storm hydrographs. The equations gave conservative estimates of peak discharge along with the shape of UH. The peak discharge prediction for the Kadavi basin using DAEMS-4 seems to be 2.25 times more than the average UH. The study is intended to provide a computationally simple and rapid procedure to use dimensional analysis techniques for empirical modeling of SUHs for small ungauged basins. This approach exhibits satisfactory performance insimulating and predicting peak discharge and shape of the hydrograph.

5.7.4 Empirical Synthesis of UHs for Basins under Study

For upper Kumbhi, Dhamani, Kadavi, and Hiranyakeshi basins, four SUH methods are tested, namely: (1) Snyder's method, (2) Mitchell's equation, (3) Commons' method, and (4) the CWC method. The peak discharges and times to peak are calculated for the basins under study using measured catchment characteristics and the regional coefficients. In the present study, two sets of values of peak discharge and time to peak are obtained for each basin. For Snyder's, Mitchell's, and Commons' methods two approaches are considered: (1) Case A: with basin lag "t_p" based on drainage area features, and (2) Case B: UH parameters based on observed "t_p" for the representative basin. By using both these approaches it is hoped to show that, SUH predictions could be improved by increasing the sampling range in a region. The comparison of SUH results using dimensional analysis approach and other synthetic methods corresponding to ARUHs with error functions (%) in prediction are summarized in Tables 5.8, 5.9, 5.10, and 5.11 for upper Kumbhi, Dhamani, Kadavi, and Hiranyakeshi basins respectively. As per Hoffmeister and Weisman (1977), the normal disparity in the prediction of Q_p of about ± 25% is considered to be probable. Figures 5.3, 5.4, 5.5, and 5.6 show the comparison of ARUHs derived from actual storm data and SUHs by the different methods for basins under study.

5.7.5 Concluding Remarks of Empirical Synthesis of UHs for Basins under Study

Snyder in his study included interflow as part of direct runoff when calculating the actual time base of the UH. This may result in a longer time base than that corresponding only to direct runoff. Hence, in the present study, the time base related to direct runoff is comparatively shorter. Snyder cautioned that lag may tend to vary slightly with flood magnitude and that SUH calculations are likely to be more accurate for fan-shaped catchments than for those of highly irregular shapes. As per Snyder, (1938), the greater the variation of any particular area from a typical fanshape, the greater is apt to be the discrepancy between the synthetic and actual unit graphs. The coefficients C_t and C_p are determined on a regional basis. The study revealed thatC_t is largely a function of catchment slope, since L and L_c are measures of length and shape of the basin. This implies that even for catchments of the same size lag is a function of slope and reflects the natural variability of catchment slopes. Basin slope has a profound effect on the velocity of overland flow. The constants C_t and C_p vary over a wide range and from region to region, and may not be equally suitable for all the regions. The differences in the SUH parameters may be due to the fact that the equations are only a general formula synthesized from many different basins; it may only give a rough approximation.

The critical factor in SUH is the lag and that good results could be expected from SUH for ungauged basins if this factor is known. In Mitchell's equation, the probable error of the peak of the SUH computed by use of the drainage formula is considerably greater than that based either on observed lag or on lag computed from time to peak. However, in most of the cases in which neither t nor t_p will be known, it becomes necessary to make the best possible use of formula involving drainage area. In the present study, Mitchell's equation predicts peak discharge within the range of

TABLE 5.8

Comparison of SUH Parameters and Error Functions Using DAEMS-4 and Other Synthetic Methods Corresponding to ARUH-S1 for the Upper Kumbhi Basin

UH Parameter	T_P (h)	Q_P (m³/s)	T_B (h)	W_{50} (h)	W_{75} (h)	T_P Error (%)	Q_P Error (%)	T_B Error (%)	W_{50} Error (%)	W_{75} Error (%)
Average UH: ARUH-S1	4.00	42.56	46	6.65	3.79	0	0	0	0	0
UH by DAEMS-4	4.03	39.56	40	5.39	3.06	−1	7	13	19	19
UH by Snyder's method Case: A	2.81	90.51	12	2.95	1.68	30	−113	74	56	56
UH by Snyder's method Case: B	3.91	59.84	19	4.60	2.62	2	−41	59	31	31
UH by Mitchell's equation Case: A	5.66	35.41	41	8.11	4.62	−42	17	11	−22	−22
UH by Mitchell's equation Case: B	3.91	35.41	41	8.11	4.62	2	17	11	−22	−22
UH by Commons'hydrograph Case: A	5.66	41.94	40	6.76	3.85	−42	1	13	−2	−2
UH by Commons'hydrograph Case: B	3.91	60.73	28	4.53	2.58	2	−43	39	32	32
UH by CWC method	6.64	49.13	19	6.27	3.72	−66	−15	59	6	2

TABLE 5.9

Comparison of SUH Parameters and Error Functions Using DAEMS-4 and Other Synthetic Methods Corresponding to ARUH-S2 for the Dhamani Basin

UH Parameter	T_P (h)	Q_p (m³/s)	T_B (h)	W_{50} (h)	W_{75} (h)	T_P Error (%)	Q_p Error (%)	T_B Error (%)	W_{50} Error (%)	W_{75} Error (%)
Average UH: ARUH-S2	3.00	51.24	47	7.12	4.06	0	0	0	0	0
UH by DAEMS-4	4.21	51.88	44	6.63	3.76	−40	−1	6	7	7
UH by Snyder's method Case: A	3.13	101.98	14	3.39	1.93	−4	−99	70	52	52
UH by Snyder's method Case: B	2.86	107.45	13	3.65	1.82	5	−110	72	49	55
UH by Mitchell's equation Case: A	6.71	45.42	41	8.11	4.63	−124	11	13	−14	−14
UH by Mitchell's equation Case: B	2.86	45.42	41	8.11	4.63	5	11	13	−14	−14
UH by Commons'hydrograph Case: A	6.71	45.36	48	8.12	4.63	−124	11	−2	−14	−14
UH by Commons'hydrograph Case: B	2.86	106.56	20	3.23	1.84	5	−108	57	55	55
UH by CWC method	8.18	56.47	22	7.05	4.18	−173	−10	53	1	−3

TABLE 5.10

Comparison of SUH Parameters and Error Functions Using DAEMS-4 and Other Synthetic Methods Corresponding to ARUH-S3 for the Kadavi Basin

UH Parameter	T_P (h)	Q_P (m³/s)	T_B (h)	W_{50} (h)	W_{75} (h)	T_P Error (%)	Q_P Error (%)	T_B Error (%)	W_{50} Error (%)	W_{75} Error (%)
Average UH: ARUH-S3	5.00	102.15	41	10.11	5.76	0	0	0	0	0
UH by DAEMS-4	4.01	229.79	43	7.29	4.14	20	−125	−6	28	28
UH by Snyder's method Case: A	4.25	197.40	20	4.96	2.83	15	−93	51	51	51
UH by Snyder's method Case: B	4.95	164.54	24	6.03	3.44	1	−61	41	40	40
UH by Mitchell's equation Case: A	13.65	125.19	41	8.11	4.63	−173	−23	0	20	20
UH by Mitchell's equation Case: B	4.95	125.19	41	8.11	4.63	1	−23	0	20	20
UH by Commons' hydrograph Case: A	13.65	61.44	98	17.50	9.98	−173	40	−138	−73	−73
UH by Commons' hydrograph Case: B	4.95	169.20	35	5.86	3.34	1	−66	15	42	42
UH by CWC method	14.29	116.10	29	9.36	5.72	−186	−14	29	7	1

TABLE 5.11

Comparison of SUH Parameters and Error Functions Using DAEMS-4 and Other Synthetic Methods Corresponding to ARUH-S4 for the Hiranyakeshi Basin

UH Parameter	T_P (h)	Q_P (m³/s)	T_B (h)	W_{50} (h)	W_{75} (h)	T_P Error (%)	Q_P Error (%)	T_B Error (%)	W_{50} Error (%)	W_{75} Error (%)
Average UH: ARUH-S4	4.00	78.16	39	7.34	4.18	0	0	0	0	0
UH by DAEMS-4	3.93	77.65	39	7.36	4.18	2	1	0	0	0
UH by Snyder's method Case: A	3.59	137.65	16	3.98	2.27	10	-76	59	46	46
UH by Snyder's method Case: B	3.91	120.32	19	4.60	2.62	2	-54	52	37	37
UH by Mitchell's equation Case: A	9.15	71.20	41	8.11	4.62	-129	9	-5	-11	-11
UH by Mitchell's equation Case: B	3.91	71.20	41	8.11	4.62	2	9	-5	-11	-11
UH by Commons'hydrograph Case: A	9.15	52.08	65	11.37	6.48	-129	33	-68	-55	-55
UH by Commons'hydrograph Case: B	3.91	122.10	28	4.53	2.58	2	-56	28	38	38
UH by CWC method	11.00	75.81	25	8.32	4.94	-175	3	36	-13	-18

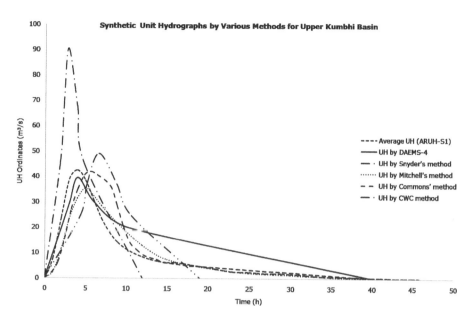

FIGURE 5.3 SUHs derived using DAEMS-4 and other synthetic methods corresponding to ARUH-S1 for the upper Kumbhi basin.

error that is probable while deriving UHs for the ungauged basins. However, the time to peak using drainage basin features is observed to be greater than the averaged value.

Commons' dimensionless hydrograph is used to determine a general shape of flood hydrographs. A relationship probably exists between hydrograph characteristics and the physical characteristics of the basin. In the present study, Commons' dimensionless hydrograph did not consistently overpredict or underpredict the peak discharge. In many cases, the time base of the dimensionless hydrograph is much greater than the averaged value, showing that the developed dimensionless hydrograph shape did not for local conditions. It is quite different in shape after the peak than the

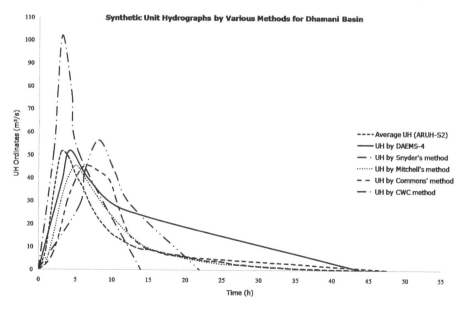

FIGURE 5.4 SUHs derived using DAEMS-4 and other synthetic methods corresponding to ARUH-S2 for the Dhamani basin.

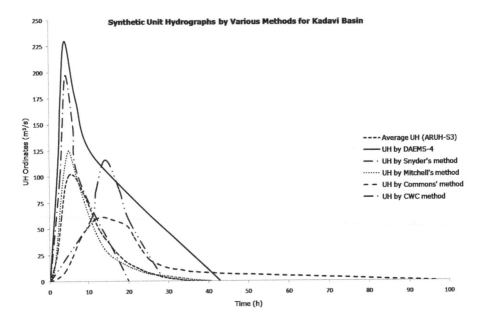

FIGURE 5.5 SUHs derived using DAEMS-4 and other synthetic methods corresponding to ARUH-S3 for the Kadavi basin.

other hydrographs. It has a very long time base on the recession portion of the hydrograph. The ability of the method to derive the complete hydrograph accurately, however, is not demonstrated in the study.

The flood estimation report of subzone 3 (h) was prepared for Krishna and Pennar Subzone 3 (h), which covers about 95% of the Krishna river basin and 65% of the area of Pennar basins covering part of the states of Maharashtra, Karnataka, and Andhra Pradesh. The UH study was carried out for 21 catchments having catchment areas ranging from 25 km² to 2,000 km²

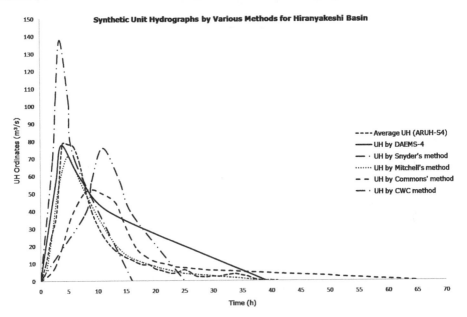

FIGURE 5.6 SUHs derived using DAEMS-4 and other synthetic methods corresponding to ARUH-S4 for the Hiranyakeshi basin.

judiciously after comparing the neighboring catchments having more or less similar characteristics. The present study reveals that the CWC method predicts peak discharge in better agreement but the shape of the hydrograph shows poor estimate as compared to average representative UHs (ARUHs). The basin lag does not necessarily improve the prediction of the shape of the UH. As per theCWC (1983), the possible limitations of the CWC method behind these inconsistencies may be: (1) the method would be applicable for reasonably free catchments with interception if any limited to 20% of the total catchment. For calculating the discharge the total area of the catchment has to be considered; (2) the generalized values of base flow and loss rate has to be assumed to hold good for the whole subzone; (3) the rainfall-runoff data of 21 catchments has been considered to develop a generalized approach of flood estimation; (4) this method has been developed for a wide range of areas, but individual site conditions may necessitate special study.

These differences in SUH parameters may be due to (1) basin characteristics such as drainage area, shape, slope; (2) hydrological and meteorological characteristics such as rainfall, intensity, duration, infiltration, runoff, storage capacity, and (3) geological characteristics such as properties relating to groundwater flow, soil type, soil erosion, porous media, sediment characteristics and sediment transport, etc. of the basins under study.

5.8 SUMMARY AND CONCLUSIONS

A number of methods for developing SUHs are reported in the literature. But these methods are applicable only to the specific regions in which they are developed based on the empirical correlations and are not applicable in general for use in other regions. It is found that the traditional synthetic unit hydrograph (SUH) models have several inconsistencies associated with them; however, these models are widely used for SUH derivation.

In the present study, from the comparison, it is observed that the results obtained from DAEMS-4 (developed for the Hiranyakeshi basin) are expected to be closer to reality than the results obtained from the other models. The DAEMS-4 proved reasonably accurate for the upper Kumbhi and the Dhamani basins producing SUHs within the range of the individual storm hydrographs. The equations gave conservative estimates of peak discharge along with the shape of UH. It is also observed that the peak discharge prediction for the Kadavi basin using DAEMS-4 is 2.25 times more than the average representative UH (ARUH-S3) of this basin. Also, it shows the poor estimate corresponding to the shape of the ARUH-S3. The probable reason behind this may be due to the large drainage area, the shape of the basin, the width of the watershed, and other influencing topographical, hydrological, and geological factors of the Kadavi basin.

Further, the results obtained by DAEMS-4 are compared with the popular synthetic methods. From this comparison, the dimensional analysis-based approach is suggested to be more suitable to derive SUHs for the basins within hydrologically similar regions. This will ensure more reliable and representative hydrographs for the catchments with more or less similar hydrological conditions. Study findings demonstrated that the use of dimensional analysis techniques to simulate SUHs offers the advantages of being a simpler and faster method of SUH analysis. The accuracy is limited by the nature of the assumptions made in their derivation, as well as how well these assumptions may apply to each specific problem.

REFERENCES

Barnes, B.S.1959. Consistency in unitgraphs, *Journal of the Hydraulics Division*, 85(8): 39–61.
Bernard, M.1935. An approach to determinate stream flow, *Transactions of the American Society of Civil Engineers*, 100: 347–362.
Bhunya, P.K., Ghosh, N.C., Mishra, S.K., Ojha, C.S.P., and Berndtsson, R.2005. Hybrid model for derivation of synthetic unit hydrograph, *Journal of Hydrologic Engineering*, 10(6): 458–467.

Bhunya, P.K., Singh, P.K., and Mishra, S.K.2009. Frechet and chi-square parametric expressions combined with Horton ratios to derive a synthetic unit hydrograph, *Hydrological Sciences Journal*, 54(2): 274–286.

Black, P.E.1972. Hydrograph response to geomorphic model watershed characteristics and precipitation variables, *Journal of Hydrology*, 17(4): 309–329.

Brater, E.F.1939. The unit hydrograph principle applied to small watersheds, *Proceedings, ASCE*, 65: 1191.

Bruen, M., and Dooge, J.C.I.1992. Unit hydrograph estimation with multiple events and prior information:II. Evaluation of the method, *Hydrological Sciences Journal*, 37(5): 445–462.

Chow, V.T., Maidment, D.R., and Mays, L.W.1988. *Applied Hydrology*, McGraw-Hill, New York, USA, pp. 201–236.

Cleveland, T.G., Xin, H., William, H.A., Xing, F., and Thompson, D.B.2006. Instantaneous unit hydrograph evaluation for rainfall-runoff modeling of small watersheds in north and south central Texas, *Journal of Irrigation and Drainage Engineering*, 132(5): 479–485.

Commons, C.G.1942. Flood hydrographs, *Civil Engineering*, 12: 571–572.

Craig, G.S., Collins, D.L., and Wilson, J.F.1969. *Study of Flood Hydrographs for Small Drainage Basins in Whyoming*, United States Department of Interior Geological Survey, Wyoming, USA, pp. 30–31.

CWC. 1983. Flood estimation report for Krishna and Pennar basins (Sub-Zone 3h): A method based on UH principle. Report No. K/6, Hydrology for Small Catchments Directorate, Central Water Commission, Government of India, New Delhi, India.

CWC. 1973. Estimation of design flood peak: A method based on unit hydrograph principle. Report No. 1/73 (Revised), Hydrology for Small Catchments Directorate, Central Water Commission, Government of India, New Delhi, India.

CWC. 2000. Flood estimation report for Krishna and Pennar Subzone-3(h) (Revised). *Central Water Commission*, Design Office Report No. KP-3(h)/R-5/45/2000.

Das, G.2010. *Hydrology and Soil Conservation Engineering*, 2nd edition, PHI Learning Pvt. Ltd., New Delhi, India, pp. 148–158.

Dooge, J.C.I.1959. A general theory of the unit hydrograph, *Journal of Geophysical Research*, 64(2): 241–256.

Dym, C.L.2004. *Principles of Mathematical Modeling*, 2nd edition, Academic Press, New York, USA, pp. 15–31.

Espey, W.H., Altman, D.G.Jr., and Graves, C.B.Jr.1977. *Nomograph for 10-Minutes Unit Hydrographs for Urban Watersheds*. Tech. Memo., 32, American Society of Civil Engineers, New York, USA.

Espey, W.H., and Altman, D.G.Jr.1978. *Nomograph for 10-Minute Unit Hydrographs for Small Watersheds*. Addendum 3 of Urban Runoff Control Planning, Report EPA-600/9-78-035, Environmental Protection Agency, Washington, DC, USA.

Genick, B.M. 2013. *Basics of Fluid Mechanics*. Last modified: Version 0.3.4.0 March 17, 2013, www.potto.org/downloads.php.

Gray, D.M.1961. Synthetic unit hydrographs for small watersheds, *Journal of Hydraulics Division. Proceedings of the American Society of Civil Engineers*, 87(4): 33–54.

Gray, D.M.1964. Physiographic characteristics and the runoff pattern. In: *Proceedings of Hydrology Symposium No. 4, Research Watersheds*, National Research Council of Canada, pp. 146–164.

Guo, J.C.Y.2006. Storm-water predictions by dimensionless unit hydrograph, *Journal of Irrigation and Drainage Engineering*, 132: 410–417.

Hoffmeister, G., and Weisman, R.N.1977. Accuracy of synthetic hydrographs derived from representative basins, *Hydrological Sciences Bulletin*, 22(2): 297–312.

Linsley, R.K., Kohler, M.A., and Paulhus, J.I.H.1988. *Hydrology for Engineers*, McGraw-Hill Book Company, New York, USA.

Maidment, D.R., Olivera, F., Calver, A., Eatherall, A., and Fraczek, W.1996. Unit hydrograph derived from a spatially distributed velocity field, *Hydrological Processes*, 10: 831–844.

Mitchell, W.D.1948. Unit Hydrographs in ILLINOIS. United States *Department of Interior Geological Survey*, Div. of Waterways, Illinois, USA.

Mujumdar, M.M., Lakshman, N., and Thorvat, A.R.2014. Development of regional flood formulas for predictions in ungauged basins of upper Krishna river basin, India, *International Journal of Global Technology Initiatives*, 3(1): A78–A95.

Narvekar, S.P., Mujumdar, M.M., and Thorvat, A.R.2013a. Innovative technique for analysis of retaining wall using dimensional analysis, *International Journal of Scientific and Research Publications*, 3: 571–576.

Narvekar, S.P., Mujumdar, M.M., and Thorvat, A.R.2013b. Use of dimensional analysis in the cement industry, *International Journal of Engineering Science and Technology*, 5(3): 500–504.

Pankhurst, R.1964. *Dimensional Analysis and Scale Factors*, Chapman and Hall, London, UK.

Patel, J.N., and Thorvat, A.R.2016. Synthetic unit hydrograph development for ungauged basins using dimensional analysis, *Journal American Water Works Association (JAWWA)*, 108(3): E145–E153, http://doi.org/10.5942/jawwa.2016.108.0014.

Pinheiro, V.B., and Naghettini, M.2013. Calibration of the parameters of a rainfall-runoff model in ungauged basins using synthetic flow duration curves as estimated by regional analysis, *Journal of Hydrologic Engineering*, 18: 1617–1626.

Ponce, V.M.1989. *Engineering Hydrology- Principles and Practices*, Prentice-Hall, Englewood Cliffs, NJ, USA, pp. 153–187.

Rajput, R.K., 2009. *Textbook of Fluid Mechanics*, S. Chand and Company Ltd., New Delhi, India, pp. 379–387.

SCS (Soil Conservation Service). 1957. *Use of Storm and Watershed Characteristics in Synthetic Hydrograph Analysis and Application*, US Department of Agriculture, Soil Conservation Service, Washington, DC, USA.

Sherman, L.K.1932. Stream flow from rainfall by the unit hydrograph method, *Engineering News Record*, 108: 501–505.

Singh, P.K., Mishra, S.K., and Jain, M.K.2014. A review of the synthetic unit hydrograph: From the empirical UH to advanced geomorphological methods, *Hydrological Sciences Journal*, 59(2): 239–261.

Singh, S.K.2000. Transmuting synthetic unit hydrographs into gamma distribution. *Journal of Hydrologic Engineering*, 5: 380–385.

Singh, S.K., 2005. Clark's and Espey's unit hydrographs vs the gamma unit hydrograph/ Les hydrogrammes unitaires de Clark et de Espey vs l'hydrogramme unitaire de forme loi gamma, *Hydrological Sciences Journal*, 50(6): 1053–1067.

Singh, V.P.1988. *Hydrologic Systems: Rainfall-Runoff Modeling*, vol. 1, Prentice-Hall, Englewood Cliffs, NJ, USA.

Singh, V.P.1994. *Elementary Hydrology*, PHI Learning Pvt. Ltd., New Delhi, India, pp. 439–566.

Snyder, F.F.1938. Synthetic unit-graphs, *Transactions, American Geophysical Union*, 19: 447–454.

Subramanya, K.2008. *Engineering Hydrology*, 3rd edition, Tata McGraw Hill Publishing Company, New Delhi, India, pp. 195–232.

Taylor, A.B., and Schwarz, H.E.1952. Unit hydrograph lag and peak flow related to basin characteristics, *Transactions, American Geophysical Union*, 33: 235–246.

Thorvat, A.R., and Patel, J.N.2015. Development of Synthetic Unit Hydrographs for Ungauged Upper Kumbhi Basin, Maharashtra (INDIA), *International Journal of Earth Sciences and Engineering (IJEE)*, 8(6): 2653–2662, ISSN: 0974-5904.

Thorvat, A.R., and Patel, J.N.2016. Empirical approach to develop synthetic unit hydrographs for ungauged dhamani basin, Maharashtra (INDIA), *International Journal of Hydrology Science and Technology (IJHST)*, 6(3): 266–284, http://doi.org/10.1504/IJHST.2016.077396.

Thorvat, A.R., and Mujumdar, M.M.2011. Design flood estimation for upper Krishna basin through RFFA, *International Journal of Engineering Science and Technology (IJEST)*, 3(6): 5252–5259.

U.S. Army Corps of Engineers (USACE). 1940. *Engineering Construction–Flood Control*, Engineering School, U.S. Army Corps of Engineers, Fort Belvoir, VA, USA.

Wesley, P.J., Phillip, W.W., and Williams, J.R.1987. Synthetic unit hydrograph. *Journal of Water Resources Planning and Management*, 113: 70–81.

Worstell, J.2014. *Dimensional Analysis: Practical Guides in Chemical Engineering*, Elsevier Inc., Oxford, UK, pp. 1–31.

Zhao, B., Tung, Y.K., Yeh, K.C., and Yang, J.C.1995. Statistical validation methods: Application to unit hydrographs, *Journal of Hydrologic Engineering*, 121:618–624.

6 Flood Forecasting
Time Series or Flood Frequency Analysis?

Priyanka Sharma, Pravin Patil, and Saeid Eslamian

CONTENTS

6.1 INTRODUCTION

6.1.1 OVERVIEW

Climate change has caused extreme rainfall events and the frequency of these events has increased; these are generally termed natural disaster floods (Lee and Kim, 2018). In recent times, it has been reported that more than one-third of the world's land is prone to natural disaster floods, affecting about 82% of the world's population (Dilley et al., 2005). In India, out of the total geographical area of 329 mha, more than 40 mha is flood-prone, mainly occurring during the monsoon season, and it is one of the most flood-prone countries in the world (Alam and Muzzammil, 2011). The most common flood-prone areas are low elevation, channels, small basins, shorelines, and alluvial fans. Pertinently, the impacts of this natural disaster flood are expected to rise due to an increase in population, economic growth, and climate change, e.g., increase in natural events and temperature rise (Tanoue et al., 2016; Bhat et al., 2019). The components of a flood forecasting and warning system

DOI: 10.1201/9780429463938-9

are shown in Figure 6.1. A flood forecasting model can be developed based on the structures and their types; classification of different flood forecasting models is depicted in Figure 6.2.

Flood forecasting only makes sense if its results reach as many of the affected people as possible in a suitable form. Flood forecasting systems involve the predetermination of flood events necessary to help the implementation of structural measures for the mitigation of flood damages, regulation, and operation of the multipurpose reservoirs with a focus on the control of incoming floods (Mutreja, 1986). The challenge of hydrometeorological flood forecasting presents a complicated task but promises significant gain when successful. Close collaboration between meteorologists and hydrologists may improve the input precipitation data and allow the running of the hydrological model with an ensemble of different numerical weather prediction datasets (Givati et al., 2016). Mutreja (1986) reported that the hydrometeorological community should redouble research and operational efforts to improve the flood alert and warning systems, especially in an era of rapid land use and climate changes.

The identification of suitable models for flood forecasting is very useful for water resources planning and management. In the last few decades, it has received tremendous attention from researchers and many types of linear and non-linear models are proposed by hydrologists for accurate forecasting of hydrologic time series (Wang et al., 2009).

Application of the process-based and data-driven models shows the need for a finer scale in the modeling process to account for extreme events as well as modeling the downstream parts of the basin and specifically the routing of wet season flow volumes along the length of a very large river with adjacent floodplains (Tshimanga et al., 2016). Due to the high degree of uncertainty in

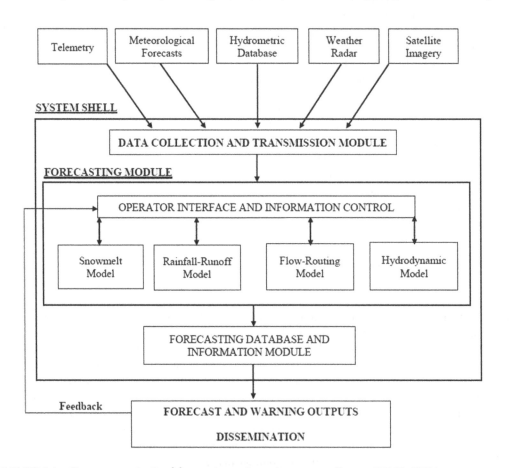

FIGURE 6.1 Components of a flood forecasting and warning system. (Source: WMO, 2011.)

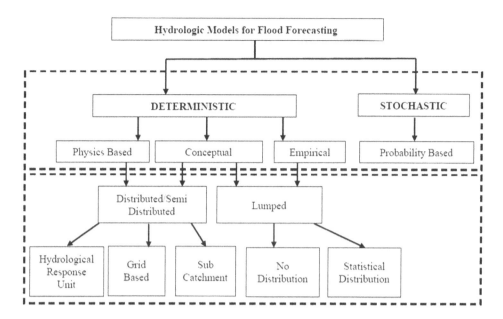

FIGURE 6.2 Classification of models used for flood forecasting based on model structure and type. (Source: WMO, 2011.)

small catchment areas, specific local forecasts accurate to the centimeter are impossible. Instead, warnings are issued for regions, if the predicted runoff from several stretches of the river exceeds a certain threshold (Demuth and Rademacher, 2016).

Time series modeling processes were mainly developed to address practical problems of data simulation, forecasting of time series either for short term or long term, and synthetic generation of data, particularly for water resources systems design and management in data-scarce situations. In time series modeling, the stochastic models are broadly classified as autoregressive moving average (ARMA) models (Box and Jenkins, 1970), disaggregation models (Valencia and Schaake, 1973), and models based on the concept of pattern recognition (Panu and Unny, 1980). According to Hipel (1985), in hydrological time series modeling a simple stochastic model may give better results than a complex deterministic model (Lohani et al., 2012), and it is a key tool to predict the effects of climate change.

Flood forecasting systems for short-term flood warnings and long-term mitigation require extensive field observations. There is considerable uncertainty about the exact conditions within the basin, and although the network is improving, even the best rainfall-runoff model will have a 5% to 10% residual error. The error in the predicted peak stage depends on the slope of the rating curve at that stage, and under average conditions, the forecast for upper reaches cannot be consistently reliable within less than 30 cm, and thus lacks reliability. Hence, it is not desirable to forecast in more precise terms than justified by the conditions; consequently, forecasts should rarely be given to less than the nearest 15 cm. Nonetheless, there is an urgent need by water managers, forecasters, policy makers, and government agencies to improve the accuracy of flood forecasting and hazard mapping by using easy and process-based tools.

Therefore, the main aim of this chapter is to facilitate the application of the accurate forecasting of floods for both long-term planning and short-term emergency warning and to raise awareness of the need for a rigorous stochastic approach to flood forecasting. As it is well known that the impacts of climate change are intensifying, there is an urgent need to understand the process to determine the extreme events, and also develop a model to improve the accuracy of forecast events. Hence, this chapter attempts to illustrate a step-by-step procedure for flood frequency analysis to estimate the magnitude of extreme flooding events and their frequency of occurrence (return period) by using flood frequency distributions, which is a key tool to characterize flood risk, and predict flood events based on the historical data (peak discharge data). It also presents the state of the art to apply stochastic modeling techniques on hydrologic time series data to forecast flood events. This process-based technique is used to explain hydrological data using statistical methods, select the best stochastic model for the data generation process, and forecast accurate hydrological data.

The capability of FFA is to interpret the forecast and flood observations, in order to provide situation updates to determine possible impacts on communities and infrastructure. Flood forecasting using time series analysis provides the capability to simulate the random and probabilistic nature of inputs and responses that given river flows, rainfall, etc. The major capability of a flood forecasting and warning system (FFWS) is to alert the general public and concerned authorities of an impending flood as much in advance, and with as much reliability, as possible (Jain et al., 2018).

The procedure for the development of stochastic models in a flow chart is presented in Figure 6.3.

6.1.2 Applications of the Flood Forecasting Model

1. It is used for predetermination of flood events necessary to help the implementation of structural measures for the mitigation of flood damages, regulation and operation of the multipurpose reservoirs with a focus on the control of incoming floods, and the evacuation of the affected people to the safer places.

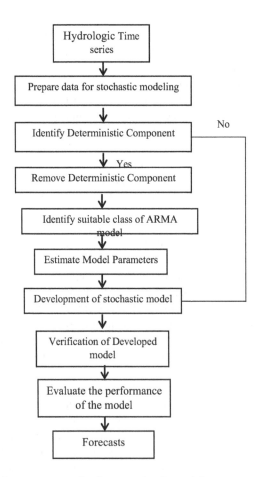

FIGURE 6.3 Flow chart for processes to develop a stochastic model.

2. It is used to quantify the real hydrological processes at the basin scale and to use the models to assess the impacts of future environmental changes in the basin (Tshimanga et al., 2016).

3. It can also help to implement the non-structural measures for mitigating floods that reduce exposure to floods, including regulation of land use and flood forecasting, among others, in flood-prone areas that are already occupied (Jain et al., 2018). It provides more reversible and less expensive mechanisms to reduce flood risk than structural protection measures (i.e., dams and levees).

4. Flood forecasting models can be implemented at global (Alfieri et al., 2013), continental (Thiemeg et al., 2015), basin (Hopson and Webster, 2010), and community scales.

6.1.3 General Guidelines for Applying Time Series Modeling to Hydrologic Time Series

1. Select the observed hydrologic time series.
2. Plot the hydrologic time series in a graph or scatter plot.
3. Observe the trend and periodicity in the time series by visual inspection.
4. If trend and periodicity are present in the time series, apply various tests to identify the trend and check the periodicity by spectral analysis (frequency domain).
5. After estimating the deterministic components (trend and periodicity), remove these components by using simple differencing, standardization, and box-cox transformation.
6. When the time series is trend-free and the periodic component is removed, the next step is to identify the tentative stochastic model by using the autocorrelation function and partial autocorrelation function in the time domain.
7. The next step is to estimate the exact parameters of the autoregressive and moving average parameters by using the method of moment, the maximum likelihood method, and the minimum least square method.
8. Conduct diagnostic checking of the residuals of the developed time series models.
9. Select the best fit model by using the performance evaluation indices.
10. Conduct forecasting.

6.1.4 Novelty of Work

This chapter gives rise to the potential for improving your stochastic model/forecasting if suitable time series data is available. Many studies have been done and have found out the probability analysis solved by the suggested methods in this chapter, giving the right direction to forecast the rainfalls, inflows, surface runoff discharge, etc. but it should be location-specific. Time series analysis is being improvised since 1970 to date. This chapter makes readers understand the concept of time series analysis and flood forecasting analysis with a given step-by-step procedure. The goal of this chapter is to understand how to obtain real-time precipitation and streamflow trends through various meteorological data, use them in a rainfall-runoff and streamflow-channel flow and reservoir routing program, and forecast flood flow rates and water levels for a period of a few hours to a few days ahead depending upon the size of the catchment in a simple procedure. In this chapter, different available models and approaches are gathered and described briefly so that readers can understand them well and use them in their research work.

6.2 METHODS OR TECHNIQUES USED FOR FLOOD FORECASTING

The methods of flood forecasting being used at present can be divided into two groups: (1) methods based on frequency analysis of the hydrological and meteorological data and (2) methods based on prediction models such as deterministic, stochastic models, etc.

6.2.1 Flood Frequency Analysis

So far, the peak discharge rates have been determined on the basis of rainfall frequency. The frequency (chance of occurrence) of the storm runoff so derived is, however, not necessarily the same as that of the rainfall due to assumptions regarding antecedent conditions and the estimate of water losses from the initial precipitation. A method of dealing with the runoff directly is called the *flood-frequency* method (Mutreja, 1986). The method, however, does not provide a hydrograph shape but gives only a peak discharge of known frequency.

Frequency studies interpret a past record of events to predict the future probabilities of occurrence. If the available runoff data are of sufficient length and reliability, they can yield satisfactory estimates. Furthermore, for the estimation of flood flows of large return periods, it is often necessary to extrapolate the magnitude outside the observed range. The accuracy of the estimate obviously reduces with the degree of extrapolation.

Basic data required:

1. Daily rainfall event data of at least the past 30 years;
2. Topographic features of the watershed;
3. Land use–land cover details of the watershed;
4. Stream number and frequency;
5. Hydrological soil group;
6. Changes in the stage–discharge relationship at the reservoir.

The following points should be taken into consideration while collecting the data:

1. The effect of man-made changes in the regime of flow should be investigated and adjustments made as required.
2. For small watersheds, a distinction should be made between daily maxima and instantaneous or momentary flood peaks.
3. Changes in the stage–discharge relationship make stage records nonhomogeneous and unsuitable for frequency analysis. It is therefore preferable to work with discharges.
4. Any useful information contained in data publications and manuscripts should be made use of after proper scrutiny.

For the statistical parameter estimation, series data should be used in the frequency analysis. Series such as annual, seasonal, monthly, or daily can be prepared, though an annual series is the most commonly used. This is a selection of the maximum event of a particular year even though this may not be higher than the maximum observed floods of each year (Mutreja, 1986) because this could be a series of maximum observed floods of each year.

6.2.1.1 Methods of Flood Frequency Analysis

Frequencies can be evaluated graphically by plotting magnitudes of hydrologic variables, say flood flows, against the frequencies with which they have been equaled or exceeded, and fitting a smooth curve through the plotted points and assuming the same as representative of future possibilities (Mutreja, 1986).

6.2.1.1.1 Frequency Analysis Using Frequency Factors

Calculating the magnitudes of extreme events by the method outlined requires that the probability distribution function be invertible, that is, given a value for T (return period) or $[F(x_T) = T/(T-1)]$, the corresponding value of x_T can be determined (Chow et al., 1988). Some probability distribution functions are not readily invertible, including the normal and Pearson type III distributions, and an alternative method of calculating the magnitudes of extreme events is required for these distributions.

The magnitude of x_T of the hydrological event is represented as the sum of mean $(\bar{x})$ plus the Δx_T of the variate from the mean, as given below:

$$x_T = \bar{x} + \Delta x_T \tag{6.1}$$

Whereas Δx_T can be taken as equal to the product of standard deviation (σ) and a frequency factor (K_T); that is $\Delta x_T = K_T \sigma$. The departure Δx_T and the frequency factor K_T are the functions of the return period and the type of probability distribution to be used in the analysis, therefore the above equation can be represented as follows:

$$x_T = \bar{x} + K_T \sigma \tag{6.2}$$

This equation was proposed by Chow (1951), and it is applicable to many probability distributions used in hydrologic frequency analysis. The frequency factor and return period relationship can be determined between the frequency factor and the corresponding return period.

6.2.1.1.2   Normal Distribution

The frequency factor from equation 6.2 can be expressed:

$$K_T = \frac{x_T - \bar{x}}{\sigma} \tag{6.3}$$

This is the same as the standard normal variable z, the value of which is corresponding to an exceedance probability of $p \left(p = \dfrac{1}{T} \right)$ which can be calculated by finding the value of an intermediate variable w:

$$w = \sqrt{\ln \frac{1}{p^2}} \quad (0 < p \leq 0.5) \tag{6.4}$$

then calculating z using the approximation

$$z = w - \frac{2.515517 + 0.802853w + 0.010328w^2}{1 + 1.432788w + 0.189269w^2 + 0.001308w^3} \tag{6.5}$$

When $p > 0.5$, $1 - p$ is substituted for p in equation 6.4 and the value of z computed by equation 6.5 is given a negative sign. The error in this formula is less than 0.00045 in z (Abramowitz and Stegun, 1965). The frequency factor K_T for the normal distribution is equal to z, as mentioned above.

6.2.1.1.3   Log-Normal Method

This is based on the log-normal probability law and assumes that the flood values are such that their natural logarithms are normally distributed. The frequency curve is derived by carrying out the following steps:

1. Compute $\bar{x} \left(\dfrac{\sum x}{n} \right)$ and $\sigma \left(\sigma = \sqrt{\dfrac{\sum X^2}{n} - \left(\dfrac{\sum X}{n} \right)^2} \right)$ from the annual series.

2. Compute the coefficient of variation C_v as $\dfrac{\sigma}{\bar{x}}$.
3. Compute C_s as equal to $(3C_v + C_v^3)$.

4. Set up a computation form similar to that in step 3 of the Gumbel method, but replace T with p, which is equal to $1/T$.
5. Enter representative p values $\bar{x}$ and σ in the computation form.
6. Select K factor corresponding to the computed value of C_s of step 3 and the selected p-value. Interpolation may be necessary. Enter the K factors in the computation form.
7. Compute the value of $K\sigma$. Compute the x value for each of the p-values by equation 6.2, then the curve is plotted.
8. Plot the x values at the appropriate p-value abscissa on the log-probability paper. Draw a straight line through the plotted points to produce the required frequency curve.

6.2.1.1.4 Log Pearson Type III

To find frequency factors for the Pearson type III (logarithmic or arithmetic) distribution, first, transform annual floods to logarithmic values ($Y = \log X$) and then find the mean, standard deviation, and skewness coefficient of the logarithms to get the value of K for the desired recurrence interval. The flood magnitude, Q, for the desired recurrence interval is now estimated from the given equation:

$$\log Q = \bar{Y} + K\sigma_y \tag{6.6}$$

It may be noted that $K = f(T, C_s)$ is a function of both the recurrence interval and skewness. Since skewness has greater variability than the mean, Beard (1962) recommends that only the average regional coefficient of skew be employed in flood analysis for a simple station unless the record exceeds 100 years. In practice, this may be impractical to attain, and the best would be to compute all parameters and compare results with any other records, experience, or regional studies. The help of logarithms is very useful in reducing the skewness of an already skewed distribution. For the best estimate of skewness coefficient $\bar{C_s}$ from small samples, Hazen (1930) recommends the following modification for Pearson type III analysis which is given in this equation:

$$\bar{C_s} = \left(1 + \frac{8.5}{N}\right)C_s \tag{6.7}$$

6.2.1.1.5 Gumbel Extreme Value Distribution

Gumbel distribution methodology was selected to perform the flood probability analysis. The Gumbel theory of distribution is the most widely used distribution for IDF analysis owing to its suitability for modeling maxima. It is relatively simple and uses only extreme events (maximum values or peak rainfalls). The Gumbel method calculates the 2, 5, 10, 50, and 100-year return intervals for each duration period and requires several calculations. Frequency precipitation PT (in mm) for each duration with a specified return period T (in years) is given by the following equation:

$$P_t = P_{avg} + K_S \tag{6.8}$$

Where K is the Gumbel frequency factor given by:

$$K = -\frac{\sqrt{6}}{\pi}\left[0.5772 + \ln\left[\ln\left[\frac{T}{T-1}\right]\right]\right] \tag{6.9}$$

Where P_{avg} is the average of maximum precipitation corresponding to a specific duration.
    In utilizing the Gumbels distribution, the arithmetic average in equation 6.8 is used:

$$P_{avg} = \frac{1}{n}\sum_{i-1}^{n}P_i \tag{6.10}$$

Where P_i is the individual extreme value of rainfall and n is the number of events or years of record. The standard deviation is calculated by equation 6.11, computed using the following relation:

$$S = \sqrt{\frac{1}{n-1}\sum_{i-1}^{n}\left(P_i - P_{avg}\right)^2}$$

(6.11)

Where S is the standard deviation of P data. The frequency factor (K), which is a function of the return period and sample size, when multiplied by the standard deviation gives the departure of a desired return period rainfall from the average. Then the rainfall intensity, I (in mm/h), for return period T is obtained from:

$$I_t = \frac{P_t}{T_d}$$

(6.12)

Where T_d is the duration in hours.

The frequency of the rainfall is usually defined by reference to the annual maximum series, which consists of the largest values observed in each year. An alternative data format for rainfall frequency studies is that based on the peak-over-threshold concept, which consists of all precipitation amounts above certain thresholds selected for different durations.

6.2.1.1.6 The Weibull Distribution

The random variable x is said to have a Weibull distribution of probability density function (pdf) is given by:

$$f(x) = \frac{a}{b}\left(\frac{x}{b}\right)^{a-1}\exp\left[-\left(\frac{x}{b}\right)^a\right], \quad a > 0,\, b > 0$$

(6.13)

Its cumulative density function (cdf) can be expressed as:

$$F(x) = \exp\left[-\left(\frac{x}{b}\right)^a\right]$$

(6.14)

In which a and b are parameters. If $a = 1$, equation 13 is an exponential pdf. Thus, the exponential distribution is a special case of the Weibull distribution. Moreover, this distribution can be generalized to the extreme value type 3 distribution by the transformation $y = x - c$, where c is the parameter.

6.2.2 Methods of Analysis of Time Series

A time series is a collection of observations, which can be recorded at a specific time t, containing data sets of a single variable which is termed as univariate. However, if the data sets of more than one variable are considered, it is termed as multivariate. In general, time-series data can be represented with two main components: deterministic and stochastic (random or without any pattern). A time series $x_1, x_2, \ldots, x_t$ is mathematically expressed as:

$$x_t = T_t + P_t + S_t$$

(6.15)

Where T is the trend component (deterministic component); P is the periodic component (deterministic component); S is the stochastic component and t represents discrete values of time 1, 2, 3, …, N, where N is the number of observations.

6.2.2.1 Deterministic Time Series

6.2.2.1.1 Basic Statistical Characteristics of Time Series

Analysis of data or time series is the first and important preliminary step for the selection of the model. This can be analyzed by using the statistical characteristics of the time series for examining

stationarity and non-stationarity in the time series. The most commonly used statistical charac-
teristics are mean, standard deviation, coefficient of variation, skewness coefficient, and kurtosis.
Details of statistical characteristics can be easily found in hydrological and statistic books, e.g.,
Machiwal and Jha (2012).

6.2.2.1.1.1 Identification of Trend   One of the most common deterministic components in any
hydrologic time series is a trend. In hydrologic time series, a trend implies the steady and regular movement through which the values are, on average, either decreasing or increasing over time
(Kottegoda, 1980). There exist various parametric and non-parametric tests for detecting a trend in
a hydrologic time series. In the parametric tests, the hydrological time series should be normal and
independent, which is usually not met in the time series data, therefore some non-parametric tests
for detecting trends have been proposed by researchers. The parametric trend identification tests
are the turning point, Kendall's phase, Kendall's rank, regression, Wald-Wolfowitz total number
of runs, sum of squared lengths, and inversion tests (Shahin et al., 1993) and non-parametric tests
are the Mann-Kendall test for a linear and/or nonlinear trend (Salas, 1993) and the modified Mann-
Kendall test (Machiwal et al., 2019). Some of the methods for estimating trends in hydrologic time
series are described in the following sections.

6.2.2.1.1.2 Turning Point Test   Before we actually estimate the trend, the first thing to find is
whether or not any trend is present at all. To check this, tests for randomness are performed on the
time series $X_1, X_2, X_3, ..., X_N$ as shown.

6.2.2.1.1.2.1 Turning Point Test for Randomness   The various tests of randomness, which are
distribution-free with minimum calculations have been developed. One test is the turning point test
in which the statistic- used is the number of turning points, T in a series. The turning point is defined
whenever $X_{i-1} < X_i > X_{i+1}$, and $X_{i-1} > X_i < X_{i+1}$, as shown tick marked in Figure 6.4.

For an independent stationary series, T is approximately normally distributed with mean $= 2(N -
2)/3$ and standard deviation $= [(16N - 29)/90]^{0.50}$.

6.2.2.1.1.3 Trend Estimation by Least Squares Method   This is used to find the equation of an
appropriate trend line, which can later be used to compute the trend value T.

First, an equation of the trend line is estimated and then the trend values are subtracted from
the time series values to remove the trend. Now, one or two types of tests are made on the residual
series to test whether or not the residual series is stationary stochastic. The other test to establish the

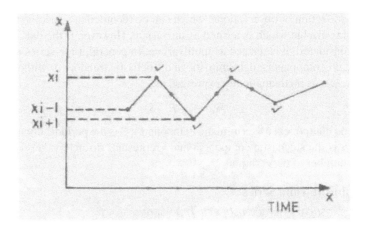

FIGURE 6.4   Tick-marked points as turning points.

presence of trend is to statistically infer whether the parameters of the fitted trend line are different from zero such as various correlation and regression parameters being significantly different from zeros.

6.2.2.1.1.4 *Trend Estimation by the Moving Average Method* The procedure of smoothening the series consists in averaging preceding and succeeding values to a given value *Xi* or a total of *2m* or *2m+I* successive members of series. The new smoothened value *X3* is at the ith position of the series. The smoothened procedure may be repeated *n* times. Generally, a polynomial is fitted to *2m+I* point.

Consider, for example, fitting a cubic equation to seven points. Let 1 = −3, −2, −1, 0, 1, 2, 3 denote the time indices of the seven points. Let a0 + a1l + a2 + a3l3 represent the cubic equation to be fitted to seven points by the least square method.

6.2.2.1.2 *Estimation of Seasonality or Periodicity*

Another important deterministic component of hydrologic time series is the periodic component, which, if it exists in the time series, should be identified and removed. The periodic component in the original hydrologic time series is due to astronomical cycles such as Earth's rotation around the sun. Periodicity can be easily identified by using the autocorrelation function or correlogram in the time domain. A correlogram will only give an idea that periodicities exist in the hydrologic time series.

A function *P(.)* is said to be periodic of period T if series of monthly precipitation, monthly run-off, as well as monthly series of many hydrologic variables have seasonal or periodic components of 12 months (T = 12) in both monthly mean and monthly standard deviation. For example, it is the effect of seasonality that the mean July discharge of river Ganga at Hardwar is always more than the mean January discharge. Experience shows that each month, day, or hour, or any multiple of these units of the year, has a different expected value (mean) and different standard deviation in hydrologic time series. A value of variable X in the year p and at the position inside the year is Xp, with p = 1, 2, ..., n and X = 1, 2, ... which is the number of discrete values in a year.

Our interest is to separate out the periodic deterministic components so as to be left with the stationary stochastic component, the properties of which, like that of distribution, etc. when studied, permit us to regenerate the new samples. In fact, the first prerequisite for reproducing properties of time series in new samples is the proper generation of samples of independent stochastic components. The fact of the matter is that generating new samples of time series can be regarded as a reversible process of decomposition of time series into its various components. The exact estimation of periodicities at the significant periodicities inherent in the time series data is better estimated by spectral analysis.

6.2.2.1.2.1 *Spectral Analysis* In this method, the spectrum of hydrologic time series shows prominent spikes, which represent the inherent periodicities in the hydrologic time series data in the frequency domain. This can be computed by the following mathematical expression given by Kite (1989):

$$x_\tau = \bar{x}_\tau + \sum_{j=1}^{\omega/2}\left[A_j\cos\left(\frac{2\pi j\tau}{\omega}\right) + B_j\sin\left(\frac{2\pi j\tau}{\omega}\right)\right] \tag{6.16}$$

Where, $\dfrac{2\pi j\tau}{\omega}$ is the circular frequency, $\tau = 1, 2, ..., \omega$; ω is the basic period, i.e., the number of seasons in a year. Hence, to identify the periodic component in a time series, the time scale, ω is to be considered less than a year (e.g., month or six months). The maximum number of harmonics, j, that can be fitted, is equal to $\omega/2$ for even values of ω and $(\omega - 1)/2$ for odd values of ω. The Fourier coefficients A_j and B_j of the jth harmonics are estimated by:

$$A_j = \frac{2}{\omega} \sum_{\tau=1}^{\omega} x_\tau \cos \frac{2\pi j\tau}{\omega} \tag{6.17}$$

$$B_j = \frac{2}{\omega} \sum_{\tau=1}^{\omega} x_\tau \sin \frac{2\pi j\tau}{\omega} \tag{6.18}$$

If var(s_j) is the variance of the hydrologic time series explained by jth harmonics, var(s_j) is given by (Lohani et al., 2012):

$$\mathrm{var}\left(s_j\right) = \frac{A_j^2 + B_j^2}{2} \tag{6.19}$$

The ratio of variance ΔP_j is estimated by the ratio of var(s_j) and original variance (s^2):

$$\Delta P_j = \frac{\mathrm{var}\left(s_j\right)}{s^2} \tag{6.20}$$

6.2.2.1.3 Removing the Deterministic Component

For stochastic modeling, the hydrologic time series needs to be stationary means it should have a constant mean and variance. When the deterministic component (trend and periodicity) are evaluated and if these components exist in the hydrologic time series, they should be removed. There are three most important methods that exist in the literature for removing deterministic components from the time series; first is non-seasonal and seasonal differencing of the time series repetitively till it becomes stationary, second is the standardization of the time series, and third is the Box-Cox transformation or logarithmic transformation. All these transformations decide the class to which the stochastic model belongs.

6.2.2.1.3.1 Differencing
Differencing is very useful for removing a trend which is done by simply to difference an original time series until it becomes stationary. This method is given by Box and Jenkins (1970):

$$y_t = x_{t+1} - x_t = \nabla x_{t+1} \tag{6.21}$$

Where z_t is the newly formed series after differencing and ∇ is the backward difference operator. This is called the first-order differencing which is generally suitable for non-seasonal time series.

6.2.2.1.3.2 Standardization of Time Series
Standardization is a very effective method and also ensures the removal of deterministic components in the hydrologic time series. The mathematical expression of standardization of the time series is given by (Kottegoda, 1980):

$$y_t = \frac{x_t - \overline{x}_\tau}{\sigma_\tau} \tag{6.22}$$

Where y_t is the standardized time series, $\overline{x}_\tau$ is the estimate of the mean streamflow of the period τ (month or ten days) to which t belongs, and σ_t is the estimate of the standard deviation of the streamflows of the period τ.

6.2.2.1.3.3 Box-Cox Transformation   Box-Cox transformation can be applied when a hydrologic time series is characterized by heteroskedasticity and it cannot be seen as the realization of an ARMA process, even after differencing, because of non-stationarity in variance. This transformation is also known as logarithmic transformation or power transformation and is given by Hipel et al. (1977):

$$z_t = \begin{cases} \dfrac{(y_t + C)^{\lambda} - 1}{\lambda} & \lambda \neq 0 \\ \ln(y_t + C) & \lambda = 0 \end{cases} \tag{6.23}$$

Where C is a constant term added to standardized data to make the complex data set to be positive and λ is another constant whose value is varied till the coefficient of skewness of the transformed series becomes near zero.

6.2.3 STEP-BY-STEP PROCESS OF STOCHASTIC MODEL DEVELOPMENT

In this chapter, the stochastic modeling process has been discussed in detail. The hydrologic time series after removal of the deterministic component (trend T_t and periodicity P_t) is the stochastic time series. In this section, a step-by-step procedure is discussed for stochastic modeling of hydrologic time series.

6.2.3.1 Model Identification

Model identification is the preliminary step for identifying the autoregressive and moving average parameters of the model. There are two ways to identify the tentative model parameters; one is the autocorrelation function (ACF) and the other is the partial autocorrelation function (PACF) (Modarres and Eslamian, 2006). These two methods are explained in the following sections.

6.2.3.1.1 Autocorrelation Function

For the hydrologic time series, the autocorrelation function or correlogram is generally used for analyzing the time series in the time domain. The autocorrelation function can be defined as the correlation between y_t with time t and y_{t+k} with time lag k. The autocorrelation function is an excellent tool for identifying the parameters of a moving average model because it is expected to "cut off" after lag q. The covariance between y_t and y_{t+k} is called autocovariance at lag k (γ_k), given by:

$$\gamma_k = \mathrm{Cov}(y_t, y_{t+k}) = E\left[(y_t - \bar{y}_t)(y_{t+k} - \bar{y}_t)\right] \tag{6.24}$$

The collection of the values of γ_k, e.g., γ_0, γ_1, … is called the autocovariance function. The mathematical expression of the autocorrelation function at lag k is given by:

$$\rho_k = \frac{E\left[(y_t - \bar{y}_t)(y_{t+k} - \bar{y}_t)\right]}{\sqrt{E\left[(y_t - \bar{y}_t)^2\right]E\left[(y_{t+k} - \bar{y}_t)^2\right]}} = \frac{\mathrm{Cov}(y_t, y_{t+k})}{\mathrm{Var}(y_t)} = \frac{\gamma_k}{\gamma_0} \tag{6.25}$$

Where ρ_k is the autocorrelation function at lag k.

6.2.3.1.2 Partial Autocorrelation Function

The partial autocorrelation function is the autocorrelation between y_t and y_{t-k} after adjusting for $y_{t-1}, y_{t-2}, …, y_{t-k+1}$. The partial autocorrelation function is given by the Yule-Walker equation in matrix notation as:

$$\begin{bmatrix} 1 & \rho_1 & \rho_2 & \cdots & \rho_{k-1} \\ \rho_1 & 1 & \rho_1 & \cdots & \rho_{k-2} \\ \vdots & \vdots & \vdots & \cdots & \vdots \\ \rho_{k-1} & \rho_{k-2} & \rho_{k-3} & \cdots & 1 \end{bmatrix} \begin{bmatrix} \phi_{k1} \\ \phi_{k2} \\ \vdots \\ \phi_{kk} \end{bmatrix} = \begin{bmatrix} \rho_1 \\ \rho_2 \\ \vdots \\ \rho_k \end{bmatrix} \tag{6.26}$$

or

$$P_k \phi_k = \rho_k \tag{6.27}$$

Where P_k is the autocorrelation matrix; ϕ_k is partial autocorrelation and ρ_k is autocorrelations. For AR(1) process, $\phi_k = 0$ for all $k > 1$. Thus, ϕ_k is nonzero for lag 1, the order of an AR(1) process. Therefore, PACF for an AR(p) process "cuts off" after the lag exceeds the order of the process. Table 6.1 summarizes the behavior of autocorrelation and partial autocorrelation functions for stochastic models.

6.2.3.2 Parameter Estimation of Model

There are several ways that can be employed to estimate the parameters of stochastic models from the tentatively identified models. These methods are the method of moment, the maximum likelihood method, and the least square method. In addition, there are many algorithms that are also available for parameter estimation, e.g., Marquardt's algorithm, given in most statistical toolboxes and the "armax" toolbox in MATLAB.

6.2.3.2.1 Fitting a Probability Distribution

A probability distribution is a function representing the probability of occurrence of a random variable. By fitting a distribution to a set of hydrologic data, a great deal of the probabilistic information in the sample can be compactly summarized in the function and its associated parameters. Fitting distributions can be accomplished by the *method of moments* or the *method of maximum likelihood*.

6.2.3.2.1.1 Method of Moments The method of moments was first developed by Karl Pearson in 1902. He considered that good estimates of the parameters of a probability distribution are those for which moments of the probability density function about the origin are equal to the corresponding moments of the sample data. If the data values are each assigned a hypothetical "mass" equal to their relative frequency of occurrence *(1/n)* and it is imagined that this system of masses is rotated about the origin $x = 0$, then the first moment of each observation x_i about the origin is the product of its moment arm x_i and its mass U_n, and the sum of these moments over all the data is expressed as:

$$\sum_{i=1}^{n} \frac{x_i}{n} = \frac{1}{n} \sum_{i=1}^{n} x_i = \bar{x} \tag{6.28}$$

TABLE 6.1

Behavior of ACF and PACF for Stochastic Models

Stochastic Model	Autocorrelation Function	Partial Autocorrelation Function
MA(q)	Cuts off after lag q	Exponential decay or tails off
AR(p)	Exponential decay or tails off	Cuts off after lag q
ARMA(p,q)	Exponential decay or tails off	Exponential decay or tails off

This is equivalent to the centroid of a body. The corresponding centroid of the probability density function is:

$$\mu = \int_{-\infty}^{\infty} xf(x)dx \tag{6.29}$$

Likewise, the second and third moments of the probability distribution can be set equal to their sample values to determine the values of parameters of the probability distribution. Pearson originally considered only moments about the origin. But later it became customary to use the variance as the second *central moment*. $\sigma^2 = E[(x - \mu)^2]$, and the coefficient of skewness as the standardized third central moment, $\gamma = E[(x - \mu)^3] \sigma^3$, to determine second and third parameters of the distribution if required.

As a matter of interest, it can be seen that the exponential probability density function, $f(x) = \lambda e^{-\lambda x}$ and the impulse response function for a linear reservoir $u(l) = (l/k)e^{-l/k}$ are identical if $x = l$ and $A = l/k$. In this sense, the exponential distribution can be thought of as describing the probability of the "holding time" of water in a linear reservoir.

6.2.3.2.1.2 Method of Maximum Likelihood The method of maximum likelihood was developed by Fisher (1922). He reasoned that the best value of a parameter of a probability distribution should be that value that maximizes the likelihood or joint probability of occurrence of the observed sample. The maximum likelihood function is given by:

$$L = \prod_{i-1}^{n} f(x_i) \tag{6.30}$$

Because many probability density functions are exponential, it is sometimes more convenient to work with the log-likelihood function:

$$\ln L = \prod_{i-1}^{n} \ln\left[f(x_i)\right] \tag{6.31}$$

The method of maximum likelihood is the most theoretically correct method of fitting probability distributions to data in the sense that it produces the most efficient parameter estimates – those which estimate the population parameters with the least average error. But, for some probability distributions, there is no analytical solution for all the parameters in terms of sample statistics and the log-likelihood function must then be numerically maximized, which may be quite difficult. In general, the method of moments is easier to apply than the method of maximum likelihood and is more suitable for practical hydrologic analysis.

6.2.4 Stochastic Model Building

6.2.4.1 Autoregressive Model

The autoregressive model regresses against the past values of the time series. Mathematically, the autoregressive model can be expressed as (Box and Jenkins, 1976):

$$z_t = \varepsilon + \varnothing_1 z_{t-1} + \ldots + \varnothing_p z_{t-p} + a_t \tag{6.32}$$

Where z_t and a_t are respectively the standardized series and residuals at time period t; ε is a constant, and ϕ_1, ϕ_2, ... ϕ_p are model parameters and p represents the order of AR model.

6.2.4.2 Moving Average Model

The moving average model uses past errors values as the explanatory variables. The mathematical expression of the MA(q) model is given by Box and Jenkins (1976):

$$z_t = \overline{z_t} + \theta_1 a_{t-1} + \ldots + \theta_p a_{t-q} + a_t \tag{6.33}$$

Where, θ_1, θ_2, ..., θ_q are the model parameters and q is the order of the MA model.

6.2.4.3 Mixed Autoregressive Moving Average Model

In order to achieve greater flexibility in developing a stochastic model of time series, it is advantageous to include both autoregressive (AR) and moving average (MA) terms in the model to form a general and useful class of time series models, known as the autoregressive moving average (ARMA) model. The mathematical expression of the mixed autoregressive moving average (ARMA) model is given by (Box et al., 2008):

$$z_t = \varnothing_1 z_{t-1} + \ldots + \varnothing_p z_{t-p} + a_t - \theta_1 a_{t-1} + \ldots + \theta_p a_{t-q} \tag{6.34}$$

Where p, q are the order of the ARMA model.

6.2.5 Verification of Fitted Stochastic Model

6.2.5.1 Testing the Goodness of Fit

Comparing the theoretical and sample values of the relative frequency or the cumulative frequency function can test the goodness of fit of a probability distribution. In the case of the relative frequency function, the χ^2 test is used. Here the value of χ^2 is given by:

$$\chi^2 = \sum_{i=1}^{m} \frac{n\left[f_s(x_i) - p(x_i) \right]}{p(x_i)} \tag{6.35}$$

Where m is the number of intervals. It may be noted that $nf_s < (x_i) = n_i$, the observed number of occurrences in interval i, and $np(x_i)$ is the corresponding expected number of occurrences in the interval.

6.3 PERFORMANCE EVALUATION INDICES

The most common measures for evaluating the performance of the model are coefficient of correlation (R), root mean square error (RMSE), Nash-Sutcliffe efficiency (NSE), index of agreement (IA), percent bias (PBIAS), and RMSE-observation standard deviation ratio (RSR). The mathematical expressions of these statistical parameters are illustrated in Table 6.2.

6.4 CONCLUSIONS

Flood forecasting is an expanding area of application of hydrologic techniques. The goal of the chapter is to understand how to obtain real-time precipitation and streamflow trend through various meteorological data, input it into a rainfall-runoff and streamflow-channel flow and reservoir routing program, and forecast flood flow rates and water levels for a period of few hours to a few days ahead depending upon the size of the catchment. However, minimum meteorological data should be available for the past 30 years. The input for this flood forecasting model includes both real-time data and the physical description of the system components that remain unchanged during a flood. The physical data such as (1) physiographic factors of stream section, roughness relationship

TABLE 6.2
Statistical Measures for Evaluating the Performance of the Model

Sr. No.	Statistical Measures	Mathematical Expression	Acceptable Range	Reference
1.	R	$$R = \frac{\sum_{i=1}^{N}\left\{\left(O_i - \overline{O_i}\right)\left(P_i - \overline{P_i}\right)\right\}}{\sqrt{\sum_{i=1}^{N}\left(O_i - \overline{O_i}\right)^2 \sum_{i=1}^{N}\left(P_i - \overline{P_i}\right)^2}}$$	−1 to +1	Butts et al., 2004
2.	RMSE	$$RMSE = \left(\frac{\sum_{i=1}^{N}\left(O_i - P_i\right)^2}{N}\right)^{1/2}$$	Minimum	Wang et al., 2009
3.	NSE	$$NSE = 1 - \left[\frac{\sum_{i=1}^{N}\left(O_i - P_i\right)^2}{\sum_{i=1}^{N}\left(O_i - \overline{O_i}\right)^2}\right]$$	0.0 to 1.0	Nash and Sutcliffe, 1970
4.	IA	$$IA = 1 - \left[\frac{\sum_{i=1}^{N}\left(P_i - O_i\right)^2}{\sum_{i=1}^{N}\left(\left|P_i - \overline{O_i}\right| - \left|O_i - \overline{O_i}\right|\right)^2}\right]$$	0.5 to 1.0	Willmott (1981)
5.	PBIAS	$$PBIS = \left[\frac{\sum_{i=1}^{N}\left(O_i - P_i\right)\times 100}{\sum_{i=1}^{N} O_i}\right]$$	If 0.0 is optimal value, +ve = underestimation, ve = overestimation	Gupta et al., 1999
6.	RSR	$$RSR = \frac{RMSE}{\sigma_i} = \frac{\left[\sum_{i=1}^{N}\sqrt{\left(O_i - P_i\right)^2}\right]}{\left[\sum_{i=1}^{N}\sqrt{\left(O_i - \overline{O_i}\right)^2}\right]}$$	0.0 is optimal value	Legates and McCabe, 1999

Note: O_i = observed streamflow; P_i = predicted streamflow; $\overline{O_i}$ = mean observed streamflow; $\overline{P_i}$ = mean predicted streamflow; N = number of observations.

(Froude number), and so on; (2) characteristics of reservoir structure (gauge), (3) the catchment area information (geomorphological and topographic characteristics, soil type, land use/land cover, etc.), and (4) hydrologic parameters are necessary for estimating the rainfall-runoff relationship. Similarly, real-time data such as (1) streamflow data and headwater and tailwater elevation at each dam, (2) rainfall data at recording rain gauges, and (3) reservoir operations (spillover discharge data and discharge data for irrigation, etc.) are needed. Since the time to make a forecast is too small, more time and effort should be devoted to the development phase of the model which is employed for the forecast so as not to waste time during the forecast operation. Therefore, worksheets based on the rainfall-runoff relationship should be prepared in advance and used at the time of forecast computation. To grasp the subject fully, it is suggested that the reader refer to standard textbooks on flood estimation, statistics, and time-series analysis.

REFERENCES

Abramowitz, M., and Stegun, I.A. 1965. *Handbook of Mathematical Functions: With Formulas, Graphs, and Mathematical Tables*, Vol. 55, Courier Corporation, p. 1046.

Alam, J., and Muzzammil, M. 2011. Flood Disaster Preparedness in Indian Scenario, *International Journal on Recent Trends in Engineering and Technology*, 5(3): 33–38.

Alfieri, L., Burek, P., Dutra, E., Krzeminski, B., Muraro, D., Thielen, J., and Pappenberger, F. 2013. GloFAS-global ensemble streamflow forecasting and flood early warning, *Hydrology and Earth System Sciences*, 17(3): 1161.

Beard, L.R. 1962. *Statistical Methods in Hydrology, Civil Works Investigation Projects CW151*, US Army Corps of Engineers, Sacramento, CA, p. 111.

Bhat, M.S., Alam, A., Ahmad, B., Kotlia, B.S., Farooq, H., Taloor, A.K., and Ahmad, S. 2019. Flood frequency analysis of river Jhelum in Kashmir basin, *Quaternary International*, 507: 288–294.

Box, G.E.P., and Jenkins, G.M. 1970. *Time Series Analysis: Forecasting and Control*, Holden-Day, Oakland, CA, USA.

Box, G.E.P., and Jenkins, G.M. 1976. *Time Series Analysis: Forecasting and Control*, Holden-Day, San Francisco, p. 575.

Box, G.E.P., Jenkins, G.M., and Reinsel, G.C. 2008. *Time Series Analysis: Forecasting and Control*, 4th edition, Wiley Series in Probability and Statistics, Wiley, Hoboken, NJ, USA.

Butts, M.B., Payne, J.T., Kristensen, M., and Madsen, H. 2004. An evaluation of the impact of model structure on hydrological modelling uncertainty for streamflow simulation, *Journal of Hydrology*, 298(1–4): 242–266.

Chow, V.T. 1951. A general formula for hydrologic frequency analysis, *Eos, Transactions American Geophysical Union*, 32(2): 231–237.

Chow, V.T., Maidment, D.R., and Mays, L.W. 1988. *Applied Hydrology*, McGrow-Hill Book Company, New York, USA, p. 149.

Demuth, N., and Rademacher, S. 2016. Flood forecasting in Germany—Challenges of a federal structure and transboundary cooperation. In: *Flood Forecasting*, Academic Press, pp. 125–151.

Dilley, M., Chen, R.S., Deichmann, U., Lerner-Lam A.L., and Arnold M. 2005. *Natural Disaster Hotspots: A global Risk Analysis*, International Bank for Reconstruction and Development/The World Bank and Columbia University, Washington, DC, USA.

Fisher, R.A. 1922. On the mathematical foundations of theoretical statistics, *Philosophical Transactions of the Royal Society of London, Series A, Containing Papers of a Mathematical or Physical Character*, 222(594–604): 309–368.

Givati, A., Fredj, E., and Silver, M. 2016. Chapter 6 Operation flood forecasting in Isreal, Flood forecasting. In: *A Global Perspective*, pp. 153–167.

Gupta, H.V., Sorooshian, S., and Yapo, P.O. 1999. Status of automatic calibration for hydrologic models: Comparison with multilevel expert calibration, *Journal of Hydrologic Engineering*, 4(2): 135–143.

Hazen, A. 1930. Flood flows: A study of frequencies and magnitudes. In: *Flood Flows: A Study of Frequencies and Magnitudes*, John Wiley & Sons, New York, USA.

Hipel, K.W. 1985. Time series analysis in perspective 1, *Journal of the American Water Resources Association*, 21(4): 609–623.

Hipel, K.W., Mcleod, A.I. and Lennox, W.C. 1977. Advances in Box Jenkins modelling: 1. Model construction, *Water Resources Research*, 13: 567–575.

Hopson, T., and Webster, P. 2010. A 1–10-day ensemble forecasting scheme for the major river basins of Bangladesh: Forecasting severe floods of 2003–07, *Journal of Hydrometeorology*, 11(3): 618–641.

Jain, S.K., Mani, P., Jain, S.K., Prakash, P., Singh, V.P., Tullos, D., Kumar, S., Agarwal, S.P., and Dimri, A.P. 2018. A brief review of flood forecasting techniques and their applications, *International Journal of River Basin Management*, 16(3): 329–344.

Kottegoda, N.T. 1980. *Stochastic Water Resources Technology*, McMillan & Co. Ltd., London, UK.

Kite, G. 1989. Use of time series analysis to detect climatic change, *Journal of Hydrology*, 111(1–4): 259–279.

Lee, E.H., and Kim, J.H. 2018. Development of a flood-damage-based flood forecasting technique, *Journal of Hydrology*, 563: 181–194.

Legates, D.R., and McCabe, Jr., G.J. 1999. Evaluating the use of "goodness-of-fit" measures in hydrologic and hydroclimatic model validation, *Water Resources Research*, 35: 233–241.

Lohani, A.K., Kumar, R., and Singh, R.D. 2012. Hydrological time series modeling: A comparison between adaptive neuro-fuzzy, neural network and autoregressive techniques, *Journal of Hydrology*, 442: 23–35.

Machiwal, D., and Jha, M.K. 2012. *Hydrologic Time Series Analysis: Theory and Practice*, Springer, Germany and Capital Publishing Company, New Delhi, India, p. 303.

Machiwal, D., Islam, A., and Kamble, T. 2019. Trends and probabilistic stability index for evaluating groundwater quality: The case of quaternary alluvial and quartzite aquifer system of India, *Journal of Environmental Management*, 237: 457–475.

Modarres, R., and Eslamian, S.S. 2006. Streamflow time series modeling of Zayandehrud River, *Iranian Journal of Science and Technology*, 30(B4): 567–570.

Mutreja, K.N. 1986. *Applied Hydrology*, Tata McGraw Hill Publication Cooperative Ltd., New Delhi, India, pp. 40–109.

Nash, J.E., and Sutcliffe, J.V. 1970. River flow forecasting through conceptual models, part I-A discussion of principles, *Journal of Hydrology*, 10(3): 282–290.

Panu, V.S., and Unny, T.E. 1980. Extension and application of feature prediction model for synthesis of hydrologic records, *Water Resources Research*, 16(1): 77–96.

Salas, J.D. 1993. Analysis and modeling of hydrologic time series. In: *Handbook of Hydrology*, McGraw-Hill, Inc., New York, USA, p. 484.

Shahin, M., Van Oorschot, H.J.L., and De Lange, S.J. 1993. *Statistical Analysis in Water Resources Engineering*, A.A. Balkema, Rotterdam, The Netherlands, p. 394.

Tanoue, M., Hirabayashi, Y., and Ikeuchi, H. 2016. Global-scale river flood vulnerability in the last 50 years, *Scientific Reports*, 6: 36021, https://doi.org/10.1038/srep36021.

Thiemig, V., Bisselink, B., Pappenberger, F., and Thielen, J. 2015. A pan-African medium-range ensemble flood forecast system, *Hydrology and Earth System Sciences*, 19(8): 3365–3385.

Tshimanga, R.M., Tshitenge, J.M., Kabuya, P., Alsdorf, D., Mahé, Gil., Kibukusa, G., and Lukanda, V. 2016. A regional perceptive of flood forecasting and disaster management systems for the Congo river basin. In: Adams, T.E., Pagano T.C. (eds.). *Flood Forecasting: An International Perspective*. Cambridge: Academic Press, pp. 87–124. ISBN 978-0-12-801884-2

Valencia, R.D., and Schaake, J.C. 1973. Disaggregation processes in stochastic hydrology, *Water Resources Research*, 9(3): 58–585.

Wang, W.C., Chau, K.W., Cheng, C.T., and Qiu, L. 2009. A comparison of performance of several artificial intelligence methods for forecasting monthly discharge time series, *Journal of Hydrology*, 374(3–4): 294–306.

Willmott, C.J. 1981. On the validation of models, *Physical Geography*, 2: 184–194.

WMO. 2011. *Manual on flood forecasting and warning, WMO No. 1072*, World Meteorological Organization, Geneva, Switzerland.

7 Analysis of Stable Channel Design Using HEC-RAS
A Case Study of Surat City

Darshan Mehta, S. M. Yadav, Sahita Waikhom,
Keyur Prajapati, and Saeid Eslamian

CONTENTS

7.1   INTRODUCTION

Stable channel design is a term aimed at minimizing flooding and sedimentation-induced river channel deformation (Mehta et al., 2020). For stable channel design, the Copeland method is widely used (Shields et al., 2003). To compute cross-section geometry, including bed slope, channel width, and flow depth, this approach combines the flow continuity and resistance equations with a sediment transport capacity equation (Firenzi et al., 2000). The Copeland modeling approach is applied in the software SAM hydraulic packages and HEC-RAS (Shields et al., 2003). Major causes of floods in India include inadequate capacity within riverbanks to contain high flows, riverbank erosion, and silting of riverbeds; it can be prevented if the rivers are managed properly, especially in densely populated and flat areas (Mehta et al., 2013). Effective flood warning systems can help take timely action during natural calamities and may save lives (Mehta et al., 2020).

The Copeland method also has the drawback of determining channel stability solely based on the equilibrium between upstream sediment supply and local sediment transport capability (Shelly and Parr David, 2009). This equilibrium, however, does not ensure stability since a channel's geometry can always be adjusted by the contact between the bed and the banks (Bahmani et al., 2012). Goitom and Zeller (1989) described how the combined constraints of cross-section shape and plan form are reflected in the channel slope along the axis of flow. Mitigation strategies can significantly reduce the effects of floods, give time for people to migrate to safer locations, and stock up essential utility items for the future (Lane, 1995). When dealing with flood mitigation measures, we have two types of mitigation measures: structural measures and non-structural measures (Patel et al., 2018). Park

DOI: 10.1201/9780429463938-10

and Datta-Gupta (2011) used an optimization technique to approximate the single-channel roughness value for open channel flow, with boundary conditions as constraints. The channel roughness
is a variable parameter that varies along the river based on changes in channel characteristics as
the flow progresses (Hey and Thorne, 1986). The aim of the chapter is to analyze the stable channel
design of the lower Tapi River reach, 7.75 km long between Sardar and Magdalla Bridges, using
hydraulic design functions HEC-RAS software. Following are the broad goals of the work:

- To collect bed material samples from the river reach.
- To carry out sieve analysis for the samples collected.
- To carry out analysis of stable channel design using the HEC-RAS software for the 1968
  flood event.

7.2   STUDY AREA

Surat metropolis is situated in the state of Gujarat; it is acknowledged and recognized for its diamond and textile business across the world. Surat is settled on the bank of River Tapi. The Ukai
Dam is constructed on the Tapi River which is 100 kilometers away from Surat city. It then flows
through Maharashtra, Gujarat, and ultimately meets the Arabian Sea which is approximately 20 km
west of Surat. Surat city is divided into seven zones, i.e., Central, East, North, West, South, South
West, and South East zones. The average annual rainfall of Surat city is 1,192 mm. Surat has faced
catastrophic floods in the years 2013, 2007, 2006, 2002, 1998, 1994, 1968, 1959, 1949, 1945, 1944,
1942, 1933, and 1884 (Mehta et al., 2020). It has been estimated that the single flood occasion,
which happened from August 7 to 14, 2006 in Surat, brought about the deaths of 300 humans and
property damage worth INR 21,000 crores (Patel et al., 2018). Surat city was hit with a massive
flood in the year 1968. The main reason for this disaster was the release from the Ukai Dam. During
this disaster, around 70–75% of the area of Surat city was flooded (Mehta et al., 2013).

The study reach, located between Sardar Bridge and Magdalla Bridge, approximately 7.75 km
long with 31 cross-sections, is shown in Figure 7.1. Surat, being a coastal city, has been prone to
large floods and endured large damages in the past. The reason for selecting the river reach for
the study is important as 80% of Surat's population lives on either side of the river (Agnihotri and
Patel, 2011). As a result, the drainage system on the left bank of the Tapi River is more extensive

FIGURE 7.1   Cross-sections in the study reach, Tapi River, Surat.

than on the right side (Mehta et al., 2013). The Purna and Girna, the two major left-bank tributaries, collectively account for nearly 45 % of the Tapi River's entire catchment area (Mehta et al., 2020).

7.3   HEC-RAS MODELING CONCEPTS

HEC-RAS 6.0.0 (Hydraulic Engineering Center – River Analysis System) is a freely available open-source software that models the hydraulics of water flow through natural rivers and other channels (Brunner, 2002). The US Department of Defense developed the model, which was made public in 1995 by the Army Corps of Engineers. HEC-RAS performs 1D and 2D computations using the St. Venant equation of conservation of mass and momentum (Mehta et al., 2013). Whereas 1D models solve the St. Venant equation along one dimension, a 2D model solves the St. Venant equation on two dimensions (Alaghmand et al., 2012). Major new features have been added to HEC-RAS 6.0.0 since version 5.0.7, containing many additional features/upgrades such as HEC-RAS Mapper editing tools, breach time series plot, 3D graphics/animation, terrain modification tools, 1D finite volume solver, non-Newtonian fluids option for 1D and 2D, and many more, among which the HEC-RAS Mapper is very important to effectively create geometry and view analysis results (Mehta et al., 2020). It's used to model water flowing through the systems of open channels and assess the water surface profiles. This model may be used in a wide application in flood management research.

7.3.1   HEC-RAS Input Parameters for Steady Flow Analysis

The model is intended for computing the water surface profiles for steady gradually varied flow. The steady flow component can represent water surface profiles in the subcritical, supercritical, and mixed flow regimes (Neary and Korte, 2002). It may also be used to examine changes in water surface profiles caused by levees. The state variables for the numerical scheme in this mathematical model are flow and stage, which are computed and stored at each cross-section. As a result, different parameters are used for analysis. Using the different parameters on the developed model in HEC-RAS, steady flow simulation has been carried out at certain flood discharges. The following parameters are used for analysis:

- Manning's n = 0.022 (Subramanya, 2009);
- Contraction coefficient = 0.1;
- Expansion coefficient = 0.3;
- Downstream reach length c/s to c/s (left bank, channel, and right bank);
- Average bed slope = 0.00842.

7.3.2   Geometric Data

Geometric data consists of the number of cross-sections, reach length, energy loss coefficient as contraction and expansion coefficient, and Manning's n value. Geometric data was collected from the Surat Municipal Corporation in the form of an AutoCAD (.dwg) file. There is no curvature in the study reach. Therefore, the effect of the meandering is neglected by providing the contraction coefficient as 0.3 and the expansion coefficient as 0.1. The Manning's n values are used primarily for calibration purposes. The study reach consists of 31 cross-sections. Surat Municipal Corporation (SMC) and Surat Irrigation Circle (SIC) have provided the detailed topographic features of the study reach (Mehta et al., 2020).

7.3.3   Cross-Sectional Data

The cross-section is ultimately perpendicular to the centerline of the river reach or flow path of river water. Cross-sections extend up to the floodplain area of the river (Mehta et al., 2020). The carrying capacity of the river, bed slope, velocity of water flow, Manning's value, and sedimentations of the river vary with the location of the river (Agnihotri and Patel, 2011). It is identified by a reach and river station label. The average distance between two adjacent cross-sections is 200 m to 250 m. The cross-section is defined as the station (in m) and elevation (in m) from left overbank (LOB) to right overbank (ROB), i.e., from upstream to downstream.

7.3.4 FLOOD CONVEYANCE PERFORMANCE

For evaluation of flood performance, the past flood data collected from Flood Cell, Surat, were used. Major flood events took place in the years 1883, 1884, 1942, 1944, 1945, 1949, 1959, 1968, 1994, 1998, 2006, 2007, 2012, and 2013 (Mehta et al., 2013). The summary of floods is given in Table 7.1.

7.4   METHODOLOGY

The following steps have been carried out for the design of a stable channel using HEC-RAS software:

Step 1: Field reconnaissance.
   - Perform a cross-sectional survey at a stable upstream point.
   - Determine the slope of the upstream section from cross-section to cross-section, i.e., slope = 0.00842.
   - Take a bed sediment sample from the study reach to determine the sediment gradation consisting of the d_{16}, d_{50}, and d_{84} sieve sizes, for which 16%, 50%, and 84% of the material is finer by weight (Garde and Raju, 2000).

Step 2: Computation of uniform flow to check design bankfull flow using HEC-RAS.

Step 3: Choose a value for Manning's "n" for the channel side.

Step 4: Calculate a Copeland stable channel design from the hydraulic design function tab in HEC-RAS main window.

Step 5: Enter the bottom width (b), bank height (y), energy slope (channel slope), and side slopes (m), calculated earlier for the upstream representative channel. If the upstream section is a meandering channel and the channel design is a meandering channel, the same n-value for the side slopes should be used in both the upstream and the design section. When the upstream section is stable but non-meandering and a meandering channel will be designed downstream or when a straight channel is designed downstream from a meandering channel, then the n-values should be different as shown in Figure 7.2.

TABLE 7.1
Flood History of Surat City, India

Sr. No.	Year	Discharge (cumecs – m³/s)	Sr. No.	Year	Discharge (cumecs – m³/s)
1	1882	2,095.5	7	1959	36,642
2	1883	2,845.8	8	1968	43,924.2
3	1884	23,956	9	1998	29,817.27
4	1944	33,527	10	2006	25,788
5	1945	28,996	11	2012	9,508.04
6	1949	23,843	12	2013	13,178

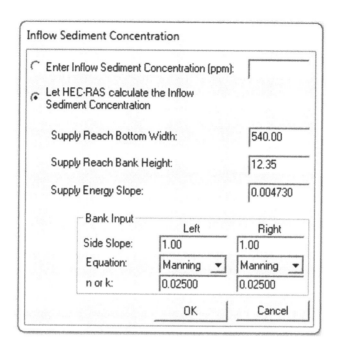

FIGURE 7.2   Inflow sediment.

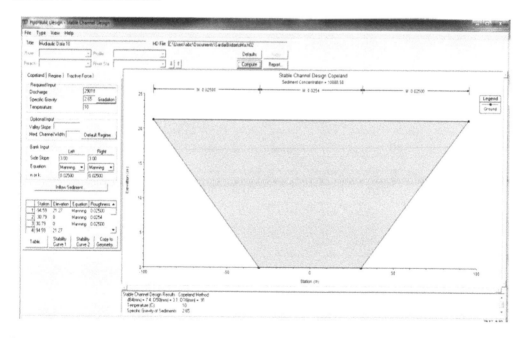

FIGURE 7.3   Stable channel design.

Step 6: After completing all the above steps, with the help of given parameters and by computing, HEC-RAS will give a trapezoidal shape using the Copeland method as shown in
Figure 7.3.

After doing this, compare the previous section with the existing section and check the following:
- Whether the section is sufficient or not; if it is not sufficient then at that particular section,
cutting and filling can be done.

7.5   RESULT AND DISCUSSIONS

7.5.1   Bed Material Analysis

As discussed in methodology, bed material samples were collected from the study reach. The following graphs (Figures 7.4–7.7) were obtained after doing the sieve analysis of the bed material sample.

After plotting the semi-logarithmic curve between the grain size and percentage finer, the following values are obtained:

$D_{16} = 0.98$ mm, $D_{50} = 4$ mm, $D_{60} = 5.5$ mm, $D_{84} = 7.8$ mm

7.5.2   Geomorphic Channel Design and Analysis

In 1D, steady-state, gradually varied flow analysis, the following assumptions are made:

- Dominant velocity is in the flow direction.
- Hydraulic characteristics of flow remain constant for the time interval under consideration.
- Streamlines are practically parallel and, therefore, hydrostatic pressure distribution prevails over the channel section (Chow, 1959).

The Copeland method (2001) was employed to estimate the preferred hydraulic geometry of the bankfull channel. The Chang method, which is rational, was developed especially for active gravel-bed channels, as opposed to the threshold channel that has no significant transport at the bankfull discharge. It predicts the width B, depth D, and slope S of the bankfull channel for a given water discharge Q, particle size D_{50}, and side slope z. The water discharge Q = 43,924 m³/s, the particle

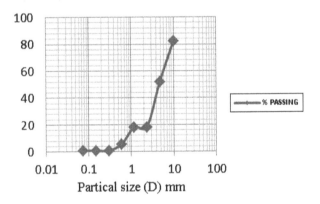

FIGURE 7.4   Grain size distribution curve.

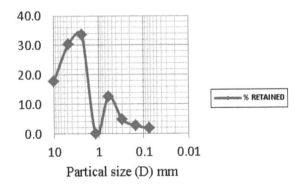

FIGURE 7.5   Frequency curve.

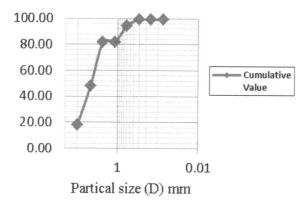

FIGURE 7.6 Cumulative frequency curve.

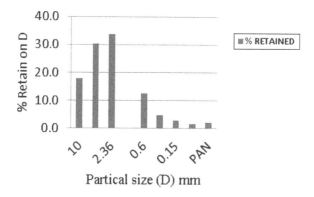

FIGURE 7.7 Histogram.

size $D_{50} = 4$ mm, and the side slope z = 1. The resulting geometry channel is a trapezoidal section with a side slope z = 1, a top width B = 800 m, bottom width b = 540 m, and depth D = 2.9 m. Comparison of an idealized section with surveyed cross-sections in the project channel indicated very good agreement in the upper portion, but poor agreement in the enlarged cut-off reach downstream, which indicates that this lower enlarged reach will be aggrading over time. The enhancement design channel includes the recommendations for the following:

- The channel plan form;
- The longitudinal profile riffle-pool morphology and the average slope;
- The cross-sectional profile, including the flow channel, a bankfull channel, and a flood conveyance channel.

In the present study, analysis of stable channel design is carried out using the Copeland method as explained in the methodology and the existing section is compared with the computed section using 1968 flood data. By following the procedure described in the methodology, the computed and existing sections are summarized in Table 7.2 for cross-sections 1 to 31 using the 1968 flood data.

From Table 7.2, the sections which are not sufficient should be modified.

7.6 CONCLUSIONS

- Based on the available data and the study presented, it is strongly recommended that the sections be found critical in the present study and require restoration work/design which needs to be raised as per requirement.

TABLE 7.2
Summary of Design and Comparison of Sections of Flood 1968

| Sr. No. | Cross-Section | Computed Section | | | | | | |
		B	T	D	Side Slope 1 in	Area (m²)	Area of Existing (m²)	Comparison
1	CS-1	470	650.33	8.21	200	3,746.23	3,800.56	Sufficient
2	CS-2	550	636.74	9.41	200	3,568.12	3,620.12	Sufficient
3	CS-3	650	751.57	9.9	200	3,720.12	3,425.12	Insufficient
4	CS-4	630	846.43	15.4	200	5,010.25	5,210.12	Sufficient
5	CS-5	730	923.34	14.33	200	6,578.25	6,100	Insufficient
6	CS-6	750	910.22	13.66	200	6,643.92	6,200.32	Insufficient
7	CS-7	650	972.36	14.33	200	7,621.03	7,007.03	Insufficient
8	CS-8	700	908.54	13.06	200	6,291.72	7,000.02	Sufficient
9	CS-9	820	1,065.77	11.96	200	4,979.99	5,479.56	Sufficient
10	CS-10	850	1,051.61	11.53	200	7,337.32	6,788.32	Insufficient
11	CS-11	230	589.82	12.09	200	3,525	4,735.36	Sufficient
12	CS-12	230	539.72	12.08	200	4,598	4,136	Insufficient
13	CS-13	230	456.13	10.64	200	3,504.45	3,600.12	Sufficient
14	CS-14	200	444.71	10.45	200	4,289.57	3,215.12	Insufficient
15	CS-15	190	488.52	11.8	200	3,635.32	3,215.12	Insufficient
16	CS-16	146.1899	453.32	11.58	200	3,202.25	3,265.1	Sufficient
17	CS-17	190	364.3	10.47	200	2,603.2	2,900	Sufficient
18	CS-18	220	480.3	12.41	200	4,012.3	3,500.71	Insufficient
19	CS-19	200	514.93	13.3	200	3,723.85	4,000.25	Sufficient
20	CS-20	100	555.77	13.76	200	4,936.23	4,756.23	Insufficient
21	CS-21	130	568.77	14.52	200	4,531.6	5,012.17	Sufficient
22	CS-22	290	577.24	14.12	200	5,700.25	5,725.85	Sufficient
23	CS-23	250	514.53	13.96	200	5,828.73	5,625.7	Insufficient
24	CS-24	200	630.3	13.94	200	6,404.25	6,305	Insufficient
25	CS-25	250	640.31	14.24	200	5,602.62	6,102	Sufficient
26	CS-26	300	637.32	14.7	200	6,897.23	6,690	Insufficient
27	CS-27	380	682.36	15.88	200	7,400.25	7,412.25	Sufficient
28	CS-28	400	639.84	14.78	200	7,236.78	7,393.25	Sufficient
29	CS-29	420	750.88	15.06	200	8,498.65	8,505.25	Sufficient
30	CS-30	450	756.42	14.89	200	8,000.65	8,336.21	Sufficient
31	CS-31	520	729.18	14.87	200	8,245.65	8,367.03	Sufficient

- It is strongly recommended that no new construction be allowed in the floodplain area.
- It is strongly recommended that the width of the river in no case be encroached upon, as already the sections are sensitive to high floods, and encroachment will result in the flooding of the study region.
- It is strongly recommended that from either side of the bank of the river no encroachment be made up to 300 m.

REFERENCES

Agnihotri, P. G., and Patel, J. N. (2011). Modification of channel of Surat city over Tapi river using HEC-RAS software. *International Journal of Advances in Engineering and Technology*, 2, 231–238.

Alaghmand, S., Bin Abdullah, R., Abustan, I., and Eslamian, S. (2012). Comparison between capabilities of HEC-RAS and MIKE11 hydraulic models in river flood risk modelling (a case study of Sungai Kayu Ara River basin, Malaysia). *International Journal of Hydrology Science and Technology*, 2(3), 270–291.

Bahmani, R., Eslamian, S. S., Naderi-Bani, M., and Fahhian, F. (2012). Investigating maximum rainfall intensity on peak discharge using IDF curves and HEC-HMS model. In *Ninth International Conference on River Engineering*, Ahvaz, Iran.

Brunner, G. W. (2002). HEC-RAS (River Analysis System). In *North American Water and Environment Congress and Destructive Water*, pp. 3782–3787, North American Water and Environment Congress & Destructive Water, ASCE, USA.

Chow, V. T. (1959). *Open-Channel Hydraulics*, pp. 507–510. McGraw-Hill Book Company, New York.

Copeland, R. R., McComas, D. N., Thorne, C. R., Soar, P. J., and Jonas, M. M. (2001). *Hydraulic Design of Stream Restoration Projects*. Coastal and Hydraulicslab, Engineer Research and Development Center, Vicksburg, MS.

Firenzi, A. L., Watson, C. C., and Bledsoe, B. P. (2000). Stable channel design for mobile gravel bed rivers. In *Building Partnerships, Proceedings EWRI of ASCE Joint Conference "Water Resource Engineering and Water Resource Planning & Management"*, Minneapolis, MN, July 30–August 2, 2000. Section 91. Chapter 2, pp. 1–9. doi: 10.1061/40517(2000)356.

Garde, R. J., and Raju, K. R. (2000). *Mechanics of Sediment Transportation and Alluvial Stream Problems*. Taylor & Francis, New York.

Goitom, T. G., and Zeller, M. E. (1989). Design procedures for soil-cement grade-control structures. In *Hydraulic Engineering*, pp. 1053–1059, ASCE, United States.

Hey, R. D., and Thorne, C. R. (1986). Stable channels with mobile gravel beds. *Journal of Hydraulic Engineering*, 112(8), 671–689.

Jani, M., Baloothiya, K., Mewar, P., and Patel, D. (2018, April). Flood potential estimation of poorly gauged Varekhadi watersheds using HEC-HMS model-A case of Lower Tapi Basin, India. In *EGU General Assembly Conference Abstracts*, 8th–13th April 2018, Vienna, Austria, p. 7326.

Lane, E. W. (1955). Design of stable channels. *Transactions of the American society of Civil Engineers*, 120(1), 1234–1260.

Mehta, D. J., and Kumar, V. (2020). Water productivity enhancement through controlling the flood inundation of the surrounding region of Navsari Purna River, India. *Water Productivity Journal*, 1(2), 11–20. doi: 10.22034/wpj.2021.264752.1024.

Mehta, D. J., Ramani, M. M., and Joshi, M. M. (2013). Application of 1-D HEC-RAS model in design of channels. *Methodology*, 1(7), 4–62.

Mehta, D. J., and Yadav, S. M. (2020). Analysis of scour depth in the case of parallel bridges using HEC-RAS. *Water Supply*, 20(8), 3419–3432.

Mehta, D. J., and Yadav, S. M. (2020). Hydrodynamic simulation of river ambica for riverbed assessment: A case study of Navsari Region. In AlKhaddar, R., Singh, R., Dutta, S., and Kumari M. (eds), *Advances in Water Resources Engineering and Management*, pp. 127–140, Springer, Singapore.

Mehta, D. J., Yadav, S. M., and Waikhom, S. (2013). Geomorphic channel design and analysis using HEC-RAS hydraulic design functions. *Global Journal for Research Analysis*, 2(4), 90–93.

Mehta, D. J., Yadav, S. M., Waikhom, S., and Prajapati, K. (2020). Stable channel design of Tapi River using HEC-RAS for Surat region. In Singh, R., Shukla, P., and Singh, P. (eds), *Environmental Processes and Management*, pp. 25–36, Springer, Cham, Germany.

Neary, V. S., and Korte, N. (2002). Preliminary channel design: Blue River reach enhancement in Kansas City. In *Global Solutions for Urban Drainage*, Ninth International Conference on Urban ASCE pp. 1–16.

Park, H. Y., and Datta-Gupta, A. (2011). Reservoir management using streamline-based flood efficiency maps and application to rate optimization. In *SPE Western North American Region Meeting*, May, OnePetro.

Patel, S. B., Mehta, D. J., and Yadav, S. M. (2018). One dimensional hydrodynamic flood modeling for Ambica River, South Gujarat. *Journal of Emerging Technologies and Innovative Research*, 5(4), 595–601.

Shelly, J., and Parr David, A. (2009). Hydraulic design functions for Geomorphic channel design and analysis using HEC-RAS. *World Environmental and Water Resources Congress*, 2(3), 41–50.

Shields Jr, F. D., Copeland, R. R., Klingeman, P. C., Doyle, M. W., and Simon, A. (2003). Design for stream restoration. *Journal of Hydraulic Engineering*, 129(8), 575–584.

Subramanya, K. (2009). *Flow in Open Channels*. Tata McGraw-Hill Education.

8 Hydro-Morpho Dynamics of River Junctions
Characteristics and Advanced Modeling

Hydar Lafta Ali, Badronnisa Yusuf,
Thamer Ahamed Mohammed, Yasuyuki Shimizu,
Mohd Shahrizal Ab Razak, and Balqis Mohamed Rehan

CONTENTS

DOI: 10.1201/9780429463938-11

8.1   INTRODUCTION

The flow in river junctions, whether confluence or branching, is the essential component in irrigation and drainage systems, referring to the fluvial systems which produce a complex hydro-morpho dynamic environment (Al Omari et al., 2018; Riley et al., 2014). There are two important aspects associated with the flow in river junctions, and these are erosion and deposition zones. An erosion zone is a morphological process that usually occurs at the beds and outer banks of confluence and branching rivers and is known as a scouring hole, while the deposition zone usually occurs at the inner banks (opposite to the location of erosion) and is recognized as points bars or islands caused by sediment deposition. The scour hole zone is a region formed in the bed sediment erosion and is considered one of the major morphological features of channel junction (Guillén-Ludeña et al., 2016; Alomari et al., 2018). The scour hole is associated with sediment transport caused by the increased flow turbulence and velocity intensities, which led to creating secondary vortexes. These vortexes play a significant role in changing the bed morphology at the junctions (Herrero et al., 2015; Leite Ribeiro et al., 2012; Rhoads et al., 2009). In contrast, the deposition can be recognized explicitly in the separation zone created under low pressure and flow recirculation. The separation zone at channel junctions exerts a direct influence on the flow dynamics and also morphological features (Birjukova et al., 2014; Thanh et al., 2010; Best and Rhoads, 2008; Ramamurthy et al., 2007). The separation zone has a direct effect on loss of capacity in irrigation channels or threatens mechanical elements of power plants' water circuits.

The size and location of the scour hole and the separation zone mainly depend on the junction angle and discharge ratio (Goudarzizadeh et al., 2010; Rhoads et al., 2009; Best, 1988; Best and Reid., 1984). Recent experimental work shows that the minimum scour hole and separation zone occurred at a 30° of junction angle (Alomari et al., 2018). Early studies have been focused on hydrodynamics features with rigid boundaries which means that their experimental works were without sediment transport and movable bed (Taylor, 1944; Grace and Priest, 1958). This approach still received attention for a vast explanation with different parameters and geometry forms (Ramamurthy et al., 2007; Mignot et al., 2013; Seyedian et al., 2014; Herrero et al., 2016). Also, there are studies under natural conditions to assess the flow with the morphological process (Szupiany et al., 2009; Riley et al., 2014; Casas, 2013; Redolfi et al., 2016; Yuill et al., 2016). There are many examples of natural river junctions that suffer from hydro-morpho dynamic issues, and one of the published examples is the confluence of the Wahei and Xianjiapu Rivers in southwest China. This confluence is exposed to flooding due to the sedimentation issue that reduces the capacity of the main river (Wang et al., 2019).

Another published example is the branching channel of the Ohio River that supplies water to the electric power project. The flow in this case clearly shows the effect of the sediment deposition that creates a sand bar at the entrance of the branching channel, which reduces the water supply to a side project (Neary et al., 1999).

Flooding and sedimentation that are recognized at the channels junctions may be found at many other locations worldwide, and a summary of the related studies shows that the topic is important due to having relevance to real-life problems. More investigations are required to mitigate and control the problems that occurred in channel junctions.

The scope of the present chapter covers reviewing the exiting work on hydro-morpho dynamics of river junctions, while the limitations are:

1. Characterize the main features of hydro-morpho dynamics in confluence and branching rivers that are obtained from the field sites, laboratories, and numerical simulations under various conditions.

2. Conduct studies that employ structures for controlling and managing the scouring and deposition zones in rivers with confluence and branching.

8.2 MATERIALS AND METHODS

8.2.1 Hydro-Dynamic Features of a Confluence

Taylor (1944) was the first to address the topic of open channel confluence, focusing on the water depth ratio and confluence angle. Confluence flow dynamics have been widely investigated (Best, 1987; Bradbrook et al., 1998; Mosley, 1976; Rhoads and Kenworthy, 1995; Schindfessel et al., 2015). The basic hydraulic characteristics of channel confluences are described in Figure 8.1, including the flow separation zone downstream following confluence, the flow stagnation zone, the flow of deflection zone, the maximum velocity area, the shear layers between two combining flows, and the gradual flow recovery area downstream from the separation zone.

Moreover, further hydrodynamic features were described by Best (1987) and Mosley (1976), who stated the presence of two counter-rotating helical flow cells downstream from the confluence, created by converging flows and divided by a shear layer. This secondary circulation has been extensively investigated in the field (Ashmore et al., 1992; Rhoads and Kenworthy, 1995; Rhoads and Sukhodolov, 2001; Riley et al., 2014), and laboratory (Liu et al., 2013; McLelland et al., 1996; Mosley, 1976; Weber et al., 2001), as well as by means of numerical models (Bradbrook et al., 1998, 2000; Lane et al., 2000). Recent studies on large natural rivers' confluences have determined that no evidence exists for the classical model of the secondary flow formulation (Parsons et al., 2008; Szupiany et al., 2009).

8.2.1.1 Flow Separation Zone

The flow separation zone is an important area formed downstream from channel confluences, under low pressure and flow recirculation. The separation zone at channel confluences exerts a direct influence on the flow dynamics features, increasing the complexity of the hydrodynamic features (Birjukova et al., 2014; Thanh et al., 2010). Best and Reid (1984) studied the nature of the separation zone in the laboratory under different flow rates, momentum ratios, and junction angles, in order to simplify the evolution calculations thereof, the results of which are detailed in Figure 8.2.

Huang et al. (2002) employed a 3D numerical model for channel confluences and determined that the separation zone disappears at the low junction angle of 30°, while its size is increased with a higher junction angle, and their results are illustrated in Figure 8.3.

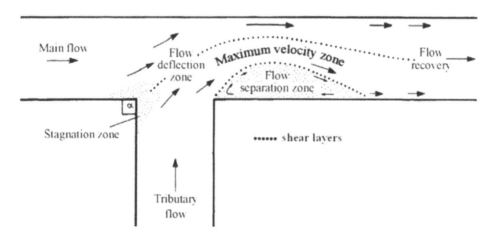

FIGURE 8.1 Characteristics of flow dynamics at channel confluences. (Source: Zhang et al., 2015.)

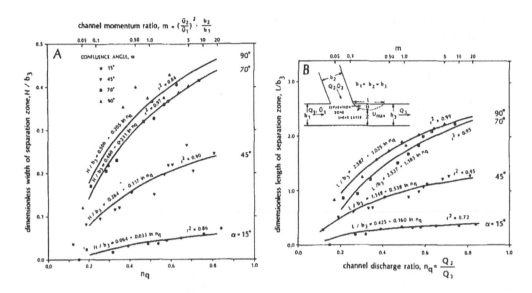

FIGURE 8.2 Calculations of separation zone (a) showing the relation between width and discharge ratio, and (b) between length and confluence angles.

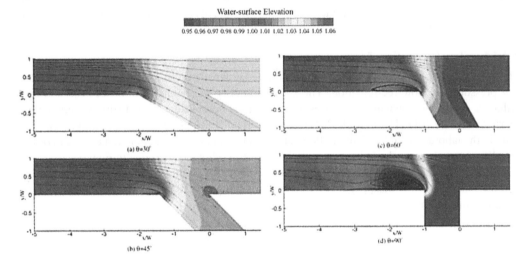

FIGURE 8.3 Variation of separation zone with confluence angle. (Source: Huang et al., 2002.)

The flow separation zone at natural river confluences exhibits a variety of profiles and at times does not appear, because in reality, it is dependent on the junction rounding and bank directions (Ashmore and Ferguson, 1992; Sukhodolov and Rhoads, 2001).

Zhang et al. (2015) implemented a physical scale for natural river confluence at the Yellow River and studied the effects of hyper-concentration on the separation zone. The authors determined that a significant amount of sediment deposition occurred, which affects the bars at the separation zone configuration.

In summary, the flow separation zone formation has been clearly demonstrated in laboratory experiments. The width-to-length dimensions increase with an increase in the confluence angle and the other flow parameters. However, at natural river confluences, it is complicated to estimate the separation zone formation and dimensions for all natural river conditions, owing to the numerous variables controlling the separation zone formulation, and further research is required for its management. Controlling the flow separation zone can be considered as an acute gap in the channel confluence research.

8.2.1.2   Flow Stagnation Zone

The stagnation zone is located at the upstream corner of the confluence and is generated by low velocities or low flow recirculation (Rhoads and Kenworthy, 1995; Rhoads and Sukhodolov, 2004). An alternative explanation for the establishment of this area in the confluence hydrodynamics zone is the negative pressure gradient arising near the upstream corner, which encourages lower velocities in this zone (Constantinescu et al., 2011; Biron et al., 2002; Sukhodolov and Rhoads, 2001). Riley et al. (2014) studied the location of the stagnation zone at the natural confluence of the Wabash and Ohio Rivers and found that increasing penetration of the tributary flow led to most of the stagnation area being shifted from the junction into the outer cross-section of the Ohio River. Furthermore, when the momentum flux ratio Mr < 1, the zone extends into the tributary channel, while when Mr > 1, the flow stagnation zone is replaced towards the main rivers. However, understanding of the stagnation zone dynamics for all types of natural confluences remains incomplete, and further research is required under different conditions in order to determine the shape of the zone and the manner in which to tackle this area.

8.2.1.3   Flow Deflection Zone

The flow deflection zone is one of the hydrodynamic confluence features formed when the tributary flow deviates the main channel stream into its left bank (Best, 1987). This area introduces streamline curvature, which has direct effects on the hydro-morpho dynamic features of channel confluence. Flow deflection manages the locations of high velocities of the main channel and also produces a helical motion over the left portion of the downstream rivers, based on the results derived from the numerical model by Roberts (2004). Best (1987), Best and Roy (1991), McLelland et al. (1996), and Sukhodolov and Rhoads (2001) have emphasized the influence of the deflection area on the flow changing movable bed motion at junctions, particularly with discordant bed confluence. Moreover, flow deflection has been found to enhance the stagnation zone surrounding the junction corner, which is illustrated by the low or negative velocity attitude near the upstream confluence corner (Riley et al., 2014). It has been determined that the stagnation zone size is responsive to the flow deflection curvature location, and increases when the Mr and confluence angle values increase (Bradbrook et al., 2000).

8.2.1.4   Maximum Velocity Zone

The maximum velocity zone, also known as the flow acceleration zone, is formed at the downstream confluence between the separation zone and main channel external bank. This area is created by decreasing the actual flow section caused by the separation zone, as well as the flow intensity at the mainstream channel left bank (Best, 1987). Best and Reid (1984) discussed the manner in which the flow separation area promotes the maximum velocity zone, and found that the velocity value near the bed at a 90-degree confluence is 1.3 times larger than at a 15-degree confluence. The maximum velocity zone has a direct effect on the bed shear stress and the potential for scour holes occurring (Guillén-Ludeña et al., 2016; Leite Ribeiro et al., 2012; Rhoads and Kenworthy, 1995; Rhoads et al., 2009). Shaheed (2016) simulated a shallow river confluence using 3D numerical modeling and found that the maximum velocity (acceleration zone) was located at the main channel outer bank, as illustrated in Figure 8.4.

Guillén-Ludeña et al. (2015, 2016) found that the origin of the scour holes is the strong flow acceleration confined along the outer bank downstream confluence of the main channel. The flow acceleration location leads to increasing availability of shear stress, which consequently results in increasing lateral bed dune length at the downstream confluence.

The maximum velocity zone formulation in natural channel confluence differs from one location to another and is not always presented in a consistent form and location, owing to its complexity

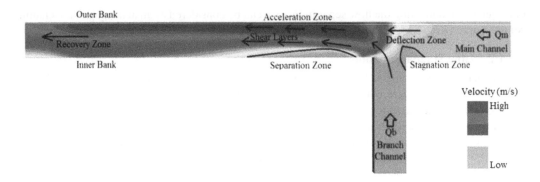

FIGURE 8.4 Maximum velocity zone (acceleration zone). (Source: modified after Shaheed, 2016.)

in natural confluence (Ashmore and Ferguson, 1992; Szupiany et al., 2009; Wallis et al., 2008). Further research in nature is required in order to elucidate the maximum velocity zone formulation.

8.2.1.5 Shear Layer Zone

The shear layer zone is another critical area of channel confluence formed between the flows originating from the tributary and the main channel at the downstream confluence. Best (1987) determined a shear layer zone is occurring twice at the main channel, caused by high-velocity gradients: one is between the surrounding flow and recirculation zone, while the second is located in the buffer zone between the flows from the tributary and main channel (Guillén-Ludeña, 2015). The shear layer zone with the other hydrodynamic features is clearly illustrated in Figure 8.5.

The second location of the shear layer, which originates from the stagnation zone and distortion in the flow recovery zone was highlighted by Best (1987) and recognized as an external shear layer described by high turbulence and shear stress. Boyer et al. (2006) and Serres et al. (1999) confirmed the relation between the momentum flux ratio (Mr) and external shear layer location and found that, at a low momentum flux ratio (Mr < 1) the shear layer position is close to the main channel inner bank, while a high momentum flux ratio (Mr > 1) results in its location being shifted to the main channel outer bank. These results were obtained from natural rivers confluences in Canada. Numerous studies have confirmed the location of this area by means of either laboratory experiments under different conditions (Guillén-Ludeña et al., 2016; Leite Ribeiro et al., 2012; Weber et al., 2001) or numerical simulation using 3D models (Shaheed, 2016; Sirdari, 2013; Shakibainia et al., 2010; Biron et al., 2004; Huang et al., 2002; Lane et al., 1999). Boyer et al. (2006) noted that the maximum bedload transport could be found at the shear layer edge, owing to sharp variations in

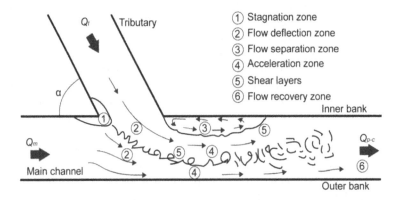

FIGURE 8.5 Hydrodynamic features at channel confluences. (Source: developed by Guillén-Ludeña et al., 2015.)

turbulence intensity. Szupiany et al. (2009) remarked that the shear layer distortion was more essential than the presence of helical cells and scour holes, owing to the flow separation zone formulation at discordant junctions, which distorts the vortices at the mixing layer and promotes faster flow mixing (Best and Roy, 1991; Biron et al., 2004; Sukhodolov and Rhoads, 2001).

8.2.1.6   Flow Recovery Zone

The final zone recognized from channel confluences occurs when the flows from the tributary and main channel tend to be near steady downstream from the main channel (Best, 1987). Numerous studies have confirmed the flow recovery zone formulation, whether in the laboratory or by means of natural or numerical simulation (Boyer et al., 2006; Tonghuan Liu et al., 2015; Sukhodolov and Rhoads, 2001; Wang et al., 2019). The flow recovery zone position in terms of the distance from channel confluences differs and is dependent on flow mixing speed, downstream morphology (Gaudet and Roy, 1995; Szupiany et al., 2009), and discharge ratios and confluence angles (Guillén-Ludeña et al., 2016; Tonghuan Liu et al., 2015).

8.2.2   Morpho-Dynamic Confluence Features

Best and Rhoads (2008) summarized a morphological model for describing the bed morphology elements of open channel confluences by means of five zones: (1) the scour hole, which is linked to increasing flow turbulence and velocities within the junction as well as sediment transport paths; (2) tributary-mouth bars, formed near the channel mouths; (3) a mid-channel bar or bars formed after the confluence channel; (4) deposition in the flow separation zone owing to flow recirculation near the inner bank upstream confluence, appearing as a lateral bar attached to the right bank; and (5) deposition in the flow stagnation zone near the corner banks of the upstream confluence. Figure 8.6 illustrates the typical bed morphology at flow confluences.

8.2.2.1   Scour Hole Zone

The scour hole zone is a region formed in the bed sediment erosion, resulting from the confluence of two flow cells originating from two channels, and is considered as one of the major morphological features of channel confluence. The classical model presented by Best (1988) explains the scour hole extension location from the tributary inflows into the left bank of the mainstream channel, as shown in Figure 8.6, where the alignment of this scour approximately divides the junction angle. The scour hole has been associated with sediment transport caused by increasing flow turbulence and velocity

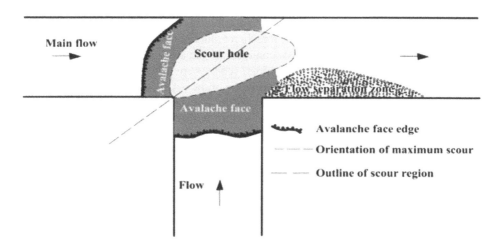

FIGURE 8.6   Typical bed morphology at flow confluences. (Source: Zhang et al., 2015.)

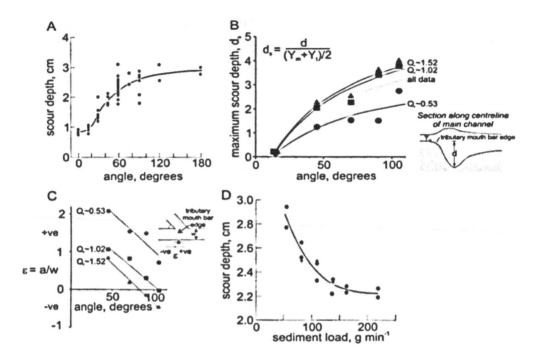

FIGURE 8.7 Factors controlling the scour hole formation resulting from laboratory experiments on channel confluence. (Source: redrawn by Best and Rhoads, 2008.)

intensities through the confluence. Best and Rhoads (2008) summarized the factors associated with the scour hole formulation, as illustrated in Figure 8.7.

Figures 8.7a to d were derived by Mosley (1976) and illustrate the functions of the confluence angles and total sediment load passing through the channel confluence on the scour hole depth. The relationships between these are nonlinear and demonstrate that increasing the degree of junction angles leads to an increase in the scour hole depth, while an increase in the total sediment load results in the scour hole depth either decreasing or being maintained. Best (1988) confirmed that the maximum scour depth location reacts to the discharge ratio and angles. Furthermore, it was found that the scour depth increases the inflow discharge from the tributaries to junctions, owing to the higher discharge ratios (Mosley, 1976). Another factor affecting the scour hole is the junction plan form shape. Bryan and Kuhn (2002) discussed the symmetrical (Y-shaped) and asymmetrical confluences and found that the plan form has a more direct influence on the scouring bed than the angle of the junction. In symmetrical (Y-shaped) confluences, the scour hole is at the central junction, while in the asymmetrical junctions, more complex scours were identified owing to bank erosion in front of the tributary inflow, resulting in the evolution of the asymmetrical junction plan form and consequently the higher confluence angles compared to the original channel plan forms.

Examinations of scouring at natural river confluences have been conducted in small rivers (Biron et al., 1993; Boyer et al., 2006; Rhoads and Kenworthy, 1995; Rhoads et al., 2009; Serres et al., 1999). Analyses of field and experimental studies have identified the junction angle, momentum flux, and discharge ratios as crucial parameters in the bed morphology of open channel confluences. Rhoads et al. (2009) studied the effects of different discharge ratio conditions on the scour hole formulation, and found that the formulation was located at the central junction with a low discharge ratio, while the scour hole location was shifted to the mainstream channel left bank owing to the erosion process with a high discharge ratio (Figure 8.8). However, the scour hole formulation was not strongly recognized at the discordant confluence and with low confluence angles (Roy et al., 1988; Biron et al., 1993).

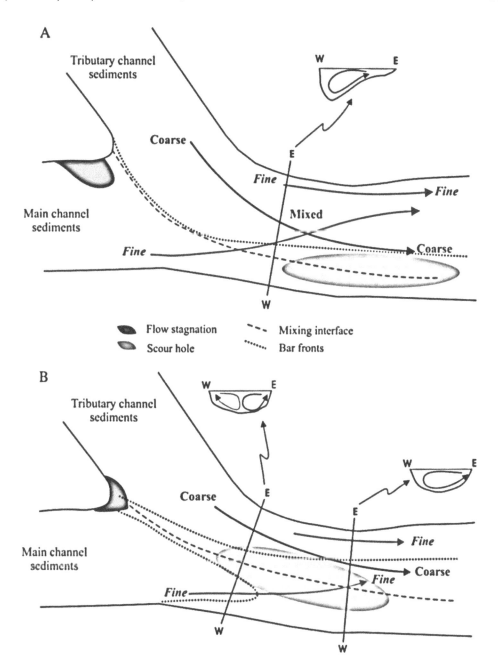

FIGURE 8.8   Scour Hole position with different discharge ratios: (a) high discharge ratio and (b) low discharge ratio. (Source: Rhoads et al., 2009.)

The scour hole in large natural river confluences was described by Best and Ashworth (1997), Parsons et al. (2007), Riley et al. (2014), and Szupiany et al. (2009). They found that the scour hole was the important feature recognized in large natural river confluences. One of the largest river confluences in the world is that of the Jamuna and Ganges rivers in Bangladesh. These confluences were studied by Best and Ashworth (1997), where the data of five bathymetric surveys were taken at different times, demonstrating that the scour depth was around 30 m at the confluence, which is equal to five times the average depth of the confluent channels. Furthermore, the slopes of the bed dipping into the scour hole at the convergence of the Jamuna and Ganges Rivers are typically less than five degrees and the avalanche faces have not been found. Moreover, a comparison of the results of small natural river confluences with widths of less than 100 m and previous laboratory investigations indicates certain scale differences in junction morphology, with the scouring depths typically ranging from two to four times the incoming tributary channel depths. Similar scour findings were provided by Parsons et al. (2007) and Szupiany et al. (2009) at the confluence of the Rio Paraná rivers in Argentina. The scour depth was approximately two to three times the pre-confluence average river depth, and reached over 22 m at the confluence, as shown in Figure 8.9.

8.2.2.2 Deposition Bars in Separation and Stagnation Zones

Additional vital bed morphology features are the depositions in the separation and stagnation areas owing to the sediment inflow, which may cause changes in the channel grade, bed and bank erosion,

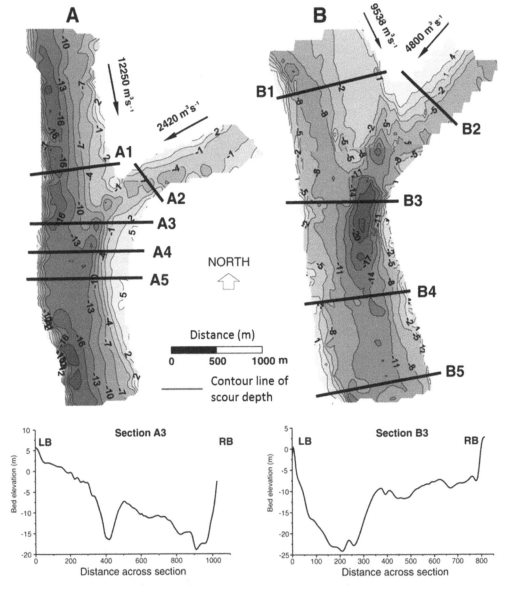

FIGURE 8.9 Scouring depths at the confluence of Rio Paraná river, Argentina, observed at two different times. (Source: Szupiany et al., 2009.)

and thalweg movement (Ettema, 2008). Flow at channel confluences is merged within the curve plan form and bar formations at the upstream confluence, associated with the flow stagnation zone, while downstream it is associated with the separation zone, and the deposition and accumulation of sediment particles increases in these zones owing to reduce the flow velocities (Best and Reid, 1984; Best and Rhoads, 2008). Furthermore, Ribeiro et al. (2012) defined the separation zone as an area of lower pressure and recirculating flow, which promotes sediment deposition. The same was found to be true for the lower velocities in the stagnation region at the upstream corner of numerous confluences. Further research is required to quantify the sediment deposition and its accumulation to form bars and the manner in which these bars are presented and changed under different channels confluence conditions for both the separation and stagnation zones.

8.2.2.3 Mid-Channel Bars

The formulation of this area is associated with high sediment deposition rates alongside the scour hole zone downstream from the confluence. Mosley (1976) was the first to identify this zone for a confluence angle higher than 60°. Best (1986) explained the combination of scour hole erosion and sediment routing contributing to the deposition conditions in this zone, and the location of this area is mostly observed at symmetrical confluences, as shown in Figure 8.10. Ashmore and Gardner (2008) found that the construction of a mid-channel bar related to sediment transport accumulates slightly downstream from the maximum bedload transport zone. Parsons et al. (2008) determined that symmetrical confluence in a sizeable natural river is an essential factor for determining a mid-channel bar and its shape, owing to its influence on flow divergence downstream from the scour hole. However, further efforts and studies are required to establish the factors affecting the mid-channel bar formulations, which are linked to the sediment erosion and deposition patterns.

8.2.3   HYDRO-DYNAMIC OF THE BRANCHING CHANNEL

8.2.3.1   Separation Zone

Flow at the separation zones occurs in the areas of water recirculation and low velocities (Neary et al., 1999; Ramamurthy et al., 2007). As a result of the recirculation and low velocities, sedimentation areas appeared in these zones (Barkdoll et al., 1999; Shamloo and Pirzadah, 2007b). These areas cause a loss of capacity in irrigation channels which consequently reduce the inflow discharge passes through the branching channel which also threatens the operation of power plants (Mignot et al., 2014). In branching channels, two main separation zones have resulted from the flow, and these zones are shown in Figure 8.11.

Zone (1) is developed in the branching channel when the flow in the main channel is discharged toward the branching channel, while Zone (2) is formed near the right bank in the main channel and downstream of the branching junction as shown in Figure 8.11. Zone (2) may not form all of

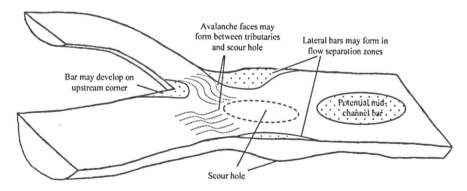

FIGURE 8.10   Location of potential mid-channel bar. (Source: Tancock, 2014.)

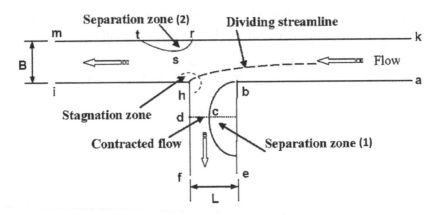

FIGURE 8.11 Characteristics of flow dynamics at branching channel. (Source: Ramamurthy et al., 2007.)

the time; it usually formed when the discharge in the branching channel took a significant portion of the discharge in the main channel. This flow condition makes the streamlines curve towards the branching channel where the separation zone expands (Ramarnurthy et al., 2007; Shamloo and Pirzadah, 2007b). The size and location of the zone (1) mainly depend on the discharge ratio and branching angle. The term discharge ratio in a branching river refers to the percentage of the branching river discharge relative to the main river discharge. The discharge ratio is considered as one of the most relevant parameters in the analysis of branching study (Alomari et al., 2018; Herrero et al., 2015; Goudarzizadeh et al., 2010; Hager, 1992; Ingle and Mahankan, 1990; and others). It can be concluded that when the discharge ratio is increased, it leads to a decrease in the size of a zone. Another essential factor that affects the formulation of the separation zone is the branching angle. Most of the branching angle studies have been done experimentally (Alomari et al., 2018; Keshavarzi and Habibi, 2005; Lama et al., 2003). The branching separation zone often occurs in the upstream side bank of the branching river, and the location, length, and size of this zone vary alongside banks of the branching depending on the junction angle. Keshavarzi and Habibi (2005) conducted a laboratory study with the different branching channel angles between 45° to 90° and found that the optimum branching angle is 55° based on the size of the separation zone. Another laboratory experiment conducted recently by Alomari et al. (2018) observed that increasing branching angle from 30° to 90° leads to moving the separation zone toward downstream in the branch channel and the smallest separation zone has occurred when the branching angle was 30°.

8.2.3.2 Stagnation Zone

Another hydrodynamic feature found at the corner downstream of the junction is a stagnation point in which the highest pressure and flow depth were found (Ramarnurthy et al., 2007). In the branching channel system, one stagnation zone is recognized at the corner junction edge as shown in Figure 8.11.

8.2.3.3 Contraction Zone

Owing to a separation zone (deposition zone) at the beginning of the branching channel, the area of the flow will be contracted slightly and the highest contracted will be formed at the middle of the flow separation zone as shown in Figure 8.11. In this area, maximum velocity was generated which is attributed to the location of the flow separation. Ramamurthy et al. (1996) found a linear relationship between the contraction area and discharge ratio (the area increases when the discharge ratio increase). Lama et al. (2002) found that the contraction flow width at the bottom of the zone is wider than at the surface.

8.2.4 MORPHO-DYNAMIC OF THE BRANCHING CHANNEL

The formation of a branching channel geometry was developed due to the behavior of the flow that comes from the main river and is distributed toward the branching river, and this geometry form causes changes in the hydraulic flow condition of the junction area. These changes lead to the appearance of the scouring and deposition regions owing to the erosion and sedimentation process in the junction zone (Kleinhans et al., 2012; Allahyonesi et al., 2008). Vortexes and secondary currents parameters are formed due to the streamlines being curved towards the branching channel. These parameters were vital for the erosion and deposition issues at the branching channel system (Moghadam and Keshavarzi, 2010).

8.2.4.1 Branching Scour Zone

One of two important aspects of morphodynamic features is the erosion process that formed at the branching junction region resulting from sediment movement caused by changes in hydraulic flow conditions. The scour hole is produced by secondary vortexes created in the junction region (Dehghani et al., 2009; Allahyonesi et al., 2008). These vortexes played a significant role in changing the bed morphology in the main and diversion channels. This zone is usually associated with maximum velocity zone and from reviewing many studies, it is found that the size and location depend on the discharge of water and sediment, the parameters of diversion angle, the roundness of the edge upstream corner of the diversion, and width ratio between the main and branching channels. These were considered as main factors that have a direct influence on the branching channel behavior (Kleinhans et al., 2012; Moghadam and Keshavarzi, 2010; Rezapour et al., 2009). Barkdoll et al. (1999) and Herrero et al. (2015) observed a scour hole in the main and diversion channels beds at the downstream junction when the diversion angle is 90°. Alomari et al. (2018) designated that the branching angle should be decreased as much as possible in order to decrease the scour depth. He conducted experiments with different diversion angles (30°, 45°, 60°, 75°, and 90°) and observed that the minimum scour depth is associated with a diversion angle of 30°.

8.2.4.2 Branching Deposition Zones

One of the most frequently stated problems in the branching channel system is sediment deposition due to it reducing the efficiency of the branching system. This zone is associated with the low and recalculation flow (separation zone). Early experiments by Bulle (1926) recognized sedimentation problems at the branching channel. Herrero (2013) developed a conceptual framework for the sedimentation zone in 90° branching as shown in Figure 8.12. He found that the deposition zone was formed at the upstream wall of the branching channel.

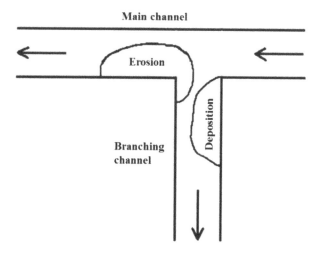

FIGURE 8.12 Deposition and erosion zones. (Source: observed by Herrero, 2013.)

As mentioned above in the separation zone, the size and location of the deposition zone are mainly depending on the discharge ratio and branching angle. However, when the branching angle is decreased from 90° to 30° it led to a move and a decrease in the size of the deposition zone. The smallest deposition zone was observed when the branching angle is 30° (Alomari et al., 2018).

8.2.5 Sediment Transport at River Junctions

Studying sediment transport is challenging, particularly in natural junctions, owing to the different parameters affecting its calculations. There are three aspects of sediment transport, namely bedload, suspended load, and total load. The total load can be obtained directly by the empirical equations or indirectly as the summation of the bed and suspended load equations. The bedload and suspended load categories are not rigid because they depend on the flow velocity and not on the sediment particle sizes. For instance, in high-velocity or very turbulent water, gravel and sediment with large sizes have mostly traveled in suspension. However, in very low velocity or turbulence, small-sized sediment particles such as silt and clay have been moved entirely in the bedload (Yang, 1996).

In channel confluences, a number of studies have explained sediment transport and its effects on channel morphology, particularly in laboratory experiments. Mosley (1976) studied the sediment transport in channel confluences and explained the earlier concepts of sediment transport patterns in symmetrical laboratory confluences. In his experiments, he found that sediment was moved predominantly along the scour hole sides, owing to the existence of helical flow cells, which move the sediment outwards from the scour. Downstream from the confluence, sediment transport was observed to converge and yielded the highest transport rates in the channel center. Moreover, he related the scour hole depths to the total sediment load and concluded that the scour depths decrease as the total sediment load increases. This finding was confirmed by Rezaur et al. (1999). Another important study on the sediment transport patterns in asymmetrical confluences was conducted by Best (1988), in which segregation was identified between the sediment loads provided by each confluent channel, and this affected the junction angle and mutual flow deflection at the confluence. These observations are opposed to those by Mosley (1976). A lack of sediment transport has been observed in the post-confluence channel center at the high junction angles (Best, 1988). Furthermore, the author confirmed his laboratory works by means of particle-tracking experiments at natural asymmetrical confluences of the Ure and Widdale Beck rivers in the United Kingdom, where sediment was moved along the bank-attached bar and scour hole flanking. Roy and Bergeron (1990) documented the gravel particle movement through the scour hole and the mixing of sediment pathways at a small asymmetrical confluence in Canada, the results of which are in contrast with those of Best (1988). Rhoads (1996) described the segregation at the natural confluence of the Kaskaskia and Cooper Slough rivers, with high sediment transport through the scour and sediment mixing downstream from the confluence when Mr < 1. Later, Rhoads et al. (2009) observed the different sediment transport patterns for Mr > 1. Under these conditions, it was found that the sediments moved in the scour hole flanking and mixed well within the confluence. For the natural discordant confluences of the Bayonne-Berthier rivers, Biron et al. (1993) observed a sharp change in the alignment of migrating ripples between the main channel bed and the tributary-mouth bar, revealing the influence of bed discordance on sediment transport patterns. Furthermore, this natural discordant confluence has been characterized by a lack of marked scour holes and dynamic protuberance of the tributary-mouth bar, which affects the patterns of sediment transport (Boyer et al., 2006), in addition to the highest bedload transport rates being quantified near the shear layer edges and impinging on the tributary-mouth bar, promoting high sediment transport rates. Moreover, Boyer et al. (2006) associated the sediment transport rates with variations in the momentum flux ratio (Mr), with these variations affecting the shear layer position and its interface with the bed morphology.

In the branching channel, the sediment transport will be divided as apportion toward the branching channel, and other quantities of sediment will continue with the flow of the main channel.

The branching channel is receiving sediment concentration higher than the original flow sediment concentration (Herrero, 2013; Omidbeigi ct al., 2009; Neary and Odgaard, 1993). The reason for this phenomenon is referred to the dividing flow streamlines width in the main channel at the upstream junction increases from the surface towards the bottom, and the water diverts from the main channel lower layers, these containing more bedload than the upper layer (Neary et al., 1999). Concurrently, the sediment concentration in lower layers is more significant than in the upper layers (Barkdoll, 2004). Secondary currents and the vortex in the branching junction have affected the amount of the branching sediment. An increase in bed roughness will lead to reducing the secondary currents, which will decrease the branching sediment. These secondary currents are depending on many factors, such as branching angle, velocity ratio, and bed width ratio. The sediment ratio of branching to the total sediment of the main channel increased with an increase in the discharge ratio and water depth of the main channel upstream junction (Moghadam and Keshavarzi, 2010). Herrero (2013) found that the pattern of sediment distribution in the main and branching channels strongly depends on the total discharge. The total discharge of his experiment was ranged from 3.5 to 6 L/s, and it was noted that when the total discharge below 4.75 L/s, about 30% of total sediment, was associated with 10% of the total discharge that entered the branching channel. Discharge more than 4.75 L/s, around 90% of the total sediment, was associated with 30% to 40% of the total discharge that entered the branching channel. However, quantification of studying sediment transport remains incomplete, particularly in natural river junctions.

8.2.6 STUDIES ON CONTROLLING EROSION AND DEPOSITION ZONES IN RIVER JUNCTIONS

Studies on controlling morphological features at the junction region had been achieved experimentally by using the submerged vanes in front of a few studies related to the practical application. These studies try to manage the movement of sediment accumulation at the branching channel section. Submerged vanes are considered as one of the most common ways to control the sediment transport at the branching channel (Wuppukondur and Chandra, 2017a; Odgaard and Wang, 1991; Neill et al., 1997; Michell et al., 2006; Odgaard, 2009; Beygipoor et al., 2013).

In confluence junctions, there are few studies on controlling erosion and deposition features (Odgaard and Wang, 1991; Wuppukondur and Chandra, 2017a, 2017b). An early study on controlling scouring and deposition zones by using a set of submerged vanes in 90° flume confluence adopted a typical vane layout for enhancing the navigation of river confluences (Odgaard and Wang, 1991). The layout was designed to decrease the velocity in the erosion zone (scouring zone) which consequently lead to a decrease in water depth; this process occurred due to redirecting the flow by vane layout from the erosion zone toward the deposition zone. Increasing the flow in the deposition zone was leading to increasing the velocity, thereby leading to the removal of a large portion of sediment from the bar that formed in the deposition zone. Two recent studies were conducted on a laboratory scale with 60° and 90° confluences (Wuppukondur and Chandra, 2017a, 2017b), in which a set of the vanes and piles were proposed to control bed erosion. The results of these studies show that the maximum percentage of reducing the scour depth is 55% and 43% with 90° and 60° confluences, respectively, through the use of a set of piles; the reduction is less than 10% from the above percentage when a set of vanes is used. Furthermore, the studies placed the vanes at angles of 15°, 30°, and 60° and found that increasing the vane angle by more than 30° leads to an increase in the scour depth.

In the branching channel, submerged vanes are installed in the form of rows which created a scour trench to reduce the bedload transport from the main channel to the branching channel (Odgaard and Spoljaric, 1986). The scour trench traps the sediment from the main channel and redirects it toward downstream of the main channel (Ouyang et al., 2008; Wang et al., 1996). Size, angle, and number of submerged vanes are considered the most effective parameters in controlling the bed sediment movement (Odgaard and Wang, 1991a). Numerous rows of vanes arrangement schemes have been investigated to evaluate their performance in preventing the bed sediment transport from entering the branching channel. Barkdoll et al. (1999) investigated two schemes of the submerged vanes arrangement and found that these vanes could reduce about 40% of the branching sediment discharge when the discharge ratio (the percentage of the branching channel discharge relative to the main channel discharge) is more than 20%. Another different vanes arrangement has been tested in different boundary conditions such as parallel, regular, and zigzag installed in the different number of rows with the different angles, Table 8.1 summarizes several studies that have been conducted experimentally and practical applications in different sites.

The studies related to the practical application to control the sediment that enters the branching channel are Wang et al. (1996), Nakata and Ogden (1998), and Michell et al. (2006). Some limitations and conditions were found in real practical cases like Wang et al. (1996), who found that the vanes are an effective way for controlling sediment at the branching channel when the flow is small enough, and the velocity at the benching channel is less than 20% of the main flow velocity at the upstream. The other researchers (Nakata and Ogden, 1998; Michell et al., 2006) combined the submerged vanes with other structures such as skimming or barrier walls in order to get a more effective way to control the sediment movement. However, it can be concluded that the effectiveness

TABLE 8.1

Summary of the Studies That Used Submerged Vanes for Controlling Sediment Dynamics at the Branching Channel

Authors	Main Channel Width (m)	Branching Channel Width (m)	Branching Angle	Submerged Vanes Arrangement and Number	Vane Angle
Wang et al. (1996)	230	25	90°	Parallel / 40	20°
Study 1	25	10		6	
Study 2					
Nakata and Ogden (1998)	Missouri River	22	90°	Zigzag	19.5°
Study 1		27.6		13	20°–45°
Study 2		8.1		5	19.5°
Study 3		48		17	22°
Study 4		27		11	22°
Study 5				17	
Barkdoll et al. (1999)	1.5	0.61	90°	Parallel with four cases, 3 rows / 54	20°
				3 rows / 21	
				2 rows / 20	
				3 rows / 18	
Michell et al. (2006)	100	26.9	90°	Zigzag with two rows / 13	22°
Allahyonesi et al. (2008)	1.5	0.6	60°	Regular / 24 and zigzag / 24	20°
AbdelHaleem et al. (2008)	0.6	0.2	90°	Single row / 4	10°, 20°, 30°, 40°, 50°
Moghadam and Keshavarzi (2010)	–	–	55°	Zigzag with two rows / 10	10°, 20°, 30°, 40°
				Parallel with three rows / 15	
Mirzaei et al. (2014)	1	0.4	90°	Parallel with two rows / 10	22°

of using submerged vanes was limited based on flow conditions in which the practical approach is with discharge ratio up to 20–30%, and in some cases may have adversely affected the navigation of the main channel. Especially these groups of the submerged vanes produce non-uniformity of the velocity distribution near its location (Barkdoll et al., 1999), and also are potentially costly due to their numbers. The limitation of using submerged vanes to control and manage the hydro-morpho dynamics displays the importance of investigating different types of obstacles and approaches.

8.2.7 Recent Works in Laboratory, Field, and Numerical

8.2.7.1 Laboratory Experiments in Rivers Confluences

Yuan et al. (2016) observed the hydrodynamics and turbulent structure of the distorted shear layer in laboratory work on a 90-degree confluence with a smaller depth-to-width ratio. The results demonstrate several essential characteristics, such as the generation of a stronger helical cell at the junction and extension downstream. The helical cell length is dependent on the tributary discharge inflow, the three main forces acting on it, and the shear layer distortion. These forces are the Reynolds shear stress, maximum turbulent kinetic energy, and occurrence possibilities of ejection and sweep events, which are distributed more in the central water depth than near the water surface. The increase in the occurrence possibilities of sweep and ejection forces within the shear layer leads to shear layer distortion, while a decrease in the occurrence possibilities of sweep and ejection actions outside the shear layer is associated with turbulence, producing vortices caused by the wall. Finally, increased flow rates from both channels lead to an increase in certain parameters, such as turbulence, velocity kinetic energy, and Reynolds shear stress at the various velocities, while decreased flow rates of each channel result in the shear layer being distorted to a greater extent. For future studies, it is suggested that studying the hydrodynamics, turbulence characteristics, and morpho-dynamics of sediment transport with a distorted shear layer at natural channels confluences determine the different features that are essential for improved understanding, and critical for the design of water intakes and sewage outlets.

Guillén-Ludeña et al. (2016) investigated the development of the hydrodynamics and morphology of mountain river confluences in the laboratory, using different discharge ratios (0.11, 0.15, and 0.23) and confluence angles (90° and 70°). The experiments were conducted under movable bed conditions with a continuous sediment feeder to both flumes, and measurements of the water surface and topography surveys were performed at different instances during the experiment as well as at equilibrium. The results demonstrate that the discharge ratio and confluence of the angle parameters exert the main control over mountain river confluence dynamics. Furthermore, certain patterns have been presented based on the evolution of bed morphology and flow dynamics for varying the discharge ratios and junction angles, which are in contrast with those described for lowland confluences. Different flow regime patterns adopted by the tributary for various junction angles, such as decreasing height of the bank-attached bar, are associated with increasing discharge ratios. Moreover, the abundant sediment load from the tributary plays a major role in the dynamics of mountain river confluences, and this load results in bed discordance in the bend affecting the flow dynamics and bed morphology. However, further laboratory studies are recommended with different sediment discharge rates in order to determine the influence of this parameter on the confluence dynamics.

Wang et al. (2019) focused on the stage-discharge relationship at mountain rivers confluences in southwest China. Laboratory flume and physical model experiments were designed and performed carefully in order to examine and analyze the complex flow structures and features of river confluences with different hydraulic and configuration conditions. The relationships of the flow confluence effects on the rating curve (stage-discharge) with other hydrodynamic features were analyzed and the results clearly indicate that channel confluences have a significant impact on the stage-discharge relationship and hydraulic features. The velocity distributions were recognized in the high and low regions, which divided the stream of open channel confluences, while the backwater and flow separation zone were identified as having complex hydrodynamic characteristics at river confluences in earlier studies. The analyses explain the potential influences of tributary river inflows on the flow structures and water levels in the confluence district, and the flooding risk may be significantly increased at the junction zone. In particular, the model test results imply that the flooding risk is underestimated for the current Boluo Power Station at this river confluence area between the Xianjiapu and Wahei rivers, and engineering project development (such as embankment elevation) is recommended at this critical location.

Zhang et al. (2015) implemented a physical scale experiment of sediment transport, including bed morphology, for natural river confluences of the Xiuliugon hyper-concentrated tributary into the upper Yellow River. The laboratory experiment results indicated four hydraulic zones comprising: (1) the backwater area above the upstream junction corner, (2) the separation flow zone, (3) the maximum velocity region, and (4) the post-confluence area downstream from the confluence corner. Furthermore, four basic bed morphology elements were identified, namely: (1) a sandbar in the backwater zone, (2) a bar at separation flow zone, (3) thalweg of flow transmission and sediment transport, and (4) bars in the reach after the separation zone. A significant amount of sediment deposition occurs from the hyper-concentrated tributary and affects the bars in the separation and backwater zones. The relationship between these is linear and tends to increase as the sediment load increases, which may lead to the main channel becoming blocked. The field data were used to derive an equation to compute the sediment deposition amount downstream from the hyper-concentration confluence based on the continuity equation and balance.

Tonghuan Liu et al. (2015) analyzed the sediment flow interactions at channel confluences under different discharge ratios and confluence angle conditions through a series of flume experiments. The sediment transport rate and bed topography were measured, and the results were compared among the different conditions. By the time the tributary flow reaches the mainstream, sediment transport is intermittent and fluctuates, with its rate increasing with an increase in discharge ratio and confluence angle. A central scour hole is formed along the shear plane and develops near the right bank, and the scour hole depth increases as the discharge ratio and confluence angle increase. These features are shaped with no sediment feeder upstream from the flume, while with the supply of heavy sediment, deposition occurs upstream from the confluence area and downstream from the separation zone. Moreover, the greatest channel confluence deposition is mainly dependent on the discharge ratio and confluence angle.

Nazari-giglou and Jabbari-sahebari (2014) assessed the laboratory work in order to understand the sediment transport and bed morphology mechanisms in open channel confluences and computed the influence of various flow and geometrical features on the sediment threshold movement, owing to its significance in river engineering designs and fluvial studies. Different parameters were used in the experimental setup, including discharge ratios (0.25, 0.3, 0.4, 0.5, 0.6, 0.75, and 0.8), confluence angles (50°, 70°, and 90°), width ratios (0.5, 0.75, and 1), and sediment size d50 (mm) (0.25, 0.5, 0.97, 1, 1.5, and 2.5), while the Froude number was less than (1). These parameters play important roles in sediment threshold movement. Moreover, mathematical relations between the above parameters and the mean threshold velocity have been derived using 57 sets of experimental data. The results demonstrate a strong correlation, which requires further studies with other conditions to reach a comprehensive conclusion. Further investigations have been conducted in order to characterize the effects of different dimensionless features, such as discharge and width ratios, and bed material size, on the development of morphological features (local scouring and deposition areas). The experiments confirmed the formation of a scour hole at the downstream confluence, and deposition zones in the separation area and further downstream from the channel. However, the outcomes indicate that the width ratio decreases when the scour hole depth increases, which consequently leads to an increasing discharge ratio. Furthermore, in order to reduce and control the scour hole, chamfer was used at the downstream corner of the confluence point, which resulted in a narrow scour hole that stretched toward downstream from the main channel.

Liu et al. (2013) conducted an experimental study in the laboratory and focused on the flow pattern and sediment transport evolutions at a 90° open channel confluence. Different discharge rates were studied in order to characterize the development of generated features, owing to its importance for the network in river engineering design. The experimental results indicate that the water surface profile at the channel confluence is influenced by the tributary mixing action. The vertical distribution of the time-averaged flow velocities is correlated with the discharge ratio and flow regime. The time-averaged velocity upstream from the convergence gradually declines when the discharge ratio increases, in both the high and low-velocity zones, and a backflow separation region exists in the downstream cross-sections, which are generated and influenced by the changing the flow discharge ratio. Moreover, the separation region development is restricted once the discharge ratio becomes relatively large, and the bedload motion is intermittent and fluctuates, based on scouring experiments with no supplementary sediment, while the bedload discharge primarily increases and then declines monotonically. The largest local scour appears near the shearing surface downstream from the confluence, associated with clear water from the tributary. The bedload transport rate and its accumulation in terms of quantity were recorded and analyzed for different discharge ratios in each case. Furthermore, 2D mapping of the bed morphology for each case was presented, and the results indicate that the features of the bedload transport and local scoured bed rely on the flow and sediment conditions from the tributary and main rivers.

8.2.7.2 Field Studies in Rivers Confluences

Martín-Vide et al. (2015) studied the bedload transport in a natural confluence of the Toltén river and its tributary, the Allipén river, located in the south of Chile, in order to understand the bedload distribution in terms of space and texture, as well as the balance between the tributary and main river. Fieldwork was conducted and the maximum bedload transport was found to reach 5,000 t/d, with a maximum discharge of 900 m^3/s. The results indicate that two-thirds of the total bedload volumes are transported through the deeper zone and gravel is predominant (64%). Average bedload volumes at the confluence appear unbalanced, and the main river bedload transport is predominantly below capacity, while that of the tributary is at capacity.

Rhoads et al. (2009) investigated the bed morphology and bed material texture under different hydrological conditions of the asymmetrical natural confluence of the Kaskaskia and Copper Slough rivers in east Illinois, USA. Cross-section surveys, a sampling of bed material, and recorded survey data over a long period illustrate the response between the channel morphology and hydrological conditions at natural asymmetrical confluences. The results demonstrate that the discharge ratio has a direct effect on the bed morphology, with high discharge ratios creating a narrow scour hole near the outer bank and a large bar complex after the downstream junction corner. However, a low discharge ratio initiates the relocation of the scour hole to the confluence midpoint, while substantial erosion of the bar structure is produced along the inner bank downstream channel. Furthermore, incoming sediment loads are observed to be segregated around the central scouring region before matching in the downstream channel. The express hydrological response of the lateral tributary is compared to that of the main channel, and the results indicate that the outer embankment erosion is a result of frequent high discharge ratio flows. Consequently, the outer embankment is shifted away from the lateral tributary mouth, leading to widening the downstream channel.

8.2.7.3 Numerical Simulation in Rivers Confluences

The continuity, energy, and momentum equations have resulted from the three principles governing fluid motion: conservation of mass, energy, and momentum. These equations have been replaced by empirical approximations and mathematical calculations with numerical models, which may be presented in different forms, such as one, two, or three dimensions, under different flow conditions and steady or unsteady circumstances (Toombes and Chanson, 2011). All numerical models in real situations have been considered as computational fluid dynamics (CFD) models, even for a simple solution to a backwater equation. The majority of the recent studies on channel confluences have considered the 3D models for explaining the complexity of river interactions (Shakibainia et al., 2010; Song et al., 2012). However, the numerical simulation of channel confluences is time-consuming, owing to the hydro-morpho dynamic interactions, particularly when building or generating the grid (Sirdari et al., 2014).

Schindfessel et al. (2015) tested the new flow patterns in an open channel confluence with an increasingly major tributary inflow under a systematic study with a fixed concordant bed. Large-eddy simulations were used to investigate the 3D compound of the flow patterns for three different discharge ratios, and the results indicate that the tributary flow influences the opposing bank when it becomes adequately dominant. This causes a recirculating eddy in the upstream confluence channel, which changes the incoming velocity distribution and consequently creates stronger helical cells at the downstream channel, along with upwelling flow regions. In turn, the changing flow patterns affect the mixing layer and flow recovery. Finally, the intermittence of the stronger upwelling flow is characterized.

Yang et al. (2013) developed a numerical simulation to track the surface position hydrodynamics of open channel confluences. A new mesh technique was employed to follow the free surface position, and the results indicate that the simulation adopted exhibits strong agreement with the experimental data. Furthermore, the impact of tracing the free-surface boundary was investigated using the different turbulence models combined with dynamic meshes, a rigid lid, or the volume of the fluid method. The free surface position, vector field, and velocity distribution were simulated and compared to the data collected in the flume test, and the results of the numerical simulations for confluence flow presented higher accuracy associated with the dynamic mesh method than the volume of fluid or rigid-lid methods.

Sirdari (2013) applied an application of a 3D numerical model of bedload transport in small river confluences of the Ara and Kurau rivers in Perak state, west of Malaysia. The Sediment Simulation in Intakes with Multiblock option (SSIIM2) program, which is associated with the CFD software, was used for modeling bedload transport and morphology, and the results indicate a strong agreement of the measured data, such as bedload, bed level, and water level, with the computed data. Furthermore, certain results confirmed similar findings to those of the previous researcher, such as the shear layer being found in the middle of the confluence when the momentum ratio (Mr) is less than 1, and the bedload being mostly traveled near the shear layer edge on the left side of the downstream confluence. Moreover, the bedload transport during high flow (flood level) was found to increase, as expected, but not remarkable regarding the amount, which is attributed to the fact that a large proportion of bed sediment loads are transported in suspension rather than in the bedload. However, the authors recommended further studies on the different sizes and shapes of natural river confluences with suspended sediment transport.

8.2.7.4   Laboratory Experimental Studies in Branching Channels

The physical models for branching junctions have been investigated in different geometry, flow conditions, and boundaries under the rigid and movable bed, depending on the purpose of the study. For instance, most of the laboratory studies achieved at branching angle of 90° (Herrero et al., 2015; Kubit and Ettema, 2001; Neary and Odgaard, 1993) while other studies under different angles were conducted by Al Omari et al. (2018) and Lama et al. (2002) in order to study the effect of the branching angle on the flow or to investigate the effect of the other variables on the flow in different angles.

Ramamurthy et al. (1996) investigated the various energy loss coefficients and contraction coefficients for a range of discharge ratios and three conduit width ratios. Also, bed condition plays a significant role in the branching flow studies (rough or smooth, movable or rigid boundary).

Kerssens and Van Urk (1986) studied the sediment transport in the branch channel, the physical model was built as a movable sand bed in order to study the effect of branching flow on the bed and water level at the up and downstream of channel junction.

Al Omari et al. (2018) implemented a laboratory study with different branching angles of 30°, 45°, 60°, 75°, and 90°, and bed with ratios of 30%, 40%, and 50% under five total discharge of

7.25, 8.5, 9.75, 11, and 12.25 L/s with movable bed material of 0.4 mm. The results indicated that the branching angle of 30° gave the best solution in terms of increasing the discharge through the branching channel, reduced the sediment concentration, and decreasing the scour depth by 46.7% compared with sour depth for the 90° branching angle.

8.2.7.5   Field Studies in Branching Channels

Wang et al. (1996) studied two projects in Rivers Rock and Cedar, USA, using submerged vanes as a tool to control the branching sediment in order to evaluate the performance of these vanes. The bed topography has been compared before and after installing the vanes. Also, a comparison with a theoretical model of the vanes was utilized early by Odgaard and Wang (1991). They found that the vanes are an effective way to control the sediment in branching river systems when the flow is small enough, and the velocity in the branching channel is less than 20% of the main flow velocity upstream

Nakato and Ogden (1998) investigated five hydraulic models to reduce the branching load sediment entering the branching channel in the case of a sand bed river along the Missouri River, between Sioux City and Lows St., Missouri. In order to restore the wetlands located in Plaquemines Parish, Louisiana on the west bank of the Mississippi River.

Michell et al. (2006) employed the submerged vanes and a skimming wall to resolve sediment accumulation at a thermal power station intake due to its withdrawing water from an alluvial river.

Miller (2004) implemented a design project and constructed a branching in the bank to help to branch water and sediment from the Mississippi River to the wetlands.

8.2.7.6   Numerical Studies in Branching Channels

Ghostine et al. (2013) studied the comparison between 1D and 2D mathematical simulations for a 90° branching channel with super-, trans-, and subcritical flow. The simulation results were validated with the previous experimental data. For the 1D simulation, they considered that the lateral flow is flowing over a zero-height side weir, while for the 2D simulation they used 2D St. Venant equations. The numerical performed the two approaches with a second-order Runge-Kutta discontinuous Galerkin (RKDG) scheme. The results of the 2D approach gave results similar to the experimental data for all types of flow, while the simulation of the 1D approach showed satisfactory results for the subcritical flow and became increasingly significant for the trans- and supercritical flow.

Omidbeigi et al. (2009) studied velocity, bed shear stresses, and turbulence flow in the branching channel. The study was implemented with the different turbulence methods and showed that using the κ-ω sub-model was better than using the κ-ε sub-model. Each sub-model has advantages over the other and can be concluded that there was no universal turbulence sub-model that can be used for solving all of the branching flow conditions.

Meselhe et al. (2016) used the 3D modeling software of Delft 3D to simulate water and sediment flow and morphology for a proposed sediment diversion from the lower Mississippi River, USA. The results showed that the sedimentation in the main flow depends on the discharge ratio and branching sediment amount.

Yuill et al. (2016) also used the delft 3D model to simulate hydro-morpho dynamics in the constructed branching channel from the west bank of the Mississippi River, USA. In order to evaluate the morphological development during the initial construction between 2004 to 2014.

8.2.8   Numerical Model

Modeling is a tool used by engineers to simulate complex hydro-morphological dynamics phenomena such as scouring and deposition zones that are usually formed at river systems (Bazrkar et al., 2017). Most of the available 2D models were designed to simulate many flow cases including river junctions. But the method of meshing of such models can only be done by using rectilinear and curvilinear methods. These types of meshing could affect the simulation results of the complicated shape like the flow in river junctions. However, the Mflow_02 model is a tool with a flexible method of meshing use to simulate the unsteady flow of river junctions. The original version of Mflow_02 was based on the program developed by Tomitokoro et al. (1985), and recently the program of the solver was improved by iRIC (2014). The improvement included adding certain functions, such as a moving boundary model and riverbed variation calculation. Thus, the program can calculate 2D plane unsteady flow and riverbed variation by using the unstructured meshes of the finite element method in an orthogonal coordinate system (Cartesian coordinate system). Subsequent development resulted in the model being able to reproduce exactly the structure and shape of complicated landforms, particularly in distributaries and confluences which can give more accurate results (Ali et al., 2019a; Nones et al., 2018).

8.2.8.1 Flow Field Model

The characteristics of the flow field model are as follows:

1. A Galerkin finite element sub-model (a type of weighted residual method) is used to discretize continuity and momentum equations.
   - Continuity equation:

$$\frac{\partial h}{\partial t} + \frac{\partial (hu)}{\partial x} + \frac{\partial (hv)}{\partial y} = 0 \tag{8.1}$$

   - Momentum equations:

$$\frac{\partial (uh)}{\partial t} + u\frac{\partial (uh)}{\partial x} + v\frac{\partial (vh)}{\partial y} - f(vh)$$

$$= -gh\frac{\partial H}{\partial x} + 2\frac{\partial}{\partial x}v_{xx}\frac{\partial (uh)}{\partial x} + \frac{\partial}{\partial y}v_{xy}\frac{\partial (uh)}{\partial y} \tag{8.2}$$

$$+ \frac{\partial}{\partial y}v_{xy}\frac{\partial (vh)}{\partial x} + \frac{\tau_s}{\rho} - \frac{\tau_{bx}}{\rho} - \frac{\tau_{tx}}{\rho}$$

$$\frac{\partial (vh)}{\partial t} + u\frac{\partial (vh)}{\partial x} + v\frac{\partial (vh)}{\partial y} + f(uh)$$

$$= -gh\frac{\partial H}{\partial y} + \frac{\partial}{\partial x}v_{xy}\frac{\partial (vh)}{\partial x} + \frac{\partial}{\partial x}v_{xy}\frac{\partial (uh)}{\partial y} \tag{8.3}$$

$$+ 2\frac{\partial}{\partial y}v_{yy}\frac{\partial (vh)}{\partial x} + \frac{\tau_s}{\rho} - \frac{\tau_{by}}{\rho} - \frac{\tau_{ty}}{\rho}$$

In which:

$$\frac{\tau_{bx}}{\rho} = C_f u\sqrt{u^2 + v^2} \qquad \frac{\tau_{by}}{\rho} = C_f v\sqrt{u^2 + v^2} \tag{8.4}$$

$$\frac{\tau_{tx}}{\rho} = C_t u\sqrt{u^2 + v^2} \qquad \frac{\tau_{ty}}{\rho} = C_t v\sqrt{u^2 + v^2} \tag{8.5}$$

Where:
t: time

u, v: flow velocity component of x, y direction

g: gravity acceleration

h: depth

H: water level (depth+ground elevation: h + z0)

f: Coriolis parameter

$v_{xx}, v_{xy}, v_{yx}, v_{yy}$: cinematic eddy viscosity

τ_{bx}, τ_{by} : bottom shear stress component of x, y direction

τ_{tx}, τ_{ty} : vegetation shear stress component of x, y direction

τ_s : shear stress on water surface

C_f: riverbed friction coefficient

C_t: coefficient of vegetation resistance

ρ: water density

- Discretization of basic equations:

$$h = N_i\,h_i$$

$$u = N_i\,u_i \tag{8.6}$$

$$v = N_i\,v_i$$

The basic equations (8.1) to (8.3) are discretized in space variable due to the Galerkin finite element method, a kind of weighted residual method. The definition is shown below, using linear triangular prism element N_i as shape function:

$$N_i = a_i + b_i x + c_i y \tag{8.7}$$

$$a_i = \left(x_j y_k - x_k y_j\right)2s$$

$$b_i = \left(y_j - y_k\right)/2s \tag{8.8}$$

$$c_i = \left(x_k - x_j\right)/2s$$

Where i, j, k are the vertexes of the triangular element, and s is the area of triangular in i, j, k.

By substituting the shape function defined in equation (8.6) into equations (8.1) to (8.3), and multiple N_i as the weight function, it is integrated it into the definition domain of the weight function.

$$\frac{\partial h_j}{\partial t}\int_s N_i N_j\,ds + \int_s N_i\left(N_j u_j\frac{\partial N_k}{\partial x}h_k + \frac{\partial N_j}{\partial x}u_j N_k h_k\right)ds$$

$$+\int_s N_i\left(N_j v_j\frac{\partial N_k}{\partial y}h_k + \frac{\partial N_j}{\partial y}v_j N_k h_k\right)ds = 0 \tag{8.9}$$

$$\frac{\partial u_j}{\partial t} \int_s N_i N_j \, N_k h_k \, ds$$

$$+ \int_s N_i \left\{ N_j u_j \frac{\partial N_k}{\partial x} u_k N_l h_l \right.$$

$$\left. + N_k u_k \left(\frac{\partial N_j}{\partial x} u_j N_l h_l + N_j u_j \frac{\partial N_l}{\partial x} h_l \right) \right\} ds$$

$$+ \int_s N_i \left\{ N_j v_j \frac{\partial N_k}{\partial y} u_k N_l h_l \right.$$

$$\left. + N_k u_k \left(\frac{\partial N_j}{\partial y} v_j N_l h_l + N_j v_j \frac{\partial N_l}{\partial y} h_l \right) \right\} ds \qquad (8.10)$$

$$= -g \int_s N_i N_j \, h_j \frac{\partial N_k}{\partial x} \left(h + z \right)_k ds$$

$$+ n_x \int_l N_i \left(2 v_{xx} \frac{\partial N_j}{\partial x} u_j + v_{yx} \frac{\partial N_j}{\partial y} v_j \right) N_k h_k \, dl$$

$$+ n_y \int_l N_i v_{yx} \frac{\partial N_j}{\partial y} u_j \, N_k h_k \, dl$$

$$- \int_l \left(2 v_{xx} \frac{\partial N_i}{\partial x} \frac{\partial N_j}{\partial x} u_j + v_{yx} \frac{\partial N_i}{\partial x} \frac{\partial N_j}{\partial y} v_j \right.$$

$$\left. + v_{yx} \frac{\partial N_i}{\partial x} \frac{\partial N_j}{\partial y} u_j \right) N_k h_k \, ds + \int_l N_i \left(\frac{\tau_s}{\rho} - \frac{\tau_{bx}}{\rho} - \frac{\tau_{tx}}{\rho} \right) ds$$

$$\frac{\partial v_j}{\partial t} \int_s N_i N_j \, N_k h_k \, ds$$

$$+ \int_s N_i \left\{ N_j u_j \frac{\partial N_k}{\partial x} v_k N_l h_l \right.$$

$$\left. + N_k v_k \left(\frac{\partial N_j}{\partial x} u_j N_l h_l + N_j u_j \frac{\partial N_l}{\partial x} h_l \right) \right\} ds \qquad (8.11)$$

$$+ \int_s N_i \left\{ N_j v_j \frac{\partial N_k}{\partial y} v_k N_l h_l \right.$$

$$\left. + N_k u_k \left(\frac{\partial N_j}{\partial y} v_j N_l h_l + N_j v_j \frac{\partial N_l}{\partial y} h_l \right) \right\} ds$$

$$= -g \int_s N_i N_j \, h_j \frac{\partial N_k}{\partial y} \left(h + z \right)_k ds$$

$$+ n_x \int_l N_i v_{yx} \frac{\partial N_j}{\partial y} v_j \, N_k h_k \, dl$$

$$(8.11)$$

$$+ n_y \int_l N_i \left(v_{xy} \frac{\partial N_j}{\partial x} u_j + 2 v_{yy} \frac{\partial N_j}{\partial y} v_j \right) N_k h_k \, dl$$

$$- \int_s \left(v_{xy} \frac{\partial N_i}{\partial x} \frac{\partial N_j}{\partial x} u_j + v_{xy} \frac{\partial N_i}{\partial x} \frac{\partial N_j}{\partial y} u_j \right.$$

$$\left. + 2 v_{yy} \frac{\partial N_i}{\partial y} \frac{\partial N_j}{\partial y} v_j \right) N_k h_k \, ds + \int_s N_i \left(\frac{\tau_s}{\rho} - \frac{\tau_{by}}{\rho} - \frac{\tau_{ty}}{\rho} \right) ds$$

$$i = i, j, k \quad j = i, j, k \quad k = i, j, k \quad l = i, j, k$$

l: the edge length of the triangle element

n_x, n_y: x, y components of unit normal vector which create the boundary of the triangular element.

It's better to consider the boundaries only, in order to cancel all the terms, including n_x, n_y inside triangular elements. In addition, each factor can be easily calculated by the formula as below:

$$\int_s N_i^a N_j^b N_k^c \, ds = \frac{a! \, b! \, c!}{(a+b+c+2)!} 2s \qquad (8.12)$$

$$\int_l N_i^a N_j^b \, dl = \frac{a! \, b!}{(a+b+1)!} 2l \qquad (8.13)$$

2. Open boundary conditions (upstream and downstream boundaries, etc.), enable the setting up of various conditions, such as time series of flow discharge, time series of water level, and together, water level with discharge.
3. The friction of the river bed can be set by using the Manning roughness coefficient. This coefficient can be represented in the model as a polygon for the entire area or for each element (cell), thus providing spatial distribution of roughness. The following equation expresses the riverbed friction coefficient (equation 8.14).

$$C_f = \frac{g \, n^2}{h^{1/3}} \qquad (8.14)$$

4. Three sub-models are available for computing the turbulence field (flows with large and small eddies). These are the zero equation sub-model, simple k-ε, and direct input of kinematic eddy viscosity. Ali et al. (2019a, 2019b) found that using a zero equation sub-model was more stable among other methods in calculating the large and small eddies of the flow in river junctions. The kinematic eddy viscosity, υ is expressed as the product of the turbulence representative velocity v_t and the representative length l.

$$v = v_t \, l \qquad (8.15)$$

In the flow fields where the water depth and roughness are changing gradually in the cross-sections of the river junctions, the kinematic eddy viscosity in horizontal and vertical directions is assumed to be in the same order, which is mainly considering the momentum transport specified by bottom friction velocity, u_* and water depth, h. The kinematic eddy viscosity, v can be expressed as below:

$$u_* = n\sqrt{g(u^2 + v^2)} \Big/ h^{\frac{1}{6}} \qquad (8.16)$$

$$v = a u_* h \qquad (8.17)$$

where a is a proportional constant.

The value of a is related to the momentum transport in the vertical direction, according to experiments by Fischer (1973) and Webel and Schatzmann (1984), is around 0.07. The eddy viscosity coefficient v is expressed by using the von Kàrmàn coefficient k (0.4).

$$v = \frac{k}{6} u_* h \qquad (8.18)$$

5. Other effects, such as the effect of wind on the water surface, are also available in this solver.

8.2.8.2 Riverbed Variation Model

The characteristics of riverbed variation are as follows:

1. The riverbed variation associated with the flow field model is calculated. This model can calculate the flow field only or together with riverbed variation.
2. The riverbed material can be selected from the uniform and mixed grain diameters. If a mixed grain diameter is selected, then a variation in grain distribution can be assumed for the deep directions and multiple layers. The accumulation curve is divided into n hierarchies, representative grain size diameter d_k, and the possibility of grain size existence P_k (Figure 8.13). Median diameter d_m is defined by the following equation:

$$d_m = \sum_{k=1}^{n} P_k d_k \qquad (8.19)$$

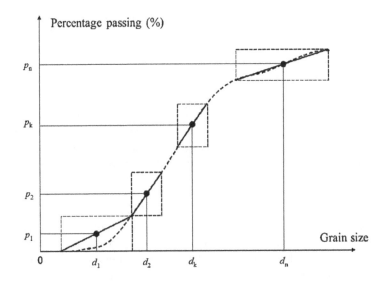

FIGURE 8.13 Handling of grain size distribution.

In addition, it is necessary to consider the shelter effect when calculating the non-dimensional critical tractive force of each grain diameter. The Ashida-Michiue formula (1971) is employed as a modification of the Egiazaroff formula (1965).

$$\frac{\partial \tau_{ck}}{\partial \tau_{cm}} = \begin{cases} 0.85 & \dfrac{d_k}{d_m} \leq 0.4 \\[2ex] \left\{ \dfrac{\log_e(19)}{\log_e\left(19\,d_k/d_m\right)} \right\} \dfrac{d_k}{d_m} & \dfrac{d_k}{d_m} > 0.4 \end{cases} \tag{8.20}$$

τ_{ck} : non-dimensional critical tractive force of each grain diameter
τ_{cm} : non-dimensional critical tractive force of central grain diameter
The formula below presents the conservation of volume for all of the sediments of entire grain diameter:

$$(1-\lambda)\left(\frac{\partial E_b}{\partial t} + \frac{\partial z}{\partial t}\right) + \frac{\partial}{\partial x}\left(\sum_{k=1}^{n} p_k\, q_{xk}\right) + \frac{\partial}{\partial y}\left(\sum_{k=1}^{n} p_k\, q_{yk}\right)$$

$$- \sum_{k=1}^{n}\left(E_{sk} - C_{ak}\right)w_{0k} = 0 \tag{8.21}$$

E_b: thickness of exchange layer
q_{xk}, q_{yk} : (x, y) direction component of sediment discharge of each grain diameter

E_{sk}: hoisting speed of each grain diameter
C_{ak}: concentration of suspended sediment at the referential level of each grain diameter
W_{0k}: sedimentation speed of each grain diameter

3. Three methods for calculating the total bedload q_b of depth-averaged flow velocity are available in the Mflow_02 solver, and these are the Meyer-Peter and Müller formula (1948), the Ashida and Michiue formula (1972), and the Engrlund-Hansen formula (1974). The Meyer-Peter and Müller formula was expressed as an example for computing the total bedload:

$$q_b = 8\sqrt{\left(\frac{\sigma}{\rho} - 1\right)g\,d^3}\ \left(\tau_*' - \tau_{*c}\right)^{1.5} \tag{8.22}$$

d: grain diameter, ρ: water density, σ: gravel density, τ_{*c} : critical tractive force (calculated by Iwagaki formula, 1956), τ_*' : calculated by Kishi and Kuroki in the formula below:

$$\frac{\bar{U}}{u_*} = \begin{cases} 7.66\left(\dfrac{h}{2d}\right)^{1/6}\left(\dfrac{\tau_*'}{\tau_*}\right)^{2/3} & \dfrac{h}{2d} < 500 \\[3ex] 11.59\left(\dfrac{h}{2d}\right)^{1/10}\left(\dfrac{\tau_*'}{\tau_*}\right)^{3/5} & \dfrac{h}{2d} \geq 500 \end{cases}$$

$$\tag{8.23}$$

$$\tag{8.24}$$

$\bar{U}$: Vertical average flow velocity in the flow direction
The total sediment discharge set by the Meyer-Peter and Müller formula is converted to the normal direction (n) and tangential direction (s) of streamline, in consideration for

the effect of secondary flow and riverbed slope which is caused by streamline curvature of depth-averaged flow velocity (Watanabe et al., 2001):

$$q_s = q_b \left(\frac{v_b}{V_b} - \sqrt{\frac{\tau_{*c}}{\mu_s \, \mu_k \, \tau_*}} \, \frac{\partial z}{\partial s} \right) \tag{8.25}$$

$$q_n = q_b \left(\frac{u_b}{V_b} - \sqrt{\frac{\tau_{*c}}{\mu_s \, \mu_k \, \tau_*}} \, \frac{\partial z}{\partial n} \right) \tag{8.26}$$

q_s: (s) direction component of sediment discharge near riverbed
q_n: (n) direction component of sediment discharge near riverbed
V_b: absolute value of velocity near riverbed
v_b: (s) direction component of flow velocity near riverbed
u_b: (n) direction component of flow velocity near riverbed
μ_s: static friction factor
μ_k: kinetic friction factor
z: height of riverbed

The flow velocity near the riverbed is calculated by streamline curvature of the depth-averaged flow velocity as below:

$$\frac{1}{r} = \frac{1}{\left(u^2 + v^2 \right)^{3/2}} \left\{ u \left(u \frac{\partial v}{\partial x} - v \frac{\partial u}{\partial x} \right) + v \left(u \frac{\partial v}{\partial y} - v \frac{\partial u}{\partial y} \right) \right\} \tag{8.27}$$

r: curvature radius of streamline

4. The scour limit of the riverbed, secondary flow coefficient, and morphological factor can be set accordingly.

8.3   CONCLUSIONS

River junctions are considered to exhibit a very complex flow, owing to different characteristics such as discharge, velocity, sediment properties, and river junctions geometry. The main controlling factors affecting a river junction are the discharge ratio and junction angle. Numerous features produced at the junctions are related to hydrodynamics, such as separation zone, stagnation zone, and maximum velocity zone. On top of that, the other vital features are related to the bed morphology, such as the erosion at the scour hole and deposition in the separation zone. Several researchers have demonstrated the results and the scales of discharge ratio and junction angle influences on the hydro-morpho dynamics features in the laboratory, and in general, increasing these factors leads to enhancing the problems of scouring and deposition zones at a river junction. However, this chapter has highlighted the flow and morphological characteristics for different cases under different geometry and boundary conditions. Furthermore, many of the junctions' flows have been reviewed in the laboratory, natural, physical, and numerical studies, and found that the main critical issues are the erosion and deposition occurring at the river junctions which continuously reduce the hydraulic capacity of rivers.

The main consequences of the scouring and deposition are restriction in navigation, flooding, and the effect on the safety of structures constructed near river banks. Also, the related studies were mainly the parametric type for prediction, characterization, and simulation of scouring and deposition zones. Only a few studies on controlling and managing these morphological features were focused on using submerged vanes in the different strategies (Odgaard and Wang, 1991; Neill et al., 1997; Michell et al., 2006; Beygipoor et al., 2013; Wuppukondur and Chandra, 2017a, 2017b).

Disadvantages of using submerged vanes are limited to discharge ratio, producing non-uniform velocity distribution due to its numbers (Barkdoll et al., 1999), and affecting the navigation in rivers.

On the other hand, the dimensions, arrangement, and inclination angle of the rehabilitation structures that can be used to control the deposition and scouring zones at the river junctions were proposed by Odgaard and Wang (1991), Mirzaei et al. (2014), Wuppukondur and Chandra (2017a, 2017b), and others. But there was a lack of studies on the best effective dimensions, arrangement, and inclination angle. Alternatively, rehabilitation works can be used, but these works are costly since dredging equipment should be used to remove and maintain sediment accumulation repeatedly from time to time. Thus, managing hydro-morpho dynamics at the channel junction is still considered challenging (Ali et al., 2019a, 2019b; Kalathil et al., 2018; Wuppukondur and Chandra, 2017a).

Although many previous investigations were made on river junctions, it is important to continue doing research in order to find suitable solutions to the problems associated with the river junctions. The new studies should focus on gaps that are not covered previously and can be summarized as:

- Studies on controlling and managing the flow dynamic with a movable bed in rivers with confluence and where branching is limited to submerged vanes.
- Studies on numerical simulation of rivers' confluences with the various scenarios that consider using various arrangements of control structures.
- Studies on introducing obstacles in rivers' junctions to control erosion at the scour hole and deposition at the separation zone and the related criteria that give the best location and configuration of the obstacles.

REFERENCES

Ali, H.L., Yusuf, B., Mohammed, T.A., Shimizu, Y., Ab Razak, M.S., and Rehan, B.M. 2019a. Enhancing the flow characteristics in a branching channel based on a two-dimensional depth-averaged flow model, *Water*, 11(9): 1863.

Ali, H.L., Yusuf, B., Mohammed, T.A., Shimizu, Y., Ab Razak, M.S., and Rehan, B.M. 2019b. Improving the hydro-morpho dynamics of a river confluence by using vanes, *Resources*, 8(1): 9.

Allahyonesi, H., Omid, M.H., and Haghiabi, A.H. 2008. A study of the effects of the longitudinal arrangement sediment behavior near intake structures, *Journal of Hydraulic Research*, 46(6): 814–819.

Alomari, N.K., Yusuf, B., Mohammad, T.A., and Ghazali, A.H. 2018. Experimental investigation of scour at a channel junctions of different diversion angles and bed width ratios, *CATENA*, 166: 10–20.

Ashida, K., and Michiue, M. 1972. Study on hydraulic resistance and bedload transport rate in alluvial streams. In: *Proceedings of the Japan Society of Civil Engineers, Japan*, pp. 59–69.

Ashida, K., and Michiue, M. 1971. Studies on bedload transportation for nonuniform sediment and river bed variation, *Disaster Prevention Research Institute Annuals*, 14.

Ashmore, P.E., Ferguson, R.I., Prestegaard, K.L., Ashworth, P.J., and Paola, C. 1992. Secondary flow in anabranch confluences of a braided, gravel-bed stream, *Earth Surface Processes and Landforms*, 17: 299–311, https://doi.org/10.1002/esp.3290170308.

Ashmore, P.E., and Gardner, J.T. 2008. Unconfined confluences in braided rivers. In: Rice, S.P., and Rhoads, B. (Eds.), *River Confluences, Tributaries and the Fluvial Network*, John Wiley and Sons, Chichester, UK.

Barkdoll, B.D. 2004. Discussion of "subcritical 90° equal width open channel dividing flow" by Chung-Chieh Hsu, Chii-Jau Tang, Wen-Jung Lee, and Mon-Yi Shieh, *Journal of Hydraulic Engineering*, 130: 171–172.

Barkdoll, B.D., Ettema, R., and Odgaard, A.J. 1999. Sediment control at lateral diversions: Limits and enhancements to vane use, *Journal of Hydraulic Engineering*, 125(8): 862–870.

Bazrkar, M.H., Adamowski, J., and Eslamian, S. 2017. Water system modeling. In: Furze, J.N., Swing, K., Gupta, A.K., McClatchey, R., and Reynolds, D. (Eds.), *Mathematical Advances Towards Sustainable Environmental Systems*, Springer International Publishing, Switzerland, pp. 61–88.

Best, J.L. 1986. The morphology of river channel confluences, *Progress in Physical Geography*, 10: 157–174.

Best, J.L. 1987. *Flow Dynamics at River Channel Confluences: Implications for Sediment Transport and Bed Morphology. Recent Developments in Fluvial Sedimentology*, SEPM (Society for Sedimentary Geology), pp. 27–35, https://doi.org/10.2110/pec.87.39.0027.

Best, J.L. 1988. Sediment transport and bed morphology at river channel confluences, *Sedimentology*, 35: 481–498, https://doi.org/10.1111/j.1365-3091.1988.tb00999.x.

Best, J.L., and Ashworth, P.J. 1997. Scour in large braided rivers and the recognition of sequence stratigraphic boundaries, *Nature*, 387(6630): 275–277, https://doi.org/10.1038/387275a0.

Best, J.L., and Reid, I. 1984. Separation zone at open-channel junctions, *Journal of Hydraulic Engineering*, 110(10): 1588–1594, https://doi.org/10.1061/(ASCE)0733-9429(1984)110:11(1588).

Best, J.L., and Rhoads, B.L. 2008. Sediment transport, bed morphology and the sedimentology of river channel confluences. In: *River Confluences, Tributaries and the Fluvial Network*, John Wiley & Sons, Ltd., Chichester, UK, pp. 45–72, https://doi.org/10.1002/9780470760383.ch4.

Best, J.L., and Roy, A.G. 1991. Mixing-layer distortion at the confluence of channels of different depth, *Nature*, 350(6317): 411–413, https://doi.org/10.1038/350411a0.

Beygipoor, G., Bajestan, M.S., Kaskuli, H.A., and Nazari, S. 2013. The effects of submerged vane angle on sediment entry to an intake from a 90 degree converged bend, *Advances in Environmental Biology*, 7(9): 2283–2292.

Birjukova, O., Guillen, S., Alegria, F., and Cardoso, A. 2014. Three dimensional flow field at confluent fixed-bed open channels, *Proc. River Flow*, http://infoscience.epfl.ch/record/202026/files/2014-989.

Biron, P.M., Ramamurthy, A.S., Asce, F., and Han, S. 2004. Three-dimensional numerical modeling of mixing at river confluences, *Journal of Hydraulic Engineering*, 130(3): 243–253, https://doi.org/10.1061/(ASCE)0733-9429(2004)130:3(243).

Biron, P.M., Richer, A., Kirkbride, A.D., Roy, A.G., and Han, S. 2002. Spatial patterns of water surface topography at a river confluence, *Earth Surface Processes and Landforms*, 27(9): 913–928, https://doi.org/10.1002/esp.359.

Biron, P., Roy, A., Best, J.L., and Boyer, C.J. 1993. Bed morphology and sedimentology at the confluence of unequal depth channels, *Geomorphology*, 8(2–3): 115–129, https://doi.org/10.1016/0169-555X(93)90032-W.

Boyer, C., Roy, A.G., and Best, J.L. 2006. Dynamics of a river channel confluence with discordant beds: Flow turbulence, bedload sediment transport, and bed morphology, *Journal of Geophysical Research: Earth Surface*, 111(4): 1–22, https://doi.org/10.1029/2005JF000458.

Bradbrook, K.F., Biron, P.M., Lane, S.N., Richards, K.S., and Roy, A.G. 1998. Investigation of controls on secondary circulation in a simple confluence geometry using a three dimensional numerical model, *Hydrological Processes*, 1396: 1371–1396, https://doi.org/10.1002/(SICI)1099-1085(19980630)12:8<1371::AID-HYP620>3.0.CO;2-C.

Bradbrook, K.F., Lane, S.N., and Richards, K.S. 2000. Numerical simulation of three-dimensional, time-averaged flow structure at river channel confluences, *Water Resources Research*, 36(9): 2731–2746, https://doi.org/10.1029/2000WR900011.

Bryan, R.B., and Kuhn, N.J. 2002. Hydraulic conditions in experimental rill confluences and scour in erodible soils, *Water Resources Research*, 38(5), https://doi.org/10.1029/2000WR000140.

Bulle, H. 1926. Untersuchungen über die geschiebeableitung bei der spaltung von Wasserläufen, *Forschungsarbeiten auf dem Gebiete des Ingenieurwesens*, 282: 57–84 (in German).

Casas, A.H. 2013. Experimental and theoretical analysis of flow and sediment transport in 90-degree fluvial diversions. Doctoral dissertation, University Politecnica of Catalunya, Barcelona, Spain.

Constantinescu, G., Miyawaki, S., Rhoads, B., Sukhodolov A., and Kirkil, G. 2011. Structure of turbulent flow at a river confluence with momentum and velocity ratios close to 1: Insight provided by an eddy-resolving numerical simulation, *Water Resources Research*, 47: W05507, https://doi.org/10.1029/2010WR010018.

Dehghani, A.A., Ghodsian, M., Suzuki, K., and Alaghmand, S. 2009. Local scour around lateral intakes in 180 degree curved channel. In: *Advances in Water Resources and Hydraulic Engineering: Proceedings of 16th IAHR-APD Congress and 3rd Symposium of IAHR-ISHS*. Springer Berlin Heidelberg, Berlin, Heidelberg, pp. 821–825.

Egiazaroff, I.V. 1965. Calculation of nonuniform sediment concentrations, *Journal of the Hydraulics Division*, 91(4): 225–247.

Engelund, F. 1974. Flow and bed topography in channel bends, *Journal of the Hydraulics Division*, 100(11): 1631–1648.

Ettema, R. 2008. River confluences, tributaries and the fluvial network - Google books. In: *Management of Confluences*, John Wiley & Sons, Chichester, UK, pp. 93–118.

Fischer, Hugo B. 1973. Longitudinal dispersion and turbulent mixing in open-channel flow, *Annual Review of Fluid Mechanics*, 5(1): 59–78.

Gaudet, J.M., and Roy, A.G. 1995. Effect of bed morphology on flow mixing length at river confluences, *Nature*, https://doi.org/10.1038/373138a0.

Ghostine, R., Vazquez, J., Terfous, A., Rivière, N., Ghenaim, A., and Mosé, R. 2013. A comparative study of 1D and 2D approaches for simulating flows at right angled dividing junctions, *Applied Mathematics and Computation*, 219(10): 5070–5082.

Goudarzizadeh, R., Hedayat, N., and Jahromi, S.M. 2010. Three-dimensional simulation of flow pattern at the lateral intake in straight path, using finite-volume method, *World Academy of Science, Engineering and Technology*, 47: 656–661.

Grace, J.L., and Priest, M.S. 1958. *Division of Flow in Open Channel Junctions*, Engineering Experiment Station, Alabama Polytechnic Institute, USA.

Guillén-Ludeña, S., Franca, M.J., Cardoso, A.H., and Schleiss, A.J. 2015. Hydro-morphodynamic evolution in a 90° movable bed discordant confluence with low discharge ratio, *Earth Surface Processes and Landforms*, 40: 1927–1938, https://doi.org/10.1002/esp.3770.

Guillén-Ludeña, S., Franca, M.J., Cardoso, A.H., and Schleiss, A.J. 2016. Evolution of the hydromorphody-namics of mountain river confluences for varying discharge ratios and junction angles, *Geomorphology*, 255: 1–15, https://doi.org/10.1016/j.geomorph.2015.12.006.

Hager, W.H. 1992. Discussion of "Dividing flow in open channels" by amruthur S. ramamurthy, duc minh tran, and luis B. carballada (march, 1990, vol. 116, no. 3), *Journal of Hydraulic Engineering*, 118(4): 634–637.

Herrero, A., Bateman, A., and Medina, V. 2015. Water flow and sediment transport in a 90° channel diversion: An experimental study, *Journal of Hydraulic Research*, 53(2): 253–263.

Herrero, H.S., García, C.M., Pedocchi, F., López, G., Szupiany, R.N., and Pozzi-Piacenza, C.E. 2016. Flow structure at a confluence: Experimental data and the bluff body analogy, *Journal of Hydraulic Research*, 54(3): 263–274.

Herrero Casas, A. 2013. Experimental and theoretical analysis of flow and sediment transport in 90-degree fluvial diversions, Doctoral Dissertation, Universitat Politècnica De Catalunya, Spain.

Huang, J., Weber, L.J., and Lai, Y.G. 2002. Three-dimensional numerical study of flows in open-channel junctions, *Journal of Hydraulic Engineering*, 128(3): 268–280, https://doi.org/10.1061/(ASCE)0733-9429(2002)128:3(268).

iRIC. 2014. *Mflow_02 Solver Manual*. Produced by Mineyuki Gamou, Released July 10, 2014, at http://i-ric.org/en/software/?c=19.

Ingle, R.N., and Mahankal, A.M. 1990. Discussion of "Division of flow in short open channel branches" by amruthur S. ramamurthy and mysore G. satish (April, 1988, vol. 114, no. 4), *Journal of Hydraulic Engineering*, 116(2): 289–291.

Iwagaki, Y. 1956. Hydrodynamical study on critical tractive force, *Journal of Japan Society of Civil Engineering*, 41: 1–21.

Kalathil, S.T., Wuppukondur, A., Balakrishnan, R.K., and Chandra, V. 2018. Control of sediment inflow into a trapezoidal intake canal using submerged vanes, *Journal of Waterway, Port, Coastal, and Ocean Engineering*, 144(6): 04018020.

Kerssens, P.J., and Van Urk, A. 1986. Experimental studies on sedimentation due to water withdrawal, *Journal of Hydraulic Engineering*, 112(7): 641–656.

Keshavarzi, A., and Habibi, L. 2005. Optimizing water intake angle by flow separation analysis, *Irrigation and Drainage*, 54: 543–552. http://dx.doi.org/10.1002/ird.207.

Kleinhans, M.G., Ferguson, R.I., Lane, S.N., and Hardy, R.J. 2012. Splitting rivers at their seams: Bifurcations and avulsion, *Earth Surface Processes and Landforms*, 38(1): 47–61.

Kubit, O., and Ettema, R. 2001. Debris-and ice-skimming booms at riverside diversions, *Journal of Hydraulic Engineering*, 127(6): 489–498.

Lama, S.K., Kudoh, K., and Kuroki, M. 2003. Study of flow characteristics of junction flow with free flow condition at branch channel, *Annual Journal of Hydraulic Engineering, JSCE*, 47: 601–606.

Lama, S.K., Kuroki, M., and Hasegawa, K. 2002. Study of flow bifurcation at the 30° open channel junction when the width ratio of branch channel to main channel is large, *Annual Journal of Hydraulic Engineering, JSCE*, 46: 583–588.

Lane, S.N., Bradbrook, K.F., Richards, K.S., Biron, P.M., and Roy, A.G. 1999. Time-averaged flow structure in the central region of a stream confluence: A discussion, *Earth Surface Processes and Landforms*, 24(4): 361–367, https://doi.org/10.1002/(SICI)1096-9837(199904)24:4<361::AID-ESP982>3.0.CO;2-5.

Lane, S.N., Bradbrook, K.F., Richards, K.S., Biron, P.M., and Roy, A.G. 2000. Secondary circulation cells in river channel confluences : Measurement artefacts or coherent flow structures ? *Hydrological Processes*, 14: 2047–2071, https://doi.org/10.1002/1099-1085(20000815/30)14:11/12<2047::aid-hyp54>3.0.co;2-4.

Leite Ribeiro, M., Blanckaert, K., Roy, A.G., and Schleiss, A.J. 2012. Flow and sediment dynamics in channel confluences, *Journal of Geophysical Research: Earth Surface*, 117(1), https://doi.org/10.1029/2011JF002171.

Liu, T., Chen, L., and Fan, B. 2013. Experimental study on flow pattern and sediment transportation at a 90 ° open-channel confluence, *International Journal of Sediment Research*, 27(2): 178–187, https://doi.org /10.1016/S1001-6279(12)60026-2.

Liu, T., Fan, B., and Lu, J. 2015. Sediment – Flow interactions at channel confluences: A flume study, *Advances in Mechanical Engineering*, 7(6): 1–9, https://doi.org/10.1177/1687814015590525.

Martín-Vide, J.P.P., Plana-Casado, A., Sambola, A., and Capapé, S. 2015. Bedload transport in a river confluence, *Geomorphology*, 250: 15–28, http://doi.org/10.1016/j.geomorph.2015.07.050.

McLelland, S., Ashworth, P., and Best, J. 1996. The origin and downstream development of coherent flow structures at channel junctions. Retrieved from http://eprints.brighton.ac.uk/id/eprint/11588.

Meselhe, E.A., Sadid, K.M., and Allison, M.A. 2016. Riverside morphological response to pulsed sediment diversions, *Geomorphology*, 270: 184–202.

Meyer-Peter, E., and Müller, R. 1948. Formulas for bed-load transport. In: *Proceedings of the 2nd Meeting of the International Association for Hydraulic Structures Research*, Stockholm, Sweden, vol. 2, pp. 39–64.

Michell, F., Ettema, R., and Muste, M. 2006. Case study: Sediment control at water intake for large thermal-power station on a small river, *Journal of Hydraulic Engineering*, 132(5): 440–449.

Mignot, E., Doppler, D., Riviere, N., Vinkovic, I., Gence, J.N., and Simoens, S. 2014. Analysis of flow separation using a local frame axis: Application to the open-channel bifurcation, *Journal of Hydraulic Engineering*, 140(3): 280–290.

Mignot, E., Zeng, C., Dominguez, G., Li, C.W., Rivière, N., and Bazin, P.H. 2013. Impact of topographic obstacles on the discharge distribution in open-channel bifurcations, *Journal of Hydrology*, 494: 10–19.

Miller, G. 2004. Mississippi river-west bay sediment diversion. In: *Critical Transitions in Water and Environmental Resources Management*, ASCE, Reston, VA, pp. 1–7.

Mirzaei, S.H.S., Ayyoubzadeh, S.A., and Firoozfar, A.R. 2014. The effect of submerged-vanes on formation location of the saddle point in lateral intake from a straight channel, *American Journal of Civil Engineering and Architecture*, 2(1): 26–33.

Moghadam, M.K., and Keshavarzi, A.R. 2010. An optimised water intake with the presence of submerged vanes in irrigation canals, *Irrigation and Drainage*, 59(4): 432–441.

Mosley, M. 1976. An experimental study of channel confluences. *The Journal of Geology*. Retrieved from http://www.jstor.org/stable/30066212.

Nakato, T., and Ogden, F.L. 1998. Sediment control at water intakes along sand-bed rivers, *Journal of Hydraulic Engineering*, 124(6): 589–596.

Nazari-giglou, A., and Jabbari-sahebari, A. 2014. An experimental study of sediment transport in channel confluences, *International Journal of Sediment Research*, https://doi.org/10.1016/j.ijsrc.2014.08.001.

Neary, V., Sotiropoulos, F., and Odgaard, A. 1999. Three-dimensional numerical model of lateral-intake inflows, *Journal of Hydraulic Engineering*, 125(2): 126–140.

Neary, V.S., and Odgaard, A.J. 1993. Three-dimensional flow structure at open-channel diversions, *Journal of Hydraulic Engineering*, 119(11): 1223–1230.

Neill, C.R., Evans, B.J., Odgaard, A.J., and Wang, Y. 1997. Discussion and closure: Sediment control at water intakes, *Journal of Hydraulic Engineering*, 123(7): 670–671.

Nones, M., Archetti, R., and Guerrero, M., 2018. Time-lapse photography of the edge-of-water line displacements of a sandbar as a proxy of riverine morphodynamics, *Water*, 10(5): 617.

Odgaard, A.J. 2009. *River Training and Sediment Management with Submerged Vanes*, ASCE Press, Virginia, USA.

Odgaard, A.J., and Spoljaric, A. 1986. Sediment control by submerged vanes, *Journal of Hydraulic Engineering*, 112(12): 1164–1180.

Odgaard, A.J., and Wang, Y. 1991. Sediment management with submerged vanes. I: Theory, *Journal of Hydraulic Engineering*, 117: 267–283.

Omidbeigi, M.A., Ayyoubzadeh, S.A., and Safarzadeh, A. 2009. Experimental and numerical investigations of velocity field and bed shear stresses in a channel with lateral intake. In: *33rd IAHR Congress*, Vancouver, Canada, pp. 1284–1291.

Ouyang, H.T., Lai, J.S., Yu, H., and Lu, C.H. 2008. Interaction between submerged vanes for sediment management, *Journal of Hydraulic Research*, 46(5): 620–627.

Parsons, D.R., Best, J.L., Lane, S.N., Kostaschuk, R.A., Hardy, R.J., and Orfeo, O. 2008. River confluences, tributaries and the fluvial network - Google books. In: *Large River Channel Confluences*, John Wiley & Sons, Chichester, pp. 73–91, https://doi.org/10.1002/9780470760383.

Parsons, D.R., Best, J.L., Stuart, N.L., Oscar Orfeo, Richard J., and Hardy, R.K. 2007. High spatial resolution data acquisition for the geosciences: Kite aerial photography, *Earth Surface Processes and Landforms*, 34: 155–161. https://doi.org/10.1002/esp.

Ramamurthy, A., Zhu, W., and Carballada, B. 1996. Dividing rectangular closed conduit flows, *Journal of Hydraulic Engineering*, 122(12): 687–691.

Ramamurthy, A.S., Qu, J., and Vo, D. 2007. Numerical and experimental study of dividing open-channel flows, *Journal of Hydraulic Engineering*, 133: 1135–1144, http://doi.org/10.1061/(ASCE)0733-9429(2007)133:10(1135).

Redolfi, M., Zolezzi, G., and Tubino, M. 2016. Free instability of channel bifurcations and morphodynamic influence, *Journal of Fluid Mechanics*, 799: 476–504.

Rezapour, S., Moghadam, K.F., and Omid Naceni, S.T. 2009. Experimental study of flow and sedimentation at 90° openchannel diversion. In: *Proceedings of the 33rd IAHR Congress*, Vancouver, Canada, pp. 2979–2986.

Rezaur, R.B., Jayawardena, A.W., and Hossain, M.M. 1999. Factors affecting confluence scour. In: Jayawardena, A.W., and Lee, J.H.W. (Eds.), *River Sedimentation: Theory and Applications*, Balkema, Rotterdam, The Netherlands, pp. 187–192.

Rhoads, B., and Kenworthy, S. 1995. Flow structure at an asymmetrical stream confluence, *Geomorphology*, 11(4): 273–293, http://www.sciencedirect.com/science/article/pii/0169555X94000694.

Rhoads, B.L. 1996. Mean structure of transport-effective flows at an asymmetrical confluence when the main stream is dominant. In: Ashworth, P.J., Bennett, J.S.J., Best, L., and McLelland, S.J. (Eds.), *Coherent Flow Structures in Open Channels*, Wiley, Chichester, UK, pp. 491–517.

Rhoads, B.L., Riley, J.D., and Mayer, D.R. 2009. Geomorphology response of bed morphology and bed material texture to hydrological conditions at an asymmetrical stream confluence, *Geomorphology*, 109(3–4): 161–173, https://doi.org/10.1016/j.geomorph.2009.02.029.

Rhoads, B.L., and Sukhodolov, A.N. 2004. Spatial and temporal structure of shear layer turbulence at a stream confluence, *Water Resources Research*, 40: 1–13.

Rhoads, B.L., and Sukhodolov, N. 2001. Field investigation of three-dimensional flow structure, *Water Resources Research*, 37(9): 2393–2410, https://doi.org/10.1029/2001WR000316.

Ribeiro, M.L., Blanckaert, K., Boillat, J., Schleiss, A.J., and Polytechnique, E. 2012. Hydromorphological implications of local tributary widening for river rehabilitation, *Water Resources Research*, 48(10): 1–19, https://doi.org/10.1029/2011WR011296.

Riley, J.D., Rhoads, B.L., Parsons, D.R., and Johnson, K.K. 2014. Influence of junction angle on three-dimensional flow structure and bed morphology at confluent meander bends during different hydrological conditions, *Earth Surface Processes and Landforms*, 40: 252–271, https://doi.org/10.1002/esp.3624.

Roberts, M. 2004. Flow dynamics at open channel confluent-meander bends. Retrieved from http://ethos.bl.uk/OrderDetails.do?uin=uk.bl.ethos.414170.

Roy, A., and Bergeron, N. 1990. Flow and particle paths at a natural river confluence with coarse bed material, *Geomorphology*, 3(2): 99–112, https://doi.org/10.1016/0169-555X(90)90039-S.

Roy, A.G., Roy, R., and Bergeron, N. 1988. Hydraulic geometry and changes in flow velocity at a river confluence with coarse bed material, *Earth Surface Processes and Landfomrs*, 13: 583–598, https://doi.org/10.1002/esp.3290130704.

Schindfessel, L., Creëlle, S., and De Mulder, T. 2015. Flow patterns in an open channel confluence with increasingly dominant tributary inflow, *Water*, 7(9): 4724–4751, https://doi.org/10.3390/w7094724.

de Serres, B., Roy, G., Biron, P.M., and Best, J.L. 1999. Three-dimensional structure of flow at a confluence of river channels with discordant beds, *Geomorphology*, 26(4): 313–335, https://doi.org/10.1016/S0169-555X(98)00064-6.

Seyedian, S.M., Bajestan, M.S., and Farasati, M. 2014. Effect of bank slope on the flow patterns in river intakes, *Journal of Hydrodynamics*, 26: 482–492, http://dx.doi.org/10.1016/S1001-6058(14)60055-X.

Shaheed, R. 2016. 3D numerical modelling of secondary current in shallow river bends and confluences. Master Thesis, Department of Civil Engineering, University of Ottawa, Ottawa, Canada.

Shakibainia, A., Reza, M., Tabatabai, M., and Zarrati, A.R. 2010. Three-dimensional numerical study of flow structure in channel confluences, *Canadian Journal of Civil Engineering*, 37(5): 772–781, https://doi.org/10.1139/L10-016.

Shamloo, H., and Pirzadeh, B. 2007b. Numerical investigation of velocity field in dividing open-channel flow. In: *Proceedings of the 12th WSEAS International Conference on APPLIED MATHEMATICS*, Cairo, Egypt, December 29–31, pp. 194–198.

Sirdari, Z.Z. 2013. Bedload transport for small rivers in Malaysia. Retrieved from https://www.academia.edu/4147500/Bed_Load_Transport_of_Small_Rivers_in_Malaysia.

Sirdari, Z.Z., Ab Ghani, A., and Hassan, Z.A. 2014. Bedload transport of small rivers in Malaysia, *International Journal of Sediment Research*, 29(4): 481–490, https://doi.org/10.1016/S1001-6279(14)60061-5.

Song, C.G., Seo, I.W., and Do Kim, Y. 2012. Analysis of secondary current effect in the modeling of shallow flow in open channels, *Advances in Water Resources*, 41: 29–48, https://doi.org/10.1016/j.advwatres.2012.02.003.

Sukhodolov, A.N., and Rhoads, B.L. 2001. Field investigation of three-dimensional flow structure at stream confluences, *Water Resources*, 37(9): 2411–2424, https://doi.org/10.1029/2001WR000316.

Sukhodolov, A.N., and Rhoads, B.L. 2001. Field investigation of three-dimensional flow structure at stream confluences: 2. Turbulence, *Water Resources Research*, 37(9): 2411–2424, https://doi.org/10.1029/2001WR000317.

Szupiany, R.N., Amsler, M.L., Parsons, D.R., and Best, J.L. 2009. Morphology, flow structure, and suspended bed sediment transport at two large braid-bar confluences, *Water Resources Research*, 45: W05415, https://doi.org/10.1029/2008WR007428.

Tancock, M.J. 2014. The dynamics of upland river confluences. Durham theses, Durham University. Available at Durham E-Theses Online: http://etheses.dur.ac.uk/10527/.

Taylor, E.H. 1944. Flow characteristics at rectangular open-channel junctions, *Transactions of the American Society of Civil Engineers*, 109(1): 893–902.

Thanh, M., Kimura, I., Shimizu, Y., and Hosoda, T. 2010. Depth-averaged 2D models with effects of secondary currents for computation of flow at a channel confluence. In: *Proceedings of the River Flow*, pp. 137–44.

Tomidokoro, G., Araki, M., and Yoshida, H. 1985. Three-dimensional analysis of open channel flows, *Annual Journal of Hydraulic Engineering*, 29: 727–732.

Toombes, L., and Chanson, H. 2011. Numerical limitations of hydraulic models. In: *Proceedings of 10th Hydraulics Conference 34th IAHR World Congress - Balance and Uncertainty, 33rd Hydrology and Water Resources Symposium*. 26 June - 1 July, Brisbane, Australia.

Wallis, E., Mac Nally, R., and Lake, P.S. 2008. A Bayesian analysis of physical habitat changes at tributary confluences in cobble-bed upland streams of the Acheron River basin, Australia, *Water Resources Research*, 44(11): 1–10, https://doi.org/10.1029/2008WR006831.

Wang, X., Yan, X., Duan, H., Liu, X., and Huang, E. 2019. Experimental study on the influence of river flow confluences on the open channel stage–discharge relationship, *Hydrological Sciences Journal*, 64(16): 2025–2039.

Wang, Y., Odgaard, A.J., Melville, B.W., and Jain, S.C. 1996. Sediment control at water intakes, *Journal of Hydraulic Engineering*, 122(6): 353–356.

Watanabe A., Fukuoka, S., Yasutake, Y., and Kawaguchi, H. 2001. *Method for Arranging Vegetation Groins at Bends for Control of Bed Variation, Collection of Papers on River Engineering*, vol. 7, pp. 285–290.

Webel, G., and Schatzmann, M. 1984. Transverse mixing in open channel flow, *Journal of Hydraulic Engineering*, 110(4): 423–435.

Weber, L.J., Schumate, E.D. and Mawer, N. 2001. Experiments on flow at a 90 open-channel confluence, *Journal of Hydraulic Engineering*, 127(5): 340–350, https://doi.org/10.1061/(ASCE)0733-9429(2001)127:5(340).

Wuppukondur, A., and Chandra, V. 2017a. Methods to control bed erosion at 90° river confluence: An experimental study, *International Journal of River Basin Management*, 15(3): 297–307.

Wuppukondur, A., and Chandra, V. 2017b. Control of bed erosion at 60° river confluence using vanes and piles, *International Journal of Civil Engineering*, 16(6): 619–627.

Yang, C.T. 1996. *Sediment Transport Theory and Practice*, McGraw Hill Companies, New York, USA.

Yang, Q.Y., Liu, T.H., Lu, W.Z., and Wang, X.K. 2013. Numerical simulation of confluence flow in open channel with dynamic meshes techniques, *Advances in Mechanical Engineering*, 5: 860431, https://doi.org/10.1155/2013/860431.

Yuan, S., Tang, H., Xiao, Y., Qiu, X., Zhang, H., and Yu, D. 2016. Turbulent flow structure at a 90-degree open channel confluence: Accounting for the distortion of the shear layer, *Journal of Hydro-Environment Research*, 12: 130–147, https://doi.org/10.1016/j.jher.2016.05.006.

Yuill, B.T., Khadka, A.K., Pereira, J., Allison, M.A., and Meselhe, E.A. 2016. Morphodynamics of the erosional phase of crevasse-splay evolution and implications for river sediment diversion function, *Geomorphology*, 259: 12–29.

Zhang, Y., Wang, P., Wu, B., and Hou, S. 2015. An experimental study of fluvial processes at asymmetrical river confluences with hyperconcentrated tributary flows, *Geomorphology*, 230: 26–36, https://doi.org/10.1016/j.geomorph.2014.11.001.

9 Shallow Water Flow Modeling

Franziska Tügel, Ilhan Özgen-Xian, Franz Simons,
Aziz Hassan, and Reinhard Hinkelmann

CONTENTS

9.1   INTRODUCTION

In water resources management, surface flow models are, for example, used to improve process understanding (e.g., Thompson et al., 2010; Khosh Bin Ghomash et al., 2019), to design restoration structures (e.g., Lange et al., 2015), to carry out case studies for decision support (e.g., Matta et al., 2018; Marafini et al., 2018), and to predict flood inundation areas (e.g., Abily et al., 2016; Amann et al., 2018; Kobayashi et al., 2015; Tügel et al., 2020b). A more recent application field deals with rainfall-runoff simulations in small catchments (e.g., Simons et al., 2014; Caviedes-Voullième et al., 2012; Mügler et al., 2011; Viero et al., 2014; Liang et al., 2016).

The origins of shallow water flow modeling can be traced back to the field of computational hydraulics, a branch of computational fluid dynamics (Abbott and Mins, 1998) with application

DOI: 10.1201/9780429463938-12

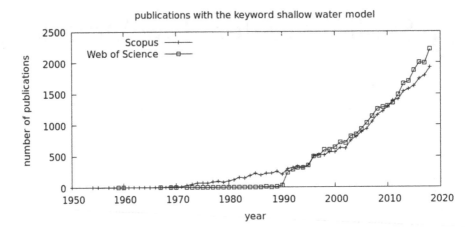

FIGURE 9.1 The number of publications per year with the keyword "shallow water model" is increasing, indicating an increasing interest in the subject. (Source: the plot is based on queries of databases at Scopus and Web of Science, accessed on June 21, 2019.)

to river flow modeling and coastal engineering (Bazrkar et al., 2017). In recent years, there is an increasing interest in computational hydraulics and shallow water flow modeling, which can be seen in the number of publications per year on this subject, which is increasing every year; see Figure 9.1. This can be explained for example by: (1) the availability of more and more high-resolution (HR) data, (2) a partial replacement of classical hydrological models by shallow water models as hydrological processes such as infiltration can be also included here, and (3) the increasing interest in investigating flooding areas in urban areas, which are generated on the one hand by river floods but also by pluvial floods due to extreme rainfalls, whose frequency and intensities might even increase due to climate change. The frontier of research is currently focusing on (1) enhancing robustness and efficiency of models by developing innovative numerical methods (Navas-Montilla and Murillo, 2016; Kesserwani et al., 2019; Xia et al., 2013), (2) fully exploiting and harnessing the latest high-performance computing (Lacasta et al., 2015; Smith and Liang, 2013; Ginting and Mundani, 2019), and (3) exploring new application fields, e.g., simulation of multiple processes (e.g., hydro-sediment, morphodynamics, debris flow, biological processes) at a catchment/basin scale, and real-time flood forecasting (Khosh Bin Ghomash et al., 2019; Viero et al., 2014; Zhao et al., 2019).

Shallow water equations (SWE) are a simplification of the Navier-Stokes equations for incompressible, free-surface flows over small bottom inclinations, which are dominated by horizontal processes. These assumptions are usually valid for cases where the wavelength is significantly larger than the water depth (by a factor ≥ 20). In such cases, the vertical acceleration and velocity can be neglected, and hydrostatic pressure distribution can be assumed over the depth. This enables averaging quantities over the water depth, yielding the two-dimensional shallow water equations that can be solved on the horizontal plane. The resulting depth-averaged shallow water equations are a system of nonlinear hyperbolic partial differential equations (PDEs) (Toro, 2001). Analytical solutions for this system are only available for very simple cases. For more complex cases, which are usually of interest in real-world applications, numerical methods have to be used to approximate the solution of the SWE. The most commonly used numerical methods are the finite-difference method, the finite-element method, and the finite-volume method (Hinkelmann, 2005). For all methods, the spatial domain of the system is subdivided into a certain number of finite cells or elements. This process is called *spatial discretization*. Depending on the chosen method, the solutions of the unknown variables in terms of water depth and unit discharge are determined either for each node or center of each element or edge.

Overland flow generated by rainfall is generally associated with very small water depths over complex topography and may feature discontinuities in the flow field. Furthermore, the propagation of wet and dry fronts, as well as transcritical flow states make the numerical modeling of this process challenging. Shock waves and other discontinuities in the flow field are usually difficult to resolve with numerical methods. For example, the classical finite-difference method is based on the assumption of smooth functions and may generate spurious oscillations in the presence of steep gradients or shock waves. These oscillations are related to the stability of the numerical method. Another difficulty in the presence of shock waves is the artificial smearing of the discontinuity over one or more grid cells, referred to as numerical diffusion (LeVeque, 2002). This is related to the accuracy of the numerical method. In order to prevent this artificial smearing, numerical methods with a higher order of accuracy must be used. However, the stability of such higher-order methods usually requires additional treatment. In the worst case, these methods may yield spurious oscillations, which can lead to partly unphysical solutions or even cause the development of non-linear instabilities making the numerical solution completely useless (Toro, 2001). These issues lead to additional difficulties to simulate flow processes appropriately. In shallow water flow modeling, the term robust numerical model has been coined to refer to numerical methods that ensure both the stability in the presence of wet and dry fronts, shock waves, and transcritical flows as well as the accuracy to an extent where numerical diffusion is considered acceptable with regard to the application.

The preferred method for such robust shallow water flow models is a second-order accurate finite-volume method (FV) (see, e.g., the models reported in Simons et al., 2014; Liang et al., 2016; Ginting and Mundani, 2019; and many others. The FV method solves the integral form of the PDEs in each computational cell. A flux function is defined across each cell interface, which drives the change of the solution in time. Hence, the solution is locally conserved in each cell, and since the integral form of the PDEs is used, discontinuous solutions are allowed (Toro, 2001; Hinkelmann, 2005). In the FV method, defining the flux function is non-trivial. There are many methods in the literature to calculate the flux function (LeVeque, 2002). A particularly established way is to solve a local Riemann problem across the cell edge. The Riemann problem is defined as an initial-value problem with piecewise constant values separated by a single discontinuity. Solving this problem exactly or approximately to define the flux function across the cell interface gives the so-called Godunov-type method (Godunov, 1959). The order of accuracy of the FV method is usually increased by means of a total variation diminishing (TVD) method, which avoids spurious oscillations. A common scheme applied in the literature is the TVD-MUSCL scheme (van Leer, 1979).

In recent years, the discontinuous Galerkin method (DG) has been explored in the context of shallow water flow modeling (Kesserwani et al., 2019; Khan and Lai, 2017). The DG method is a finite-element method (FE) with discontinuous base functions. In contrast to the classical FE methods, e.g., the streamline-upwind Petrov-Galerkin method (SUPG) (Hinkelmann, 2005), the DG method calculates a local flux across cell interfaces and is therefore locally conservative. The main advantage of the DG method is that its extension to a higher order of accuracy has a compact stencil, compared to the FV method. In this chapter, only the FV method is introduced.

Considerable progress in survey technologies results in the availability of more and more accurate data with high spatial resolutions. As the topography is often the main driving force of flow, especially accurate and high-resolution data about the topography is crucial to generate appropriate solutions with 2D shallow water models. Using high-resolution numerical meshes increases the computational effort significantly, as the mesh consists of more cells and therefore more computations per time step have to be carried out. If data with a resolution in the range of decimetres is used for a domain of several square kilometres, the computational mesh with the same resolution would have a huge cell number coming along with an unfeasible high CPU-time.

The issue of computational cost is currently approached in three ways: (1) high-performance parallel computing on shared or distributed memory including the usage of graphics processing units (GPUs) (Smith and Liang, 2013; Lacasta et al., 2015; Kobayashi et al., 2015), (2) adaptive mesh refinement (Kesserwani et al., 2019), and (3) scaling approaches (Viero et al., 2014). The third approach reduces the computational effort by taking topographic details conceptually into account instead of explicitly discretizing them with high-resolution meshes. There exist different possibilities to realize the conceptional consideration of subgrid-scale information.

This chapter aims to give an introductory overview of the physical-mathematical background of shallow water flow models from an engineering perspective. As a consequence, not all numerical methods and application areas of shallow flow models are covered here. The discussion of numerical methods is limited to the finite volume method, which is currently considered the state of the art for flow-dominated problems. Other methods such as the finite difference method, the finite element method, the discontinuous Galerkin method, the Lattice Boltzmann method, and smoothed-particle hydrodynamics are not discussed. Two presented application cases consider rural catchments in different climate regions. Interesting applications in urban areas also exist, but these applications are not part of the current chapter. Section 9.2 of this chapter gives an overview of the governing equations, namely the 2D shallow water flow equation and the Green-Ampt infiltration equation. Section 9.3 introduces robust numerical methods. In this context, a cell-centered finite volume MUSCL scheme with explicit time discretization is presented. Section 9.4 shows some basic benchmark tests to verify the numerical code and validate its performance. Section 9.5 presents two case studies. The first one considers flash floods in an arid region in Egypt, where the impact of infiltration on the flood extent is studied and different flood mitigation strategies are assessed. The second case considers a small alpine catchment in Austria, where automatic parameter calibration is applied to match discharge data. A friction-law-based upscaling approach, which uses a modified friction law to take into account subgrid-scale microtopography, is derived.

9.2   GOVERNING EQUATIONS

9.2.1   GENERAL FORM OF CONVERSATION LAW

The general form of the 2D conversation law can be expressed as:

$$\frac{\partial \mathbf{q}}{\partial t} + \frac{\partial \mathbf{f}}{\partial x} + \frac{\partial \mathbf{g}}{\partial y} = \mathbf{s} \tag{9.1}$$

where q is the vector of conserved state variables, t is the time, f and g denote the vectors of advective and diffusive fluxes in x- and y-direction, respectively and the vector s represents the source terms. This equation describes mathematically, that a temporal change of the conserved variables in the control volume can only be caused by a net flux over the surface of the control volume and/or by sinks/sources within the control volume.

9.2.2   2D SHALLOW WATER MODEL

Inserting the following vectors in the general conservation law (Equation 9.1) yields in the SWE:

$$q = \begin{bmatrix} h \\ uh \\ vh \end{bmatrix}, \; f = \begin{bmatrix} uh \\ uuh + \dfrac{1}{2} gh^2 - v_t \dfrac{\partial uh}{\partial x} \\ uvh - v_t \dfrac{\partial vh}{\partial x} \end{bmatrix},$$

$$g = \begin{bmatrix} vh \\ vuh - v_t \dfrac{\partial uh}{\partial y} \\ vvh + \dfrac{1}{2} gh - v_t \dfrac{\partial vh}{\partial y} \end{bmatrix}, \; s = \begin{bmatrix} r \\ -gh \dfrac{\partial z_B}{\partial x} - s_{f,x} - f_x \\ -gh \dfrac{\partial z_B}{\partial y} - s_{f,y} - f_y \end{bmatrix} \tag{9.2}$$

Here, the first row of each vector contains the mass balance equation and the second and third rows represent the momentum balance equations in x- and y-direction, respectively. h is the water depth, u and v are the velocity vector components in x- and y-direction, respectively, and z_B is the bottom elevation above datum. uh and vh represent the specific discharge in x- and y-direction, respectively, and could be replaced by q_x and q_y. r is a mass source/sink term accounting for precipitation, infiltration or injection/abstraction of water, g denotes the gravitational acceleration, ν_t the turbulent viscosity, and ρ the density of water. The turbulent viscosity ν_t is not a property of the fluid, but strongly depends on the local turbulence structure. As the turbulent viscosity ν_t is usually much higher than the physical viscosity ν, the latter one is commonly neglected. In applications, where turbulence is of minor importance, the turbulent viscosity ν_t is also set to zero, and therefore the diffusive fluxes of the flow variables are completely neglected. Besides the bottom slope term $\left(gh\dfrac{\partial z_B}{\partial x} \right)$, the momentum source/sink terms also include the momentum sinks due to bottom friction $s_{f,x}$ and $s_{f,x}$ as well as external source terms f_x and f_y, which can include the Coriolis force or wind shear stress. The bottom friction is considered with a sink term in depth-averaged models, as it is not directly considered with a boundary condition. Different laws can be used to determine the bottom friction. The Manning's law given in equation 9.3 is one of the most common ones and is also used in the studies presented later in this chapter:

$$s_{f,x} = \frac{g \cdot n^2}{h^{1/3}} u\sqrt{u^2 + v^2}, s_{f,y} = \frac{g \cdot n^2}{h^{1/3}} v\sqrt{u^2 + v^2} \tag{9.3}$$

where n denotes the Manning roughness coefficient and is the reciprocal value of the also well-known Strickler coefficient $K_{St} = \dfrac{1}{n}$.

9.2.3 INFILTRATION

One commonly used and physically based model concept to consider infiltration processes is the Green-Ampt model, where the cumulative infiltration and the infiltration rate are calculated with the following equations:

$$F(t) = Kt + (h_0 - \psi)\Delta\theta \ln\left(1 + \frac{F(t)}{(h_0 - \psi)\Delta\theta}\right) \tag{9.4}$$

$$\frac{dF}{dt} = f(t) = K\left(1 + \frac{(h_0 - \psi)\Delta\theta}{F(t)}\right) \tag{9.5}$$

where F(t) denotes the cumulative depth of infiltration, f(t) the infiltration rate, K the hydraulic conductivity at residual air saturation, which is assumed to be 50% of the saturated hydraulic conductivity K_s (Whisler and Bouwer, 1970). ψ the wetting front soil suction head, h_0 the ponding water depth, and $\Delta\theta$ the soil moisture deficit, which is the difference between the effective porosity n_{eff} and the initial moisture content θ_i. The effective porosity, wetting front soil suction head and hydraulic conductivity are called Green-Ampt parameters. Rawls et al. (1983) analyzed different soils and determined average Green-Ampt parameters based on soil texture classes, where four of them are shown in Table 9.1.

9.3 ROBUST NUMERICAL METHODS

As mentioned before, the Godunov-type finite volume method is an appropriate method to build up a robust 2D shallow water model, and is implemented in our in-house shallow water model called

TABLE 9.1

Green-Ampt Parameters

Soil Texture Class	Effective Porosity n_{eff} (–)	Soil Suction Head ψ (cm)	Hydraulic Conductivity K (m/s)
Sand	0.417	4.95	$3.27 \cdot 10^{-5}$
Loamy sand	0.401	6.13	$8.31 \cdot 10^{-6}$
Sandy clay loam	0.330	21.85	$4.17 \cdot 10^{-7}$
Clay loam	0.309	20.88	$2.87 \cdot 10^{-7}$

Source: after Rawls et al. (1983).

the Hydroinformatics Modeling System (hms). These methods are further described in the following paragraphs.

9.3.1 SPATIAL DISCRETIZATION

Within the cell-centered finite-volume method (FVM), the differential equations in the conservative form are integrated over all control volumes, and therefore local conservation is ensured. By using FVM, the model domain is divided into several control volumes of finite size. These control volumes or cells can be either of rectangular shape organized on a structured grid, or of triangular or other shapes organized on an unstructured mesh. The conserved state variables can be determined for the center of each control volume representing the averaged values for the whole cell. Typically, a finer mesh results in more accurate results, but with an increasing cell number, the computational effort and therefore the computing time also increase. To find the optimum numerical mesh regarding the accuracy and computational effort, a so-called *convergence test* is carried out. Here, the cell sizes are gradually reduced to increase the mesh resolution and the results between one mesh and the next finer one are compared. If no differences between one mesh and the next finer one can be observed, mesh convergence is achieved.

In the following, the basic steps to derive the solution of partial differential equations with the finite-volume method are explained for the general form of the balance equation (equation 9.1). The first step is to integrate the conservative form of the equations over the control volume (equation 9.6) and by using the Green-Gauss theorem, the volume integral of the flux terms is then transformed into a surface integral (equation 9.7) (Simons, 2020; Hinkelmann, 2005):

$$\int_{\Omega} \frac{\partial \mathbf{q}}{\partial t} d\Omega + \int_{\Omega} \left(\frac{\partial \mathbf{f}}{\partial x} + \frac{\partial \mathbf{g}}{\partial y} \right) d\Omega = \int_{\Omega} \mathbf{s} d\Omega \qquad (9.6)$$

$$\int_{\Omega} \frac{\partial \mathbf{q}}{\partial t} d\Omega + \oint_{\Gamma} \mathbf{F} \mathbf{n} d\Gamma = \int_{\Omega} \mathbf{s} d\Omega \qquad (9.7)$$

Here, Ω denotes the domain and Γ the surface of the control volume, n is an outward pointing normal vector, and F a flux term consisting of the diffusive and advective fluxes in x- and y-direction. Integrating the surface integral over a two-dimensional cell leads to:

$$\oint_{\Gamma} \mathbf{F} \mathbf{n} d\Gamma = \sum_{k=1}^{n_b} \mathbf{F}_k \mathbf{n}_k l_k \qquad (9.8)$$

where n_b is the number of cell edges, k is the index of a cell edge and l_k is the cell length. With the assumption that the temporal change of the state variables as well as the source term is constant over the cell, equation 9.7 can be written as:

$$\frac{\partial \mathbf{q}}{\partial t} A + \sum_{k=1}^{n_b} \mathbf{F}_k \mathbf{n}_k l_k = \mathbf{s} A \tag{9.9}$$

where A denotes the cell area (Hinkelmann, 2005; Simons, 2020).

9.3.2 TEMPORAL DISCRETIZATION

The simulation time is subdivided into several constant or variable time steps Δt, where the solution functions are determined. Approximating the temporal derivative in equation 9.9 can be done by forward differencing, which can be derived by a Taylor series expansion and skipping the higher-order terms leading to equation 9.10:

$$\frac{\partial \mathbf{q}}{\partial t} \approx \frac{q^{n+1} - q^n}{\Delta t} \tag{9.10}$$

with n + 1 denoting the new time level and n denoting the current time level. Inserting equation 9.10 into equation 9.9 and rearranging it to have the unknowns q^{n+1} on the left-hand side leads to:

$$\mathbf{q}^{n+1} = \mathbf{q}^n - \frac{\Delta t}{A} + \sum_{k=1}^{n_b} \mathbf{F}_k^n \mathbf{n}_k l_k - \Delta t \mathbf{s}^n \tag{9.11}$$

Equation 9.11 shows the simplest case of an *explicit* method, which is first-order accurate in time and is called the forward Euler method. Here, the state variables on the new time level n + 1 are determined only by flux and source values of the current time level n. Therefore, the solution is relatively simple and for each equation, there is only one unknown, which leads to a low computational effort per time step. On the other hand, the time step size is limited with the Courant-Friedrichs-Lewy criterion (CFL) to keep the simulation stable. This criterion defines, that the maximum propagation velocity must be equal or smaller than the characteristic cell distance Δx divided by the time step size Δt:

$$Cr = \frac{|v| + c}{\Delta x / \Delta t} \leq 1 \tag{9.12}$$

Here, Cr denotes the Courant number, |v| the magnitude of the flow velocity and $c = \sqrt{gh}$ is the wave velocity. As these velocities are variable during the simulation, instead of choosing a fixed time step size it is more efficient to adapt the time step size during the simulation depending on the maximum velocities in the current time step. Therefore, equation 9.12 is rearranged to calculate the time step size depending on the Courant number, the cell distance Δx, and the maximum propagation velocity:

$$\Delta t \leq \frac{Cr \cdot \Delta x}{|v| + c} \tag{9.13}$$

Explicit methods are specifically useful for simulations with large wetting and drying areas, such as flood plains during flood simulations. Therefore, the explicit time discretization was chosen to be implemented in our in-house software Hydroinformatics Modeling System (hms).

Another possibility is to use a *fully implicit* time discretization, where the state variables at the new time level n + 1 depend on the flux and source values on the new time level n + 1. Obviously, this approach leads to higher computational effort per time step, because the unknowns depend on each other, and therefore, a system of equations has to be solved. The advantage of implicit methods compared to explicit ones is that there are no stability problems limiting the time step size. This scheme is called the backward Euler method and is first-order accurate in time.

In general, the temporal accuracy of a scheme can be increased by taking more than only one term of the Taylor series. Further methods for the time discretization are for example the Crank-Nicholson scheme or predictor-corrector schemes, which are described in Hinkelmann (2005), Simons (2020), etc.

9.3.3 Godunov's Method

As in the finite-volume method the values in the cell center are representative of the whole cell, the values at a cell face are not unique, but contain the values of the cells to the right and left of the face. Therefore, it is a non-trivial problem which values should be used for the calculation of the numerical fluxes, and different approaches can be chosen to determine the fluxes.

The first-order upwind method (FOU) is the easiest way to determine the numerical fluxes and is of first-order accuracy in space. Here, the flux is separated into a transporting variable and a transported variable, for example, the flow velocity u and the water depth h in the mass balance equation in the x-direction. The value of the transporting variable u at the cell edge k is determined by linear interpolation, and the transported variable h is taken from that cell, where the flow is coming from. Although FOU is a robust method to calculate the numerical flux at a cell edge, it introduces high numerical diffusion leading to smeared solutions of sharp gradients (Simons, 2020).

When applying Godunov's method (Godunov, 1959) to solve hyperbolic equations with FVM together with piecewise constant data, the following steps are carried out: firstly, the values at the cell faces have to be reconstructed from the averaged values in the cell center. In first-order schemes, the values at the cell faces will be the same as in the cell center. As the cells at the left and right sides of a cell interface have different values, a discontinuity at the cell interface occurs. Therefore, the numerical fluxes are solved with this initial data and by applying the theory of the Riemann problem. Afterward, the cell values are updated by summing up the numerical fluxes and finally, new cell averages are determined (LeVeque, 2002; Simons, 2020).

The solution of the Riemann problems contains a wave structure, which describes the propagation of the discontinuity (Figure 9.2). Three different wave types are distinguished, where the middle one is a contact discontinuity or shear wave with a constant water depth and momentum in flow direction and a jump in the momentum perpendicular to the flow direction. The left and right waves are either rarefaction or shock waves, according to the present flow conditions. In contrast to a shock wave, a rarefaction wave describes a smooth gradient in the variable. In the most left and most right regions of the solution structure, the unchanged initial states are preserved. Across the left and right waves, the water depth and normal flow velocity change, the tangential flow velocity stays constant. The region between the left

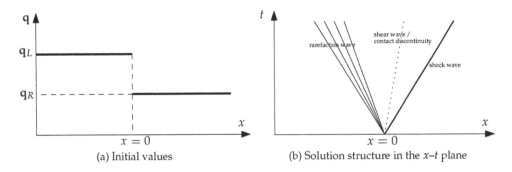

(a) Initial values   (b) Solution structure in the x–t plane

FIGURE 9.2   Riemann problem. (Source: Simons, 2020.)

and right waves is called the star region, where a new constant water depth and normal flow velocities are obtained. The star region is divided by the middle wave (contact discontinuity or shear wave) across which the tangential flow velocity will change discontinuously. The aim is to find the unknown water depth and flow velocity in the star region and therefore, two functions to connect the state of the star region to the left and right state, are derived. The derivation of the connecting functions differs depending on whether the left and right waves are shock or rarefaction waves. Further information can be found in Simons (2020) and the complete derivation is given by Toro (1992).

As the exact solution of the Riemann problem is computationally costly and for the flux computation in the FVM method only the solution at the sampling point x/t = 0 is needed, approximate Riemann solvers have been developed, where the Harten-Lax-van Leer (HLL), the Harten-Lax-van Leer-Contact (HLLC), which in contrast to the HLL solver also restores the contact wave, and the Roe solver are the most common ones. In the presented studies, the HLLC Riemann solver is used; a detailed derivation of the HLLC solver can be found in Simons (2020).

To avoid negative water depths and preserve a well-balanced state including wet-dry fronts, hydrostatic reconstruction is an appropriate approach. Well-balanced in that context means that a motionless water surface is ensured when simulating standing water or steady-state conditions even on complex topography. Before the Riemann problem is solved, the bottom elevation and water depths are modified according to the assumption of hydrostatic conditions. In a first step, the bed elevation at the midpoint of the considered cell edge k is determined by taking the maximum value of the bed elevations in the cells left and right from the cell edge. Then the water depths in the left and right cell are reconstructed by taking the maximum of zero, and the water elevation in the considered cell minus the bed elevation at cell edge k, which was determined in the first step. Finally, the modified specific discharges on both sides of edge k are calculated under consideration of the reconstructed water depths (Audusse et al., 2004; Hou et al., 2013; Simons, 2020). When simulating sheet-like overland flow, the water depth can be smaller than the variation in bottom topography (Simons et al., 2014) (Figure 9.3a). Applying the hydrostatic reconstruction in first-order schemes to such kinds of problems, e.g., sheet-like flow on a sloped plane, reduces this problem to piecewise flux calculations between wet and dry cells (Figure 9.3b). This leads to the calculation of a wrong flow field. However, high-order schemes discretize the considered flow situation as a continuous problem (Figure 9.3c), and therefore very shallow water flow on coarse grids with steep bottom gradients can be calculated more appropriately.

9.3.4 MONOTONIC UPWIND-CENTERED SCHEMES FOR CONSERVATION LAWS (MUSCL)

As mentioned before, it is very challenging to exactly resolve a discontinuity as a perfect jump with numerical methods, as the discontinuity is usually spread over several cells leading to a smearing

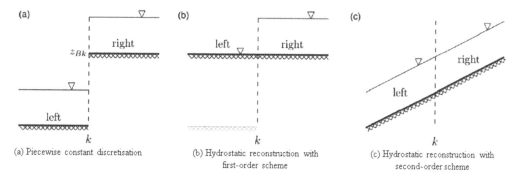

(a) Piecewise constant discretisation

(b) Hydrostatic reconstruction with first-order scheme

(c) Hydrostatic reconstruction with second-order scheme

FIGURE 9.3 Influence of hydrostatic reconstruction on the first-order scheme and result of the second-order reconstruction of the face values for water depths smaller than the variation of the bottom topography. (Source: Simons et al., 2014.)

of the sharp front. The ability to capture a discontinuity as exact as possible is called the *resolution* of a scheme. Schemes, which are based on the consideration of constant values within each cell and therefore having the same values at the cell face and in the cell center, have a first-order accuracy. A piecewise linear approximation of the cell values results in a higher-order scheme achieving higher accuracy. However, directly at discontinuities or sharp gradients, these linear approximations can lead to new extrema without physical meaning due to over- or undershooting slopes, resulting in spurious oscillations. To avoid this, total variation diminishing methods (TVD) are used to preserve monotonicity in higher-order Godunov-type schemes. Here, second-order approximations are used at smooth gradients, while the accuracy is reduced to first-order accuracy at discontinuities. Van Leer introduced this method as a monotonic upwind-centered scheme for conservation laws (MUSCL) (van Leer, 1979; LeVeque, 2002; Simons, 2020). Different TVD limiter functions have been developed; some examples are the van Leer, minmod, superbee, QUICK, and MC limiter functions. While the minmod limiter function has the highest limitation and lowest accuracy, the superbee limiter function has the lowest limitations. While the minmod function chooses the smaller of two neighboring slopes leading to the stronger smearing of sharp gradients, the superbee function tends to make smooth transitions near reflection points steeper and squared off, because of choosing the larger of two neighboring slopes. The MC limiter function seems to give good results for a wide range of problems (LeVeque, 2002). Due to its high numerical stability, the minmod limiter function is often used, although it has the lowest accuracy (Simons, 2020).

Due to very small water depths, bottom friction has a significant influence on surface runoff applications. This could lead to numerical instabilities, and therefore the splitting point-implicit method, which is described, e.g., in Bussing and Murman (1988) and Simons (2020), is used for the friction source term.

If the water depth in a cell reaches a threshold value, e.g., of 10^{-6} or 10^{-8} m, the cell is considered dry and the water depth is set to zero for the flux calculation. To ensure mass conservation, the value of the state variable itself is not changed. If only one of the two cells next to an edge is dry, the value at the cell edge is determined by first-order reconstruction.

9.4 BENCHMARKS

Two different benchmarks are presented in the following sections to verify the correctness and accuracy of the numerical schemes, which are implemented in our in-house modeling software Hydroinformatics Modeling System (hms). Both benchmarks are included in the Ph.D. thesis of Simons (2020).

9.4.1 1D Dam Break

The simulation of a quasi-one-dimensional dam break is a typical benchmark to test if the numerical scheme can deal with discontinuities in the initial conditions. In the following, two test cases with initially wet and dry downstream conditions are represented. The model domain of the presented test cases had an extension of 20 m × 2 m and was discretized with a cell size of 0.05 m. No friction was considered. For both cases the initial upstream water depth was set to 4 m, and the downstream water depth was 0 m in the dry case and 1 m in the wet case. All boundaries were closed, and the simulations were stopped before the waves reached the upstream or downstream boundary. The results of three different schemes are compared to the exact Riemann solver: first-order upwind, first-order HLLC, second-order HLLC. The CFL criterion was set to 0.3 and to reach second-order spatial accuracy, the minmod TVD limiter was used.

9.4.1.1 Dry Bed Initial Conditions

Figure 9.4 shows the temporal development of the water elevation (a) and flow velocity (b) profiles calculated with the different schemes under dry bed initial conditions. In general, all schemes

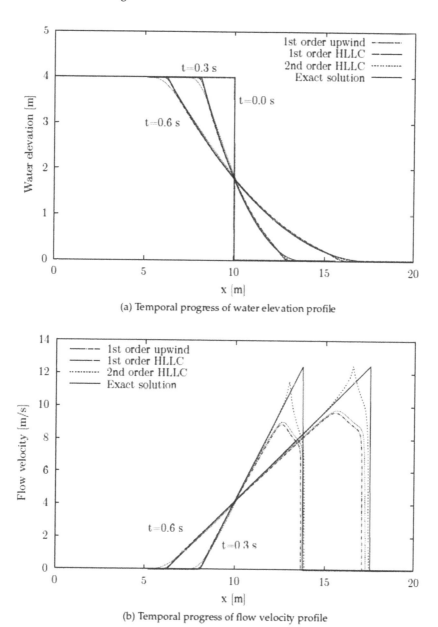

(a) Temporal progress of water elevation profile

(b) Temporal progress of flow velocity profile

FIGURE 9.4 Comparison of exact and numerical solutions of first- and second-order schemes for dam break on dry bed test case. (Source: Simons, 2020.)

performed well and could deal with the initial discontinuity and dry bed. The results of all schemes agree well with the exact Riemann solver and only small differences can be observed between the schemes. Looking at the velocity profiles, it becomes obvious that the HLLC solver outperformed the two others, as its results agree much better with the solution of the exact Riemann solver, while the two other schemes clearly underestimated the flow velocity at the propagating front.

9.4.1.2   Wet Bed Initial Conditions

Figure 9.5 shows the temporal development of the water elevation (a) and flow velocity (b) profiles under wet bed initial conditions. Here, a shock wave propagates to the right and a rarefaction wave

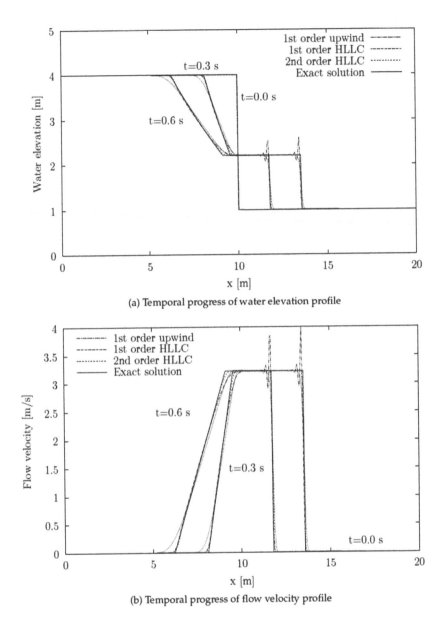

FIGURE 9.5 Comparison of exact and numerical solutions of first- and second-order schemes for dam break on wet bed test case. (Source: Simons, 2020.)

to the left. While both HLLC solvers could capture the shock front without instabilities, the first-order upwind scheme produced spurious oscillations at the shock front. The first-order HLLC solver showed the highest numerical diffusion, and thus the second-order HLLC solver outperformed the two others in representing the shock wave.

9.4.2 2D RAINFALL-RUNOFF SIMULATION

In this test case, rainfall is simulated over an idealized V-shaped catchment to analyze the accuracy of different schemes in simulating shallow surface runoff generated by rainfall. The catchment geometries after Di Giammarco et al. (1996) are shown in Figure 9.6. The channel depth varied

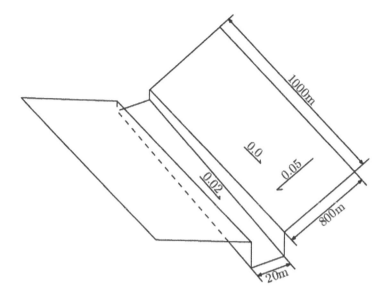

FIGURE 9.6   Geometry of the idealized V-shaped catchment. (Source: Simons et al., 2014.)

between 1 m at the upstream end and 20 m at the downstream end. The roughness for the river was set to a Manning coefficient of $n = 0.15$ sm$^{-1/3}$ and for the hillsides to $n = 0.015$ sm$^{-1/3}$. The domain was discretized by rectangular cells of 10 m × 10 m and the CFL criterion was set to 0.1. Critical flow depth was assumed as a boundary condition for the outlet, while all other boundaries were considered closed. A constant rainfall intensity of 10.8 mm/h with a duration of one hour was simulated over the hillsides.

The results of a first-order and second-order HLLC scheme were compared to the analytical solution presented in Overton and Brakensiek (1970). To obtain second-order accuracy, the minmod TVD limiter function was used. Figure 9.7 shows the simulated hydrographs at one hillside (a) and at the channel outlet (b). It can be observed that the first-order HLLC scheme did not perform well, as the results did not agree well in the ascending and descending phase of the hydrograph. This can be explained by the fact, that the first-order scheme assumes piecewise constant bottom topography, where the water depth can be smaller than the variation of the bottom topography. This leads to artificial detention and therefore a delayed ascending and descending of the hydrograph. In contrast, by using the second-order accurate scheme the sloped domain is discretized as a continuous flow problem, which results in a better representation of thin water flow on coarse grids with steep bottom slopes.

9.5   CASE STUDIES

9.5.1   Flash Flood Simulations in El Gouna

9.5.1.1   Study Area

The area of this case study is located in the eastern desert of Egypt, where an extremely arid climate with almost no rainfall is present all year. However, from time to time strong rainfall events in the mountainous areas can generate very fast and devastating flash floods, as the usually dry wadi systems turn into rivers with very fast flows. Many settlements, infrastructure, and cities have been constructed inside those usually dry wadi catchments, mostly close to the Red Sea coast or river Nile. During flash flood events, those areas are strongly affected and the flooding can cause severe damages to the environment, infrastructure, and properties, and in the worst case endanger human lives. The consequences are economic, ecological, and social problems. El Gouna is a

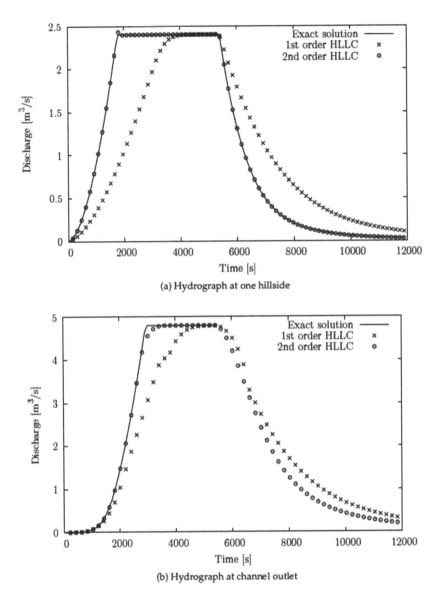

FIGURE 9.7 Comparison of analytical and numerical solution of the first- and second-order scheme. (Source: Simons et al., 2014.)

touristic town, which is located inside the Wadi Bili catchment (Figure 9.8), and during the flash flood event in March 2014, discharge measurements could have been carried out and were published in Hadidi (2016).

The two-dimensional shallow water model Hydroinformatics Modeling System (hms) is used to determine the flooding areas in El Gouna during the flash flood event in March 2014. Afterward, different mitigation measures to reduce the damages to infrastructure and urban areas are investigated. The model could be further used to study the effects of even stronger rainfall events, which might occur in the future. Infiltration excess or so-called Hortonian overland flow, which occurs when the rainfall intensities exceed the infiltration capacity of the soil, is the common type of runoff generation that occurs during heavy rainfalls in arid areas rather than saturation excess (Blöschl and Sivapalan, 1995). As the study area is covered mostly with natural soils, infiltration cannot be neglected and is included in the model with the well-known Green-Ampt approach, which

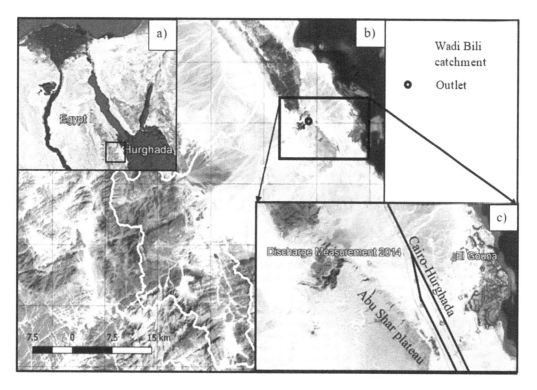

FIGURE 9.8 Location of Wadi Bili catchment in Egypt (a), catchment boundary and outlet (b), and location of discharge measurements in 2014 and the city of El Gouna (c) (background: Data SIO, NOAA, U.S. Navy, NGA, GEBCO ©2017 ORION-Me, Image ©2017 CNES/Airbus, Image©2017 Digital Globe). (Source: Tügel et al., 2020a.)

also includes the temporal development of infiltration depending on the occurring rainfall and the propagation of the flood wave. Several cases are simulated to study the effect of infiltration, also in combination with mitigation measures.

9.5.1.2 Model Set-Up

The model domain of about 11 km × 8 km is discretized by cells of 15 m × 15 m. The topography is represented by the digital surface model (DSM) ALOSWORLD 3D (AW3D) with a horizontal resolution of 30 m, which was interpolated to a cell size of 10 m (Figure 9.9). The DSM includes the big buildings of El Gouna, and as the model domain is hardly covered by plants, it was assumed that the DSM will not give a wrong topography due to plant canopies. The friction was set to 0.025 s•m$^{-1/3}$. All boundaries were considered open except for 30 m at the western boundary, where the inflow from the upstream wadi catchment enters the coastal plain. Here, the hydrograph measured by Hadidi (2016) was implemented as a boundary condition (Figure 9.10). Furthermore, a constant rain intensity of 4.25 mm/h over the first eight hours of simulation time was imposed.

9.5.1.3 Infiltration

Figure 9.11 shows the spatial distribution of water depth after nine hours of simulation time for different scenarios. While the left one represents the reference case without any infiltration, the middle and right ones show the effect of infiltration for two different soils using the Green-Ampt model and tabled Green-Ampt parameters after Rawls et al. (1983) (see Table 9.1). Under consideration of sandy clay loam (middle), the whole rainfall directly infiltrated and no surface runoff generated from rainfall occurred. The incoming flood wave from the wadi catchment just reached the city area after approximately nine hours. In the case of clay loam (right), some surface runoff was generated

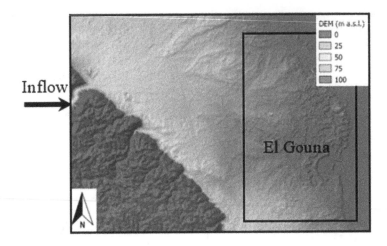

FIGURE 9.9   Model domain and bottom elevation. (Source: AW3D30 © JAXA, visualized with QGIS.)

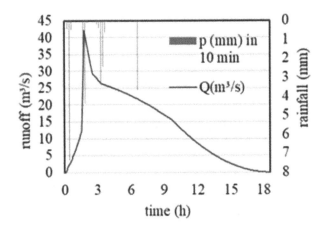

FIGURE 9.10   Rainfall and runoff on March 9, 2014. (Source: according to Hadidi, 2016.)

also by the rainfall and the flood wave from Wadi Bili propagated already a bit further compared to the case of sandy clay loam. Figure 9.12 shows the temporal development of water depth at the location of maximum water depth in the reference case (circle in Figure 9.11). The consideration of infiltration strongly dampens the hydrograph and leads to a delay of the peak water depths. While the peak water depth in the reference case was 0.82 m and occurred after about nine hours, it was only 0.28 m occurring after 14 hours in the case of clay loam, and 0.17 m after 16.5 hours, when the Green-Ampt parameters for sandy clay loam were considered. These values correspond to a reduction of the peak water depth of 66 % and 79 %, respectively. In Figure 9.12 it becomes also obvious that under consideration of loamy sand, all water infiltrated and no surface runoff occurred at the considered location. As loamy sand would be the best presentation of the dominating soil type in that area (based on own field investigations), one can conclude that the infiltration with tabled Green-Ampt parameters are strongly overestimated, as no runoff was simulated for this case, while in reality, water depths and surface runoff were reported for this location during the event in 2014. Reasons for this could be surface clogging, which can occur during heavy rainfall events and reduce the infiltration rate, as mentioned e.g., in Mein and Larson (1971), as well as subgrid-scale rill flow. More investigations should be carried out to achieve a better representation of infiltration in the model domain.

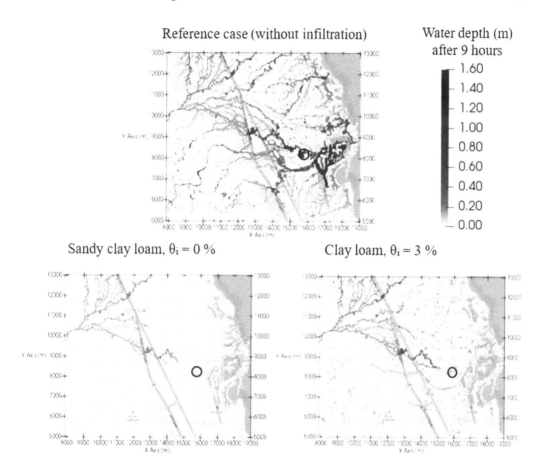

FIGURE 9.11 Spatial distribution of water depth after nine hours of simulation time for different scenarios without and with infiltration, plotted on OSM © OpenStreetMap contributors; the circle indicates the location of the maximum water depth in the reference case. (Source: Tügel et al., 2020b.)

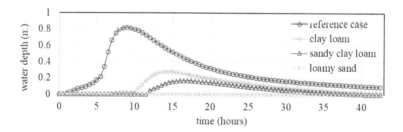

FIGURE 9.12 Temporal development of water depth for different scenarios without and with infiltration, at location of maximum water depth in the reference case (circle in Figure 9.12). (Source: Tügel et al., 2020b.)

9.5.1.4 Mitigation Measures

Preliminary studies about different mitigation strategies to reduce the water depths and flooding areas in the city of El Gouna have been carried out within the scope of a master thesis and are published in Marafini et al. (2018). One of the strategies was further developed and is represented in Figure 9.13. Here, one bigger retention basin is proposed behind the valley, where the flood wave from the wadi catchment reaches the coastal plain and spreads into several streams. This basin should capture the incoming flood wave from upstream. Furthermore, two smaller retention basins close to the city area are aimed to capture the surface runoff, which is generated by rainfall.

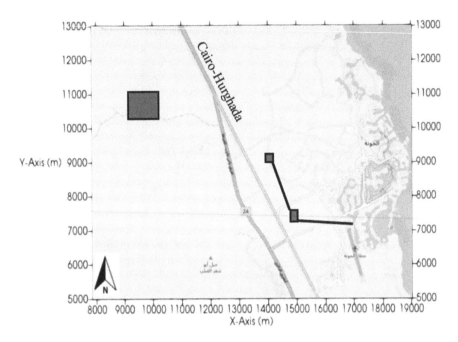

FIGURE 9.13 Possible mitigation strategy with one big and two small retention basins as well as two drainage channels, plotted on OSM © OpenStreetMap contributors. (Source: Tügel et al., 2020b.)

In addition, two drainage channels should release the water from the smaller basins to one of the lagoons and into the Red Sea, so that the basins do not overflow.

Figure 9.14 shows the spatial water depth distribution inside the model domain for the case of mitigation but without infiltration, and under consideration of mitigation measures and infiltration at the same time, again for the two soil types, sandy clay loam and clay loam. The mitigation measures could significantly reduce the water depths inside the city area. In the case of mitigation measures and infiltration for sandy clay loam, no surface runoff occurs in the city area, as the rainfall directly infiltrated and the flood wave from the wadi catchment was completely captured within the bigger retention basin. In the case of mitigation measures and infiltration for clay loam, only very small water depths occurred in the city area.

Figure 9.15 represents the temporal development of water depths at the location of maximum water depth in the reference case (circle in Figure 9.13) for three different cases: (1) the reference case with neither mitigation measures nor infiltration, (2) with mitigation measures and without infiltration, and (3) with mitigation measures combined with infiltration for clay loam.

The mitigation measures reduced the peak water depth by 74% from 0.82 to 0.21 m. When infiltration for clay loam was considered additionally, the water depths were reduced to 0.02 m, which corresponds to a total reduction of 97.5%.

9.5.2 Heumöser Slope

9.5.2.1 High-Resolution Rainfall-Runoff Simulation

Heumöser slope is a natural catchment located in the Vorarlberg Alps, Austria. In this case study, rainfall-runoff in a 100,000 m² (0.1 km²) large sub-catchment of Heumöser slope is simulated. The topography data is provided by the Austrian Department for Torrent and Avalanche Control in the form of a 1 m resolution digital elevation model (Figure 9.16). The model domain was discretized with 147,400 rectangular cells with a size of 1 m × 1 m. All boundaries were defined as open with a free outflow condition. The catchment response to a five-day-long rainfall event in July 2008 with a

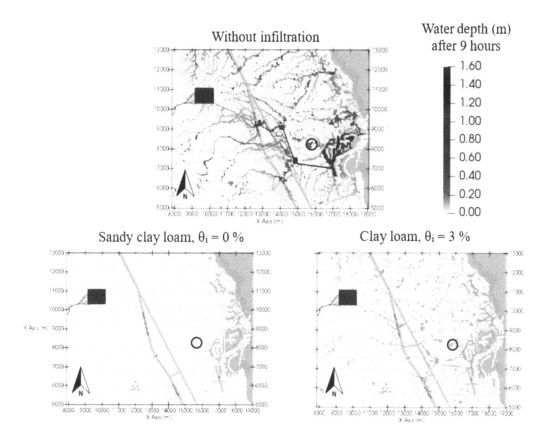

FIGURE 9.14 Spatial distribution of water depth after nine hours of simulation time with mitigation measures and different cases without and with infiltration, plotted on OSM © OpenStreetMap contributors. (Source: Tügel et al., 2020b.)

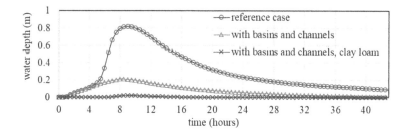

FIGURE 9.15 Temporal development of water depth for different scenarios without and with mitigation measures and infiltration, at location of maximum water depth in the reference case (circle in Figures 9.13 and 9.15). (Source: Tügel et al., 2020b.)

total amount of approximately 160 mm is simulated. Measurements of discharge are available from a measurement weir located at the outlet of the sub-catchment; see Figure 9.16.

Infiltration was considered with the simple approach of a constant runoff coefficient. Here, the total rainfall is reduced by a coefficient to consider only the *effective rainfall*, which generates surface runoff, while the rest is considered as losses due to infiltration. While one part of the infiltrated water will percolate to deeper soil layers into the saturated zone and contribute to the groundwater recharge, another part will contribute to the so-called *interflow*, which flows in the upper soil layers following the natural topography, partially exfiltrates and leads to a delayed contribution to the total

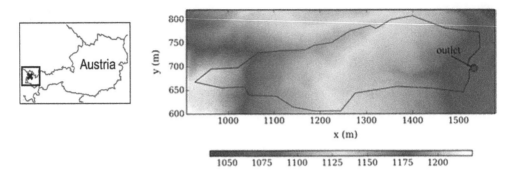

FIGURE 9.16 Location in Austria and digital elevation model (1 m resolution) of the sub-catchment of Heumöser slope. The line denotes the watershed, the dot at the outlet denotes the position of the measurement weir. (Source: Özgen et al., 2015.)

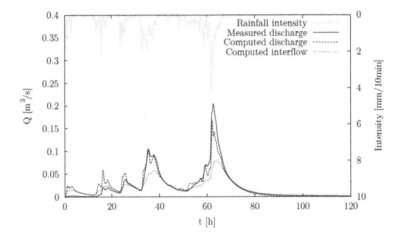

FIGURE 9.17 Temporal distribution of rainfall, simulated interflow, and discharge compared with the measured discharge hydrograph at catchment outlet. (Source: Simons et al., 2014.)

discharge at the catchment outlet. Investigations showed, that this process cannot be neglected in the studied catchment due to its steep slopes. When this process was not considered, the discharge was strongly underestimated, as the infiltrated water did not contribute to the discharge at the measurement point. Therefore, interflow processes were considered with a simple linear reservoir model to take into account the retention effect when the water is flowing through the upper soil layers and consequently reaching the outlet with some delay. The shape of the interflow hydrograph is therefore dampened and widened compared to the one generated by direct surface runoff. The average residence time, or reservoir constant, which controls the shape of the interflow hydrograph, was determined through calibration. Much better results could be achieved when interflow was taken into account. Further information about the approach to consider interflow, as well as a sensitivity analysis to investigate the impact of runoff coefficient and Manning roughness coefficient can be found in Simons (2020). Figure 9.17 shows the temporal distribution of rainfall as well as the results of discharge and interflow at the catchment outlet compared to the measured discharge. Here, a runoff coefficient was of $\Psi = 0.3$, a Manning coefficient of 0.067 sm$^{-1/3}$, and a reservoir constant of 6 h was used. While the first peaks were overestimated by the model, the middle one was captured very well and the last one was underestimated. A good agreement of the overall hydrograph shape, base flow, and arrival times of peak discharges could be achieved.

9.5.2.2 Optimization

The results presented in the last section were based on a manual calibration, where the runoff coefficient and Manning coefficient were determined by comparing only a few variations and choosing the parameter set with the best agreement of simulation results and measurements. To optimize the calibration process, an automated calibration can be carried out, where the chosen calibration parameters are adapted automatically until no further improvement of the results is taking place. There exist many different optimization techniques, in the study of the Heumöser slope, automated calibration of model optimization is carried out by means of:

- Basin-Hopping method, which is one of the most progressing global optimization techniques primarily based on the Monte Carlo method, and it is mandatory to use a searching method for a local minimum (e.g., Nelder-Mead) (Wales and Doye, 1997).
- Nelder-Mead method, which is a simplex-type algorithm that converges to a local minimum, which is ideally the global minimum, of a given objective function (Nelder and Mead, 1965).
- Objective function root mean squared error, which is chosen to measure the deviations between the model predictions for the candidate parameters (sim$_i$) and observed data (obs$_i$), shown in equation 9.14:

$$\text{RMSE} = \sum_{i=0}^{n} \left(\text{sim}_i - \text{obs}_i \right)^2 \tag{9.14}$$

The Basin-Hopping and the Nelder-Mead method have been chosen because of their high success rates to find the global optimum and their higher performance compared with other methods in terms of computing time. The optimization methods and techniques, which are utilized to implement automatic calibration, were performed by the SciPy optimization package, written in Python. The optimization algorithm was applied to two parameters, namely the roughness coefficient and the runoff coefficient. As many simulations had to be carried out to find the optimum, a grid resolution of 10 m was used here to reduce the overall simulation time. In the end, 117 simulations were conducted and the parameter values were more accurate (Figure 9.18 and Table 9.2) and the attainment of results was faster than by hand calibration.

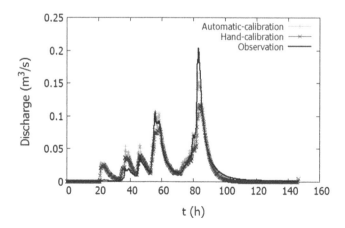

FIGURE 9.18 Hand and automatic calibration of model for rainfall-runoff simulation. (Source: Hassan et al., 2018.)

TABLE 9.2

Values of the Model Parameters for Rainfall-Runoff Simulation

Parameter	Manual Calibration	Automated Calibration
Ψ [–]	0.30	0.14
K_{st} [m$^{1/3}$/s]	35	24
RMSE [m^3/s]	1.44	1.11

Source: Hassan et al. (2018).

9.5.2.3 Upscaling

High-resolution models for whole catchments need huge computational effort and therefore the computing time is often too long to generate a real-time flood-forecasting. Besides the usage of high-performance parallel computing techniques, *upscaling methods* have been developed that can enable real-time forecasting. *Upscaling*, in this sense, is the approximation of a system of PDEs by another system of PDEs that can be solved with fewer computational resources (Farmer, 2002). In the literature, the SWEs have been upscaled by artificially increasing the roughness coefficient to account for subgrid-scale effects (Liang et al., 2016; Néelz and Pender, 2007). However, modifying the roughness closure formulation itself is also used, e.g., in Mügler et al. (2011) to account for vegetation, in Razafison et al. (2012) to account for furrows, and in Bellos et al. (2018) for application in flood modeling. Here, the modified roughness formulation from Özgen et al. (2015) is used to account for unresolved topography. The idea is to coarsen the resolution of the computational mesh to reduce the computational cost, taking the unresolved high-resolution data into account by means of the roughness formulation, and to finally obtain similar results on the coarse grid when compared to the fine one.

The proposed friction law reads:

$$s_f = -\left(\frac{g}{C_0} + K\right)|\mathbf{v}|v \tag{9.15}$$

where C_0 is the Chezy coefficient (m$^{1/2}$s^{-1}), which can be related to the Manning coefficient by:

$$C_0 = h^{1/6}n^{-1} \tag{9.16}$$

and K is calculated as:

$$K = \alpha_0 \exp\left(-\alpha_1(\Lambda - 1)\right) \tag{9.17}$$

with the free parameters α_0 and α_1, that account for the geometry of the unresolved topography, and the inundation ratio Λ, calculated as:

$$\Lambda = \frac{h}{(1-I)k} \tag{9.18}$$

where k is the characteristic roughness length, and I is the slope of the cell. For k, the standard deviation of the microtopography is used.

The model is calibrated by minimizing the root mean square deviation between upscaled model results and a high-resolution (1 m × 1 m) reference model result. The Limited-memory Broyden, Fletcher, Goldfarb, and Shanno algorithm (L-BFGS-B) (Byrd et al., 1995) was used to

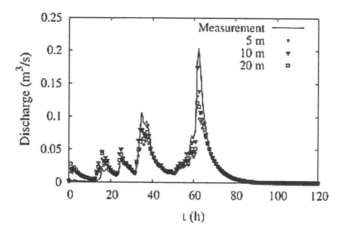

FIGURE 9.19 Good agreement with measurement data is obtained by calibrating the upscaled model. The calibration used a 10 m resolution model, hence it shows the best agreement of all. (Source: Özgen et al., 2015.)

optimize the free parameters. The calibration results in $n = 0.15$ s•m$^{-\frac{1}{3}}$, $\alpha_0 = 28.57$, and $\alpha_1 = 7.26$, with a root mean square deviation of 0.007 m^3/s. The overall agreement between the upscaled model and measurement data is good. The peaks at the beginning of the simulation are overshot, while the last peak is underestimated. Yet, the arrival times of the peaks are all captured accurately; see Figure 9.19. The 5 m resolution simulation is 56 times faster, the 10 m resolution simulation is 336 times faster, and the 20 m resolution simulation is 2,520 times faster than the 1 m resolution reference simulation.

The upscaled models are not able to reproduce the high-resolution solution exactly; however, in general the upscaling approach is found to be a good trade-off between accuracy and computational cost.

9.6 CONCLUSIONS

The basics of shallow water flow models as well as robust numerical methods to handle specific complex flow problems and the propagation of wet-dry fronts have been represented. Here, a cell-centered Godunov-type finite-volume scheme with second-order accuracy in space, explicit time discretization with first-order accuracy, and the TVD methods to avoid spurious oscillations were chosen and the main numerical concepts were introduced. The scheme was verified with two different benchmarks and the results showed that the scheme is capable of simulating shallow flow generated by rainfall over varying topography as well as handling discontinuities in the initial conditions and at wet-dry-fronts, as well as capturing important features such as sharp fronts.

In two case studies, the scheme was applied to real-world problems, where the first one was a flash flood simulation in the eastern desert of Egypt, and the second one dealt with rainfall-runoff simulations in a small alpine catchment in Austria. For the first case study, different scenarios of infiltration and mitigation measures were represented for a flash flood event in March 2014. It was shown, that infiltration has a significant impact on the flash flood propagation, as the soils are mainly natural and dominated by sand. Different reasons which lead to an overestimation of infiltration need to be further investigated. Furthermore, the proposed mitigation measures of retention basins and drainage channels at appropriate locations could significantly reduce the flooding areas and water depths in the city area. The model could be used to investigate stronger rainfall events or further mitigation measures to find optimum solutions.

Within the second real-world application in the Heumöser catchment in Austria, high-resolution rainfall-runoff simulations were carried out for a five-day-long rainfall event in July 2008. Under consideration of interflow processes, a good agreement between simulated and measured runoff hydrograph could be achieved. Different optimization techniques were used to carry out an auto-mated calibration, and it could be shown, that the automated calibration could reach better results than the hand calibration. By using the friction-based coarse grid approach as an upscaling method, the measured runoff hydrograph could be successfully reproduced to an acceptable amount of accu-racy, while the computing time was two to three orders of magnitude shorter compared to the high-resolution model.

ACKNOWLEDGMENTS

The simulations were partially computed on the supercomputers of Norddeutscher Verbund für Hoch- und Höchstleistungsrechnen (HLRN), as well as on the High-Performance-Computing-Cluster of TU Berlin, Germany.

REFERENCES

Abbott, M.B., and Mins, A.W. 1998. *Computational Hydraulics*, Ashgate Publishing, Aldershot, Hamshire, UK.

Abily, M., Bertrand, N., Delestre, O., Gourbesville, P., and Duluc, C.-M. 2016. Spatial global sensitivity anal-ysis of high resolution classified topographic data use in 2D urban flood modelling, *Environmental Modelling and Software*, 77: 183–195.

Amann, F., Özgen-Xian, I., Abily, M., Zhao, J., Liang, D., Kobayashi, K., Oishi, S., Gourbesville, P., and Hinkelmann, R. 2018. Integral porosity shallow water model at district scale – Case study in Nice. In: *E3S Web of Conferences, 40, 06018, River Flow 2018*.

Audusse, E., Bouchut, F., Bristeau, M.-O., Klein, R., and Perthame, B. 2004. A fast and stable well-balanced scheme with hydrostatic reconstruction for shallow water flows, *SIAM Journal on Scientific Computing*, 25(6): 2050–2065.

Bazrkar, M.H., Adamowski, J., and Eslamian, S. 2017. Water System Modeling, in *Mathematical Advances Towards Sustainable Environmental Systems*, edited by Furze, J.N., Swing, K., Gupta, A.K., McClatchey, R., Reynolds, D., Springer International Publishing, Switzerland, pp. 61–88.

Bellos, V., Nalbantis, I., and Tsakiris, G. 2018. Friction modeling of flood flow simulations, *Journal of Hydraulic Engineering*, 144: 04018073.

Blöschl, G., and Sivapalan, M. 1995. Scale issues in hydrological modelling: A review, *Hydrological Processes*, 9: 251–290.

Bussing, T.R.A., and Murman, E.M. 1988. Finite-volume method for the calculation of compressible chemi-cally reacting flows, *AIAA Journal*, 26(9): 1070–1078.

Byrd, R., Lu, P., Nocedal, J., and Zhu, C. 1995. A limited memory algorithm for bound constrained optimiza-tion, *SIAM Journal on Scientific Computing*, 16: 1190–1208.

Caviedes-Voullième, D., Murillo, J., and Garcia-Navarro, P. 2012. Influence of mesh structure on 2D full shal-low water equations and SCS Curve Number simulation of rainfall/runoff events, *Journal of Hydrology*, 448–449: 39–59.

Di Giammarco, P., Todini, E., and Lamberti, P. 1996. A conservative finite elements approach to overland flow: The control volume finite element formulation, *Journal of Hydrology*, 175(1–4): 267–291.

Farmer, C.L. 2002. Upscaling: A review, *International Journal for Numerical Methods in Fluids*, 40: 63–78.

Ghomash, S.K.B., Caviedes-Voullieme, D., and Hinz, C. 2019. Effects of erosion-induced changes to topogra-phy on runoff dynamics, *Journal of Hydrology*, 573: 811–828.

Ginting, B.M., and Mundani, R.-P. 2019. Parallel flood simulations for wet-dry problems using dynamic load balancing concept, *Journal of Computing in Civil Engineering*, 33: 1–18.

Godunov, S.K. 1959. A finite difference method for the computation of discontinuous, *Mat. Sb.*, 47(89): 271–306.

Hadidi, A. 2016. *Wadi Bili Catchment in the Eastern Desert - Flash Floods, Geological Model and Hydrogeology*, Dissertation. Fakultät VI – Planen Bauen Umwelt der Technischen Universität Berlin, Berlin, Germany.

Hassan, A., Özgen, I., and Hinkelmann, R. 2018. Using a simplex-type optimization method to calibrate a hydrodynamic model for rainfall-runoff simulations. In: *Proceedings of the 5th IAHR Europe Congress - New Challenges in Hydraulic Research and Engineering*, 12–14 June, Trento, Italy.

Hinkelmann, R. 2005. *Efficient Numerical Methods and Information-Processing Techniques for Modeling Hydro- and Environmental Systems*, 21st edition: Lecture Notes in Applied and Computational Mechanics, Springer-Verlag, Berlin Heidelberg, Germany.

Hou, J., Simons, F., Liang, Q., and Hinkelmann, R. 2013. An improved hydrostatic reconstruction method for shallow water model, *Journal of Hydraulic Research*, 52(3): 432–439.

Kesserwani, G., Shaw, J., Sharifian, M.K., Bau, D., Keylock, C.J., Bates, P., and Ryan, J.K. 2019. (Multi)wavelets increase both accuracy and efficiency of standard Godunov-type hydrodynamic models, *Advances in Water Resources*, 129: 31–55.

Khan, A.A., and Lai, W. 2017. *Modeling Shallow Water Flows Using the Discontinuous Galerkin Method*, CRC Press, Boca Roton, FL.

Kobayashi, K., Kitamura, D., Ando, K., and Ohi, N. 2015. Parallel computing for high-resolution/large-scale flood simulation using the K supercomputer, *Hydrological Research Letters*, 9: 61–68.

Lacasta, A., Morales-Hernández, M., Murillo, J., and García-Navarro, P. 2015. GPU implementation of the 2D shallow water equations for the simulation of rainfall/runoff events, *Environmental Earth Sciences*, 74: 7295–7305.

Lange, C., Schneider, M., Mutz, M., Haustein, M., Halle, M., Seidel, M., Sieker, H., Wolter, C., and Hinkelmann, R. 2015. Model-based design for restoration of a small urban river, *Journal of Hydro-environment Research*, 9(2): 226–236.

LeVeque, R.J. 2002. *Finite Volume Methods for Hyperbolic Problems*, Cambridge University Press, Cambridge, UK.

Liang, Q., Xia, X., and Hou, J. 2016. Catchment-scale high-resolution flash flood simulation using the GPU-based technology, *Procedia Engineering*, 154: 975–981.

Marafini, E., Tügel, F., Özgen, I., Hinkelmann, R., and La Rocca, M. 2018. Flash flood simulations based on shallow water equations to investigate protection measures for El Gouna, Egypt. In: *13th International Conference on Hydroinformatics*, Palermo, Italy.

Matta, E., Koch, H., Selge, F., Simshäuser, M.N., Rossiter, K., Nogueira da Silva, G.M., Gunkel, G., and Hinkelmann, R. 2018. Modelling the impacts of climate extremes and multiple water uses to support water management in the Icó-Mandantes Bay, Northeast Brazil, *Journal of Water and Climate Change*, 10(4): 893–906.

Mein, R.G., and Larson, C.L. 1971. *Modeling the Infiltration Component of the Rainfall-Runoff Process*, Water Resources Research Center, Minneapolis, MN.

Mügler, C., Planchon, O., Patin, J., Weill, S., Silvera, N., Richard, P., and Mouche, E. 2011. Comparison of roughness models to simulate overland flow and tracer transport experiments under simulated rainfall at plot scale, *Journal of Hydrology*, 402: 25–40.

Navas-Montilla, A., and Murillo, J. 2016. Asymptotically and exactly energy balanced Augmented flux-ADER schemes with application to hyperbolic conservation laws with geometric source terms, *Journal of Computational Physics*, 317: 108–147.

Néelz, S., and Pender, G. 2007. Sub-grid scale parameterisation of 2D hydrodynamic models of inundation in the urban area, *Acta Geophysica*, 55: 65–72.

Nelder, J.A., and Mead, R. 1965. A simplex method for function minimization, *Computer Journal*, 7: 308–313.

Overton, D., and Brakensiek, D. 1970. A kinematic model of surface runoff response. In: *Proceedings of the Wellington Symposium*, Unesco/IAHS, Paris, pp. 100–112.

Özgen, I., Teuber, K., Simons, F., Liang, D., and Hinkelmann, R. 2015. Upscaling the shallow water model with a novel roughness formulation, *Environmental Earth Sciences*, 74: 7371–7386.

Rawls, W., Brakensiek, D., and Miller, N. 1983. Green-Ampt infiltration parameters from soils data, *Journal of Hydraulic Engineering*, 1: 62–70.

Razafison, U., Cordier, S., Delestre, O., Darboux, F., Lucas, C., and James, F. 2012. A shallow water model for the numerical simulation of overland flow on surfaces with ridges and furrows, *European Journal of Mechanics - B/Fluids*, 31: 44–52.

Simons, F. 2020. *A High Resolution Hydrodynamic Numerical Model for Surface Water Flow and Transport Processes Within a Flexible Software Framework*. PhD thesis. Technische Universität Berlin, Germany.

Simons, F., Busse, T., Hou, J., Özgen, I., and Hinkelmann, R. 2014. A model for overland flow and associated processes within the hydroinformatics modelling system, *Journal of Hydroinformatics*, 16(2): 375–391.

Smith, L.S., and Liang, Q., 2013. Towards a generalised GPU/CPU shallow-flow modelling tool, *Computers and Fluids*, 88: 334–343.

Thompson, S., Katul, G., and Porporato, S. 2010. Role of microtopography in rainfall-runoff partitioning: An analysis using idealized geometry, *Water Resources Research*, 46: W07520.

Toro, E.F. 1992. Riemann problems and the WAF method for solving twodimensional shallow water equations, *Philosophical Transactions: Physical Sciences and Engineering*, 338(1649): 43–68.

Toro, E.F. 2001. *Shock-Capturing Methods for Free-Surface Shallow Flows*, John Wiley and Sons, Ltd., Chichester, UK.

Tügel, F., Abdelrahman, A., Özgen-Xian, I., Hadidi, A., and Hinkelmann, R. 2020a. Rainfall-runoff modeling to investigate flash floods and mitigation measures in the Wadi Bili catchment, Egypt. In: *Advances in Hydroinformatics*, Springer Water, Springer, Singapore.

Tügel, F., Özgen-Xian, I., Marafini, E., Hadidi, A., and Hinkelmann, R. 2020b. Investigation of flash flood mitigation measures for an Egyptian city and the impact of infiltration, *Urban Water Journal*, 17(5): 396–406.

van Leer, B. 1979. Towards the ultimate conservative difference scheme V: A second order sequel to Godunov's method, *Journal of Computational Physics*, 32: 101–136.

Viero, D.P., Peruzzo, P., Carniello, L., and Defina, A. 2014. Integrated mathematical modeling of hydrological and hydrodynamic response to rainfall events in rural lowland catchments, *Water Resources Research*, 50: WR014293.

Wales, D.J., and Doye, J.P. 1997. Global optimization by basin-hopping and the lowest energy structures of lennard-jones clusters containing up to 110 atoms, *The Journal of Physical Chemistry A*, 101(28): 5111–5116.

Whisler, F., and Bouwer, H. 1970. Comparison of methods for calculating vertical drainage and infiltration for soils, *Journal of Hydrology*, 10: 1–19.

Xia, X., Liang, Q., Pastor, M., Zou, W., and Zhuang, Y. 2013. Balancing the source terms in a SPH model for solving the shallow water equations, *Advances in Water Resources*, 59: 25–38.

Zhao, J., Özgen-Xian, I., Liang, D., Wang, T., and Hinkelmann, R. 2019. A depth-averaged non-cohesive sediment transport model with improved discretization of flux and source terms, *Journal of Hydrology*, 570: 647–665.

Part IV

Floods, River Restoration, and Climate Change

10 Sediment Transport and Changes in the River Bottom Topology Downstream of the Jeziorsko Reservoir

Hämmerling Mateusz, Kałuża Tomasz, and Zaborowski Stanisław

CONTENTS

10.1 INTRODUCTION

A retention reservoir is an element of the river network that significantly affects its hydrological regime, primarily by leveling extreme flows. It decreases maximum flow rates by reducing high water or accumulation and increases minimum flows, supporting the watercourse in low-flow periods (Song et al., 2020). Partitioning of a given riverbed and creating a dam reservoir consequently change the hydromorphological conditions both below and above the partition. The processes occurring during the operation of damming facilities are particularly important to the transformation of river valleys (Sindelar et al., 2017). When the continuity of sediment flow is disturbed, i.e., water flowing out of the reservoir is deprived of bedload and floating material, which results in intensification of the process of bottom erosion downstream of the damming structure (Sun et al., 2016). It manifests in two ways: as local erosion directly below the structure and as linear erosion, which successively lowers the bottom of the riverbed (Hammerling et al., 2019). Bidorn et al. (2016) observed characteristic sediment transport of the Ping River downstream of the Bhumibol Dam. It was found that sediment characteristics of the Ping River, by decreasing suspended sediment loads downstream of the dam, consequently caused an increase in *bed-to-suspended-sediment* ratio.

The literature provides numerous examples of simple and more complex types of sediment division. However, the most commonly used criterion is based on the method the material is transported. Nittrouer [6] distinguishes floating, dragged (bedload), suspended, and dissolved sediment. Sediment transport in the river is analyzed by direct field measurements and calculations using empirical equations (Sun at al., 2016; Camenan and Larson, 2005; Latosinski et al., 2017; Liébault et al., 2016). Also, more and more elaborate numerical models are being applied by researchers.

Sediment transport is a complex process, therefore its description by means of equations may involve some simplifications and assumptions. These assumptions have been used by researchers to develop the equations using one or two dominant factors such as water discharge, average flow velocity, energy slope, and shear stress (Yang et al., 2009).

Influential parameters for sediment transport according to (Sinnakaudan et al., 2010) are dimensionless unit stream power, time rate of potential energy expenditure per unit weight in an alluvial channel, dimensionless flow depth (water depth ratio), relative roughness on bed, roughness on the bed, fall velocity Reynolds number, relative sediment particle size, flow parameter, ratio of shear velocity and fall velocity, dimensionless grain size, sediment mobility number, motion of near-bed particles, hiding-exposure function, particle densimetric Froude number for initiation of motion, and sheltering factor.

Tayfur and Singh (Tayfur and Singh, 2012) investigate transport capacity models based on six different dominant variables – shear stress, stream power, unit stream power, discharge, velocity, and slope in the modeling of unsteady and non-equilibrium sediment waves conceptualized as kinematic waves in alluvial channels. Some of them are presented by Wyżga (Wyżga et al., 2017).

Bohorquez and Ancey (Bohorquez and Ancey, 2015) analyze, among others, parameters critical to the beginning of particles' movement and their rest, such as stresses acting on the river bottom or the Shields parameter.

Fischer-Antze (Fischer-Antze et al., 2008) used a 3D computational fluid dynamics model to compute bed changes in a section of the Danube River. Forecasting of sediment transport with the use of mathematical models is presented in (Tayfur and Singh, 2012; Lotsari et al., 2014; Pinto et al., 2012; Recking et al., 2016; Siviglia et al., 2013; Neupane and Yager, 2013), which used a one-dimensional (1D) bedload transport model for gravel-bedded river networks. Juez et al. (2016) present a 2D depth-averaged model to the application of sediment replenishment in open channels. Carrivick et al. (2010) applied Delft3D software for fluid dynamics.

The literature offers a very interesting approach to sediment transport calculations which uses artificial neural networks. The implementation of machine learning techniques in this context has produced very good results. For instance, the models derived from ANNs and ANFIS performed equally well, better than SR (Kitsikoudis et al., 2014). Yang et al. (2009) show that the ANN model with four input nodes V, S, D, and d50 is able to estimate the sediment transport rate accurately. However, it is worthy of mention that the ANN model with three input data estimates the sediment load as accurately as some well-known formulas.

The paper contains the analyses of changes caused by the process of bottom erosion downstream of the Jeziorsko reservoir in the Warta River (Ptak et al., 2019), located in the central part of Poland. The effect of erosion is changing over time, albeit systematic lowering of the river bottom, which after years of operation may cover tens of kilometers of the watercourse (Zheng et al., 2018). At the same time, the phenomenon is accompanied by a decrease in the water table, reduction in longitudinal gradient, changes in sediment particle size, and modifications to the groundwater system (Dysarz et al., 2017). The authors used the HEC-RAS 5.0.1 computer program to model changes in the river bottom topology downstream of the Jeziorsko reservoir, referring the obtained results to the results of field measurements

10.2 METHODOLOGY

The object of the presented research is the section of the Warta River from the frontal dam of the Jeziorsko reservoir to the town of Uniejów (km 483 + 657 – km 465 + 850), which is 16,578 km long. The test section includes four stabilizing thresholds constructed at the turn of 1993–1994 with threshold no. 1 and threshold no. 2 located at a distance of 270 m and 380 m downstream of the frontal dam of the reservoir. They were constructed as it was necessary to raise the water table downstream of the weir and the hydropower plant and to reduce the intensity of the erosion process occurring below the reservoir. Due to the necessity to weaken the process of longitudinal erosion in the section below the Jeziorsko reservoir, another two stabilizing thresholds were built in 2005 (threshold no. 3 and threshold no. 4). Figure 10.1 shows the map of the Warta River catchment area.

FIGURE 10.1 Map of the Warta River catchment area (Laks et al., 2018).

The Warta River is the largest right-bank tributary of the Odra, to which it flows at km 617 + 600 (Malinger et al., 2020). The most important tributaries within the Warta catchment area are Noteć, Obra, Prosna, Ner, and Wełna (Kałuża et al., 2020). In addition to the well-developed river network, the hydrography of the Warta River catchment area is characterized by the presence of numerous artificial reservoirs and lakes of various sizes. Significant changes in the riverbed took place after the construction of the Jeziorsko reservoir in 1986. The reservoir was created as a result of partitioning the riverbed in its middle course, on the border of Greater Poland and the Łódź province. The decision to build the facility was justified, among others, by the need to ensure an open cooling cycle for the Konin–Pątnów–Adamów power plant. Currently, it is the fourth dam reservoir in terms of total capacity in Poland. The basic components of the reservoir are:

- Frontal dam;
- Weir;
- Hydroelectric power plant;
- Pichna side dam with Pęczniew pump station;
- Teleszyna side dam with Miłkowice and Jeziorsko pump stations;
- Glinno backwater dam with pump station;
- Proboszczowice backwater dam with pump station;
- Fish ponds.

The frontal dam with a length of 2,732 m, erected at 484 + 300 km of the river along the line of Skęczniew and Siedlątków closes the catchment area of 9,012.6 km². Its body is made of medium-grained sands. The sealing is a tight reinforced concrete screen, which at the same time serves as protection of the upstream slope against its exposition to wave motion. The three-span weir, which is the main component of the reservoir, is located in the axis of the dam. It is a drain and overflow facility made as a dock structure. It has three 12 m overflow openings and four bottom outlets (3.3 × 2.0 m). The outlets are placed in pairs in the extreme spans of the weir and equipped with steel segments driven hydraulically. The closures of the upper overflows are shell flaps. The total discharge of overflows and outlets allows for passing $Q_{0,02\%} = 890$ m^3/s of test water at a maximum damming of 121.50 masl. The allowable outflow below the weir is $Q_{doz} = 110$ m^3/s, and the permissible outflow equals $Q_{dop} = 310$ m^3/s (Kałuża et al., 2017).

The hydropower function at the barrage is performed by a run-of-river hydroelectric power plant commissioned in 1995. The power plant building is situated on the right bank downstream and is equipped with two vertical shaft Kaplan turbines. The chamber of valves and pipelines as well as the water intake for the hydroelectric power plant are fitted in the body of the frontal dam. The average annual electricity production is 20 million kWh, provided by two turbine sets with a maximum installed power of 4.89 MW.

In order to recognize the morphology, granulometric composition, and hydraulic conditions in the analyzed section of the Warta River, a series of field measurements were made in 1975–2015. The scope of work included measurements of the water table topology and cross-sections over the period from 1989 to 2013 and the sieve analyses of bottom sediment samples taken between 1995–2010. Figure 10.2 shows the distribution of selected cross-sections with the location of field measurements.

Granulometric measurements consisting in taking the samples of bottom material were carried out in 1995–2010. Each time, the material was obtained directly from the bottom surface and considered as samples with a disturbed structure. The measurement points were located in the axis of the riverbed in stabilized cross-sections. The characteristic diameters of bottom sediment: d_{10}, d_{35}, d_{50}, d_{90}, and d_{95} were determined on the basis of granulometric analyses of the samples.

10.2.1 Modeling

The HEC-RAS 5.0.1 mathematical model was used for calculations. The program was applied to create a spatial model of the analyzed section of the Warta River, and then simulate its steady motion and the transport of sediment in quasi-unsteady motion.

Modeling of the water table topology was performed as a one-dimensional model. Under the conditions of steady motion, the water table topology was determined by solving the Bernoulli equation describing the energy balance between two successive cross-sections.

The prognostic studies of the erosion process were performed using a one-dimensional model describing the two-phase movement of water and bottom material. Modeling consists in combining the parameters of transported sediment with hydraulic parameters. Calculations can be made using the concept of steady and quasi-unsteady motion.

Quasi-unsteady motion assumes the division of simulation time into time steps in which flow rate parameters remain unchanged over time. Then, based on the Bernoulli equation, hydraulic calculations are made for each time step. The model simplifies the hydrodynamics of the system – the hydrograph of flow rates is divided into individual steady flows. The obtained solution is more stable than the Saint Venant equation solution (which is the basis of unsteady motion) and allows for shorter computation time.

10.2.2 Formulas Describing the Discharge of Sediment Transport

The calculation algorithm of sediment is solved using the equation of sediment continuity that can be described by the Exner equation and expressed with the formula (10.26):

$$\left(1 - \lambda_P\right) \cdot B \cdot \frac{\delta \eta}{\delta t} = -\frac{\delta Q_s}{\delta x} \tag{10.26}$$

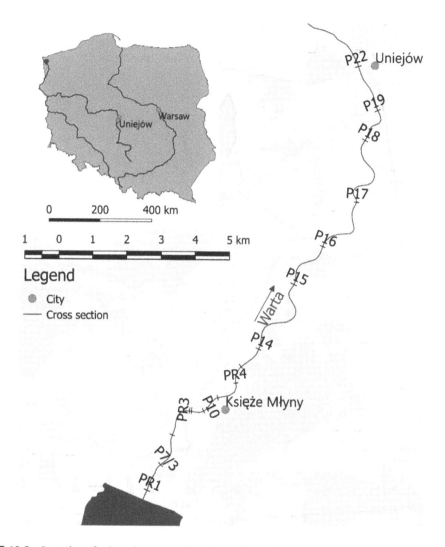

FIGURE 10.2 Location of selected cross-sections in the Warta River.

gdzie:

B – bottom width [m],

η – riverbed ordinate [m],

λ_P – porosity of the bottom sediment layer,

t – time [s],

x – distance [m],

Q_s – volumetric discharge of sediment transport [m³/s].

The key element in forecasting sediment transport and assessing erosion processes is to select an appropriate formula that will enable to obtain the required compliance of calculations with field measurement results. The Yang formula is an equation that describes the total load of transported sediment. It uses the concept of unit stream power and fall velocity (Radecki-Pawlik, 2014). The Meyer-Peter and Muller formula, known under the abbreviated name of MPM, uses the assumption that part of stream energy is used to overcome additional resistance to motion caused by the irregularities in the riverbed (Radecki-Pawlik, 2014; Gao, 2012). On the other hand, the Engelund-Hansen formula is used to determine the total discharge of sediment transport (Radecki-Pawlik, 2014; Choi and Lee, 2015; Okcu et al., 2016). The calculations also used the van Rijn formula, the results of which were compared with the results obtained from the mathematical model (10.27; 10.28).

Unit discharge of bedload sediment can be calculated from the formula:

$$\frac{s_b}{Uh} = C_{rb} \left[\frac{U - U_c}{\left[g D_{50} \left(\frac{\rho_s - \rho}{\rho} \right) \right]^{0,5}} \right]^{2,5} \left(\frac{D_{50}}{h} \right) \tag{10.27}$$

$$\frac{s_u}{Uh} = C_{ru} \left[\frac{U - U_c}{\left[g D_{50} \left(\frac{\rho_s - \rho}{\rho} \right) \right]^{0,5}} \right]^{2,4} \left(\frac{D_{50}}{h} \right) D_*^{-0,6} \tag{10.28}$$

The average critical velocity of water flow Uc can be calculated from the formula (10.29; 10.30):

$$U_c = 0,19 \left(D_{50} \right)^{0,1} \log \left(\frac{12 R_h}{3 D_{90}} \right) \quad 0.1 \leq D_{50} \leq 0,5 \, \text{mm} \tag{10.29}$$

$$U_c = 8,5 \left(D_{50} \right)^{0,6} \log \left(\frac{12 R_h}{3 D_{90}} \right) \quad 0.5 \leq D_{50} \leq 2,0 \, \text{mm} \tag{10.30}$$

10.2.3 Assumptions for the Model

For the purpose of the study, the effect of adopted boundary conditions and applied sediment transport formulas on the results of calculations was analyzed. The simulations were conducted for the period of one year (June 25, 2009–June 24, 2010), and their results were compared with the results of bottom measurements made between June 22–26, 2010.

The characteristics of granulometric composition of bottom material in cross-sections were determined based on the particle size curves prepared in 2009. Calculations for sediment transport were made using the Thomas sorting method (Exner 5), and for particle fall velocity with the Ruby method. After taking into account the applicability of formulas, the Meyer Peter Muller (MPM), Engelund-Hansen, and Yang functions were selected for the simulation. Modeling of the flow rate in the concept of quasi-unsteady motion required the introduction of a hydrograph of daily flows. Other equations determining sediment transport were also used for detailed calculations.

The shape of cross-sections measured in the section of the Warta River downstream of the Jeziorsko reservoir (km 483 + 300) to the town of Uniejów (km 465 + 850) was used as a starting point for the mathematical model and its further verification.

The model was supplemented with adopted boundary conditions for the inflow cross-section by including a flow hydrograph for the period from June 25, 2009 to June 24, 2010 and for the outflow cross-section by including the value of normal depth, provided in the form of water table gradient ($i = 0.32\%$).

Modeling of sediment transport in quasi-unsteady motion was performed assuming different calculation variants. The interruption of sediment flow continuity is the effect of partitioning the Warta riverbed with the dam (km 484 + 300). This has led to a reduction in the mass of material flowing into the sections below the dam. For the selected formulas describing sediment transport, there were made calculations with variable boundary conditions set in the upper calculation cross-section. The following configurations were adopted:

- Variant A – upper boundary condition: "Equilibrium load";
- Variant B – upper boundary condition: hydrograph of sediment discharge $Q_r = 0$.

10.3   RESULTS

The results of field measurements allowed for analyzing the variability in sediment diameters over time. In detailed analyses, the diameter D90 was used, which was the most important in terms of sediment transport. Figures 10.3–10.5 show changes in the diameter D90 in the tested section.

Based on the presented results, it can be observed that in 1995 the variability in characteristic diameters D90 (Figures 10.3–10.5) ranged from 0.75 mm (km 482 + 136) to 9.2 mm (km 483 + 390). In 2004, it ranged from 0.7 mm to 32 mm, and in 2010 from 0.34 mm to 46 mm. The highest value in the analyzed period was recorded in 2007 at the P5 cross-section (km 483 + 514) and P5/1 cross-section (km 483 + 220), and it was 85 mm.

The analysis of results showed that in 2008 there was a significant drop in particle size in the section from km 479 + 148 (P12/2) to km 465 + 850 (P22) (Figure 10.4). The smallest changes of

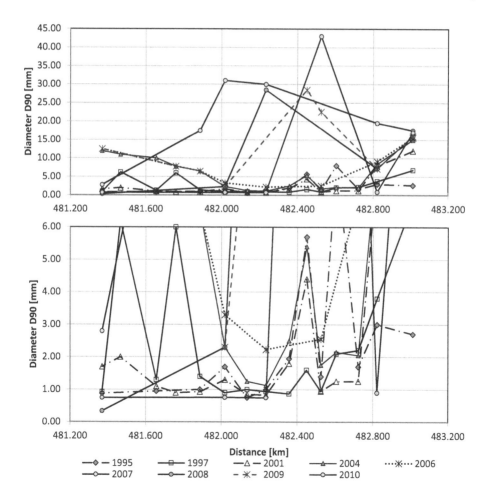

FIGURE 10.3   Changes in the diameter of particle size D90 in the section of approx. km 481 + 369 (P8/4) to km 483 + 012 (P6) in 1995–2010.

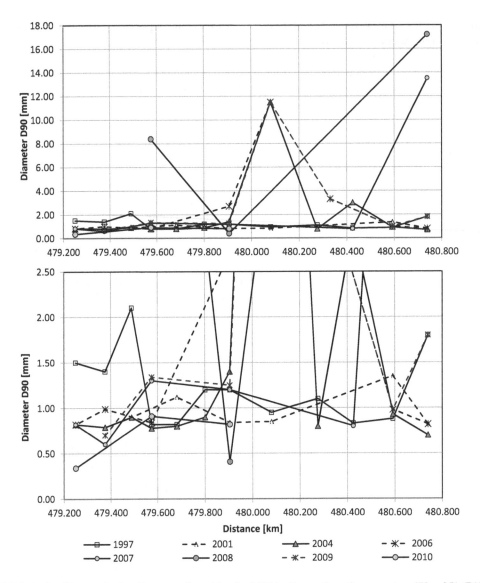

FIGURE 10.4 Changes in the diameter of particle size D90 in the section of approx. km 479 + 251 (P12/1) to km 480 + 740 (P10) in 1997–2010.

0.47 mm were observed at the P10/1 cross-section (km 480 + 592), and the largest (84.35 mm) at P5 (km 483 + 514). Until 2010, so after a 27-year service life of the reservoir, the value of characteristic diameters D90 was within the following range:

- D90 = 0.34–46 mm in the section from km 483 + 012 (P6 cross-section) to km 481 + 369 (P8/4 cross-section).
- D90 = 0.34–17.20 mm, in the section from km 480 + 740 (P10 cross-section) to km 479 + 251 (P12/1 cross-section).
- D90 = 0.3–18 mm, in the section from km 479 + 148 (P12/2 cross-section) to km P8/4 465 + 850 (P22 cross-section).

When compared, the results of measurements carried out from 1995 to 2010 show that in the analyzed section of the Warta River there is a systematic, but time-varying increase in the diameters of

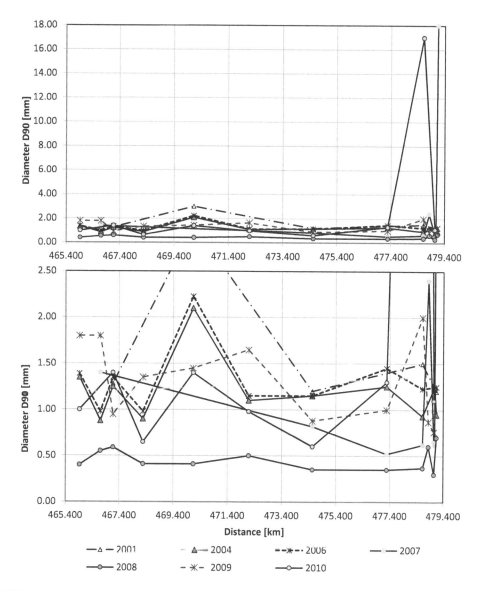

FIGURE 10.5 Changes in the diameter of particle size D90 in the section from km 465 + 850 (P22) to km 479 + 148 (P12/2) in 2001–2010.

bottom sediment. It results from the progressive process of bottom erosion and the accompanying phenomenon of material segregation. When water flows out of the reservoir, it scours finer fractions, leading to an increase in characteristic diameters and bottom paving. This particularly applies to the section of the river closest to the dam (from km 483 + 012 to km 480 + 850). It is worth noting that both the size and the variability in characteristic diameters are greater in the cross-sections closest to the dam (Figures 10.3–10.5). Referring to individual cross-sections, it can be noted that the diameters of bottom sediment in the further section (from km 479 + 148 to km 465 + 850) are characterized by a lower variability in particle size. A local reduction in the diameters was observed in 2008 in the section from the P12/2 cross-section (479 + 920) to the P22 cross-section (km 465 + 850). It was most likely caused by the construction of stabilizing thresholds no. 3 and no. 4, which decreased the dynamics of longitudinal erosion in the studied section of the river. The hydrological characteristics of the downstream side of the Jeziorsko reservoir were made on the basis of

TABLE 10.1

Characteristic Flows of the Second Order in the Multi-Year Period (1993–2013) Downstream of the Jeziorsko Reservoir

Characteristic Flow	Designation	Flow Rate [m³/s]
Highest flow rate of maximum annual values	WWQ	360.00
Average flow rate of maximum annual values	SWQ	124.88
Average flow rate of average annual values	SSQ	51.15
Average flow rate of minimum annual values	SNQ	23.87
Lowest flow rate of minimum annual values	NNQ	15.00

observations of daily water outflows. The characteristic flows of the second-order downstream of the reservoir covering the period of 21 years are presented in Table 10.1.

Calculations of the water table topology at the characteristic flows of the second order were made for the geometry of the riverbed measured in 2009 and 2010. The simulation was carried out in steady motion. The water table elevations read from the flow rate curve in the Uniejów stream gauge profile (km 465 + 850) were assumed as the lower boundary condition. Figure 10.6 shows

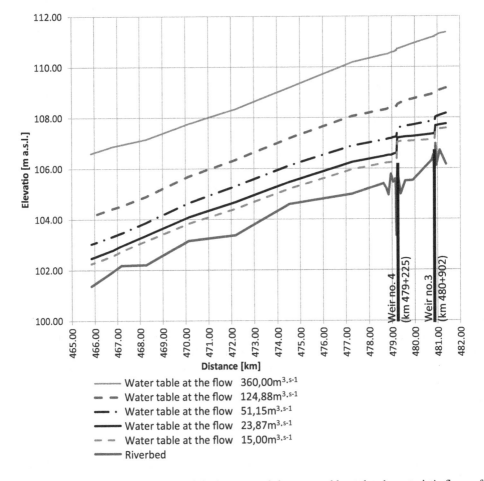

FIGURE 10.6 Changes in the topology of the bottom and the water table at the characteristic flows of the second order in 2009.

changes in the water table topology at the characteristic flows of the second order for the river bottom in 2009.

The analysis of results showed that the stabilizing thresholds exert a significant effect on the formation of the water table in the studied section (threshold no. 3 and threshold no. 4). Differences in the water table levels noticeable within threshold 4, at the flow NNQ = 15.00 m³/s, range from 0.74 m (2009) to 0.72 m (2010). Wherein, the impact of thresholds decreases with an increase in flow values, and it is 0.32 m (2009) and 0.42 m (2010) at the flow SSQ = 51.05 m³/s. Threshold no. 3 at the minimum flow records 0.39 m (2009) and 0.36 m (2010), whereas at the flow SSQ = 23.87 m³/s the values are 0,31 m (2009) and 0.28 m (2010), respectively. Under the conditions of high-water transition (WWQ = 360 m³/s), the flow along the entire section occurs outside the main channel, thus the above-mentioned structures do not have a significant impact on the water table topology.

When examining the longitudinal gradients, it can be noticed that they are variable in the section, and at the maximum flow (WWQ = 360 m³/s) they are as follows:

- In the section from km 481 + 369 (P8/4) to km 480 + 822 (PR3/11):
  - I = 0.35 ‰ in 2009;
  - I = 0.27 ‰ in 2010.
- In the section from km 480 + 740 (P10) to km 479 + 138 (P12/2):
  - I = 0.34 ‰ in 2009;
  - I = 0.29 ‰ in 2010.
- In the section from km 479 + 092 (P12/3) to km 465 + 850 (P22):
  - I = 0.30 ‰ in 2009;
  - I = 0.31 ‰ in 2010.

At the lowest observed flow (NNQ = 15 m³/s), the longitudinal gradients range from 0.8 ‰ (from P8/4 cross-section at km 481 + 369 to P3/11 cross-section at km 480 + 800) to 0.3‰ (from P10 cross-section at km 480 + 824 to P12/2 cross-section at km 479 + 138). For the analysed flow rates, the gradients in the section from km 481 + 369 to km 465 + 850 are always within the range of 0.30–0.32‰.

10.3.1 THE RESULTS OF SEDIMENT TRANSPORT MODELING

The simulations for sediment transport were made for the section of the Warta River from km 481 + 369 (P8/4 cross-section) to km 465 + 850 (P22 cross-section). Water flow modeling was carried out with the assumption of the concept of quasi-unsteady motion using the HEC-RAS program.

The curves of sediment discharge as a function of water flow rate for the lower calculation cross-section (P22 at km 465 + 850) are presented in Figure 10.7.

The highest values of sediment discharge at the P22 cross-section (km 486 + 850) were obtained from the Engelund-Hansen formula. Significantly lower values were calculated from the Ackers-White and Laursen equations. The values calculated using the Van Rijn formula were the closest to the values obtained from the Yang equation. However, when analyzing the range of flows 10–30 m³/s, the flow rate curve was the closest to the values obtained from the MPM equation. Whereas in the flows ranging 40–130 m³/s, to the Ackers-White formula. For further analyses, due to the number of obtained results, three equations describing sediment transport were selected, i.e., Engelund-Hansen, Meyer, Peter Muller, and Yang.

In the 38 tested cross-sections, the difference in the averaged bottom elevations between the first and the last day of the simulation was determined. Then, on the basis of these values, the number of cross-sections in which the processes of material erosion or accumulation occurred were estimated and compared with the results of field measurements (Table 10.2).

Certain trends can be observed in the variability of the presented results. They are evident both in the analysis of preferable variants and adopted calculation formulas. The results of field

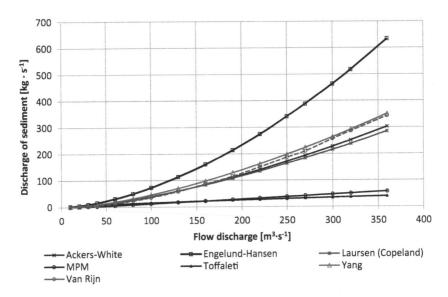

FIGURE 10.7 Changes in sediment discharge depending on the flow rate in the cross-section P22 (km 465 + 850) for adopted sediment transport formulas.

TABLE 10.2

Analysis of Bottom Accumulation and Erosion Processes in the Section from km 481 + 369 to km 465 + 850

Calculation Variant	Calculation Formula	Erosion	Accumulation	No Changes to Bottom Elevations
		Number of Tested Cross-Sections		
A	Engelund-Hansen	9	28	1
	Meyer Peter Muller	8	29	1
	Yang	10	27	1
B	Engelund-Hansen	29	9	0
	Meyer Peter Muller	11	26	1
	Yang	20	18	0
Results of measurements		14	19	5

measurements indicate that in 2009–2010 the accumulation process was predominant in the studied section, and it was recorded in 19 cross-sections. In five of the analyzed cross-sections, the bottom elevations did not change.

Assuming variant A, the calculation results for each used equation take similar values. The number of cross-sections with sediment accumulation was 27 to 29, while erosion was recorded in 8 to 10 cross-sections. For variant B, the tendency to erosion or accumulation was strongly dependent on the method used to calculate sediment transport. The application of the Engelund-Hansen formula resulted in the ultimate bottom topology logically consistent with the dominant erosion process. The opposite results were obtained using the MPM method. The results of calculations with the Yang equation show a similar number of cross-sections with a tendency toward the accumulation and erosion of the riverbed. On the basis of conducted analyses, it was found that the results obtained with the use of variant A reflect the natural state of the riverbed more accurately than the assumption of a complete lack of inflow (variant B). Figure 10.8 shows the variability in the bottom elevations modeled with the use of various equations in relation to the results of field measurements.

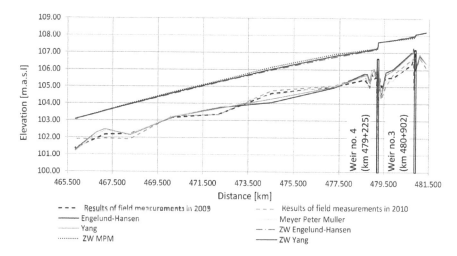

FIGURE 10.8 Calculation results according to variant A.

Regardless of the adopted boundary condition, each time the results closest to the real ones were obtained using the MPM equation. The calculations of sediment discharge were made using a set of formulas available in the HEC-RAS program. The results are presented in the form of the dependence of sediment discharge on water flow rate.

In the studied section, the mass transport of bottom material occurs during the transition of medium and high (flood) flows, with higher values recorded at the P22 cross-section (km 465 + 850).

The course of changes in sediment mass flowing in and out of cross-sections was also analyzed. In order to illustrate more efficiently the effect of the variability in sediment mass on the shaping of the riverbed, Figures 10.9 and 10.10 show changes in the averaged bottom elevations during the simulations.

Figure 10.9 shows the variability in sediment mass flowing in and out of the P8/4 cross-section. The analysis of results showed that in the initial simulation period, the mass of material flowing out of the cross-section exceeded the mass of inflowing sediment, which led to a lowering of the river bottom.

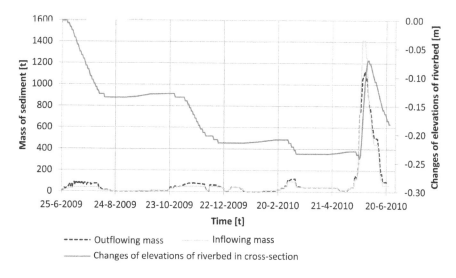

FIGURE 10.9 Effect of sediment mass flowing in and out of P8/4 cross-section (km 481 + 369) on changes in the bottom elevations.

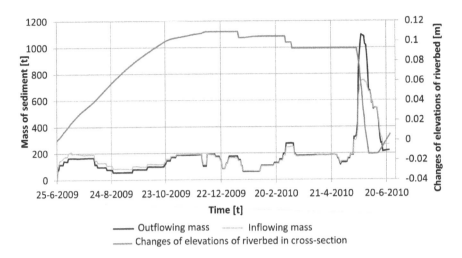

FIGURE 10.10 Effect of sediment mass flowing in and out of P22 cross-section (km 465 + 850) on changes in the bottom elevations.

From September 2009 to October 2009, there was an equilibrium in sediment transport, which resulted in the stabilization of the bottom elevations. In the analyzed period, on May 25–27, 2010, the highest mass of sediment brought to the cross-section (1,410.58 t) was recorded at the flow of 360 m^3/s.

The results of the variability in sediment mass at km 465 + 850 shown in Figure 10.10 indicate that in the initial simulation period (June 25, 2009–December 7, 2009), the cross-section was dominated by the process of material accumulation. It was manifested by a systematic increase in the averaged bottom ordinate. With the slight variation in flow rates (28–64 m^3/s), both the mass of sediment flowing in and out of the cross-section remained constant. The effect of the variability in sediment mass on the formation of the river bottom was most clearly visible during the transition of the flood wave (May 20–June 9, 2010). At the climax (May 24, 2010), 1,095 t of sediment was washed out during one day, whereas 748 t/d was washed in. As a consequence, the bottom elevations dropped sharply by 0.10 m. During this period, 3,174.46 tons of material eroded (151 t/d on average).

The mass of transported sediment is presented as the sum of the mass of all fractions of bottom sediment. The Meyer Peter Muller formula was used in the calculations, and the results are presented for three cross-sections: P8/4 (km 481 + 369), P14 (km 477 + 284), and P22 (km 465 + 850). The sum curves of sediment mass flowing into the above-mentioned sections are marked as "IN," while the mass of sediment flowing out of as "OUT" (Figure 10.11).

The mass of sediment flowing into the P8/4 cross-section (km 481 + 369) and flowing out of P22 cross-section (km 465 + 850) was determined as the sum of sediment transported during one year. Based on the difference between the mass of sediment flowing in and out of, it can be concluded that during one year in the section from P8/4 cross-section (km 481 + 369) to P22 cross-section (km 465 + 850), 36,295 t of material was washed away, which gives 99.4 t/d. In the observed period, the section of 4.085 km (from P8/4 cross-section to P14 cross-section) eroded 8,485 t. The section from P14 to P22 was characterized by a slightly higher value. For one kilometer of the riverbed, it was 2,444.69 t/year. By comparing the amount of sediment flowing in and out of the P14 cross-section, it can be seen that there was a systematic accumulation during the computations, which is confirmed by the variability in the bottom elevations (Figure 10.7).

After the entire simulation period, 181 t of material was accumulated at the P22 cross-section, and 5671 t at P14 cross-section (236.3 t/d on average). An inverse relationship was observed at the P8/4 cross-section, with the noticeable process of bottom erosion, which resulted in 1,520 t of sediment being washed out during the year under study. Figure 10.12 presents the variability in sediment mass flowing out over time.

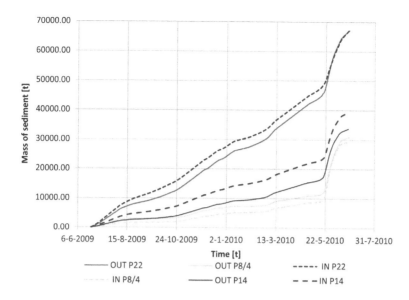

FIGURE 10.11   Sum curves of sediment mass flowing in and out of selected cross-sections.

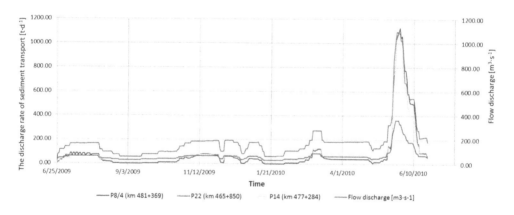

FIGURE 10.12   Variability in the mass of flowing sediment for selected measurement cross-sections.

By comparing the sediment flow at P8/4, P14, and P22 cross-sections, it can be concluded that the P22 cross-section is characterized by higher values of sediment discharge. The amount of sediment transport for P8/4 and P14 cross-sections was very similar for the flow rates and approximately equal to the flow rates with average values. Based on the analysis of results, it can be stated that a six-fold increase in the water flow rate causes a 20-fold increase in the discharge of sediment transport. The analysis of results indicated that high water, which intensifies the processes of material erosion and accumulation, plays an important role in the formation of the river bottom.

10.4   DISCUSSIONS

It is very important to determine the dynamics of sediment transport in freely flowing rivers as well as in rivers transformed by damming water. The transport of sediment is influenced by many factors, one of them is the speed of water flow which is determined by the fall of the water table (Hachoł et al., 2017). Particularly in the latter case, understanding the processes of bottom changes is vital. This is confirmed by the works of other researchers, indicating, as was the case in the studied section of the Warta, that local bottom obstacles and flow rate variabilities are important.

For example, Kabir et al. (2012) used as the case study the Abukuma River for evaluating bedload transport formulae and for simulating riverbed level variations. The model results have also reasonably illustrated the possible riverbed level variations. For example, deposition has been found to have occurred mostly at the confluence of different branches, whereas channel sections having steeper gradients have been subjected to substantial erosion. Recking et al. (2012) tested 16 common bedload transport equations with several data sets corresponding to different measurement periods. In all cases, similarly to the analyzed case of the Warta River, the formulas hide large temporary or seasonal fluctuations that may affect the average values calculated. However, even when formulas are assumed to be efficient, it might be difficult to predict bedload transport with precision considering the uncertainty associated with input parameters.

Determination of the mass of transported sediment is very important, particularly in the context of the operation of facilities interrupting the continuity of sediment transport. The analyses revealed a very strong influence of flow rate on the intensity of sediment transport. This was particularly evident during the passage of the flood that occurred in the spring of 2010 (Figure 10.12). Similar correlations presented Erskine and Saynor, Angelis et al., and López et al. (Erskine and Saynor, 2013; Angelis et al., 2013; López et al., 2014) in the form of regression relationships simultaneously describing the measured water flow rate and sediment transport in the Ngarradj River (Northern Australia), the Nestos River (Greece), and the Ebro River (Spain).

The determination of sediment transport intensity is closely related to the equation used. As is the case in the analyses for the downstream side of the Jeziorsko reservoir, Bidorn et al. (2016), based on their studies, claimed that the Acker and White, England Hansen, and Yang functions over-estimate total sediment loads for most flow conditions. The Toffaletti and Laursen-Copeland functions give either over- or under-prediction of total sediment loads.

Choi and Lee (2015) said that DuBoys type formula such as MyereMuller and Peter's relationship predicted no bedload if the shear stress is less than its critical value although a significant bedload is transported in reality. However, Camenene Larson's formula appropriately predicted the lateral distribution of the bedload. On the basis of the obtained data, it was found that the largest sediment flows in individual cross-sections were obtained using the Engelund-Hansen equation, the smallest using Toffaleti. Similar results were obtained from the analysis of predictions of changes in the bedload system of the Flint River under the influence of applied deflectors, as presented in (Szałkiewicz et al., 2019). Calculated in this paper intensity of sediments transport using an empirical van Rijn equation was within the limits of values obtained by the largest number of equations (Figure 10.7). It was also confirmed by the results of field measurements.

The results presented by Kaless et al. (2015) show the downstream trend of erosion/deposition, with positive differences indicating siltation and negative values erosion. In there is a general trend of downstream siltation that is higher for predictions than observations. They certify that the approach sub-reach (cross-sections between x = 0 m and x = 300 m) predictions are lower than observations, and there are overestimations in the final sub-reach (cross-sections between x = 1,250 m and x = 1,400 m). Erwin et al. (2012)ured all components of a sediment mass balance for Jordanelle Dam on the Provo River. Based on transport rate measurements, sediment input to the reach exceeded outputs, producing a net sediment accumulation of approximately 290 m^3 (95% CI is 180–489). The difference between the value of erosion and deposition was also positive (472 m^3), but uncertainty in the topographic differencing (61,344 m^3) was larger than the observed net change in storage. The calculations indicated that in cross-section P22 more than 67,000 tons of sediments during the year which is approximately 2,800 m^3 of sediments. Analysis of Figure 10.11 does not show any difference between the mass of sediments flowing into and out of cross-section P22 at the end of the simulation. It is caused by the passage of a surge wave, because in periods of average flows the difference between the mass of inflowing and outflowing sediments was significant. Through cross-section P22 during the passage of the flood wave flowed 18,000 tons of sediments. It was about 25% of the total mass that flowed during the year.

This can refer to the results Liu et al. (2011) carried out very detailed research of the Urumqi River and certify that the 25-year average values are ~17 t × km^{-2} × yr^{-1} for chemical weathering and ≈29 t × km^{-2} × yr^{-1} for mechanical erosion. This gives a total of 46 t × km^{-2} × yr^{-1} of erosion on the upper catchment.

10.5 CONCLUSIONS

The ability to forecast the amount of transported material plays a vital role in describing changes in the morphological characteristics and stability of riverbeds. However, in order to obtain reliable results, a number of factors determining the conditions of its transport must be taken into account (Dalezios et al., 2018).

The purpose of the study was to analyze changes in the Warta riverbed caused by the interruption of the continuity of sediment movement. The sediment movement was calculated with the HEC-RAS program. The presented sediment transport simulation for the section of the Warta River from km 481 + 369 to km 465 + 850 covered the period of one year (June 24, 2009–June 25, 2010).

The author analyzed the effect of adopted boundary conditions and a number of applied computational formulas on the obtained results. The best calculation variant was selected by comparing the final changes in the geometry of the riverbed resulting from the simulation with the results of field measurements made in 2010. The conditions most similar to those prevailing in the studied section were obtained by specifying *equilibrium load* and the MPM equation as the boundary condition.

The analysis of calculated bottom levels with the use of various sediment transport equations showed the suitability of the model for sediment movement calculations. When comparing the values of sediment discharge calculated with various formulas, differences in the obtained results were found. The lowest intensity of sediment transport was obtained with the Toffaleti equation and the highest with the Engelund-Hansen equation. By comparing the values of sediment discharge obtained with the use of mathematical modeling and the van Rijn empirical equation, it can be concluded that the obtained results were similar.

On the basis of field investigation, it was found that the flood in 2010 caused large changes in the cross-section geometry of the river channel. During field surveys carried out after the flood, washed sand deposits were observed in the floodplains.

On the basis of analyses carried out with the HEC-RAS model, the highest transport of sediment was observed at the end of May and the beginning of June 2010. This period was characterized by flows six times higher than the average flow. The results showed an over fivefold increase in the discharge of sediment transport as compared to the average flows. Analyzing daily values of sediment transport in the research year it can be concluded that during near average flows higher expenditure characterize cross-section P22. Whereas during surge flows daily sediment flows were similar for all cross-sections.

During the flood, the transport capacity of the river was so large that at the P22 cross-section, the mass of inflowing sediment was much smaller than the mass of outflowing sediment. Analysis of bottom changes during the study year in cross-section P22 showed a slow increase in elevations over time until high flows occurred, which caused a lowering of the bottom level in the analyzed cross-section.

The prognostic assessment of sediment transport prepared on the basis of the selected variant indicated that mass transport of bottom material can be observed during flood flows. Its discharge during the transition of the flood wave (May 20–June 9, 2010) at P8/4 cross-section (km 481 + 369) was 788 t/d on average. The dynamics of the phenomenon are evidenced as this value at medium and low flows in the studied period was 41.4 t/d on average.

Conducted analysis proves the possibility of using a computer program to predict bedload transport below the reservoir on a lowland river. Using computer software it is also possible to estimate the dynamics of the bedload transport process both in terms of its quantity and variability of bed elevations. Determination of erosion dynamics is particularly important in the context of erosion below the reservoir. Too large changes in the bottom may pose a threat to structures located along the river.

REFERENCES

Angelis, I., Metallinos, A., and Hrissanthou, V. 2013. Regression analysis between sediment transport rates and stream discharge for the Nestos River, Greece, *Global NEST Journal*, 14: 362–370.

Bidorn, B., Kish, S.A., Donoghue, J.F., Bidorn, K., and Mama, R. 2016. Sediment transport characteristic of the Ping River Basin, Thailand, *Procedia Engineering*, 154: 557–564.

Bohorquez, P., and Ancey, C. 2015. Stochastic-deterministic modeling of bed load transport in shallow water flow over erodible slope: Linear stability analysis and numerical simulation, *Advances in Water Resources*, 83: 36–54.

Camenen, B., and Larson, M. 2005. A general formula for non-cohesive bed load sediment transport, *Estuarine, Coastal and Shelf Science*, 63: 249–260.

Carrivick, J.L., Manville, V., Graettinger, A., and Cronin, S.J. 2010. Coupled fluid dynamics-sediment transport modelling of a Crater Lake break-out lahar: Mt. Ruapehu, New Zealand, *Journal of Hydrology*, 388: 399–413.

Choi, S.-U., and Lee, J. 2015. Assessment of total sediment load in rivers using lateral distribution method, *Journal of Hydro-environment Research*, 9: 381–387.

Dalezios, N.R., Eslamian, S., Ostad-Ali-Askari, K., Rabbani, S., and Saeidi-Rizi, A. 2018. Sediments. In: Bobrowsky, P., and Marker, B. (eds) *Encyclopedia of Engineering Geology, Encyclopedia of Earth Sciences Series*, Springer.

Dysarz, T., Szałkiewicz, E., and Wicher-Dysarz, J. 2017. Long-term impact of sediment deposition and erosion on water surface profiles in the Ner river, *Water*, 9: 168.

Erskine, W.D., and Saynor, M.J. 2013. Hydrology and bedload transport relationships for sand-bed streams in the Ngarradj Creek catchment, northern Australia, *Journal of Hydrology*, 483: 68–79.

Erwin, S.O., Schmidt, J.C., Wheaton, J.M., and Wilcock, P.R. 2012. Closing a sediment budget for a reconfigured reach of the Provo river, Utah, United States, *Water Resources Research*, 48: 2011WR011035.

Fischer-Antze, T., Olsen, N.R.B., and Gutknecht, D. 2008. Three-dimensional CFD modeling of morphological bed changes in the Danube River: 3-D CFD modeling bed changes in the Danube, *Water Resour Research*, https://doi.org/10.1029/2007WR006402.

Gao, P. 2012. Validation and implications of an energy-based bedload transport equation: An energy-based bedload transport equation, *Sedimentology*, 59: 1926–1935.

Hachoł, J., Hämmerling, M., and Bondar-Nowakowska, E 2017. Applying the Analytical Hierarchy Process (AHP) into the effects assessment of river training works, *Journal of Water and Land Development*, 35: 63–72.

Hammerling, M., Walczak, N., Nowak, A., Mazur, R., and Chmist, J. 2019. Modelling velocity distributions and riverbed changes using computer code SSIIM belowSills stabilizing the riverbed, *Polish Journal of Environmental Studies*, 28: 1165–1179.

Juez, C, Battisacco, E., Schleiss, A.J., and Franca, M.J. 2016. Assessment of the performance of numerical modeling in reproducing a replenishment of sediments in a water-worked channel, *Advances in Water Resources*, 92: 10–22.

Kabir, M.A., Dutta, D., Hironaka, S., and Pang, A. 2012. Analysis of bed load equations and river bed level variations using basin-scale process-based modelling approach, *Water Resources Management*, 26: 1143–1163.

Kaless, G., Mao, L., Moretto, J., Picco, L., and Lenzi, M.A. 2015. The response of a gravel-bed river planform configuration to flow variations and bed reworking: A modelling study: Response of a gravel-bed river to flow variations, *Hydrological Processes*, 29: 3812–3828.

Kałuża, T., Sroka, Z., and Lewandowska, J. 2017. Impact of decreasing the normal damming level of the Jeziorsko reservoir on low flows in the Warta river, *Acta Scientiarum Polonorum Formatio Circumiectus*, 2: 107–122.

Kałuża, T., Sojka, M., Wróżyński, R., Jaskuła, J., Zaborowski, S., and Hämmerling, M. 2020. Modeling of river channel shading as a factor for changes in hydromorphological conditions of small lowland rivers, *Water*, 12: 527.

Kitsikoudis, V., Sidiropoulos, E., and Hrissanthou, V. 2014. Machine Learning Utilization for bed load transport in gravel-bed rivers, *Water Resources Management*, 28: 3727–3743.

Laks, I., Kałuża, T., and Zawadzki, P. 2018. Impact of a weir damage located on a polder on flood wave transformation - A case study of the Golina polder/ Poland, *WasserWirtschaft*, 108: 21–26.

Latosinski, F.G., Szupiany, R.N., Guerrero, M., Amsler, M.L., and Vionnet, C. 2017. The ADCP's bottom track capability for bedload prediction: Evidence on method reliability from sandy river applications, *Flow Measurement and Instrumentation*, 54: 124–135.

Liébault, F., Jantzi, H., Klotz, S., Laronne, J.B., and Recking, A. 2016. Bedload monitoring under conditions of ultra-high suspended sediment concentrations, *Journal of Hydrology*, 540: 947–958.

Liu, Y., Métivier, F., Gaillardet, J., Ye, B., Meunier, P., Narteau, C., Lajeunesse, E., Han, T., and Malverti, L. 2011. Erosion rates deduced from seasonal mass balance along the upper Urumqi river in Tianshan, *Solid Earth*, 2: 283–301.

López, R., Vericat, D., and Batalla, R.J. 2014. Evaluation of bed load transport formulae in a large regulated gravel bed river: The lower Ebro (NE Iberian Peninsula), *Journal of Hydrology*, 510: 164–181.

Lotsari, E., Wainwright, D., Corner, G.D., Alho, P., and Käyhkö, J. 2014. Surveyed and modelled one-year morphodynamics in the braided lower Tana River: Surveyed and modelled one-year morphodynamics, *Hydrological Process*, 28: 2685–2716.

Malinger, A., Kałuża, T., and Dysarz, T. 2020. LIDAR data application in the process of developing a hydrodynamic flow model exemplified by the Warta river reach. In: Kalinowska, M.B., Mrokowska, M.M., and Rowiński, P.M. (eds) *Recent Trends in Environmental Hydraulics*, Springer International Publishing, Cham, Germany, pp. 159–170.

Neupane, S., and Yager, E.M. 2013. Numerical simulation of the impact of sediment supply and streamflow variations on channel grain sizes and Chinook salmon habitat in mountain drainage networks: Sediment supply and hydrograph impacts on grain size and habitat, *Earth Surface Processes and Landforms*, 38: 1822–1837.

Nittrouer, J.A., Shaw, J., Lamb, M.P., and Mohrig, D. 2012. Spatial and temporal trends for water-flow velocity and bed-material sediment transport in the lower Mississippi River, *Geological Society of America Bulletin*, 124: 400–414.

Okcu, D., Pektas, A.O., and Uyumaz, A. 2016. Creating a non-linear total sediment load formula using polynomial best subset regression model, *Journal of Hydrology*, 539: 662–673.

Pinto, L., Fortunato, A.B., Zhang, Y., Oliveira, A., and Sancho, F.E.P. 2012. Development and validation of a three-dimensional morphodynamic modelling system for non-cohesive sediments, *Ocean Modelling*, 57–58: 1–14.

Ptak, M., Sojka, M., Kałuża, T., Choiński, A., and Nowak, B. 2019. Long-term water temperature trends of the Warta River in the years 1960–2009, *Ecohydrology and Hydrobiology*, 19: 441–451.

Radecki-Pawlik, A. 2014. Hydromorfologia rzek i potoków górskich. In: *Polish: Hydromorphology of Rivers and Mountain Streams*, Wydawnictwo Uniwersytetu Rolniczego w Krakowie, Kraków, Poland.

Recking, A., Liébault, F., Peteuil, C., and Jolimet, T. 2012. Testing bedload transport equations with consideration of time scales: Bedload Modelling, *Earth Surface Processes and Landforms*, 37: 774–789.

Recking, A., Piton, G., Vazquez-Tarrio, D., and Parker, G. 2016. Quantifying the morphological print of bedload transport: Morphological print, *Earth Surface Processes and Landforms*, 41: 809–822.

Sindelar, C., Schobesberger, J., and Habersack, H. 2017. Effects of weir height and reservoir widening on sediment continuity at run-of-river hydropower plants in gravel bed rivers, *Geomorphology*, 291: 106–115.

Sinnakaudan, S.K., Sulaiman, M.S., and Teoh, S.H. 2010. Total bed material load equation for high gradient rivers, *Journal of Hydro-environment Research*, 4: 243–251.

Siviglia, A., Stecca, G., Vanzo, D., Zolezzi, G., Toro, E.F., and Tubino, M. 2013. Numerical modelling of two-dimensional morphodynamics with applications to river bars and bifurcations, *Advances in Water Resources*, 52: 243–260.

Song, X., Zhuang, Y., Wang, X., Li, E., Zhang, Y., Lu, X., Yang, J., and Liu, X. 2020. Analysis of hydrologic regime changes caused by Dams in China, *Journal of Hydrologic Engineering*, 25: 05020003.

Sun, X., Li, C., Kuiper, K.F., Zhang, Z., Gao, J., and Wijbrans, J.R. 2016. Human impact on erosion patterns and sediment transport in the Yangtze River, *Global and Planetary Change*, 143: 88–99.

Szałkiewicz, E., Dysarz, T., Kałuża, T., Malinger, A., and Radecki-Pawlik, A. 2019. Analysis of in-stream restoration structures impact on hydraulic condition and sedimentation in the Flinta river, Poland, *Carpathian Journal of Earth and Environmental Sciences*, 14: 275–286.

Tayfur, G., and Singh, V.P. 2012. Transport capacity models for unsteady and non-equilibrium sediment transport in alluvial channels, *Computers and Electronics in Agriculture*, 86: 26–33.

Wyżga, B., Mikuś, P., Zawiejska, J., Ruiz-Villanueva, V., Kaczka, R.J., Czech, W. 2017. Log transport and deposition in incised, channelized, and multithread reaches of a wide mountain river: Tracking experiment during a 20-year flood, *Geomorphology*, 279: 98–111.

Yang, C.T., Marsooli, R., and Aalami, M.T. 2009. Evaluation of total load sediment transport formulas using ANN, *International Journal of Sediment Research*, 24: 274–286.

Zheng, S., Xu, Y.J., Cheng, H., Wang, B., Xu, W., and Wu, S. 2018. Riverbed erosion of the final 565 kilometers of the Yangtze River (Changjiang) following construction of the Three Gorges Dam, *Scientific Reports*, 8: 11917.

11 Impact of Climate Change on Flooding

Conrad Wasko

CONTENTS

11.1 INTRODUCTION

Despite design flood hydrology generally being based on the assumption of stationarity, climate change is impacting hydrology and flooding in a way that often violates this assumption (François et al., 2019; Wasko et al., 2021b; Milly et al., 2008). This chapter presents the evidence of non-stationarity for flooding based on historical data and understanding of natural systems (Mujere and Eslamian, 2014). Non-stationarity in flooding is an active area of research, and has been so for a long time (Schwarz, 1977). For example, even on geological time scales, using a 7,000-year record of flooding for the upper Mississippi, during the warmer drier periods floods were relatively smaller, but during the cooler wetter periods, floods were larger with an order of magnitude change in the recurrence interval of the flood (Knox, 1993). The order of magnitude changes in recurrence interval were associated with mean changes in temperature of 1–2° C and in annual precipitation changes of less than 10–20% (Knox, 1993). This suggests that the changes to the climate we observing currently may have large consequences on our ecosystems and economy (Alley et al., 2003).

Although large changes can be expected in flooding due to climatic change, observed changes in flooding and understanding the mechanisms causing the change remain poorly understood. The two major hurdles to this are discussed in Hall et al. (2014). The first hurdle is that flood measurements are subject to large errors. Flood discharges are measured indirectly as a height or level with this height being converted to a discharge based on a rating curve. This conversion of a flood level to a discharge is often based on a limited number of observations, performed on a logarithmic scale and generally an extrapolation because streamflow recording or measurement can rarely be performed in a flood situation (Kuczera, 1996). This method of flood measurement will also fail to incorporate the uncertainties due to the flood itself changing the channel geometry and hence the rating curve (Mcmillan and Westerberg, 2015). The second hurdle, related to the first, is the small signal-to-noise ratio (Hall et al., 2014). One can hardly imagine a noisier climatic signal in nature than flooding. The variability in flooding is a result of coupling random (aleatory) uncertainty in the precipitation causing the flood (Apel et al., 2004) with the catchment conditions prior to the flood event, which will be influenced by where in the catchment the rain occurred, as well as the temporal and spatial pattern of the rainfall.

Another problem in identifying changes in flooding is a lack of consistent terminology. The terminology often varies from adopting percentiles or probabilities of non-exceedance which generally correspond to the events that occur on a frequency of multiple times a year, to annual recurrence intervals or annual exceedance probabilities which implies that the events occur with frequency less than once a year. Further, often qualitative terminology is used such as "heavy" or "extreme" with the meaning of these terms varying between the background of the author and the target audience. Although much of this terminology is generally reserved for precipitation, it is also often adopted for streamflow and demonstrates the difficulty in reconciling published literature. This is further compounded by the fact that streamflow is generally reported as a mean daily flow, whereas, for engineering design, an instantaneous maximum is used. Although transfer functions exist from mean daily flow to the instantaneous peak flow (Fill and Steiner, 2003), this remains an added complication when interpreting the literature.

Finally, it is a standard practice to investigate trends in flooding using a series of annual maxima, but this in itself may not be particularly extreme (Bennett et al., 2018). For example, the design of infrastructure such as housing is generally planned around the 1-in-100-year flood level. As streamflow record lengths are generally only multiple decades in length (Do et al., 2018), either extrapolation is required or inferences have to be made using changes in the climatic drivers of flooding to understand changes in flooding relevant to engineering design (Villarini and Wasko, 2021). With this in mind, this chapter begins by discussing the physical basis of why flooding is changing with climate change, before discussing the changes in these physical drivers, and finally focusing on the observed changes in flooding itself. Additional considerations such as other anthropogenic factors are also presented. For brevity, the discussion of changes focuses on annual maxima, consistent with the majority of literature investigating historical trends in precipitation extremes and flooding.

11.2 PHYSICAL BASIS

Much of the physical basis for changed flooding is described in Sharma et al. (2018). As a change in extreme precipitation will naturally imply a change in pluvial flooding (Seneviratne et al., 2012), the discussion of the physical basis for changes in flooding begins with a discussion of changes in precipitation. A necessary (but not sufficient) condition of changes in pluvial flooding is a change in extreme precipitation. Pendergrass (2018) summarizes the different mechanisms for changing precipitation intensity. The first mechanism is dependent on the moisture-holding capacity of the atmosphere. Different literature will use differing terminology, but in brief, as global temperatures increase so the saturation vapor pressure increases also, and hence the amount of water vapor at saturation in the atmosphere will increase (Trenberth et al., 2003; Trenberth, 2011). This is described as the Clausius-Clapeyron scaling rate, after the thermodynamic relationship which governs this behavior, and is generally approximately to 6–7% per degree centigrade (Westra et al., 2014; Wasko, 2021). The second mechanism is often referred to as super Clausius-Clapeyron scaling (Lenderink and van Meijgaard, 2008; Wasko and Sharma, 2015; Trenberth, 2011) whereby latent heat release strengthens the storm's dynamics resulting in increases in precipitation extremes above 7% per degree centigrade, particularly for short duration (hourly) precipitation extremes (Fowler et al., 2021). The final mechanism is often more related to average precipitation changes. Warmer temperatures are associated with increased atmospheric stability and weakened circulations reducing precipitation events intensities. Indeed the overall intensity of the hydrologic cycle is controlled not by the availability of moisture but "by the availability of energy; specifically, the ability of the troposphere to radiate away latent heat released by precipitation" (Allen and Ingram, 2002). Hence although average precipitation is expected to increase with climatic change, the rate of increase is expected to be much less than the changes in extreme precipitation.

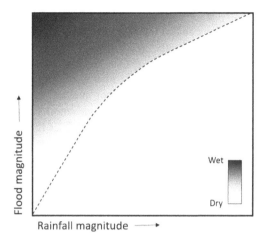

FIGURE 11.1 Conceptualization of the relationship between flood magnitude, rainfall magnitude, and the antecedent moisture conditions preceding the flood event assuming events are sampled on the basis of extreme rainfall. The shading represents the antecedent moisture condition state with a darker shading being a wetter moisture condition. A large flood magnitude can result from both a large rainfall magnitude and a small rainfall magnitude as long as the moisture conditions prior to the rainfall onset are wet.

It can be expected that the changes to pluvial flooding in highly urban areas may be driven by changes in precipitation extremes as there is little impervious area, but this simple transfer function will not hold in general (Wasko et al., 2019). Ivancic and Shaw (2015) presented the probability of observing the 99th percentile discharge given the 99th percentile of precipitation across the contiguous United States. Aggregated across all sites analyzed, only 36% of the precipitation events resulted in a discharge of corresponding extremity. When the precipitation was conditioned on the catchment being wet before the start of the event, that is, with a soil moisture wetness above the median, 62% of events achieved a discharge commensurate with the probability of exceedance of the precipitation. In contrast, when the state of the catchment was below the median wetness state only 13% of the events achieved the 99th percentile of discharge (Ivancic and Shaw, 2015). Figure 11.1 conceptualizes this point and is used to demonstrate the importance of antecedence in considering flood discharge magnitude (Wasko and Nathan, 2019). As mentioned, precipitation is required for flooding and a large precipitation magnitude will correspondingly cause a large flood event. But this flood event will not be as large if it does not fall onto a wet catchment. Conversely, small precipitation can also cause a large flood event, but a wet catchment is necessary for this to occur. The point here is two-fold and cannot be stressed strongly enough. At no point is the transfer function between a percentile of precipitation to a percentile of discharge equal to one as antecedent moisture conditions modulate the resultant streamflow (Bennett et al., 2018; Boughton and Droop, 2003; Pathiraja et al., 2012; Kuczera et al., 2006; Stephens et al., 2018a). But also, in the context of non-stationarity, if there is a shift to a drier or wetter catchment state then indeed the streamflow, for a given level of exceedance, will be modulated accordingly (Sharma et al., 2018).

The last primary factor driving changes in flooding discussed here is snow and snowmelt. Rising temperatures will have a two-fold impact on snowfall. First, precipitation will fall preferentially as rain and not snow (Trenberth, 2011); and second, snowmelt can be expected to occur earlier and possibly at a greater rate, albeit from a lower snowpack (Hamlet and Lettenmaier, 2007). These two factors can compete creating both increases and decreases in the resultant flood. Although it appears conclusive that rainfall will become the more dominant flood generating mechanism as compared to snow with climatic change (Vormoor et al., 2016) the resultant impact on flooding will be mixed. Naturally decreased snowpack should result in decreased flooding. However, if an anomalous larger-snow pack occurs then higher temperatures are likely to melt the snow more rapidly causing a larger flood. Alternatively, an earlier snow melt could result in increased soil moisture and once again, for a given precipitation cause a larger flood (Hamlet and Lettenmaier, 2007). Changes in flooding due to snow are also complicated by changes in rain on snow events. As temperatures increase, it has been shown that rain on snow events are less frequent at lower elevations due to snowpack declines, but at higher elevations where snow cover persists the risk of flooding due to rain on snow becomes more frequent due to a shift from snowfall to rain (Musselman et al., 2018).

Of course, there are a multitude of other factors that could cause the changes in flooding. Increases in temperatures will change the land cover (Liu et al., 2015) affecting transmission losses, such as those related to evapotranspiration (Huntington, 2006), as well as runoff generation processes (Ajami et al., 2017). Urbanization likewise will change winds and temperatures changing precipitation characteristics (Shepherd et al., 2002), and will impact the pervious fraction of the catchment, again changing runoff characteristics. However, the impact of changes in the above factors on flooding is difficult to quantify. For example, evaporation is expected to increase with increasing temperatures, but trends in evaporation are mixed (Johnson and Sharma, 2010; Stephens et al., 2018b; McMahon et al., 2013). In addition, for the duration of a flood event, the hydrologic losses due to evaporation volumetrically are much less than the magnitude of the precipitation causing flooding and the losses due to the antecedent moisture conditions of the catchment. For these reasons, and consistent with the published literature, the focus of this chapter remains on understanding how changes in precipitation, antecedent moisture, and snowmelt relate to changes in flooding (Berghuijs et al., 2016).

11.3   CHANGES IN CLIMATIC DRIVERS OF FLOODING

11.3.1   PRECIPITATION

One of the primary drivers of changes in flooding is the changes in precipitation, and in particular, the changes in extreme precipitation. Despite the strong physical basis for increases in extreme precipitation with increased temperatures due to climatic change, trends are difficult to identify due to the large amount of natural variability, sparse gauging networks, and the lack of long records. Indeed, it has been concluded in the UK that natural variability appears to dominate the currently observed trends in short-duration precipitation extremes (Kendon et al., 2018).

Global mean precipitation changes are mixed. Although global average precipitation trends demonstrate an increase in the order of 2 mm/year per decade (Hartmann et al., 2013), these trends are not spatially homogenous. There is a suggestion that in recent years, most of the increases have occurred over tropical areas with decreases elsewhere (Beck et al., 2019). But this is an oversimplification and much of this generalization only applies to the oceans, whereas for precipitation to cause flooding it needs to occur over land. Summarizing the results in the latest IPCC report (Hartmann et al., 2013), since 1950, mean precipitation has decreased across Spain and southern Europe but increased elsewhere through Europe. In Australia, the tropics have experienced increases, but all coastal areas in the extratropics have experienced a large decrease of over 10 mm/year/decade. Across Northern America and Southern America, the observed increases have often been more than 10 mm/year/decade. As a continent, Africa has experienced decreases as has much of East Asia. Across Russia, strong increases have been observed.

There are limited studies that investigate the changes in extreme precipitation on a global scale. Here the focus is on manuscripts that investigate the trends in annual daily maxima (Table 11.1). Using the GCOS Surface Network (GSN), the European Climate Assessment (ECA), and the daily Global Historical Climatology Network (GHCN-Daily) data sets, Alexander et al. (2006) developed a set of gridded climatic indices. For a grid resolution of 2.5 degrees latitude by 3.75 degrees longitude, the trend in the maximum annual daily precipitation (termed RX1day) was investigated. Of the 2,291 grid cells, at the field significance level of 5% (Wilks, 2006), 7.0% of land grid points showed a statistically significant increase in precipitation extremes while only 2.7% showed

TABLE 11.1

Summary of Manuscripts Investigating Changes in Annual Daily Precipitation Maxima on a Global Scale

Reference	Data	Conclusion
(Alexander et al., 2006)	HadEX Rx1day (1951–2003)	7.0% of land grid points show statistically significant increases while 2.7% show statistically significant decreases using a field significance level of 5%
(Donat et al., 2013)	HadEX2 Rx1day (1901–2010)	21.9% of land grid points show statistically significant increases while 7.1% show statistically significant decreases using at site significance level of 5%
(Westra et al., 2013)	Rx1day at gauge locations used in HadEX2	8.6% of gauge stations had significant increasing trends, while 2.0% had significant decreasing trends at a field significance level of 5%
(Sun et al., 2021)	Rx1day using multiple data sets largely based on HadEX2 locations	9.1% of gauge stations had significant increasing trends, while 2.1% had significant decreasing trends at a field significance level of 5%

a statistically significant decrease. This dataset was expanded in 2013 to include additional data sources and lengthened to incorporate the years from 1901 to 2010 (Donat et al., 2013). From 420 land-based grid boxes, 21.9% of sites showed statistically significant increases at the 5% level while 7.1% showed decreases (Donat et al., 2013). Using the annual maximum precipitation values from the 11,391 land-based observing stations used in Donat et al. (2013), a Mann-Kendall test showed that 8.6% of stations had significant increasing trends, while 2.0% had significant decreasing trends (Westra et al., 2013). This study was subsequently updated and expanded to 14,796 land-based observing stations and obtained similar conclusions (Sun et al., 2021).

Here the results from Westra et al. (2013) are replicated using the Global Historical Climatology Network – Daily (GHCN-Daily) data set (Menne et al., 2015). The results are limited to the stations with more than 50 years of annual maxima records resulting in 18,281 sites being used. Of the sites analyzed, 8.7% of sites show an at-site statistically significant increase in the annual daily maxima while 3.4% show a decrease at the 5% level (Figure 11.2). These results are comparable with those presented above and allow comment on the spatial pattern of the changes. The spatial coverage is clearly not uniform (Figure 11.2a), but where sites exist increases in the annual maxima are generally observed. Throughout the United States, Australia, south-eastern Brazil, South Africa, and India, increases are observed (Figure 11.2b). Although Europe has a dense gauge network, the number of sites with increases appears to be proportionally less than the other regions across the world. In comparison to the number of sites with increases in the annual maxima precipitation, there are very few sites in Europe or elsewhere with decreases (Figure 11.2c). The exception to this is Australia but again, this is a function of the dense gauging network and proportionally more sites show an increase than a decrease.

The overall evidence for increasing precipitation extremes globally is compelling. There are many local studies that examine the changes in precipitation extremes, but their results are not necessarily universal due to the examining of smaller regions. Of the studies that consider annual maxima across continental regions, increases in US annual maxima daily precipitation from the period of 1950 to 2011 were found to be commensurable with the Clausius-Clapeyron relationship of approximately 7%/°C when the increase was compared to the change in global temperature. But for hourly precipitation, which is expected to intensify more than daily precipitation, the converse was true, with a sensitivity of approximately 4%/°C (Barbero et al., 2017). Across the east coast of Australia, an average sensitivity of 5.6%/°C per degree of Australian land surface temperature was

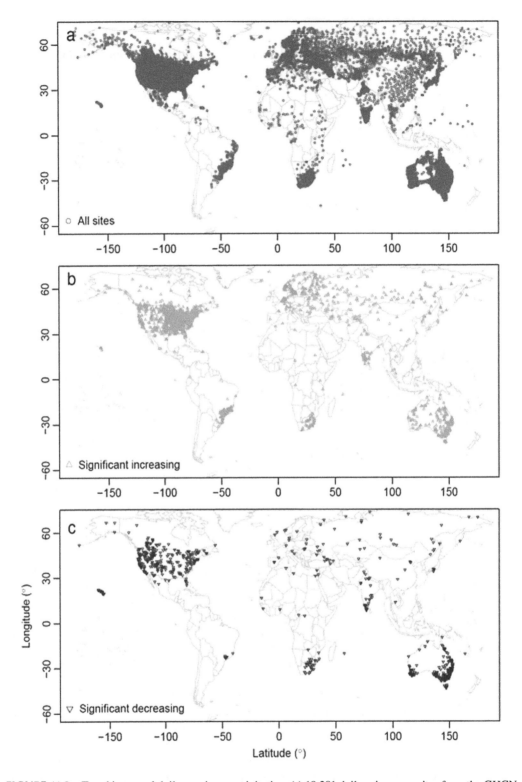

FIGURE 11.2 Trend in annual daily maxima precipitation. (a) 18,281 daily rain gauge sites from the GHCN data set (Menne et al., 2015) that contain more than 50 years of data; (b) sites showing a statistically significant positive linear trend at the 5% level; (c) sites showing a statistically significant decreasing linear trend at the 5% level. Of the sites analyzed, 8.7% of sites show a statistically significant increase in the annual daily maxima while 3.4% show a decrease.

found for precipitation of a six-minute duration with this rate of increase declining with increasing storm duration (Westra and Sisson, 2011). This is in stark contrast to a recent study across Australia which compared hourly precipitation from the period 1990–2013 and 1966–1989 for the various levels of exceedance and found the rarest events increased on average at a rate well in excess (close to double) the Clausius-Clapeyron relation (Guerreiro et al., 2018). Globally, Westra et al. (2013), through the application of a non-stationary generalized extreme value distribution, found an association to the daily annual maxima change with the changes in global mean temperature at a rate of between 5.9% and 7.7% per degree Kelvin, in line with the Clausius-Clapeyron relationship, with a recent update of this study finding similar conclusions (Sun et al., 2021).

11.3.2 Soil Moisture

There have been very few studies directly examining trends in antecedent moisture conditions. Woldemeskel and Sharma (2016) investigated the trend in precipitation occurring prior to an extreme precipitation event and the results are summarized in Table 11.2. Despite positive trends for extreme precipitation almost universally there was little evidence of a consistent trend in the antecedent precipitation. This suggests a disconnect between flood-causing precipitation and the antecedent moisture content. However, this study did not use a direct measure of soil moisture. Of studies that have investigated changes in antecedent soil moisture directly, decreases in flood magnitude have been attributed to drier soils before rainfall events (Wasko and Nathan, 2019; Tramblay et al., 2019).

If storm events occur independently of whether the soil moisture is wet or dry, as suggested in previous studies (Ivancic and Shaw, 2015), then the antecedent conditions will be independent of the precipitation. Hence mean soil moisture trends should be equivalent to antecedent soil moisture trends. Although there is evidence to suggest that precipitation occurrence is not independent of soil moisture (Holgate et al., 2019), the following presents a review of studies investigating changes in the mean soil moisture.

Using the CCI global soil moisture data obtained from the European Space Agency, Feng and Zhang (2015) examined 30 years of records (1979–2013) to calculate the trends in soil moisture. Despite spatial variability, 30% of global land experienced robust moisture trends with 22% trending to drier and 7% trending to wetter, pointing to an overall drying trend globally. Regionally, the results suggest drying in the extra-tropics of the southern hemisphere and wetting in the tropics, consistent with the expansion of the Hadley cell (tropical) expansion (Seidel et al., 2008). In the northern hemisphere, the trend is mixed with drying trends in the western United States, but wetting trends in the east. Europe has drying trends in the north but wetting trends in the south. As direct measures of soil moisture are difficult to find, Greve et al. (2014) investigated changes using the land surface water balance and the aridity index from 1948 to 2005 and found comparable results. Much of Africa and Australia was found to be drying, with South America wetting. In the US, the

TABLE 11.2
Annual Maximum Precipitation and Antecedent Precipitation Trends

Region	Extreme Precipitation Trend	Antecedent Precipitation Trend (API)
North America	Positive	No trend
South America	Positive	No trend
Africa	Positive	Positive
Eurasia	Positive	Negative
Australia	Negative	Positive

Source: Woldemeskel and Sharma, 2016.

east coast was wetter, and the west coast was dryer. Through Europe, the south was drying and the north was wetting, with most of Asia drying except for the northern extremes.

Another global trends analysis for 1988–2010 (Albergel et al., 2013) suggests similar trends. The study used three remotely sensed products were used: ECMWF Interim Re-Analysis (ERA-Interim; ERA-Land); a revised version of the Modern-Era Retrospective Analysis for Research and Applications (MERRA) reanalysis from NASA (MERRA-Land); and a microwave-based multi-satellite surface soil moisture dataset (SM-MW). The results were mixed but also pointed to an overall drying trend. Although the coverage of the products varies, 72% of the significant trends were drying trends for the ERA-Land data set and 73% were drying from the SM-MW data set. But only 41% were drying from the MERRA data set, suggesting a possible overall wetting. This may be consistent with the overall small wetting trend in global soil moisture consistent with the increases in mean precipitation found using a terrestrial hydrologic simulation (Sheffield and Wood, 2008). The spatial trends from the three products also did not necessarily match and exhibited spatial variability (Albergel et al., 2013). Broadly, although Australia shows overall drying trends, the east coast of Australia shows wetting trends, with most of Europe tending to wetter conditions, but Asia to drying conditions. In this case, the majority of the US and South America also showed drying (Albergel et al., 2013).

Jung et al. (2010) also analyzed the soil moisture trends from the Tropical Rainfall Measuring Mission's (TRMM) for 1998–2008 to explain the trends in a global land-based estimate of evapotranspiration and concluded that since 1998 a decrease in evapotranspiration was primarily driven by moisture limitations in the Southern Hemisphere, particularly Africa and Australia. This generally coincided with the above studies with a large part of Asia and the US also showing the drying trends and wetting trends appearing around the equatorial region of Mexico and Central America.

It appears that trends in soil moisture are very mixed, with no overall consistent "drying" or "wetting" trend for any time period considered, across either the globe, latitude bands, or continents. Although shorter trends can correlate with interannual variability and human influence, by far the biggest driver of changes in soil moisture is precipitation (Rodell et al., 2018), hence it may be that changes in average precipitation are a good indicator of soil moisture (Wasko et al., 2021a). This is not surprising as a simple antecedent moisture index based on maximum daily temperature and precipitation has been found to be very well correlated to measured soil moisture (in situ) soil moisture (Holgate et al., 2016).

11.3.3 Snowfall and Snowmelt

Changes in snowmelt form a large driver for changing streamflow and hence flooding, but generally, the manuscripts published in this domain focus on shifts in the timing of the snowmelt and resulting streamflow (Trenberth, 2011). For example, by examining the center date of flow volume for sites that receive at least 30% of their annual precipitation as snow across the United States, it was found that the center date had shifted by 8.2 days earlier for the years 1940–2014 and 8.6 days earlier for the period 1960–2014 (Dudley et al., 2017). Similar shifts to earlier snowmelt have been observed across Europe (Blöschl et al., 2017) and Asia (Shen et al., 2018).

Examining snowfall from 440 stations across the US, Kunkel et al., (2009) examined the frequency of exceedance station wise for the 90th and 10th percentile. Although looking over the past century did not demonstrate a trend, during the period from 1950–1951 to 2006–2007 the United States as a whole showed a significant decrease in years experiencing high extreme snowfall, and the converse was found for the low extreme years. Using 76 stations ranging up to a century in length across Switzerland, it was found the relative days of snow versus days of rain have been decreasing, and this trend is correlated to temperature increases (Serquet et al., 2011). Generally, it can be concluded that the snowfalls are decreasing (Hartmann et al., 2013), and as a result, snow cover extent has decreased over the past century with most of the decrease occurring since 1980 (Brown and Robinson, 2011) particularly at lower elevations or in warmer areas (Vaughan et al., 2013).

Trends in the magnitude of streamflow resulting from snowmelt are more complex with, as discussed, the possibility of both increased and reduced streamflow as a result of changes in the snow (Hamlet and Lettenmaier, 2007; Barnhart et al., 2016). Though global pictures are not possible, localized studies suggest reductions in snowpack and snowmelt due to higher temperatures (Hamlet and Lettenmaier, 2007). There does appear, in the western United States at least, that declines in snowpack and earlier snowmelt are resulting in decreased streamflow (Barnhart et al., 2016). Using a hydrologic model for the western US driven by precipitation and temperature, reductions in snowpack over the 20th century were demonstrated (Hamlet and Lettenmaier, 2007). Using a Variable Infiltration Capacity model and calculated streamflow anomalies from 1950 to 2013, a reduced snowmelt rate and hence decreased streamflow was also found across the western United States (Barnhart et al., 2016). Over Norway, decreases in snowmelt are likely the cause of a reduced magnitude and frequency in flood events (Vormoor et al., 2016).

11.4   CHANGES IN FLOOD TIMING

As discussed at the beginning of the chapter, identifying trends in flood magnitude with climatic change is difficult. As an alternative, a number of studies attempt to identify a climate change signal by investigating the shifts in the flood timing (Schwarz, 1977). A global study of changes to flood timing showed mixed trends (Wasko et al., 2020b) hinting at changes in the flood mechanisms discussed earlier. For example, as temperatures have increased, greater snowmelt coupled with less precipitation falling as snow (Lettenmaier and Gan, 1990) and lower snowpack (Hamlet and Lettenmaier, 2007) have caused a shift to earlier snowmelts (Trenberth, 2011). However, this is complicated by changes in the interaction between the rain and snow in so-called rain-on-snow events (Musselman et al., 2018). In non-snow-dominated environments, shifts in the flood seasonality were found to be closely linked to shifts in soil moisture seasonality.

Using 84 natural rivers in Canada, a Mann-Kendall test found a shift to an earlier peak for the spring snowmelt runoff event at 44% of the sites at the 10% level whereas no sites exhibited a trend in the opposite direction, consistent with anthropogenic climatic change. A similar shift to earlier flood peak occurrence was observed across Scandinavia (Matti et al., 2017). However, mixed patterns of shifting flood peak timing were found in China (Zhang et al., 2017) and Europe (Blöschl et al., 2017). Using 4,262 streamflow gauges from 1960–2010, it was found that shifts in flooding vary across Europe, for example, to earlier in the year in western Europe and along the Atlantic coast due to earlier soil moisture maximum, and earlier in the year in northeastern Europe due to earlier snowmelt. But in other regions, such as the North Sea, later winter storms are resulting in later flooding (Blöschl et al., 2017). Using 221 streamflow gauges across Australia it was found changes in the flood timing are more linked to changes in the soil moisture timing than precipitation maxima timing, with shifts to earlier flood maxima in the north of the continent and later flood maxima in the south of the continent (Wasko et al., 2020a).

11.5   TRENDS IN FLOOD MAGNITUDE

There are a plethora of local investigations which study the changes in historical flooding. Despite the creation of larger data sets and increased computing power allowing global studies, the fact that converting rainfall to streamflow is a local scale process often necessitates that changes in flooding due to the climatic change need to be evaluated considering each catchment individually. Most streamflow gauges are concentrated in Australia, the USA, and Europe. For example, in Australia, using instantaneous maxima from 491 small to medium-sized catchments with record lengths from 30 to 97 years, a Mann-Kendall test showed that the number of stations showing statistically significant decreasing trends (26%) exceeded those showing increasing trends (6%) when the entire record at each site was considered (Ishak et al., 2013). The proportion of sites showing decreases exceeded the proportion showing increases regardless of the study period chosen. Using a high-quality data set of 222 stations across Australia and daily average streamflow annual maxima, again 44% of sites showed the decreasing trends and only 8% showed an increase at the 10% level (Zhang et al., 2016). It has been demonstrated that the decreases in streamflow annual maxima across Australia are linked to drier antecedent moisture conditions with increases for the rarest floods still observed (Wasko and Nathan, 2019).

Across the US, Lins and Cohn (2011) analyzed trends in annual instantaneous peak stream-flow for 1,491 stations from the USGS Hydro-Climatic Data Network (HCDN) for the 60-year period from 1948 to 2007. No spatially consistent trend across the US was identified, and the results appeared mixed based on the histogram of the observed magnitude of the changes. However, there were more sites showing statistically significant decreasing trends than increasing trends. Using 50 stations with at least 100 years of record across the USA, a set of non-parametric tests concluded that no monotonic temporal trend in the annual maximum instantaneous peak discharge could be detected (Villarini et al., 2009a). Using 395 stations across the US for the period 1944–1993, again the number of stations with decreasing trends (21) outnumbered those increasing (14) at the 5% level using a Mann-Kendall test (Lins and Slack, 1999). Though these proportions differed depending on the start and end years and the length of record, the authors concluded their results indicated that the conterminous US is getting "less extreme." A similar study using data from 400 sites across the conterminous Unites States from 1941 to 1999 suggested these trend tests could be affected by a significant step change (McCabe and Wolock, 2002). Using Kendall's Tau, trends using varying windows of a minimum length of ten years were analyzed. A large number of sites with increasing trends were found for periods with a start year before 1970 and an end year after 1970. This would suggest a step increase around 1970. A uniform trend was not identified post-1970. However, the most recent study for the US using 774 stream gauge stations across the United States for the common period 1962–2011 found that 20% (158) of stations showed statistically significant changes in annual maxima with the majority of sites (101 out of 158) showing increases (Mallakpour and Villarini, 2015).

A review of flooding in Europe equally presents mixed flood trends (Hall et al., 2014). There were decreasing trends in the magnitude of annual maxima in Spain (Mediero et al., 2014) regard-less of the time period analyzed. Using 195 gauges across France, although no consistent trend was found, some increases in flood peaks in the northeast were identified (Renard et al., 2008). Using 145 gauges across Germany with a common period from 1951 to 2002, increases were found at 28% of sites for the maximum discharge for each hydrological year (Petrow and Merz, 2009), whereas only 1% of sites had a statistically significant decreasing trend (based on a Mann-Kendall test with 10% significance). But further east, using 39 sites across Poland from the years 1921–1990, it was found the streamflow annual maxima had decreased (Strupczewski et al., 2001). No evidence of consistent changes was found across the Russia and Arctic (Shiklomanov et al., 2007). The results across the UK for the instantaneous streamflow annual maxima (23 sites over the years 1959–2003) showed that, at the 10% level, 8.7% of sites have positive trends and zero have negative trends, point-ing to a wetting trend consistent with the precipitation trends in this region (Hannaford and Marsh, 2008). However, many parts of Europe continue to have no clear signal for changes in flooding (Hall et al., 2014), for example, Sweden (Lindström and Bergström, 2004).

A global investigation of trends in annual flood maxima using the 9,213 stations from the Global Runoff Data Centre (GRDC, 2015) split this data set into three subsets of varying size and duration and found that on average, the number of sites showing statistically decreasing trends (based on a Mann-Kendall test) was greater than the number of sites showing statistically significant increases (Do et al., 2017). This analysis is repeated here using the GSIM compilation of indices (Do et al., 2018; Gudmundsson et al., 2018). Only stations with more than 30 years of record are included and the results of linear regression on all the 14,882 sites are presented in Figure 11.3. Consistent with Do et al. (2017), the proportion of at site statistically significant decreases (14.1%) is greater than those showing increases (10.4%) which is the reverse of the extreme precipitation trend presented in Figure 11.2. Spatially homogeneous regions are difficult to observe but the overall trends are

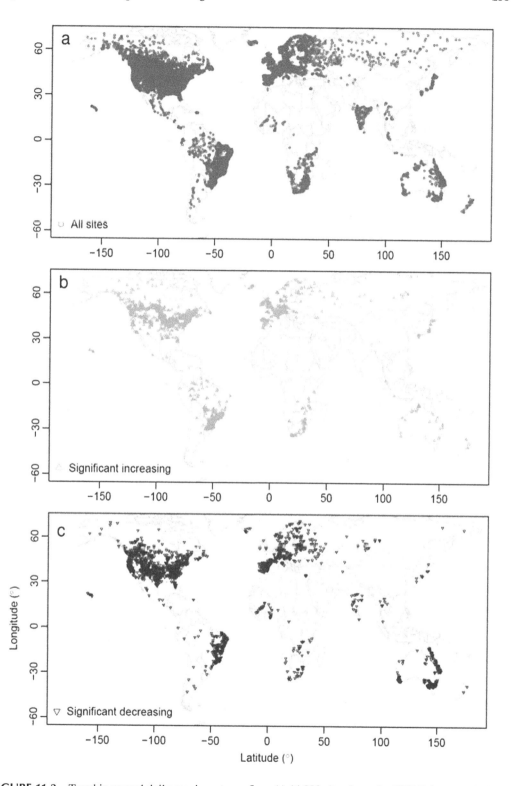

FIGURE 11.3 Trend in annual daily maxima streamflow. (a) 14,882 sites from the GRDC data set (GRDC 2015) with more than 30 years of data; (b) sites showing a statistically significant positive linear trend at the 5% level; (c) sites showing a statistically significant decreasing linear trend at the 5% level. Of the sites analyzed 10.4% of sites show a statistically significant increase in the annual daily maxima while 14.1% show a decrease.

consistent with a recent summary of global flood trends (Wasko, 2021). There are increases in flooding in the colder regions of the US. The central US and UK also exhibit increases, as do southern Brazil and the coast of South Africa. However, decreases are observed throughout the majority of the US, southern Europe, and in particular Spain. Brazil and Australia also on average show statistically significant decreases.

The conflicting evidence for trends in flooding for data-rich parts of the world shows the difficulty in identifying trends in flooding, how the generalization of results is not possible (Archfield et al., 2016), and the need to understand the mechanisms behind the flood trend (Sharma et al., 2018). Figure 11.4 presents a conceptualization of the sensitivity of precipitation and streamflow (varying with catchment size) with an increase in temperature for a location where antecedent soil moisture is decreasing with increasing temperatures. The change in flooding depends on changes to precipitation and soil moisture and interacts with the event magnitude considered (Wasko and Nathan, 2019) and the size of the catchment (Wasko and Sharma, 2017; Bennett et al., 2018). Indeed Do et al. (2017) noted, the proportion of sites showing statistically significant positive trends, increases as the catchment area decreases. Or in other words, the smaller the catchment, the more likely an increase in the annual maxima is observed.

As temperatures increase, precipitation extremes will increase and are likely to do so at an increasing rate the more extreme the event (Pendergrass, 2018). However, some of the precipitation increase will be captured by the soil moisture storage in the catchment. If the soil moisture storage has decreased as a result of climatic change then not all of the increase in precipitation will be translated to an increase in flow, resulting in a possible decrease in streamflow (Wasko et al., 2019). However, this relationship is dependent on how rare an event is. As an event becomes rarer the impact of antecedent moisture conditions matters less (Ivancic and Shaw, 2015) and so the change in streamflow will approach the change in precipitation. Indeed the rarest of floods are expected to increase with climatic change (Knox, 2000; Milly et al., 2002; Berghuijs et al., 2017).

The flood response in a changing climate also depends on the catchment area (Wasko and Sharma, 2017; Do et al., 2017). With a smaller catchment size, there is an increased chance a storm will cover the entire catchment and hence lead to soil moisture saturation, with more of the precipitation contributing to the streamflow response, and so the streamflow increase will approach the precipitation increase more closely. This demonstrates that changes in the size of storm events

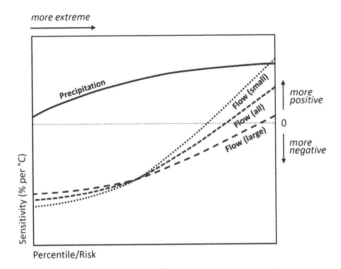

FIGURE 11.4 Conceptualized sensitivity of precipitation and streamflow to changes in temperature for undeveloped catchments with a tendency to drier antecedent moisture conditions as presented in Sharma et al. (2018). The sensitivity of the streamflow is stratified nominally on catchment size.

(Chang et al., 2016; Wasko et al., 2016) and a possible trend to smaller convective rainfall events with climatic change (Molnar et al., 2015) will also affect the resultant flood response.

There will almost certainly be an increase in urban flood risk as a result of increased rainfall due to the impervious nature of the catchment (Hettiarachchi et al., 2018). Gauging of urban catchments is rare, and indeed as urban gauges are subject to increases in streamflow due to further urbanization, this conflates any observation of changes in streamflow that may result from climatic changes. Of the few studies that have looked at flooding in urban regions, Vogel et al. (2011) considered three sets of streamflow gauges across the US; 14,893 nonregulated sites, 4,537 regulated sites, and 1,588 sites included in the USGS Hydro-Climatic Data Network (HCDN). All groups exhibited approximately the same proportion of sites with a positive trend, however, those non-HCDN sites had a much higher range of results, and included sites with development, suggesting development does increase flood magnitude. Finally, it should be noted that, although it is generally thought that changes in antecedent moisture conditions will not have an impact in urban areas as most of the catchment is considered impervious, there is evidence that changes in the storage in urban areas will modulate flood increases (Hettiarachchi et al., 2019). As a result, there is the possibility of reduced stream flows and nuisance flooding with climatic change in urban areas despite increases in precipitation extremes.

11.6 ADDITIONAL ANTHROPOGENIC FACTORS

The focus of this chapter has been on the impact of hydroclimatic changes on flooding. However, it is worth commenting on additional considerations to the nature of changed flooding. Aside from climate-driven changes, the other primary anthropogenic change causing the changes to flooding is changing catchment land use. Given the difficulty in identifying trends in flooding, it is almost impossible to distinguish between the changes in flooding as a result of hydroclimatic changes and human-induced changes due to such factors as the urbanization of catchments which make the catchment itself non-stationary. However, a physical understanding of the drivers of pluvial flooding can lead to some general inferences which are now discussed.

Firstly, urbanization will increase runoff and flooding as a result of decreased infiltration, increasing runoff volumes, and reduced conveyance time of rainfall through the catchment (Shuster et al., 2005). Of the very few studies available that investigate non-stationarity in urban catchments, they all point to the increases in flooding due to increased urbanization. For example, small floods may be increased by a factor of ten due to urbanization, with the effect diminishing in relative terms as the flood recurrence interval increases (Hollis, 1975). Villarini et al. (2009b) fitted a non-stationary exceedance probability model to Little Sugar Creek, a highly urbanized catchment in North Carolina from 1924 to 2007, and found annual flood peaks increased, in particular during the period of rapid urbanization and population increase from the 1960s. Increases in flooding in urban areas can also be expected from the heat island effect, where the construction of paved surfaces results in increased local temperatures, increased aerosols in the atmosphere, and obstructed airflow, causing precipitation intensification with the intensification being relative to city size (Huff and Changnon, 1973).

Analogously, deforestation can also result in increased runoff and increased flooding in rural catchments. As vegetation is lost, the rainfall intercepted through the canopy or evapotranspiration is reduced, and the rate at which flow can occur both overland and through the soil increases. Using remotely sensed flood data from 1990 to 2000 from 56 developing countries a negative correlation between flood frequency and forest cover was found (Bradshaw et al., 2007). In paired catchment studies reductions in vegetation have increased streamflow, but the majority of catchments are small and less than 10 km^2 (Peel, 2009) with the impact scale-dependent and unlikely to be as significant in large catchments (Peña-Arancibia et al., 2012).

Finally, reservoir construction can cause reductions in flooding. Not only can a reservoir reduce flood flow by physically obstructing flow and creating storage for volume reduction of a flood event, but diversions will also take place from the storages for example for irrigation and urban water supply further reducing flow (Vörösmarty and Sahagian, 2000). Dams will affect all aspects of streamflow, including the timing and magnitude. Although one may see a regulated river system as having less variability, the controlled release can actually increase the variability in the streamflow characteristics including maximum monthly streamflow, however, this will depend on the size of the catchment (Lajoie et al., 2007). Looking specifically at the impact of dams on flooding, annual maxima have been quantified across the US as a function of catchment charactersitics as well as population density and dam storage (FitzHugh and Vogel, 2011). The authors conclude that the percent of rivers with greater than a 25% reduction in the median annual flood is 55% for large rivers, 25% for medium rivers, and 10% for small rivers due to dam storage.

11.7   CONCLUSIONS

This chapter summarized the current scientific understanding of changes in flooding with climatic change on a global scale. As the changes in flooding are difficult to identify a large part of the chapter focused on the hydroclimatic drivers of flooding. These were identified as being extreme precipitation, antecedent moisture conditions, and snowmelt. The interaction of these variables and the resultant impact on flooding will be dependent on both how extreme the event is and the catchment size.

To the best available scientific knowledge, it can be argued that precipitation extremes are increasing, and for daily precipitation extremes, this rate of increase is consistent with the Clausius-Clapeyron relation (Fowler et al., 2021). However, despite more evidence for increases in precipitation extremes than decreases, historical floods do not have the same pattern of increase, with more sites showing decreases rather than the increases in flood magnitude. The evidence presented suggests that a decrease in flood magnitude is likely the result of drying antecedent moisture conditions due to increasing temperatures. However, this relationship is highly dependent on the local catchment characteristics and the definition of what is meant by an "extreme event." As the event becomes more extreme (in excess of an annual recurrence interval) it becomes increasingly likely that floods will increase due to greater rainfall volume dominating any changes to the antecedent moisture conditions (Wasko and Nathan, 2019). Likewise, for smaller, and more urban catchments, a greater chance of increased flooding from increased precipitation extremes is expected due to less impervious surfaces and less flow attenuation.

It should be noted this simplified conceptualization does not account for dependencies that might exist such as those between the spells of extreme precipitation (Pui et al., 2012) due to the possible changes in the frequency of storm events (Lu et al., 2013; Molnar et al., 2015) which will subsequently also affect the antecedent conditions (Wasko et al., 2015a). To limit the literature examined in this chapter, there was a specific emphasis on annual maxima and changes in their timing and magnitude. In addition, changes in the frequency of flooding were not detailed here (Hirsch and Archfield, 2015; Mallakpour and Villarini, 2015). Although this chapter had a global focus, flooding remains a local-scale phenomenon and will depend very much on the changes in the local climatic conditions as well as the catchment characteristics. Hence, although the governing principles presented in this chapter are likely to be translatable to the catchment-specific conditions, catchment-scale investigations will be required for calculating the future changes to the magnitude, frequency, and timing of flooding.

ACKNOWLEDGMENTS

The precipitation data used in this chapter is freely available from www.ncdc.noaa.gov/ghcnd-data -acccs. The streamflow data used in this chapter is freely available from www.bafg.de/GRDC/EN/ Home/homepage_node.html. This chapter was edited by Professor Saeid Eslamian.

REFERENCES

Ajami, H., Sharma, A., Band, L.E., Evans, J.P., Tuteja, N.K., Amirthanathan, G.E., and Bari, M.A. 2017. On the non-stationarity of hydrological response in anthropogenically unaffected catchments: An Australian perspective, *Hydrol. Earth Syst. Sci.*, 21: 281–294, http://www.hydrol-earth-syst-sci-discuss.net/hess-2016-353/.

Albergel, C., Dorigo, W., Reichle, R.H., Balsamo, G., de Rosnay, P., Muñoz-Sabater, J., Isaksen, L., de Jeu, R., and Wagner, W. 2013. Skill and global trend analysis of soil moisture from reanalyses and microwave remote sensing, *J. Hydrometeorol.*, 14: 1259–1277, http://journals.ametsoc.org/doi/abs/10.1175/JHM-D-12-0161.1.

Alexander, L.V., Zhang, X., Peterson, T.C., Caesar, J., Gleason, B., Klein Tank, A.M.G., Haylock, M., Collins, D., Trewin, B., Rahimzadeh, F., and Tagipour, A. 2006. Global observed changes in daily climate extremes of temperature and precipitation, *J. Geophys. Res.*, 111: D05109, http://doi.wiley.com/10.1029/2005JD006290.

Allen, M.R., and Ingram, W.J. 2002. Constraints on future changes in climate and the hydrologic cycle, *Nature*, 419: 224–232, http://www.ncbi.nlm.nih.gov/pubmed/12226677.

Alley, R.B., Marotzke, J., Nordhaus, W.D., Overpeck, J.T., Peteet, D.M., Pielke, Jr. R.A., Pierrehumbert, R.T., Rhines, P.B., Stocker, T.F., Talley, L.D., and Wallace, J.M. 2003. Abrupt climate change, *Science*, 299(80): 10, https://www.jstor.org/stable/3833703.

Apel, H, Thieken, A.H., Merz, B., and Blöschl, G. 2004. Natural hazards and earth system sciences flood risk assessment and associated uncertainty, *Nat. Hazards Earth Syst. Sci.*, 4: 295–308, https://www.nat-hazards-earth-syst-sci.net/4/295/2004/nhess-4-295-2004.pdf.

Archfield, S.A., Hirsch, R.M., Viglione, A., and Blöschl, G. 2016. Fragmented patterns of flood change across the United States, *Geophys. Res. Lett.*, 43(10): 232–239, http://doi.wiley.com/10.1002/2016GL070590.

Barbero, R., Fowler, H.J., Lenderink, G., and Blenkinsop, S. 2017. Is the intensification of precipitation extremes with global warming better detected at hourly than daily resolutions? *Geophys. Res. Lett.*, 44: 974–983, http://doi.wiley.com/10.1002/2016GL071917.

Barnhart, T.B., Molotch, N.P., Livneh, B., Harpold, A.A., Knowles, J.F., and Schneider, D. 2016. Snowmelt rate dictates streamflow, *Geophys. Res. Lett.*, 43: 8006–8016.

Beck, H.E., Wood, E.F., Pan, M., Fisher, C.K., Miralles, D.G., van Dijk, A.I.J.M., McVicar, T.R., and Adler, R.F. 2019. MSWEP V2 global 3-hourly 0.1 precipitation: Methodology and quantitative assessment, *Bull. Am. Meteorol. Soc.*, 100: 473–500, http://journals.ametsoc.org/doi/10.1175/BAMS-D-17-0138.1.

Bennett, B., Leonard, M., Deng, Y., and Westra, S. 2018. An empirical investigation into the effect of antecedent precipitation on flood volume, *J. Hydrol.*, 567: 435–445, https://doi.org/10.1016/j.jhydrol.2018.10.025.

Berghuijs, W.R., Aalbers, E.E., Larsen, J.R., Trancoso, R., and Woods, R.A. 2017. Recent changes in extreme floods across multiple continents, *Environ. Res. Lett.* 12: 114035, http://iopscience.iop.org/10.1088/1748-9326/aa8847.

Berghuijs, W.R., Woods, R.A., Hutton, C.J., and Sivapalan, M. 2016. Dominant flood generating mechanisms across the United States, *Geophys. Res. Lett.*, 43: 4382–4390.

Blöschl, G., Hall, J., Parajka, J., Perdigão, R.A.P., Merz, B., Arheimer, B., Aronica, G.T., Bilibashi, A., Bonacci, O., Borga, M., Čanjevac, I., Castellarin, A., Chirico, G.B., Claps, P., Fiala, K., Frolova, N., Gorbachova, L., Gül, A., Hannaford, J., Harrigan, S., Kireeva, M., Kiss, A., Kjeldsen, T.R., Kohnová, S., Koskela, J.J., Ledvinka, O., Macdonald, N., Mavrova-Guirguinova, M., Mediero, L., Merz, R., Molnar, P., Montanari, A., Murphy, C., Osuch, M., Ovcharuk, V., Radevski, I., Rogger, M., Salinas, J.L., Sauquet, E., Šraj, M., Szolgay, J., Viglione, A., Volpi, E., Wilson, D., Zaimi, K., and Živković, N. 2017. Changing climate shifts timing of European floods, *Science*, 357: 588–590, http://www.sciencemag.org/lookup/doi/10.1126/science.aan2506.

Boughton, W., and Droop, O. 2003. Continuous simulation for design flood estimation—A review, *Environ. Model. Softw.*, 18: 309–318, http://linkinghub.elsevier.com/retrieve/pii/S1364815203000045.

Bradshaw, C.J.A., Sodhi, N.S., Peh, K.S.H., and Brook, B.W. 2007. Global evidence that deforestation amplifies flood risk and severity in the developing world, *Glob. Chang. Biol.*, 13: 2379–2395.

Brown, R.D., and Robinson, D.A. 2011. Northern Hemisphere spring snow cover variability and change over 1922–2010 including an assessment of uncertainty, *Cryosphere*, 5: 219–229.

Chang, W., Stein, M.L., Wang, J., Kotamarthi, V.R., and Moyer, E.J. 2016. Changes in spatiotemporal precipitation patterns in changing climate conditions, *J. Clim.*, 29: 8355–8376, http://journals.ametsoc.org/doi/10.1175/JCLI-D-15-0844.1.

Do, H.X., Gudmundsson, L., Leonard, M., and Westra, S. 2018. The global streamflow indices and metadata archive (GSIM)-Part 1: The production of a daily streamflow archive and metadata, *Earth Syst. Sci. Data*, 10: 765–785.

Do, H.X., Westra, S., and Leonard, M. 2017. A global-scale investigation of trends in annual maximum streamflow, *J. Hydrol.*, 552: 28–43, http://linkinghub.elsevier.com/retrieve/pii/S0022169417304171.

Donat, M.G., Alexander, L.V., Yang, H., Durre, I., Vose, R., Dunn, R.J.H., Willett, K.M., Aguilar, E., Brunet, M., Caesar, J., Hewitson, B., Jack, C., Klein Tank, A.M.G., Kruger, A.C., Marengo, J., Peterson, T.C., Renom, M., Oria Rojas, C., Rusticucci, M., Salinger, J., Elrayah, A.S., Sekele, S.S., Srivastava, A.K., Trewin, B., Villarroel, C., Vincent, L.A., Zhai, P., Zhang, X., and Kitching, S. 2013. Updated analyses of temperature and precipitation extreme indices since the beginning of the twentieth century: The HadEX2 dataset, *J. Geophys. Res. Atmos.*, 118: 2098–2118, http://doi.wiley.com/10.1002/jgrd.50150.

Dudley, R.W., Hodgkins, G.A., McHale, M.R., Kolian, M.J., and Renard, B. 2017. Trends in snowmelt-related streamflow timing in the conterminous United States, *J. Hydrol.*, 547: 208–221, http://dx.doi.org/10.1016/j.jhydrol.2017.01.051.

Feng, H., and Zhang, M. 2015. Global land moisture trends: Drier in dry and wetter in wet over land, *Sci. Rep.*, 5: 18018, http://dx.doi.org/10.1038/srep18018.

Fill, H.D., and Steiner, A.A. 2003. Estimating instantaneous peak flow from mean daily flow data, *J. Hydrol. Eng.*, 8: 365–369, http://ascelibrary.org/doi/10.1061/%28ASCE%291084-0699%282003%298%3A6%28365%29.

FitzHugh, T.W., and Vogel, R.M. 2011. The impact of dams on flood flows in the United States, *River Res. Appl.*, 27: 1192–1215, http://doi.wiley.com/10.1002/rra.1417.

Fowler, H.J., Lenderink, G., Prein, A.F., Westra, S., Allan, R.P., Ban, N., Barbero, R., Berg, P., Blenkinsop, S., Do, H.X., Guerreiro, S., Haerter, J.O., Kendon, E.J., Lewis, E., Schaer, C., Sharma, A., Villarini, G., Wasko, C., and Zhang, X. 2021. Anthropogenic intensification of short-duration rainfall extremes, *Nat. Rev. Earth Environ.*, 2: 107–122, http://www.nature.com/articles/s43017-020-00128-6.

François, B., Schlef, K.E., Wi, S., and Brown, C.M. 2019. Design considerations for riverine floods in a changing climate – A review, *J. Hydrol.*, 574: 557–573.

GRDC. 2015. The Global Runoff Data Centre, https://grdc.com.au/resources-and-publications/all-publications/bookshop/2015.

Greve, P., Orlowsky, B., Mueller, B., Sheffield, J., Reichstein, M., and Seneviratne, S.I. 2014. Global assessment of trends in wetting and drying over land, *Nat. Geosci.*, 7: 716–721, http://www.nature.com/articles/ngeo2247.

Gudmundsson, L., Do, H.X., Leonard, M., and Westra, S. 2018. The global streamflow indices and metadata archive (GSIM)-Part 2: Quality control, time-series indices and homogeneity assessment, *Earth Syst. Sci. Data*, 10: 787–804.

Guerreiro, S.B., Fowler, H.J., Barbero, R., Westra, S., Lenderink, G., Blenkinsop, S., Lewis, E., and Li, X.-F. 2018. Detection of continental-scale intensification of hourly rainfall extremes, *Nat. Clim. Chang.*, 8: 803–807, http://www.nature.com/articles/s41558-018-0245-3.

Hall, J., Arheimer, B., Borga, M., Brázdil, R., Claps, P., Kiss, A., Kjeldsen, T.R., Kriaučiūnienė, J., Kundzewicz, Z.W., Lang, M., Llasat, M.C., Macdonald, N., McIntyre, N., Mediero, L., Merz, B., Merz, R., Molnar, P., Montanari, A., Neuhold, C., Parajka, J., Perdigão, R.A.P., Plavcová, L., Rogger, M., Salinas, J.L., Sauquet, E., Schär, C., Szolgay, J., Viglione, A., and Blöschl, G. 2014. Understanding flood regime changes in Europe: A state-of-the-art assessment, *Hydrol. Earth Syst. Sci.*, 18: 2735–2772, http://www.hydrol-earth-syst-sci.net/18/2735/2014/.

Hamlet, A.F., and Lettenmaier, D.P. 2007. Effects of 20th century warming and climate variability on flood risk in the western U.S. *Water Resour. Res.*, 43: W06427, http://doi.wiley.com/10.1029/2006WR005099.

Hannaford, J., and Marsh, T.J. 2008. High-flow and flood trends in a network of undisturbed catchments in the UK, *Int. J. Climatol.*, 28: 1325–1338, http://cdiac.esd.ornl.gov/oceans/GLODAP/glodap_pdfs/Thermohaline.web.pdf

Hartmann, D.L., Klein Tank, A.M.G., Rusticucci, M., Alexander, L.V., Broennimann, S., Charabi, Y., Dentener, F.J., Dlugokencky, E.J., Easterling, D.R., Kaplan, A., Soden, B.J., Thorne, P.W., Wild, M., and Zhai, P.M. 2013. Observations: Atmosphere and surface. In: *Climate Change 2013: The Physical Science Basis. Contribution of Working Group I to the Fifth Assessment Report of the Intergovernmental Panel on Climate Change*, edited by T. Stocker, D. Qin, G.-K. Plattner, M. Tignor, S.K. Allen, J. Boschung, A. Nauels, Y. Xia, V. Bex, and P.M. Midgley, Cambridge University Press, Cambridge, UK and New York, NY, USA, 159–254.

Hettiarachchi, S., Wasko, C., and Sharma, A. 2018. Increase in flood risk resulting from climate change in a developed urban watershed – The role of storm temporal patterns, *Hydrol. Earth Syst. Sci.*, 22: 2041–2056, https://doi.org/10.5194/hess-22-2041-2018.

Hettiarachchi, S., Wasko, C., and Sharma, A. 2019. Can antecedent moisture conditions modulate the increase in flood risk due to climate change in urban catchments? *J. Hydrol.*, 571: 11–20, https://linkinghub.elsevier.com/retrieve/pii/S0022169419301064.

Hirsch, R.M., and Archfield, S.A. 2015. Flood trends: Not higher but more often, *Nat. Clim. Chang.*, 5: 198–199, http://dx.doi.org/10.1038/nclimate2551.

Holgate, C., De Jeu, R.A.M., van Dijk, A.I.J., Liu, Y., Renzullo, L.J., Vinodkumar, D.I., Parinussa, R.M., Van Der Schalie, R., Gevaert, A., Walker, J., McJannet, D., Cleverly, J., Haverd, V., Trudinger, C.M., and Briggs, P.R. 2016. Comparison of remotely sensed and modelled soil moisture data sets across Australia, *Remote Sens. Environ.*, 186: 479–500, http://dx.doi.org/10.1016/j.rse.2016.09.015.

Holgate, C.M., Van Dijk, A.I.J.M., Evans, J.P., and Pitman, A.J. 2019. The importance of the one-dimensional assumption in soil moisture - Rainfall depth correlation at varying spatial scales, *J. Geophys. Res. Atmos.*, 124: 2964–2975, http://doi.wiley.com/10.1029/2018JD029762.

Hollis, G.E. 1975. The effect of urbanization on floods of different recurrence interval, *Water Resour. Res.*, 11: 431–435.

Huff, F.A., and Changnon, S.A. 1973. Precipitation modification by major urban areas, *Bull. Am. Meteorol. Soc.*, 54: 1220–1232.

Huntington, T.G. 2006. Evidence for intensification of the global water cycle: Review and synthesis, *J. Hydrol.*, 319: 83–95, http://linkinghub.elsevier.com/retrieve/pii/S0022169405003215.

Ishak, E.H., Rahman, A., Westra, S., Sharma, A., and Kuczera, G. 2013. Evaluating the non-stationary of Australian annual maximum flood, *J. Hydrol.*, 494: 134–145, http://linkinghub.elsevier.com/retrieve/pii/S0022169413003156.

Ivancic, T.J., and Shaw, S.B. 2015. Examining why trends in very heavy precipitation should not be mistaken for trends in very high river discharge, *Clim. Change*, 133: 681–693, http://link.springer.com/10.1007/s10584-015-1476-1.

Johnson, F., and Sharma, A. 2010. A comparison of australian open water body evaporation trends for current and future climates estimated from class a evaporation pans and general circulation models, *J. Hydrometeorol.*, 11: 105–121, http://dx.doi.org/10.1175/2009JHM1158.1.

Jung, M., Reichstein, M., Ciais, P., Seneviratne, S.I., Sheffield, J., Goulden, M.L., Bonan, G., Cescatti, A., Chen, J., de Jeu, R., Dolman, A.J., Eugster, W., Gerten, D., Gianelle, D., Gobron, N., Heinke, J., Kimball, J., Law, B.E., Montagnani, L., Mu, Q., Mueller, B., Oleson, K., Papale, D., Richardson, A.D., Roupsard, O., Running, S., Tomelleri, E., Viovy, N., Weber, U., Williams, C., Wood, E., Zaehle, S., and Zhang, K. 2010. Recent decline in the global land evapotranspiration trend due to limited moisture supply, *Nature*, 467: 951–954, http://www.nature.com.ezproxy.library.wisc.edu/nature/journal/v467/n7318/full/nature09396.html%5Cnhttp://www.nature.com.ezproxy.library.wisc.edu/nature/journal/v467/n7318/pdf/nature09396.pdf.

Kendon, E.J., Blenkinsop, S., and Fowler, H.J. 2018. When will we detect changes in short-duration precipitation extremes?, *J. Clim.*, 31: 2945–2964.

Knox, J.C. 1993. Large increases in flood magnitude in response to modest changes in climate, *Nature*, 361: 430–432.

Knox, J.C. 2000. Sensitivity of modern and Holocene floods to climate change, *Quat. Sci. Rev.*, 19: 439–457.

Kuczera, G. 1996. Correlated rating curve error in flood frequency inference, *Water Resour. Res.*, 32: 2119–2127.

Kuczera, G., Lambert, M., Heneker, T., Jennings, S., Frost, A., and Coombes, P. 2006. Joint probability and design storms at the crossroads, *Aust. J. Water Resour.*, 10: 63–79, http://search.informit.com.au/documentSummary;dn=232986079825698;res=IELENG.

Kunkel, K.E., Palecki, M.A., Ensor, L., Easterling, D., Hubbard, K.G., Robinson, D., and Redmond, K. 2009. Trends in twentieth-century U.S. extreme snowfall seasons, *J. Clim.*, 22: 6204–6216, http://journals.ametsoc.org/doi/abs/10.1175/2009JCLI2631.1.

Lajoie, F., Assani, A.A., Roy, A.G., and Mesfioui, M. 2007. Impacts of dams on monthly flow characteristics. The influence of watershed size and seasons, *J. Hydrol.*, 334: 423–439.

Lenderink, G., and van Meijgaard, E. 2008. Increase in hourly precipitation extremes beyond expectations from temperature changes, *Nat. Geosci.*, 1: 511–514, http://www.nature.com/doifinder/10.1038/ngeo262.

Lettenmaier, D.P., and Gan, T.Y. 1990. Hydrologic sensitivities of the Sacramento-San Joaquin, *Water Resour. Res.*, 26: 69–86.

Lindström, G., and Bergström, S. 2004. Runoff trends in Sweden 1807–2002/Tendances de l'écoulement en Suède entre 1807 et 2002, *Hydrol. Sci. J.*, 49: 69–83, http://www.informaworld.com/openurl?genre =article&doi=10.1623/hysj.49.1.69.54000&magic=crossref%7C%7CD404A21C5BB053405B1A640AF FD44AE3.

Lins, H.F., and Cohn, T.A. 2011. Stationarity: Wanted dead or alive?, *J. Am. Water Resour. Assoc.*, 47: 475–480.

Lins, H.F., and Slack, J.R. 1999. Streamflow trends in the United States, *Geophys. Res. Lett.*, 26: 227–230, http://doi.wiley.com/10.1029/1998GL900291.

Liu, Y.Y., Van Dijk, A.I.J.M., De Jeu, R.A.M., Canadell, J.G., McCabe, M.F., Evans, J.P., and Wang, G. 2015. Recent reversal in loss of global terrestrial biomass, *Nat. Clim. Chang.*, 5: 470–474.

Lu, M., Lall, U., Schwartz, A., and Kwon, H. 2013. Precipitation predictability associated with tropical moisture exports and circulation patterns for a major flood in France in 1995, *Water Resour. Res.*, 49: 6381–6392.

Mallakpour, I., and Villarini, G. 2015. The changing nature of flooding across the central United States, *Nat. Clim. Chang.*, 5: 250–254, http://dx.doi.org/10.1038/nclimate2516.

Matti, B., Dahlke, H.E., Dieppois, B., Lawler, D.M., and Lyon, S.W. 2017. Flood seasonality across Scandinavia-Evidence of a shifting hydrograph?, *Hydrol. Process.*, 31: 4354–4370, http://doi.wiley.com/10.1002/hyp .11365.

McCabe, G.J., and Wolock, D.M. 2002. A step increase in streamflow in the conterminous United States, *Geophys. Res. Lett.*, 29: 38-1–38-4, http://doi.wiley.com/10.1029/2002GL015999.

McMahon, T.A., Peel, M.C., Lowe, L., Srikanthan, R., and McVicar, T.R. 2013. Estimating actual, potential, reference crop and pan evaporation using standard meteorological data: A pragmatic synthesis, *Hydrol. Earth Syst. Sci.*, 17: 1331–1363.

Mcmillan, H.K., and Westerberg, I.K. 2015. Rating curve estimation under epistemic uncertainty, *Hydrol. Process.*, 29: 1873–1882.

Mediero, L., Santillán, D., Garrote, L., and Granados, A. 2014. Detection and attribution of trends in magnitude, frequency and timing of floods in Spain, *J. Hydrol.*, 517: 1072–1088, http://doi.org/10.1016/j .jhydrol.2014.06.040.

Menne, M.J., Durre, I., Korzeniewski, B., McNeal, S., Thomas, K., Yin, X., Anthony, S., Ray, R., Vose, R.S., Gleason, B.E., and Houston, T.G. 2015. *Global Historical Climatology Network - Daily (GHCN-Daily) Version 3.22*, NOAA National Climatic Data Center. Available at http://doi.org/10.7289/V5D21VHZ.

Milly, P.C.D., Betancourt, J., Falkenmark, M., Hirsch, R.M., Kundzewicz, Z.W., Lettenmaier, D.P., and Stouffer, R.J. 2008. Climate change: Stationarity is dead: Whither water management?, *Science*, 319: 573–574, http://www.ncbi.nlm.nih.gov/pubmed/18239110.

Milly, P.C.D., Wetherald, R.T., Dunne, K.A., and Delworth, T.L. 2002. Increasing risk of great floods in a changing climate, *Nature*, 415: 514–517, http://www.ncbi.nlm.nih.gov/pubmed/11823857.

Molnar, P., Fatichi, S., Gaál, L., Szolgay, J., and Burlando, P. 2015. Storm type effects on super Clausius–Clapeyron scaling of intense rainstorm properties with air temperature, *Hydrol. Earth Syst. Sci.*, 19: 1753–1766, http://www.hydrol-earth-syst-sci.net/19/1753/2015/.

Mujere, N., and Eslamian, S. 2014. Climate change impacts on hydrology and water resources. In: S. Eslamian (Ed.), *Modeling, Climate Changes and Variability*, Taylor and Francis, CRC Group, USA, 113–126.

Musselman, K.N., Lehner, F., Ikeda, K., Clark, M.P., Prein, A.F., Liu, C., Barlage, M., and Rasmussen, R. 2018. Projected increases and shifts in rain-on-snow flood risk over western North America, *Nat. Clim. Chang.*, 8: 808–812, http://doi.org/10.1038/s41558-018-0236-4.

Pathiraja, S., Westra, S., and Sharma, A. 2012. Why continuous simulation? The role of antecedent moisture in design flood estimation, *Water Resour. Res.*, 48: W06534, http://www.agu.org/pubs/crossref/2012 /2011WR010997.shtml.

Peel, M.C. 2009. Hydrology: Catchment vegetation and runoff, *Prog. Phys. Geogr.*, 33: 837–844.

Peña-Arancibia, J.L., van Dijk, A.I.J.M., Guerschman, J.P., Mulligan, M., (Sampurno) Bruijnzeel, L.A., and McVicar, T.R. 2012. Detecting changes in streamflow after partial woodland clearing in two large catchments in the seasonal tropics, *J. Hydrol.*, 416–417: 60–71.

Pendergrass, A.G. 2018. What precipitation is extreme?, *Science*, 360: 1072–1073, http://www.sciencemag.org /lookup/doi/10.1126/science.aat1871.

Petrow, T., and Merz, B. 2009. Trends in flood magnitude, frequency and seasonality in Germany in the period 1951–2002, *J. Hydrol.*, 371: 129–141, http://doi.org/10.1016/j.jhydrol.2009.03.024.

Pui, A., Sharma, A., Santoso, A., and Westra, S. 2012. Impact of the El Niño–Southern Oscillation, Indian Ocean Dipole, and Southern Annular Mode on Daily to Subdaily Rainfall Characteristics in East Australia, *Mon. Weather Rev.*, 140: 1665–1682, http://journals.ametsoc.org/doi/abs/10.1175/MWR-D-11 -00238.1.

Renard, B., Lang, M., Bois, P., Dupeyrat, A., Mestre, O., Niel, H., Sauquet, E., Prudhomme, C., Parey, S., Paquet, E., Neppel, L, and Gailhard, J. 2008. Regional methods for trend detection: Assessing field significance and regional consistency, *Water Resour. Res.*, 44: W08419, http://doi.wiley.com/10.1029 /2007WR006268.

Rodell, M., Famiglietti, J.S., Wiese, D.N., Reager, J.T., Beaudoing, H.K., Landerer, F.W., and Lo, M.H. 2018. Emerging trends in global freshwater availability, *Nature*, 557: 651–659.

Schwarz, H.E. 1977. Climate change and water supply: How sensitive is the Northeast? In: *Climate, Climatic Change, and Water Supply*, National Academy of Sciences, Washington, DC, 111–120.

Seidel, D.J., Fu, Q., Randel, W.J., and Reichler, T.J. 2008. Widening of the tropical belt in a changing climate, *Nat. Geosci.*, 1: 21–24, http://www.nature.com/articles/ngeo.2007.38.

Seneviratne, S., Nicholls, N., Easterling, D., Goodess, C., Kanae, S., Kossin, J., Luo, Y., Marengo, J., McInnes, K., Rahimi, M., Reichstein, M., Sorteberg, A., Vera, C., and Zhang, X. 2012. Changes in climate extremes and their impacts on the natural physical environment. In: C.B. Field, V. Barros, T.F. Stocker, D. Qin, D.J. Dokken, K.L. Ebi, M.D. Mastrandrea, K.J. Mach, G.-K. Plattner, S.K. Allen, M. Tignor, and P.M. Midgley (Eds.), *Managing the Risk of Extreme Events and Disasters to Advance Climate Change Adaptation, A Special Report of Working Groups I and II of the Intergovernmental Panel on Climate Change (IPCC)*, Cambridge University Press, Cambridge, UK, and New York, 109–230.

Serquet, G., Marty, C., Dulex, J.-P., and Rebetez, M. 2011. Seasonal trends and temperature dependence of the snowfall/precipitation-day ratio in Switzerland, *Geophys. Res. Lett.*, 38, http://doi.wiley.com/10.1029 /2011GL046976.

Sharma, A., Wasko, C., and Lettenmaier, D.P. 2018. If precipitation extremes are increasing, why aren't floods?, *Water Resour. Res.*, 54: 8545–8551, http://doi.wiley.com/10.1029/2018WR023749.

Sheffield, J., and Wood, E.F. 2008. Global trends and variability in soil moisture and drought characteristics, 1950–2000, from observation-driven simulations of the terrestrial hydrologic cycle, *J. Clim.*, 21: 432–458, http://journals.ametsoc.org/doi/abs/10.1175/2007JCLI1822.1.

Shen, Y., Shen, Y., Fink, M., Kralisch, S., Chen, Y., and Brenning, A. 2018. Trends and variability in streamflow and snowmelt runoff timing in the southern Tianshan mountains, *J. Hydrol.*, 557: 173–181, https:// doi.org/10.1016/j.jhydrol.2017.12.035.

Shepherd, J.M., Pierce, H., and Negri, A.J. 2002. Rainfall modification by major urban areas: Observations from Spaceborne Rain Radar on the TRMM Satellite, *J. Appl. Meteorol.*, 41: 689–701, http://journals .ametsoc.org/doi/abs/10.1175/1520-0450%282002%29041%3C0689%3ARMBMUA%3E2.0.CO%3B2.

Shiklomanov, A.I., Lammers, R.B., Rawlins, M.A., Smith, L.C., and Pavelsky, T.M. 2007. Temporal and spatial variations in maximum river discharge from a new Russian data set, *J. Geophys. Res. Biogeosciences*, 112, http://doi.wiley.com/10.1029/2006JG000352.

Shuster, W.D., Bonta, J., Thurston, H., Warnemuende, E., and Smith, D.R. 2005. Impacts of impervious surface on watershed hydrology: A review, *Urban Water J.*, 2: 263–275.

Stephens, C.M., Johnson, F.M., and Marshall, L.A. 2018a. Implications of future climate change for event-based hydrologic models, *Adv. Water Resour.*, 119: 95–110, https://doi.org/10.1016/j.advwatres.2018.07 .004.

Stephens, C.M., McVicar, T.R., Johnson, F.M., and Marshall, L.A. 2018b. Revisiting pan evaporation trends in australia a decade on, *Geophys. Res. Lett.*, 45(11): 11–164.

Strupczewski, W.G., Singh, V.P., and Mitosek, H.T. 2001. Non-stationary approach to at-site flood frequency modelling. III. Flood analysis of Polish rivers, *J. Hydrol.*, 248: 152–167.

Sun, Q., Zhang, X., Zwiers, F., Westra, S., and Alexander, L.V. 2021. A global, continental, and regional analysis of changes in extreme precipitation, *J. Clim.*, 34: 243–258, https://journals.ametsoc.org/jcli/article /354601/A-global-continental-and-regional-analysis-of.

Tramblay, Y., Mimeau, L., Neppel, L., Vinet, F., and Sauquet, E. 2019. Detection and attribution of flood trends in Mediterranean basins, *Hydrol. Earth Syst. Sci.*, 23: 4419–4431, https://hess.copernicus.org/ articles/23/4419/2019/.

Trenberth, K.E. 2011. Changes in precipitation with climate change, *Clim. Res.*, 47: 123–138, http://www.int -res.com/abstracts/cr/v47/n1-2/p123-138/.

Trenberth, K.E., Dai, A., Rasmussen, R.M., and Parsons, D.B. 2003. The changing character of precipitation, *Bull. Am. Meteorol. Soc.*, 84: 1205–1217, http://journals.ametsoc.org/doi/abs/10.1175/BAMS-84 -9-1205.

Vaughan, D.G., Comiso, J.C., Allison, I., Carrasco, J., Kaser, G., Kwok, R., Mote, P., Murray, T., Paul, F., Ren, J., Rignot, E., Solomina, O., Steffen, K., and Zhang, T. 2013. Observations: Cryosphere. In: T. Stocker, D. Qin, G.-K. Plattner, M. Tignor, S.K. Allen, J. Boschung, A. Nauels, Y. Xia, V. Bex, and P.M. Midgley

(Eds) *Climate Change 2013 - The* Physical Science Basis, Cambridge University Press, Cambridge, UK, and New York, 317–382, https://www.cambridge.org/core/product/identifier/CBO9781107415324A020/type/book_part.

Villarini, G., Serinaldi, F., Smith, J.A., and Krajewski, W.F. 2009a. On the stationarity of annual flood peaks in the continental United States during the 20th century, *Water Resour. Res.*, 45: W08417, http://doi.wiley.com/10.1029/2008WR007645.

Villarini, G., Smith, J.A., Serinaldi, F., Bales, J., Bates, P.D., and Krajewski, W.F. 2009b. Flood frequency analysis for nonstationary annual peak records in an urban drainage basin, *Adv. Water Resour.*, 32: 1255–1266, http://dx.doi.org/10.1016/j.advwatres.2009.05.003.

Vogel, R.M., Yaindl, C., and Walter, M. 2011. Nonstationarity: Flood magnification and recurrence reduction factors in the united states, *J. Am. Water Resour. Assoc.*, 47: 464–474.

Vormoor, K., Lawrence, D., Schlichting, L., Wilson, D., and Wong, W.K. 2016. Evidence for changes in the magnitude and frequency of observed rainfall vs. snowmelt driven floods in Norway, *J. Hydrol.*, 538: 33–48, http://doi.org/10.1016/j.jhydrol.2016.03.066.

Vörösmarty, C.J., and Sahagian, D. 2000. Anthropogenic disturbance of the terrestrial water cycle, *Bioscience*, 50: 753, https://academic.oup.com/bioscience/article-lookup/doi/10.1641/0006-3568(2000)050[0753:ADOTTW]2.0.CO;2

Wasko, C. 2021. Review: Can temperature be used to inform changes to flood extremes with global warming? *Philos. Trans. R. Soc. A Math. Phys. Eng. Sci.*, 379: 20190551, https://royalsocietypublishing.org/doi/10.1098/rsta.2019.0551.

Wasko, C., and Nathan, R. 2019. Influence of changes in rainfall and soil moisture on trends in flooding, *J. Hydrol.*, 575: 432–441, https://linkinghub.elsevier.com/retrieve/pii/S0022169419304998.

Wasko, C., Nathan, R., and Peel, M.C. 2020a. Changes in antecedent soil moisture modulate flood seasonality in a changing climate, *Water Resour. Res.*, 56: e2019WR026300, https://onlinelibrary.wiley.com/doi/abs/10.1029/2019WR026300.

Wasko, C., Nathan, R., and Peel, M.C. 2020b. Trends in global flood and streamflow timing based on local water year, *Water Resour. Res.*, 56: e2020WR027233, http://doi.wiley.com/10.1029/2020WR027233.

Wasko, C., Pui, A., Sharma, A., Mehrotra, R., and Jeremiah, E. 2015a. Representing low-frequency variability in continuous rainfall simulations: A hierarchical random Bartlett Lewis continuous rainfall generation model, *Water Resour. Res.*, 51: 9995–10007, http://doi.wiley.com/10.1002/2015WR017469.

Wasko, C., Shao, Y., Vogel, E., Wilson, L., Wang, Q.J., Frost, A., and Donnelly, C. 2021a. Understanding trends in hydrologic extremes across Australia, *J. Hydrol.*, 593: 125877, https://doi.org/10.1016/j.jhydrol.2020.125877.

Wasko, C., and Sharma, A. 2015. Steeper temporal distribution of rain intensity at higher temperatures within Australian storms, *Nat. Geosci.*, 8: 527–529, http://www.nature.com/doifinder/10.1038/ngeo2456.

Wasko, C., and Sharma, A. 2017. Global assessment of flood and storm extremes with increased temperatures, *Sci. Rep.*, 7: 7945, http://www.nature.com/articles/s41598-017-08481-1.

Wasko, C., Sharma, A., and Lettenmaier, D.P. 2019. Increases in temperature do not translate to increased flooding, *Nat. Commun.*, 10: 5676, http://www.nature.com/articles/s41467-019-13612-5.

Wasko, C., Sharma, A., and Westra, S. 2016. Reduced spatial extent of extreme storms at higher temperatures, *Geophys. Res. Lett.*, 43: 4026–4032, http://doi.wiley.com/10.1002/2016GL068509.

Wasko, C., Westra, S., Nathan, R., Orr, H.G., Villarini, G., Villalobos Herrera, R., and Fowler, H.J. 2021b. Incorporating climate change in flood estimation guidance, *Philos. Trans. R. Soc. A Math. Phys. Eng. Sci.*, 379: 20190548, https://royalsocietypublishing.org/doi/10.1098/rsta.2019.0548.

Westra, S., Alexander, L., and Zwiers, F. 2013. Global increasing trends in annual maximum daily precipitation, *J. Clim.*, 26: 3904–3918.

Westra, S., Fowler, H.J., Evans, J.P., Alexander, L., Berg, P., Johnson, F., Kendon, E.J., Lenderink, G., and Roberts, N.M. 2014. Future changes to the intensity and frequency of short-duration extreme rainfall, *Rev. Geophys.*, 52: 522–555, http://onlinelibrary.wiley.com/doi/10.1029/88EO01108/abstract.

Westra, S., and Sisson, S.A. 2011. Detection of non-stationarity in precipitation extremes using a max-stable process model, *J. Hydrol.*, 406: 119–128, http://linkinghub.elsevier.com/retrieve/pii/S0022169411004112.

Wilks, D.S. 2006. On "field significance" and the false discovery rate, *J. Appl. Meteorol. Climatol*, 45: 1181–1189.

Woldemeskel, F., and Sharma, A. 2016. Should flood regimes change in a warming climate? The role of antecedent moisture conditions, *Geophys. Res. Lett.*, 43: 7556–7563, http://doi.wiley.com/10.1002/2016GL069448.

Zhang, Q., Gu, X., Singh, V.P., Shi, P., and Luo, M. 2017. Timing of floods in southeastern China: Seasonal properties and potential causes, *J. Hydrol.*, 552: 732–744, http://doi.org/10.1016/j.jhydrol.2017.07.039.

Zhang, X.S., Amirthanathan, G.E., Bari, M.A., Laugesen, R.M., Shin, D., Kent, D.M., MacDonald, A.M., Turner, M.E., and Tuteja, N.K. 2016. How streamflow has changed across Australia since the 1950s: Evidence from the network of hydrologic reference stations, *Hydrol. Earth Syst. Sci.*, 20: 3947–3965, http://www.hydrol-earth-syst-sci.net/20/3947/2016/.

12 Sea-Level Rise Due to Climate Change

Brij Bhushan and Abhishek Sharma

CONTENTS

DOI: 10.1201/9780429463938-16

12.1   INTRODUCTION

The people living on this planet are adding huge quantities of many gases into the atmosphere. These gases, called greenhouse gases (GHGs), are mainly carbon dioxide (CO_2), ozone (O_3), chlorofluorocarbons (CFCs), nitrous oxide (N_2O), methane (CH_4), and sulfur dioxide (SO_2). These gases tend to make the earth and the atmosphere warmer as they retard the process of Earth's radiation going into space in a similar way as a greenhouse. These gases absorb outgoing radiation and keep the atmosphere warm. The roof of a greenhouse does not allow the infrared radiation of land to go out of the greenhouse. This process keeps the interior of the greenhouse warmer than ambient open space. Trees, crops, and other vegetations absorb CO_2, help counter CO_2 emission with photosynthesis, and produce oxygen (O_2).

Meteorological services all over the world have been continuously observing Earth's temperature almost since 1880. These data have steadily grown into large datasets and, today, the temperatures are recorded manually and automatically at several thousand locations on land, at sea, in the depths of oceans, and at different layers of the atmosphere, and also through satellite remote sensing. Satellite remote sensing, since the 1970s, has boosted monitoring of the environment. Many types of sensors, on hundreds of satellites, have mapped the changes in the temperature of the land, ocean, and atmosphere. Research organizations, including the NASA Goddard Institute for Space Studies, the Hadley Centre for Climate Change, the Japan Meteorological Agency, and NOAA's National Climate Data Center have used these data sets to determine global temperature changes.

12.2   OBSERVATIONS ON GLOBAL WARMING

Analysis of the observations shows that Earth's average temperature has increased by more than 0.8° C in the 20th century. Most of the increase occurred during 1975–2010. A temperature rise of 0.8° C in 100 years may not seem much if compared with daily or seasonal fluctuation, but it is significant when there is thought to be a permanent increase across the Earth.

A number of indicators of global warming have been found. These are: (1) heat wave frequency has increased, (2) cold snaps are becoming shorter and milder, (3) snow and ice cover are getting reduced in the Northern Hemisphere, (4) glaciers and ice caps are melting, and (5) many plant and animal species are entering into colder latitudes or higher elevations. This is a clear and consistent indication that the Earth is getting warmer.

In the 1820s, the importance of certain gases in regulating the temperature of the atmosphere was established. The CO_2, CH_4, N_2O, and water vapor, which are the constituents of GHGs, act as a blanket in the atmosphere. These gases retain heat in the lower atmosphere. The GHGs are only a very small fraction of the Earth's atmosphere. They are critical for keeping the planet warm to support life. Some of the Sun's energy coming to Earth is reflected back to space and the remaining is absorbed by the land and oceans. The Earth also emits heat as infrared rays. In the absence of GHGs, this heat would escape to space, and the Earth's average temperature would be below freezing. The GHGs absorb, retain, and redirect some of this energy back to the Earth. As the concentration of the GHGs increases, the atmosphere's greenhouse effect is enhanced. It results in an increase in the temperatures of the Earth.

12.3   DISRUPTION OF CARBON CYCLE AS CAUSE OF GLOBAL WARMING

The factors responsible for warming or cooling the atmosphere are known as climate forcing agents. GHGs are climate forcing agents because they can change the Earth's energy balance. The GHGs differ in their forcing power, e.g., a CH_4 molecule has about 25 times the warming capacity of a CO_2 molecule. The CO_2 has a larger warming effect than CH_4 because the CO_2 is more abundant and stays in the atmosphere for a longer time. Some climate forcing agents work towards cooling, offsetting some of the heating effects of GHGs. Aerosols like tiny liquid or solid particles suspended in air and most of the visible air pollutants have a cooling effect because they scatter a portion of incoming sun rays back into space. Human activities like the burning of fossil fuels have increased the concentration of aerosols, especially over and around the major cities and industries. Change in land use and land cover also influence Earth's climate. Deforestation is responsible for 10–20% of the excess CO_2 injected into the atmosphere every year. Agriculture adds N_2O and CH_4 into the atmosphere.

Earth's reflectivity depends on land use and land cover. The more reflective a surface, the more energy is sent back into space. Cropland is more reflective than a forest. The urban areas reflect less energy than land. Baron land reflects more energy than tar roads. Considering all human and natural forcing agents, it was found that the net climate forcing during 1750–2005 is warming the Earth by about 1.6 watts per square meter of Earth's surface. This energy is about 800 trillion watts per year and it is about 50 times the power produced in the world (Keeling, 1960).

An important GHG is water vapor. The amount of water vapor in the atmosphere increases as the atmosphere warms up. An increase of air temperature by 1° C increases water vapor content in the air by about 7%. This increase of water vapor results in further warming of the atmosphere as water vapor is a GHG.

There is an inherent time lag in the warming process caused by climate forcing. This lag occurs because it takes time for oceans to respond. Even if all climate forcing agents are held at the present level, the Earth will continue to warm beyond the level of 0.8° C already reached.

The sun's output has an influence on Earth's temperature. The records of solar activity were analyzed to determine if the changes in solar output are responsible for global warming. The direct measurements of solar output are satellite observations, available since 1979 has not shown a net increase during the past 30 years and thus the sun is not responsible for global warming. Prior to the satellite era, solar output was estimated by indirect methods, such as data of the number of sunspots in a year. There was a slight increase in solar energy reaching Earth during the first few decades of the 20th century. This increase may have contributed to global temperature increases during that period, but it does not explain the warming in the latter part of the century. Current warming is not a result of solar changes as shown by the temperature trends in the layers of the atmosphere. The upper air data from meteorological observatories and satellite data show a warming trend in the troposphere and a cooling trend in the stratosphere. This is the vertical pattern of temperature changes expected from the increased GHGs, which trap energy closer to the Earth's surface. If an increase in solar output was responsible for the warming, the vertical pattern of warming would have been more uniform in the troposphere and the stratosphere.

The IPCC AR4, Chapter 5 analyzes the global and regional GHG emission trends up to the recent past and the main drivers causing warming trends. It also analyzed the past GHG emissions trends, including aggregate emissions flows, per capita emissions, cumulative emissions, sectoral emissions, and territory-based vs. consumption-based emissions. The GHG emissions increased from 27 ± 3.2 to 49 ± 4.5 Gt CO_2 eq/yr (+80%) during 1970–2010 (Gt = gigaton). GHG emissions grew on average by 1 Gt CO_2 eq (2.2%) per year during 2000–2010. It was 0.4 Gt CO_2 eq (1.3%) per year in 1970–2000. CO_2 emissions from fossil fuel and industrial processes contributed about 78% of the total GHG emission increase in 1970–2010. Fossil fuel-related CO_2 emissions increased consistently during 1966–2005 reaching 32 ± 2.7 Gt CO_2/yr, i.e., 69% of global GHG emissions in 2010. It increased further by about 3% in 2010–2011 and by about 1–2% during 2011–2012.

Agriculture, deforestation, and other land-use changes were the second-largest contributors to GHG emissions, which has reached 12 Gt CO_2 eq/yr, 24% of global GHG emissions in 2010. Since 1970, CO_2 emissions increased by about 90%, CH_4 increased by about 47%, and N_2O by 43%. Fluoride gases emitted continue to be less than 2% of GHG emissions. Out of 49 (± 4.5) Gt CO_2 eq/yr total GHG emissions in 2010, the CO_2 remains the major GHG accounting for 76% (38 ± 3.8 Gt CO_2 eq/yr) of total GHG emissions in 2010. CH_4 contributed 16%, i.e., 7.8 ± 1.6 Gt CO_2 eq/yr and the N_2O about 6.2% (3.1 ± 1.9 Gt CO_2 eq/yr) and fluorinated gases contributed 2.0% (1.0 ± 0.2 Gt CO_2 eq/yr).

During 1970–2010, GHG emissions have risen in every region other than the economies in transition, where trends were dissimilar in the different regions. In Asia, GHG emissions grew by 330% reaching 19 Gt CO_2 eq/yr in 2010, in the Middle East and Africa by 70%, in Latin America by 57%, in the member countries of the Organisation for Economic Co-operation and Development (OECD-1990) by 22%, and in economies in transition (EIT) by 4%. Though small in quantity, GHG emissions from international transport are growing rapidly.

Cumulative fossil CO_2 emissions since 1750 have had more than tripled from 420 Gt CO_2 in 1970 to 1,300 ($\pm$8%) Gt CO_2 by 2010. Cumulative CO_2 emissions associated with agriculture, deforestation, and other land-use changes (AFOLU) have increased from about 490 Gt CO_2 in 1970 to 680 Gt CO_2 in 2010. Cumulative CO_2 emissions in 1750–2010 in the OECD-1990 region were a major contributor with 42%; Asia with 22% is increasing its share. The findings are summarised in Table 12.1.

Production-related and consumption-related emissions of the CO_2 gap show that a considerable share of CO_2 emissions from fossil fuels combustion in developing countries is released in the production of goods exported to developed countries. By 2010, the developing country group has overtaken the developed country group in terms of annual CO_2 emissions from fossil fuel and industrial processes from the production and consumption perspectives.

The carbon cycle exchanges CO_2 between the atmosphere, ocean, biosphere, and land on different timescales. In the short term, CO_2 is exchanged continuously among vegetation and

TABLE 12.1
Growth of GHG Emissions

S. No.	Name of the GHG	Year	Quantity CO2 Eq Gigatons	Year	Quantity CO2 Eq Gigatons	Growth	% of Total GHG
1	All GHG	1970	27 ($\pm$ 3.2)	2010	49 ($\pm$ 4.5)	80%	
2	GHG growth	2000		2010		1.0 Gt/yr, 2.2%	
3	GHG growth	1970		2010		0.4 Gt/yr, 1.3%	
4	Fossil fuel- related CO2	1966		2005	32 ($\pm$ 2.7)		78
		2010		2011		3%	
		2011		2012		1–2%	
5	Agriculture, deforestation			2010	12 Gt		24
6	CO2	1970		2010		90%	76
7	CH4			2010	7.8 $\pm$ 1.6	47%	16
8	N2O			2010	3.1 $\pm$ 1.9	43%	
9	Fluorides			2010	1.0 $\pm$ 0.2	76%	2.0
10	GHG (Asia)	1970		2010	19 Gt	330%	
11	GHG (Middle East, Africa)					70%	
12	GHG (Latin America)					57%	
13	OECD-1990					22%	
14	EIT					4%	
	Cumulative CO2						
15	Fossil CO2	1750	420	2010	1300 $\pm$ 8%	300%	
16	CO2-AFOLU	1970	490	2010	680		
17	OECD-1990	1750		2010		42%;	
18	CO2 (Asia)	1750		2010		22%	

humans/animals. Humans and animals inhale oxygen for respiration and release CO_2. The vegetations consume CO_2 and by the process of photosynthesis fix carbon in the plants and release oxygen. The weathering of rocks and the formation of fossil fuels are very slow processes occurring over thousands of years. Most of the world's oil and coal reserves were formed when the remains of plants and animals were buried in the sediment at the bottom of shallow seas or in the earth millions of years ago, and then subjected to heat and pressure for many millions of years. A small amount of carbon is released into the atmosphere by volcanoes. Human activities, like the burning of coal, oil, and natural gas, are disrupting the natural carbon cycle by adding a large amount of CO_2 to the air in a short time. When humans began digging up coal and oil and burning them for energy, additional CO_2 began entry into the atmosphere more rapidly than in the natural carbon cycle. Other human activities, like cement production and deforestation, also add CO_2 into the atmosphere.

Till the 1950s, it was believed that oceans absorb most of the excess CO_2. Revelle and Suess (1957) and Bolin and Eriksson (1959) led to the conclusion that the oceans could not absorb all the excess CO_2 emitted into the atmosphere. Keeling (Keeling, 1960) collected air samples at the Mauna Loa Observatory to find variations in the CO_2 concentration. Today, there are thousands of such observatories around the globe. The observations reveal a steady growth of atmospheric CO_2. To determine the CO_2 variations prior to these measurements, scientists have studied the composition of air bubbles trapped in ice cores extracted from Greenland and Antarctica. These data show that, for at least 2,000 years before the Industrial Revolution, atmospheric CO_2 concentrations were steady and then began to rise sharply in the late 1800s. Today, the atmospheric CO_2 is at 390 parts per million (ppm), about 40% more than the pre-industrialization period, and, according to ice core data, higher than at any point in the past 800,000 years.

Human activities have increased the concentrations of the other GHG as well. The CH_4 produced by the burning of fossil fuels, rearing livestock, decay of landfill wastes, and production and transport of natural gas has increased sharply through the 1980s to about 250% of its pre-industrialization value. N_2O has increased about 15% since 1750 predominantly because of high fertilizer use. Chlorofluorocarbons (CFCs), not naturally produced, act as potent GHGs.

Through the data of atmospheric CO_2, detailed records of coal, oil, and natural gas burned every year were estimated. It was also estimated how much CO_2 is absorbed, on average, by the oceans and the land. About 45% of the CO_2 emitted by humans remains in the atmosphere. The atmospheric CO_2 is increasing and will remain high for many centuries.

12.4 DRIVERS OF GHG EMISSIONS

The IPCC AR4 (IPCC, 2007) described (1) per capita production and consumption growth, (2) reductions in the energy intensity, (3) population growth, (4) technological innovation and diffusion, (5) infrastructural choices, (6) energy use, technological choices, lifestyles, and consumption preferences, and (7) policies addressing fossil fuel use as seven major GHG emission drivers. Details of these drivers can be found in the IPCC AR4.

12.5 CLIMATE CHANGE'S IMPACTS

Increased concentrations of GHGs have been warming the land and the oceans. The 1951–2000 witnessed winter warming across parts of Canada, Alaska, northern Europe, and Asia. Summer warming was strong across the Mediterranean and Middle East and parts of the western United States. Heatwaves and record high temperatures have increased in many regions, cold snaps have decreased, and record low temperatures have increased. Global warming has reduced ice across the Arctic. The average annual extent of Arctic sea ice has dropped by about 10% per decade as observed with satellites. The melting was strong in late summer when the Arctic Ocean had no ice for weeks. Water from melting glaciers, ice sheets, and icecaps adds to SLR. Many of the world's glaciers and ice sheets are seen melting, and long-term average winter snowfall and snowpack have declined in many regions.

12.5.1 IMPACT ON SALINITY

Because CO_2 gets dissolved in water the acidification of the oceans is another outcome of more CO_2 emissions. The oceans have absorbed 25–33% of the excess CO_2 released making oceans nearly 30% more acidic than as in the pre-industrial period. This change has taken place in a very short timeframe, and evidence shows its potential to alter the marine ecosystems, and the health of coral reefs, shellfish, and fisheries. The saline water through higher waves and surges on the elevated sea may enter into freshwater resources in the coastal regions.

12.5.2 IMPACT ON PRECIPITATION

Changes are observed in the frequency and distribution of precipitation. Total precipitation in the United States has increased by about 5% over the past 50 years. This is not geographically uniform. It is wetter in the northeast, drier in the southeast, and much drier in the southwest. Warmer air can hold more water vapor, which has led to a measurable increase in the intensity of precipitation. In the United States, the total precipitation falling in the heaviest 1% of rainstorms has increased by about 20% in the last century. The northeastern states have experienced an increase of 54%. This change has increased the risk of flooding.

12.5.3 IMPACT ON ECOSYSTEMS

All organisms attempt to acclimate to a changing environment or else move to a location where they find it more comfortable. Climate change threatens some species beyond their capacity to adapt or move. Special stress is on cold-adapted species on mountains and at high latitudes. The ability of species to move and adapt is also affected by infrastructural barriers like roads, rails land use, and interaction with other species (Eslamian, 2014).

As the climate is changing, many species have shifted poleward and to higher altitudes. The timing of different seasonal activities is also changing. Several plant species are blooming earlier in spring, and some birds, mammals, fish, and insects are migrating earlier, while other species are altering their seasonal breeding patterns. Global analyses show these behaviors occurred on an average of five days earlier per decade in 1970–2000. Such changes can disrupt feeding patterns, pollination, and other interactions between species, and they also affect the timing and severity of insects, disease outbreaks, and other disturbances. In the western United States, climate change has increased the population of some forest pests like pine beetles.

12.5.4 IMPACT ON SNOW COVER

Snow cover has decreased in most regions, especially in spring. Satellite observations of the Northern Hemisphere snow from 1966 to 2005 show a decrease in every month except in November and December, with a stepwise drop of 5% in the annual mean in the late 1980s. There is either a decrease or no changes during 1971–2010 in the Southern Hemisphere.

12.5.5 IMPACT ON PERMAFROST

Permafrost and seasonally frozen ground in most regions have shown large changes in recent decades. Temperature rise at the top of the permafrost layer of up to 3° C since the 1980s was observed. Permafrost warming has also been observed with variable magnitudes in the Canadian

Arctic, Siberia, the Tibetan Plateau, and Europe. The permafrost base has been thawing at a rate ranging from 0.02 m/yr on the Tibetan Plateau to 0.04 m/yr in Alaska.

12.5.6 GREENLAND ICE SHEET (GIS)

The Greenland ice sheet shrunk during 1993–2003, with the thickening in central regions being more than offset by increased melting in coastal regions (IPCC, 2007). An assessment of the data showed a mass balance for the GIS of –50 to –100 Gt/year during 1993–2003, with even larger losses in 2005. The estimated range in mass balance for the GIS for 1961–2003 is between growth of 25 Gt/yr and shrinkage of 60 Gt/yr, i.e., –0.07 to +0.17 mm/yr SLE.

12.5.7 ANTARCTIC ICE SHEET (AIS)

Using measurements of time-variable gravity of the Gravity Recovery and Climate Experiment (GRACE) satellites, Velicogna and Wahr (2006) found that the Antarctic ice sheet mass has decreased at a rate of 152 ± 80 km/yr of ice, equivalent to 0.4 ± 0.2 mm/yr of global SLR, during 2002–2005, with most of the loss from the WAIS. The EAIS exhibits the smallest range of variability among recent mass balance estimates. Losses from the WAIS were more than offset by any growth occurring in the EAIS, leading to a net loss for the AIS as a whole. Thus most of the dynamic changes in the AIS are driven by losses in the WAIS. It is estimated that the overall AIS mass balance ranges from growth of 100 Gt/yr to shrinkage of 200 Gt/yr. This is equivalent to –0.27 to +0.56 mm/yr of SLR during 1961–2003 and also +50 to –200 Gt/yr, i.e., –0.14 to +0.55 mm/yr of SLR during 1993–2003 (IPCC, 2007).

Rignot et al. (2008), with Advanced Land Observation System Phased-Array Synthetic-Aperture Radar (ALOS PALSAR) data, has shown that pronounced regional warming in the Antarctic Peninsula has triggered an ice shelf collapse, in turn, leading to a ten-fold increase in glacier flow and rapid ice sheet retreat. In West Antarctica, the Pine Island Bay sector is draining far more ice into the ocean than is stored upstream from snow accumulation. This sector alone would raise sea levels by 1 m and trigger widespread retreat of ice in West Antarctica. Its neighboring Thwaites Glacier is widening and may double its width when the weakened eastern ice shelf breaks up. Acceleration in this sector may be caused by glacier un-grounding from ice shelf melting by the ocean that warmed up by 0.3° C. The glaciers buffered from oceanic change by large ice shelves have only small contributions to SLR.

At present, many glaciers in East Antarctica are close to a state of mass balance, but sectors grounded well below sea level, such as the Cook Ice Shelf, Ninnis/Mertz, Frost, and Totten glaciers, are thinning and losing mass. Thus East Antarctica is not immune to climate change.

Vaughan (2006) found that long-term records from meteorological stations on Antarctica show strong rising trends in the annual duration of melting. Observation of 1950–2000 from the Faraday/Vernadsky station showed a 74% increase in the number of positive degree-days. A simple parameterization of the likely effects of the warming on the rate of snowmelt suggests an increase on the Antarctic Peninsula ice sheet from 28 ± 12 Gt/yr in 1950, to 54 ± 26 Gt/yr by 2000 which may reach 100 ± 46 Gt/yr in next 50 years.

Rignot et al. (2008) have shown 85% of Antarctica's coastline to estimate the total mass flux into the ocean during 1992–2006. The mass fluxes from large drainage basins were compared with interior snow accumulation for 1980–2004. According to these estimates, in East Antarctica, small glacier losses in Wilkes Land and glacier gains at the mouths of the Filchner and Ross ice shelves combine to a loss of 4 ± 61 Gt/yr. In West Antarctica, losses along the Bellingshausen and Amundsen seas increased the ice sheet loss by 59% in ten years to 132 ± 60 Gt/yr in 2006. The losses increased by 140% to 60 ± 46 Gt/yr in 2006 in the Peninsula.

Climate change impacts also include:

- Regional CLIMATE CHANGE could change forests, crop yields, and water supplies.
- In the 21st century, the projected 2° C temperature rise could shift the ideal range of many forest tree species by about 300 km towards the north in the United States.
- Climate change may alter grazing activity and wild habitats due to a shift in water bodies.
- Human and animal health may be affected by more heat-induced deaths. There may be increases in vector-borne diseases like plague, malaria, encephalitis, and yellow fever. Cold-related deaths may reduce.
- Loss of habitat may affect populations of many of the wildlife, including birds, reptiles, mammals, fish, and sea life. An increased rate of extinction of rare species is expected.
- Agriculture may be affected in many ways due to changes in rainfall and temperatures. Farming will require the adaptation of new crop varieties and new agricultural practices.
- The snow lines may shift poleward in both the hemispheres.
- Snowmelt into the rivers may result in more water in the rivers resulting in more floods.
- Flooding may cause increased land erosion, increased coastal erosion, and destruction of coastal habitats of fish and other aquatic species.
- There may be a significant increase in the SLR-related extremes.
- Mean significant wave heights will increase in the Southern Ocean as a result of enhanced wind speeds. Southern Ocean-generated swells are likely to affect the heights, periods, and directions of waves.
- Islands and coastal cities with low height above MSL may get submerged in the sea.
- Coastal cities may face more threats of high tide, waves, and surges on elevated sea chocking the storm drainages. Concurrent heavy rains may result in severe floods.

12.6 PROJECTIONS BY CLIMATE MODELS

Climate models simulate the Earth's climate. These models are based on mathematical equations representing the basic laws of physics that govern behavior of the atmosphere, oceans, land, and other parts of the climate system, and also interactions therein. Climate models represent the past, present, and future climates. These variations are caused by many processes that can occur over a large range of timescales ranging from warm summers or snowy winters to changes in millions of years. These models predicted that for undisturbed Earth without human-produced GHGs, there would have been negligible warming, or even a slight cooling, in the 20th century. When GHG emissions and other activities were included in the models, the estimated temperature changes resemble the observations.

The future global warming is projected by future CO_2 and other GHGs emissions. It will depend on (1) how energy will be produced and used, (2) national and international policies be implemented to control emissions, and (3) new technologies. Scientists try to account for these uncertainties by developing different scenarios of future emissions. Each of these scenarios is based on the estimates of how different socioeconomic, technological, and policy factors including population growth, economic activity, energy-conservation practices, energy technologies, and land use will change over time. The most comprehensive suite of modeling experiments to project climate changes was completed in 2005. It included 23 models, each of which used the same set of GHG emission scenarios. The modeling experiments were part of the World Climate Research Program's Coupled Model Inter-Comparison Project phase 3 (CMIP3). The observed and projected mean global SLR is given in Table 12.2. The values are mean SLR in mm/year and values in brackets are the range.

Dangendorf et al. (2017) estimated the mean GSLR as 3.1 ± 1.4 mm·y^{-1} from 1993 to 2012. Some of the projected changes by the climate models are as follows.

TABLE 12.2

Observed and Estimated Global Mean SLR

Source ↓	Period→	1901–1990	1971–2010	1993–2010
Observed contributions to global mean SLR (mm/yr)				
Thermal expansion		–	0.8 [0.5 to 1.1]	1.1 [0.8 to 1.4]
Glaciers except in Greenland and Antarctica		0.54 [0.47 to 0.61]	0.62 [0.25 to 0.99]	0.76 [0.39 to 1.13]
Glaciers in Greenland		0.15 [0.10 to 0.19]	0.06 [0.03 to 0.09]	0.10 [0.07 to 0.13]
Greenland ice sheet		–	–	0.33 [0.25 to 0.41]
Antarctic ice sheet		–	–	0.27 [0.16 to 0.38]
Land water storage		–0.11 [–0.16 to –0.06]	0.12 [0.03 to 0.22]	0.38 [0.26 to 0.49]
Total of contributions		–	–	2.8 [2.3 to 3.4]
Observed GMSL rise		1.5 [1.3 to 1.7]	2.0 [1.7 to 2.3]	3.2 [2.8 to 3.6]
Modeled contributions to global mean SLR (mm/yr)				
Thermal expansion		0.37 [0.06 to 0.67]	0.96 [0.51 to 1.41]	1.49 [0.97 to 2.02]
Glaciers except in Greenland and Antarctica		0.63 [0.37 to 0.89]	0.62 [0.41 to 0.84]	0.78 [0.43 to 1.13]
Glaciers in Greenland		0.07 [–0.02 to 0.16]	0.10 [0.05 to 0.15]	0.14 [0.06 to 0.23]
Total with land water storage		1.0 [0.5 to 1.4]	1.8 [1.3 to 2.3]	2.8 [2.1 to 3.5]
Residual		0.5 [0.1 to 1.0]	0.2 [–0.4 to 0.8]	0.4 [–0.4 to 1.2]

Source: Church et al. 2013.

12.6.1 PROJECTED TEMPERATURE CHANGE

Models have projected global the mean temperature change during the 21st century for future emissions scenarios (1) high, (2) medium, and (3) low rise in the range 1.1–6.1° C for the low to the high emissions scenario. Warming will be maximum in the high latitudes of the Northern Hemisphere having more land than ocean. The Hotter Days Model suggests that, relative to the 1960s and 1970s, the number of days with a heat index above 100° F will increase by 60 to 90 additional days in the United States. Heatwaves may last longer. The global temperature rise would increase the risk of heat-related illness and deaths. The cold extremes may decrease, and it would reduce the cold-related deaths.

12.6.2 FUTURE PRECIPITATION CHANGES

Global warming is expected to intensify the exiting regional contrasts in rain and snowfall. Dry areas are expected to get drier and the wet areas wetter. This is because the warmer temperatures tend to increase the evaporation from oceans, lakes, plants, and soil, which will boost the water vapor in the atmosphere by about 7% per 1° C of warming. Enhanced evaporation may increase the precipitation in some areas, as it dries out the land faster, it may intensify droughts. The subtropics, where most of the world's deserts are located, are likely to see 5–10% reductions in precipitation for each 1° C of global warming. The sub-polar and polar regions are expected to see increased precipitation, especially in winter. The overall pattern of change in the continental United States is somewhat complicated, as it lies between the drying subtropics of Mexico and the Caribbean and the sub-polar regions.

Observations in many parts of the world show an increase in the intensity of rainstorms. Climate models indicate that this trend will continue as Earth warms, even in subtropical regions where overall precipitation will decrease. The projections show an increase in dry days between the rainstorms with the average rainfall going down. The rainstorms are likely to intensify by 5–10% for each 1° C of global warming, with maximum intensification in the tropics.

12.6.3 Impact on Surface Hydrology

Global climate models show that future runoff is likely to decrease throughout most of the United States, except for parts of the northwest and northeast, with sharp drops in the southwest. A decrease in runoff of 5–10% per 1° C of warming is expected in some basins, including the Arkansas and the Rio Grande. This decrease would be due to increased evaporation because of the high temperatures, which will not be offset by an increase in rainfall. Streamflow in many temperate river basins outside Eurasia is likely to decrease, especially in arid and semiarid regions. Rivers depending on snowmelt may have more severe flooding.

12.6.4 Effect on Agriculture

An increase in the CO_2 in the atmosphere favors the growth of many plants. But it may not translate into more food. Crops tend to grow more quickly in high temperatures, leading to shorter growing periods and needing less time to ripen. Rising temperatures may result in greater water stress and the risk of higher temperature peaks that can quickly damage crops. Moderate warming, the increases in CO_2, and changes in precipitation are expected to benefit the crop and pasture lands in the mid to high latitudes but decrease yield in seasonally dry and low latitudes. In California, climate change is projected to decrease the yields of almonds, walnuts, avocados, and grapes by up to 40% by 2050. Assessments for other parts of the world consistently show that climate change presents a serious risk to staple crops in sub-Saharan Africa and in places depending on irrigation from ice melt. For each 1° C of warming, yields of corn in the United States and Africa, and wheat in India may drop by 5–15% threatening food security.

12.6.5 Impact on Forest Fire

Rising temperatures, increased evaporation, and more frequent and intense droughts may result in an increased risk of fire. The forests that are already fire-prone are likely to become more vulnerable to fire. The average area burnt by wildfire per year in parts of the western United States relative to 1971–2001 is expected to increase annually by 2–4 times per 1° C of warming. Areas dominated by shrubs and grasses may experience a reduction in fires as warmer temperatures may cause shrubs and grasses to die. Fewer fires would be countered by the loss of existing ecosystems. Fire damage in the northern Rocky Mountains is expected to more than double annually for each 1° C of global warming.

12.6.6 Projected Effect on Sea Ice and Snow Cover

Many forms of ice on the Earth will decrease in extent, thickness, and duration in the coming years due to global warming. Snow in Antarctica has, on average, expanded during the past decades. This increase may be linked to the ozone hole over the Antarctic, which developed because of the use of ozone-depleting chemicals in refrigerants and spray cans. The ozone hole allows damaging the UV light to enter the lower atmosphere and may result in lower temperatures as more heat escapes to space. This effect is expected to disappear as ozone returns to normal by later this century, due to the success of the Montreal Protocol that banned the use of ozone-depleting chemicals. Antarctic sea ice may decrease at a lower rate than the Arctic, as in the Southern Ocean heat penetrates to greater depths than the Arctic. In many areas of the world, snow cover is expected to diminish, with delayed snowpack building in the cold season and melting earlier in the spring reducing the duration of the snowpack in a year. In places such as Siberia, parts of Greenland, and Antarctica, where temperatures are low enough to support snow for long periods, the amount of snowfall may increase even if the season shortens, due to the increased amount of water vapor in the atmosphere.

12.7 SEA-LEVEL RISE (SLR)

SLR is occurring due to the processes of (1) melting of ice and (2) thermal expansion of the seawater. The ice melt flows into the ocean and adds a huge volume of water to the ocean. The second-largest contributor to SLR is the expansion of water volume due to an increase in temperature. There are other processes like visco-elastic deformation of the Earth, changing the ocean floor height and local sinking of land, and also changing the regional ocean currents that play role in a local SLR. Variations are due to circulation, variations in temperature and salinity, and static equilibrium caused by mass redistributions changing gravity and Earth's rotation. These sea-level variations form unique spatial patterns and there are a few observations for these patterns. These influences cause the hot spots that have shown a higher than average local SLR, e.g., the East and Gulf Coasts of the United States (Sallenger et al., 2012).

The IPCC AR4 shows that if emissions continue to at the worst IPCC scenarios, the global average SLR by 2100 could be about 1 meter from 0.52–0.98 meters of 1986–2005 baseline and it would be 0.28–0.6 meters for the lowest emissions scenario. Other estimates suggest that for the same period, the global mean SLR could be 0.2–2.0 meters, relative to MSL in 1992.

The Third National Climate Assessment (NCA), 2014, projected an SLR of 0.3–1.20 meters by 2100. Marine Isotope Substage 5e data showed that SLR could accelerate in the next few decades. The ice sheets in contact with the ocean are vulnerable to disintegration due to warming. Paleoclimate data reveal that subsurface ocean warming causes ice shelf melt and ice sheet discharge into the ocean. Holland G. of NCAR found that the SLR in the IPCC is low, and the fact lies somewhere in between estimates of IPCC and of Hansen (2006).

The IPCC AR4 identified the factors that contribute to SLR are: (1) changes in glaciers and ice-caps, (2) glacial melt from the Greenland and Antarctica ice sheets, (3) ocean thermal expansion, and (4) contributions from terrestrial storage, snow on land, and permafrost. It is estimated that each 1° C of temperature rise would contribute about 2.3 meters SLR for the next 2,000 years. Warming beyond 2° C would lead to rates of SLR dominated by ice loss from Antarctica. CO_2 emissions from fossil fuels may result in additional tens of meters of SLR in the next 1,000 years. This may eliminate the entire Antarctic ice sheet, causing about 58–70 meters of SLR. If the Greenland ice sheet were to melt completely, it would result in an average SLR of about 7 meters. There is continued evidence of ice sheet growth in eastern Antarctica but the losses of ice in the west and Greenland may result in an increased SLR.

12.7.1 MELTING OF GLACIERS AND SNOW

Glaciers melt each summer and grow in winter. As temperatures rise, the ice growth in winter is often less than the ice melt in summer. Almost all the world's glaciers, ice caps, and the GIS are losing ice, adding water to the oceans, and causing global SLR. The ice melts also reduce the temperature and salinity of water by adding freshwater to the sea. The pace of ice loss from small glaciers and large ice sheets has accelerated in the past few decades.

Although it is difficult to predict the ablation that will become runoff, a calculation based on an established criterion for runoff indicates that the contribution from the Antarctic Peninsula, as a direct and immediate response to warming, is equivalent to 0.008–0.055 mm/yr of global SLR. Given future warming, this may increase in the next 50 years. This contribution due to increased runoff could lead to a dynamic imbalance in the glaciers draining the ice sheet.

IPCC projections have been criticized by many, even by some members of the IPCC (Solomon et al., 2008). Much of the debate was on the exclusion of dynamic changes in the ice sheets which tend to produce a much larger impact on estimates. Hansen et al. (2016) was against the IPCC for its neglect of potential ice sheet disintegration in projections of SLR, which have remained relatively small until the past few years. Due to the nonlinearity of the ice disintegration, it is difficult to accurately predict the SLR, however, the process is bounded by thresholds, which once crossed, may trigger many meters of SLR by 2100.

The largest uncertainties are in multiple feedbacks occurring on and under the ice and in the nearby oceans that accelerate the process of ice sheet disintegration. For example, a key feedback of the ice sheets is the albedo flip that occurs when snow and ice begin to melt. Snow-covered ice reflects most of the sunlight back to space however as warming causes increased melting on the surface, the darker wet ice absorbs more solar energy. Most of the resulting melt burrows through the ice sheet, lubricates its base, and speeds up the discharge of icebergs into the ocean. Satellite images of the Greenland ice sheet have revealed the appearance of large meltwater pools on the ice surface in recent years.

Even with these dynamic uncertainties, recent satellite observations point to accelerating trends in the rates of change among the contributors of SLR. Rahmstorf et al. (2007) noted that since 1990 the sea level has been rising faster than that projected by most climate models. Satellite data show a linear trend in SLR of 3.3 mm/year over the period 1993–2006, whereas previous predictions by the IPCC (2007) projected an estimated rise of less than 2 mm/year.

An alternative to the IPCC AR4 SLR approach pioneered by Rahmstorf (2007) estimates SLR indirectly from changes in the global average near-surface temperature, one of the more accurate variables in CGCMs. This semi-empirical technique estimates SLR based on the changes in global average temperature and sea level between 1880 and the present. A proportionality constant of 3.4 mm/yr/°C was computed from data on the global SLR and global mean surface temperature for the 20th century. This proportionality constant applied to IPCC AR4 emission scenarios resulted in a projected SLR in 2100 of 0.5 to 1.4 meters above the 1990 level. Using a wider range of CGCM models, Horton et al. (2008) updated Rahmstorf's results and produced a broader range of the SLR projections.

Pfeffer et al. (2008) explored the kinematic scenarios of glacier contributions to SLR in the 21st century by setting 2 m and 5 m SLR targets and compared current loss rates to see whether these objectives can be achieved. A total SLR of about 2 meters by 2100 may occur under physically possible glaciological conditions but only if all variables are quickly accelerated to extremely high limits. More plausible but still accelerated conditions could lead to a total sea-level rise by 2100 of about 0.8 m. They suggested that this estimate be a starting point for the forecast refinements, but the range of 0.8–2.0 m is still possible with the inclusion of ice flow dynamics.

There is an absence of a thorough understanding of the processes that control ice sheet behavior and limited observations on how glaciers and ice sheets respond to climate change. Paleo-climatic data showed that the warming during 1951–2000 was unusual in the past 1,300 years. The last time the Polar Regions were significantly warmer than the present for an extended period of about 125,000 years ago. Otto-Bliesner et al. (2006) used a global climate model, a dynamic ice sheet model, and paleo-climatic data to evaluate the warming in the Northern Hemisphere and its impact on the Arctic ice fields during the Last Interglaciation (LIG), ~116,000–130,000 years ago. Their simulated climate matches paleo-climatic observations of past warming, and the combination of physically based climate and ice-sheet modeling with ice-core constraints indicated that the GIS and other circum-Arctic ice fields contributed 2.2 to 3.4 meters of SLR during the LIG. Overpeck et al. (2006) using a similar method concluded that under a present scenario, similar areas of Greenland could melt causing an SLR of several meters by 2100.

12.7.2 Thermal Expansion of Oceans

Water expands when it warms above 4° C. The oceans have absorbed 85% of the excess heat trapped by the atmosphere since 1880 (Church et al., 2013; IPCC, 2013). The thermal expansion of seawater also adds to SLR. The mass components explain the majority of SLR over the 20th century, but the thermal expansion has increasingly contributed to SLR, starting from 1910 onward and in 2015 accounting for 46% of the total simulated sea-level change (Slangen et al., 2017). The share of thermal expansion in GSLR has declined in recent decades as the shrinking of land ice has accelerated.

Observations revealed that the world's oceans have warmed since 1955–2006. During 1961–2003, the layer 0–3,000 m depth of ocean has absorbed up to 14.1×10^{22} Joules of heat. This is equivalent to an average heating rate of 0.2 watts/m2. During 1993–2003, the corresponding rate of warming in the shallow 0–700 m ocean layer was about 0.5 ± 0.18 W/m2. Hence, relative to 1961–2003, the period 1993–2003 had much higher rates of warming, especially in the upper 700 m layer of the global ocean. The thermal expansion has not only caused SLR but will increase the number and intensity of storms in the warmer sea.

12.7.3 MEASURING LOCAL SLR

The height of the ocean surface at a location is measured (1) relative to the Earth's surface, called relative sea level (RSL), taking some reference points on the Earth or (2) taking Earth's center as a reference point to give geocentric sea level (GSL). RSL is more relevant when studying the impacts of SLR in the coastal regions, and it has been measured using tide gauges. GSL is now measured using satellite altimetry since 1975. A temporal average of sea level for a given location, known as mean sea level (MSL), is used to remove the short time variability. The spatially averaged MSL is known as global mean sea level (GMSL).

Tide gauges, landmarks, and satellite data are used in specific coastal regions to estimate the local SLR. The local sea level is influenced by factors like weather, ocean currents, geologic factors, surface and groundwater flows, dams, drilling, dredging, geography, silts, exploration, and construction. The SLR is found accelerating globally and regionally in many places. During 1993–2008, the GSRL has been 2.8 to 3.3 mm/y. This acceleration is mainly due to ocean warming. The coastline from Nova Scotia to the Gulf of Mexico faced the world's fastest rates of SLR in the 20th century from 2.5 mm/y in Boston to 9.6 mm/y in Louisiana.

Satellite altimetry reveals a higher response of SLR to climate change. The global average rate of SLR during 1993–2003 is 3.1 ± 0.7 mm/yr, which is close to the estimated total of 2.8 ± 0.7 mm/yr for the climate-related contributions. Due to thermal expansion (1.6 ± 0.5 mm/yr) and changes in land ice (1.2 ± 0.4 mm/yr). On average, thermal expansion and melting of ice each accounted for about 50% of the observed SLR during this period.

12.7.4 THE SLR AS OBSERVED

As described in the IPCC AR4 (Bindoff et al., 2007) statistical analyses of tide-gauge data have shown an increase in observed sea level extremes worldwide. Using particle size analysis of cores collected in the Mackenzie Delta in the Arctic, Vermaire et al. (2013) inferred increased storm surge in the region during 1961–2010, which was linked to the annual mean temperature anomaly in the Northern Hemisphere and decrease in summer sea ice extent. SLR, as observed by NOAA and displayed on the website, is given in Figure 12.1.

Some important findings on SLR include:

1. Satellite altimetry since 1993 produced clear evidence of regional variability of SLR. The largest SLR since 1992 has taken place in the Western Pacific and Eastern Indian Oceans.
2. Nearly all of the Atlantic Ocean showed SLR during 1996–2005. The sea levels in Eastern Pacific and Western Indian Oceans are found going down (IPCC, 2007).
3. IPCC (2007) supports that the warming of 1951–2000 is unusual in at least the last 1,300 years.
4. GHGs found in the atmosphere include CO_2, CH_4, N_2O, HFCs, perfluorocarbons (PFCs), sulfur hexafluoride (SF_6), hydrochlorofluorocarbons (HCFCs), chlorofluorocarbons

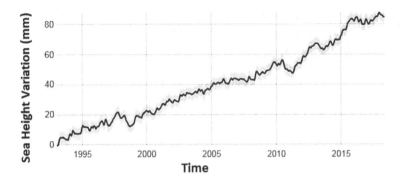

FIGURE 12.1 Satellite sea-level observations from NASA Goddard Space Flight Center. (Source: https://climate.nasa.gov/vital-signs/sea-level/.)

(CFCs), aerosols, sulfur dioxide (SO_2), carbon monoxide (CO), nitrogen oxides (NO_x), and non-methane volatile organic compounds.

5. Polar Regions were significantly warmer than at present for a period about 125,000 years ago. The reductions in polar ice resulted in 4–6 meters of SLR at that time.

6. The contribution of the groundwater, lakes, and reservoirs is difficult to estimate with confidence as data is not available for many regions, and the interactions between SLR, aquifers, lakes, and reservoirs are largely unknown (IPCC, 2007).

7. Floating ice does not contribute to SLR. Ice melt exactly fills the space vacated by the ice.

8. Changes in the mass of floating ice shelves can affect the adjacent non-floating ice, and it can affect SLR indirectly.

9. The sea-level equivalent of sea ice, ice shelves, and seasonally frozen land is near zero.

10. Approximately 360 gigatons of ice melt are equivalent to 1 mm of GSLR.

11. Many of the AOGCM experiments do not include the influence of Mt. Pinatubo, the omission of which may reduce the projected rate of thermal expansion during the early 21st century.

12. A study of snow accumulation areas of glaciers and icecaps has shown that the observed accumulation areas are small, forcing glaciers to lose 27% of their volume to attain equilibrium with the climate. At least 18.4 ± 3.3 cm of SLR by 2100 are forced by mass loss of the world's glaciers and ice caps even if assumed that climate does not warm.

13. If the climate continues to warm at the current rate, a minimum of 37.3 ± 2.1 cm of SLR by 2100 is expected from the glaciers and ice melt (Bahr and Radić, 2012).

14. Antarctic mass variability is difficult to estimate because of the ice sheet's size and complexity. Standard estimates use techniques with intrinsic limitations and uncertainties.

15. The Pine Island Glacier loss accelerated 38% since 1975, and most of this occurred during 1996–2007. This has increased to 42% and the glacier has gone ungrounded.

16. The Smith glacier loss has accelerated by 83% and is now ungrounded.

17. ENVISAT detected new rifts on the Wilkins ice shelf in November 2008. The Wilkins ice shelf has been stable for most of the last century and began retreating in the 1990s. In February 2008, an area of about 400 km² broke off. In May–July 2008, the ice shelf saw further disintegration and lost 1,350 km². The new rifts on the ice shelf could lead to the opening of an ice bridge which prevented the ice shelf from breaking away (*Science Daily*, 2008).

18. If the ice shelf breaks from the peninsula of Antarctica, it will not cause SLR as the ice therein is floating, but it may allow the ice sheets on land to slip and add water to the sea.

19. The size of ice bodies is due to equilibrium in snowfall and ice melt. The oldest ice has disappeared, and 58% of multiyear ice is 2–3 years old compared to 35% in the mid-1980s.

20. Arctic ice coverage in the summer of 2007 reached a record minimum, with ice extent declining by 42% compared to that in the 1980s.

12.8   GREENLAND ICE SHEET (GIS) AND ANTARCTIC ICE SHEET (AIS)

The average annual snowfall on the GIS and AIS is equivalent to 6.5 mm of SLR. This is approximately balanced by loss from melting and calving (IPCC, 2001). The balance of these processes is not the same for the two ice sheets. Antarctic temperatures are so low that there is no runoff. The reduction in size results from ice discharge into the ocean and the formation of icebergs. GIS experiences summer temperatures are high enough to cause widespread melting.

Satellite and in situ observations have shown increased melting and accelerated ice flow around the periphery of Greenland during 1986–2010. A few simulations of long-term ice sheets suggest that the GIS will significantly decrease in volume and area over the coming centuries if a warmer climate continues (Gregory et al., 2004). A threshold of annual mean warming of 1.9° C to 4.6° C in Greenland may eliminate the GIS (Gregory and Huybrechts, 2006).

Observations of accelerated ice streams in the Amundsen Sea sector of the WAIS, the rapidity of propagation of this signal upstream, and the acceleration of glaciers that fed the Larsen B Ice Shelf after its collapse have raised the concern of a collapse of the WAIS. It is possible that the presence of ice shelves tends to stabilize the ice sheet, at least regionally. Thus, a weakening or collapse of ice shelves, caused by melting on the surface or at the bottom due to a warmer ocean, may destabilize the WAIS.

12.9   IMPACTS OF SLR

SLR is a serious threat to countries with high concentrations of population and economic activity in coastal regions. A World Bank estimate suggests that 1 meter SLR in coastal countries of the developing world would submerge 194,000 square km of land, and displace at least 56 million people (Dasgupta and Meisner, 2009). The biophysical effects of SLR include (1) inundation, flood, and storm damage, (2) wetland loss, (3) erosion, (4) saline water intrusion, (5) coral bleaching, (6) ocean productivity changes, and (7) species migration. These effects on the natural system have a range of socio-economic impacts, including (1) increased loss of property and coastal habitats, (2) increased flood risk and potential loss of life, (3) damage to coastal protection works and other infrastructure, (4) loss of renewable and subsistence resources, and (5) loss of tourism, recreation, and transportation functions.

12.9.1   IMPACT OF SLR ON WEATHER SYSTEMS

A warmer sea is likely to create changes in weather patterns. The climate model gave feedback that slows Antarctic bottom water formation and increases the ocean temperature near ice shelf is cooling the ocean surface and increasing sea ice cover and water column stability. The ocean surface cooling in the North Atlantic and the Southern Ocean are increasing tropospheric horizontal temperature gradients, eddy flow, and baroclinicity. This may result in more powerful storms. The land may experience more rains and so more intense floods. The intensity and number of storms may increase. There is likely to be an overall trend of increased precipitation and evaporation, blizzards, more and more intense cyclones, and also increased drier soils which may increase the intensity of droughts.

The potential destructiveness of hurricanes shows an upward trend since the mid-1970s, with a trend towards longer lifetimes and greater intensity, and such trends are correlated with higher sea surface temperatures. Based on the model projections, the IPCC AR4 states that it is likely that future cyclones will become more intense, with larger peak wind speeds and heavier precipitation due to ongoing increases of tropical sea surface temperatures. An IWTC statement has also asserted that if the projected SLR due to global warming occurs, then the vulnerability to storm surge flooding would increase. It is likely that some increase in tropical cyclone peak wind speed and rainfall will occur if the climate continues to warm. Model studies and theory project a 3–5% increase in wind speed per 1° C increase in sea surface temperature and so the storms will be more intense with the stronger winds and higher surge.

12.9.2   IMPACT OF SLR ON FLOODING

Some coastal wetlands and populated areas already flood regularly particularly during high tides over the elevated sea. The surge may be higher due to stronger winds. A rise of 60 cm above the present sea level would risk more than $1 trillion of property and structures in the United States alone. Also, the saline water from higher waves and surges could reach further into coastal groundwater and surface water resources, increasing the salinity of freshwater.

Some of the most populated and beautiful cities are at low elevations, making SLR a serious threat. There are some danger zones particularly in large urban areas on coastal deltas, including those of the Mississippi, Nile, Ganges, and Mekong. In densely populated cities there are unplanned human settlements where drainage is not proper. In some coastal cities, drainage outlets get blocked at the time of high tide. These locations will be more vulnerable to floods. This situation will enhance the intensity of the flood and the associated damages.

If the average SLR is 0.50 meters relative to a 1990 baseline, coastal flooding could affect 5 to 200 million people. Due to the melting of ice and permafrost, four million people may have to be permanently relocated. Relocation has already started in towns on the coast of Alaska. Sea waves may erode the shoreline of more than 250,000 square kilometers of land.

12.9.3   IMPACTS OF SLR ON SURGE

The storms often cause surges as high winds push water inland. With rising seas, storm surges occur on top of an elevated sea. Tides, sea waves, and surges will cause more damages when they travel more inland on the elevated sea. Thus a storm today could create more extensive flooding than an identical storm in 1900. In the future, with higher sea levels, surge penetration may be further inland. SLR increases the potential for erosion by allowing waves to penetrate further inland, even during calm weather. The rate of land loss from erosion can be 100 times more than the SLR itself. This is also forcing investments in expensive measures such as repeated beach replenishment and construction of barriers.

Communities on the Bering Sea coast and in the Arctic are experiencing increased damage from storm surges, shoreline loss to winter storms, and saltwater intrusion into freshwater resources, sanitation lagoons, and fish and wildlife habitats. Indications are that this is being made worse by a gradual SLR.

12.9.4   IMPACTS OF SLR ECOSYSTEMS

The ocean circulation changes may result in impacts on ecosystems. Satellite data shows that warm surface waters are mixing less with deeper cold waters. It is separating near-surface marine life from the nutrients below and reducing the amount of phytoplankton, which is part of the ocean food web. Climate change will exacerbate this in the tropics and subtropics. In temperate and polar waters, vertical mixing of waters may increase because of higher losses in sea ice. Ocean warming will continue to push many marine species toward the poles.

Warmer waters may result in a decline in dissolved oxygen. Acidification, due to excess CO_2 will threaten many species over time, especially mollusks and coral reefs. But not all life forms will suffer. Some of the phytoplankton and other photosynthetic organisms may grow better with higher CO_2. Also, the growth of phytoplankton is likely to retard due to the higher ocean surface temperatures. This will create a greater distance between warmer surface waters and cooler deep waters, separating upper layer marine life from nutrients found in the deep sea.

12.9.5 IMPACTS OF SLR ON GLACIERS AND ICECAPS

The loss of protective sea ice makes the impact of storm waves and surges on the shoreline more severe. The decline of sea ice increases the distance the waves can travel uninterrupted. This increases wave height also. The decline of the ice on the shore leaves the coast unprotected from waves. As ice coverage decreases so does the damping effect it has on the waves. The effects of SLR and storm surges are thus becoming more destructive than they were otherwise. Rapid erosion has threatened homes and forced emergency evacuations in coastal towns of Alaska. One hundred and sixty villages have been identified as threatened by climate-related erosion.

Thawing permafrost increases the coast's susceptibility to erosion. When melted water in the soil drains off, it leaves the soil soft and porous, so when it is hit by waves it easily erodes. In some places, the shoreline has receded by as much as 100 feet in a single storm.

12.10 REDUCING GLOBAL WARMING BY CONTROLLING GHG EMISSIONS

Global warming can be reduced by controlling emissions of GHGs. It is required to limit emissions of the most abundant GHG, CO_2, to control warming. The United States is producing about 50% of human-produced CO_2 and currently accounts for about 20% of global CO_2 emissions, though having only 5% of the world's population. The CO_2 emissions from the United States are likely to decline in the coming decades but emissions from fast-developing countries like China and India will increase. Some suggestions to reduce the GHG emissions are:

1. Reduce demand for goods and services that require more energy.
2. Create awareness and incentives to influence consumer behavior and preferences.
3. Curtail sprawling development patterns that increase our dependence on petroleum.
4. Use fuel-efficient cars, hydrogen-operated cars, or electric cars to reduce emissions.
5. Reduce electricity consumption. High electricity demand causes coal-based power generation.
6. Use more efficient methods for insulating, heating, cooling, and lighting
7. Avoid buildings having only glass on outer walls as they need a large amount of energy for cooling.
8. Use energy-efficient equipment.
9. Upgrade industrial equipment and processes to energy-efficient systems.
10. Encourage buying efficient home appliances and vehicles.
11. Expand the use of low- and zero-carbon energy sources. Switch from coal and oil to natural gas, and expand the use of nuclear power and renewable energy sources.
12. Capture and sequester CO_2 from power plants and factories.
13. Capture and sequester CO_2 directly from the atmosphere.
14. Manage forests and soils to enhance CO_2 intake.
15. Develop methods to scrub CO_2 directly from ambient air.
16. Trees are natural CO_2 absorbers. Plant as many native trees as possible.
17. Buy products in reusable or recyclable packaging. Recycle all of your home's waste.
18. Walk, bicycle, carpool, or take the bus instead of using a car if you are going alone.
19. Avoid burning as far as possible. Biomass should be composted instead of burning.
20. Avoid firewood and dung-based stoves.
21. Cleaning and bringing water to a place of requirement consumes a lot of energy. Harvesting water and avoiding its contamination can help in energy saving in a big way.

These actions to reduce emissions will depend on government policies, private investments, and the behavioral and consumer choices of the people. Governments have a large role to play in influencing the stakeholders through policies, incentives, and public awareness. Most economists and policy makers think that putting high taxes on CO_2 emissions and increasing them over time is the least costly path to reduce GHG emissions.

The only solution to prevent global warming is a reduction in the use of fossil fuel–based energy wherever possible, driving less, and using public transport. Making significant improvements will require worldwide changes in society and the economies of the world. It will take decades or even centuries for the current trends of global warming and SLR to be reversed.

12.10.1 To Reduce Warming Agents Other Than GHGs

There are opportunities to reduce GHG emissions of other than CO_2, like CH_4, N_2O, and HFCs. These gases are generally much stronger climate forcing agents than CO2. Reducing methane leaks from oil and gas systems, coal mining, and landfills is cost-effective because there is a market for CH_4. The largest source of CH_4 emissions is agriculture. The CH_4 is produced by livestock in digesting food, and N_2O and CH_4 are from manure and nitrogen fertilizer. These emissions can be reduced in many ways, including by the use of precision agriculture that helps farmers to cut over-fertilization, and by improving livestock waste management. The gas produced from dung can be used for cooking. It will convert CH_4 to CO_2 which is a low heat absorber and will have a reduced greenhouse effect.

Some short-lived pollutants are not GHGs but cause warming. Black carbon, or soot, emitted from the burning of fossil fuels, biofuels, and biomass, the dried dung used in cooking stoves in many developing countries can be reduced. Black carbon can cause strong local or regional-scale warming where it is emitted. It can also amplify warming in some regions by leaving a heat-absorbing coating on reflective surfaces like arctic ice and snow. Reducing emissions of these short-lived warming agents could reduce the warming and reduce the SLR.

12.10.2 Preparedness for the Climate Change

Disaster preparedness plans in all sectors of economy and development at almost all of the levels in a large number of countries all over the world contain climate change response plans. Adaptation planning and response efforts are underway in a number of countries and communities. At the same time, many countries do not have mechanisms for monitoring climate change and the SLR. Most of the experience of disaster managers in protecting vulnerable people, resources, and infrastructure is based on the historic record of rapid-onset disasters. The records of climate variability during a time of relatively stable climate are helpful in situations of slow-onset disasters. Adaptation efforts are hampered by a lack of information about benefits, costs, potential, and limits of different types of responses. Adaptation of easily deployable actions includes low-cost strategies. In the longer term, high-cost responses may be required.

Even though there are uncertainties in SLR it should not be a reason for inaction, and there are many things already known about climate change that can be acted upon. Reasons for taking action include the following:

- The sooner the efforts to reduce GHG emissions are initiated, the lower the cost and lower the risks of climate change and the SLR. It will avoid more rapid and expensive actions required later.
- Some climate change impacts like SLR, once manifested, will persist for hundreds or even thousands of years and it will be difficult to undo the damages.
- Major investments are being made in equipment and infrastructure that can control GHG emissions. Incentives and policies will provide crucial guidance for these investments.

- Many actions that are needed to reduce vulnerability to climate change and SLR are simple investments that will provide protection against SLR and extreme events.

The challenge for society is to evaluate the risks and benefits and make wise choices even knowing that there are uncertainties in predictions of climate and the SLR. A valuable framework for supporting climate choices is an iterative risk management approach. This refers to a process of systematically identifying risks and possible response options.

12.11 CONCLUSIONS

Excess GHG emission has been the root cause of global warming and so that of the SLR through the process of ice melting and thermal expansion of water in the oceans. SLR is found to be accelerating. People living in coastal regions all over the world will face difficult situations in responding to the increased threat of SLR. More and more coastal communities will need to assess the costs and risks of the SLR, retreating from there, or protecting the properties and infrastructure with different kinds of measures.

It is a common belief that since SLR is slow and gradual, society has enough time to adapt to these challenges, and thus immediate alarm is not warranted. Even small changes in SLR have a profound implication on the impact of storm surges, which occur annually and typically with devastating consequences in coastal areas. SLR acts as the baseline reference to which storm surge height is added.

The best options for the coastal communities would be to share experiences and coordinate policies and actions across local, state, regional, and national jurisdictions. As we work to adapt to unfolding impacts of global warming and those of SLR, reduction in GHG emissions is one of the best ways to limit the magnitude and pace of SLR and reduce the costs of protective measures to prevent the damages. The simplest way is to live in elevated places and not disrupt the natural processes and climate systems like ocean, land, and atmosphere.

An estimated 25 billion tons of soil, which is not replenished, enter the oceans every year. Climate change is causing increasing trends in the frequency and intensity of natural disasters and thus increased quantities of the soil are moving with the water into the oceans through erosions caused by SLR, more severe cyclones, and floods. This process will increase the sea bed height and lower the height of the land in the long run. Rising river bed height due to silt deposition has also been one cause of flooding and thus causes more soil to be taken to oceans. This is an indirect impact of climate change on the sea bed, thus causing SLR. This aspect of SLR has not been included in this chapter.

REFERENCES

Bahr, D.B., and Radić, V. 2012. Significant contribution to total mass from very small glaciers, *Cryosphere*, 6: 763–770.

Bindoff, N.L., Willebrand, J., Artale, V., Cazenave, A. Gregory, J M., Gulev, S. Hanawa, K. Le Quéré, C. Levitus, S. Nojiri, Y. and Shum, C.K. 2007. Observations: Oceanic climate change and sea level. In: Solomon, S., Qin, D., Manning, M., Chen, Z., Marquis, M., Averyt, K.B., Tignor, M., and Miller, H.L. (eds.) *Climate Change 2007: The Physical Science Basis. Contribution of Working Group I to the Fourth Assessment Report of the Intergovernmental Panel on Climate Change*, Cambridge University Press, Cambridge, UK and New York, pp. 747–845.

Bolin, B., and Eriksson, E. 1959. *The Atmosphere and Sea in Motion*, edited by Bolin, B., Rockefeller Institute Press, New York.

Church, J.A., Clark, P.U., Cazenave, A., Gregory, J.M., Jevrejeva, S., Levermann, A., Merriield, M.A., Milne, G.A., Nerem, R.S., Nunn, P.D., Payne, A.J., Pfeffer, W.T., Stammer, D., and Unnikrishnan, A.S. 2013. Sea level change. In: *Climate Change 2013, The Physical Science Basis. Contribution of Working Group I to the Fifth Assessment Report of the IPCC*.

Dangendorf, S., Marcos, M., Woppelmann, G., Conrad, C.P., Frederikse, T., and Riva, R. 2017. Reassessment of 20th century global mean sea level rise, *Proc. Nat. Acad. Sci.*, 114: 5946–5951.

Dasgupta, S., and Meisner, C. 2009. *Climate Change and Sea Level Rise, A Review of the Scientific Evidence*, The World Bank Environment Department, Paper Number 118, Climate Change Series.

Eslamian, S. 2014. *Handbook of Engineering Hydrology, Vol. 2: Modeling, Climate Change and Variability*, Taylor and Francis, CRC Group, USA, p. 646.

Gregory, J.M., and Huybrechts, P. 2006. Ice-sheet contributions to future sea-level change, *Philos. Trans. R. Soc. Ser. A*, 364: 1709–1731.

Gregory, J.M., Huybrechts, P., and Raper, S.C.B. 2004. Threatened loss of the Greenland ice-sheet, *Nature*, 428: 616.

Hansen, J. 2006. Can we still avoid dangerous human made climate change? Presentation to the American Geophysical Union in San Francisco, California, http://www.columbia.edu/~jeh1/newschool_text_and _slides.pdf.

Hansen, J., et al. 2016. Ice melt, sea level rise and superstorms: Evidence from paleoclimate data, climate modeling, and modern observations that 2°C global warming could be dangerous, *Atmos. Chem. Phys.*, 16: 3761–3812, https://www.atmos-chem-phys.net/16/3761/2016/.

Horton, R., Herweijer, C., Rosenzweig, C., Liu, J., Gornitz, V., and Ruane, A.C. 2008. Sea level rise projections for current generation CGCMs based on the semi-empirical method, *Geophys. Res. Lett.*, 35: L02715.

IPCC. 2007. *Climate Change 2007, Impacts, Adaptation and Vulnerability*, Cambridge University Press, UK.

IPCC. 2013. *IPCC Fifth Assessment Report: Climate Change*.

Keeling, C.D. 1960. Concentration and isotopic abundance of carbon dioxide in the atmosphere, *Tellus*, 12: 200–203.

Otto-Bliesner, B., Marshall, S., Overpeck, J., Miller, G., and Hu, A. 2006. Climate Change Series 21 References ablation zone of the Greenland ice sheet Science 321, *Atmos. Chem. Phys.*, 16: 3761–3812.

Overpeck, J., Otto-Bliesner, B., Miller, G., Muhs, D., Alley, R., and Kiehl, J. 2006. Paleoclimatic evidence for future ice-sheet instability and rapid sea-level rise, *Science*, 311: 1747–1750.

Pfeffer, W.T., Harper, J.T., and O'Neel, S. 2008. Kinematic constraints on glacier contributions to 21st-century sealevel rise, *Science*, 321: 1340–1343.

Rahmsdorf, S. 2007. A semi-empirical approach to projecting future sea-level rise, *Science*, 308: 368–370.

Rahmsdorf, S., Cazenave, A., Church, J., Hansen, J., Keeling, R., Parker, D., and Somerville, R. 2007. Recent climate observations compared to projections, *Science*, 316: 709.

Revelle, R., and Suess, H.E. 1957. Carbon Dioxide Exchange between Atmosphere and Ocean and the question of an increase of Atmospheric CO2 during past Decades, *Tellus*, 9: 18.

Rignot, E., Bamber, J.L., Van Den Broeke, M.R., Davis, C., Li, Y.H., Van De Berg, W.J., and Van Meijgaard, E. 2008. Recent Antarctic ice mass loss from radar interferometry and regional climate modelling, *Nat. Geosci.*, 1: 106–110.

Sallenger, A.H., Doran, K.S., and Howd, P.A. 2012. Hotspot of accelerated sea-level rise on the Atlantic coast of North America, *Nat. Clim. Change*, 2: 884–888.

Science Daily. 2008. *Antarctica: Wilkins Ice Shelf Under Threat*, http://www.sciencedaily.com/releases/2008 /11/081128132029.htm.

Slangen, A.B., Meyssignac, B., Agosta, C., Champollion, N., Church, J.A., Fettweis, X., Ligtenberg, S.R., Marzeion, B., Melet, A., Palmer, M.D., and Richter, K. 2017. Evaluating model simulations of twentieth-century sea level rise. Part 1: Global mean sea level change, *J. Climate*, 30: 8539–8563.

Solomon, S., Alley, R., Gregory, J., Lemke, P., and Manning, M. 2008. A closer look at the IPCC report science, *Science*, 319: 409–410.

Vaughan, D.G. 2006. Recent trends in melting conditions on the antarctic peninsula and their implications for ice-sheet mass balance and sea level, *Arct. Antarct. Alp. Res.*, 38(1): 147–152.

Velicogna, I., and Wahr, J. 2006. Measurements of time-variable gravity show mass loss in Antarctica, *Science*, 311: 1754–1756.

Vermaire, J.C., Pisaric, M.F.J., Thienpont, J.R., Mustaphi, C.J.C., Kokelj, S.V., and Smol, J.P. 2013. Arctic climate warming and sea ice declines lead to increased storm surge activity, *Geophys. Res. Lett.*, 40: 1386–1390.

Part V

Flood Optimization and Simulation

13 Real-Time Operation of Reservoirs during Flood Conditions Using Optimization-Simulation with One- and Two-Dimensional Modeling

Hasan Albo-Salih, Larry W. Mays, and Daniel Che

CONTENTS

13.1   INTRODUCTION

Floods are natural disasters that affect millions of people and their properties all over the world. Over the years, humans somehow succeeded in diminishing the effect of floods by constructing many dams on the riverways, and later they developed these structures to control the flood water by adding more facilities. Flood management includes both planned management and real-time management. The modeling approach described herein focuses on real-time management of river-reservoir systems. Real-time flood management requires real-time inflow data to determine the releases from the control facilities. Many times inflow data are not available at the event time, so forecasting the required data for the short-term depending on the availability may be the best solution. This capability of the model focuses on the flooding caused by large rainfall events that require immediate response (Eslamian et al., 2018)

Flood forecasting studies endeavor to produce as accurate as possible future estimates of discharges from a reservoir-river system based on the present state, forecasted rainfall, and the past

DOI: 10.1201/9780429463938-18

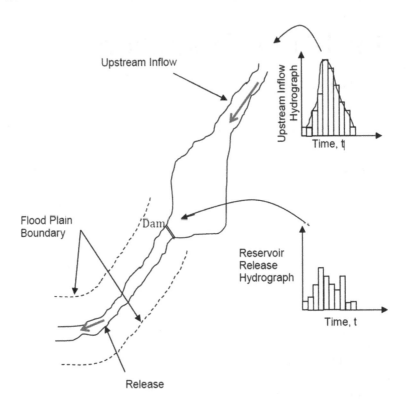

FIGURE 13.1   Schematic of a river-reservoir system showing reservoir inflows and releases.

behavior of the river-reservoir system. Bálint (2002) defined flood forecasting as "an operational, result-oriented activity and as such pays less attention to the modeled system than to the output of the forecasting procedure." The outputs are peak stage and/or flood crest, flood flow, and stage or discharge hydrographs, and flood volume. Flood forecasting and warning processes provide timely and reliable information to operators. The entire forecasting and warning process should be done in enough lead time to allow the decision-makers to take the possible measures to prevent or minimize the prospective flooding.

Flooding is the inundation of land downstream of a river and/or reservoir system by the overflow and rise of water level resulting from extreme rainfall, dam breach, or snowmelt, exceeding the capacity of channels of the river system, lake, or the way in which it runs. Figure 13.1 is a schematic of a simple river-reservoir system with reservoir inflows and reservoir releases. The flooding downstream of the reservoir and the flooding within the reservoir are dependent on the reservoir inflow and the reservoir releases. The combined flow of several tributaries can cause even more flooding along rivers and their floodplains. Despite the extensive studies that many previous researchers have performed to analyze the floods and the management of flow, many flooding problems still occur causing tremendous devastation in life and properties in both the short and the long terms. As of late, new approaches and systems created using the GIS techniques, allow more productive storage and processing of information and joint examination of various datasets.

13.2   REAL-TIME FLOOD FORECASTING

One of the most effective measures for flood management is real-time flood forecasting. As a focused activity in the hydro-meteorological sector, flood forecasting is a relatively recent development, which might indicate a growing seriousness of flood impacts. Formerly, in many different European countries, response focused on flood forecasting and warning through the global meteorological forecasting of severe weather. The estimations of flood forecasts regarding quantity and time have grown with the acknowledgment of the consequence of flood warnings as a contribution to flood management. That implies that the conventional methods of simple or basic extrapolation of forecasts from the gauged sites are no longer sufficient (Moore et al., 2006). However, the occurrence of numerous severe events has resulted in numerous national flood forecasting and warning centers. These warning centers contributed to enhancing the development of monitoring networks particularly for flood forecasting and warning purposes. Hydrological systems consist of instruments that have electronic facilities for data storage and transmission (rain gauges and water surface elevations), and meteorological effort has focused on the collection and delivery of satellite and radar data. Due to the nature of hydrological and meteorological phenomena, flood forecasting and warning are exposed to uncertainty, as they are somehow based on the principles of probability (World Meteorological Organization, 2011).

The process of any hydrologically related forecast is an estimation of the future state of a hydrological event, such as flow rate, the volume of water, water level, areas that would be affected or inundated by water, and velocity of flow. The lead time for such a forecast can be defined as the interval of time initiating the forecast to the future point at a time for which the forecast is applied. Determining a lead-time requires many constraints to be considered; the main one of them is the size of the catchment within a particular region or even country. However, for instance, a short-term forecast of lead time between 2 and 48 hours is considered in the United States of America, while between two and ten days is classified as a medium-term forecast, and a long-term forecast could be more than ten days (World Meteorological Organization, 2011).

To formulate a real-time flood forecasting approach for a river-reservoir system, meteorological and flow data are observed and then transmitted to the determining station through a different method of information correspondences. The meteorological and stream data obtained in real-time are then used in the flood forecasting model to estimate the movement of the flood and the corresponding water levels. The lead-time mentioned above ranges from a few hours to days depending on the catchment size and aim of the forecast. The modeling framework should not have excessive input requirements, but at the same time, the forecasted flood should be as accurate as possible.

In many cases, modeling based upon one-dimensional flow is problematic for accurately modeling floodplain water surface elevations, whereas a two-dimensional model is more appropriate. Usually, modeling flow in a network of channels can be performed using one-dimensional modeling. While the modification in a direction is considered as a part of the solution in two-dimensional models (Beffa and Connell, 2001).

Diffusion of flow in a floodplain includes many issues to be considered, especially in complex topography. During flooding conditions, water at a particular time allocating stream flow can exceed the flood level and then propagate horizontally onto the floodplain in different directions, so it is going to be challenging to model in one direction. Starting with a dry floodplain is another critical feature of two-dimensional solutions. Based on the topography of the flooding area, water spreads out in the floodplain in different directions at the beginning of modeling. Two-dimensional flow models were first developed and applied to the estuaries flows in 1967 by Leendertse (see Beffa and Connell, 2001). They succeeded in implementing the finite difference method to solve the problem of subcritical flow regime. Later, in the 80s and 90s, that type of modeling become widely applied in the simulation of tidal flows and lowlands flows. However, the use of these schemes receded because of their inability to predict the critical and supercritical flows regimes accurately. In other words, they cannot model flow in the steep slope channels.

One-dimensional modeling results in one water surface elevation for a cross-section at a particular time. So, the fluctuations across the section will not happen in the model as they do in the real event. However, the one-dimensional analysis can predict good results for river reaches. This research will be using both one- and two-dimensional unsteady flow routing for river reaches. The river segments are modeled using one-dimensional flow equations, while floodplains will be modeled using the two-dimension analysis. The basic equations that describe the one-dimensional unsteady flow (propagation of a flood wave) are the Saint-Venant equations represented by following the continuity and momentum equations,

Continuity equation:

$$\frac{\partial Q}{\partial x} + \frac{\partial s_{co}(A + A_0)}{\partial t} - q = 0 \tag{13.1}$$

Momentum equation:

$$\frac{\partial(s_m Q)}{\partial x} + \frac{\partial(\beta Q^2/A)}{\partial x} + gA\left[\frac{\partial h}{\partial x} + S_f + S_e + s_i\right] - L + W_f B = 0 \tag{13.2}$$

Where,

Q discharge.

A cross-sectional area of flow.

A_0 inactive off-channel cross-sectional area.

h water surface elevation.

s_{co} and s_m sinuosity factors which vary with h.

q lateral inflow or outflow per lineal distance.

x distance along the river.

t time.

β momentum correction coefficient.

g acceleration of gravity.

L momentum effect of lateral flow.

W_f surface wind resistance.

B top width of the channel.

S_f slope of the energy grade line derived from Manning's equation.

S_e contraction/expansion slope.

S_i additional friction slope associated with internal viscous dissipation of non-Newtonian fluids.

In practice, two-dimensional unsteady flow simulation models are one of the approaches for streamflow and floodplain forecasting as well. For a given set of operation policies, a two-dimensional unsteady flow simulation model can be used to simulate the flow rates, water surface elevations, and velocities in both X and Y directions at various locations for specified time steps. The basic equations that describe the two-dimensional unsteady flow (propagation of a wave) in an open channel and floodplain are the Saint-Venant equations represented by continuity and momentum equations in both the X and Y directions:

Two-dimensional conservation of mass:

$$\frac{\partial h}{\partial t} + \frac{\partial hu}{\partial x} + \frac{\partial hv}{\partial y} = 0 \tag{13.3}$$

X-direction momentum:

$$\frac{\partial u}{\partial t} + u\frac{\partial u}{\partial x} + v\frac{\partial u}{\partial y} + g\frac{\partial(h+z)}{\partial x} + \frac{gn^2 u\sqrt{u^2+v^2}}{h^{4/3}} - \frac{v}{h}\left(2\frac{\partial^2 hu}{\partial x^2} + \frac{\partial^2 hu}{\partial x^2} + \frac{\partial^2 hv}{\partial x \partial y}\right) = 0 \tag{13.4}$$

Y-direction momentum:

$$\frac{\partial v}{\partial t} + u\frac{\partial v}{\partial x} + v\frac{\partial v}{\partial y} + g\frac{\partial(h+z)}{\partial y} + \frac{gn^2 v\sqrt{u^2+v^2}}{h^{4/3}} - \frac{v}{h}\left(\frac{\partial^2 hv}{\partial x^2} + 2\frac{\partial^2 hv}{\partial y^2} + \frac{\partial^2 hu}{\partial x \partial y}\right) = 0 \tag{13.5}$$

Where,

   u and v horizontal velocity components in the X and Y direction.
   h water surface elevation.
   z bed elevation.
   x and y horizontal distances in the x and y directions respectively.
   t time.
   g acceleration due to gravity.
   n Manning's coefficient of roughness.

There are various types of one, and two-dimensional unsteady flow models, most of which are commercial models and used in practice such as the HEC-RAS. The two-dimensional unsteady flow equations solver in HEC-RAS uses the implicit finite volume algorithm. The implicit solution algorithm allows for larger computational time steps than explicit methods. The finite volume method provides an increment of improved stability and robustness over traditional finite difference and finite element techniques. The wetting and drying of 2D cells are very robust allowing the 2D flow areas to start completely dry, even with a sudden rush of water into the area. Additionally, the algorithm can handle subcritical, supercritical, and mixed flow regimes (flow passing through critical depth, such as a hydraulic jump) (Brunner, 2016a).

Mays and Tung (1992) defined the lead time as the interval of time between the issuing of a forecast and the expected arrival of the forecasted event. In flood forecasting, both the location and time are necessary. For instance, a relatively short lead time for a short river reach may become a long lead time for locations much further downstream. Considering the scenario depicted in Figure 13.2 having three urban areas: A, B, and C receiving a significant rainfall in the upper region of the watershed. A short lead time is required for urban area A, with a longer time for urban area B, whereas urban area C has the longest lead time. Due to the travel time of the flood down the river, longer lead times are needed. The flood hydrographs in urban areas A, B, and C are illustrated in Figure 13.3. In this example, the lead time for urban area A is relatively short, whereas the lead time for urban area C is longer. Moreover, the beginning of the flood hydrograph at urban area C occurs approximately at the same time the rainfall ends. This example also illustrates that, in order to forecast a flood hydrograph from urban area A, precipitation forecasts are required, whereas, for urban area C, the precipitation will be observed throughout the rainfall event in order to properly forecast.

Previous optimization-simulation models for real-time operation of river-reservoir systems have been developed by Unver et al. (1987), Unver and Mays (1990), Ahmed and Mays (2013), Che and Mays (2015, 2017), and Albo-Salih and Mays (2021). These optimization-simulation models have been formulated as optimal control problems, the mathematics of which have been discussed in detail by Mays (1997).

13.3 RAINFALL-RUNOFF MODELING

Reliable estimation of a river flow generated from the watershed is required as part of data set that helps in decision making for planning and managing river-reservoir systems. The characteristic of the time series of a river flow that can affect modeling, simulation, and planning of river-reservoir system can include the sequencing of flows hourly and longer time steps, the spatial and temporal changes in flows, season allocation, and flow characteristics. The better prediction of streamflow would be expected to come from water level observations at gauging stations, converted to the flow estimates using a stage-discharge relationship. However, such observations are only available for a limited number of gauging locations and for the relatively short time range. Estimates for ungauged locations and a much more extended period are required for contemporary water management, and ways to make the estimates for possible future conditions are required as well (Vaze et al., 2012).

A variety of methods are available to determine the runoff from catchments, using either observed or forecasted data wherever possible, or estimating by experimental and statistical techniques, and

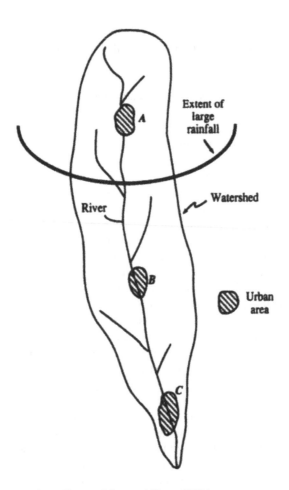

FIGURE 13.2 Effect of lead time. (Source: Mays and Tung, 1992.)

more commonly using the rainfall-runoff models. The methodology of modeling used to estimate the streamflow depends on the purpose of the modeling, time, and skills and instruments are available within the organization. With increasing the levels of inter-agency collaboration in water planning and management, the development of a best practice approach to rainfall-runoff modeling is desirable to provide a consistent process and improve the interpretation and acceptability of the modeling results.

The hydrologic modeling system (HEC-HMS) was developed by the US Army Corps of Engineers (USACE) Hydrologic Engineering Center (HEC) (Feldman, 2000). HEC-HMS simulates the rain-runoff processes of dendritic watershed systems. It is designed to be applicable in a wide range of geographic areas for solving a wide range of problems. The model may include large river basin water sources and flood hydrology to urban or natural watershed runoff. The program produces hydrographs that can be used in different ways with different software for water management studies, reservoir spillway design, urban drainage, future urbanization impact, flow forecasting, flood damage reduction, floodplain regulation, wetlands hydrology, and systems operation (Fleming and Brauer, 2016).

For precipitation-runoff-routing simulation, the program provides the following components (Feldman, 2000):

- Precipitation-specification options that can describe an observed (historical) precipitation event, a frequency-based hypothetical precipitation event, or an event that represents the maximum limit of precipitation likely at a given location.

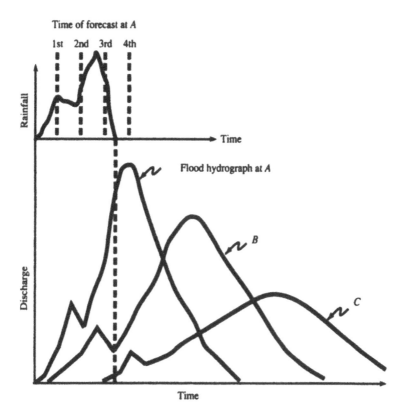

FIGURE 13.3   Flood hydrograph at the downstream location in a watershed. (Source: Mays and Tung, 1992.)

- Loss models that can estimate the volume of runoff, given the precipitation and properties of the watershed.
- Hydrologic routing models that account for storage and energy flux as water moves through stream channels.
- Direct runoff models that can account for storage, overland flow, and energy losses as water runs off a watershed and into the stream channels.
- Models of naturally occurring confluences and bifurcations.
- Models of water-control measures, including diversions and storage facilities.

HEC-HMS is used to model the rainfall-runoff as the first component of the optimization simulation model. Figure 13.4 shows the watershed scale rainfall-runoff process represented by HEC-HMS.

13.4   ONE- AND TWO-DIMENSIONAL UNSTEADY FLOW MODELING

Open channel flow can be modeled as steady or unsteady (such as natural streams, drainage channels, and even storm sewers). These unsteady flow variations with time are significant in river-reservoir systems, especially during and after a storm event. In practice, for flood studies, sometimes the steady flow equations use to determine the maximum flow depths in a stream channel, assuming the flow is steady at peak discharge. Nevertheless, this approach is conservative, since it does not account for the attenuation of flood waves due to the storage effect of the channel. Prediction of how a flood wave propagates in a channel is possible only through unsteady flow as flood routing or channel routing calculations (Akan, 2006).

The unsteady one and two-dimensional Saint-Venant equations are solved by an implicit finite difference scheme and finite volume method, respectively to handle floodplain and river channel flows. The complete derivation of the Saint-Venant equations can be found in Chow et al. (1988).

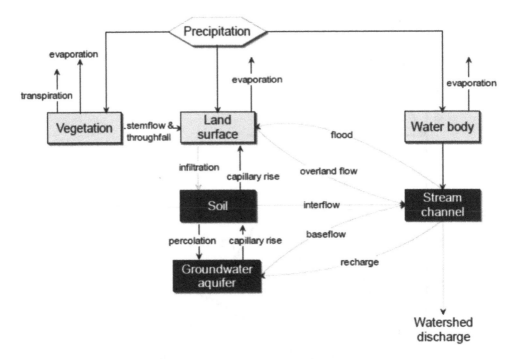

FIGURE 13.4   Watershed scale rainfall-runoff process represented by HEC-HMS. (Source: Feldman, 2000.)

13.5   HYDROLOGIC ENGINEERING CENTER'S RIVER ANALYSIS SYSTEM

One of the most popular hydraulic analysis software programs in the world is HEC-RAS (Brunner et al., 2014). It is an integrated system of software, programmed for interactive use in a multi-tasking environment. The HEC-RAS system contains the following river analysis components for (1) steady flow water surface profile computations; (2) one-dimensional and two-dimensional unsteady flow simulation; (3) quasi unsteady or fully unsteady flow movable boundary sediment transport analysis; and (4) water quality studies. All four components use a common geometric data representation and ordinary geometric and hydraulic computation routines. Besides the four river analysis components, the system contains several hydraulic design features that can be invoked once the water surface profiles are computed. HEC-RAS is designed to perform one-dimensional and two-dimensional hydraulic calculations for a full network of natural and constructed channels, overbank/floodplain areas, and levee-protected areas (Brunner, 2016a, 2016b).

HEC-RAS can be used to perform two-dimensional (2D) hydrodynamic flow routing within the unsteady flow analysis portion of HEC-RAS. Users can now perform one-dimensional (1D) unsteady flow modeling, two-dimensional (2D) unsteady flow modeling (full Saint-Venant equations or diffusion wave equations), as well as combined one-dimensional and two-dimensional (1D/2D) unsteady-flow routing. The two-dimensional flow areas in HEC-RAS can be used in several ways. The following are examples of how the 2D flow areas can be utilized to support modeling with HEC-RAS (Brunner, 2014):

- Detailed 2D channel modeling.
- Detailed 2D channel and floodplain modeling.
- Combined 1D channels with 2D floodplain areas.
- Combined 1D channels with 2D flow areas behind levees.
- Directly connect 1D reaches into and out of 2d flow areas.
- Directly connect a 2D flow area to 1D storage area with a hydraulic structure.

- Multiple 2D flow areas in the same geometry.
- Directly connect multiple 2D flow areas with hydraulic structures.
- Simplified to very detailed dam breach analyses.
- Simplified to very detailed levee breaching studies.
- Mixed flow regime. The 2D capability (in addition to the 1D) can handle the supercritical and subcritical flow, as well as the flow transitions from subcritical to supercritical and supercritical to subcritical (hydraulic jumps).

Two-dimensional flow modeling is accomplished by adding two-dimensional flow area elements into the model in the same manner as adding a storage area. A two-dimensional flow area can be added by drawing a two-dimensional flow area polygon; developing the 2D computational mesh; then linking the two-dimensional flow areas to one-dimensional model elements and directly connecting boundary conditions to the 2D areas. In this research, HEC-RAS is the second significant component of the optimization-simulation model to simulate flow routing in floodplains.

13.6   ROUTING INFLOW FLOOD THROUGH RESERVOIRS

HEC-RAS can be used to route inflow flood hydrographs through reservoirs by using one of the following methods:

- Unsteady flow routing in one-dimensional using Saint-Venant equations.
- Unsteady flow routing two-dimensional using diffusion-wave equations or Saint- Venant equations.
- Level pool reservoir routing.

The one or two-dimensional unsteady flow routing methods can capture the water surface slope through the reservoir pool. Reservoirs with long narrow pools will exhibit a greater water surface slope upstream of the dam than reservoirs that are short and wide. Thus, the most accurate modeling technique to capture the pool elevations and outflows of long narrow reservoirs is full dynamic wave (unsteady flow) routing. For wide and short reservoirs, level pool routing may be appropriate.

13.7   CONTROLLING HEC-RAS THROUGH MATLAB

A feature available in HEC-RAS, called the HEC-RAS controller, is a part of the application programming interface. Also available are specific functions and subroutines that can be employed to allow a programmer to operate HEC-RAS from outside by preparing input data, retrieving input or output data, and execution common functions such as opening and closing HEC-RAS, changing plans, running HEC-RAS, and getting output plots (Goodell, 2014).

The interfacing among the various simulation programs used in the optimization-simulation model was a major objective of this research. The interfacing and data exchange between HEC-RAS using MATLAB was a challenging task. The users of HEC-RAS often have unique applications for using this software that may include the coupling with other software to perform systems analysis such as flood risk analysis, optimization of flooding structures under uncertainty and multi-objective reservoir operation under uncertainty, multi-objective reservoir operation under uncertainty, among others (Leon and Goodell, 2016). One state-of-the-art environment for integrating proprietary software and/or open-source codes is MATLAB, which is a high-performance language for technical computing. MATLAB integrates computation, visualization, and programming in an easy-to-use environment (Mathworks, 2016). The interfacing of the various components of the optimization simulation is discussed later.

13.8   REAL-TIME RESERVOIR OPERATION

Reservoir operation for flood control is a complicated problem that involves several conflicting objectives. These objectives include some water releases from reservoirs before the arrival of flood-waters, the storage and water level in the reservoir during flood events, and ensuring reservoir gate releases during flood events will not heavily damage downstream areas. If a river basin consists of a system of reservoirs, the problem becomes even more complicated, as each of the decisions made for one reservoir would have significant impacts on the rest of the reservoirs in the system and also the flood conditions in the entire basin (Che and Mays, 2015, 2017). Usually, decision-makers of flood control reservoir operations use fixed reservoir rule curves and stage-discharge relationships to determine the reservoir releases based on the next reservoir stages. These rigid reservoir rules are based typically on past flood records. However, when facing an extreme precipitation event, traditional methods such as using reservoir stage-discharge relationships are not sufficient to achieve flood control objectives since most of these reservoir operation rules are not backed up by extreme flooding scenarios.

The principle of any reservoir operation, whether in real-time or planning operation, is governed by the conservation of mass (continuity equation). Reservoir operation in real-time fashion is a process of continuously determining the releases from the reservoir gates to keep the water surface elevations at the upstream and downstream within the desired level. Strategic planning always decides and defines the rules and operation policy. The involvement of all time-related information justifies the necessary complication of reservoir operation in real-time. While the reservoir operation in some parts is based upon forecasting information, the process is not without error. Uncertainty and inaccuracy are unavoidable in the real-time reservoir operation and even in the planning processes that come from the forecasting and then determining the net inflow into the reservoir, and that is the source of uncertainty and inaccuracy.

Another component of the operation model is the reservoir operation model. The principle that any reservoir is governed by the conservation of mass (continuity), this law is expressed mathematically in the following equation. Consider the primary control volume which is shown in Figure 13.5.

$$\frac{d(m_{cv})}{d_t} + \sum_{cs} m_{out} - \sum_{cs} m_{in} = 0. \tag{13.6}$$

Where

$d(m_{cv})$ is the accumulation of mass in the control volume.

$\sum_{cs} m_{out}$ is the total mass inflow through the control surface.

$\sum_{cs} m_{in}$ is the total mass outflow through the control surface.

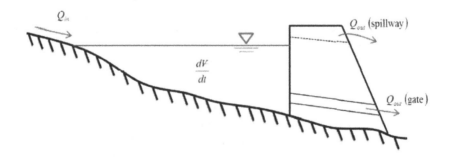

FIGURE 13.5   Reservoir inflow, outflow, and storage.

Equation (12.6) states that the accumulation rate of mass in the control volume plus the net rate of the outflow of mass through the control surface is equal to zero.

Instead of using the flow of mass rate, equation (13.6) can be written in terms of volume inflow by dividing all its terms on the density of the water ρ as the in following:

$$\frac{1}{\rho}\frac{d(m_{cv})}{d_t} + \frac{1}{\rho}\sum_{cs} m_{out} - \frac{1}{\rho}\sum_{cs} m_{in} = 0 \qquad (13.7)$$

Moreover, that means,

$$\frac{d(V_{cv})}{d_t} + \sum_{cs} Q_{out} - \sum_{cs} Q_{in} = 0 \qquad (13.8)$$

Where,

$\dfrac{d(V_{cv})}{d_t}$ is the volume change in the control volume.

$\displaystyle\sum_{cs} Q_{out}$ is the total volumetric in flow through the control surface.

$\displaystyle\sum_{cs} Q_{in}$ is the total volumetric outflow through the control surface.

Equation (13.8) is the basis of a gated spillway model. The schematic reservoir with components of flows and the storage of the reservoir is shown in Figure 13.5.

Gated spillway operations in real-time define releases determined using the optimization-simulation model. The optimization is based on a genetic algorithm (GA) in MATLAB. The optimizer interfaces with the other components of the optimization model to determine actual gate operations during the real-time operation of the river-reservoir systems. The optimization model is the next major component of the optimization/simulation model, which its complete formulation is explained next.

13.9 RESERVOIR OPERATIONAL CONSTRAINTS

The reservoir operational constraints can be assigned as greater than and less than sorts of constraints, which define the operational target, upper and lower variables bounds, limitation of the outlet and spillway gates, and the reservoir storage capacity. All the constraints mentioned above are included in the problem formulation in the optimization model.

The process of gate opening is a main operational constraint that is designed to be operated between the range of minimum and maximum allowed gate opening. The gate operation limits depend on the physical limitation of the gate operation at which the minimum and maximum allowed rate of change in gates opening are predetermined.

The optimization-simulation modeling approach is used to help make real-time operation decisions (gate operations) for a river-reservoir system during flooding conditions by incorporating real-time precipitation and streamflow data and forecasted rainfall throughout the system. The model should consist of the following:

- Forecasting model, to predict the precipitation for next the time period of simulation.
- Rainfall-runoff model, to demonstrate the hydrologic response of a watershed and then determine the discharge come off from the watershed.
- Unsteady one- or two-dimensional flow model for the reservoir system (in our case here two-dimensional model), to route the water flow in the river stream and further in the floodplain area in two-dimensional cases.

- Operation model, for operating the system dams' gates.
- Optimization model, for determining the optimal reservoir spillways' gates operation.

The real-time reservoir operation problem involves the operation of a reservoir system by making decisions about reservoir releases as information becomes available, with relatively short time intervals, ranging from several minutes to several hours. The real-time operation of multi-reservoir systems involves many considerations, such as hydrologic, hydraulic, operational, technical, and institutional considerations. That will enable engineers in the field to make critical decisions about releases from the reservoirs to control floodwaters. For an operation to be efficient, a monitoring system is essential to provide the operator of the reservoir with the flows and water levels at various locations in the river system. These include upstream flow conditions, tributaries, reservoir levels, and precipitation data for the watersheds of which output (rainfall and runoff) are not gaged. Flood forecasting in general and real-time flood forecasting, in particular, have always been significant problems in hydrologic engineering, especially when flood-control reservoir operations are involved.

The forecasting problem can be viewed as a system with inputs and outputs. The inputs of the system are inflow hydrographs at the upstream end of the river system and runoff from rainfall in other catchments converging to the system. The outputs of the system are flow rates and water levels at points of interest in a river system.

13.10   OPTIMIZATION AND SIMULATION APPROACH

A real-time optimization-simulation approach must be able to operate the river-reservoir system to determine the releases before, during, and even after the extreme flood event. Such a model could help the decision-maker to control the discharge of floodwater and consequently maintain water surface elevations within desired levels. The decision variable is the water discharge (Q), based on the reservoir gate openings that represent the control variables.

The model starts inputting actual rainfall data when it is needed to decide on how much water should be released to assure enough volume in the reservoir to accommodate the upcoming flood water wave. Meanwhile, the model generates a short-term rainfall forecast and then the floods that can be happened by using the real-time rainfall of the precipitation gages and expects the real-time flood water elevation in the river-reservoir system to avoid accumulating a headwater on the gate of the reservoir.

A methodology of projecting future rainfall within the next few minutes to hours has been developed by Che (2015) as a part of the methodology. Forecasted rainfall data will be used to simulate the watershed rainfall-runoff through the HEC-HMS model, and then to produce hydrographs as time series of the reservoir inflows that are going to be used as inputs of the optimization model to compute the releases of the reservoirs gates in a river-reservoir system. The optimization model will come up with sets of feasible solutions of how much water should be released to satisfy some of the problem constraints. Once these set of feasible solutions for the decision variable which may or might not contain the optimal sets are obtained, the obtained data then is to use as the input of one and two-dimensional unsteady models to be routed downstream and simulated through the HEC-RAS 5 model to check if the flooding will occur or not. If the answer were yes, the model would repeat the process until the target water surface elevations are achieved without or with minimum damage effect for the system or the area downstream. So, the objective of this study is to control the resales and keep the water surface elevation up and downstream under a certain level. For instance, 100-year flood elevation might be one of the targeted levels in the processed system. At this moment, the next iteration begins, and the model uses the projected real rainfall data for running the HEC-HMS model to compute the actual runoff quantities from the watershed and then the reservoir water level and consequently the releases from the reservoir and hence route it downstream as a decision for the next iteration of the operation period. These processes continue and repeat until the objective is met at all times, satisfying all the constraints for the entire period of simulation. The reason behind that model enabled to forecast and run the simulation in advance of the storm event can help to make pre-decision and take the necessary action to minimize the flood condition as much as possible. Figure 13.6 is a flowchart of the optimization-simulation model.

The first stage is to obtain real-time rainfall data from precipitation gages stations for the HEC-HMS to start simulating the rainfall-runoff model and to produce the required hydrographs to input them into one dimensional unsteady model for routing them from the watershed exit point to the

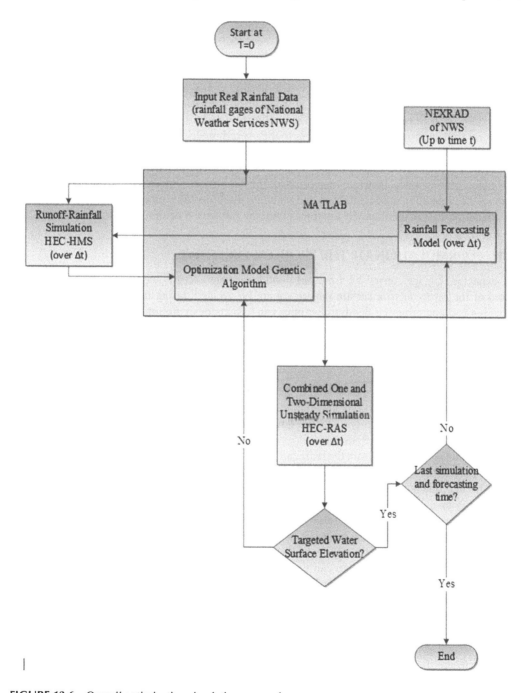

FIGURE 13.6 Overall optimization-simulation approach.

reservoir location. Now, the optimization model has the all required data and is ready to run and search for the optimal solution to make the proper decision of real-time releases from the reservoir. The optimization model starts to generate the possible operation to the reservoir and determine the gate opening. These possible solutions represent the releases from the reservoir gates, so the next step is to route these discharges downstream into a two-dimensional unsteady model by using the new version of HEC-RAS. The new generation is then simulated further downstream to the two-dimensional area of study where other constraints should be stratified and meet the objective of the targeted water level at the control cell in the study area. This stage is the most crucial operation in the model where the model begins to test the objective and repeats the process to readjust the releases from the reservoir. For example, if the releases of the first iteration after being routed in the unsteady two-dimensional model produced water level in the downstream control cell higher than the target level, the model will repeat the process and reduce the releases and vice versa until reaching the exact water level desired or close to desired.

Once the objective is met and the target water level is determined, the overall process will be repeated for the next period, t to t + Δt, in the optimization-simulation model. The process of forecasting rainfall is a complementary process to the overall model to ensure the continuity of model operation, right after they enter the actual rainfall data. The rainfall forecasting model forecasts the rainfall over the next Δt, which is used with the known rainfall up to the next simulation time. The rainfall forecasting model generates and provides the required rainfall data for the next iteration that prompts the simulation process to start and simulate the hydrologic model. The overall process of the simulation-optimization model continues until the last period of simulation.

13.11   FORMULATION OF THE PROBLEM OBJECTIVE

The expected damage caused by a natural flood event that exceeds the hydrologic design return period of the reservoir to a certain degree cannot be avoided. Using optimization techniques in managing and operating the flooding facilities can tremendously minimize the effect and damage cost. The general theoretical formulation for optimizing and simulation the releases under flood conditions from a river-reservoir system that minimizes the expected flood damage subjected to many constraints and as the following:

- Hydrologic constraints represented by a rainfall-runoff model simulated by HEC-HMS.
- Diffusion-wave or Saint-Venant equations for simulating unsteady flow modeling in a two-dimensional form which predicts the flow and its various components in a river-reservoir system solved by HEC-RAS 5.03.
- The allowable releases from the reservoir, in other words, the maximum and minimum and limits and the flow at particular locations or computational cells.
- The maximum allowable water surface elevation at specific computational cells in the floodplain and minimum water surface elevation in the river system.
- The operation rule of the reservoir gate during flooding conditions.
- The storage capacity of the reservoir.

The simulation part of the problem can be described by constraint equations (12.11), (12.12), and (12.15). The HEC-RAS 5.0 two-dimensional computational module has an option of either using the 2D diffusion wave equations, or the full 2D Saint-Venant equations (sometimes referred to as the full 2D shallow water equations) to run a simulation model. The 2D diffusion wave equation is set as the default. Generally, the 2D diffusion wave equations for flooding applications work fine as the diffusion wave equation model runs faster and is more stable. However, there are many applications and cases where the 2D full Saint-Venant equations should be used for more accuracy. The choice between the two options is a matter of selecting the equation set and one can use both in two different runs and compare the results (Brunner, 2016).

The objective function is to minimize the flow rate at a control point which minimizes the water depth at a certain cell, i, in the downstream area during all the simulation period. The objective function can be expressed in terms of discharge to minimize the total damage caused by the rising water elevation at certain locations as expressed below:

$$\text{Min } Z = \sum_{x=1,y=1}^{x,y} \sum_{t=1}^{T} \left[C_{x,y} Q_{x,y,t} \right] \tag{13.9}$$

Where,

$Q_{x,y,t}$ is the flow rate time-series at the control point (x, y), of the river-reservoir system that affects cell, i, in the downstream two-dimensional area.

C_{xy} is the penalty coefficient at control cells x, y.

x, y, and t are the spatial and temporal indices, respectively.

Because of a lack of information on the relation between the total damage and the water surface elevation, an alternative object function has been adopted to solve the problem. Minimizing the maximum downstream water surface elevation (h) at the control cell (x, y) with satisfying all the constraints can also minimize the total damage in the urban area, so the objective function could also be expressed as:

$$\text{Min } Z = \text{Min} \left[\text{maximum } h_{x,y,t} \right] \tag{13.10}$$

The objective minimizes the total damage in the entire river system at all times, t, including the upstream side, as it relieves the headwater upstream reservoir gate. In general, the constraints that governed the objective function and as it has been described earlier in this chapter can be classified into three main types:

a) Hydrologic constraints: Represented by the rainfall-runoff relationships defined by the sub-basins areas, rainfall losses due to canopy interceptions, depression storage, soil infiltration, excess rainfall transform methods, watershed runoff routing method, internal boundary conditions, and initial conditions that depict the rainfall-runoff process in different components of a watershed system and function as below:

$$h\left(P_{i,t}, L_{i,t}, Q_{i,t}\right) = 0 \tag{13.11}$$

where $P_{i,t}$ is the matrix of precipitation data in the system at location $i=x$, y; $(L_{i,t})$ is the rainfall losses matrix of the watershed system; and $Q_{i,t}$ is the discharge matrix of the system. All the hydrologic constraints are in matrix form because the problem has dimensions of space, x, y, and time, t.

b) Hydraulic constraints which are defined by the Saint-Venant equations for two-dimensional unsteady flow (equations 12.1 and 12.2), and related relationships of upstream boundary condition, downstream boundary condition, external two-dimensional flow area boundary conditions, internal two-dimensional area boundary conditions, and initial conditions that depict the flow in different components of a river-reservoir system.

$$g\left(h_{x,y,t}, Q_{x,y,t}\right) = 0 \tag{13.12}$$

where $(h_{xy,t})$ is the matrix of water surface elevations in the system; $Q_{x,y,t}$ is the discharge matrix of the system. All the hydraulic constraints are in matrix form because the problem has dimensions of space, i, and time, t.

c) The bound constraints include upper and lower discharge limits that define the maximum and minimum allowable reservoir release and flow rate at target locations:

$$\underline{Q}_{x,y} \leq Q_{x,y,t} \leq \overline{Q}_{x,y} \tag{13.13}$$

The bars above and underneath the variable denote the upper limit and lower limit for the discharge, respectively. Another significant hydraulic constraint is the water surface elevation bounds defined by the allowable upper and lower limits at specified locations in the downstream two-dimensional area, including reservoir levels:

$$\underline{h}_{x,y} \leq h_{x,y,t} \leq \overline{h}_{x,y} \tag{13.14}$$

d) Operation constraints which include the rules of reservoir operation and releases, the beginning and the end of the simulation period, reservoir storage capacities, etc., are included in the optimization-simulation model:

$$\left(Q_{x,y,t}, h_{x,y,t}\right) \leq 0. \tag{13.15}$$

13.12 REDUCED OPTIMIZATION PROBLEM

The above optimization formulation is converted to an unconstrained optimization problem by placing the bound constraints (equations 13.13 and 13.14) and the operational constraints (equation 13.15) into the objective function using penalty weights $W_{1,t} - W_{3,t}$. The hydrologic constraints (equation 13.11) are solved using the HEC-HMS and the hydraulic constraints (equation 13.12) each time these constraints are solved.

The reduced objective for the unconstrained optimization problem is

$$\text{Minimize } Z = W_{1,t} \sum_{x=1,y=1}^{X,Y} \sum_{t=1}^{T} C_{x,y} Q_{x,y,t} + W_{2,t} \sum_{i=0}^{I} \sum_{t=0}^{T} \left(\left(\max\left(0, Q_{i,t} - Q_{\max,i}\right)\right)^n + \left(\max\left(0, Q_{\min,i,t} - Q_{i,t}\right)\right)^n\right)$$

$$+ W_{3,t} \sum_{i=0}^{I} \sum_{t=0}^{T} \left(\left(\max\left(0, h_{i,t} - h_{\max,i}\right)\right)^n + \left(\max\left(0, h_{\min,i,t} - h_{i,t}\right)\right)^n\right)$$

$$\tag{13.16}$$

This unconstrained optimization problem is solved using the genetic algorithm within MATLAB. A genetic algorithm (GA) is an optimization method that is based on a heuristic solution-search, fundamentally based on the principle of Darwinian evolution through genetic selection. Genetic algorithms use an abstract version of evolutionary processes to evolve solutions to given problems. Genetic algorithms differ from the conventional optimization methods, such as gradient-based methods and the simplex method. GA does not necessarily require a distinct fitness function (objective function).

The above model formulation is a real-time optimal control problem in which the gate operations defining the reservoir releases are the decision variables. The reduced objective function (unconstrained problem), equation (13.16), is not a well-defined continuous function but is amenable to GA solutions when the constraints are solved using the simulation models (HEC-HMS and HEC-RAS) each time the constraints need to be solved.

13.13 RAINFALL FORECASTING

Rainfall forecasting is another necessary component in the optimization/simulation model. Forecasted precipitation is needed for flood forecasting since reservoir management personnel would have to make reservoir releases decision based upon the forecasted information before the actual rainfall event and floodwater arrive. A statistical regression analysis approach is used for the rainfall forecasting model.

A model for forecasting observations presented by Montgomery et al. (2012) can be applied to estimate rainfall in a period of time in the following form:

$$\hat{P}_{t+\Delta t} = \hat{\Phi} P_t + \left(1 - \hat{\Phi}\right) \hat{\beta}_0 + \hat{\beta}_1 \left[\left(t + \Delta t\right) - \hat{\Phi} t\right] \tag{13.17}$$

Where,

$\hat{P}_{t+\Delta t}$ is the vector of predicted rainfall values over time $(t+\Delta t)$.

$t+\Delta t$ is the forecasting time period.

t is the current time period.

P_t is the vector of known rainfall values at the end of the current time period, t.

Φ is an autocorrelation parameter, defined as $\Phi = \sum_{t=2}^{t} \dfrac{e_t e_{t-1}}{\sum_{t=1}^{t} e_t^2}$

e_t is the vector of prediction residuals.

$\hat{\beta}_0, \hat{\beta}_1$ are model parameters.

Using actual rainfall up to the current time of operation, t, the prediction model (equation 13.17) rainfall at time $t + \Delta t$. The projected rainfall is used in the rainfall-runoff model of the optimization-simulation approach. When the last simulation period ends, the forecasting model repeats the process by obtaining the actual rainfall up to the current time, t. A new prediction will be generated for each simulation period. The process repeats until the very last simulation period when forecasting is no longer needed.

13.14 EXAMPLE APPLICATION

Before using the methodology of the optimization simulation model presented in this chapter in a more realistic or relatable problem, a simple hypothetical model is introduced for demonstrating the application of the optimization-simulation model in a real-time manner. A built-in two-dimensional example data (Muncie project) that comes with the HEC-RAS 5.0.3 data example has been used here as a hypothetical example shown in Figure 13.7. The Muncie project is a part of the White River that flows in two forks crosses most of central and south Indiana, creating the largest watershed located entirely within the state, discharging part or all of 42 of the 92 in Indiana.

Muncie, located in east-central Indiana, is approximately 50 miles (80 km) northeast of Indianapolis (see Figure 13.8). The west fork of the White River is a relatively small creek close to the Ohio border in central Indiana and flows slightly westward. At the time it passes Muncie, however, it is a substantial river. The west fork of the White River starts in a farmer's field in Randolph County, southern Winchester. Along its first few miles, travels north, then turns west through Muncie and Anderson before flowing south through Noblesville, Indianapolis, Martinsville, and Spencer (see https://friendsofwhiteriver.org/about-white-river/ and http://www.in.gov/dnr/outdoor /4474.htm).

As a portion of the Mississippi River system, the White River Basin drains 11,350 square miles of central and southern Indiana. The average streamflow is about 12,300 cubic feet per second close

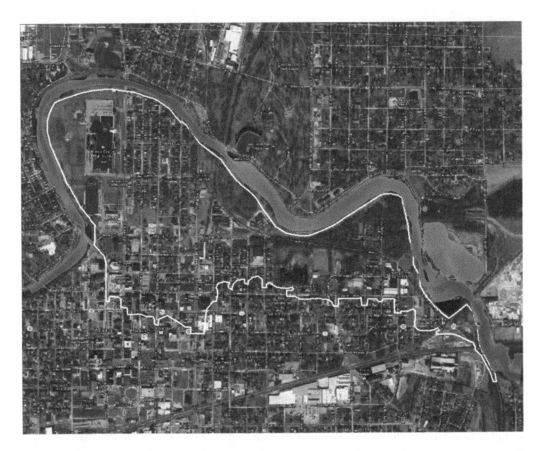

FIGURE 13.7 West fork of the White River through Muncie, Indiana. Area within the red line indicates the floodplain area modeled using 2D. (Source: Brunner et al., 2014.)

to the White River's confluence with the Wabash River in southwestern Indiana. Changes in stream-flow generally occurred seasonally and moderated. The peak flows are typically recorded in April and May, whereas the lowest flows are in summer and fall. The annual precipitation rates on average from between 40 inches in the northern part of the basin and 48 inches in the south-central part. The precipitation is evenly distributed throughout the year. The winter and early spring rainfall are generally characterized by a long-term duration, steady and of moderate intensity, while during late spring and summer rainfall seems to be of high intensity of short duration (see https://in.water.usgs.gov/nawqa/wr00002.htm).

As mentioned above, the Muncie project (Brunner, 2014), which is the start of the west fork of the White River, is one of two examples of two-dimensional unsteady flow of HEC-RAS 5.0.3 that has been used here as a hypothetical model in the optimization/simulation model. Unfortunately, there is no actual reservoir or storage area is associated with this part of the west fork of the White River and hence no gated structure to control the releases to be considered as a river-reservoir system; however, some modifications have been introduced to the model to be compatible with the problem formulation. The modifications added to the model are a reservoir pool (storage area) and an inline hydraulic structure (dam with gated spillway). This hydraulic structure connects the introduced storage area with the Muncie project as depicted in Figure 13.9. The reservoir pool is described by the elevation versus volume storage relation.

The system is modeled using a combined one- and two-dimensional modeling approach. That is the river reach is modeled using one-dimensional, while the area around it, which represents the floodplain and the probable inundation area that may be inundated with flooding water during the

FIGURE 13.8 Muncie City and White River Project. (Source: USACE.)

flooding period, modeled using two-dimensional modeling approach, Figures 13.10 and 13.11 show the river reach and the inundation area respectively.

The hypothetical reservoir (storage area) is connected to the river reach at the very first upstream cross-section see Figure 13.11, through a gated spillway inline structure with a big enough radial gate to allow passing a range of significant discharges for the simulation process of a flooding event. A radial gate type is assumed to regulate the flow from the dam to the downstream channel. The total gate width is assumed to be 30 feet long with a maximum opening of 21 feet, and the discharge coefficient of the gate is 0.98. The river channel has an initial steady flow of 1,000 cfs. The watershed receives a 35-hour storm and a simulation period is 64 hours is used. A computation interval Δt of 1.0 hour is used, starting at time $t = 0$. A rainfall forecasting time interval is one hour. The results are presented in the form of inundation maps at the city of Muncie.

To determine the best sequence of the reservoir gate openings for the targeted downstream water surface elevations, an adequate theoretical optimization scheme should be presented and the potential constraints that must be satisfied during a flooding event. These time series of gates openings should achieve downstream water surface elevations within the allowable range of elevations at the

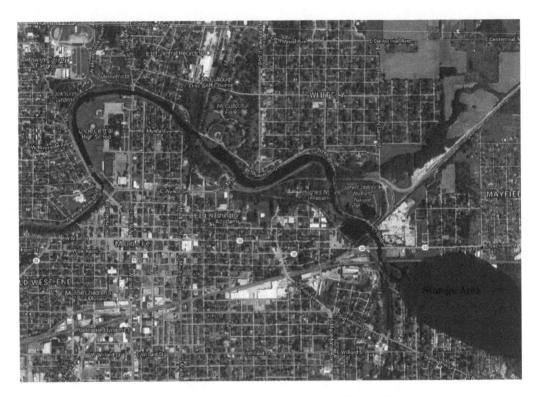

FIGURE 13.9   West fork of the White River through Muncie and the hypothetical storage area.

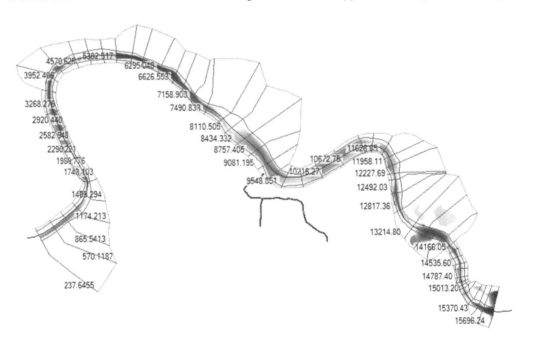

FIGURE 13.10   One-dimensional river reach.

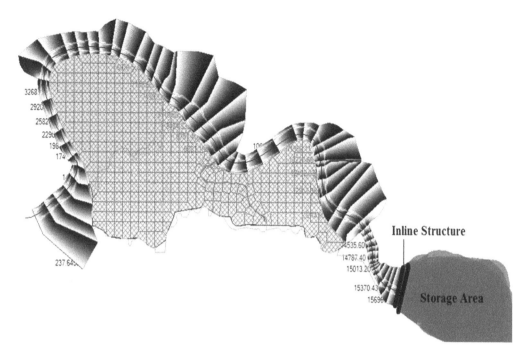

FIGURE 13.11 1D and 2D flow areas with storage area connected to river through an inline structure.

point of interest, which is here in the model represented by Muncie City. So, the problem objective is to minimize the flow rates at the city of Muncie, and then minimize the flooding area, while satisfying all hydraulic, hydrologic, and operational constraints during the entire period of flooding event simulation.

This example did not rely upon a rainfall forecasting model or a hydrologic model to determine the reservoir inflow. Instead, a hypothetical reservoir inflow shown in Figure 13.12a was used to illustrate the optimization-simulation model application. The time series of the optimal gate openings of the inline structure is depicted in Figure 13.12b. The reservoir operation model and the genetic algorithm solver succeeded to keep the reservoir storage within the desirable range, in which the storage for the reservoir was kept above the inactive storage and below the maximum flood storage, thus preventing any potential dam failure.

The results are presented in the form of inundation maps at the city of Muncie. The flooding condition (maximum inundation area downstream of the dam), assuming a fully open gate, is illustrated in Figure 13.13. Figure 13.14 shows the inundation area using the reservoir operation determined by the optimization-simulation (O/S) model. The city is kept entirely safe from the flooding. Albo-Salih and Mays (2021) present more detail on the results of the application to the Muncie example.

13.15 CONCLUSIONS

An optimization/simulation model has been developed to determine the optimal operation of gates (reservoir release schedules) before, during, and after flooding events. The proposed optimization-simulation model consists of four individual models (interfaced through MATLAB) that make up the larger simulation model created in this chapter and are listed as follows: a rainfall forecasting model, a rainfall-runoff model, a one- and/or two-dimensional unsteady flow model, and a reservoir operation model with genetic algorithm optimization model. The two-dimensional optimization-simulation approach uses the US Army Corps of Engineers Hydrologic Engineering Center HEC-HMS and the newest version of the river analysis system HEC-RAS 5.0.3. By this

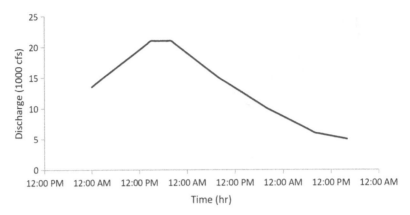

(a) Reservoir (storage area) hypothetical inflow hydrograph

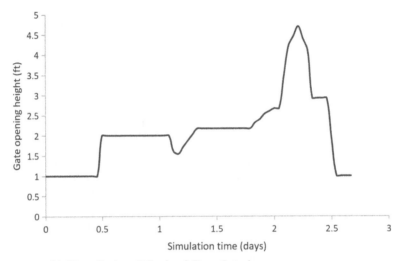

(b) Time Series of Optimal Gate Openings.

FIGURE 13.12 Inflow hydrograph and optimal gate openings.

achievement, HEC-RAS 5.0.3 has become the first two-dimensional accurate software released to the public for free of charge. These are very popular in the private and public industries. In the example application presented, the river reach was modeled using one-dimensional modeling and the floodplain was modeled using two-dimensional modeling. The reservoir operation model is optimized using the genetic algorithm within MATLAB. Although the genetic algorithm requires many function evaluations (simulation runs) to reach an optimum solution, the approach has the significant benefit of being able to interface with simulators. Recently, heuristic search methods have gained significant attention from researchers due to the advancing technology of computers which reduce computation times.

There are few limitations, if any, to the theory of the optimal control approach described herein. Limitations in the application could be the computer time required because of the number of simulations that are required for the genetic algorithm and the size of the floodplain (2D) areas, especially for very large floodplain areas on long river reaches. For 2D modeling of floodplain areas, the use of the 2D diffusion wave model in HEC-RAS could possibly save significant amounts of computation time as compared to the 2D full equation model, depending on the application.

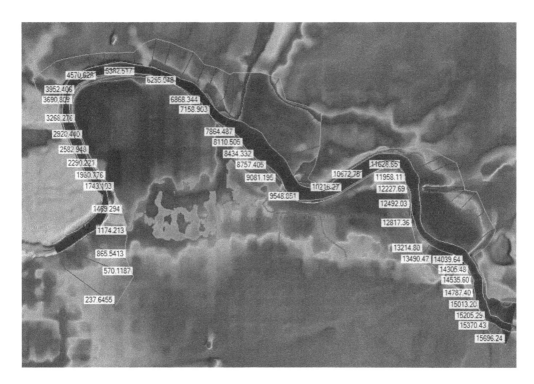

FIGURE 13.13 Inundation area (in blue) at Muncie without the O/S model.

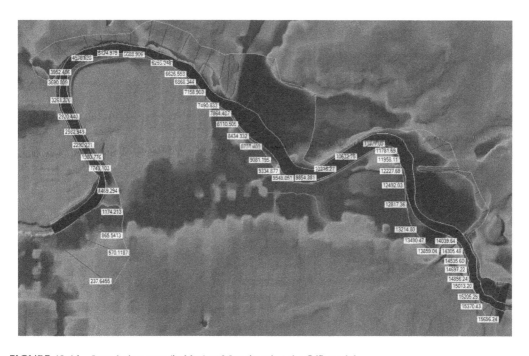

FIGURE 13.14 Inundation area (in blue) at Muncie using the O/S model.

REFERENCES

Ahmed, E.S.M.S., and Mays, L.W. 2013. Model for determining real-time optimal dam releases during flooding conditions, *Natural Hazards*, 65(3): 1849–1861, https://doi.org/10.1007/s11069-012-0444-6.

Akan, A.O. 2006. *Open Channel Hydraulics*, Elsevier Ltd., Oxford, UK.

Albo-Salih, H., and Mays, L.W. 2021. Testing of an optimization-simulation model for real-time flood operation of river-reservoir systems, *Water*, 13: 1207, https://doi.org/10.3390/w13091207.

Bálint, G. 2002. *State-of-the-Art for Flood Forecasting Modelling*. *Water Resources Research Centre*, VITUKI, Hungary.

Beffa, C., and Connell, R.J. 2001. Two-dimensional flood plain flow. I: Model description, *Journal of Hydrologic Engineering*, 6(5): 397–405.

Brunner, G.W. 2014. *Combined 1D and 2D Modeling with HEC-RAS*, Davis, CA.

Brunner, G.W. 2016a. *HEC-RAS, River Analysis System Hydraulic Reference Manual*, US Army Corps of Engineers, Hydraulic Engineering Center, Davis, CA.

Brunner, G. 2016b. *Benchmarking of the HEC-RAS Two-Dimensional Hydraulic Modeling Capabilities*, US Army Corps of Engineers, Hydrologic Engineering Center, Davis, CA.

Brunner, G., Jensen, M., Piper, S., Ackerman, C., Kennedy, A., and Chacon, B. 2014. *2D Modeling and Mapping with HEC-RAS 5.0*, US Army Corp of Engineers, Davis, CA.

Che, D., and Mays, L.W. 2015. Development of an optimization/simulation model for real-time flood-control operation of river-reservoirs systems, *Water Resources Management*, 29: 3987–4005, https://doi:10.1007/s11269-015-1041-8.

Che, D., and Mays, L.W. 2017. Application of an optimization/simulation model for real-time flood-control operation of river-reservoirs systems, *Water Resources Management*, 31(7): 2285–2297, https://doi.org/10.1007/s11269-017-1644-3.

Che, D.C. 2015. *Optimization/Simulation Model for Determining Real-time Optimal Operation of River-Reservoirs Systems during Flooding Conditions*. Ph.D. Dissertation, Arizona State University, Tempe, AZ.

Chow, V.T., Maidment, D.R., and Mays, L.W. 1988. *Applied Hydrology*, McGraw-Hill, Inc, New York.

Eslamian, S., Gohari, A.R., Ostad-Ali-Askari, K., and Sadeghi, N. 2018. Reservoirs. In: Bobrowsky, P., Marker, B. (eds) *Encyclopedia of Engineering Geology, Encyclopedia of Earth Sciences Series*, Springer.

Feldman, A.D. 2000. Hydrologic modeling system HEC-HMS, Technical Reference Manual. *Technical Reference Manual*, (March), 145. https://doi.org/CDP-74B.

Fleming, M., and Brauer, T. 2016. *Hydrologic Modeling System, HEC-HMS, Quick Start Guide*. U.S. Army Corps of Engineers, Institute for Water Resources, Hydrologic Engineering Center (CEIWR-HEC), Davis, CA.

Goodell, C. 2014. *Breaking the HEC-RAS Code* (1st ed.), h2Ls, Portland, OR.

Leon, A.S., and Goodell, C. 2016. Controlling HEC-RAS using MATLAB, *Environmental Modelling and Software*, 84: 339–348, https://doi.org/10.1016/j.envsoft.2016.06.026.

Mathworks. 2016. *MATLAB*, Natick, MA.

Mays, L.W. 1997. *Optimal Control of Hydrosystems*, Marcel Dekker, Inc., New York.

Mays, L.W., and Tung, Y.-K. 1992. *Hydrosystems Engineering and Management*, McGraw-Hill, Inc., New York.

Montgomery, D.C., Peck, E.A., and Vining, G.G. 2012. *Introduction to Linear Regression Analysis* (5th ed.), John Wiley & Sons, INC, Hoboken, NJ.

Moore, R.J., Cole, S.J., Bell, V.A., and Jones, D.A. 2006. Issues in flood forecasting: Ungauged basins, extreme floods and uncertainty, *Frontiers in Flood Research*, 305: 103–122. Retrieved from http://iahs.info/redbooks/a305/305006.htm.

Unver, O.I., and Mays, L.W. 1990. Model for real-time optimal flood control operation of a reservoir system, *Water Resources Management*, 4(1): 21–46, https://doi.org/10.1007/BF00429923.

Unver O.I., Mays L.W., and Lansey K.E. 1987. Real-time flood management model for the Highland Lakes, *J. Water Resources Planning and Management*, 113(9): 620–638.

Vaze, J., Jordan, P., Beecham, R., Frost, A., and Summerell, G. 2012. *Guidelines for Rainfall-Runoff Modeling: Towards Best Practice Model Application*, eWater Ltd., Bruce, Australia.

World Meteorological Organization. 2011. *Manual on Flood Forecasting and Warning*. WMO, Geneva, Switzerland.

14 Application of Physically Based Distributed Flood Models for Large-Scale Flood Simulations

Siddharth Saksena and Venkatesh Merwade

CONTENTS

14.1 INTRODUCTION

Floods are considered to be the most damaging natural disasters in terms of both economic and human losses. One of the things that can be done to minimize these losses is to provide accurate information about future flood-related risks in the form of flood inundation extent and depth. Traditionally, this has been done through static flood inundation maps corresponding to a 100-year flow. Considering the non-stationary nature of floods, these static maps are proving to be inadequate in providing the information needed to make sound decisions. Additionally, the methodology that is typically followed to create these maps also needs to be re-evaluated to include an accurate representation of all the natural and non-natural processes that interact during high flood events. Even though the flood modeling approach has evolved significantly in the last few decades towards using complex 2D hydrodynamic models, it is still based on simulating flood hydrodynamics through point discharge measurements at discrete locations across a watershed. The 2D hydrodynamic models are capable of handling complex simulations using computationally efficient techniques such as unstructured grids, finite volume models, and parallelization to produce flood inundation maps (Carrivick, 2006; Ernst et al., 2010), but they cannot incorporate the interaction between the river reach and the subsurface, which can play a critical role in dictating the duration and extent of flood inundation.

DOI: 10.1201/9780429463938-19

Despite the significance of subsurface storage dynamics in hydrology, it has received relatively less attention within the context of flood hazards. In addition, when it comes to simulating the hydrodynamics of rivers and their floodplains, the role of subsurface storage and/or floodplain storage has been ignored. These interactions are likewise ignored in environmental management practices. For example, in the United States, 1D and 2D flood inundation models are used by the Federal Emergency Management Agency (FEMA) to generate flood insurance rate maps (FIRMs) (Saksena and Merwade, 2015). These models typically assume that the observed flow, stage, and flood extents are solely a function of the roughness parameter distribution, upstream streamflow forcing, and topography. As a result, the simulated hydrodynamics may be different than the actual field observations in many regions (Highfield et al., 2013). It is important to realize that similar rainfall depths can lead to different flow hydrographs and flood impacts on the same reach depending on the antecedent conditions, watershed characteristics, rainfall intensity, and climatic conditions. For example, saturated conditions in subsurface and floodplains can lead to a more prolonged inundation from low intensity but continuous flood events (Govindaraju and Kavvas, 1991; Pathiraja et al., 2012).

For events occurring at watershed scales instead of reach scales, the flood hydrodynamics are influenced by multiple factors, including but not limited to, topography, bathymetry, rainfall variability, soil moisture retention, infiltration, groundwater (GW) flow, evapotranspiration, reservoir storage, land use characterization, stormwater drainage, and floodplain storage (Kim et al., 2012; Meire et al., 2010). Therefore, accurate prediction of future flood risk requires incorporating the feedbacks from these physical factors affecting the hydrologic cycle in an integrated system (Viero et al., 2014). For example, the deeper groundwater layers that form the confined aquifer systems do not necessarily interact with surface water during floods; however, the surficial (or unconfined) aquifer systems can play a vital role in governing the flood hydrodynamics. The location of the water table determines the available water storage in the subsurface, and therefore, may control surficial ponding during flood events. Thus, incorporating antecedent soil conditions, including the soil moisture profile in the vadose zone and the water table location, is critical for an improved understanding of how flood hydrodynamics changes over space and time under different flow conditions. Finally, several studies have shown that the existing flood modeling approach does not incorporate spatial heterogeneities in watershed properties and non-stationarity in hydrologic response to climate and land-use change at watershed scales (Bloschl et al., 2007; Falter et al., 2016). With increasing focus on large-scale planning and allocation of resources for protection against future flood risk, it is necessary to analyze and improve the deficiencies in the existing flood modeling techniques for identifying regions that are more susceptible to flooding (Bazrkar et al., 2017). For accurate large-scale flood prediction, computationally efficient models that can capture the feedback from the subsurface during extreme events while providing a highly detailed channel and overland flow characterization for high-resolution hydrodynamic modeling are needed.

14.1.1 BACKGROUND OF DISTRIBUTED MODELING

Over the last few decades, studies have shown that it is possible to improve the understanding of hydrologic processes using physically based distributed models that incorporate observable data through assimilation for identifying hydrologic variables and simulating hydrologic fluxes using the fundamental laws of conservation of mass, energy, and momentum at multiple spatiotemporal scales (Fatichi et al., 2016; Pokrajac and de Lemos, 2014; Saksena and Merwade, 2017). These models, often referred to as integrated surface/subsurface hydrologic models (ISSHMs), are capable of solving overland flow, channel flow, and subsurface flow, under different atmospheric conditions (Ebel et al., 2008; Fatichi et al., 2016; Maxwell et al., 2014; Saksena et al., 2019).

One of the earliest blueprints for physically based hydrologic modeling by Freeze and Harlan (1969) highlighted three major issues: (1) accurate representation of physical processes through mathematical formulations; (2) accurate estimation of hydrologic parameters at high spatial resolution; and (3) computational efficiency for simulating large-scale process dynamics. Over the last few decades, these models have evolved significantly through advances in mathematical derivations of hydrologic processes that can capture non-homogeneity, anisotropy, and non-stationarity in physical processes (Fatichi et al., 2016). Similarly, the application of satellite-based products for model parameterization, data assimilation, and calibration has reduced input data uncertainties. Further, creating computationally efficient distributed models for large watersheds is now possible through parallel processing, cluster computing, and big-data analysis (Berg and Sudicky, 2018). Due to these advances, physically based distributed models are being applied for several hydrologic applications (Maxwell et al., 2014), including but not limited to, agriculture sustainability (Schoups et al., 2005), dam removal (Heppner and Loague, 2008), climate change impacts (Kollet and Maxwell, 2008a), residence time distributions (Kollet and Maxwell, 2008b), slope instability (Ebel et al., 2008), runoff generation (Kollet and Maxwell, 2006; VanderKwaak and Loague, 2001), sediment and solute transport (Heppner et al., 2006), stream-aquifer exchanges (Gunduz and Aral, 2005), and short-term forecasting and inundation mapping of natural hazards (Chen et al., 2017; Saksena et al., 2020).

In recent years, a variety of physically based distributed models with varying degrees of structural complexity and process representation have been applied across multiple spatial scales ranging from point to continental scales (Barthel and Banzhaf, 2016; Kollet et al., 2010). While significant strides have been taken by the research community in estimating streamflow, surface-subsurface volumes, and soil moisture, predicting flood depths and extents accurately across large scales remains a challenge for distributed models. Due to the large number of processes and the interconnections between them, the generic issue that has derailed the progress of distributed models revolves around estimating true and meaningful model parameters, which can be a computationally challenging step without parameter optimization (Todini, 2007). However, when parameter optimization is used, the resultant parameter values can be unrealistic due to equifinality (Beven and Freer, 2001). Therefore, a balance between realistic model parameterization and model complexity needs to be attained for optimal hydrologic simulations using distributed models.

Another issue that has derailed the application of distributed models for large-scale flood simulation is the appreciation of different computational constraints, data resolutions, and timescales across multiple river networks. While hydrologic processes can be simulated much faster due to larger timescales (minutes to days), hydrodynamic modeling requires estimating the flow of water from one location to the other across very small timescales (milliseconds to a few seconds). Hence, a smaller iterative time-step is required to maintain model stability, especially when simulating flood events as the hydraulic gradients are often steep and conveyance volumes are very high. Therefore, without optimal parallelization (where different river networks are simulated across different systems), the process of integrating flood hydrodynamics with hydrological processes becomes cumbersome.

This chapter addresses the challenges with the existing flood modeling approach while advocating for a large-scale holistic approach to flood inundation modeling and mapping. As the way forward, this chapter presents an overview of physically based distributed modeling by providing valuable insights into the essential characteristics of distributed models and highlighting the important factors that need to be considered for large-scale flood simulation. Finally, the chapter provides an application of distributed modeling at a large scale using a modeling framework that can be applied across models for flood inundation mapping.

14.2 NEED FOR LARGE-SCALE FLOOD RISK ASSESSMENT

Large-scale (area > 10,000 km^2) flood risk assessment is essential for developing mitigation and planning strategies for decision-makers (Berg and Sudicky, 2018). Recently, the importance of large-scale analysis has risen with the advent of predictive modeling and forecasting for analyzing the impact of population growth, land use, and climate change on water resources (Paniconi and Putti, 2015). Understanding the spatial and temporal variability of storm events becomes even more important at large scales as it can help in identifying areas that are more "reactive" to storm events, thus facilitating a more focused flood mitigation approach by the local agencies. Further, the hydraulics of flood storage and its role in flood hazard mitigation is complicated by the range of spatial and temporal scales involved in causing the actual flood, and the heterogeneity of the floodplains (Vivoni et al., 2007). A flash flood can result from an intense storm lasting for a few hours; a relatively non-intense storm lasting for a day preceded by a wet rainfall season can trigger prolonged flooding of an area; and sometimes significant rainfall and flooding in the upstream part of a watershed can cause flooding in downstream areas within a few hours (Govindaraju and Kavvas, 1991). The spatial and temporal variability of rainfall also impacts flood propagation across large watershed scales since high-intensity events upstream can potentially change the hyporheic exchange dynamics of the entire downstream region (Segond et al., 2007). Additionally, understanding the rainfall-runoff dynamics across entire watersheds becomes even more essential when urban areas are present within the watershed as they are highly reactive to high intensity–low duration events (Berne et al., 2004).

Large-scale water management decisions often rely on a thorough understanding of how one water resource affects the other which can only be accomplished by tracking the flow of water from the atmospheric system to the surface-groundwater system. Several water-resource issues arise at large watershed scales affecting multiple states, and therefore, national agencies such as the United States Geological Survey (USGS) and the Environmental Protection Agency (EPA) are focusing on developing large-scale water management policies (Winter et al., 1999). For example, over the last century, several natural wetlands and river floodplains that contributed to attenuating the impact of extreme rainfall were removed for constructing levees, reservoirs, agricultural growth, and urbanization, leading to increased impervious regions and an alteration of existing flow paths, thereby leading to higher runoff volumes and subsequent flooding (Hutchins et al., 2017). Due to these issues, even the US Army Corps of Engineers (USACE), which led the implementation of most structural measures in the US, is focusing on implementation of more and more nonstructural measures in the form of floodplain storage or creating wetlands for attenuating flood peaks (Brody et al., 2008). Therefore, instead of focusing on the construction of reservoirs for surface storage, there is a growing interest in addressing flood-related issues through natural alternatives.

This shift in focus from structural to nonstructural measures is driven by the goal to achieve environmental sustainability in flood damage reduction (Bednarek, 2001). For example, nonstructural or "soft" flood mitigation measures involve retaining the natural characteristics of floods, by letting the regions in the floodplain sustain the extra volume of water during a flood (Kundzewicz, 2002). In addition to providing flood protection by increasing surficial storage, restored floodplains around the riverbanks can cause the underlying shallow water table to directly transpire water to the atmosphere which can increase the lateral inflow of surface water for recharging groundwater, thereby reducing the overland storage. Therefore, the development of nonstructural measures such as restored wetlands and floodplain storage requires an understanding of the broader impacts from extreme events across large spatial scales. A key step in understanding the true impacts of nonstructural measures involves developing more holistic methods involving large-scale models for estimating flood risk. To predict flood risk accurately, we need to know the factors that are responsible for causing floods under various settings and revisit the existing assumptions of the flood modeling approach. This can be accomplished using distributed flood modeling, by incorporating feedback from multiple watershed processes during flood events.

14.3   CHARACTERISTICS OF LARGE-SCALE DISTRIBUTED MODELS

The physical factors influencing flood hydrodynamics are traditionally modeled as four separate components and are further used to create inputs for one another through model interoperability, namely: (1) hydrologic modeling involving overland and channel flow generation through rainfall-runoff partitioning; (2) hydraulic modeling for routing flow through the rivers; (3) groundwater models for estimating subsurface recharge, water table movement, and subsequent lateral seepage into the rivers; and (4) stormwater modeling for estimating drainage volumes from urban storm sewer networks into the river. In most studies, the outputs from hydrologic models are used as input variables in hydraulic/stormwater models to obtain flood inundation maps. Similarly, the subsurface outputs such as infiltration rates and soil moisture content are used as boundary conditions in groundwater models. Traditionally, stormwater effects are excluded when modeling large flood events as the impact of stormwater drainage is negligible compared to the rainfall-induced riverine flow.

In the case of physically based distributed models, the three major components (hydrologic-hydraulic-groundwater) are formulated in a single system, therefore, several degrees of complexities arise on the basis of spatial scale of the watershed, data availability, the focus of interest, objectives, and problem setting (Fatichi et al., 2016). Due to these complexities, there is a lot of structural variability across different distributed models that results in multiple classifications and numerical schemes. Therefore, distributed models differ from each other significantly. The basic steps involved in the modeling process include solving partial differential equations using the conservation of mass principle based on the shallow water (Saint-Venant) equations for overland and channel flow, Darcian subsurface flow, and a vadose zone based on an approximation of the 3D solution of the Richards equation (Park et al., 2011). Generally, the modeling schemes for distributed models are decided on the basis of five major factors: (1) what are the physical processes that are being modeled; (2) what is the modeling structure or conceptual/mathematical representation required to incorporate these processes; (3) what coupling mechanisms are used to incorporate feedbacks from one process onto the other; (4) which boundary conditions are needed to drive the modeling process; and (5) what is the optimal spatiotemporal scale and resolution required to capture process variability accurately (Paniconi and Putti, 2015). Since this chapter focuses on the application of distributed models towards flood modeling, each of these fundamental factors is discussed in the following sections.

14.3.1 Physical Process Representation

The most important consideration when choosing a distributed model involves selecting the physical processes needing to be modeled. There are several natural hydrologic and hydraulic processes that can affect water movement in the physical systems. These processes can be broadly characterized as atmospheric, land surface, and subsurface processes. Examples of these processes include rainfall, solar radiation, evapotranspiration, snowmelt, infiltration, runoff, channel flow, water table fluxes, tile drainage, GW recharge, lateral seepage, interception and depression storage, floodplain storage, and river-floodplain fluxes (Kampf and Burges, 2007; Mirus and Loague, 2013). In more complex urban systems, the movement of water is affected by the incorporation of man-made structures including, reservoirs, levees, drop structures, weirs, stormwater drainage, flows through bridges and culverts, and flows across road networks. However, with respect to flood modeling, the influence of some of these processes can be neglected depending on the physiological characteristics of the watershed.

For example, evapotranspiration is usually neglected when modeling high-intensity rainfall events. Also, snowmelt is usually ignored when simulating flood hydrodynamics, especially in tropical environments. Similarly, the movement of water in the subsurface is assumed to be in a liquid state even though it can exist in both liquid and vapor form (Govindaraju and Kavvas, 1991). Additionally, the subsurface is assumed to be incompressible. The estimation of deep-aquifer recharge may be crucial for hydrogeological applications and long-term streamflow estimation, but may not be considered for flood modeling, where only the surficial aquifer contributions are essential. Finally, the influence of groundwater processes is less significant for urban systems with a high impervious cover as there is a minimal possibility of groundwater recharge. However, for

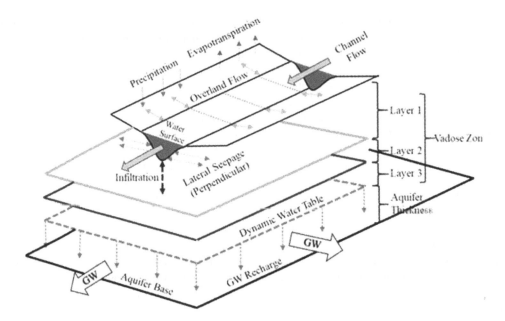

FIGURE 14.1 A 2D illustration of physical processes interacting during rainfall-induced flooding. (Source: modified from Saksena, 2019.)

large-scale flood modeling, it may be essential to incorporate all processes mentioned above as there can be spatially variable atmospheric and physiological conditions within the watershed. The process selection is also dependent on the duration of the simulation, for example, when simulating river hydrodynamics for a continuous long-duration, processes such as evapotranspiration and snowmelt can become crucial, and therefore, should not be neglected. A 2D illustration of the physical processes involved during rainfall-induced flooding is presented in Figure 14.1.

14.3.2 Model Structure

The model structure is formed by the collective mechanisms chosen for formulating the surface and subsurface processes in distributed models. The surface zone can be divided into channel flow and overland flow while the subsurface zone can be divided into unsaturated and saturated flow. These zones interact with one another through different processes, for example, infiltration, seepage, runoff, and river-floodplain fluxes. Channel flow can be simulated using several modifications of Saint-Venant equations based on the level of complexity including the energy, momentum, or diffusive wave equations. Further, channel flow can be divided into one (1D) or two (2D) dimensions, depending on the model domain. Other methods for routing channel flow include the kinematic wave or empirical iterative solutions, however, these methods are not suitable for large-scale flood modeling, where hydraulic routing plays an integral role in determining the model accuracy. The overland flow processes can be formulated using Hortonian flow approximation (Horton, 1931), where the precipitation rate exceeds the saturated hydraulic conductivity of the soil, or Dunne flow approximation, where the precipitation rate is less than the saturated hydraulic conductivity with a shallow initial water table. At large scales, a combination of both types of overland flow approximations is needed to accurately simulate surficial ponding outside the floodplain and infiltration in areas near the channel where the water table is shallow.

The complexity of the model structure is heavily determined by the type of method used for calculating infiltration. While the most optimal solution for infiltration is determined by the three-dimensional (3D) solution of the Richards equation, its application to flood modeling is limited by its computational inefficiency at large scales. Therefore, several other computationally efficient methods are used, including but not limited to, the Boussinesq equation, the 1D Green-Ampt equation, a non-iterative kinematic solution with soil redistribution, and the Dupuit-Forchheimer approximation. Although a 3D solution for the unsaturated zone is recommended for hydrologic modeling, a simpler 1D method with an assumption of a sharp wetting front is commonly used. However, simpler methods should still allow the soil moisture redistribution in the vadose zone to interact with a dynamic water table at the top of the saturated zone. The most widely used infiltration method in hydrologic models called the SCS Curve Number method is not valid for distributed modeling especially for continuous simulations with groundwater flow. Evapotranspiration (ET) is generally simulated using either the Penman-Monteith equation or the Priestley-Taylor method. These methods solve for ET using the amount of solar radiation, soil moisture in the root zone, and crop type (Zotarelli et al., 2013).

The groundwater flow through the surficial aquifer is simulated using 1D, 2D, or 3D approximations of the continuity equation. The mathematical solutions for saturated flow are based on a combination of numerical schemes, for example, the finite difference, finite volume, or finite element methods. While the finite difference and finite volume methods provide a semi- or fully-discrete solution at each grid point in the model domain (Qu and Duffy, 2007), the 2D or 3D finite element methods use continuous functions to simulate subsurface flow across complex geometries, thereby, providing a more realistic model of groundwater dynamics. Figure 14.2 provides an example of a distributed modeling framework with all the elements required for simulating flood hydrodynamics.

14.3.3 Coupling Mechanisms

The two-way interactions between the surface and subsurface zones are established using coupling mechanisms. These coupling mechanisms need to be devised in a way that can incorporate the feedback from physical processes accurately (Dawson, 2008). When the governing equations for both the zones are solved simultaneously, the coupling is referred to as a first-order coupling (Kampf and Burges, 2007). However, with first-order coupling, the time-step required to simultaneously solve all processes together can create a greater computational burden at a large scale. Therefore, first-order coupling may not be feasible for modeling flood inundation during extreme events like Hurricane Harvey in the United States (Lindner and Fitzgerald, 2018). On the other hand, when

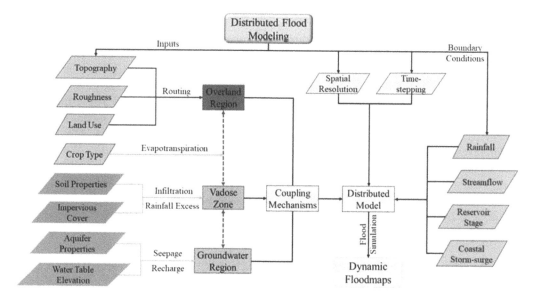

FIGURE 14.2 Example of a distributed modeling framework with specific application to flood prediction. (Source: modified from Saksena, 2019.)

the surface and subsurface equations are solved independently and then linked through an iterative water balance at a specific time-step (Park et al., 2009), the coupling mechanism is referred to as a sequential iterative coupling (SIC).

In some models, the unsaturated and saturated zones in the subsurface are not solved simultaneously, instead, these zones are further coupled sequentially (Streamline Technologies, 2018). While sequential coupling may cause uncertainties due to inaccurate scaling of model parameters across different zones, it provides a much faster alternative to first-order coupling if the model parameters are estimated accurately. Table 14.2 presents examples of widely used distributed models with information on their model structure and coupling mechanisms based on previous studies on distributed model inter-comparisons for application towards hydrologic modeling (Kampf and Burges, 2007; Maxwell et al., 2014).

14.3.4 Boundary Conditions

Boundary conditions established by stream gauge or tidal wave (coastal watersheds) inflows are traditionally not used in distributed models developed for streamflow prediction. However, these may be important for simulating flood hydrodynamics as reservoir inflows, storm-surge inflows, and river inflows are all important factors that can affect flood inundation in large watersheds. While hydrologic prediction involves rainfall-runoff modeling for an entire watershed, the model domain for flood simulation does not necessarily contain an entire watershed. Consequently, the model domain for large-scale flood prediction can also contain multiple watersheds. Therefore, it is important to consider which boundary conditions are needed to be specified depending on the watershed size, location, and influence of man-made structures. For example, when integrating groundwater dynamics with overland flow, an initial water table surface may be required to identify the location of the boundary between groundwater and vadose zone. This is because, in addition to the soil moisture content, the water table location determines the antecedent subsurface conditions which are crucial for accurate characterization of physical conditions in the watershed before flood simulation. Similarly, in some coastal watersheds, the inclusion of storm-surge and tidal boundary inputs may be required when modeling rivers that are affected due to coastal effects.

14.3.5 Spatiotemporal Scale and Resolution

The spatial discretization of distributed models can be characterized as fixed or unstructured within which several geometries can exist, including rectangular or square grids, triangulated-irregular-networks (TIN), hexagonal, or polygons with more than six edges. While a fixed-grid structure provides the same resolution across the entire model domain, it can also be computationally challenging, instead, a TIN with a flexible structure can provide better computational efficiency while compromising on the spatial resolution at certain locations within the model domain. In practice, the smallest model scale should be identified based on the resolution of the finest dataset used for model creation. For example, topographic data should be used as a benchmark for identifying the minimum resolution, since other datasets are usually available at a coarser resolution. While hydrologic applications like rainfall-runoff modeling or the GW recharge analysis may produce optimal results even with a coarse resolution spatial structure, flood prediction requires a much finer spatial resolution to reduce the likelihood of over-prediction or under-prediction of inundation.

Similarly, hydrodynamic modeling requires a much smaller time-step compared to hydrologic rainfall-runoff partitioning and groundwater flow. This is because surface, subsurface, and atmospheric processes occur at different timescales, hence the iterative timescales and temporal resolution of model outputs should be variable for flood prediction. Finally, the optimal time interval required for capturing large-scale watershed dynamics should also be considered. While the flood hydrodynamics for small-scale watersheds with low-intensity events can be captured using event-based modeling (ranging from a few hours to a few days), large-scale watersheds with a high spatio-temporal distribution of rainfall require longer (continuous) time periods ranging from a few days

TABLE 14.1

Examples of Distributed Models Suitable for Large-Scale Flood Modeling and Their Characteristics

Model	Surface Domain	Subsurface Domain	Coupling Mechanism	Sources
HGS (HydroGeoSphere)	2D	3D	FOC	Aquanty Inc., 2013
ParFlow (Parallel Flow)	2D	3D	FOC	Kollet and Maxwell, 2006
OGS (OpenGeoSys)	2D	3D	SC	Delfs et al., 2009
GSSHA (Gridded Surface/Subsurface Hydrologic Analysis)	2D	3D/2D	SC	Downer and Ogden, 2006
ICPR (Interconnected Channel and Pond Routing)	1D/2D	1D/2D	SC	Streamline Technologies, 2018
tRIBS (Triangulated Irregular Network-Based Real Time Integrated Basin Simulator)	1D	1D/2D	SC	Ivanov et al., 2004
InHM-Integrated Hydrology Model	2D	3D	FOC	VanderKwaak, 1999
CATHY (CATchment HYdrology)	1D	3D	SC	Bixio et al., 2002
MIKE-SHE	1D/2D	1D/3D	SC	Abbott et al., 1986
MODHMS	1D/2D	3D	FOC	Panday and Huyakorn, 2004
WASH123D	2D	3D	SC	Yeh et al., 2006
PAWS (Process-Based Adaptive Watershed Simulator)	2D	Quasi-3D	FOC	Shen and Phanikumar, 2010
PIHM (Penn State Integrated Hydrologic Model)	1D/2D	1D/2D	FOC	Qu and Duffy, 2007

Notes:

FOC: first-order coupling; SC: sequential coupling, which can be both iterative and non-iterative; 1D/2D options in surface domain refer to 1D channel flow and 2D diffusive wave; 3D in subsurface domain refers to both 3D vadose zone (Richard's equation) and 3D solution of groundwater continuity equation; 3D/2D in subsurface domain refers to 3D vadose zone (Richard's equation) and 2D groundwater. For more information, please refer to model sources.

to several months. Therefore, the application of distributed models for large-scale flood inundation requires that the models are capable of continuous simulations with adaptive and variable timescales for simulating different processes. Based on the characteristics highlighted in section 14.3 that are crucial for large-scale distributed flood modeling, the models that are most suitable for this application are presented in Table 14.1.

14.4 APPLICATION OF DISTRIBUTED FLOOD MODELING TO A LARGE-SCALE WATERSHED

To illustrate how physically based modeling can be used for large-scale flood simulations, the Wabash River Basin (WRB) with a drainage area of 85,320 km², is chosen as the test site (shown in Figure 14.3). The Wabash River is a major tributary of the Ohio River in the United States and forms the main basin system for the state of Indiana, while also extending into the states of Illinois and Ohio. Based on the Hydrologic Unit Code (HUC) specification, the WRB is characterized as a HUC-4 basin, a catchment located in the Ohio River Basin (HUC-2), which is one of the major river basins in the United States. This region has experienced several major floods in the past decade, with a 50-year return period event in 2013, a 25-year event in 2015, and ten-year events in 2016 and

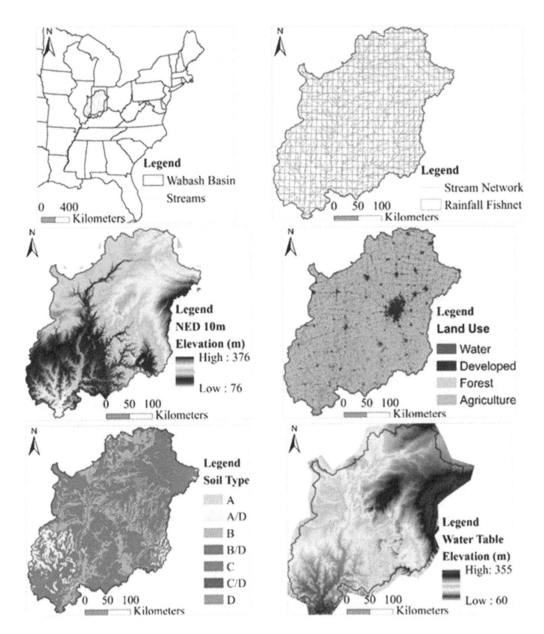

FIGURE 14.3 Geospatial representation of datasets used for distributed modeling in ICPR.

2018. According to FEMA, approximately 12% of the total area in the watershed is located inside a 100-year flood zone, signifying the high flooding potential in the region. Table 14.2 presents information on the watershed characteristics including the land and soil distribution for the Wabash River Basin. The following section provides information on the sources of data used for building the large-scale flood model for this watershed.

14.4.1 Data Acquisition

The large-scale distributed modeling is achieved by linking the unsaturated (vadose) zone with the overland flow region and the saturated zone (surficial aquifer). The creation of distributed

TABLE 14.2

Watershed Characteristics

Drainage area, km²	85,320
Land use as per NLCD 2011 (%)	**% of total area**
Open water	1.1
Low-intensity residential	5.7
High-intensity residential	2.5
Commercial/industrial	0.8
Transportation	0.3
Barren land	0.1
Deciduous forest	19.2
Evergreen forest	0.3
Mixed forest	0.1
Shrub/scrub	0.1
Grasslands	1.1
Hay/pasture	6.6
Cultivated crops	61.3
Woody wetlands	0.5
Herbaceous wetlands	0.2
Soil type as per STATSGO (%)	**% of total area**
A	0.8
A/D	5.5
B	23.8
B/D	0.1
C	68.0
C/D	0.3
D	1.5
No. of NLDAS rainfall weather stations	649

models requires information on multiple datasets, for example, topography, land use, surface roughness (Manning's n), rainfall, unsaturated soil zone parameters, and surficial aquifer parameters. Considering the critical role of topography for accurate hydrodynamic simulation of rivers (Saksena, 2015), a 10 m resolution topographic dataset obtained from the United States Geological Survey's National Elevation Dataset (NED) is used in this study (Figure 14.3). The land use data for the basin is obtained from the National Land Cover Database (NLCD; http://nationalmap.gov/viewer.html) and the values of Manning's roughness corresponding to the land use classes in NLCD are adapted from Chow (1959).

For the subsurface part, vadose zone properties such as moisture content, vertical conductivity, and soil type are extracted from the State Soil Geographic Database (STATSGO) which is available at a 1:250,000 scale. Hourly rainfall data at a spatial resolution of a one-eighth degree grid for the Wabash River basin are obtained from the North American Land Data Assimilation System (NLDAS). Since this is a grid-based dataset, a total of 649 unique grids are used to provide bias-corrected rainfall information, which results in a higher resolution of rainfall inputs compared to gauge-based techniques. Similarly, water table information in the form of contour maps is obtained from the Indiana Department of Natural Resources (IDNR) Potentiometric Surface Maps (www.in.gov/dnr/water/7256.htm), which are converted into gridded raster formats using ArcGIS. The water table raster for this basin is shown in Figure 14.3. Additionally, the surficial aquifer properties are extracted using spatial datasets obtained from the Indiana Geographic Information Council's (IGIC) Indiana Map server (http://maps.indiana.edu/index.html). Stream network features including the river and its tributaries that are required for hydraulic routing are extracted from the National Hydrography Dataset (NHD) and corrected using the NED 10 m digital elevation model (DEM) and aerial imagery in a GIS framework.

14.4.2 MODEL DESCRIPTION

The Interconnected Channel and Pond Routing (ICPR) model, a two-dimensional (2D) integrated surface water-groundwater modeling tool based on distributed model parameterization procedures, is used for this application (Streamline Technologies, 2018). ICPR is a FEMA-approved distributed model with capabilities of combining precipitation, stormwater infrastructure, evapotranspiration, 1D/2D surface flow, pond storage, and groundwater simulations for both event-based and continuous storm events (Streamline Technologies, 2018). Using the ICPR Linux simulation engine and Purdue University's supercomputers, a high-resolution hydrologic model for the Wabash River Basin (85,320 km^2) is created. ICPR is based on the solution of the fundamental watershed mass balance equation at every time-step, and therefore, can provide a more physical and dynamic representation of hydrologic and hydraulic fluxes by simultaneously incorporating watershed processes in a single system.

ICPR uses a finite volume approach for surface flow using unstructured mesh networks. Momentum equations (with options for energy and diffusive wave equations) are lumped along triangle edges, and the mass balance equations are lumped at the triangle vertices where irregularly shaped polygons are formed establishing local control volumes. This set of polygons is referred to as the surface honeycomb mesh. The mass balance equations used in ICPR for every time-step are presented in equations 14.1 to 14.3. These equations are applied across each individual honeycomb structure at every time step to evaluate the volume of water from individual processes such as seepage, rainfall excess, and streamflow generation.

$$dz = \left(\frac{(Q_{in} - Q_{out})}{A_{surface}} \right) dt \tag{14.1}$$

$$Q_{in} = \sum Q_{link_{in}} + \sum Q_{excess} + \sum Q_{external} + \sum Q_{seepage} \tag{14.2}$$

$$Q_{out} = \sum Q_{link_{out}} + \sum Q_{irrigation} \tag{14.3}$$

Where, dz = incremental change in stage (L); dt = computational time-step (T); Q_{in} = total inflow rate (L^{3}T^{-1}); Q_{out} = total outflow rate (L^{3}T^{-1}); $A_{surface}$ = "wet surface area" (L^2); $\sum Q_{link_{in}}$ = sum of all link flow rates entering a control volume (L^{3}T^{-1}); $\sum Q_{link_{out}}$ = sum of all link flow rates leaving the control volume (L^{3}T^{-1}); $\sum Q_{excess}$ = sum of rainfall excess rates for polygons in control volume (L^{3}T^{-1}); $\sum Q_{external}$ = sum of all inflows from external sources such as streamflow gauges (L^{3}T^{-1}); $\sum Q_{seepage}$ = sum of lateral seepage inflow from groundwater model (L^{3}T^{-1}); $\sum Q_{irrigation}$ = sum of water pulled out of the system for irrigation.

The spatial heterogeneity in the watershed characteristics for WRB is represented using approximately one million triangular elements that are used to build a honeycomb mesh that contains structural elements that act as an individual sub-basin (or control volumes) for hydrologic and hydraulic simulation. Even within each element of the honeycomb mesh, the geospatial datasets highlighted in Figure 14.3, for example, soil, land use, and rainfall fishnet are intersected to form sub-polygons containing unique intersections of these datasets. The sub-polygons are used to extract rainfall, soil, and land-use properties from the input datasets, and they function as individual hydrologic units within a honeycomb. Therefore, even though approximately one million triangular elements are used to characterize the watershed, the number of sub-polygons exceeds one and a half million, thereby providing a high-resolution characterization of watershed properties.

As mentioned earlier, ICPR uses the one-dimensional form of St. Venant equations and average 2D ground slopes to move water between control volumes along the edges of the triangles. Equation 14.4 presents the momentum equation for overland flow including inertial terms for local and convective acceleration. In this study, a simpler version of the momentum equation, called the diffusive wave equation, is used to simulate the flow between triangular edges (2D surface water routing) as shown in equation 14.5. The diffusive wave equation is based on eliminating the local and convective acceleration, as well as the pressure terms from the momentum equation. The second equation used to formulate the flow alongside the diffusive wave equation is obtained from the continuity equation of flow, presented in equation 14.6. Equation 14.6 is used in conjunction with equations 14.7 and 14.8 to evaluate the flow through the triangular edges, where the conveyance (equation 14.8) is calculated using the Manning's roughness, which is further used to evaluate the discharge (flow) using the friction slope.

$$\frac{\partial Q}{\partial t} + \frac{\partial \left(\frac{Q^2}{A} \right)}{\partial x} + gA \frac{\partial Z}{\partial x} + gA(S_0 - S_f) = 0 \tag{14.4}$$

$$Q = \left(\frac{Z_1 - Z_2}{\Delta x C_f} \right)^{\frac{1}{2}} \tag{14.5}$$

$$\frac{\partial Q}{\partial x} + \frac{\partial A}{\partial t} = 0 \tag{14.6}$$

$$Q = K \times S_f^{0.5} \tag{14.7}$$

$$K = \sum_{i=1,p} \frac{C R_i^{2/3} A_i}{n_i} \tag{14.8}$$

Where, Q=channel discharge or flow rate (L³T⁻¹); A=cross-sectional area (L²); t=time (T); g=acceleration due to gravity (LT⁻²); Z=gravitational head (L); x=distance in direction of flow (L); S_f=friction slope; S_0=slope of the channel bed; K=channel conveyance (L³T⁻¹); p=number of segments with unique Manning's roughness; n_i=Manning's roughness for segment i; A_i=cross-sectional area of ith segment (L²); R_i=hydraulic radius of the ith segment (L); C_f=coefficient of friction for head loss calculation; Δx=distance between adjacent cross-sections.

ICPR allows the application of variable surface roughness for shallow and deep channels. During large flood events, the channel roughness changes over time when the depth of the water in the floodplain increases due to persistent flooding (Arcement Jr. and Schneider, 1984). Therefore, ICPR uses an exponential decay function (shown in equations 14.9 and 14.10) dependent on surface depth to incorporate the variability in surface roughness, which improves the overland flow routing significantly.

$$n = n_{shallow} e^{(k)(d)} \tag{14.9}$$

$$k = \frac{\ln \left(\frac{n_{deep}}{n_{shallow}} \right)}{d_{max}} \tag{14.10}$$

Where, n=Manning's roughness at depth d; n_{shallow}=Manning's roughness at ground surface; n_{deep}=Manning's roughness at depth ≥ 3 feet; k=exponential decay factor (L^{-1}); d=depth of flow (L); d_{max}=user-specified maximum depth for transitioning to n_{deep} (L).

The Brooks-Corey soil water retention–hydraulic conductivity relationship (Rawls and Brakensiek, 1982), shown in equation 14.11, is used to determine unsaturated conductivities based on initial soil moisture contents. The soil moisture accounting, and subsequent GW recharge, is computed using a non-iterative kinematic approach which distributes the unsaturated soil zone into three layers of non-homogeneous soil properties (Table 14.3). These layers are further divided into multiple cells to enable tracking of hydrologic fluxes through each individual cell in both upward and downward directions. The moisture contents are updated based on these fluxes, followed by rebalancing the cells from the bottom, back to the surface, to ensure that the moisture content in any cell does not exceed saturation.

The deeper 2D groundwater simulations are based on the groundwater continuity equation using finite element approximations. A 2D subsurface discretization allows the distributed model to account for the non-uniform distribution of the hydrologic properties in the vadose zone and the surficial aquifer. A finite element approach with a six-point quadratic triangular element is used in ICPR to solve the continuity equation for unsteady phreatic two-dimensional groundwater flow (MARTÍNEZ, 1989; Streamline Technologies, 2018), and the system of equations (14.12 and 14.13) is solved using the Cholesky method (Kuiper, 1981). A detailed description of the modification of the continuity equation for finite element formulation can be accessed from ICPR's Technical Reference Manual (Streamline Technologies, 2018). Finally, the seepage rates along a sloping ground surface, riverbank, or seepage faces on a hill are calculated using equation 14.14.

$$\frac{K(\theta)}{K_s} = \left(\frac{\theta - \theta_r}{\varphi - \theta_r}\right)^n \tag{14.11}$$

$$n\frac{\partial h}{\partial t} = -\frac{\partial(uh)}{\partial x} - \frac{\partial(vh)}{\partial y} \tag{14.12}$$

$$u = -K \cdot \frac{\partial h}{\partial x}; \text{ and, } v = -K \cdot \frac{\partial h}{\partial y} \tag{14.13}$$

$$Q_{\text{seepage}} = \frac{(h_1 - h_2) \times (A) \times \varphi_b}{dt_{gw}} \tag{14.14}$$

Where, θ=current moisture content; θ_r=residual moisture content; φ=saturated moisture content; $K(\theta)$=unsaturated vertical conductivity at θ; K_s=saturated vertical conductivity; $n = 3 + \dfrac{2}{\lambda}$; and

λ=pore size index n is the fillable porosity (or specific yield); h is the GW elevation (piezometric head); u, v are the velocity vector components; t is time; x, y are the Cartesian coordinates; K is the permeability (conductivity) of the porous media; Q_{seepage}=seepage rate (L^{3}T^{-1}); h_1=calculated water table elevation (L); h_2=ground surface elevation at node (L); A=groundwater control volume surface area (L^2); φ_b =below ground fillable porosity; and dt_{gw}=groundwater computational time increment (T).

14.4.3 MODEL CALIBRATION AND SIMULATION

The previous section describes the model formulations and process characterizations using ICPR. For more information on the model building process, please refer to ICPR's technical reference

TABLE 14.3

ICPR Subsurface Parameters

Vadose Zone	Soil Type	K_v (mm/hr)	Saturated MC	Residual MC	Initial MC	Field Capacity MC	Wilting Point MC	PSI	Ψ (cm)
Layer 1	A	15.24	0.277	0.069	0.128	0.128	0.107	0.518	38.3
150 cm	A/D	8.01	.320	0.057	0.220	0.220	0.160	0.540	30.7
	B	6.20	0.298	0.061	0.200	0.200	0.138	0.620	25.5
	B/D	2.60	0.420	0.046	0.310	0.310	0.190	0.226	99.8
	C	2.34	0.458	0.051	0.300	0.300	0.225	0.296	59.2
	C/D	0.80	0.361	0.050	0.240	0.240	0.141	0.270	108.2
	D	1.40	0.328	0.035	0.240	0.240	0.118	0.161	197.9
Layer 2	A	8.38	0.200	0.040	0.080	0.125	0.063	0.296	59.2
90 cm	A/D	4.05	0.280	0.057	0.220	0.220	0.160	0.540	30.7
	B	3.10	0.280	0.070	0.170	0.220	0.135	0.316	67.5
	B/D	1.29	0.380	0.046	0.310	0.310	0.190	0.226	99.8
	C	1.17	0.320	0.078	0.220	0.220	0.155	0.270	106.8
	C/D	0.39	0.300	0.050	0.240	0.240	0.141	0.270	108.2
	D	0.80	0.300	0.040	0.120	0.200	0.090	0.161	197.9
Layer 3	A	2.10	0.120	0.030	0.090	0.090	0.060	0.540	30.7
60 cm	A/D	1.01	0.240	0.057	0.220	0.220	0.160	0.540	30.7
	B	0.77	0.200	0.040	0.100	0.100	0.080	0.226	99.8
	B/D	0.33	0.320	0.046	0.310	0.310	0.190	0.226	99.8
	C	0.29	0.180	0.045	0.120	0.120	0.075	0.161	168.4
	C/D	0.13	0.260	0.050	0.240	0.240	0.141	0.270	108.2
	D	0.20	0.170	0.030	0.010	0.010	0.080	0.161	197.9

GW Zone	Type	Effective Porosity, η_e	Hydraulic Conductivity, K (mm/hr)
Aquifer	A1	0.175	30.48
	B1	0.270	12.40
	C1	0.310	4.67
	D1	0.360	2.79

Notes: MC = moisture content; K_v = saturated vertical permeability, Ψ = soil matric potential; PSI = pore size index; A1 = moraine aquifer; B1 = complex aquifer; C1 = till aquifer; and D1 = outwash aquifer.

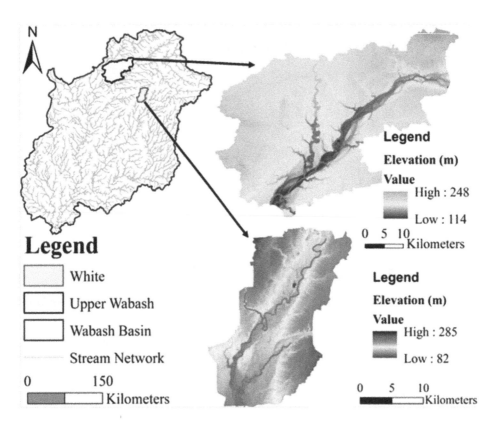

FIGURE 14.4 Small-scale sites (Upper Wabash River Basin and White River Basin) selected for evaluating optimal model parameters.

manual (Streamline Technologies, 2018). After creating the model inputs and processing watershed information from open-access datasets, the best-fit model parameters for simulating flood hydrodynamics for the entire basin are evaluated by selecting smaller sub-watersheds inside the model domain and calibrating and validating the predicted streamflow at multiple locations inside the watershed using historic flood events (Figure 14.4).

Therefore, the best-fit parameters were derived by simulating two sub-watersheds of the contributing area of 1,760 km² (Upper Wabash River Basin) and 380 km² (White River Basin) with different physical characteristics. The White River Basin is characterized by a significantly higher (86%) proportion of developed (urban) regio, when compared to the Upper Wabash Basin (10%), which has agricultural land use. Using these sub-watersheds, a 300-hour-long simulation for a 50-year storm event in April 2013 is simulated and the results are compared with observed flows at four USGS streamflow gauges namely, USGS 03328500, Eel River near Logansport, IN; USGS 03333050, Tippecanoe River near Delphi, IN; USGS 03335500, Wabash River at Lafayette, IN; and USGS 03353000, White River at Indianapolis, IN. After simulating the models, the best-fit parameters are selected for hydraulic routing (surface water) and sub-surface parameters for vadose zone and groundwater.

Tables 14.3 and 14.4 present the best-fit subsurface parameters and surface roughness parameters respectively based on the simulation across both basins. As shown in Table 14.4, a channel roughness (Manning's n) value of 0.030 is used as the deep channel roughness (depth > 0.91 m or 3 feet), while a value of 0.045 is used as the shallow channel roughness (depth < 0.91 m or 3 feet). The values of roughness outside the river channels are not calibrated and are derived from the land-use-based roughness classification (Table 14.4) presented in Chow (1959). Figure 14.5 shows the performance of the best-fit model parameters in estimating streamflow when compared with

TABLE 14.4
ICPR Surface Roughness Parameters

Surface Zone	Type of Land Use	Shallow Surface Roughness	Deep Surface Roughness
1	Open water	0.045	0.030
2	Low intensity residential	0.020	0.011
3	High intensity residential	0.015	0.011
4	Commercial/industrial	0.014	0.010
5	Transportation	0.013	0.010
6	Barren land	0.100	0.050
7	Deciduous forest	0.198	0.184
8	Evergreen forest	0.198	0.184
9	Mixed forest	0.198	0.184
10	Shrub/scrub	0.200	0.100
11	Grasslands	0.350	0.240
12	Hay/pasture	0.200	0.100
13	Cultivated crops	0.180	0.100
14	Woody wetlands	0.060	0.045
15	Herbaceous wetlands	0.060	0.035

Notes: The roughness parameters transition from shallow (maximum) to deep (minimum) based
on an exponential decay function dependent on the depth of flow described in equations
14.9 and 14.10.

observed streamflow across four USGS gauge locations. Since the overall results shown in Figure
14.5 are satisfactory, the same model parameters are used to simulate the entire Wabash River Basin
for a duration of two months, from April 1 to May 31, 2013.

14.4.4 LARGE-SCALE FLOOD SIMULATION

After simulating the flood hydrodynamics for the entire basin, the overall flood extents are extracted
at an hourly time-step. Therefore, the propagation of the flood event can be dynamically observed
throughout the watershed. Figure 14.6a shows the 100-year FEMA-derived flood extent map for
the river networks in the Wabash Basin. These flood extents only capture the flooding in the river-
floodplain region and are not available for all the streams in the basin. This is because traditional
flood modeling is designed to capture only river-floodplain fluxes, regardless of the type of hydraulic model (1D or 2D or a combination). The Federal Emergency Management Agency specifies the
quality standards for creating acceptable maps that can be used for flood estimation to maintain
consistency in the quality of the results obtained from several small-scale projects that are combined to form the map shown in Figure 14.6a. However, the combination of different county-based
surveys, models, calibration techniques, and input datasets still create several uncertainties in the
results when evaluated at large scales. Also, several lower-order streams are not included in this
map due to the high economic and computational costs associated with creating these maps.

On the other hand, Figure 14.6b shows the maximum flood extent across the entire watershed
using distributed modeling in ICPR. The flood extents suggest that the integration of rainfall and
subsurface hydrology with hydrodynamics becomes essential for watersheds with higher drainage
density in the stream network as the volumetric contributions from lower-order streams increase
at larger scales. Further, the flood extents are not only generated in the river-floodplain, but also
outside the floodplain due to surficial ponding, thereby, highlighting the advantage of this approach

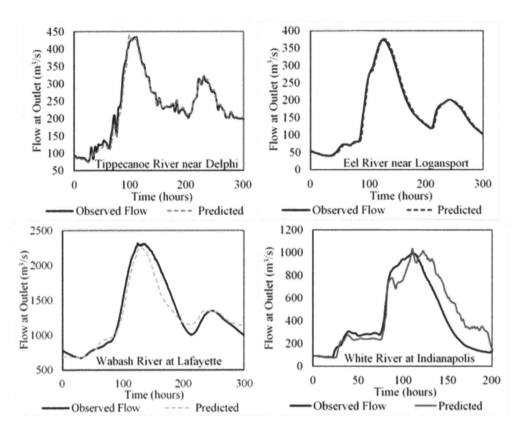

FIGURE 14.5 Streamflow comparison with observed data across four USGS gauges in Indiana. (Source: modified from Saksena, 2019.)

over the traditional approach that assumes the surface to be impervious. Finally, Figures 14.6c and 14.6d show the progression of inundation due to rainfall at time $t = 1000$ hours and $t = 1400$ hours. These results highlight the spatiotemporal variability of rainfall-induced flooding since the maximum inundation extent does not occur everywhere at the same time. Although Figure 14.6b shows the peak inundation at every location, this inundation propagates dynamically throughout the basin at different times. This spatiotemporal variability in flood inundation is captured here, highlighting the advantage of using distributed flood modeling for large-scale flood simulation.

14.5 CONCLUSIONS

This chapter focuses on providing an overview of distributed models and their application for large-scale flood modeling. The chapter also discusses the various factors that influence flood hydrodynamics and how they can be incorporated into distributed models. Finally, a prototype framework for large-scale distributed flood modeling using ICPR is presented for the Wabash River Basin. The flood extent maps in this study are generated using only precipitation as the input boundary condition, which makes this approach useful for flood forecasting because the expected rainfall can be predicted in advance through regional forecast centers and applied to the model for flood forecasting. Although only flood extents are shown in this chapter, the modeling approach can be used to generate location-specific information such as the timing, magnitude, and duration of inundation everywhere inside the watershed. All the datasets used for this modeling effort are open access, highlighting the spatial transferability of the modeling approach to other large-scale watersheds. The results presented in this chapter highlight the potential of distributed modeling in identifying

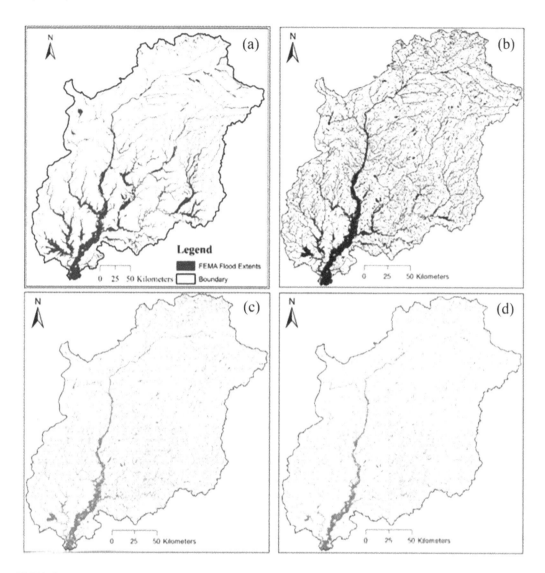

FIGURE 14.6 (a) 100-year FEMA flood extent map; (b) maximum flood extent during the simulation; (c) flood inundation extent at time t = 1,000 hours; and (d) flood inundation extent at time t = 1,400 hours.

areas that are at a higher risk of inundation due to land use, topographic-slope, or saturated subsurface conditions.

It should be noted that the hydrodynamic simulations used here are based on the 2D diffusive wave formulations for hydraulic routing of floodwaters. However, future model applications should include a 1D channel representation using the energy equation in combination with a 2D overland flow. This combination of a 1D–2D channel-floodplain discretization is expected to add more detail in the river channels compared to the 2D approach used here. This would not only result in more accurate hydraulic routing, but also improve the computational efficiency of the large-scale model. Similarly, the model presented here is devoid of channel bathymetry which may be crucial for capturing river-floodplain fluxes accurately (Dey et al., 2019), therefore, incorporation of river-network bathymetry for the entire study area should be considered for modeling flood hydrodynamics accurately.

Studies have suggested that land-use change can occur drastically with increasing urbanization and population growth (El-Khoury et al., 2015), which can have severe implications on streamflow generation (Rajib and Merwade, 2017), causing increased flood risk. Additionally, flood forecasting for unprecedented events such as Hurricane Harvey (> 1,000-year return period) in a changing climate would require integrated and dynamic flood prediction in the future. Therefore, in addition to short-term flood forecasting, future work on distributed modeling should involve evaluating the effect of climate and land use in exacerbating future flood risk.

Even though the application of distributed modeling in flood prediction is promising for decision-makers in understanding flood hydrodynamics at large scales, there are several issues that should be addressed by the scientific community for improving the model precision, accuracy, and computational efficiency. For example, optimal spatial and temporal scaling, estimation of model parameters and state variables, incorporation of accurate boundary conditions, estimation of initial physical and climatic conditions, resolution and quality of data, and climatic uncertainty are some of the challenges that the scientific community must encounter for a more reliable and spatially transferable application of distributed models in flood prediction. Finally, the issue of data visualization should be addressed for improving the applicability of distributed models for flood mapping. Hydrodynamic models (for example, HEC-RAS 2D and LISFLOOD-FP) that only simulate surface hydraulics have tools for high-resolution visualization of flood dynamics, in addition to a user-friendly interface that enables the broad applicability of these models. On the contrary, only a limited number of distributed models such as GSSHA and ICPR have such tools for high-resolution visualization and animation of flood dynamics, which has enabled their application for generating FEMA-approved flood inundation maps. Several other distributed models with a more complex model structure do not have this capability. For improving the applicability of distributed models in simulating flood hydrodynamics, the scientific community should address the issue of data visualization. In conclusion, by addressing some of the challenges highlighted above, distributed models can provide more robust and realistic flood simulations across large scales.

ACKNOWLEDGMENTS

This work was supported by grants from the Indiana Water Resources Research Center (IWRRC) and the Pathfinder Fellowship sponsored by the Consortium of Universities for the Advancement of Hydrologic Science, Inc. (CUAHSI) and the National Science Foundation (NSF). The authors would like to thank Peter Singhofen from Streamline Technologies, Winter Springs, Florida for his initial analysis and valuable feedback on this chapter. We thank Dr. Marty Frisbee from Purdue University for his help in obtaining the groundwater data. The authors also thank Kimberly Peterson from the Lyles School of Civil Engineering for proofreading the chapter.

REFERENCES

Abbott, M.B., Bathurst, J.C., Cunge, J.A., O'Connell, P.E., Rasmussen, J.1986. An introduction to the European hydrological system – Systeme hydrologique Europeen, "SHE", 1: History and philosophy of a physically-based, distributed modelling system, *J. Hydrol.* 87: 45–59.

Aquanty Inc. 2013. *HGS 2013, HydroGeoSphere User Manual*, Aquanty Inc., Waterloo, ON, Canada, p. 435.

Arcement, Jr., G.J., and Schneider, V.R. 1984. *Guide for Selecting Manning's Roughness Coefficients for Natural Channels*, Fed. Highw. Adm. FHWA-TS-84, 68.

Barthel, R., and Banzhaf, S. 2016. Groundwater and surface water interaction at the regional-scale – A review with focus on regional integrated models, *Water Resour. Manag.*, 30: 1–32, https://doi.org/10.1007/s11269-015-1163-z.

Bazrkar, M.H., Adamowski, J., and Eslamian, S. 2017. Water system modeling. In: Furze, J.N., Swing, K., Gupta, A.K., McClatchey, R., Reynolds, D. (eds) *Mathematical Advances Towards Sustainable Environmental Systems*, Springer International Publishing, Switzerland, pp. 61–88.

Bednarek, A.T. 2001. Undamming rivers: A review of the ecological impacts of dam removal, *Environ. Manage.*, 27: 803–814, https://doi.org/10.1007/s002670010189.

Berg, S.J., and Sudicky, E.A. 2018. Toward large-scale integrated surface and subsurface modeling, *Groundwater*, 57: 1–2, https://doi.org/10.1111/gwat.12844.

Berne, A., Delrieu, G., Creutin, J.D., and Obled, C. 2004. Temporal and spatial resolution of rainfall measurements required for urban hydrology, *J. Hydrol.*, 299: 166–179, https://doi.org/10.1016/j.jhydrol.2004.08.002.

Beven, K., and Freer, J. 2001. Equifinality, data assimilation, and uncertainty estimation in mechanistic modelling of complex environmental systems using the GLUE methodology, *J. Hydrol.*, 249: 11–29, https://doi.org/10.1016/S0022-1694(01)00421-8.

Bixio, A.C., Gambolati, G., Paniconi, C., Putti, M., Shestopalov, V.M., Bublias, V.N., Bohuslavsky, A.S., Kastelteva, N.B., and Rudenko, Y.F. 2002. Modeling groundwater-surface water interactions including effects of morphogenetic depressions in the Chernobyl exclusion zone, *Environ. Geol.*, 42: 162–177, https://doi.org/10.1007/s00254-001-0486-7.

Bloschl, G., Ardoin-Bardin, S., Bonell, M., Dorninger, M., Goodrich, D., Gutknecht, D., Matamoros, D., Merz, B., Shand, P., and Szolgay, J. 2007. At what scales do climate variability and land cover change impact on flooding and low flows? *Hydrol. Process.*, 21: 1241–1247.

Brody, S.D., Zahran, S., Highfield, W.E., Grover, H., and Vedlitz, A. 2008. Identifying the impact of the built environment on flood damage in Texas, *Disasters*, 32: 1–18, https://doi.org/10.1111/j.1467-7717.2007.01024.x.

Carrivick, J.L. 2006. Application of 2D hydrodynamic modelling to high-magnitude outburst floods: An example from Kverkfjoll, Iceland, *J. Hydrol.*, 321: 187–199, https://doi.org/10.1016/j.jhydrol.2005.07.042.

Chen, Y., Li, J., Wang, H., Qin, J., and Dong, L. 2017. Large-watershed flood forecasting with high-resolution distributed hydrological model, *Hydrol. Earth Syst. Sci.*, 21: 735–749, https://doi.org/10.5194/hess-21-735-2017.

Dawson, C. 2008. A continuous/discontinuous Galerkin framework for modeling coupled subsurface and surface water flow, *Comput. Geosci.*, 12: 451–472, https://doi.org/10.1007/s10596-008-9085-y.

Delfs, J.O., Park, C.H., and Kolditz, O. 2009. A sensitivity analysis of Hortonian flow, *Adv. Water Resour.*, 32: 1386–1395, https://doi.org/10.1016/j.advwatres.2009.06.005.

Dey, S., Saksena, S., and Merwade, V. 2019. Assessing the effect of different bathymetric models on hydraulic simulation of rivers in data sparse regions, *J. Hydrol.*, 575: 838–851, https://doi.org/10.1016/j.jhydrol.2019.05.085.

Downer, C.W., and Ogden, F.L. 2006. *Gridded Surface Subsurface Hydrologic Analysis (GSSHA) User's Manual, Version 1.43 for Watershed Modeling System 6.1*, US Army Corps Eng. Eng. Res. Dev. Cent, USA.

Ebel, B.A., Loague, K., Montgomery, D.R., and Dietrich, W.E. 2008. Physics-based continuous simulation of long-term near-surface hydrologic response for the Coos Bay experimental catchment, *Water Resour. Res.*, 44: 1–23, https://doi.org/10.1029/2007WR006442.

El-Khoury, A., Seidou, O., Lapen, D.R.L., Que, Z., Mohammadian, M., Sunohara, M., and Bahram, D. 2015. Combined impacts of future climate and land use changes on discharge, nitrogen and phosphorus loads for a Canadian river basin, *J. Environ. Manage.*, 151: 76–86, https://doi.org/10.1016/j.jenvman.2014.12.012.

Ernst, J., Dewals, B.J., Detrembleur, S., Archambeau, P., Erpicum, S., and Pirotton, M. 2010. Micro-scale flood risk analysis based on detailed 2D hydraulic modelling and high resolution geographic data, *Nat. Hazards*, 55: 181–209, https://doi.org/10.1007/s11069-010-9520-y.

Falter, D., Dung, N.V., Vorogushyn, S., Schröter, K., Hundecha, Y., Kreibich, H., Apel, H., Theisselmann, F., and Merz, B. 2016. Continuous, large-scale simulation model for flood risk assessments: Proof-of-concept, *J. Flood Risk Manag.*, 9: 3–21, https://doi.org/10.1111/jfr3.12105.

Fatichi, S., Vivoni, E.R., Ogden, F.L., Ivanov, V.Y., Mirus, B., Gochis, D., Downer, C.W., Camporese, M., Davison, J.H., Ebel, B., Jones, N., Kim, J., Mascaro, G., Niswonger, R., Restrepo, P., Rigon, R., Shen, C., Sulis, M., and Tarboton, D. 2016. An overview of current applications, challenges, and future trends in distributed process-based models in hydrology, *J. Hydrol.*, 537: 45–60, https://doi.org/10.1016/j.jhydrol.2016.03.026.

Freeze, R.A., and Harlan, R.L. 1969. Blueprint for a physically-based ddigitally-simulated hydrologic response model, *J. Hydrol.*, 9: 237–258.

Govindaraju, R.S., and Kavvas, M.L. 1991. Dynamics of moving boundary overland flows over infiltrating surfaces at hillslopes, *Water Resour. Res.*, 27: 1885–1898, https://doi.org/10.1029/91WR00689.

Gunduz, O., and Aral, M.M. 2005. River networks and groundwater flow: A simultaneous solution of a coupled system, *J. Hydrol.*, 301: 216–234, https://doi.org/10.1016/j.jhydrol.2004.06.034.

Heppner, C.S., and Loague, K. 2008. A dam problem: Simulated upstream impacts for a Searsville-like watershed, *Ecohydrology*, 1: 408–424, https://doi.org/10.1002/eco.34.

Heppner, C.S., Ran, Q., VanderKwaak, J.E., and Loague, K. 2006. Adding sediment transport to the integrated hydrology model (InHM): Development and testing, *Adv. Water Resour.*, 29: 930–943, https://doi.org/10.1016/j.advwatres.2005.08.003.

Highfield, W.E., Norman, S.A., and Brody, S.D. 2013. Examining the 100-year floodplain as a metric of risk, loss, and household adjustment, *Risk Anal.*, 33: 186–191, https://doi.org/10.1111/j.1539-6924.2012.01840.x.

Horton, R.E. 1931. The field, scope, and status of the science of hydrology, *Trans. Am. Geophys. Union*, 12: 189, https://doi.org/10.1029/TR012i001p00189-2.

Hutchins, M.G., McGrane, S.J., Miller, J.D., Hagen-Zanker, A., Kjeldsen, T.R., Dadson, S.J., and Rowland, C.S. 2017. Integrated modeling in urban hydrology: Reviewing the role of monitoring technology in overcoming the issue of 'big data' requirements, *Wiley Interdiscip. Rev. Water*, 4: e1177, https://doi.org/10.1002/wat2.1177.

Ivanov, V.Y., Vivoni, E.R., Bras, R.L., and Entekhabi, D. 2004. Catchment hydrologic response with a fully distributed triangulated irregular network model, *Water Resour. Res.*, 40: 1–23, https://doi.org/10.1029/2004WR003218.

Kampf, S.K., and Burges, S.J. 2007. A framework for classifying and comparing distributed hillslope and catchment hydrologic models, *Water Resour. Res.*, 43, https://doi.org/10.1029/2006WR005370.

Kim, J., Warnock, A., Ivanov, V.Y., and Katopodes, N.D. 2012. Coupled modeling of hydrologic and hydrodynamic processes including overland and channel flow, *Adv. Water Resour.*, 37: 104–126, https://doi.org/10.1016/j.advwatres.2011.11.009.

Kollet, S.J., and Maxwell, R.M. 2008a. Capturing the influence of groundwater dynamics on land surface processes using an integrated, distributed watershed model, *Water Resour. Res.* 44, https://doi.org/10.1029/2007WR006004.

Kollet, S.J., and Maxwell, R.M. 2008b. Demonstrating fractal scaling of baseflow residence time distributions using a fully-coupled groundwater and land surface model, *Geophys. Res. Lett.* 35: 1–6, https://doi.org/10.1029/2008GL033215.

Kollet, S.J., Maxwell, R.M., Woodward, C.S., Smith, S., Vanderborght, J., Vereecken, H., and Simmer, C. 2010. Proof of concept of regional scale hydrologic simulations at hydrologic resolution utilizing massively parallel computer resources, *Water Resour. Res.*, 46: 1–7, https://doi.org/10.1029/2009WR008730.

Kollet, S.S.J., and Maxwell, R.R.M. 2006. Integrated surface – Groundwater flow modeling : A free-surface overland flow boundary condition in a parallel groundwater flow model, *Adv. Water Resour.*, 29: 945–958, https://doi.org/10.1016/j.advwatres.2005.08.006.

Kuiper, L.K. 1981. A comparison of the incomplete Cholesky-Conjugate Gradient Method with the strongly implicit method as applied to the solution of two-dimensional groundwater flow equations, *Water Resour. Res.*, 17: 1082–1086, https://doi.org/10.1029/WR017i004p01082.

Kundzewicz, Z.W. 2002. Non-structural flood protection and sustainability, *Water Int.*, 27: 3–13, https://doi.org/10.1080/02508060208686972.

Lindner, J., and Fitzgerald, S. 2018. *Immediate Report – Final Hurricane Harvey - Storm and Flood Information*. Harris Cty. Flood Control Dist., Houston, TX, USA.

Martínez, J.B. 1989. *Simulación Matemática de Cuencas Subterráneas, Flujo Impermanente Bidimensional*, CIH, ISPJAE, Ciudad La Habana, Cuba.

Maxwell, R.M., Putti, M., Meyerhoff, S., Delfs, J.-O., Ferguson, I.M., Ivanov, V., Kim, J., Kolditz, O., Kollet, S.J., Kumar, M., Lopez, S., Niu, J., Paniconi, C., Park, Y.-J., Phanikumar, M.S., Shen, C., Sudicky, E.A., and Sulis, M. 2014. Surface-subsurface model intercomparison: A first set of benchmark results to diagnose integrated hydrology and feedbacks, *Water Resour. Res.*, 50: 1531–1549, https://doi.org/10.1002/2013WR013725.

Meire, D., Doncker, L., Declercq, F., Buis, K., Troch, P., and Verhoeven, R. 2010. Modelling river-floodplain interaction during flood propagation, *Nat. Hazards*, 55: 111–121, https://doi.org/10.1007/s11069-010-9554-1.

Mirus, B.B., and Loague, K. 2013. How runoff begins (and ends): Characterizing hydrologic response at the catchment scale, *Water Resour. Res.*, 49: 2987–3006, https://doi.org/10.1002/wrcr.20218.

Panday, S., and Huyakorn, P.S. 2004. A fully coupled physically-based spatially-distributed model for evaluating surface/subsurface flow, *Adv. Water Resour.*, 27: 361–382, https://doi.org/10.1016/j.advwatres.2004.02.016.

Paniconi, C., and Putti, M. 2015. Physically based modeling in catchment hydrology at 50: Survey and outlook, *Water Resour. Res.* 51: 7090–7129, https://doi.org/10.1002/2015WR017780.

Park, Y.J., Sudicky, E.A., Brookfield, A.E., and Jones, J.P. 2011. Hydrologic response of catchments to precipitation: Quantification of mechanical carriers and origins of water, *Water Resour. Res.*, 47: 1–11, https://doi.org/10.1029/2010WR010075.

Park, Y.J., Sudicky, E.A., Panday, S., and Matanga, G. 2009. Implicit subtime stepping for solving nonlinear flow equations in an integrated surface-subsurface system, *Vadose Zo. J.*, 8: 825–836, https://doi.org/10.2136/vzj2009.0013.

Pathiraja, S., Westra, S., and Sharma, A. 2012. Why continuous simulation? the role of antecedent moisture in design flood estimation, *Water Resour. Res.*, 48: 1–15, https://doi.org/10.1029/2011WR010997.

Pokrajac, D., and de Lemos, M.J.S. 2014. A coupled surface-subsurface model of overbank flood flow and air entrapment in a permeable floodplain. In: *Proc. eight IAHR River Flow Conf.*, Lausanne, Switzerland, pp. 591–596.

Qu, Y., and Duffy, C.J. 2007. A semidiscrete finite volume formulation for multiprocess watershed simulation, *Water Resour. Res.*, 43: 1–18, https://doi.org/10.1029/2006WR005752.

Rajib, A., and Merwade, V. 2017. Hydrologic response to future land use change in the Upper Mississippi River Basin by the end of 21st century, *Hydrol. Process.*, 31: 3645–3661, https://doi.org/10.1002/hyp.11282.

Rawls, W.J., and Brakensiek, D.L. 1982. Estimating soil water retention from soil properties, *J. Irrig. Drain. Div.*, 108: 166–171.

Saksena, S. 2015. Investigating the role of DEM resolution and accuracy on flood inundation mapping. In: *World Environmental and Water Resources Congress 2015: Floods, Droughts, and Ecosystems - Proceedings of the 2015 World Environmental and Water Resources Congress*, pp. 2236–2243, https://doi.org/10.1061/9780784479162.220.

Saksena, S. 2019. *Integrated Flood Modeling for Improved Understanding of River-Floodplain Hydrodynamics: Moving beyond Traditional Flood Mapping*, Purdue University Graduate School, https://doi.org/10.25394/PGS.8984219.v1.

Saksena, S., Dey, S., Merwade, V., and Singhofen, P.J. 2020. A computationally efficient and physically based approach for urban flood modeling using a flexible spatiotemporal structure, *Water Resour. Res.*, 56: 1–22, https://doi.org/10.1029/2019WR025769.

Saksena, S., and Merwade, V. 2015. Incorporating the effect of DEM resolution and accuracy for improved flood inundation mapping, *J. Hydrol.*, 530: 180–194, https://doi.org/10.1016/j.jhydrol.2015.09.069.

Saksena, S., and Merwade, V. 2017. Integrated modeling of surface-subsurface processes to understand river-floodplain hydrodynamics in the Upper Wabash River Basin. In: *World Environmental and Water Resources Congress 2017. American Society of Civil Engineers*, Reston, VA, pp. 60–68, https://doi.org/10.1061/9780784480595.006.

Saksena, S., Merwade, V., and Singhofen, P.J. 2019. Flood inundation modeling and mapping by integrating surface and subsurface hydrology with river hydrodynamics, *J. Hydrol.*, 575: 1155–1177, https://doi.org/10.1016/j.jhydrol.2019.06.024.

Schoups, G., Hopmans, J.W., Young, C.A., Vrugt, J.A., Wesley, W., Tanji, K.K., and Panday, S. 2005. Sustainability of irrigated agriculture in the San Joaquin Valley, California, *Proc. Natl. Acad. Sci. U. S. A.*, 102: 15352–15356.

Segond, M.L., Wheater, H.S., and Onof, C. 2007. The significance of spatial rainfall representation for flood runoff estimation: A numerical evaluation based on the Lee catchment, UK, *J. Hydrol.*, 347: 116–131, https://doi.org/10.1016/j.jhydrol.2007.09.040.

Shen, C., and Phanikumar, M.S. 2010. A process-based, distributed hydrologic model based on a large-scale method for surface-subsurface coupling, *Adv. Water Resour.*, 33: 1524–1541, https://doi.org/10.1016/j.advwatres.2010.09.002.

Streamline Technologies. 2018. *ICPR4 Technical Reference Manual*. Streamline Technol. Inc., Winter Springs, Florida, www.streamnologies.com/misc/ICPR4_DOCS.zip.

Te Chow, V. 1959. *Open Channel Hydraulics*, McGraw-Hill B. Company, Inc., https://doi.org/10.1016/B978-0-7506-6857-6.X5000-0.

Todini, E. 2007. Hydrological catchment modelling: Past, present and future, *Hydrol. Earth Syst. Sci.*, 11: 468–482, https://doi.org/10.5194/hess-11-468-2007.

VanderKwaak, J.E. 1999. Numerical Simulation of Flow and Chemical Transport in Integrated Surface-Subsurface Hydrologic Systems, Univ. Waterloo, Ontario, Canada.

VanderKwaak, J.E., and Loague, K. 2001. Hydrologic-response simulations for the R-5 catchment with a comprehensive physics-based model, *Water Resour. Res.*, 37: 999–1013, https://doi.org/10.1029/2000WR900272.

Viero, D.P., Peruzzo, P., Carniello, L., and Defina, A. 2014. Integrated mathematical modeling of hydrological and hydrodynamic response to rainfall events in rural lowland catchments, *Water Resour. Res.*, 50: 5941–5957, https://doi.org/10.1002/2013WR014293.

Vivoni, E., Entekhabi, D., Bras, R.L., and Ivanov, V. 2007. Controls on runoff generation and scale-dependence in a distributed hydrologic model, *Hydrol. Earth Syst. Sci.*, 11: 1683–1701, https://doi.org/10.5194/hessd-4-983-2007.

Winter, T.C., Harvey, J.W., Franke, O.L., and Alley, W.M. 1999. *Ground Water and Surface Water: A Single Resource*, U.S. Geological Survey Circular 1139, USA.

Yeh, G.-T., Huang, G., Cheng, H.-P., Zhang, F., Lin, H.-C., Edris, E., and Richards, D. 2006. *A First-Principle, Physics-Based Watershed Model: WASH123D. Watershed Model*, edited by Singh, V.P., Frevert, D.K., CRC Press, Boca Raton, FL, pp. 210–244, https://doi.org/10.1201/9781420037432.ch9.

Zotarelli, L., Dukes, M.D., Romero, C.C., Migliaccio, K.W., and Morgan, K.T. 2013. *Step by Step Calculation of the Penman-Monteith Evapotranspiration (FAO-56 Method)*, Univ. Florida AE459, USA, pp. 1–10.

15 Continuous Large-Scale Simulation Models in Flood Studies

*Rouzbeh Nazari, Md Golam Rabbani Fahad,
Maryam Karimi, and Saeid Eslamian*

CONTENTS

15.1 INTRODUCTION

Flooding, a major calamity during extreme storm events, is a natural hydrological event that is becoming more frequent due to global warming, changes in precipitation patterns, and sea-level rise (Fahad et al., 2018; Hirabayashi et al., 2013; Tebaldi et al., 2012). Coastal and river floods cause significant damage to public properties, infrastructures, and services (Arrighi et al., 2013; Hatzikyriakou and Lin, 2017; Selvanathan et al., 2018). Elevated water levels and waves generated due to hurricanes as well as nor'easters are the leading cause of coastal flooding in the Mid-Atlantic United States (Schwartz, 2007). Two hundred and ninety-four North Atlantic hurricanes have originated since 1851 producing hurricane-force winds in 19 states along the Atlantic coast (Landsea and Franklin, 2013), causing extensive damage to the infrastructures as well as loss of lives. Recent hurricanes such as Irene, Sandy, Matthew, Harvey, Maria, Irma, Florence, Michael, etc. highlighted the devastating impact of flooding during extreme storms.

Studies show that about 6.6 million properties in the USA, combining an economic value of nearly $1.5 trillion, are at risk from extreme storm events (Botts et al., 2015). This significant exposure to losses is expected to increase in the future (Blake et al., 2013; Hirabayashi et al., 2013). For example, coastal regions such as New Jersey and New York City in the US are expected to face a 400-year flooding event similar to Hurricane Sandy to a ~130-year event by the end of the 21st century (Lin et al., 2016). Considering this extensive and increasing risk, fine-scale and accurate assessment of flooding due to storm surge is significant. Efficient and sustainable flood management heavily relies on the holistic assessments of flood events and their consequences on the biological and built environment.

Over the past few decades, numerous efforts have been made in the field of flood risk assessment focusing on varying spatial scales and final objectives. Access to high-performance computing,

DOI: 10.1201/9780429463938-20

availability of fine-scale data, and innovative visualization techniques enable modern methods of flood risk management at different spatial scales (i.e., macro, meso, and micro levels). Developing national or regional disaster mitigation policy necessitates large-scale flood risk analysis as well as comprehensive risk assessments for the insurance industry. National scales models have been developed in different parts of the world utilizing large-scale flood modeling (Jongman et al., 2012). The FLEMO model developed in Germany has been used for scientific flood risk analyses from the local to national scale (Apel et al., 2004; Vorogushyn et al., 2012). The Rhine Atlas damage model (RAM) (ICPR, 2001) was developed to identify flood risk and increase flood awareness in the Rhine basin, Germany. Jonkman et al. (2008) developed an integrated hydrodynamic model using SOBEK 1D-2D (Delft Hydraulics, 2003) and an economic model to assess the flood damage in the Netherlands. To assess flood risk under climate change, a damage model has been developed for pan-European regions (Huizinga, 2007). Alfieri et al. (2014) have developed another pan-European flood hazard map at 100 m resolution using Lisflood-ACC. In China, powerful spatial analytic tools have been utilized in developing the GIS-based risk assessment models for flood disasters. A common approach for large-scale flood risk assessment involves the discharge calculation for the river network based on a spatially uniform T-year return period. A few examples of practical implementation of this methodology include the damage assessment for flooding in the river Rhine (Thieken et al., 2015), peak discharge analysis for a 26,000 km river in Austria as discussed by Merz et al. (2008), national scale floodplain mapping from Bradbrook et al. (2005), etc.

The spatially homogeneous return period assumption is certainly useful for small-scale hazard assessment but falls short for the large-scale flood risk assessment where the non-uniform nature of weather parameters, variable landscape, and river characteristics plays a crucial role in flood origin and propagation. The current alternative approach for large-scale flood risk assessment involves coupling the storm surge and riverine flooding through different sets of hydrologic, hydraulic, and storm surge modeling tools driven by the observed climate data or scenarios. Flooding in coastal regions is largely dominated by the wave action and storm surge generated during extreme storm events. Continuous modeling of flood hazards involves a two-step process where, in the first step, a storm surge modeling will be applied to generate the possible water level along the coast which works as a boundary condition for the second step which is inland flood modeling (Bazrkar et al., 2017).

Table 15.1 shows a classification of the hydraulic models considering the dimensionality of a solution algorithm. Although flow characteristics in a river can be explained by one-dimensional (1D) modeling, spatial variation of flood propagation, inundation depth, and flow velocity can be adequately represented by two-dimensional (2D) hydrodynamic modeling. In recent years, the popularity of the 2D hydrodynamic model has increased substantially and TUFLOW (Syme, 2006; Huxley, 2004; Lhomme et al., 2008; Phillips et al., 2005) is one of the most applied models in this area.

Considering the coupled effect of storm surge and riverine flooding, ADvanced CIRCulation (ADCIRC) along the east coast of the US and a 2D hydrodynamic model (TUFLOW) were implemented for a robust flood hazard assessment tool.

15.2 STUDY AREA, DATA, AND METHODOLOGY

To evaluate the proposed approach, in this study, the Brick Township in northeastern Ocean County in the state of New Jersey (NJ), USA, was selected (Figure 15.1).

According to the United States Census, Brick Township is one of the largest municipalities and ranks third in terms of population in Ocean County, NJ (US Census Bureau, 2010). Brick Township consists of 25.71 square miles of land and 6.60 square miles of water, totaling an area of 32.31 square miles. Although most of this township is situated on the mainland, three ocean beaches are located on the Barnegat Peninsula separating Barnegat Bay from the Atlantic Ocean. The township is surrounded by five major watershed areas (i.e., Manasquan and Metedeconk River, Beaver Dam, Kettle and Reedy

TABLE 15.1

Summary of the Numerical Tools for Flood Modeling and Their Potential Application

Method	Description	Software Examples	Potential Application
0D	No physical laws	ArcGIS, Delta mapper	Broad-scale assessment of flood extents and flood depths
1D	Solution of the 1D equations	Mike 11, HEC-RAS	Design scale modeling, which can be of the order of tens to hundreds of km depending on catchment size
1D+	1D plus a flood storage cell approach flow	Mike 11, HEC-RAS	Design scale modeling, which can be of the order of tens to hundreds of km depending on catchment size, also has the potential for broad-scale application if used with sparse cross-sectional data
2D–	2D minus the law of conservation of momentum for the floodplain flow	LISFLOOD-FP, CA model	Large-scale modeling or urban inundation depending on cell dimensions
2D	Solution of the 2D shallow wave equations	TUFLOW, Mike 21, TELEMAC, DIVAST	Design scale modeling of the order of tens of km. May have the potential for use in broad-scale modeling if applied with coarse grids
2D+	2D plus a solution for vertical velocities using continuity only	TELEMAC 3D	Predominantly coastal modeling applications where 3D velocity profiles are important. Has also been applied to reach scale river modeling problems in research projects

Source: Teng et al., 2017.

Creek watersheds). Brick Township has a high density of waterfront property in New Jersey and has access to several major state highways, traveling northeast to southwest through the central portion of the township. This township was severely impacted during Hurricane Sandy in 2012, and it was on mandatory evacuation for all residents during the event (SRPR, 2012; Fahad et al., 2020).

Being an ocean-front township, Brick is a place for tourism and recreational activity, containing the largest amount of waterfront property of any township in New Jersey. Development patterns within the township are mostly suburban oriented with single-family detached residential properties being the most common type of housing. During Hurricane Sandy, Brick Township experienced severe structural damage or destruction due to the direct impact of wave energy and storm surge (SRPR, 2012).

15.3 MODEL FORMULATION IN ADCIRC

The current trend in coastal ocean tidal modeling by utilizing the larger computational domains has been demonstrated by previous research (Westerink et al., 1994, 1995).

Studies conducted by Flather (1988), Vincent and Le Provost (1988), Hagen and Parrish (2004), and recently Cialone et al. (2017) and Bacopoulos and Hagen (2017) have all implemented tidal and/ or storm surge models considering a large portion of the North Atlantic region. These studies concluded that precise tidal predictions could be simulated using large computational domains through hydrodynamic modeling. Utilizing a large computational domain allows an accurate specification of boundary conditions in the deep ocean where flow is mostly linear, and tidal constituents can be defined precisely. The domain of the Western North Atlantic (WNAT) model encloses the Western North Atlantic Ocean, the Gulf of Mexico, and the Caribbean Sea (Figure 15.2a).

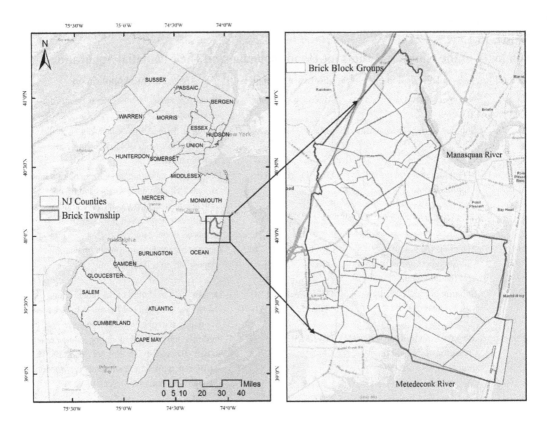

FIGURE 15.1 Map of study area, main river channels, and block groups (BG) for Brick Township, NJ, USA.

The open ocean boundary for the WNAT model domain lies along the 60° W meridian extending from the area of Glace Bay, Nova Scotia, Canada, to the vicinity of Corcoran Island in eastern Venezuela, and is situated entirely in the deep ocean (Figure 15.2a). This large computational domain covers an area of approximately 8.4 million km². An unstructured mesh was implemented to ensure a fine-scale resolution in shallow water regions where steep bathymetry and rapid change in gradient leads to complex geometry while allowing coarser but still adequate resolution further deep in the ocean. The grid consists of approximately 65,000 nodes (Figure 15.2b). The topography within the WNAT domain includes the continental shelf with a depth varying from 0 m to 130 m and the continental rise and deep ocean (depths from approximately 3,000 m to almost 8,300 m) as shown in Figure 15.2a. The required shoreline was obtained from the Global Self-Consistent, Hierarchical, High-Resolution Geography Database (GSHHG) (Wessel and Walter, 1996). The data was obtained from the National Centers for Environmental Information (NCEI) in ESRI shapefile format with WGS84 geographic horizontal datum (source: www. ngdc. noaa. gov/mgg/shorelines/).

Bathymetry data (ETOPO1) from the National Geophysical Data Center (NGDC) was available for the whole WNAT study region. The ETOPO1 1 arcminute (2.5 km) bathymetric dataset is a global relief model of Earth's surface that integrates land topography and ocean bathymetry (Amante and Eakins, 2009). NOAA's VDATUM (Parker et al., 2003) was used to convert the bathymetry data to the common vertical datum NAVD88.

Hourly observed water level data from NOAA was collected from October 21 to November 1, 2012, for 13 stations as shown in Figure 15.3 and used to validate the results from the ADCIRC simulation. The eight most important tidal constituents (M2, S2, N2, K2, K1, P1, O1, and Q1) were used as a tidal forcing along the open ocean boundary. The time step for the ADCIRC model was

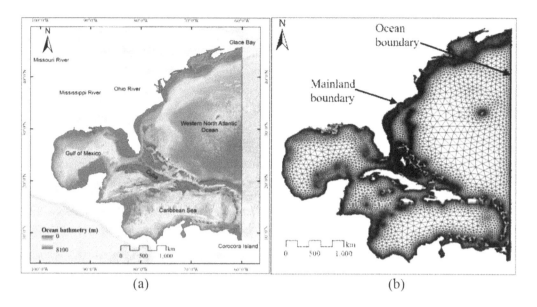

FIGURE 15.2 (a) WNAT model domain with ocean bathymetry and (b) unstructured mesh generated for ADCIRC simulation.

Stations	Latitude	Longitude
Montauk	41.05	-71.96
Kings Point	40.81	-73.77
Sandy Hook	40.47	-74.01
Atlantic City	39.36	-74.42
Cape May	38.97	-74.96
Ship John Shoal	39.31	-75.38
Brandywine Shoal Light	38.99	-75.11
Lewes	38.78	-75.12
Ocean City Inlet	38.33	-75.09
Wachapreague	37.61	-75.69
Kiptopeke	37.17	-75.99
Sewells Point	36.95	-76.33
Chesapeake	36.97	-76.11

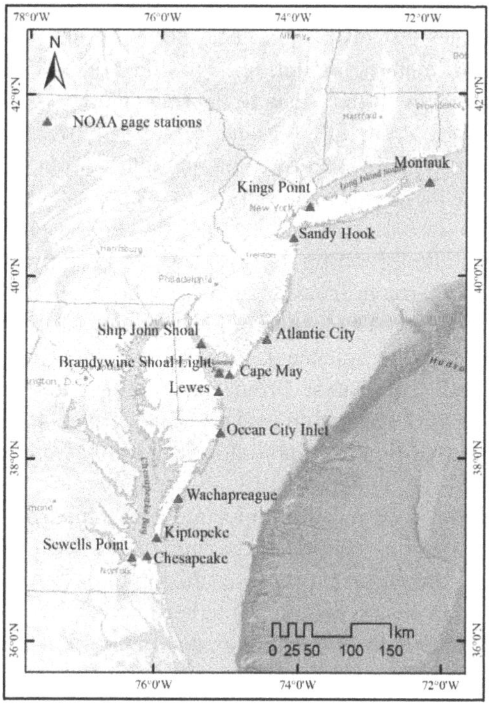

FIGURE 15.3 Locations of observed data from NOAA tidal gages for calibration and validation of ADCIRC model.

set to 5.0 seconds to achieve computational stability. The required Hurricane Sandy best track file was obtained from the National Hurricane Centre (NHC, 2018) database. The best track file was obtained in ATCF format, which contains the information on atmospheric pressure, radius of influence, and coordinates of the hurricane tracks as well as the timing of landfall.

15.4   MODEL FORMULATION IN SMS-TUFLOW

Fine-scale hydrodynamic modeling of a floodplain requires a tool, with the capability for large-scale flood modeling, to ensure sufficient detail concerning the depth and extent of a flood event. For this study, SMS-TUFLOW was utilized due to its capacity for large-scale flood modeling at a very high resolution, performing flood hazard analysis at a two-dimensional level, and support for spatial data processing and viewing capabilities. SMS, which is primarily a GIS-based system for developing, running, and processing water surface models using a wide variety of river and coastal hydraulics models. The two-dimensional solution algorithm for TUFLOW was based on Stelling (1983).

The SWE algorithms are based on the Navier-Stokes equations for the motion of fluid in a two-dimensional horizontal conserving the laws of mass and linear momentum in a Cartesian coordinate system (WBM TUFLOW, 2008). The SWE solution algorithm can adequately model the different hydrodynamic phenomena such as gravitational wave propagation, momentum transportation during advection, the effect of bottom friction, the Coriolis effect due to earth's rotation, changes in atmospheric pressure, etc. The main three components of data for hydrodynamic modeling include the topographic, land-use/land cover (LU/LC), and water level vs. time. The bathymetry data for the study area was obtained from Coastal National Elevation Database (CoNED) Project. Figure 15.4a shows the seamless topobathymetric data used for hydrodynamic modeling. Overall, Brick Township has a low-lying topography because of the adjacency to the ocean to the east. The west and northwest parts of Brick Township have higher elevations than the eastern region (Figure 15.4a). The topobathymetry used in the study also represents the accurate channel delineation within the study area as shown in Figure 15.4b for three different cross-sections (CS). There were ten high water marks (HWMs) within the model domain collected and verified by USGS, which were used to calibrate and validate the model simulated flooding during Hurricane Sandy. Once the topobathymetry data was prepared, the next major processing involves the LU/LC data. The land-use data was gathered from the New Jersey Department of Environmental Protection (NJDEP) by the Bureau of Geographical Information Science (GIS). The data was gathered in 2012 and used the categories: agriculture, barren land, forest, urban land, water, and wetlands. A vector clipping and projection conversion were conducted in ArcGIS according to the extent of our study area. The required boundary condition for inland flood modeling was extracted from the ADCIRC simulation. All of the DEM, bathymetry, and land-use data were imported into the SMS interface for the hydrodynamic modeling. An important step in hydrodynamic modeling is to define the mesh size of the 2D domain. The mesh size must fulfill two important criteria: (1) being fine enough to reproduce the physical processes during flooding, thus ensuring a stable model, and (2) reducing runtime to ensure computational efficiency. It is also recommended to provide fine meshing across the channel where complex flow conditions might occur during flooding. Considering all these criteria, a mesh size of 7.5 m was implemented to ensure the model stability and computational efficiency for the model domain. The model was also discretized by a three-second timestep to ensure numerical stability. An appropriate Manning's friction factor was implemented so that the flood flow reflects the real-world scenario in the channel as well as in the floodplain. Model calibration was performed by changing Manning's friction factor and the initial water level in the channel. We have utilized the Graphics Processing Unit (GPU) for parallel computing to achieve faster computation. The GPU had 2,560 CUDA cores, which runs well over 100 times faster than utilizing only a central processing unit (CPU) for hydrodynamic modeling.

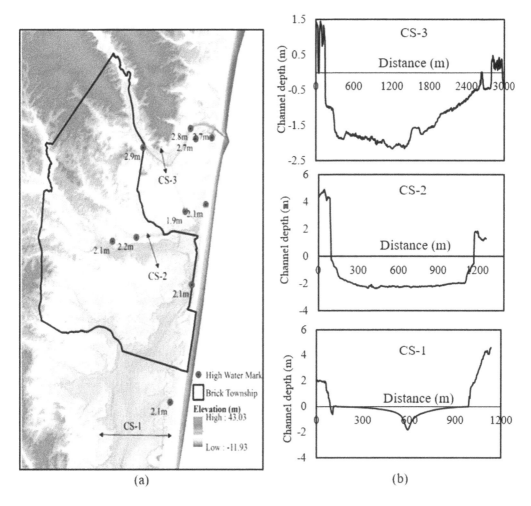

FIGURE 15.4 (a) Elevation and observed high water marks (HWM) during hurricane Sandy; (b) channel cross-section (CS) at three different locations in the study area.

15.5 RESULTS AND DISCUSSION

15.5.1 VALIDATION OF STORM SURGE MODELING

Model validation was performed to ensure that ADCIRC adequately predicts the hydrodynamics of the study area. The model accuracy is influenced by the accuracy of the forcing functions applied in the open ocean boundaries, accurate representation of the geometry of the study area (i.e., coastline and coastal bathymetry), and the values selected for model parameters such as wave continuity, bottom stress, etc. A satisfactory agreement between the predictions and measurements in the validation procedure ensures confidence that modeling represents the pertinent hydrodynamic process. The results for the model validation in this study were accomplished by comparing the observed data with the ADCIRC simulated data for 13 tidal stations from NOAA. In order to quantify the accuracy of the observed and model-simulated data, different statistical parameters such as Nash-Sutcliffe efficiency (NSE), root mean squared error (RMSE), mean absolute error (MAE), coefficient of determination (R-square), Pearson correlation coefficient, and ratio of standard deviations were used (Table 15.2). The purpose of using the various parameters to determine the model's efficiency was to ensure that results from the ADCIRC simulation were able to capture various

TABLE 15.2

Statistical Parameters to Compare the Observed and ADCIRC Simulation Results

	NSE	RMSE	MAE	R^2	Pearson Correlation	Ratio of Standard Deviations
Montauk	0.779	0.207	0.162	0.873	0.934	0.955
Kings Point	0.776	0.491	0.357	0.866	0.931	1.034
Sandy Hook	0.727	0.359	0.273	0.851	0.923	0.925
Atlantic City	0.743	0.290	0.206	0.833	0.913	0.958
Cape May	0.610	0.389	0.274	0.796	0.892	0.901
Ship John Shoal	0.370	0.516	0.391	0.706	0.840	1.032
Brandywine Shoal Light	0.779	0.274	0.193	0.795	0.891	0.856
Lewes	0.542	0.396	0.272	0.742	0.861	0.852
Ocean City Inlet	0.745	0.213	0.157	0.769	0.877	0.879
Wachapreague	0.747	0.266	0.203	0.784	0.885	0.946
Kiptopeke	0.510	0.302	0.227	0.626	0.792	0.899
Sewells Point	0.416	0.371	0.318	0.565	0.751	0.832
Chesapeake	0.805	0.212	0.179	0.919	0.959	0.953

characteristics of a flow such as an accurate representation of flood peaks, flow variability, as well as combined dispersion between the observed and model-simulated results.

Figure 15.5 exhibits the performance of the ADCIRC simulation in comparison with the observed water level data. ADCIRC simulation was able to represent the observed tidal pattern. The simulated normal tide was found to be slightly underestimating than that of observed data in high tide, which could be a consequence of slight overestimation of bottom friction in the ADCIRC model. The peak water level during Sandy was also well captured by the ADCIRC model.

The results show a good agreement between the observed peak and decay of water level in Atlantic City station near which Sandy made the landfall. More refined mesh and better bathymetric data near the shoreline could have solved the issue for those three stations mentioned above. Although the R^2 and Pearson correlation manifested that the ADCIRC model simulated water level accurately represents the tidal pattern, peak, and rise of the observed stations during the period of simulation concluding a better fit between the observed one and the model. The ratio of standard deviations also suggests that the variability of the observed one was well represented by the ADCIRC model, as they are close to one (Table 15.2).

15.5.2 Validation of 2D Hydrodynamic Modeling

The study considered the Hurricane Sandy event. The dynamic flood simulation contains information on flood depth at every 15-minute interval. Among all these time intervals, the maximum flooding was extracted from the results in terms of maximum flood depth and extent of flooding. This maximum flooding information from hydrodynamic modeling of Hurricane Sandy was compared to the observed Hurricane Sandy impact analysis from the official FEMA Modeling Task Force, also known as FEMA-MOTF (2014). This high-resolution (3 m) product was created based on mission-assigned, field-verified USGS HWM data, USGS surge sensor data, and three-meter USGS DEM. A simulated inundation map was masked to remove the physical waterbodies (i.e., rivers) to match with the USGS reanalysis product (Figure 15.6). The results indicated that the model predicts the spatial variation of inland flooding very well (Figure 15.6) with a small degree of overestimation (~0.61 m) near the river channel. The overall difference in the observed and model-simulated flooding information was calculated by subtracting and plotting them as a bias between the observed and

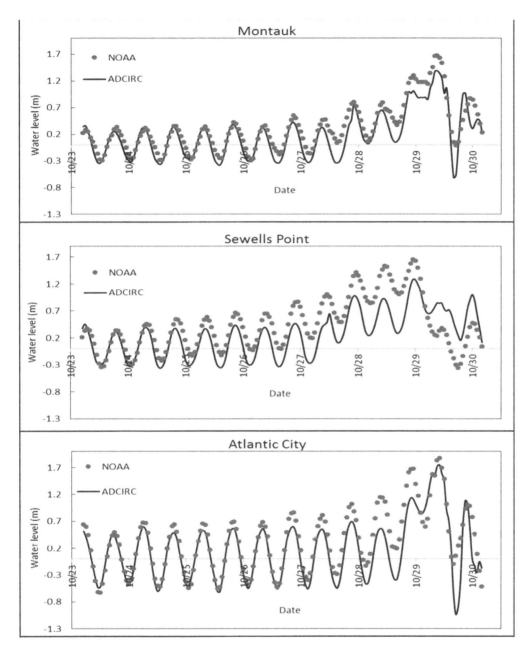

FIGURE 15.5 Example comparison of water level (m) from ADCIRC simulation with the observed data from October 23 to October 30, 2012.

simulated data. The model reasonably predicts the inland flooding with a bias up to approximately 0.2 m except for a small portion at the top east where it underestimates the inundation but also an area outside of Brick Township. Comparisons were also conducted by utilizing ten USGS HWMs that were collected and verified for the Sandy event. The correlation coefficient, RMSE, and ratio of the standard deviations were 0.79, 0.71, and 1.41, respectively. These indicate the good agreement between the observed and estimated model simulated flooding.

Several predictive skill tests were conducted to quantify similarity or agreement between the observed and model-simulated flood inundation extent (Saleh et al., 2017; Sampson et al., 2015;

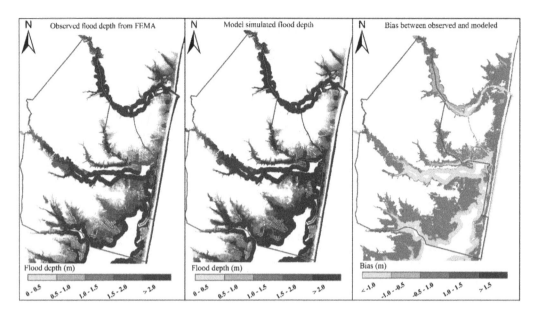

FIGURE 15.6 Comparison of observed and model-simulated inundation depth for Hurricane Sandy. The solid line outlines the admin boundary of Brick Township, NJ.

Schubert and Sanders, 2012; Bates and Roo, 2000). Similarities between two datasets were determined by the critical coefficient of similarity (F_a) as defined by equation 15.1:

$$F_a = \frac{E_p \cap E_m}{E_p \cup E_m} \tag{15.1}$$

Where E_p indicates the model simulated or predicted flood extent and E_m indicates the observed flood inundation extent. Dividing the intersection ($\cap$) and union ($\cup$) between to dataset shows the similarity of flooding in terms of inundation extent. A value of $F_a = 1$ corresponds to a perfect similarity between the predicted and observed inundation extent. Likewise, the metric for the overpredicted (F_{op}) and underpredicted (F_{up}) was quantified as follows (equations 15.2 and 15.3),

$$F_{op} = \frac{E_p - E_p \cap E_m}{E_p \cup E_m} \tag{15.2}$$

$$F_{up} = \frac{E_m - E_p \cap E_m}{E_p \cup E_m} \tag{15.3}$$

$F_{op} = F_{up} = 0$ indicates a perfect model validation with no over or under prediction for the observed inundation extent. Furthermore, three other metrics were derived, including the probability of detection (POD), false-alarm ratio (FAR), and critical success index (CSI), which were also used to further assess a model accuracy in replicating flooding during the Hurricane Sandy event (Saleh et al., 2017; Bhatt et al., 2017).

$$POD = \frac{hits}{hits + misses} \tag{15.4}$$

$$FAR = \frac{false\ alarms}{(hits + false\ alarms)} \tag{15.5}$$

$$CSI = \frac{hits}{hits + misses + false\ alarms} \tag{15.6}$$

POD (equation 15.4) describes a pixel-by-pixel analysis by comparing a simulated inundation map overlaid to the FEMA MOTF reanalysis product at 3 m spatial resolution. On the other hand, FAR (equation 15.5) indicates the fraction of simulated flood extent over the predicted by the model. CSI (equation 15.6), which is similar to the similarity metric, indicates the model performance when compared to the FEMA MOTF product. Table 15.3 summarizes the results of different metrics used in the study to validate model performance.

A random sampling analysis was performed to analyze the spatial variation of flood depth. Sixty random samples of flood depth were created over the FEMA MOTF inundation map as well as the model-simulated map using ArcGIS.

The values were compared to the model simulated flood depth corresponding to the randomly sampled points. The results indicated a satisfactory agreement ($R^2 = 0.9308$) between the two datasets in terms of flood depth (Figure 15.7a). Figure 15.7b exhibits the spatial distribution of performance

TABLE 15.3

Summary of Performance Metrics by Comparing Model Simulated Results with Observed FEMA MOTF Inundation Product

Metrics	F_a	F_{up}	F_{op}	POD	FAR	CSI
Value	96.49%	2.61%	0.89%	97.38%	0.91%	96.5%

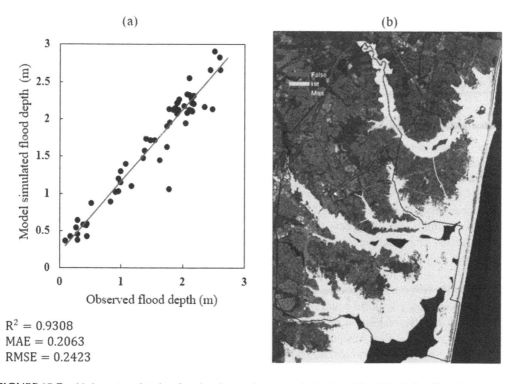

FIGURE 15.7 (a) A scatter plot showing the observed vs. model-simulated flood depth for 60 random points sampled within the boundary of the study area. (b) Spatial distribution of performance metrics compared with FEMA MOTF reanalysis product.

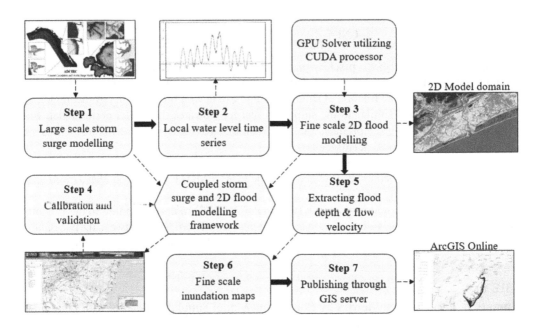

FIGURE 15.8 Implementation framework for high-resolution flood resiliency assessment through continuous simulation of extreme storm events.

metrics in terms of false alarms, hit, and miss from a pixel-by-pixel perspective. The analysis yielded a good agreement in predicted flood extent supported by POD = 97.38%, FAR = 0.91%, and CSI = 96.5% respectively.

Figure 15.8 represents the operational framework for this study. The process starts with the large-scale storm surge modeling in ADCIRC, followed by a high-resolution 2D hydrodynamic modeling to simulate the inland flooding. The results were disseminated using online geospatial analytics to develop a user-friendly decision-making framework.

15.6 CONCLUSIONS

A pair of state-of-the-art hydrodynamic models were used to study the hydrodynamic response to a major storm in the Western North Atlantic Domain, USA, as well as inland. The models reasonably reproduced the tides, storm surges, and large waves compared with a tide gauge. The resolution of the coastal bathymetry and shoreline is an important factor to represent the actual near coast characteristics through the ADCIRC modeling. A high-resolution unstructured grid, which greatly affects the hydraulics of tidal storm surge, is essential for capturing the actual storm scenarios. Although increasing the amount of mesh near the shoreline cloud increases the accuracy of the model results, it also requires an extensive computational capacity. The inland flood model also predicted the spatial and temporal variation of flooding quite well. However, a more accurate representation of topography, proper model parameterization, and boundary condition could certainly improve the models' performance in reducing the uncertainty. Coarse-resolution models may not exhibit satisfactory accuracy for micro-scale inundation mapping. Fine-resolution 2D hydrodynamic modeling could provide more accurate mapping of floods, flood levels, and flood hazards. Flood modeling at high resolution can reduce the uncertainties in predicting the flood behavior by accurate representation of flood depth, velocity, duration, and warning time especially in urban areas and floodplains where the flow patterns are complex. The capacity of high-resolution flood modeling in terms of temporal resolution could be utilized to build a robust framework for evacuation planning as well.

ACKNOWLEDGMENTS

This work was supported through the New Jersey Department of Community Affairs (NJDCA) from the Superstorm Sandy Community Development Block Grant Disaster Recovery (CDBG-DR); the Center for Advanced Infrastructure and Transportation under grant number 69A3551847102 from the US Department of Transportation, Office of the Assistant Secretary for Research and Technology (OST-R); and the National Science Foundation under award number NSF DUE-1610911. We also acknowledge the World Climate Research Programme's Working Group on Coupled Modeling, which is responsible for CMIP. For CMIP the US Department of Energy's Program for Climate Model Diagnosis and Inter-Comparison provides coordinating support and led the development of software infrastructure in partnership with the Global Organization for Earth System Science Portals.

REFERENCES

Alfieri, L., Salamon, P., Bianchi, A., Neal, J., Bates, P., and Feyen, L. 2014. Advances in pan-European flood hazard mapping, *Hydrological Processes*, 28(13): 4067–4077.

Amante, C., and Eakins, B.W. 2009. ETOPO1 1 arc-minute global relief model: Procedures, data sources and analysis. NOAA Technical Memorandum NESDIS NGDC-24. National Geophysical Data Center, NOAA, 10, V5C8276M, USA.

Apel, H., Thieken, A.H., Merz, B., and Bloschl, G. 2004. Flood risk " assessment and associated uncertainty, *Natural Hazards and Earth System Sciences*, 4: 295–308, https://doi.org/10.5194/nhess-4-295-2004.

Arrighi, C., Brugioni, M., Castelli, F., Franceschini, S., and Mazzanti, B. 2013. Urban micro-scale flood risk estimation with parsimonious hydraulic modelling and census data, *Natural Hazards and Earth System Sciences*, 13(5): 1375–1391.

Bacopoulos, P., and Hagen, S.C. 2017. The intertidal zones of the South Atlantic Bight and their local and regional influence on astronomical tides, *Ocean Modelling*, 119: 13–34.

Bates, P.D., and De Roo, A.P.J. 2000. A simple raster-based model for flood inundation simulation, *Journal of Hydrology*, 236(1–2): 54–77.

Bazrkar, M.H., Adamowski, J., and Eslamian, S. 2017. Water system modeling. In: Furze, J.N., Swing, K., Gupta, A.K., McClatchey, R., Reynolds, D. (eds) *Mathematical Advances Towards Sustainable Environmental Systems*, Springer International Publishing, Switzerland, pp. 61–88.

Bhatt, C.M., Rao, G.S., Diwakar, P.G., and Dadhwal, V.K. 2017. Development of flood inundation extent libraries over a range of potential flood levels: A practical framework for quick flood response, *Geomatics, Natural Hazards and Risk*, 8(2): 384–401.

Blake, E.S., Kimberlain, T.B., Berg, R.J., Cangialosi, J.P., and Beven Ii, J.L. 2013. *Tropical Cyclone Report: Hurricane Sandy*, vol. 12, National Hurricane Center, USA, pp. 1–10

Botts H, Jeffery T, Du W, and Suhr L 2015. *CoreLogic Storm Surge Report*, CoreLogic, California, USA.

Bradbrook, K., Waller, S., and Morris, D. 2005. National floodplain mapping: Datasets and methods–160,000 km in 12 months, *Natural Hazards*, 36(1–2): 103–123.

Cialone, M.A., Grzegorzewski, A.S., Mark, D.J., Bryant, M.A., and Massey, T.C. (2017). Coastal-storm model development and water-level validation for the North Atlantic Coast Comprehensive Study, *Journal of Waterway, Port, Coastal, and Ocean Engineering*, 143(5): 04017031.

Delft Hydraulics. 2003. *Delft Hydraulics HIS Overstromings Module*, Delft, the Netherlands, www.sobek.nl.

Fahad, M.G.R., Nazari, R., Motamedi, M.H., and Karimi, M.E. 2020. Coupled hydrodynamic and geospatial model for assessing resiliency of coastal structures under extreme storm scenarios, *Water Resources Management*, 34(3): 1123–1138.

Fahad, M.G.R., Saiful Islam, A.K.M., Nazari, R., Alfi Hasan, M., Tarekul Islam, G.M., and Bala, S.K. 2018. Regional changes of precipitation and temperature over Bangladesh using bias-corrected multi-model ensemble projections considering high-emission pathways, *International Journal of Climatology*, 38(4): 1634–1648.

FEMA Modeling Task Force. 2014. FEMA MOTF hurricane sandy impact analysis. *Federal Emergency Management Agency Modeling Task Force*, accessed 14 December 2018. Available online at https://data.femadata.com/MOTF/.

Flather, R.A. 1988. A numerical model investigation of tides and diurnal-period continental shelf waves along Vancouver Island, *Journal of Physical Oceanography*, 18(1): 115–139.

Hagen, S.C., and Parrish, D.M. 2004. Meshing requirements for tidal modeling in the western North Atlantic, *International Journal of Computational Fluid Dynamics*, 18(7): 585–595.

Hatzikyriakou, A., and Lin, N. 2017. Simulating storm surge waves for structural vulnerability estimation and flood hazard mapping, *Natural Hazards*, 89(2): 939–962.

Hirabayashi, Y., Mahendran, R., Koirala, S., Konoshima, L., Yamazaki, D., Watanabe, S., ... and Kanae, S. 2013. Global flood risk under climate change. *Nature Climate Change*, 3(9): 816–821.

Huizinga, H.J. 2007. Flood damage functions for EU member states, HKV Consultants, Implemented in the framework of the contract# 382442-F1SC awarded by the European Commission–Joint Research Centre. Tech. Rep., European Commission-Joint Research Center.

Huxley, C.D. 2004. *TUFLOW Testing and Validation*, Griffith University, Brisbane, Australia.

ICPR. 2001. *Atlas of Flood Danger and Potential Damage due to Extreme Floods of the Rhine*, International Commission for the Protection of the Rhine, Koblenz, Germany.

Jongman, B., Kreibich, H., Apel, H., Barredo, J.I., Bates, P.D., Feyen, L., ... and Ward, P.J. 2012. Comparative flood damage model assessment: Towards a European approach, *Natural Hazards and Earth System Sciences*, 12(12): 3733–3752.

Jonkman, S.N., Bočkarjova, M., Kok, M., and Bernardini, P. 2008. Integrated hydrodynamic and economic modelling of flood damage in the Netherlands, *Ecological Economics*, 66(1): 77–90.

Landsea, C.W., and Franklin, J.L. 2013. Atlantic hurricane database uncertainty and presentation of a new database format, *Monthly Weather Review*, 141(10): 3576–3592.

Lhomme, J., Sayers, P., Gouldby, B., Wills, M., and Mulet-Marti, J. 2008. Recent development and application of a rapid flood spreading method, *Flood Risk Management: Research and Practice*, 15–24, https://doi.org/10.1201/9780203883020.ch2.

Lin, N., Kopp, R.E., Horton, B.P., and Donnelly, J.P. 2016. Hurricane Sandy's flood frequency increasing from year 1800 to 2100, *Proceedings of the National Academy of Sciences*, 113(43): 12071–12075.

Merz, R., Blöschl, G., and Humer, G. 2008. National flood discharge mapping in Austria, *Natural Hazards*, 46(1): 53–72.

National Hurricane Center. 2018. Costliest US tropical cyclones tables updated. NOAA Tech. Memo. NWS NHC-6, National Hurricane Center, https://www.nhc.noaa.gov/news/UpdatedCostliest.pdf.

Parker, B., Milbert, D., Hess, K., and Gill, S. 2003. National VDatum–The implementation of a national vertical datum transformation database. In: *Proceeding from the US Hydro, Conference*, Biloxi, MS, USA, pp. 24–27.

Phillips, B.C., Yu, S., Thompson, G.R., and De Silva, N. (2005, August). 1D and 2D Modelling of Urban Drainage Systems using XP-SWMM and TUFLOW. In: *10th International Conference on Urban Drainage*, pp. 21–26. Copenhagen, Denmark.

Saleh, F., Ramaswamy, V., Wang, Y., Georgas, N., Blumberg, A., and Pullen, J. 2017. A multi-scale ensemble-based framework for forecasting compound coastal-riverine flooding: The Hackensack-Passaic watershed and Newark Bay, *Advances in Water Resources*, 110: 371–386.

Sampson, C.C., Smith, A.M., Bates, P.D., Neal, J.C., Alfieri, L., and Freer, J.E. 2015. A high-resolution global flood hazard model, *Water Resources Research*, 51(9): 7358–7381.

Schubert, J.E., and Sanders, B.F. 2012. Building treatments for urban flood inundation models and implications for predictive skill and modeling efficiency, *Advances in Water Resources*, 41: 49–64.

Schwartz, R. 2007. *Hurricanes and the Middle Atlantic States*, Blue Diamond Books, Springfield, VA, USA.

Selvanathan, S., Sreetharan, M., Lawler, S., Rand, K., Choi, J., and Mampara, M. 2018. A framework to develop nationwide flooding extents using climate models and assess forecast potential for flood resilience, *JAWRA Journal of the American Water Resources Association*, 54(1): 90–103.

Stelling, G.S. 1983. *On the Construction of Computational Methods for Shallow Water Flow Problems*. *Rijkswaterstaat Communications*, vol. 35, Government Printing Office, The Hague, The Netherlands.

Strategic Recovery Planning Report (SRPR). Post Sandy Planning Assistance Grant Program. New Jersey Department of Community Affairs (NJDCA). 2012. Available at https://www.nj.gov/dca/services/lps/SR PRs/Brick_SRPR.pdf.

Syme, B. 2006, July. 2D or not 2D? –An Australian perspective. In: *Proceedings of the Defra Flood and Coastal Risk Management Conference*, pp. 4–6, York, UK.

Tebaldi, C., Strauss, B.H., and Zervas, C.E. 2012. Modelling sea level rise impacts on storm surges along US coasts, *Environmental Research Letters*, 7(1): 014032.

Teng, J., Jakeman, A.J., Vaze, J., Croke, B.F., Dutta, D., and Kim, S. 2017. Flood inundation modelling: A review of methods, recent advances and uncertainty analysis, *Environmental Modelling and Software*, 90: 201–216.

Thieken, A.H., Apel, H., and Merz, B. 2015. Assessing the probability of large-scale flood loss events: A case study for the river Rhine, Germany, *Journal of Flood Risk Management*, 8(3): 247–262.

U.S. Census Bureau. 2010. Profile of general population and housing chracterstics: 2010 for Brick Township, Ocean County, NJ, USA. Retrieved from https://factfinder.census.gov/faces/tableservices /jsf/pages/pro-ductview.xhtml?src=bkmk.

Vincent, P., and Le Provost, C. 1988. Semidiurnal tides in the northeast Atlantic from a finite element numeri-cal model, *Journal of Geophysical Research: Oceans*, 93(C1): 543–555.

Vorogushyn, S., Lindenschmidt, K.E., Kreibich, H., Apel, H., and Merz, B. 2012. Analysis of a detention basin impact on dike failure probabilities and flood risk for a channel-dike-floodplain system along the river Elbe, Germany, *Journal of Hydrology*, 436: 120–131.

WBM, BMT. 2008. *TUFLOW User Manual-GIS Based 2D/1D Hydrodynamic Modelling*. Report. Australia.

Wessel, P., and Smith, W.H. 1996. A global, self-consistent, hierarchical, high-resolution shoreline database, *Journal of Geophysical Research: Solid Earth*, 101(B4): 8741–8743.

Westerink, J.J., Luettich Jr, R.A., Blain, C.A., and Scheffner, N.W. 1994. ADCIRC: An advanced three-dimen-sional circulation model for shelves, coasts, and estuaries. Report 2. User's Manual for ADCIRC-2DDI (No. WES/TR/DRP-92-6-2). Army Engineer Waterways Experiment Station, Vicksburg, MS, USA.

Westerink, J.J., Luettich, R.A., Blain, C.A., and Hagen, S.C. 1995. Surface elevation and circulation in conti-nental margin waters. In: *Finite Element Modeling of Environmental Problems*, pp. 39–59, Wiley.

Part VI

Flood Software

16 Riverine and Flood Modeling Software

Mustafa Goodarzi and Saeid Eslamian

CONTENTS

16.1 INTRODUCTION

The increasing trend in mortality and financial losses caused by flooding over the past decades in the world has led water engineers and other relevant experts to rely on modern tools to take control of this natural phenomenon. On the other hand, it is clear that full control of floods is desirable, but it is impossible to do so, and it is only possible to minimize the damage caused by them. The most basic steps in flood management, flood control, flood damages estimation, and flood insurance premiums are to precisely determine the flood boundaries or floodplain zoning, which cannot be achieved except by hydraulic analysis. In this regard, computer models play an important role in these analyses, and by using these models, it is possible to simply determine the surface water profiles along the river and the floodway, for different flow intensities.

Humankind needs to find and analyze many of these natural phenomena according to their needs. In this regard, it is necessary to use the proper tool to maintain the realities of the natural phenomenon and made it possible to analyze by the human mind, this device is called a model. The term "model" can be referenced to any representation of a system. By this definition, even an image of a phenomenon will be a model of it. However, what is being inferred today from the model concept is a system that emulates or simulates the behavior of a real system. When the models are used to analyze a process, choosing the appropriate type of model is important. Choosing the wrong type of model can result in wasting costs and time or gaining inaccurate information. In general, models can be categorized into different types based on different perspectives. Based on one of the most common divisions, models are categorized into four groups: physical models, induction models, mathematical models, and computer models (Goodarzi and Absalan, 2014).

In fact, computer models are the same as mathematical models, but they are called computer models because these types of models are analyzed by a computer and a user interface software. Today, the use of these models is widespread, the reasons for which are the simplicity and low

DOI: 10.1201/9780429463938-22

cost of these models (Goodarzi and Absalan, 2014). Given the many advances made in computer sciences, today, a variety of computer models have been developed to solve the various problems in water engineering, including flood simulation. Due to the precise calculations and flexibility in changing flow parameters and lower cost than physical models, computer models have attracted more attention from researchers over the past few years, which has led to the development of different software for simulating the flood and low flow in the river basin. The present study attempts to introduce the flood modeling software to show the importance of their application in forecasting and estimating flood and river flows, their application in watersheds, and emphasis on their use more than before for flood management. Therefore, in this chapter, due to the importance of the application of flood and river analysis software, the most popular modeling software in the field of flood and river flow simulation have been introduced, and their effectiveness is discussed.

16.2 HEC-RAS SOFTWARE

The HEC-RAS software, or the River Analysis System (RAS) developed by the US Army Corps of Engineers, is a toolkit that allows the user to perform hydrodynamic calculations of the river in a steady unsteady flow. The HEC-RAS models the hydraulics of water flow through natural rivers and other channels. The ability of this software to model the various types of structures and the capability to analyze the flow in an unsteady state, as well as its powerful graphics, have made this software one of the most applicable types of water engineering software. The software consists of three components of one-dimensional hydraulic analysis for the calculation of the water surface profile in steady-state flow, simulation of unsteady flow, and the sediment transport calculations at the river bed (Goodarzi and Abessalan, 2014). The general computational process in steady flow is based on the solution of the one-dimensional energy equation. However, the momentum equation is also used in situations where the water surface profile changes rapidly, such as hydraulic jumps, hydraulic bridges, and river confluences. In unsteady flow, the software solves the full dynamic 1D Saint Venant equation, using an implicit finite difference method. This software has the ability to model a network of channels, a dendritic system, or only a river reach. This software is able to simulate subcritical, supercritical, and mixed flow regimes along with the effects of various structures such as bridges, culverts, weirs, etc. (Brunner, 2010). The HEC-RAS software has special applications in floodplain management and flood insurance studies to assess floodway encroachments. This software can be used for floodplain zoning, river bed floodway determination, river basin management, reservoir organization, and other similar uses. One of the important applications of the HEC-RAS software is floodplain zoning in order to determine the land use around the river and the division of the floodplain area into different risk zones. Floodplain zoning maps are also widely used in urban management studies. Therefore, by using the HEC-RAS software, it is possible to determine how high the water surface will be in the floodplain, and which areas are affected by it (Rezaei, 2018) (Figure 16.1).

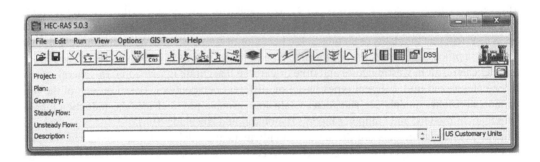

FIGURE 16.1 The startup window of the HEC-RAS software.

In general, for the purpose of floodplain zoning in the HEC-RAS software, the following basic steps are taken:

- Collecting hydrologic and hydraulic data (roughness coefficient, river status, etc.), topographic information (river bed and floodway profiles), flood discharge (flood hydrograph, discharge rating curve, etc.).
- Entering the required information in the HEC RAS software: the most important cases are including the introduction of river and floodway boundary plan, the riverbed and floodway roughness coefficient, the upstream and downstream boundary conditions, entering the required flow rate for flow analysis, introducing various structures in the flow path, and choosing the desired flow regime (subcritical, supercritical, or mixed).
- Running the model and determination of areas affected by the flood flow and providing the floodplain zoning maps (Figure 16.2).

The floodplain flow can be modeled for different return periods (5, 10, 20, 25, 50, 100, 200, 500, 1000, and PMF) by HEC-RAS software. Finally, after computations and modeling by the software, the output results maps can be entered into the ArcGIS environment, and by identifying the topographic level of the area, the inundated area of the river basin is determined. The produced maps could be used for various applications such as flood insurance premiums, rehabilitation of structures for floods, river bed boundaries, permitting for various land uses in the river basin, and so on.

16.3   HEC-GEORAS ADD-ON

The HEC-GeoRAS add-on is a set of methods, tools, and utilities for processing geospatial data on the ArcGIS environment using a friendly graphical user interface. This add-on creates a link between the ArcGIS and the HEC-RAS software, and is specially designed for the processing of geospatial data for use in modeling with the HEC-RAS and interchangeably for processing the HEC-RAS results in the ArcGIS environment (Hejazi et al., 2019). This add-on gives the user specific, logical, and simple access to the Geographical Information System (GIS), and the user can focus on the principles of hydraulic modeling in the HEC-RAS software. This add-on uses ArcGIS to create the input data for the HEC-RAS model from the digital terrain model (DTM) data and other GIS databases (ESRI). After the modeling results have been prepared in the HEC-RAS, they can be processed in the HEC-GeoRAS and create different outputs such as depth of flood, floodplain mapping, modeled velocity speed distribution map, an ice depth map, and a sediment transport map. The processing of geometric data and the other GIS data in the ArcGIS software allows the HEC-GeoRAS to create and export a special file for analysis in HEC-RAS. This file includes the stream centerline, flow path centerlines (optional), main channel banks (optional), cross-section cut

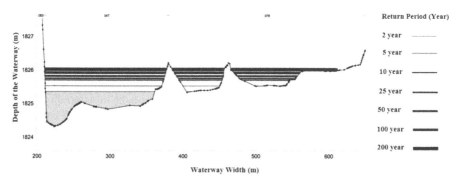

FIGURE 16.2   Cross-section profile and flood levels with different return periods determined by HEC-RAS software at a section of Taleghan River in Iran. (Source: Yamani et al., 2012.)

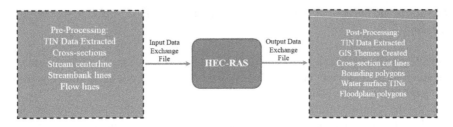

FIGURE 16.3 Evans' pre- and post-processing for HEC-RAS. (Source: Tate et al., 2002.)

lines, station information, cross-sections, Manning roughness coefficients, and more (Hejazi et al., 2019). The basic steps in working with this add-on as shown in Figure 16.3 include the construction of a regional DTM file, the introduction of the river banks and cross-sections, introducing the river's hydraulic characteristics, and providing inputs for the HEC-RAS program (Salajegheh et al., 2009; Tate and Maidment, 1999; Tate et al., 2002).

So far, the HEC-RAS has been successfully used in various flood investigations around the world (Knebl et al., 2005; Yamani et al., 2012; Lian et al., 2013; Mohammadi et al., 2014; Moya Quiroga et al., 2016; Rezaei, 2018; and Hejazi et al., 2019). For example, Rezaei et al. (2014) by using the combination of GIS and the HEC-RAS, while determining the floodplain extent in a part of the Moraghak River in Iran, identified reaches of the river with the highest flood extent for the return periods of 25 years. Rezaei (2018), in research using the combination of GIS and the HEC-RAS, simulated the hydraulic behavior of the Goharud River in Iran and determined the floodplain area of the river. Based on the results of this study, the study area is divided into three zones with low, moderate, and high risk.

Hejazi et al. (2019) in a study using the HEC-RAS software, determined the floodplain surface level for the 25 and 50 years return periods along the Varkesh Chai River in East Azarbaijan, Iran. Based on the results, 100 square kilometers of the entire catchment area are affected by floods with a return period of 50 years and 63 kilometers from floods with a return period of 25 years. Napradean and Chira (2006) used the combination of the Wetspa and HEC-RAS software to identify floodplain zones in the catchment areas near the Usturoi Valley, Romania. For this, the WetSpa software outputs were used as inputs for the HEC-RAS software and identified the floodplain zones and obtained a flood risk map. Moya Quiroga et al. (2016) simulated a Bolivian Amazonian flood event in February 2014, used the HEC-RAS software, and compared the results with the satellite imagery of the flood event, and the results showed good performance and applicability of the HEC-RAS model.

16.4 HEC-HMS SOFTWARE

The HEC-HMS software is designed to simulate the precipitation-runoff processes in a catchment area. In other words, this software simulates the response of the surface runoff in a catchment to a specific rainfall event. The software is designed to simulate a wide range of geographic areas, as well as the ability to analyze the rivers of large river basin water supply, reservoirs, flood hydrology, and runoff from natural and urban catchment areas. The hydrograph resulting from the HEC-HMS software analysis is used either directly in designing hydraulic structures or is used in conjunction with the other software for further hydrological studies, urban drainage, flow prediction, reservoir weir design, flood damages reduction, floodplain regulation, systems operation, and so on. This model shows the watershed as an integrated system with hydrological and hydraulic components. Each model component simulates an aspect of the runoff process within a part of the basin, which is usually considered as a sub-basin. In other words, various components are used to simulate the physical system of the basin, and each component represents one of the factors of rainfall transformation into the runoff in the basin. The combined effect of these factors will be the final hydrograph of the flood (Goodarzi and Absalan, 2014).

To simulate in the HEC-HMS model, the basin model is firstly created manually or using the DEM map of the study area, by the HEC-GeoHMS add-on in the ArcGIS environment. In the next step, by introducing the position of the basin outlet station, the watershed computational domain is determined for the model. Then different parts of the basin model should be created from the components menu. These components include six main parts: basin models, meteorological models, control specifications, time-series data, paired data, and grid data. Once the information required for simulation has been entered into the model, it should be calibrated and validated by using several flood events in the basin. The output results in this software include graphs, summary tables, and flood time series tables (Goodarzi and Absalan, 2014).

16.5   HEC-GEOHMS ADD-ON

The HEC-GeoHMS add-on, like HEC-GeoRAS, is a set of methods, tools, and utilities for processing geospatial data on the ArcGIS environment using a friendly graphical user interface. This add-on creates a link between the ArcGIS and the HEC-HMS software and is specially designed for the processing of geospatial data for use in modeling with the HEC-HMS. Also, some of the parameters required for modeling, such as basin area, length and slope of the river, slope of the basin and the gravity center of the basin, the time of concentration, and the lag time of the basin, are calculated by this software.

Shieh et al. (2007) used the HEC-HMS and HEC-RAS software to simulate the flow in the basin and investigate the effect of the construction of a check dam in Taiwan's Tsengwen basin. In a study for evaluation of the impact of climate change on the hydrology of the Tunga- Bhadra river basin in India, conducted by Meenu et al. (2012), the HEC-IIMS model was used to simulate the runoff.

Verdhen et al. (2013) examined the capability of the HEC-HMS model to simulate the snowmelt runoff in one of the Himalayan sub-basins using the temperature index and spatial and temporal analysis of the model parameters. The results of the simulation with a coefficient of determination of 0.7 indicated that this method is suitable. They also concluded that the HEC-HMS snow melting model has a high sensitivity to the ATI Cold Melt Rate Functions. Choudhari et al. (2014) simulated 24 rainstorm events between 2010 and 2013 in the Indian Balijore Nala watershed using the HEC-HMS model and obtained satisfactory results, indicating that this model is highly capable of simulating the rainfall-runoff process.

Oleyiblo and Li (2010) used the HEC-HMS and HEC-GeoHMS models to predict the floods in Misai and Wan'an Catchments in China. The results of this study showed that this model has high accuracy in flood forecasting at the studied basins. Farrokhzadeh et al. (2018), investigated the effects of climate change and land-use change on surface runoff in the Balighlo Chai Watershed in Iran, using the HEC-HMS model. The results showed an increase in peak flow and flood volume in April while it will decrease in March, May, and June. Also, if land-use change occurs with climate change, this increase will be intensified. Jahanbakhsh Asl et al. (2018) explored the ability to continuously simulate rainfall-runoff using the combination of the HEC-HMS and HEC-GeoHMS at the Urmia Basin in Iran. The results of this study showed that the HEC-HMS model was able to simulate the hydrologic behavior of the Shahrchay basin.

16.6   HEC-FIA SOFTWARE

The HEC-FIA software allows the users to estimate flood damages and monitor the benefits associated with flood control projects. The analysis periods can be a single flood event or a long recorded period of flood events. The program calculates the amount of agricultural damage, urban damage, human casualties, and benefits for an event, or annually, and summarizes the results. The HEC-FIA can also investigate the dam failure events, and conduct flood risk assessments for dams. All damage assessments in HEC-FI are computed on a structure-by-structure basis using inundated area depth and arrival grids, or hydrograph data. In general, due to the proper link of this software to hydraulic software (such as HEC-RAS), having flexible features, and the ability to properly illustrate with the help of the GIS, it could be a suitable choice for modeling and evaluating damages in different flood scenarios (Hasanzadeh Nafari, 2013).

The required data for simulation by this software include the central flow line, the precise DEM of the area, the agricultural and urban structure information, the raster map of inundated areas (derived from HEC-RAS simulation), the flood-affected area, and the computation points. The results of the simulation by this software provide flood damage estimates for the agricultural and urban sectors separately and in tabular form. First, in order to determine the extent of the inundated area, it is necessary to simulate the flow depth by hydrologic software like HEC-RAS. Then, the computations will be done by the HEC-FIA for the affected areas. The HEC-FIA model is an international authentic model and requires a lot of information about floodplain land use and applications. Therefore, more complete data and information about flooding in the region will result in more accurate outputs, and the analysis will be closer to reality (HEC-FIA, 2015; Lehman and Light, 2016) (Figure 16.4).

Daliran Firouz et al. (2016) assessed the flood damage in Ghamsar and Ghohrood Watershed in Iran using HEC-FIA software. In this study, HEC-RAS and HEC-GeoRAS were used to produce inundation maps for different return periods of floods, and flood damages to regional agricultural land, buildings, and human and financial losses were estimated in the HEC-FIA model. The benefits of this model are the direct estimates of economic and financial losses from floods and the potential future occurrence of floods, which can be helpful in managing the watershed, flood insurance, and flood risk management. Lehman and Light (2016) described the capability of the HEC-FIA model to determine the indirect economic losses caused by floods in the catchment areas. Prakash et al. (2018) in a study using the HEC-FIA software, assessed flood damage in the Guadalupe river basin in the United States. For this purpose, the HEC-RAS software was used to determine the depth of the flood in different areas of the basin and, by analyzing its results in the HEC-FIA model, investigated the flood damage.

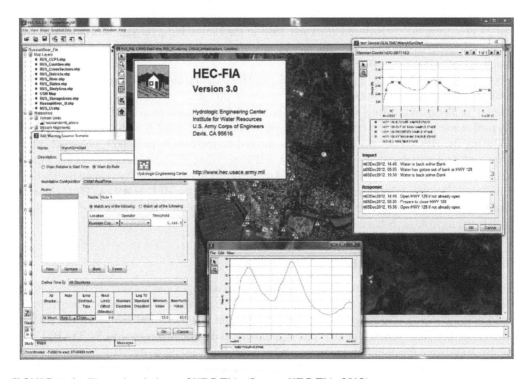

FIGURE 16.4   The main windows of HEC-FIA. (Source: HEC-FIA, 2015.)

16.7 MIKE FLOOD SOFTWARE

This software is a powerful tool for professional simulation of floods in 1D and 2D. Using this software, various flood-related issues, such as floods in the river, plains, urban areas, drainage networks, coastal areas, levee and dyke breaches, or any combination of these, can be modeled. The Mike Flood software can be used at any scale from a single parking lot to regional modeling of the entire catchment area. The software applications range from classical flood extent and risk mapping to environmental impact assessments of severe flood events. The Mike Flood software is ideal for flood forecasting, management and mitigation of flood risks, flood risk analysis, flood vulnerability mapping, climate change impact assessment, dam failure, and integrated flood modeling in urban, river, and coastal areas (DHI, 2017).

The Mike Flood model is a dynamically coupled modeling system, consisting of a one-dimensional river model, Mike 11, and a two-dimensional river model, Mike 21, that is capable of hydrodynamic simulation of the combined flows in the river and floodplains. To model the flood events, a Mike 11 and a Mike 21 model must be created and introduced into the Mike Flood model. In Figure 16.5, the Mike Flood Editor window is shown. In this window, the various combinations of the model connection can be applied (DHI, 2017). This model simulates and solves the equations of energy, continuity, and momentum (Navier-Stokes equations) in one- and two-dimensional modes. In order to simulate the flood in the Mike Flood model, the topographic data of the study area are processed in the ArcGIS environment and converted into a suitable format for the river bed topography in the Mike Flood software. Then, by calibrating the model based on one or more flood events, the validated model can be used for simulating the flood events in the region (Ghasemi et al., 2018).

The capabilities of this model have been evaluated in various studies around the world. Ghasemi et al. (2018) investigated the dam break and its effects on the downstream using the Mike Flood model for Mirzai Shirazi Storage Dam in Iran. Vanderkimpen et al. (2008) examined the flood risk for a section of the Belgian coastal plain using the Mike Flood and SOBEK 1D2D software. The

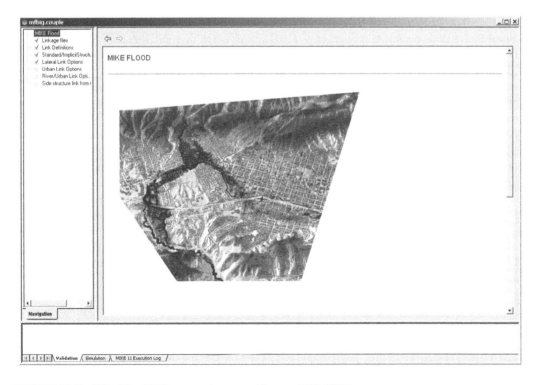

FIGURE 16.5 Mike Flood Editor; opening page. (Source: DHI, 2017.)

results showed that both models offer almost identical results, and uncertainty in the selection of each software package is negligible. Patro et al. (2009) used the Mike Flood model to simulate the flood inundation extent and flooding depth in the delta region of the Mahanadi River basin in India. Fayazi et al. (2010) investigated the accuracy of the Mike Flood and Mike 11 software in floodplain zoning in the Kashkan River basin in Iran. The results of this study showed that there are no differences between the two models in the V-shaped valleys, but in the plains, there is a large difference between the two models, and the accuracy of the Mike Flood model is higher. Kavianpour et al. (2015), in a study using the Mike Flood model and ArcGIS software, investigated the reduction of flood damage in alluvial fans in the Jamhash River in Iran. The results of this study showed that the use of longitudinal walls and spillway significantly reduces the damage and losses from the flood.

16.8   WMS SOFTWARE

The WMS model is a comprehensive watershed modeling system. This model includes hydraulic and hydrological modeling systems related to the watershed. The inputs of this model include meteorological data, digital elevation model (DEM), land use, and soil type, which are used to calculate physical parameters of watershed and model precipitation-runoff in the catchment area (Sen and Kahya, 2017). In general, with the help of the WMS model, one can simulate a watershed and a variety of related issues. The software supports a number of hydraulic and hydrologic models that can be used to create drainage basin simulations. The software provides tools to automate the various basic and advanced delineations, calculations, and modeling processes. By using these tools, which are taken from authentic models, both water quantity and water quality in the watershed could be modeled, easily. The most important models supported in this software are HEC-1, HEC-RAS, HEC-HMS, TR-20, TR-55, NFF, Rational, MODRAT, HSPF, CE-QUAL-W2, GSSHA, and SMPDBK (WMS, 2018).

In 2008, Sadrolashrafi et al. evaluated a flash flood in the Dez basin in Iran using the WMS software. Also, they used the WMS software for integrated modeling for flood hazard mapping. In this study, both HEC-1 and HEC-RAS models were used to simulate the main flood scenario to control future flood events. Ghaswari et al. (2011), in research on rainfall-runoff simulation in the Shohr River basin in Iran by WMS software, after mapping the channel networks with a TOPAZ sub-model and hydrologic simulation of the basin by the HEC-HMS sub-model, showed that in the estimation of peak discharge and flood volume, the sensitivity of the model to the CN changes is greater than the initial infiltration and the lag time changes. Abu Sharkh (2009) simulated a runoff process in the Wadi watershed in Palestine using the WMS software, the results of which showed that the software could well simulate the flood in the basin. Nozari et al. (2017), in a study, identified and prioritized the potential flood inundation areas of the Dez basin in Iran using the WMS model. For this purpose, the rainfall-runoff process was simulated in the WMS model and the sub-basins that had the most impact on the flood formation were identified (Figure 16.6).

Eshagh Teymori et al. (2012), in a study by using the WMS model, simulated rainfall-runoff in low data basins and as a case study in the Chalus watershed in Iran. The results showed that the SCS method has higher accuracy due to the use of in-basin parameters, and also, upstream sub-basins have the highest potential for runoff production. Karimi and Aref (2018) used the WMS model and SDSM software to predict runoff in the Khansar watershed in Iran under climate change conditions. Ahmadzadeh et al. (2015) used the WMS model to investigate and zone the areas prone to floods, especially in urbanized watersheds.

16.9   CCHE2D SOFTWARE

The CCHE2D model is an integrated numerical model for simulation and analysis of unsteady turbulent flow, sediment transport, and morphological processes in open channels. This model is a two-dimensional hydrodynamic model, and additionally includes a mesh generator (CCHE-MESH) and

FIGURE 16.6 Example floodplain mapping by WMS.

a Graphical User Interface (CCHE2D-GUI), which will help to use the CCHE2D model more easily and efficiently. As illustrated in Figure 16.7, the CCHE-MESH provides meshes for CCHE2D-GUI and CCHE2D numerical models, while the CCHE2D GUI provides a graphical interface to handle the data input and visualization for the CCHE2D numerical model (Zhang, 2009). Some of the capabilities of this model are the simulation of flood flow and sediment transport along the rivers, simulation of dam failure, studying the effect of flood flow patterns on river changes, and preparing the floodplain zoning maps.

The CCHE2D model is based on water flow governing equations and uses the finite-volume numerical method to solve the mass conservation and momentum equations. In this model, the simulation of water flow is based on solving the Navier-Stokes average depth equations. The turbulent shear stress is calculated using the estimation of the Boussinesq equations, and three different turbulence models can be used to calculate the turbidity vortex viscosity (Khosravi, 2018). Initial data required for running this model include the meshed data of land surface, regional coordinates, initial water level of the river before the floods, Manning roughness coefficient of the riverbed,

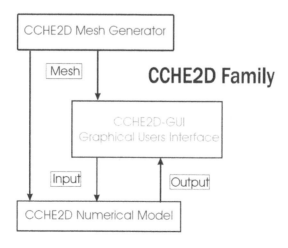

FIGURE 16.7 Process of CCHE2D model simulations. (Source: Zhang, 2009.)

computation range and limits, and model control parameters such as the total simulation time and the time difference of the output files. The most important outputs of the CCHE2D model are the depth and speed of the flood flow wave along the river. Output files, define the depth and velocity parameters in each cell of the computing grid in units of meters and meters per second (Kardan et al., 2018). Niknam et al. (2018) investigated the effects of floods in the Kor River in Iran using the CCHE2D model. The results of this study indicated that the CCHE2D model has the ability to simulate the flood and sediment transport in the river.

Kardan et al. (2018) simulated and evaluated urban flood flows with the CCHE2D model in the city of Aq Qala in Iran. The results of this study showed that the city of Aq Qala is severely vulnerable to 100-year floods, and the city's evacuations are necessary. Khosravi (2018) studied the turbulent flow and sediment transport pattern using the CCHE2D model downstream of the Minab Dam in Iran. The results of this study indicate high accuracy and low error of the model in predicting flow and sediment parameters, and also the erosion and sedimentation conditions of the region indicate erodibility of the studied area against possible floods. Abu Hasan et al. (2007) using the CCHE2D model in a reach of the Muda River in Malaysia, evaluated the flow and sediment pattern in the river and stated that this model is useful and can be used to analyze flow behavior, both in the river channel and the floodplains. Abu Hasan et al. (2007), by examining the two-dimensional CCHE2D model and one-dimensional HEC-RAS in the Muda River, concluded that although the HEC-RAS model was widely used, it was unable to analyze some of the hydraulic characteristics of the flow, such as the flow and sediment pattern in meanders, determining the locations susceptible to longer inundation, erosions and deposition in floodplains, and velocity distribution, and therefore requires a more sophisticated 2D model and modeling techniques like CCHE2D. Malekabbaslu et al. (2012), in a study, evaluated the CCHE2D model on floodplain zoning in rivers on alluvial fans, and as a case study in the Rudan River in Iran. They concluded that the floodplain area obtained from the two-dimensional model is highly matching to the river bed determined from the field visits and aerial photos, while the results of a one-dimensional model are unreliable. Arzanlou et al. (2017), using the two-dimensional CCHE2D model, investigated the Shahrchay earthfill dam break due to overtopping in Iran, and then prepared a floodplain zoning map downstream by ArcGIS.

16.10   SWAT SOFTWARE

The SWAT software is a large-scale or sub-watershed model developed by Arnold for the US Agricultural Research Service and has been developing steadily since its introduction in the early 1990s. This model is used to simulate the hydrological processes such as the quality and quantity of surface and groundwater and predict the environmental impact of land use, land management practices, and climate change (Gholami and Nasiri, 2015). Also, all computations can be simulated on a daily, monthly, or annual basis. One of the important features of this model is that hydrological modeling can be done on a sub-daily basis, a feature that is unique in hydrological models, and few models are able to simulate hydrological processes at hourly time-steps, which is especially very important for flash-floods modeling. The implementation of this model in the ArcGIS environment makes it easy to use and enhances the functionality of this model. The base maps required for the simulation in this software include a digital elevation model (DEM), land use map, soil map, and vegetation map, all of which must be presented in a raster form (Neitch et al., 2009). Other information, including meteorological data, water quality data, factors affecting surface and subsurface flow, groundwater, water extraction, land management, comprehensive water quality information, reservoirs, and the other required data with respect to the aim of the study, should be entered in the model (Neitch et al., 2009).

In Figure 16.8, inputs and outputs of the SWAT model are schematically presented. This model simulates very large watersheds or varies the management strategy and enables users to simulate long-term modeling without spending extra cost or time. The SWAT software is a physically based distributed hydrological model that uses the Soil Conservation Service (SCS) Runoff Curve Number

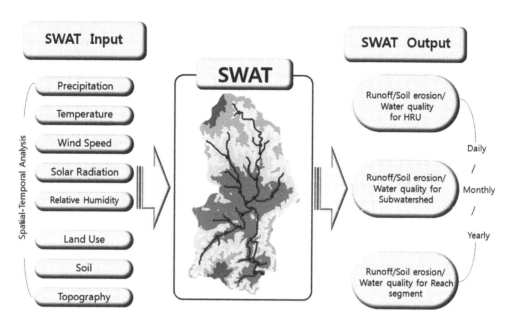

FIGURE 16.8   Overview of the SWAT model – model input/output parameters. (Source: Heo et al., 2008.)

(CN) method to calculate surface runoff, and the degree-day factor method to calculate snowmelt runoff (Duan et al., 2018). In general, this software is a comprehensive model for assessing flow discharge, long-term effects of management operations on water, sediment, and agricultural chemicals in large watersheds. In order to use this model properly, parameters which the output of the model is more sensitive to their accuracy must be specified, and the model is calibrated based on the observation data of a given period (e.g., measured discharge flow data from the river in the watershed) (Goodarzi, 2016).

Jodar-Abellan et al. (2019) in a study using the SWAT model, assessed the effect of land-use change on a flash flood in five Mediterranean watersheds. For this purpose, different land-use change scenarios were studied using the SWAT model, and the results showed that this model has a good performance in predicting the flood, and in the studied areas, with increasing urban areas, the frequency of flash floods increases. Boithias et al. (2017) simulated the flash floods of the Têt watershed in southern France at hourly time-step with the SWAT software. The results of this study showed that the SWAT model is an appropriate tool for modeling and predicting flash flood events. Duan et al. (2018) simulated snow melting and flood processes with the SWAT model in the Tizinafu watershed in China. In this study, the importance of considering snow melting in the occurrence of floods, especially the spring floods, has been determined by using the SWAT model. Also, they applied a series of changes in the model, which increased the accuracy of the model in the flood forecast. Goodarzi et al. (2016) evaluated the efficiency of the SWAT model in the simulation of the rainfall-runoff process at the Gharesou watershed in Iran. In this study, the 1992–1996 rainfall-runoff observation data were used to calibrate the model. The results showed that by using optimized parameters, the model simulates the stream flow with high accuracy.

16.11   SWMM SOFTWARE

The SWMM model or Storm Water Management Model was developed by the US Environmental Protection Agency. The SWMM model is a dynamic rainfall-runoff simulation model with the ability to consider different hydrological phenomena, including evaporation, snowmelt, infiltration, deep percolation, interception, depression storage, and subsurface flow. This model can be used for a single event or continuous simulation of runoff quantity and quality from urban areas. In this model, flood estimation is carried out using the kinematic and dynamic wave method and the combination of ground and channel flow elements. Also, this model has the ability to be linked with the other models and provides satisfactory results in small basins (Rossman, 2009). The SWMM has been widely used throughout the world for planning, analysis, and design related to stormwater runoff, combined sewers, sanitary sewers, and other drainage systems in urban areas (Rossman, 2009).

The most common uses of the SWMM model are design and sizing of drainage system components for flood control, sizing of detention facilities and their appurtenances for flood control and water quality, floodplain mapping of natural channel systems, and determining the effects of land-use change runoff and flooding in urban areas (Rossman, 2009). The required data for flood modeling in this software are including the DEM map of the area, the roughness coefficients of the existing surfaces, the characteristics of the joints and the runoff entry point of each sub-basin, the characteristics of existing ducts and channels in the basin, the measured rainfall and runoff data for calibration of the model, the land use map, and the curve number of sub-basins. The main window of the SWMM software is shown in Figure 16.9.

Over the past few years, researchers have used the capabilities of the SWMM model in various study areas around the world. The results in four areas in Korea showed that this model can improve errors, such as estimating less peak flow and a longer peak time for the post-expansion of urban areas, and evaluate the hydrological effects of urban development very well (Janga et al., 2007). The model of SWMM and ArcGIS was used to simulate the runoff in the Macau metropolitan area in Korea. The results showed that the use of GIS is very useful in obtaining some important parameters of the SWMM model, and also, the SWMM model has high flexibility in simulating runoff when sufficient parameters are available (Dongquan, et al., 2009). The SWMM model was used

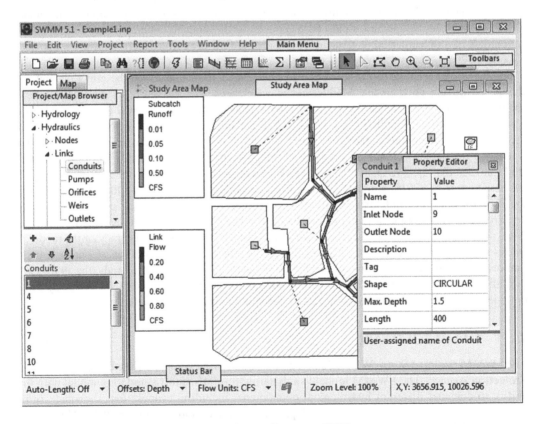

FIGURE 16.9   The SWMM main windows. (Source: Rossman, 2009.)

to manage the urban runoff in order to reduce the risk of flood and urban inundation in Mahdasht city in Iran. After sensitivity analysis, the model was calibrated and evaluated. Also, in this study, the concentration time of the area was estimated using different experimental methods. The results showed a good performance of the SWMM model in estimating the urban runoff in the study area (Shahbazi, 2013).

Mikovits et al. (2014), in evaluating the impact of urban development with SWMM, found that increasing the impenetrable surfaces causes a two-fold increase in the flood volume. Einloo et al. (2016) investigated the effect of urban development on changes in runoff volume in the Zanjan urban watershed in Iran using the SWMM model. The results showed that urban development and land conversion to impenetrable surfaces increased the amount of runoff so that during the ten-year period the runoff volume increased by 65% in the studied area.

Tavasoli et al. (2017) used the SWMM model to determine the best locations for rainwater harvesting reservoirs in Tehran's urban area in Iran, and in addition to preventing flood damage, identified the potential of produced runoff to meet the irrigation requirement of urban landscapes. Izanloo and Sheikh (2017), using the SWMM model, simulated the infiltration trench system for urban flood control in Bojnord, Iran. The results showed that the implementation of the infiltration trench system reduced the runoff production volume by about 6%. Khalghi (2010) evaluated the performance of the SWMM model in simulating the flow hydrograph of the urban river of Shiraz in Iran. The calibration and validation results of the model showed that there is a good correlation between the simulated and observed flow and the values of the performance indicators of the model are acceptable. Rostami Khalaj (2012) used the SWMM and HEC-RAS models in order to urban flood hazard zonation in Mashhad, Iran. For this purpose, the simulated runoff in the SWMM model entered the HEC-RAS model and identified the flood map inundation of the city. Badiezadeh et al. (2015) assessed the capability of the SWMM model in urban flood management through surface runoff simulation in Gorgan city, Iran. The results of this study showed that the SWMM model has the accuracy required for urban runoff simulation, and this model can be used for urban runoff management projects and the design of the surface water collection and drainage networks.

16.12 CONCLUSIONS

In recent years, computer models have turned out to be more important because of the precise computations, the flexibility to change flow parameters, and lower costs than the other models. As mentioned before, several models have been developed to predict flood and flow in rivers in the world, which are divided into several categories in terms of complexity and ease of use. The results of the studies have shown that these types of software are extremely valuable for hydrological simulations in watersheds. Regarding the use of the computer modeling system as confirmed in various studies, it can be admitted that using these models, due to the high capability and access to different methods for evaluation, would be very useful in flood control and hydrological studies, provided that logical calibration parameters are included. The most important factor in hydrologic simulation is access to complete and correct information about the watershed. Therefore, before choosing a software hydrological simulation, it is absolutely necessary to ensure the complete availability of the required information based on the selected model. Hydrological software is diverse based on the degree of complexity, the processes involved, the data required for calibration, and how the model is used. There is no model that is the best model for all applications, but the most suitable model for a specific application should be selected according to the goal and furthermore based on the characteristics of the watershed to be investigated. In a general conclusion, it could be said that in flood management and damage analysis by using computer models, many studies have been done around the world, and according to the regional conditions, it is possible to choose the appropriate software from different available models, and after calibration of the model, it could be utilized for riverine and flood management in the studied area.

REFERENCES

Abu Sharkh, M.S. 2009. Estimation of runoff for small watershed using watershed modeling system (WMS) and GIS. In: *13th International Water Technology Conference, IWTC*, March 2009, Hurghada, Egypt, pp. 1185–1200.

Ahmadzadeh, H., Saed Abadi, R., and Noori, A. 2015. A study and zoning of the areas prone to flooding with an emphasis on urban floods (case study: City of Maku), *Hydrogeomorphology*, 1(2): 1–24.

Arzanlou, A., Hassanzadeh, Y., Kardan, N., and Hasanzadeh, M. 2017. Investigation of shahrchay earthfill dam break due to overtopping and flood mapping, *Journal of Irrigation and Water Engineering*, 8(1): 1–17.

Badiezadeh, S., Bahremand, A.R., Dehghani, A.A., and Noura, N. 2015. Urban flood management by simulation of surface runoff using SWMM model in Gorgan city, Golestan Province-Iran, *Journal of Water and Soil Conservation*, 22(4): 155–170.

Boithias, L., Sauvage, S., Lenica, A., Roux, H., Abbaspour, C.K., Larnier, K., Dartus, D., and Sánchez-Pérez, M.J. 2017. Simulating flash floods at hourly time-step using the SWAT model, *Water*, 9(929): 1 25.

Brunner, W.G. 2010. *HEC-RAS River Analysis System, Hydraulic Reference Manual*, US Army Corps of Engineers, Hydrologic Engineering Center, USA.

Choudhari, K., Panigrahi, B., and Paul, C.J. 2014. Simulation of rainfall-runoff process using HEC-HMS model for Balijore Nala watershed, Odisha, India, *International Journal of Geomatics and Geosciences*, 5(2): 253–265.

Daliran Firouz, H., Mokhtari, F., Soltani, S., and Mousavi, S.A. 2016. Flood damage assessment in Ghamsar and Ghohrood watershed basins using HEC-FIA, *Journal of Water and Soil Science*, 19(74): 63–76.

Danish Hydraulic Institute (DHI). 2017. *Mike Flood Modeling, User Manual, Scientific Documentation*, Kopenhagen, Denmark.

Dongquan, Z., Jining, C., Haozheng, W., Qingyuan, T., Shangbing, C., and Zheng, S. 2009. GIS- based urban rainfall-runoff modeling using an automatic catchment-discretization approach: A case study in Macau, *Environmental Earth Sciences*, 59(2): 465–472.

Duan, Y., Liu, T., Meng, F., Luo, M., Frankl, A., Maeyer, D.P., Bao, A., Kurban, A., and Feng, X. 2018. Inclusion of modified snow melting and flood processes in the SWAT model, *Water*, 10(1715): 1–24.

Einloo, F., Salajegheh, A., Malekian, A., and Ahadnejad, M. 2016. Evaluation of urbanization effect on runoff volume by using stormwater management model (case study: Zanjan City Watershed), *Iranian Journal of Watershded Management Sciences*, 10(33): 37–46.

Eshagh Teymori, M.A., Habib Nejad, M., Kaviyan, A., and Shahedi, K. 2012. Estimation of Process rainfall-Runoff in low data basin with use of WMS model (case study; Chalous Basin), *Iranian Irrigation and Water Engineering*, 9(3): 12–25.

Farrokhzadeh, B., Choobeh, S., Nouri, H., and Goodarzi, M. 2018. Study of climate change and land use changes impacts on surface runoff: Balighlo Chai Watershed in Ardebil, *Watershed Engineering and Management*, 10(3): 318–331.

Fayazi, M., Bagheri, A., Sedghi, H., Keyhan, K., and Kaveh, F. 2010. Flood plains simulation of Kashkan River, Lorestan, Iran with MIKE11and MIKE FLOOD. In: *Eighth International River Engineering Conference*, Shahid Chamran University, Ahvaz, Iran.

Ghasemi, K., Zonemat Kermani, M., and Samareh Hashemi, M. 2018. Investigating the hydraulics of dam break and its effects on the downstream using MIKE-FLOOD (case study: Reservoir dam of Mirza Shirazi), *Dam and Hydroelectric Powerplant*, 4(15): 35–48.

Ghaswari, M.J., Bustani, F., and Rostami Rawari, A. 2011. Calibration and sensitivity analysis of WMS/HEC-HMS model for estimating flood hydrograph properties in the Jahrom Shoor River basin. In: *National Conference on Civil Engineering and Sustainable Development*, Estahban, Fars, Iran.

Gholami, S., and Nasiri, M. 2015. Simulation of Atrak River monthly discharge using SWAT model, Case study: Maraveh Tappeh watershed, Gholestan province, *Watershed Engineering and Management*, 7(2): 126–135.

Goodarzi, M. 2016. *Evaluation of Climate Change Impacts on Groundwater Resources by Using a Combination of MODFLOW Model and Thornthwaite and Mather's Method*. PhD Thesis, Department of Water Engineering, College of Agriculture, Isfahan University of Technology, Isfahan, Iran.

Goodarzi, M., and Absalan, S. 2014. *Introduction to Water Engineering Softwares*, Kankash Publication, Isfahan, Iran.

Goodarzi, M., Zahabiyoun, B., and Massah Bavani, A. 2016. Simulation of rainfall-runoff for gharesou watershed using SWAT model, *International Journal of Environmental Science and Technology*, 18(2): 12–20.

Hasan, A.Z., Ghani, A., and Zakaria, N. 2007. Application of 2-D modeling for muda river using CCHE2D. In: *2nd International Conference on Managing Rivers in the 21 Century: Solution Towards Sustainable River Basins*, June 2007, Kuching, Sarawak, Malaysia.

Hasanzadeh Nafari, R. 2013. *Flood Damage Assessment with the Help of HEC-FIA Model*. Master Thesis, Faculty of Civil and Environmental Engineering, Politecnico Di Milano, Italy.

HEC-FIA. 2015. *Flood Impact Analysis User's Manual*, US Army Corps of Engineers, Hydrologic Engineering Center, USA.

Hejazi, A., Khodaie Geshlag, F., and Khodaie Geshlag, L. 2019. Zoning the villages at flood risk in the Varkesh-Chai drainage basin by GIS and HEC - RAS software and HEC- GEO - RAS extension, *Researches in Geographical Sciences*, 19(53): 137–155.

Heo, S., Kim, N., Park, Y., Kim, J., Kim, S.J., Ahn, J., Kim, K.S., and Lim, K.J. 2008. Evaluation of effects on SWAT simulated hydrology and sediment behaviors of SWAT watershed delineation using SWAT ArcView GIS extension patch, *Journal of Korean Society on Water Quality*, 24(2): 147–155.

Izanloo, R., and Sheikh, V. 2017. Feasibility study and modeling with SWMM of infiltration trench system in urban flood control, a case study in Bojnord, Iran, *Iranian Journal of Rainwater Catchment Systems*, 5(16): 51–62.

Jahanbakhsh Asl, S., Rezaee Banafshe, M., Rostamzadeh, H., and Aalinejad, M.H. 2018. Continuous simulation of rainfall-runoff of shahrchay basin of urmia using HEC-HMS model, *Hydrogeomorphology*, 4(16): 101–118.

Janga, S., Chob, M., Yoonc, J., Yoond, Y., Kime, S., Kimf, G., Kimg, L., and Aksoyh, H. 2007. Using SWMM as a tool for hydrologic impact assessment, *Desalination*, 212(1): 344–356.

Jodar-Abellan, A., Valdes-Abellan, J., Pla, C., and Gomariz-Castillo, F. 2019. Impact of land use changes on flash flood prediction using a sub-daily SWAT model in five Mediterranean ungauged watersheds (SE Spain), *Science of the Total Environment*, 657: 1578–1591.

Kardan, N., Hassanzadeh, Y., and Arzanlou, A. 2018. 2D numerical simulation of urban floods using CCHE2D (case study: Aghghala City), *Iranian Journal of Murine Technology*, 4(4): 25–36.

Karimi, M., and Aref, M. 2018. Coupling SDSM downscaling climate and WMS hydrologic models in Khansar basin for runoff prediction under future climate change conditions, *Geographic Exploration of Desert Areas*, 6(1): 75–90.

Kavianpour, M.R., Nikrou, P., and Pourhasan, M.A. 2015. Reduction of flood losses in alluvial fans case study: Jamash River in Bandar Abbas, *Iran-Water Resources Research*, 11(1): 87–91.

Khalghi, A. 2010. *Simulation of Flow Hydrograph Using SWMM Model and Predict the Effects of Watershed Management Practices in Dry River Shiraz*. The Master's Thesis, Gorgan University of Agricultural Sciences and Natural Resources, p. 128.

Khosravi, Q. 2018. Hydrodynamic investigation of turbulence flow and sediment transport pattern by CCHE2D (case study: Of Downstream Meander of Minab Dam), *Iran-Watershed Management Science and Engineering*, 12(40): 23–39.

Kncbl, M.R., Yung, Z.L., Hutchison, K., and Maidment, D.R. 2005. Regional scale flood modelingusing NEXRAD rainfall, GIS, and HEC-HMS/RAS: A case study for the San Anto-nio River Basin Summer 2002 storm event, *Journal of Environmental Management*, 75(4 Special Issue): 325–36.

Lehman, W., and Light, M. 2016. Using HEC-FIA to identify indirect economic losses. In: *3rd European Conference on Flood Risk Management*, Lyon, France.

Lian, J.J., Xu, K., and Ma, C. 2013. Joint impact of rainfall and tidal level on flood risk in a coastal city with a complex river network: A case study of Fuzhou City, China, *Hydrology and Earth System Sciences*,17(2): 679–89.

Malekabbaslu, A. Haji Kennedy, H., and Pirestani, M. 2012. Two-dimensional modeling of flow pattern in the river alluvial fans using two dimensional model CCHE2D. In: *The Ninth International Seminar on River Engineering*, Shahid Chamran University, Ahvaz, Iran.

Meenu, R., Rehana, S., and Mujumdar, P.P. 2012. Assessment of hydrologic impacts of climate change in Tunga-Bhadra river basin, India with HEC-HMS and SDSM, *Hydrological Processes*, 27(11): 1572–1589.

Mikovits C., Rauch W., and Kleidorfer M. 2014. Dynamics in urban development, population growth and their influences on urban water infrastructure, *Procedia Engineering*, 70: 1147–1156.

Mohammadi, S.A., Naziriha, M., and Mehrdadi, N. 2014. Flood damage estimate (quan-tity), using HEC-FDA model. Case study: The Neka river, *Procedia Engineering*, 70: 1173–1182.

Moya Quiroga, V., Kure, S. Udo, K., and Mano, A. 2016. Application of 2D numerical simulation for the analysis of the February 2014 Bolivian Amazonia flood: Application of the new HEC-RAS version 5V. *RIBAGUA - Rev Iberoam Agua*, http://dx.doi.org/10.1016/j.riba.2015.12.001.

Napradean, I., and Chira, R. 2006. The hydrological modeling of the Usturoi Valley - Using two modeling programs - WetSpa And HecRas, *Carpathian Journal of Earth and Environmental Sciences*, 1(1): 53–62.

Neitch, S.L, Arnold, J.G., Kiniry, J.R., and Williams, J.R. 2009. *Soil and Water Assessment Tool Theoretical Documentation*, Blackland Research Center, Temple, TX, USA, p. 618.

Niknam, A., Nohegar, A., Jafarpoor, A., and Taghi Avand, M. 2018. Evaluating the effect of flood flow patterns on river trends using CCHE2D model (case study: The Kor River, Abbas Abad Bridge to Doroodzan Dam Distance), *Hydrogeomorphology*, 4(16): 23–41.

Nozari, H., Marofi, S., and Edirsh, M. 2017. Identification and prioritize of potential areas to flood inundation in the Dez basin using WMS, *Journal of Range and Watershed Management*, 70(3): 805–820.

Oleyiblo, O.J., and LI, Z.J. 2010. Application of HEC-HMS for flood forecasting in Misai and Wan'an catchments in China, *Water Science and Engineering*, 3(1): 14–22.

Patro, S., Chatterjee, C., Mohanty, S., Singh, R., and Raghuwanshi, N.S. 2009. Flood inundation modeling using MIKE FLOOD and remote sensing data, *Journal of the Indian Society of Remote Sensing*, 37(1): 107–118.

Prakash, O., Avance, A., Schwind, M., and Curtis, D. 2018. Flood damage assessment for the guadalupe river basin using HEC-FIA model. In: *Floodplain Management Association (FMA) Conference*, September 04–07, Reno, NV, USA.

Rezaei, P. 2018. Determining the flooding zone using GIS and HEC-RAS hydraulic model, case study: Goharrood River, Rasht, *Geography and Environmental Hazards*, 7(27): 27–41.

Rezai, P., Tajdari, K., and Mirghasemi, S.E. 2014. Determining the flood prone areas of morghak river using HEC-GeoRAS, *Journal of Spatial Analysis Environmental Hazards*, 1(2): 29–45.

Rossman, A.L. 2009. *Storm Water Management Model User's Manual Version 5.0*, Water Supply and Water Resources Division National Risk Management Research Laboratory, Cincinnati, OH, USA.

Rostami Khalaj, M. 2012. *Urban Flood Hazard Zoning Combining Hydrologic and Hydraulic Model Study, the Two Mashhad Municipality Watershed*, Master's Thesis, Agriculture and Natural Resources Faculty, University of Tehran, p. 116.

Sadrolashrafi, S.S., Mohamed, A.T., Mahmud, R.B.A., Kholghi, K.M., and Samadi, A. 2008. Integrated modeling for flood hazard mapping using watershed modeling system, *American Journal of Engineering and Applied Science*, 1(2): 149–156.

Salajegheh, A., Bakhshaei, M., Chavoshi, S., Keshtkar, A.R., and Najafi Hajivar, M. 2009. Floodplain mapping using HEC-RAS and GIS in semi-arid regions of Iran, *Desert*, 14(1): 83–93.

Sen, O., and Kahya, E. 2017. Determination of flood risk: A case study in the rainiest city of Turkey, *Environmental Modelling and Software*, 93: 296–309.

Shahbazi, A. 2013. *Urban Runoff Management to Reduce Risks Using SWMM Model: Case Study: Mahdasht Town*, Master Thesis, Agriculture and Natural Resources Faculty, University of Tehran, p. 144.

Shieh C.L., Guh, Y.R., and Wang, S.O. 2007. The application of range of variability approach to the assessment of a check dam on riverine habitat alteration, *Environmental Geology*, 52(3): 427–435.

Tate, E., and Maidment, D.R. 1999. *Floodplain Mapping Using HEC-RAS and ArcView GIS*. Master Thesis, Department of Civil Engineering, University of Texas at Austin, USA.

Tate E., Maidment, D.R., Olivera, F., and Anderson, D.J. 2002. Creating a terrain model for floodplain mapping, *Journal of Hydrology Engineering*, 7: 100–108.

Tavasoli, A., Hoseinnia, A., and Shahbazi, A. 2017. Site selection for rainwater harvesting reservoirs in urban areas using the SWMM model (Case Study: Tehran's first district), *Iranian Journal of Rainwater Catchment Systems*, 5(15): 13–27.

Vanderkimpen, P. Melger, E., and Peeters, P. 2008. Flood modeling for risk evaluation: A MIKE FLOOD vs. SOBEK 1D2D benchmark study. In: *Flood Risk Management: Research and Practice*, Taylor and Francis group, pp. 77–84.

Verdhen, A., Chahar, B., and Sharma, O. 2013. Snowmelt runoff simulation using HEC-HMS in a Himalayan Watershed. In: *World Environmental and Water Resources Congress*, May 2013, Cincinnati, Ohaio, United States of America, pp. 3206–3215.

WMS User Manual. 2018. *The Watershed Modeling System*, Aquaveo Company, Utah, United States of America.

www.esri.com.

Yamani, M., Toorani, M., and Chezghe, S. 2012. Determination of the flooding zones by using HEC-RAS model (case study: Upstream the Taleghan Dam), *Geography and Environmental Hazards*, 1(1): 1–16.

Zhang, Y. 2009. *CCHE-GUI – Graphical Users Interface for NCCHE Model User's Manual – Version 3.0*, Technical Report No. NCCHE-TR-2009-01, Mississippi University, MS, USA.

17 Flood Modeling Using Open-Source Software

Thomas J. Scanlon and Saeid Eslamian

CONTENTS

17.1 INTRODUCTION

Given the now-established link between extreme weather events and climate change (Blöschl et al., 2019; Robins and Lewis, 2019), an increasing likelihood of severe flooding exists worldwide. A significant number of inundation events have occurred over the last 20 years in the UK with the current flood damage costs estimated at around £1.3 billion each year (ECIU, 2019). This is in addition to the human psychological damage where one in six properties in the UK are exposed to significant flooding risk (CCRA, 2019). Although technical and economic barriers mean that the complete elimination of flood risk is impractical, comprehensive flood management plans must be employed to mitigate such adverse weather events. These strategies can be informed by the appropriate flood modeling techniques and this is the focus of this book chapter.

Several commercial flood models are available (MIKE, 2019; TUFLOW, 2019; ISIS, 2019; SOBEK, 2019; INFOWORKS_ICM, 2019) and in 2013 such codes were invited by the UK government Department of Environment, Food, and Rural Affairs (DEFRA) to take part in a series of flood modeling exercises (DEFRA, 2013). The goal of this project was to generate a series of benchmark cases for flood modeling and gauging the performance of each code with a view to establishing the best practice when applied to the flooding events. Among the codes tested in the DEFRA exercise was the open-source code ANUGA (ANUGA, 2019) developed by the Australian National University and Geoscience Australia. The code performed well in comparison with its commercial counterparts, however, it was only applied to a select number of benchmark cases. shallowFoam (Mintgen, 2017) is the open-source flood model solver used in this book chapter and represents an alternative code to ANUGA. The code has been released open-source via the software repository GitHub (shallowFoam, 2019). Developed within the framework of the open-source computational fluid dynamics (CFD) solver OpenFOAM (OpenFOAM, 2019), shallowFoam solves the shallow water equations (SWE) in a finite volume meshing environment. With an estimated user base of 10,000 (OpenORG, 2019), OpenFOAM is a parallelized CFD code widely employed in academic and industrial communities. shallowFoam has been successfully applied to the case of a dam break (Zeng et al., 2018), however, this work necessitated the use of radial basis functions in the generation of the digital elevation map (DEM). Other flood modeling comparison exercises have been made (Alaghmand et al., 2012). In the book chapter that is presented, modeling is made exclusively of free-to-download software in the generation of the DEM and does not rely on any external mathematical manipulation.

A new methodology for flood modeling using exclusively free-to-download, open-source software has been presented. The geographical information system software QGIS has been applied to a LiDAR image to produce a 3D file in stereolithography format. This file was then converted to a digital elevation model in the computational fluid dynamics package OpenFOAM. Finally, the shallow water equations were solved within the framework of OpenFOAM using the flood model shallowFoam to produce the maps of water height and velocity. In comparison with the benchmark results for a range of commercial codes, the shallowFoam solutions compared well and produced satisfactory results for key flooding parameters. The final test case showed how the methodology could be applied to a practical flooding problem with the mitigation measures successfully implemented. Future work will include the incorporation of 1D–2D links for features such as culverts and sluice gates and the implementation of a shock-capturing scheme for supercritical flows. The availability of such computational technologies means that effective open-source flood modeling software is now accessible for the wider academic, industrial, and citizen science communities.

17.1.1 AIMS AND KEY OBJECTIVES

This chapter aims to propose an effective novel methodology for flood modeling using exclusively free-to-download, open-source software. The key objectives are as follows:

- To characterize the new modeling methodology and demonstrate its use in terms of practical flood applications;
- To develop a combined GIS and computational fluid dynamics (CFD) approach to assess flood risk;
- To assess and compare flood modeling results for different commercial and open-source software techniques; and
- To promote the new methodology as being fully accessible to the wider academic, industrial, and citizen science communities.

17.2 SHALLOWFOAM SOLVER

The shallowFoam solver solves the 2D (x, y), depth-averaged, shallow-water equations (SWE) according to the coordinate system shown in Figure 17.1. Here, h is the flow depth (m), z_b is the bottom surface level (m), and z_w is the water level (m).

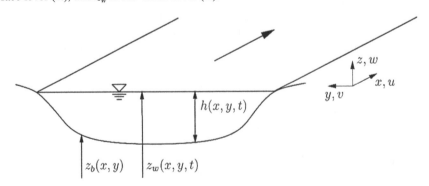

FIGURE 17.1 Coordinate system and variables for the shallow water equations. (Source: Mintgen, 2017.)

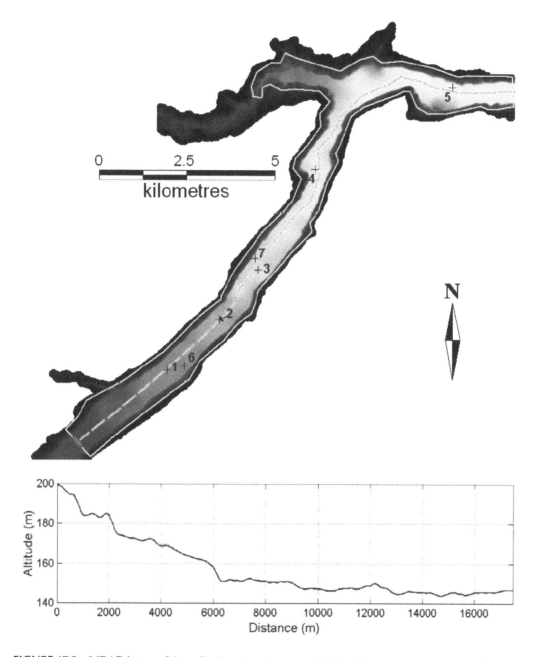

FIGURE 17.2 LiDAR image of the valley benchmark case and height-distance graph along centerline.

The SWE correspond to the transport equations for the conservation of mass and momentum according to:

$$\frac{\partial h}{\partial t} + \frac{\partial q_i}{\partial x_i} = 0 \qquad (i = 1,2),$$ (17.1)

and

$$\frac{\partial q_i}{\partial t} + \overline{u}_j \frac{\partial q_i}{\partial x_j} = -\frac{g}{2}\frac{\partial h^2}{\partial x_i} - gh\frac{\partial z_b}{\partial x_i} - \frac{\tau_{bi}}{\rho} + \frac{\partial}{\partial x_j}\left[\upsilon_t \left(\frac{\partial q_i}{\partial x_j} + \frac{\partial q_j}{\partial x_i} \right) \right] \qquad (i = 1,\ 2),$$ (17.2)

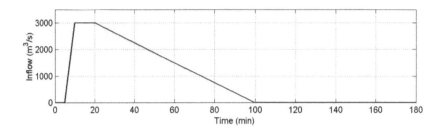

FIGURE 17.3   Inlet hydrograph for dam-break scenario.

respectively. $q_i = h\bar{u}_i$ has been introduced as the specific discharge (m^2s^{-1}), g is the gravitational acceleration ($m.s^{-2}$), and the bottom surface shear stress τ_{bi} is modeled using:

$$\frac{\tau_{bi}}{\rho} = \frac{n^2 g}{h^{\frac{1}{3}}} \bar{u}_i |\bar{u}| \tag{17.3}$$

where $|\bar{u}|$ is the velocity vector magnitude and n is Manning's roughness coefficient. Turbulence closure is via a depth-averaged parabolic eddy viscosity model where the turbulent viscosity υ_t is set as:

$$\upsilon_t = \frac{\kappa}{6} u^* h \tag{17.4}$$

with the von-Kármán constant $\kappa = 0.41$ and the shear velocity u^* calculated by:

$$u^* = \sqrt{\frac{\tau_b}{\rho}} \tag{17.5}$$

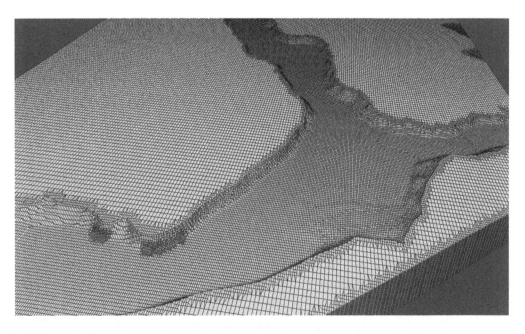

FIGURE 17.4   3D snappyHexMesh showing the lower valley zone.

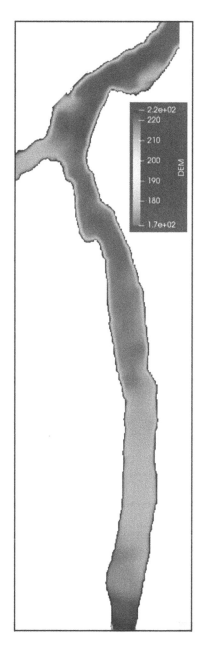

FIGURE 17.5 Valley DEM generated using OpenFOAM utility Test-WallDist.

17.3 METHODOLOGY

This section will present the general methodology for setting up, running, and post-processing a flood model case. The case presented here is Case 2, one of the four test cases considered in this book chapter. All of the other cases follow the same methodology. Case 2 is one of the DEFRA benchmark cases and considers a dam burst scenario into a valley of approximately 17 km in length. Figure 17.2 shows the LiDAR image of the valley and corresponding shapefile outlined in light blue which outlines the computational domain.

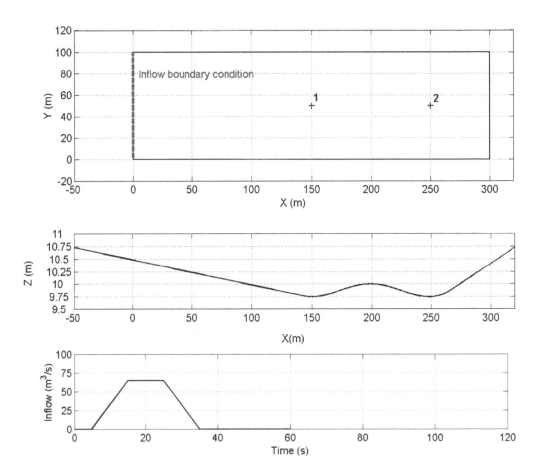

FIGURE 17.6   Plan view, elevation, and inlet hydrograph for test Case 1.

The ground slope is shown as an altitude-distance graph along the valley center-line in green and sampling points are shown as numbered crosses. Inlet flow conditions are supplied at the red line in the upper valley as a hydrograph as shown in Figure 17.3.

The first step in the methodology is to create the digital elevation map (DEM). The open-source geographical information system (GIS) software QGIS (QGIS 2019) is used to process the LiDAR .asc file supplied as part of the DEFRA benchmark case. A QGIS plug-in called DEMto3D is employed to generate a stereolithography (STL) file and this STL file is brought into OpenFOAM to generate a 3D computational mesh using the OpenFOAM meshing utility snappyHexMesh. The result of this meshing is shown in Figure 17.4.

The height to the cell face centers from a bottom reference plane is then calculated with a new OpenFOAM utility called Test-wallDist. The resulting DEM is shown in Figure 17.5.

Following the generation of the DEM, a new 2D mesh is created which covers the same computational area as the 3D one. The 2D mesh is generated using the combined OpenFOAM utilities blockMesh, which creates an orthogonal, structured mesh, and snappyHexMesh, which creates an unstructured polyhedral mesh. The OpenFOAM utility extrudeMesh is then employed to generate a 2D mesh with "*empty*" patch types at the top and bottom surfaces. Such "*empty*" patches are necessary to force the OpenFOAM solution to be two-dimensional in nature. It is also at this stage that additional STL files may be incorporated into the 2D model to provide, for example, zones of different surface roughness for rivers, woodland, grassland,

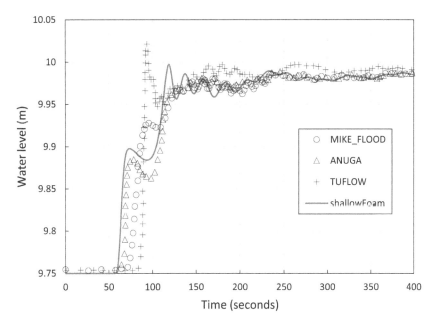

FIGURE 17.7 Water level versus time for point 1 in test Case 1 (see Figure 17.6 for point locations).

buildings, etc. – see Case 4 for an example of this. Finally, the digital elevations are mapped from the 3D solution onto the 2D mesh using the OpenFOAM utility mapFields. The solver shallowFoam is then ready to be run, taking advantage of OpenFOAM's unlimited parallel processing capability. Post-processing is via the open-source visualization application Paraview which is supplied with OpenFOAM.

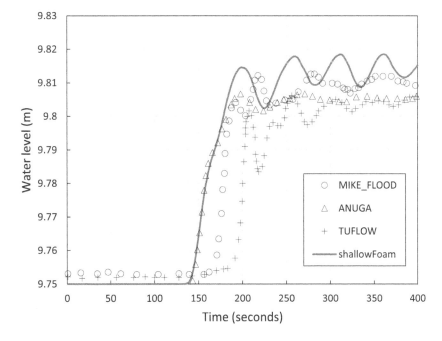

FIGURE 17.8 Water level versus time for point 2 in test Case 1 (see Figure 17.6 for point locations).

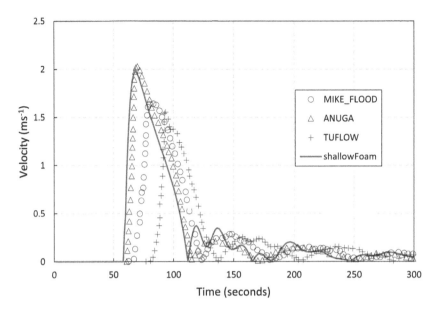

FIGURE 17.9   Water velocity versus time for point 1 in test CASE 1 (see Figure 17.6 for point locations).

17.4   RESULTS AND DISCUSSIONS

17.4.1   CASE 1

Test Case 1 consists of a sloping topography with two depressions separated by an obstruction as shown in Figure 17.6, and of width 100 m.

A time-varying inflow discharge is applied as an upstream boundary condition at the left-hand side, causing a flood wave to travel down the 1:200 slope. While the total inflow volume is only sufficient to fill the left-hand side depression at $X = 150$ m, some of this volume is expected to overtop the obstruction because of momentum conservation. The objective of the test is to assess the code's

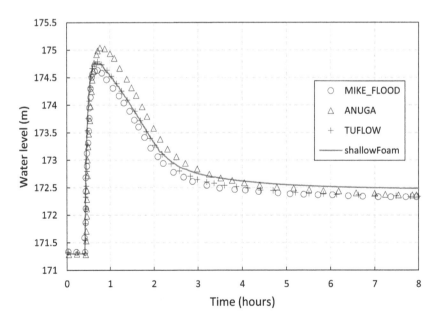

FIGURE 17.10   Water level versus time for point 1 in test Case 2 (see Figure 17.2 for point locations).

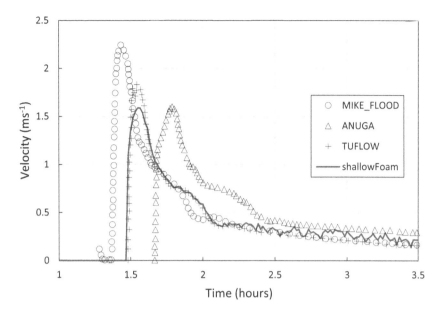

FIGURE 17.11   Water velocity versus time for point 4 in test Case 2 (see Figure 17.2 for point locations).

ability to conserve momentum over an obstruction in the topography and settle in the depression on the right-hand side at X = 250 m.

Figures 17.7–17.9 show that the shallowFoam results for water level and velocity are in satisfactory concurrence with the results produced by the commercial codes MIKE_FLOOD and TUFLOW and the open-source code ANUGA. This gave confidence in the shallowFoam code and the DEM methodology to proceed to the more challenging test Case 2.

This shallowFoam case was run on a single Intel Core i7-7820HK CPU at 2.90 GHz. A computational mesh of 1,200 cells was used, a time-step size of 1 s was employed, and the computational time to complete the run was 10 s.

17.4.2   CASE 2

The geometry and boundary conditions for test Case 2 valley dam break were outlined in the section on methodology. Figures 17.10–17.11 for the water level and velocity, respectively, show a very reasonable agreement between the shallowFoam model and the other flood codes.

A total simulation time of 30 hours was modeled in this case and Figure 17.12 shows the contours of the wetted area after two hours.

This shallowFoam case was run in parallel using three Intel Core i7-7820HK CPUs at 2.90 GHz. A computational mesh of 246,000 cells was used, a time-step of 1 s was employed, and the computational time to complete the run was 2.7 hours.

17.4.3   CASE 3

In this river and floodplain modeling test case, the site to be modeled is approximately 7 km long by 0.75 to 1.75 km wide (see Figures 17.13–17.14), and consists of a set of three distinct floodplains FP1, FP2, and FP3 in the vicinity of the village of Upton-upon-Severn, England.

In the test, the River Severn that flows through the site is modeled for a total distance of ~20 km. Boundary conditions are a hypothetical inflow hydrograph (DEFRA, 2013). The objective of the test is to assess the package's ability to simulate the fluvial flooding in a relatively large river, with floodplain flooding taking place as the result of riverbank overtopping.

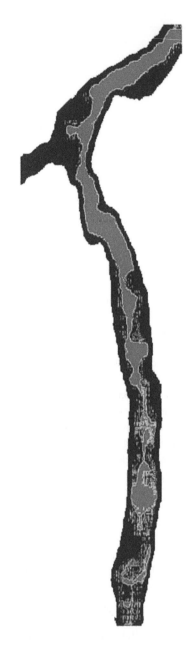

FIGURE 17.12   Contours of wetted area (red) for the shallowFoam simulation of test Case 2 after a duration of two hours.

It is evident that there is a relatively wide variation in the range of results for the codes tested in this complex benchmark case. The shallowFoam results, both qualitatively and quantitatively, appear to be in reasonable agreement across the spectrum of data presented with peak water levels and velocities corresponding well with the commercial codes as shown in Figures 17.15–17.17.

This *shallowFoam* case was run in parallel using three Intel Core i7-7820HK CPUs at 2.90 GHz. A computational mesh of 244,000 cells was used, a time-step of 1 s was employed, and the computational time to complete the run was 5.05 hours.

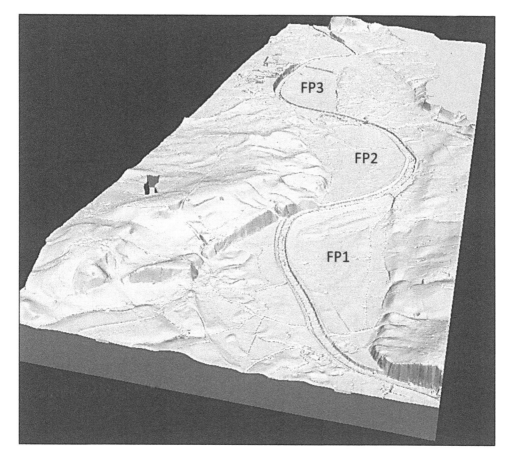

FIGURE 17.13 3D STL file generated by the QGIS plugin DEMto3D showing floodplain regions FP1, FP2, and FP3.

17.4.4 CASE 4

Case 4 is a practical example of how the methodology presented in this paper may be used to mitigate a flooding issue. In January 2015, storm Frank caused severe flooding and damage in the town of Ballater, Scotland (FRANK, 2019). The methodology has been applied to, firstly, model the flood as it occurred and suggest a possible flood defense solution. Storm Frank hydrograph inputs for the rivers Gairn/Dee and the river Muick were included as the inlet boundary conditions in the model and the shallowFoam code was run for a simulation period of 24 hours.

Figure 17.18 shows the extent of the computational domain and the proposed flood defense solution in red. Sampling points around the flood barrier are indicated by points P8 to P16. These points are used to extract the water level in front of the flood defense to give the necessary barrier height to counteract flooding during such a storm.

As discussed in the section on methodology, the different roughness parameters in terms of Manning-Strickler coefficients may be introduced into the model and these roughness zones are shown in Figure 17.19.

Figure 17.20 shows the excessive inundation occurring at the time of the peak river flow rate, while Figure 17.21 shows the mitigation of this by using flood barrier protection. The results from test Case 4 show that the methodology described in the book chapter can be applied to a practical situation and propose solutions for flood mitigation in complex topography.

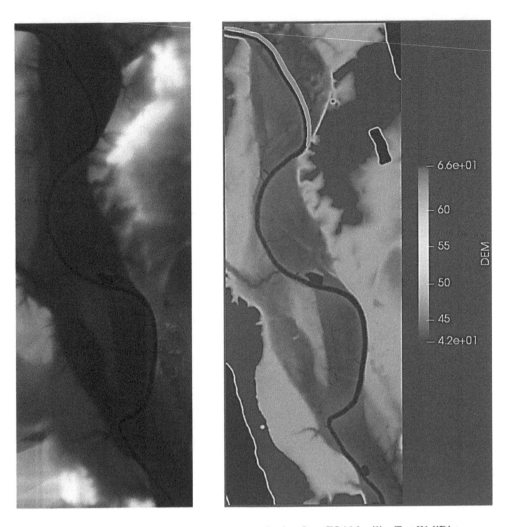

FIGURE 17.14 LiDAR image (left) and DEM generated using OpenFOAM utility Test-WallDist.

This shallowFoam case was run in parallel using three Intel Core i7-7820HK CPUs at 2.90 GHz. A computational mesh of 176,000 cells was used, a time-step of 1 s was employed, and the computational time to complete the run was 1.92 hours.

17.5 CONCLUSIONS

A new methodology for flood modeling using exclusively free-to-download, open-source software has been presented. The geographical information system software (QGIS) has been applied to a LiDAR image to produce a 3D file in stereolithography format. This file was then converted to a digital elevation model in the computational fluid dynamics package OpenFOAM. Finally, the shallow water equations were solved within the framework of OpenFOAM using the flood model shallowFoam to produce the maps of water height and velocity. In comparison with benchmark results for a range of commercial codes, the shallowFoam solutions compared well and produced satisfactory results for key flooding parameters. The final test case showed how the methodology could be applied to a practical flooding problem with the mitigation measures successfully implemented. Future work will include the incorporation of 1D–2D links for

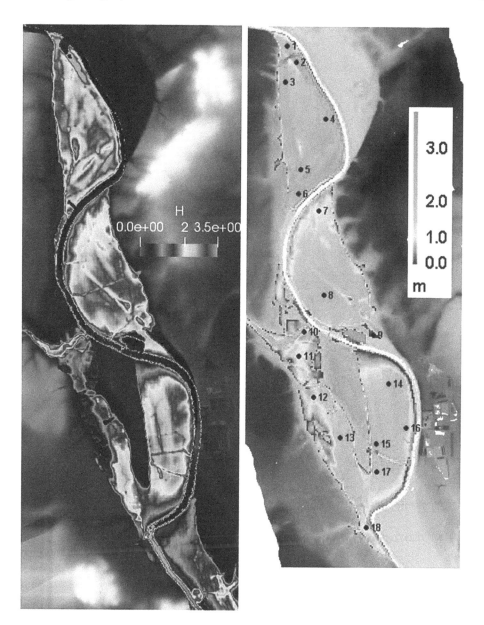

FIGURE 17.15 Peak water depths predicted by shallowFoam (left) and the commercial code ISIS-FAST (right). Sampling points P11 and P17 in floodplain FP3 are indicated on the right-hand figure.

features such as culverts and sluice gates and the implementation of a shock-capturing scheme for supercritical flows. The availability of such computational technologies means that effective open-source flood modeling software is now accessible for the wider academic, industrial, and citizen science communities.

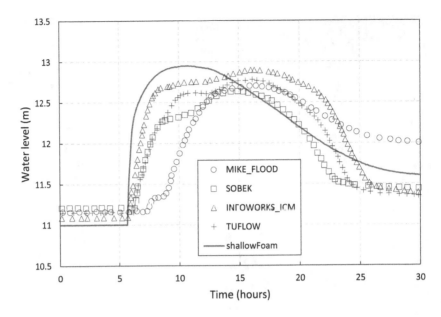

FIGURE 17.16 Water level versus time for point 11 in test Case 3 (see Figure 17.15 for point locations).

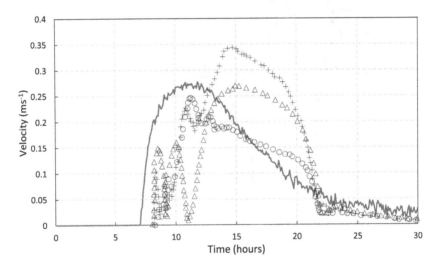

FIGURE 17.17 Water velocity versus time for point 17 in test Case 3 (see Figure 17.15 for point locations).

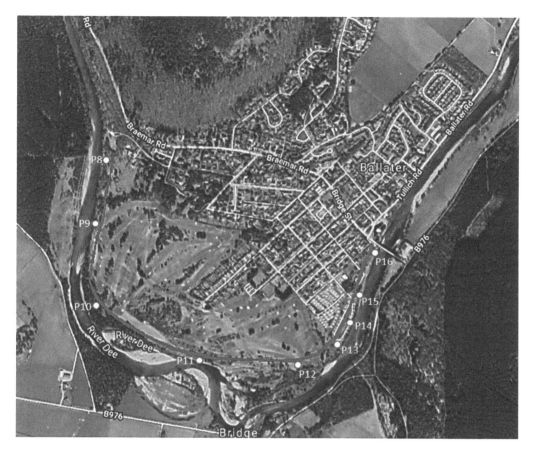

FIGURE 17.18 Extent of computational domain around the town of Ballater in test Case 4. Proposed flood defense barrier (red) and probe sampling points are also highlighted.

FIGURE 17.19 Roughness zones for different Manning-Strickler coefficients in test Case 4.

FIGURE 17.20   Contours of water height (m) at the time of maximum river flow rate (12 hours) with no flood defense.

FIGURE 17.21 Contours of water height (m) at the time of maximum river flow rate (12 hours) with flood defense in place.

REFERENCES

Alaghmand, S., Bin Abdullah, R., Abustan, I., and S. Eslamian, 2012, Comparison between capabilities of HEC-RAS and MIKE11 hydraulic models in river flood risk modeling (a case study of Sungai Kayu Ara River basin, Malaysia), *International Journal of Hydrological Science and Technology*, 2(3): 270–291.

ANUGA. https://anuga.anu.edu.au/ (accessed 25th October 2019).

Blöschl, G., Hall, J., and Živković, N. 2019. Changing climate both increases and decreases European river floods, *Nature*, 573: 108–111.

CCRA. 2019. http://ccra.hrwallingford.com/CCRAReports/downloads/CCRASummaryFloods.pdf (accessed 25th October 2019).

DEFRA.   2013.   https://www.gov.uk/government/publications/benchmarking-the-latest-generation-of-2d-hydraulic-flood-modelling-packages (accessed 25th October 2019).

ECIU. 2019. https://eciu.net/briefings/climate-impacts/flood-risk-and-the-uk (accessed 25th October 2019).

FRANK.   2019.   https://www.bbc.co.uk/news/uk-scotland-north-east-orkney-shetland-38332590   (accessed 25th October 2019).

INFOWORKS_ICM. 2019. https://www.innovyze.com/en-us (accessed 25th October 2019).

ISIS. 2019. https://www.floodmodeller.com/ (accessed 25th October 2019).

MIKE-FLOOD. 2019. https://www.mikepoweredbydhi.com/products/mike-flood (accessed 25th October 2019).

Mintgen, G. F. 2017. *Coupling of Shallow and Non-Shallow Flow Solvers - An Open Source Framework.* PhD Thesis, Technischen Universität München, https://mediatum.ub.tum.de/doc/1369622/1369622.pdf (accessed 25th October 2019).

OpenFOAM. 2019. https://www.openfoam.com/ (accessed 25th October 2019).

QGIS. 2019. https://qgis.org/en/site/ (accessed 25th October 2019).

Robins, E., and Lewis, M.J. 2019 Changing hydrology: A UK perspective, *Coasts and Estuaries - The Future*, 611–617.

ShallowFoam. 2019. https://github.com/mintgen/shallowFoam (accessed 25th October 2019).

SOBEK. 2019. https://www.deltares.nl/en/software/sobek (accessed 25th October 2019).

TUFLOW. 2019. https://www.tuflow.com/ (accessed 25th October 2019).

OpenORG. 2019. https://openfoam.org/news/funding-2019/ (accessed 25th October 2019).

Zeng, H., Grbčić, L., Lučin, I., and Kranjčević, L., 2018 Mesh creation for realistic terrain cases for shallowfoam –2D OpenFOAM solver. In: Katalinic, B. (ed) *Proceedings of the 29th DAAAM International Symposium*, pp. 1065–1070, DAAAM International, Vienna, Austria, ISBN 978-3-902734-20-4, ISSN 1726-9679, 10.2507/29th.daaam.proceedings.152.

18 Iterative Floodway Modeling Using HEC-RAS and GIS

Majid Galoie, Artemis Motamedi,
Jihui Fan, and Saeid Eslamian

CONTENTS

18.1 INTRODUCTION

The study of an inundation area (an area of land subject to flooding) and the evaluation of its water surface level are the most important parts of each flood management project. Floodplain analysis and mapping studies involve the intensive hydrologic and hydraulic modeling of catchments and rivers in order to evaluate the flood hydrographs and flow components. Floodway modeling is usually implemented and considered as the final task in determining a floodplain area. A floodway is normally a dry place located along a river where water flow through a riverbank can cover it.

Based on the Federal Emergency Management Agency (FEMA); a floodway is defined as "the channel of a river or other watercourse and the adjacent land areas that must be reserved in order to discharge the base flood without cumulatively increasing the water surface elevation by more than a designated height" (Ackerman, 2009). The base flood is a 100-year flood and the designated height of flood, which is also known as the surcharge, is 1.00 feet (0.305 m).

Since the formation of floodplain boundaries depends on topography and hydraulics of flow, the spatial distribution of terrain should be considered during floodway modeling. The Geographic Information System (GIS) is a digital layering and computerized system which is used to capture, store, retrieve, analyze, and display spatial data. This system is an effective planning tool in the various fields of sciences. Due to the nature and spatial distribution of floodway modeling, GIS

DOI: 10.1201/9780429463938-24

technologies, tools, and procedures are being used extensively in floodplain modeling and management. Basically, an encroachment analysis is required in order to evaluate the floodway of a river (Brunner, 2021a).

For floodplain modeling and visualization of input or output data, HEC-RAS was designed in order to model the hydraulic engineering properties of river flow and to import/export data to ArcGIS to facilitate the decision-making. Since the objective of floodway modeling is to keep the encroached water surface profiles within the specified surcharge value as defined by FEMA, the procedure is based on engineering judgments and it is usually an iterative process. The goal of this chapter is to provide further information about iterative floodway modeling using HEC-RAS. For this reason, a brief introduction about HEC-RAS and HEC-GeoRAS is presented first and then the floodway modeling will be discussed.

18.2   HEC-RAS

The HEC-RAS (Hydrologic Engineering Center – River Analysis System) model is currently one of the most famous engineering software in numerical modeling of various fields of hydraulics engineering, especially in flood management projects. Although HEC-RAS was developed by the US Army Corps of Engineers (USACE) in the 1990s in order to manage the rivers and improve the channels and waterways, this software in fact was built in 1964 with another name: HEC2. Table 18.1 represents a brief history of HEC-RAS.

The HEC-RAS software was designed by Gary Brunner, leader of the HEC-RAS software development team, and the first version was released in August 1995. During these years, this model has been improved in both steady and unsteady flow conditions, sediment transport interface, water quality computations, channel design and modifications, and also ice cover modeling. It should be mentioned that the different parts were added to the first version by different scientists.

TABLE 18.1
A Brief History of HEC-RAS

HEC-RAS Version	Release Date
1.0	August 1995
2.0	April 1997
2.2	October 1998
3.0	April 2003
3.1.1	May 2003
3.1.2	June 2004
3.1.3	May 2005
4.0 B	November 2006
4.0.0	March 2008
4.1	January 2010
5.0	February 2016
5.0.1	April 2016
5.0.3	September 2016
5.0.4	May 2018
5.0.5	June 2018
5.0.7	March 2019
6.0	June 2021
6.1	September 2021

Source: Wikipedia.

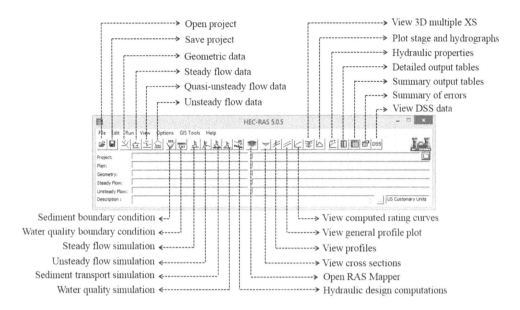

FIGURE 18.1   HEC-RAS software overview.

Today's HEC-RAS software builds upon all of the engineering advances and can be downloaded free of charge from the Hydrologic Engineering Center's home page at: www.hec.usace.army.mil/. It would enable to calculate the several variables such as water levels, flow depths, or flow velocities for the different flow conditions and various cross-sectional zones along rivers. The users can model the channels/rivers for both 1D and 2D flow simulations in various regimes such as subcritical, supercritical, and mixed flow. Significantly, the recent developments in sediment transport modeling and also water temperature modeling have been added to the software to accurately capture environmental conditions (Brunner, 2021b).

At the present time, the latest available version of HEC-RAS is version 6.1 (Figure 18.1) which includes a 2D flow analysis, allowing the consideration of much more complex rivers including lateral flows, eddies and bend losses, rain-on-grid modeling, and mapping capabilities (Brunner, 2021c).

This software can be used in the GIS program ArcView and can be read in (or used along with) AutoCAD, Civil 3D, and Map 3D drawings for the additional backgrounds. Also, it is the most appropriate one to use it in the combination with river CAD or HEC-GeoRAS (2D) to create cross-sections and other geometric data. Exporting data from HEC-RAS back into ArcGIS is possible easily and creating the flood maps with depths of flooding would be created quickly.

18.3   HEC-GEORAS

HEC-GeoRAS is an extension for ArcGIS with multiple tools for geospatial data analysis and extraction of all needed data for the HEC-RAS modeling projects. The HEC-GeoRAS output file, which is referred to as the RAS GIS import file, is a complex containing input data for HEC-RAS such as river, reach, and cross-section properties, cross-sectional cut lines, cross-sectional surface lines, cross-sectional bank stations, downstream reach lengths for the left over-bank, main channel and right over-bank; and cross-sectional roughness coefficients. Moreover, all hydraulic structures like bridges, weirs, levees, ineffective areas, and storage may be determined with this extension. HEC-GeoRAS has many capabilities in GIS visualization of inundation areas which can help and guide modelers to have a robust flood management modeling. One of the most helpful of these capabilities is publishing the KMZ output file which allows modelers to meet and animate the floodplain

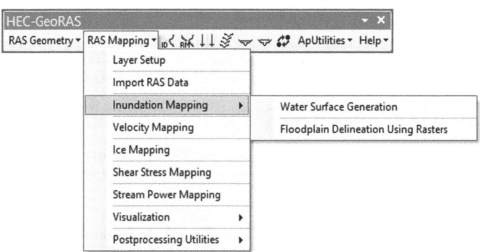

FIGURE 18.2   HEC-GeoRAS toolbar.

results in KML viewing clients like Google Earth (Ackerman, 2009). Figure 18.2 represents the HEC-GeoRAS toolbar menu in ArcGIS.

Since this extension is a geospatial analyzer, a Digital Elevation Model (DEM) is needed in order to derive the spatial data. For a successful and detailed floodplain modeling, the grid cell resolution of this DEM should not be very coarse because in this way geometric data and even river slope

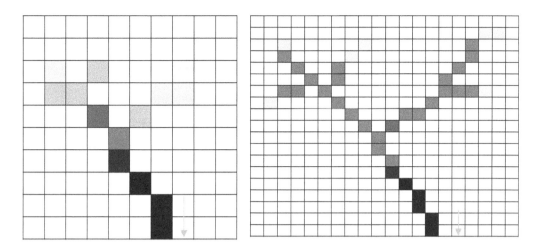

FIGURE 18.3 Derivation of river network in a given catchment using DEMs with different resolutions; left: 10 by 10 resolution cells, and right: 20 by 20 resolution cells.

may not be achieved properly. Figure 18.3 shows how a low-resolution DEM can affect the river's properties.

Since the preprocessing of data in HEC-GeoRAS is not covered by this chapter, further information and tutorials can be found in the HEC-GeoRAS user's manual (Merwade, 2010).

18.4 GENERAL EQUATIONS

In fluid dynamics, the general behavior of fluid is represented by conservation laws which are basically based on Newton's laws. In hydraulics engineering, these laws are expressed by the continuity equation, energy equation, and momentum equation. Most computational procedures in HEC-RAS are based on the one-dimensional (1D) equations, therefore only the one-dimensional general equations are represented here.

18.4.1 THE CONTINUITY EQUATION

The continuity equation (the law of conservation of mass) in hydraulics engineering which considers water as an incompressible fluid can be represented as:

$$Q = A_1V_1 = A_2V_2 \tag{18.1}$$

In which Q, A, and V represent discharge, cross-section area, and mean velocity respectively. The area of a cross-section when it is defined by the user as a polygon (Figure 18.4) can be computed by HEC-RAS using the following formula:

$$A = \frac{\sum_{i=1}^{n}\left(x_iy_{i+1} - x_{i+1}y_i\right)}{2} \qquad \left(\text{if } i = n, \text{ then } i+1 = 1\right) \tag{18.2}$$

In which n represents the number of nodes defining the cross-section boundary. The calculation must move in a sequential clockwise or counterclockwise along the cross-section boundary.

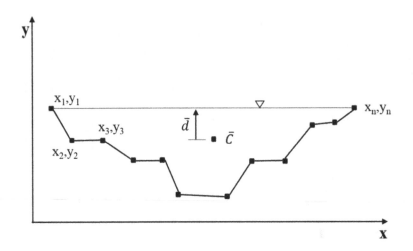

FIGURE 18.4   Definition of a cross-section in HEC-RAS.

18.4.2   THE ENERGY EQUATION

The energy equation, which is basically based on laws of thermodynamics and is also known as the Bernoulli equation, can be represented as (Figure 18.5):

$$Z_2 + y_2 \cos^2 \theta + \frac{\alpha_2 V_2^2}{2g} = Z_1 + y_1 \cos^2 \theta + \frac{\alpha_1 V_1^2}{2g} + h_{L_{1-2}} \tag{18.3}$$

In which Z, y, θ, V, g, and $h_{L_{1-2}}$ represent channel bed elevation, water depth, channel bed slope angle, mean velocity, and head loss respectively. The total head loss is the summation of all of the local and longitudinal losses along a reach.

18.4.3   THE MOMENTUM EQUATION

The 1D equation of momentum in open channel hydraulics is expressed as:

$$\frac{\beta_1 Q^2}{g A_1} + \overline{d_1} A_1 = \frac{\beta_2 Q^2}{g A_2} + \overline{d_2} A_2 \tag{18.4}$$

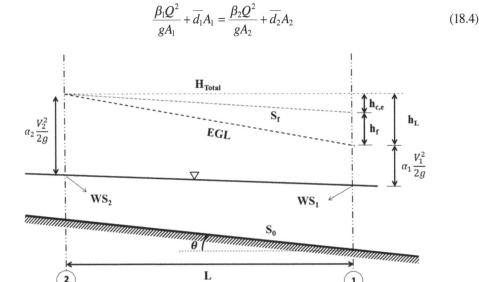

FIGURE 18.5   The energy equation parameters.

In which $\bar{d}$ is the distance between the centroid of the cross-section to the water level (Figure 18.4). This parameter can be computed by HEC-RAS using the following formula:

$$\bar{d} = \frac{\sum_{i=1}^{n} \left(x_i y_{i+1} - x_{i+1} y_i \right) \left(y_i + y_{i+1} \right)}{6A} \qquad \left(\text{if } i = n, \text{ then } i+1 = 1 \right) \tag{18.5}$$

In which n represents the number of nodes defining the cross-section boundary. The calculation must move in a sequential clockwise or counterclockwise along the cross-section boundary.

18.4.4 SAINT-VENANT EQUATIONS

In the case of modelling an unsteady flow, HEC-RAS uses a 1D dynamic wave equation (the Saint-Venant equation) in order to find the unknown parameters. The one-dimensional Saint-Venant equations in the conservative form are expressed as (Popescu, 2014):

$$\frac{\partial A}{\partial t} + \frac{\partial Q}{\partial x} = q \tag{18.6}$$

$$\frac{\partial Q}{\partial t} + \frac{2Q}{A}\frac{\partial Q}{\partial x} + \left(-\frac{Q^2}{A^2} + \frac{gA}{B} \right)\frac{\partial A}{\partial x} - gAS_0 + gAS_f = qu_q \tag{18.7}$$

Kinematic wave   ----------------------------|
Diffusion wave   ---|
Dynamic wave   --|

In which q and u_q are the lateral flow and velocity of lateral flow respectively. Equation (18.6) is a continuity equation and the equation (18.7) is a momentum equation, and both together are known as the 1D Saint-Venant equation (along x-direction). The analytical solution of this 1D equation is very difficult or impossible, except in very simplified cases (Popescu, 2014). Therefore, unsteady-state equations are usually solved by the numerical method which is briefly explained later.

18.5 BASIC COMPUTATIONAL PROCEDURE

18.5.1 CRITICAL DEPTH

HEC-RAS needs to calculate frequently many geometric data along a reach in order to evaluate and balance the basic governing equations (continuity equation, energy equation, and momentum equation). Most of these geometric data can be obtained directly from user input data and some of them must be calculated frequently during the modeling process. One of the main parameters which should be evaluated by HEC-RAS is critical depth, which has a key role in the computation of many hydraulics events. There are two methods available in HEC-RAS for computing the critical depth in a cross-section: the parabolic method and the secant method (Dyhouse et al, 2007).

Although the parabolic method is the default method for calculating the critical depth in HEC-RAS, sometimes it may give incorrect or inaccurate estimates of critical depth especially when there is more than one minimum on the total energy curve. This situation may occur when a cross-section is defined with wide and flat overbanks or ineffective flow areas. When the parabolic method is unsuccessful, HEC-RAS automatically (or by user) switches to the secant method which is computationally slower than the parabolic method. The computational procedure of critical depth in HEC-RAS is an iterative process which is briefly explained here for the secant method.

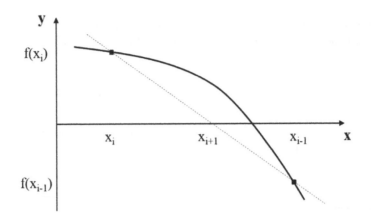

FIGURE 18.6 The secant method.

The secant method which is also known as a quasi-Newton method is an approximation method to find an iterative solution of the equation $f(x) = 0$. This method starts with two initial values x_0 and x_1 which should preferably be chosen close to the root. By connecting these two points $(x_0, f(x_0))$ and $(x_1, f(x_1))$, the secant line equation can be expressed as (Figure 18.6):

$$f(x) - f(x_1) = \frac{f(x_1) - f(x_0)}{x_1 - x_0}(x - x_1) \tag{18.8}$$

In the case of $f(x) = 0$, the equation can be rearranged and be expressed in the form of finite difference method as follows:

$$x_{i+1} = x_i - \frac{x_i - x_{i-1}}{f(x_i) - f(x_{i-1})} f(x_i) \qquad i = 2, 3, 4, \ldots \tag{18.9}$$

Equation (18.9) is used to find the critical depth in HEC-RAS iteratively. In this case, the function is:

$$f(x) = 1 - \frac{\alpha Q^2 T}{g A^3 \cos \theta} \tag{18.9}$$

In which T is the free surface width.

18.5.2 WATER LEVEL IN STEADY STATE

In order to find the water surface elevation $(z+y)$ in each cross-section and to draw the water surface profile along a reach, HEC-RAS assumes a one-dimensional steady $\left(\dfrac{\partial y}{\partial t} = 0\right)$ gradually varied flow $\left(\dfrac{V^2}{r} \approx 0,\ r \text{ is the radius of streamline curvature}\right)$ and computes it numerically (in the case of existing rapid variation in water surface profile, the momentum equation is used). Although in river networks, the water depth (y) and velocity (V) vary with respect to the time (which refers to as unsteady flow), these changes often occur very slowly and the assumption of steady flow may not alter the final results significantly even during a flood hydrograph (except in the case of the high velocities, i.e., steep mountain rivers).

There are several graphical and analytical methods available for the computation of water surface elevations in a 1D steady gradually varied flow, but two of them are most common: the direct step method and the standard step method. HEC-RAS uses the standard step method because of its applicability for both prismatic and non-prismatic channels. Furthermore, this method can be used for both sub-critical (Fr < 1) and super-critical (Fr > 1) flows. In the standard step method, water surface elevations are computed iteratively and based on the energy equation in the following form (Dyhouse et al, 2007):

$$WS_2 = WS_1 + \left(\frac{\alpha_1 V_1^2}{2g} - \frac{\alpha_2 V_2^2}{2g} \right) + \underbrace{L\bar{S}_f + C_{e,c} \left| \left(\frac{\alpha_2 V_2^2}{2g} - \frac{\alpha_1 V_1^2}{2g} \right) \right|}_{h_{L1-2}}$$
(18.10)

In which (the subtitle numbers refer to cross-section location):

WS: water surface elevation (or potential energy $= Z + y\cos^2\theta$),

$\frac{\alpha V^2}{2g}$: kinetic energy,

$h_{L_{1-2}}$: total energy loss due to the friction slope and expansion/contraction between two adjacent nodes (cross-sections) on a 1D computational domain,

L: the distance between the two adjacent nodes,

$\bar{S}_f$: the average energy slope between the two adjacent nodes,

C_e: the contraction coefficient (for supercritical flow is often taken as 0.3),

C_c: the expansion coefficient (for supercritical flow is often taken as 0.1).

All of these parameters are shown in Figure 18.5.

The iterative computational procedure of the standard step method can be started if boundary and initial conditions are available. It means that discharge, geometry, roughness values (Manning's values), and expansion/contraction coefficient should be known at each cross-section. Also, the direction of the computation depends on the flow regime in the reach; for subcritical flow, the flow control section is located at the downstream end of the reach, and computation would begin at this location and proceed upstream. For supercritical flow, the situation is vice versa. The iterative procedure for computation of water level at each cross-section along a reach under a subcritical state is shown as a flowchart in Figure 18.7.

The iterative procedure for a supercritical state is almost similar but the direction of computation is repercussive (from the most upstream cross-section to the downstream end of a reach). It should be noted that although the computational procedure of water level is based on an iterative process, it usually can be obtained by four or five iterations.

18.5.3 WATER LEVEL IN UNSTEADY STATE

As it was mentioned in section 18.3.4, HEC-RAS uses a 1D Saint-Venant equation for the numerical solution of unsteady flow in a reach. Although the capability of numerical modeling of 2D unsteady flow is added to HEC-RAS in some new versions (version 5 and higher), 1D simulation for simplification is still applied for modeling of many river network systems whose flow is under an unsteady state (Brunner, 2021a).

In order to solve the 1D Saint-Venant equation, HEC-RAS utilizes a frequently used numerical scheme known as a Preissmann scheme (also known as box scheme) which is a finite difference method and it is suitable for the numerical solution of hyperbolic types of equations (Popescu, 2014). The scheme is developed based on four nodes of the computational grid and the derivatives are not approximated at these nodes but at points inside the grid (Figure 18.8).

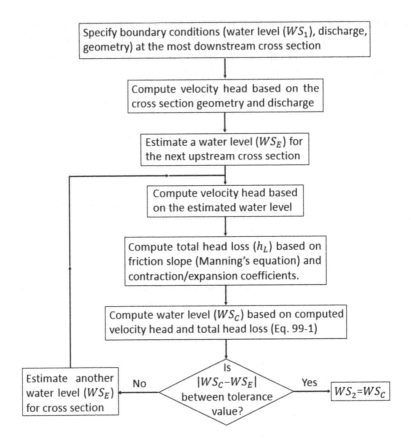

FIGURE 18.7 Computational iterative procedure of water surface elevation in a subcritical flow.

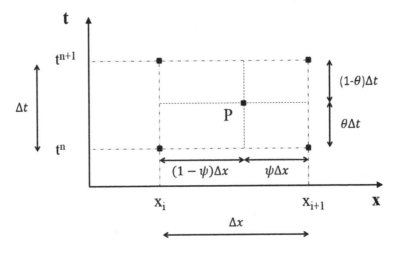

FIGURE 18.8 Priessmann numerical scheme.

The derivative of each quantity like N (with respect to time and space) can be approximated based on the following relations:

$$\frac{\partial N_P}{\partial t} = \psi \frac{N_{i-1}^{n+1} - N_{i-1}^{n}}{\Delta t} + \left(1 - \psi\right) \frac{N_{i}^{n+1} - N_{i}^{n}}{\Delta t} \qquad 0 \leq \psi \leq 1 \tag{18.11}$$

$$\frac{\partial N_P}{\partial x} = \theta \frac{N_{i}^{n+1} - N_{i-1}^{n+1}}{\Delta x} + \left(1 - \theta\right) \frac{N_{i}^{n} - N_{i-1}^{n}}{\Delta x} \qquad 0 \leq \theta \leq 1 \tag{18.12}$$

Considering these relations instead of derivatives in the 1D Saint-Venant equation leads us to the FDM form of unsteady flow in open channel flow modeling. For each node in the computation of domain, in this case, there are two unknown parameters that should be computed numerically. As $\theta > 0$, the scheme is implicit and for the M nodes in each row, the number of 2M relations for the 2M unknown parameters is obtained which should be solved by any iterative specific algorithm (i.e., double sweep algorithm).

18.6 FLOODPLAIN ANALYSIS

Since each floodplain analysis needs an encroachment analysis, the encroachment analysis in HEC-RAS is presented here.

18.6.1 THE ENCROACHMENT ANALYSIS IN HEC-RAS

An encroachment analysis will show the limits of encroachment which will change the water surface elevation until the specified value. The encroachment procedure is mainly relied on the computation of a natural profile as the first profile in a multiple profile run (Ackerman, 2009). In a run, the other profiles are computed using one of the various encroachment options as above. After this step, the floodway will be divided into the floodway fringe and the floodway. Floodway fringe is blocked by the encroachment and the floodway is the portion of floodplain in which a 100-year flood must flow without raising the water level more than the target amount (Ackerman, 2009). Therefore, a model must be performed for an existing river system for at least the 100-year flood, then it should be calibrated and validated before preparation for the encroachment analysis.

There are five options available in HEC-RAS for evaluating encroachments in a steady state (Brunner, 2021a):

1) Specify the left and right encroachment stations.

In this option, the modeler manually specifies the exact location of the encroachments on either side of the main channel, so an equal conveyance condition may not exist. Figure 18.9 shows a cross-section that defines method 1. The vertical distance between the encroached water surface and the natural water surface is the surcharge.

2) Specify floodway top width.

In this option, the modeler specifies a floodway top width for each cross-section and it is centered on the channel centerline by HEC-RAS as shown in Figure 18.10. As encroachments cannot be specified into the main channel, the narrow encroachments will be fixed automatically at the channel bank stations. This option is good when one wants to set equal widths from the centerline in floodway modeling.

3) Specify the percent conveyance reduction.

In this option, the modeler specifies a percentage reduction in the conveyance at a cross-section of interest and by default, HEC-RAS is considered, one-half of which in each

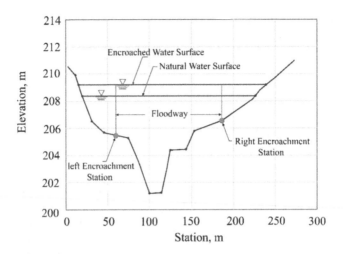

FIGURE 18.9 Specify left and right encroachment stations – method 1 (Ackerman, 2009).

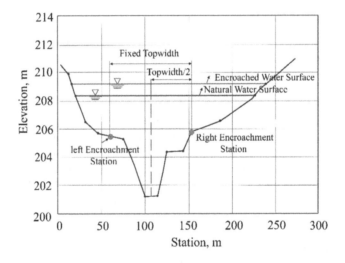

FIGURE 18.10 Specify left and right encroachment stations – method 2 (Ackerman, 2009).

side of the cross-section. HEC-RAS also allows the modeler to determine the conveyance reduction in each side of the cross-section either equal or proportional. In Figure 18.11, K represents the percent of conveyance reduction in the model.

4) Specify target surcharge to reduce conveyance equally.

In this option which is almost similar to the previous one, the modeler specifies a target surcharge and HEC-RAS computes the difference between the base 100-year flood elevation and the surcharge elevation at the cross-section. An equal conveyance reduction which is the difference between the base and surcharge conveyances is performed on each side of the cross-section (Figure 18.12).

5) Specify target surcharges for water surface and maximum change in energy.

In this option, the modeler specifies two targets: a water surface level and an energy grade elevation (Figure 18.13). In fact, this is option 4 plus an extra target for defining energy grade elevation which would be met the first target by equal conveyance. If only one of the targets is specified, HEC-RAS considers only the specified target, or else this optimization algorithm is run up to 20 iterations per cross-section in order to satisfy the

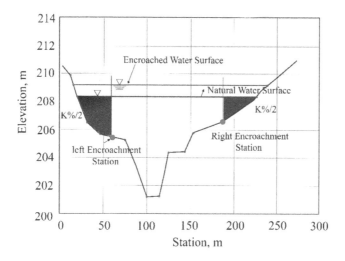

FIGURE 18.11 Specify left and right encroachment stations – method 3 (Ackerman, 2009).

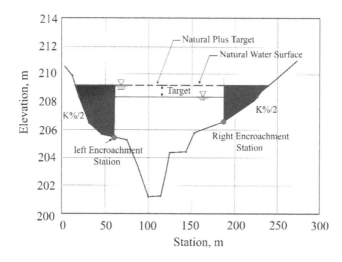

FIGURE 18.12 Specify left and right encroachment stations – method 4 (Ackerman, 2009).

two targets. If this iterative process is failed, among 20 iterations, the best values are considered.

So far for an unsteady state, only option number one is available in HEC-RAS (version 6.1).

18.7 ITERATIVE FLOODWAY MODELING

The procedure of floodway modeling using HEC-RAS is performed iteratively so that a base flood modeling together with multiple profile runs for varying floodways are implemented. Most often, options 4 (or 5) and 1 are used as encroachment analysis, and the procedure iterates until all of the cross-sections have an allowable surcharge. The process is started in GIS (HEC-GeoRAS) and an initial floodplain is developed. Then, a steady-state analysis is implemented in HEC-RAS for a 100-year flood in order to prepare the data for the encroachment analysis. During this process and by switching between HEC-RAS and ArcGIS visual environment, the map floodway boundaries are modified and improved. Finally, a floodway with a smooth boundary and base as per FEMA

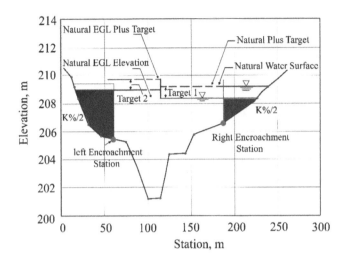

FIGURE 18.13   Specify left and right encroachment stations – method 5 (Ackerman, 2009).

descriptions will be developed in this iterative procedure which is briefly explained by a flowchart in Figure 18.14.

In the proceeding sections, an iterative floodway process is briefly presented.

18.7.1 Steady Flow Analysis and Encroachments

Floodway analysis in HEC-RAS is usually started from the steady flow analysis window (Figure 18.15). The encroachments window then can be called from the Options menu. In floodway analysis, all profiles should use the same discharge for the base flood which is the first profile in the analysis. As it was mentioned before, there are five options available in this editor for encroachment analysis which can be simply selected.

The water surface elevation for starting the analysis is very important because, for a floodway, the water surface profile should be considered as the water surface of a 100-year un-encroached profile plus the target surcharge. For method 4, the normal depth for floodway profile may be better than using the water surface profile plus the target surcharge. There are many options available for the modeler in the encroachments analysis which can be reviewed in the reference manual of HEC-RAS (Brunner, 2021a).

After the encroachment analysis, the modeler can view the results with the 3D plot tool of HEC-RAS (Brunner, 2021b). This useful tool provides a 3D view of the river in which floodway top width and boundaries can be checked. The final floodway limits should be set through method 1. Using this method, the modeler can adjust the encroachments so that the appearance of the floodway would be enhanced.

18.7.2 Excessive and Negative Surcharges

In floodway modeling, the surcharge values should be kept between 0.0 and 1.0 feet (typically 1.0 feet). After finishing the floodway analysis, the modeler should investigate the excessive and negative surcharge values from downstream cross-section to upstream. Negative surcharges or increases in water elevation (even 0.001 feet) above the allowable increase are not accepted.

If in a floodway profile the water surface elevation is less than the natural profile, then the negative surcharge occurs. By widening narrow floodways, correcting bridge modeling, narrowing wide non-optimized floodways, and inserting additional cross-sections may eliminate the negative surcharge values.

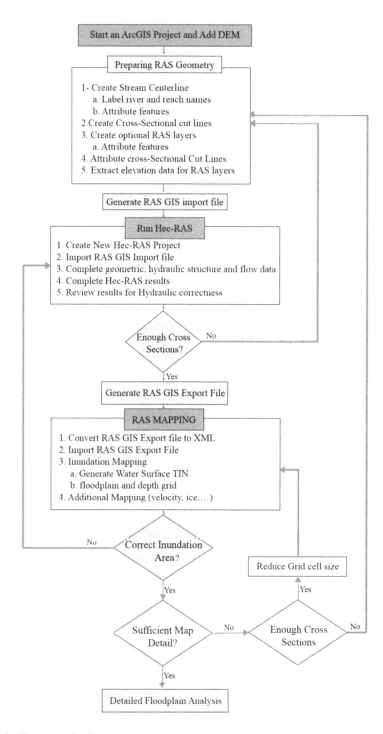

FIGURE 18.14 Flowchart for floodway modeling using HEC-GeoRAS and HEC-RAS.

Since the energy grade elevation at each cross-section is the sum of water surface elevation and velocity head, therefore, if the velocity head at a cross-section is less than that of the base profile then the water surface elevation increases to balance the energy grade elevation which is higher than the allowable surcharge value. Also, if the difference in energy grade elevation between the base profile and floodway profile is greater than the maximum allowable surcharge at a cross-section,

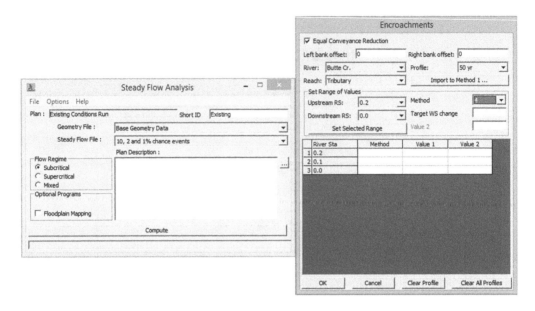

FIGURE 18.15 Steady flow analysis window and encroachment analysis options.

then the water surface elevation at the next upstream cross-section may be increased to balance the energy.

18.8 CONCLUSIONS

In this chapter, a brief guideline is provided in order to facilitate floodway modeling using the HEC-RAS software. Also, HEC-GeoRAS which is an extension for ArcGIS is introduced to easily prepare and export the input file to HEC-RAS. The floodway map, which is the final objective of every floodplain analysis project, can be achieved through an iterative process using HEC-RAS and HEC-GeoRAS. The procedure of floodway analysis is started by a steady-state analysis and encroachments analysis for a 100-year flood profile (base flood profile) and then continued with the multiple surcharge elevations. HEC-RAS provides five options for encroachments analysis which can be used individually or in combination with other options within the same floodway analysis run. Most of the time, a target surcharge is specified and HEC-RAS computes the difference between the base 100-year flood elevation and the surcharge elevation at the cross-section. An equal conveyance reduction which is the difference between the base and surcharge conveyances is performed on each side of the cross-section (option 4). In addition, sometimes, two targets may be specified: a water surface level and an energy grade elevation, which is option 4, plus an extra target for defining energy grade elevation (option 5). Practically, one of these options is used for the encroachments analysis and it will be finalized with a manual specification of the exact location of the encroachments on either side of the main channel. In addition to the floodway procedure in HEC-RAS, an intensive engineering judgment is also needed for successful and reliable results.

REFERENCES

Ackerman, C.T. 2009. *HEC-GeoRAS; GIS Tools for Support of HEC-RAS Using ArcGIS*, United States Army Corps of Engineers, Davis, USA.

Brunner, G.W. 2021a. *HEC-RAS River Analysis System. Hydraulic Reference Manual. Version 6.0*, Hydrologic Engineering Center, Davis CA, USA.

Brunner, G.W. 2021b. *HEC-RAS River Analysis System: User's Manual. Version 6.0*, US Army Corps of Engineers, Institute for Water Resources, Hydrologic Engineering Center, USA.

Brunner, G.W. 2021c. *HEC-RAS River Analysis System: 2D Modeling User's Manual. Version 6.0,* US Army Corps of Engineers, Institute for Water Resources, Hydrologic Engineering Center, USA.

Dyhouse, G., Hatchett, J., and Benn, J. 2007. *Floodplain Modeling Using HEC-RAS*, Bentley Institute Press.

Merwade, V. 2010. *Tutorial on Using HEC-GeoRAS with ArcGIS 9.3*, School of Civil Engineering, Purdue University, USA.

Popescu, I. 2014. *Computational Hydraulics*, IWA Publishing, UK.

19 Floodplain Mapping Using HEC-RAS and ArcGIS

Mehdi Vafakhah, Saeid Janizadeh,
Mohammadtaghi Avand, and Saeid Eslamian

CONTENTS

19.1 INTRODUCTION

19.1.1 Background

Flood phenomena are among the most dangerous natural disasters that threaten human societies. The frequency of floods in recent decades has caused most regions to be exposed to periodic and destructive floods and the mortality and financial losses of floods have increased dramatically (Degiorgis et al., 2012). Floodplains and adjacent rivers, which, due to special circumstances, are suitable spaces to for economic and social activities, are always threatened by the dangers of flooding. Therefore, in these areas, determining the rate of flooding advance in different return periods, as well as flood depths relative to the surface of the earth (called "the flood map"), is of great importance (Motevalli and Vafakhah, 2016). Flood zoning is one of the most sensitive stages in river management. One of the newest methods for providing a flood zoning map is the use of Geographic Information System (GIS) and its integration with hydraulic models. Flood zoning plans are widely used in floodplain management studies (Jalayer et al., 2014). Hydraulic models can be categorized according to the number of dimensions in which they represent the spatial domain and flow processes and, for particular problems, one-, two-, and three-dimensional models may be more suitable (Hunter et al., 2007). The main objectives of this chapter are: (1) to train the steps of implementing the HEC-RAS hydraulic model; (2) to apply HEC-RAS model in two case studies used in flood hazard mapping.

19.2 FLOOD ZONING MODELS

One of the branches of computational fluids dynamics is the water or hydraulic engineering science, which has recently been a success story. Computational fluid dynamic science can be successful in simulating and predicting water behavior by applying mathematical flow simulation models based on computational fluid dynamics (Villazón et al., 2013). Several flood zonings models have been developed in the world that has been categorized to varying degrees in terms of complexity and ease of use. In this regard, numerous hydrodynamic models of one-, two-, three-dimensional and dual-axle hydraulics have been developed in the field of hydraulic flood simulation (Amini et al., 2019; Bates, 2004).

The processes involved in the current model are of three-dimensional nature, which ideally should use 3D models to model them, but 3D models are time-consuming. While one-dimensional and two-dimensional models, which are obtained by various simplifications such as intermediate cross-sections, depth and width, are suitable solutions for engineering activities (Lazaridou et al., 2004). One-dimensional models are widely used in simulation studies of flood and sedimentation processes in direct drains, separating currents and at the junction of waterways. In one-dimensional models, the variations of the parameters are assumed to be small in two dimensions, and the governing equations on the flow mediocre in those two dimensions. When variations in the mean quantities of flow in the vertical direction (flow depth) are negligible, for this reason, using simpler assumptions, two-dimensional equations can be used in this design. Two-dimensional models should be used for flow analysis. In order to select the appropriate model for flow simulation in the study area, several major criteria should be considered: The purpose of the modeling is to consider the spatial dimensions of the flow (one-dimensional or two-dimensional), the time dimension of the flow (steady or unsteady) (Horritt and Bates, 2002).

Hardy et al., (1999) showed that one-dimensional models require limited computational time, but only allows the user to examine the hydraulic parameters of the river channel. While the two-dimensional models provide the ability to study the changes in the water level between the main channel and the flood of the plains and the two-dimensional flow in the flood plains. Moussa and Bocquillon (2009) showed that one of the most important advantages of one-dimensional models in comparison with two-dimensional models is the simplicity of the governing equations (due to the neglect of some parts of the equations with considering the physical condition of the justifying

TABLE 19.1

Types of Flow Analysis Models

Row	Model Name	Developer	Dimension	Model Structure
1	WMS	Aquaveo	1D	Hydraulic
2	MIKE11	Danish Hydraulic Institute (DHI)	1D	Hydraulic
3	CCHE2D	University of Mississippi	2D	Hydraulic
4	HEC-RAS	U.S. Army Corps of Engineers	2D and 1D	Hydraulic

ruler), less computational nodes, resulting in lower computational volume, shorter runtimes, and easy analysis of their results. In contrast, one-dimensional models compute the hydraulic parameters such as velocity, flow depth, water surface width, etc. in the middle, and do not provide information on flow details in the flood plains. Also, the flow paths must already be identified for them. While two-dimensional, three-dimensional or dual-mode models provide flow-flood detail (depth, velocity, etc.) without the need to define flow paths, they require more execution time due to heavy computing (Table 19.1).

19.2.1 WATERSHED MODELING SYSTEM

The Watershed Modeling System (WMS) model is a complete environment for hydraulic analysis, developed and expanded by the Bringham Hydraulic Laboratory. By integrating GIS and various hydrologic models, this software has been able to create a powerful tool for hydrologic simulation of watersheds, so that raw input data are first downloaded by the software and, after initial processing, to create the required digital platform, like are converted to Triangular Irregular Networks (TIN) or Digital Elevation Model (DEM) format. Then, WMS calculates the watersheds and sub-watersheds by extracting the drains and receiving the outlet, and finally calculates the appropriate hydrologic model of the hydrological rainfall, which can be done according to the needs of the project under study to simulate flood plains and flood degradation effects, or hydraulic calculations of structures along the waterways and in urban environments to design a flood system in the city (Mohamadi et al., 2009).

19.2.2 MIKE 11 SOFTWARE

This software was developed by the Danish Hydraulic Institute (DHI) and is capable of simulating one-dimensional flow, sediment transport and unstable water quality in rivers, estuaries and irrigation networks (Alaghmand et al., 2012). This program uses a finite difference method to solve a dimensional flow governing, sediment transport and water quality equations. The hydrodynamic model is in fact the underlying element of all of these systems (Thompson et al., 2004).

19.2.3 CCHE2D MODEL

The CCHE2D model is a numerical model for simulating the turbulent non-permanent flow and sediment transport in open channels that was developed at the International Center for the Study of Hydraulics and Computing Engineering, Faculty of Engineering, University of Mississippi. This model is a two-dimensional hydrodynamic model that has a pre-processor CCHE-MESH for geometry and field networking, and solves the flow and sediment transport field as well as graphical observation results in graphical software environment, called CCHE-GUI (Ding et al., 2016).

19.2.4   HEC-RAS Model

A complex package is a series of hydraulic analysis programs in which the user communicates with the system through the user's graphic user interface. The system has the ability to perform water-level profile calculations in stationary and non-stationary flow conditions, sediment transport calculations and several other hydraulic designs (Warner and Brunner, 2001). A project in the HEC-RAS terminology series is a collection of data files associated with a particular river system. The user can perform one or more different types of analyzes included in the HEC-RAS bundle as part of the project. The data files of a project are categorized as follows: plan data, geometric data, continuous flow data, unexplained flow data, sedimentation data and hydraulic design data (Alaghmand et al., 2012; Brunner, 2010).

The data required for the preparation of flood hazard maps with HEC-RAS model include digital terrain models, cross-sections of the river, the Manning roughness coefficient and discharge with different return periods. The scale of maps used to produce a hazard map for the river plan and its margin is a scale of 1:1,000 to 1:5,000; for longitudinal profiles, a horizontal scale of 1:1,000 to 1:5,000 and a vertical scale of 1:10 to 1:500; for horizontal profiles, a horizontal cross-sectional of 1:100 to 1:2,000 and vertical scale of 1:10 to 1:200 are used.

19.2.4.1   Steps to Create a Flood Hazard Mapping with HEC-RAS

There are six main steps in creating a flood hazard mapping with HEC-RAS:

- Create new project;
- Introduce river network and enter geometric information of sections;
- Introduce flow and boundary conditions;
- Calculate hydraulic of the model;
- View output results;
- Generate final report for entering GIS.

19.2.4.2   Creating a New Project

The first step in creating a hydraulic model with HEC-RAS is the route you want to work and enter a title for the new project.

19.2.4.3   Introducing River Network and Entering Geometric Information of Sections

The next step is to enter the geometric data required, including the river system schematic, transverse sectional data and hydraulic structures data (bridges, culverts, dams, etc.). The user creates geometric data first by plotting the schematic diagram of the river system. The river system is carried out based on a reach-by-reach basis and drawing in a reach from upstream to downstream (in a positive flow direction). The river and reach identifiers can be up to sixteen characters in lenght. During connection reach, splits are formed automatically by the interfaces. After plotting the schematic design of the river system, the user can begin entering data of transverse sections and hydraulic structures. Each cross-section has a river name, a reach name, a river station, and a description. River indicators, reach and river station are used to illustrate where this cross-section is located in the river system. The river station index should not be the actual river station (miles or kilometers), the transverse section is located in that station in the stream, but a numerical value must be assigned to it (for example, 1/1, 2, 5/3, ...). These numerical values are used to place cross-sections in an appropriate order within the range. Cross-sections are arranged within an interval from the highest river station to the lowest station at the bottom.

After entering cross-sectional data, the user can add any hydraulic structure such as bridge, culvert and overflow to the model. Data editors are similar to the cross-section data editor for all types of hydraulic structures. If there are splits of other waterways in the river system, then we will need to enter additional data for each branch. The split data editor is available through the geometric data window. These data should be stored in a file.

Another method is to use the HEC-GeoRAS software to provide geometric data. To do this, a digital earth model should be available as a raster (grid squares), triangular irregular networks (TIN), or level lines.

First, layers of the central axis of the river and the left and right sides of the river should be created in ArcGIS and then, with cross-sections perpendicular to the river axis on a digital earth model, cross-sections were created in the ArcGIS environment and entered into the HEC-RAS to make.

19.2.4.4 Entering Flow Data and Boundary Conditions

After entering geometric data, the user can enter permanent and non-permanent flow data. The type of input data depends on the type of analysis desired. Permanent flow data include the number of profiles to be calculated, flow data, and boundary conditions of the river system. At least one flow rate for each reach in the system should be included. Flow discharge can be changed at any position within the system, in addition to the river. Flow rate values for all profiles must be entered.

Boundary conditions are essential to establish the starting water surface at the ends of the river system. If a subcritical flow analysis is to be considered, then only boundary conditions will be required at the downstream, if the supercritical flow analysis is to be considered, then the definition of boundary conditions in the upstream will be necessary. If the user wishes to perform mixed-flow regime calculations, it is necessary to introduce boundary conditions upstream and downstream.

19.2.4.5 Performing Hydraulic Calculations

After geometric data and flow data were entered, the user can start the hydraulic calculations. Three types of permanent flow analysis, non-permanent flow analysis, and hydraulic design can be done with the HEC-RAS. The user can use any of the existing hydraulic analyses from the "run menu" bar on the main HEC-RAS window.

19.2.4.6 View and Print Results

After completing all calculations by the model, the user can see the results.

Several examples of output results of the model include cross-sectional diagrams, surface water profile charts, discharge rating curve charts, three-dimensional XYZ graphs, outputs in different situations as tables, outputs for multiple situations as a table, and a summary of errors, warnings, notes, and sample from a cross-sectional graph.

Graphic outputs can be printed in two different ways. Graphic diagrams can be sent directly from HEC-RAS to the printer or platform that the user has introduced on the Windows print manager. Graphic charts can also be sent to the Windows clipboard. After copying to the clipboard, it can be sent to other word processors.

Output results are available in two different ways. The first type outputs the hydraulic results table in detail in the position of a particular cross-section. The second type of output indicates a limited number of hydraulic variables for several cross-sectional sections of different profiles.

19.2.4.7 Implementation of Flood Risk Map within ArcGIS

From the HEC-RAS file menu, select Export GIS Data.

Figure 19.1 shows the most common data that must be entered into ArcGIS environment.

The message for creating this information is then given as Figure 19.2.

After adding the HEC-GeoRAS extension to ArcGIS and to ensure the presence of two spatial analyst and 3D analyst extensions needed to execute program the ArcGIS environment. The steps are then as follows: invoke the saved project file to the ArcGIS environment while making sure that the HEC-GeoRAS extension is open (see Figure 19.3).

Click the "Import RAS SDF File" icon. At this point, the SDF file is converted to an XML file. The XML file name must be the same as the original file. Then press the "OK" button (see Figure 19.4).

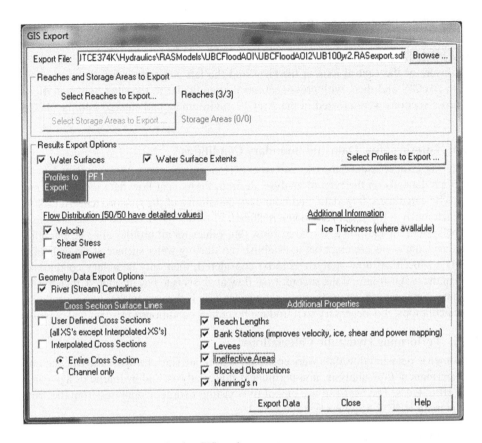

FIGURE 19.1   Data needed to enter the ArcGIS environment.

FIGURE 19.2   Creates a file for calling in the ArcGIS environment.

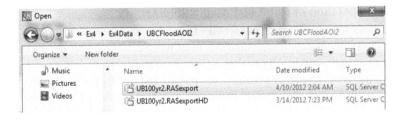

FIGURE 19.3   Project file saved to call ArcGIS environment.

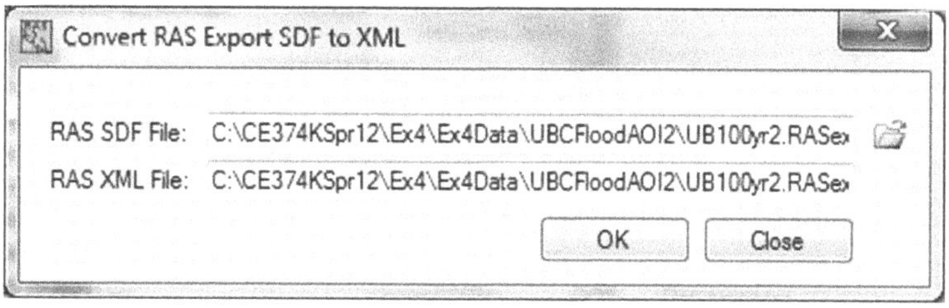

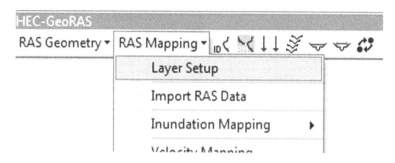

FIGURE 19.4 How to call sdf file in ArcGIS environment.

FIGURE 19.5 How to define a newly called project.

If there is a message about the XML file on the page, it shows that there is no need to run this program. Define the newly called project by executing the Layer Setup command from the RAS mapping menu on the HEC-GeoRAS extension as in Figure 19.5.

The items to be defined include the name of the new project (HermineDam). The file name should be RAS GIS) UB100yr2.RASexport.XML(; the file name of the digital earth model (ubcaoi); the path to save the output file and the size of the raster cell that is accepted by the default software. If you define the above items correctly, Figure 19.6 will be appeared.

Then press "OK". At this time, the new file will be created with the name of the original file of the new project, along with the data of the digital land model on which it was uploaded. But data is not actually called yet (the digital earth model is not selected by default), so the new data page will appear as blank. Save the project at the end. Figure 19.7 shows the file created after executing this command.

Next click on RAS Mapping -> Import RAS, you will see a series of messages as shown in Figure 19.8. The execution of this command may take several minutes depending on the complexity of the results (Figure 19.9).

Through this process, GIS reagent files, bank points, velocities, water surface extent, River2D central line, and XS cut lines have been used to derive transverse cross-sections on a DEM and a bounding polygon is around the data. Finally, Figure 19.10 will be created.

If you only select ubcaoi, Figure 19.11 will be appealed.

To generate a water level TIN, order the Water Surface Generation from the Inundation Mapping submenu from the RAS mapping menu, as shown in Figure 19.12. If you want the water level TIN to appear, check the "Draw Output Layers" option. Choose other options for combining and smoothing the floodplain polygon (these are usually selected automatically). The result of this command is given in Figure 19.13.

Flood area and depth of water can be represented by selecting "Floodplain Delineation Using Raster" from submenu Inundation Mapping from the main menu RAS Mapping run as shown in Figure 19.14. Select one or more profiles that you want to create the flood area (automatically, deep squares grid will be generated). This will only work for the profiles previously selected. If you want

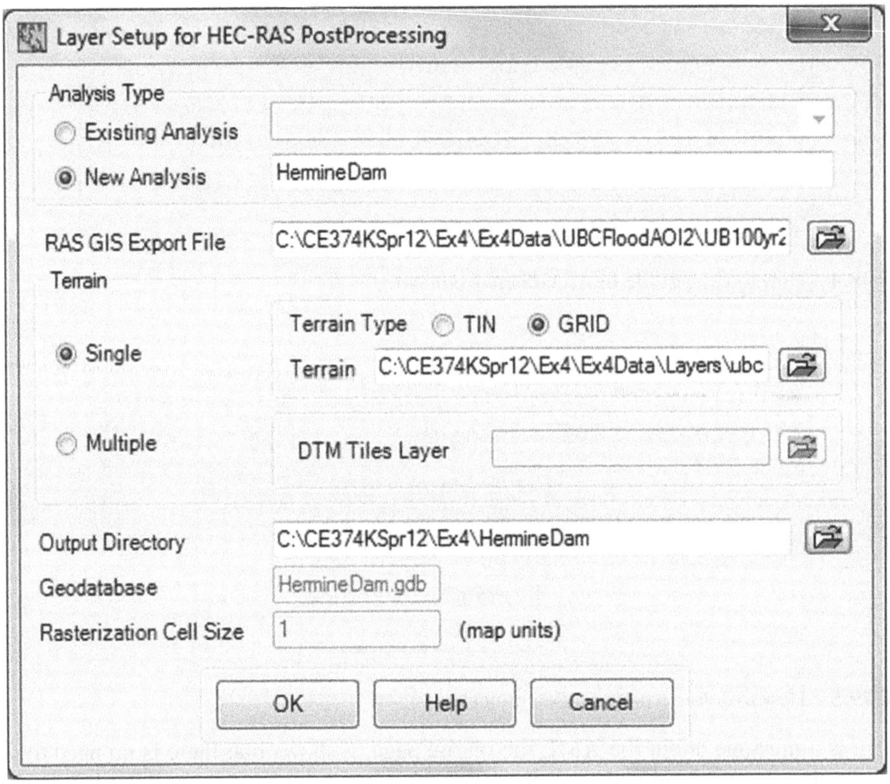

FIGURE 19.6   How to define items in the Layer Setup menu.

FIGURE 19.7   The file generated after the execution of the Layer Setup command.

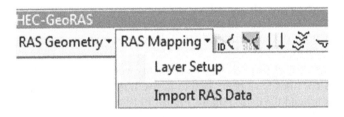

FIGURE 19.8   Executive routine Import RAS Data.

to output layers, check "Draw Output Layers". If you want to smooth the jaggedness of the flood zoning, check the "Smooth Floodplain Delineation" (but first you should not check it). For each profile, the squares of the depth and polygon class of the flood zone will be generated.

The result of this command is given in Figure 19.15. This figure (flow depth) is produced by subtracting the squares of the surface water map from the surface elevation map.

Using the tool ☉ on the dPF 1 layer, you can control the numerical value of the flow depth. You can add a background to the map using the Basemap command. To see the XS cut lines sections, you can see the cross-sections on the figure with magnification.

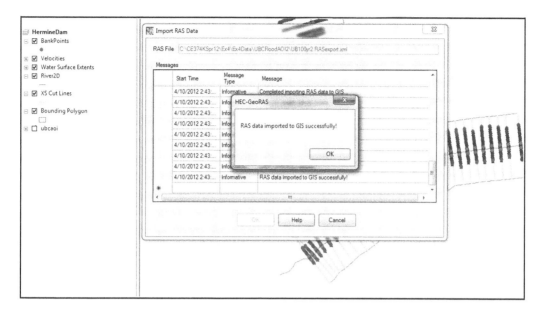

FIGURE 19.9 Information messages appearing after the execution of the Import RAS Data.

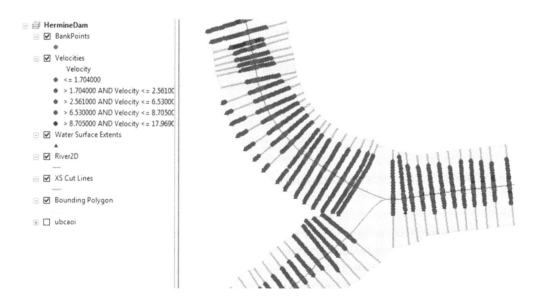

FIGURE 19.10 The production file appeared after the import RAS Data command.

19.3 CASE STUDIES

19.3.1 PREPARATION OF FLOOD RISK MAPS AT WATERSHED SCALE

19.3.1.1 The Study Area

The study area is located in Zirab City with 59,842 people (Iranian Statistic Center, https://www .amar.org.ir/), Mazandaran Province, Iran (Figure 19.16). The Zirab watershed envelopes a district of almost 92 km^2 with average elevation of 687 m. The average annual rainfall and 24-hour maximum average rainfall is 607 mm and 40.3 mm, respectively, in the study area. The maximum and minimum annual temperature varies between 8°C and 17°C. According to Emberger's and De

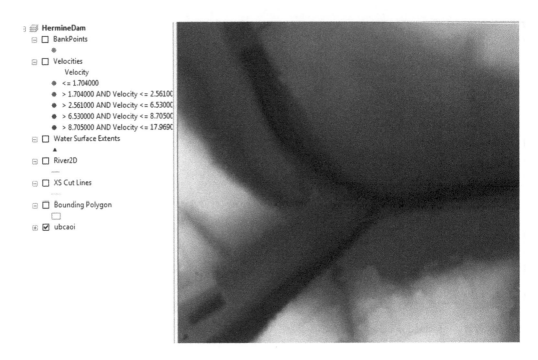

FIGURE 19.11   The file appears after selecting the ubcaoi layer.

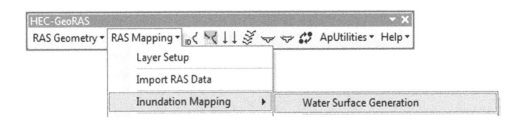

FIGURE 19.12   How to execute the command Water Surface Generation.

Martonne's climatic classifications, the study area has a semi-humid climate. The land-use types of the Zirab watershed were classified as grassland with an area of 1.02 km² (1.11% of the total area), forestland with an area of 79.1 km² (86.14%), cultivated area with an area of 4.7 km² (5.1%), orchard with an area of 1.2 km² (1.3%), residential area with an area of 5.4 km² (5.9%), and bare land with an area 0.4 km² (0.45%). The flooding of July 2015, due to heavy rainfall, damaged or destroyed buildings, roads, urban infrastructure and farmlands around the river.

19.3.1.2   Methodology

In order to calculate flood discharge with different return periods, peak discharge records were collected in Shirgah-Kasilian hydrometric station at the geographical location of 36°17' 57" north latitude and 52°53' 10" east longitude from Iran Water Resources Research Company (IWRRC) over a 54-year period. The main inputs of HEC-RAS are cross-sections, the surface roughness parameter and flow boundary conditions. In order to derive cross-sections, stream centerline, banks and cross-section cut line were created in HEC-GeoRAS, a triangular irregular network (TIN) was developed from a DEM with one-meter vertical resolution within ArcGIS for a total 19 cross-sections over a length of about 2.5 km (Figure 19.17).

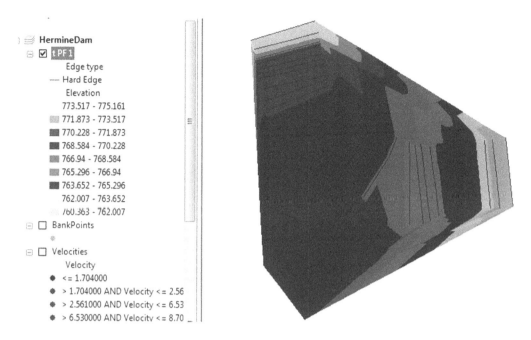

FIGURE 19.13 The result of the Water Surface Generation command.

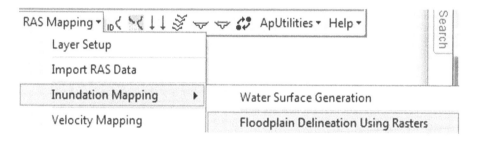

FIGURE 19.14 How to execute the command floodplain delineation using raster.

19.3.1.3 Topographic Wetness Indexes

TWI was proposed by Kirkby and Beven, (1979) to simulate the watershed response to a precipitation event. TWI for a given pixel within the hydrographic watershed is computed as following:

$$TWI = \log\left(\frac{A_s}{\tan \beta}\right) \tag{19.1}$$

where A_s is the specific watershed area expressed in meters and computed as the territorial up-slope area terrain via generic spot, per unit contour length; β is the vernacular ramp at the pixel in query expressed in degrees. TWI permits for the explanation of a contribution of a hydrographic watershed potentially exposed to flood inundation by detecting all the region determined by a TWI that exceeds a given threshold (DeRisi et al., 2015).

19.3.1.4 Maximum Likelihood Estimation of the Flood Depth-Related TWI Threshold

The probability of the accurate scheme of the flood-prone region, or the likelihood function for the flood depth related TWI threshold τ explain as L(τ|W) for different values of τ, can be calculated as follows (Jalayer et al., 2014):

$$L\left(\tau \mid W\right) = P(FP, IN\left(T_R\right) \mid \tau, W) + P(\overline{FP}, \overline{IN}\left(T_R\right) \mid \tau, W) \tag{19.2}$$

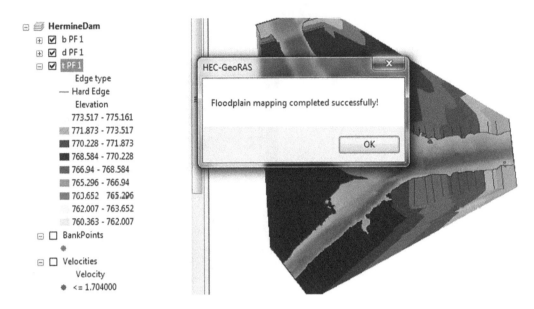

FIGURE 19.15   Water depth map.

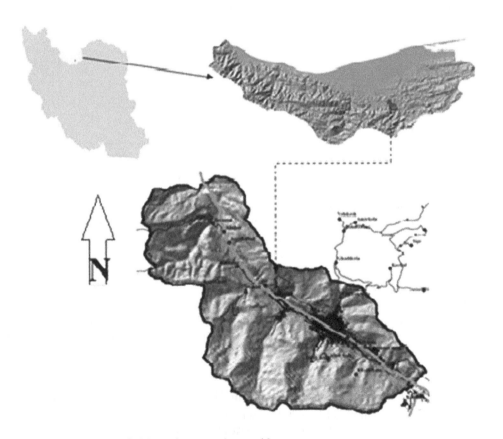

FIGURE 19.16   Study area in Mazandaran province and Iran.

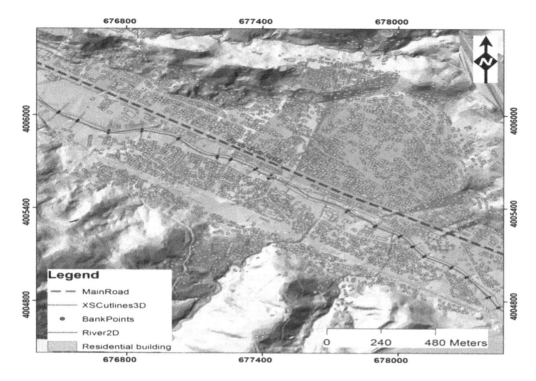

FIGURE 19.17 Triangular Irregular Network (TIN) for extraction of cross-sections.

where $P(FP, IN(T_R) | \tau, W)$ denotes the probability that a given pixel within a zone window is recognized both as flood-prone FP (using the TWI procedure) and inundated IN(TR) (based on flood inundation maps), for a certain return period TR and constitutional on a specified value of τ of the TWI threshold (Jalayer et al., 2014). Comparably, $P(\overline{FP}, \overline{IN}(T_R) | \tau, W)$ denotes the probability that a given pixel within the region of desire is neither recognized as FP nor as IN(TR), conditioned on a given value of τ of the TWI threshold. The region identified as not FP and not IN(TR)(Jalayer et al., 2014). In equation (19.2), the courses $P(FP, IN(T_R) | \tau, W)$ and $P(\overline{FP}, \overline{IN}(T_R) | \tau, W)$ can be extended, using the probability theory's product rule (Jalayer et al., 2014), as follows:

$$P(FP, IN(TR) | \tau, W) = P(FP | \tau, W) \cdot P(IN(TR) | FP, \tau, W) \qquad (19.3)$$

$$P(\overline{FP}, \overline{IN}(TR) | \tau, W) = P(\overline{FP} | \tau, W) \cdot P(\overline{IN}(TR) | \overline{FP}, \tau, W) \qquad (19.4)$$

where $P(IN(TR) | FP, \tau, W)$ explains the probability of being IN(TR) given that it is identified as FP and IN and $P(\overline{IN}(TR) | \overline{FP}, \tau, W)$ explains the probability of not being IN(TR) conditioned on not being FP, certain the threshold value τ. $P(FP | \tau, W)$ and $P(\overline{FP} | \tau, W)$ illustrate the probability of existence FP or not existence FP, respectively, given the flood depth dependent TWI threshold value τ (Jalayer et al., 2014).

19.3.1.5 Regression Between Food Depth and TWI

The flood depth-dependent TWI threshold was estimated by maximum likelihood for various return periods. Herein, a simple linear regression model is used in order to probabilistically characterize the correlation between TWI threshold and the flood depth (DeRisi et al., 2015).

19.3.1.6   Accuracy Assessment of Flood Hazard Maps

The 59 and 253 damaged points in Zirab City and Zirab's watershed, respectively, were used to evaluate the accuracy of the flood hazard maps obtained at different return periods. The receiver operating characteristics (ROC) curve is a fruitful manner for demonstrating the quality of definitive and likely detection and predicting systems (Swets, 1988). The aim of an ROC curve analysis is to characterize the cutoff value. The ROC curve is a graph of sensitivity (y-axis) versus 1 – specificity (x-axis). The area under the ROC curve (AUC) defines the state of a predict system by explaining the system's capability to predict the valid incidence or non-incidence of pre-determined course of events. The illustrious procedure has a curve with the largest AUC that equals 1, if the model does not predict the incidence of the event any better than chance, the AUC would equal 0.5. A ROC curve of 1 illustrates thorough prediction. The qualitative-relevance between AUC and prediction precision can be classified as follows; 0.9–1, excellent; 0.8–0.9, very good; 0.7–0.8, good; 0.6–0.7, average; and 0.5–0.6, poor (Swets, 1988).

19.3.1.7   Results and Discussion

The results of a TWI map for Zirab's watershed are given in Figure 19.18b. As can be seen from Figure 19.18b, the TWI values change between 3.42 and 16.19. In particular, largest TWI values can be spotted around the natural water channels. Based on Kolmogorov–Smirnov test, a Wakeby distribution was used for the estimation of flood peak discharge with various return periods. In order to calibrate the TWI threshold for Zirab watershed, the flood inundation maps for different return periods have been performed for the Zirab City (Figure 19.18a). It is obvious from Figure 19.18a that the flood depth values vary between 0.01 and 3.202 m. As shown in Figure 19.19a, pink points denote the probability that a given pixel within zone Window is recognized both as FP (using the TWI procedure) and IN(TR) (based on the HEC-RAS maps), for a certain return period TR and constitutional on a specified value of τ of the TWI threshold $\left(P(FP, IN\,(T_R)\,\middle|\,\tau, W \right)$ value in equation 19.3). Comparably, pink points in Figure 19.19b denote the probability that a given pixel within the region of desire is neither recognized as FP nor as IN (TR), conditioned on a given value of τ of the TWI threshold ($P(\overline{FP}, \overline{IN}\,(T_R)\,\middle|\,\tau, W)$ value in equation 19.4). The summation of pink points ($P(FP, IN\,(T_R)\,\middle|\,\tau)$ in Figures 19.6a and b resulted in a maximum likelihood interval for a certain return period TR.

The resulting likelihood function $L(\tau|W)$ for a return period of TR = 100 years and for a flood depth h(TR) larger than 0, is drawn in Figure 19.20 as a function of the TWI threshold τ. Therefore,

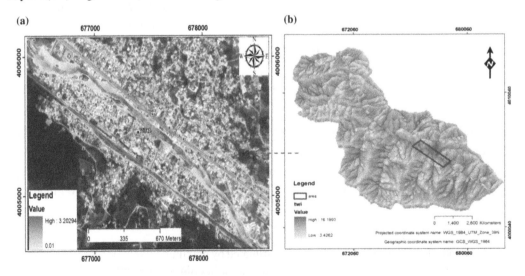

FIGURE 19.18   The flood depth values for return period of 100 years and b the TWI map.

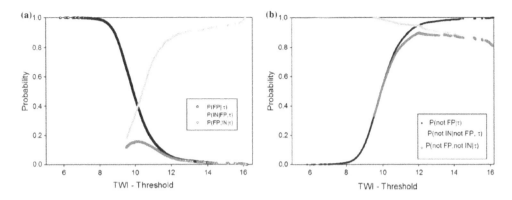

FIGURE 19.19 a Probability of essence FP and IN given s and b probability of essence FP and IN given s for TR = 100 years.

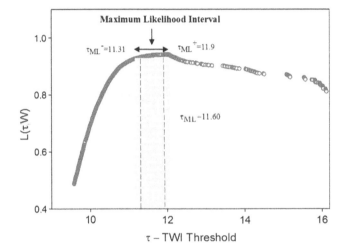

FIGURE 19.20 The likelihood function for TWI thresholds for flood depth values larger than 0.

the maximum likelihood estimate for τ can be identified as $\tau = 11.60$. In addition, by knowing the τ values, maximum likelihood value, it is feasible to determine a maximum likelihood interval, that changes between $\tau_{ML}^- = 11.31$ and $\tau_{ML}^+ = 11.90$.

The resulting same procedure for a flood depth h(TR) larger than 0 for various return periods is presented in Table 19.2. As can be seen from Table 19.2, with a large increase flood, the maximum likelihood interval was less. The reason for this decrease is that fewer TWI values necessary to exposed to flooding, because the characteristics and competence of flow in flood conditions are more dominant than those of morphometric factors. As for the low return periods, TWI needs more to be placed in flood prone areas.

In general, handling the same method for flood depth values larger than h(TR) = h, the maximum likelihood assessment for flood depth related TWI threshold τ can be established, conditioned on a prescribed return period TR. Figure 19.21 demonstrates the linear regression of flood depth versus corresponding maximum likelihood τ assessments, for certain return periods TR. Then, the relationship between the depth of the flood and the threshold of the topographic humidity index was obtained in different return periods. For example, Figure 19.21 shows this relation for the period of return of 100 years. These relations were applied to generalize the depth of the flood to the watershed surface on the Figure 19.18.

TABLE 19.2

Threshold Value of Topographic Wetness Index in Different Return Periods

Return Period	τ_{ML}^-	τ_{ML}^+	τ_{ML}
2	11.82	12.11	12.07
10	11.63	12.30	11.75
25	11.46	12.20	11.65
50	11.41	12.20	11.62
100	11.31	11.90	11.60

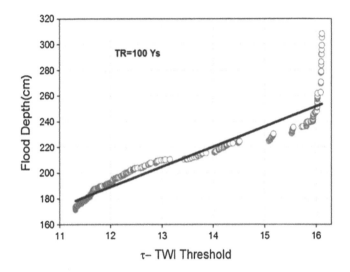

FIGURE 19.21 The linear regression between flood depth (h*) and corresponding maximum likelihood estimates, for 100-year return period.

In order to qualitatively check of the accuracy of the zonation of the flood prone areas, the inundation partition (obtained from the hydraulic routine) and the TWI map for threshold values larger than the maximum likelihood estimate $\left(\tau_{ML} \right)$ was overlaid. On the other hand, the inundation partition and those obtained by mapping the TWI layer using the correlation prediction, for various values of flood depth given TR = 100 years was overlaid. Figure 19.22 depicts the result of overlaying the hydraulic routine (IN) and the TWI map (FP) for the three values of flood depth equal to 0.0, 1.0, and 2.0 m corresponding to TR = 100 years and TR = 10 years.

As Figure 19.22 shows, TWI is able to provide a good description of flood-prone areas because a good agreement between the inundation map (IN) (obtained from the hydraulic routine) and the TWI contours (FP) appears.

19.3.1.8 Accuracy Evaluation of Flood Hazard Maps

A quantitative evaluation of the accuracy of the results can be carried out by ROC curve. The consequences of the ROC curve examination are illustrated in Figure 19.23.

As shown in Figure 19.23a, flood hazard map for Zirab City with return period of 100 years had a good accuracy of 72% (AUC = 0.72), while the accuracy of flood hazard maps for other return periods was evaluated poor. The AUC values show that the accuracy increased with increasing return period. As shown in Figure 19.24b, flood hazard map for the Zirab watershed with return period of 100 years had a good accuracy of 65% (AUC = 0.65), while the accuracy of the flood hazard map

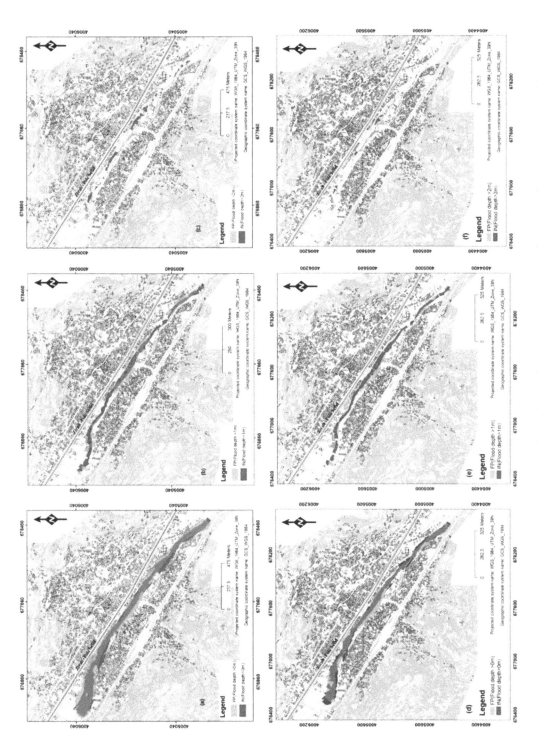

FIGURE 19.22 Graphical representation of overlapping the layers of flood inundation and TWI depends on flood depth for return period of 100 years a 0 m, b 1 m, c 2 m and flood depth for return period of 10 years d 0 m, e 1 m, f 2 m.

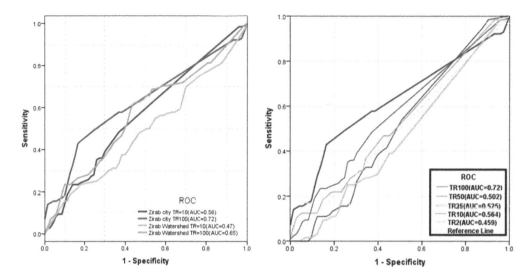

FIGURE 19.23   The ROC curves for (a) Zirab City and (b) Zirab watershed.

for ten years return period was evaluated as poor (AUC = 0.47). For the Zirab City area, the poor performance of the procedure at short return periods may be due to the effect of urban sewer system (DeRisi et al., 2015). Figure 19.24 shows the flood hazard maps for return periods of ten and 100 years based on hazard zoning thresholds set as: 0.0–1.0 m, 1.0–2.0 m, and larger than 2.0 m.

19.3.2   Flood Zoning Using HEC-RAS Hydraulic Model

19.3.2.1   The Study Area

The studied area is west of Iran and the center of Lorestan province located between 47°55' to 48°50' eastern longitude and 32°40' to 34°20' north latitude, which in terms of hydrological divisions is one of the sub-basins of Karkheh watershed with an area of 2501 km^2 (Mazidi and Kushki, 2015) (Figure 19.25).

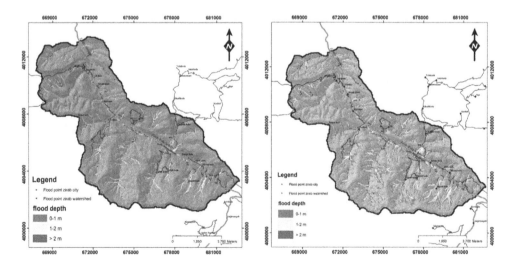

FIGURE 19.24   Flood hazard maps at watershed scale for (a) TR = 10 years and (b) TR = 100 years.

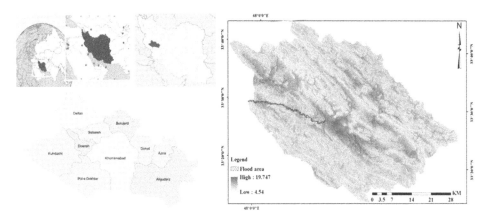

FIGURE 19.25 Location of Khorramabad Watershed in Lorestan province and Iran.

19.3.2.2 Materials and Methods

First, the longitudinal sections of the maps and the maximum instantaneous discharge in Duab Veysian hydrometric station with elevation of 965 m at the geographical location of 33°26'00" north latitude and 47°58'00" east longitude from the Regional Water Authority of Lorestan province were obtained. Then, given that the data used should have three conditions of adequacy, correctness and relevance (Jaydari et al., 2012; Zeraatkar et al., 2015), then maximum instantaneous discharge in Duab Veysian hydrometric station controlled, corrected and reconstructed. To determine the maximum instantaneous discharge with different return periods, after the homogeneity test, using the easy fit software (to select the best statistical distribution) and the Kolmogorov–Smirnov test (to detect normalization of the data), the three-parameter normal log distribution was identified as the most appropriate statistical distribution function. Then, using this distribution, discharge was obtained with different return periods.

There are several methods for calculating roughness coefficients. One of the most practical methods for determining the Manning coefficient is Cowan's method. First, using the existing table, the Manning coefficient is estimated and then according to other conditions i.e., irregularity, character of variations of size and shape cross-section, obstructions, vegetation, and meandering is modifies (Cowan, 1956).

Manning roughness coefficients of the main channel, left bank, and right bank were obtained by using the Cowan method and a field survey (Arcement and Schneider, 1989; McCuen, 2016). After entering geometric and flow data, hydraulic calculations can be calculated using three methods: steady flow analysis, unsteady flow analysis, and hydraulic design functions. In this study, hydraulic calculations were carried out using steady flow analysis. It is also possible to perform computations for subcritical, supercritical, and complex flow regimes. In this study, the boundary conditions flow was considered to be mixed.

After introducing the boundary conditions of the flow, the flood discharge with different return periods, cross-sections and their distance, and Manning roughness coefficient for each cross-sections, the water surface profiles in different return periods were computed using HEC-RAS (Warner and Brunner, 2001). After the water surface profiles computation, the output results were imported to ArcGIS and flood zoning maps with different return periods were prepared.

19.3.2.3 Results and Discussions

The flood discharge with different return periods and Manning roughness coefficients of the main channel, left bank, and right bank in the study reach are given in Tables 19.3 and 19.4, respectively.

The Khorramabad River, with a length of 40 km, was considered in this study. Figure 19.26 shows the schematic of the river and the cross-sections in HEC-RAS.

TABLE 19.3

Flood Discharge with Different Return Periods at Duab Veysian Hydrometric Station

Return Period	Discharge(m³/s)
2	145.125
5	253.254
10	326.496
25	418.775
50	486.727
100	553.781

TABLE 19.4

Manning Roughness Coefficients of the Main Channel, Left Bank, and Right Bank in the Study Reach

Cross-section Number	Manning Roughness Coefficient		
	Right Bank	Main Channel	Left Bank
1–11	0.128	0.54	0.128
12–25	0.13	0.54	0.13
26–186	0.091	0.54	0.091
187–238	0.109	0.45	0.109
239–302	0.106	0.45	0.106
303–421	0.11	0.45	0.11

Figure 19.27 shows flow depth in the return periods of 2, 5, 10, 25, 50, and 100 years.

By studying the cross-sections in the study reaches, it is observed that flow will be entered into floodplain during flood at 83 cross-sections and will be caused damages. These critical points are observed throughout the river, but the largest critical points are in the middle of the river. For example, cross-section 18872.41 is given in Figure 19.28.

Water surface profiles with different return periods are presented in Figure 19.29.

Figures 19.30 and 19.31 show the flood zone with the return period of 100 years.

The results obtained from flood zoning showed that flood prone area in the return period of two years with 145.125 $m^3 s^{-1}$ and the return period of 100 years with 553.781 $m^3 s^{-1}$ effect 8.63 and 10 km^2 in area, respectively. So that about 4.4 km^2 of the total rain-fed farming area, 2.4 km^2 of total rangeland area and 1.4 km^2 of total residential, 1.6 road area, and 0.2 km^2 abandoned effect by flood in the return period of 100 years. Similarly, for other flood return periods was also observed that the most flood prone areas are related to rain-fed farming, rangeland, road, residential area and abandoned the land.

19.4 CONCLUSIONS

Rivers are considered to be the main source of water for humans and other creatures and, sometimes, this source of life causes destruction and irreparable damage. Therefore, it is necessary to determine the safe zones for human activities around the river by studying the hydraulic properties of the stream and the riparian river. Investigating the determination of the riparian and river bed means the exact definition of the sections of its studies and the correct relationships between these

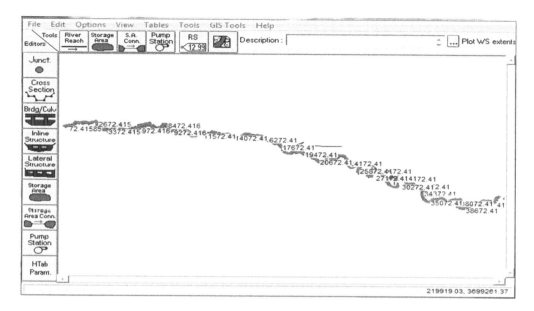

FIGURE 19.26 The schematic of the river and the cross-sections in HEC-RAS.

components that can play a key role in the correct estimation of the riparian rivers. Considering the importance of determining the flood zones with different return periods to prevent the risks of floods and organizing and modifying the river route and identifying the status of existing facilities in adjacent rivers, the need for river flow hydraulic studies and the determination of flood zonation with the different return periods.

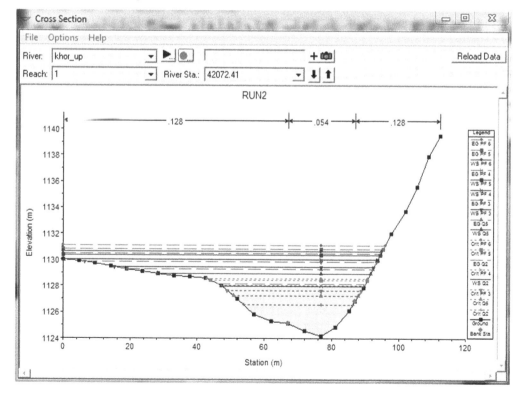

FIGURE 19.27 Flow depth in different return periods at Duab Veysian hydrometric station.

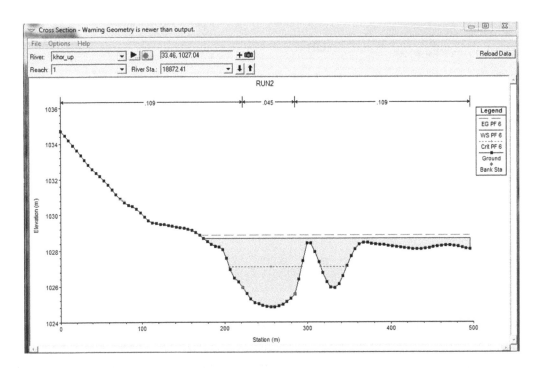

FIGURE 19.28   Flow depth in cross-section 18872.41.

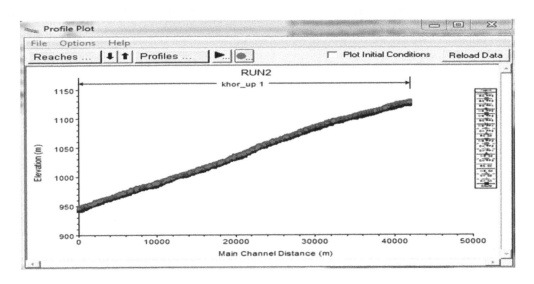

FIGURE 19.29   Longitudinal profile of the river.

Several mathematical models have been developed to study the flow characteristics of rivers flow and flood zoning in the world, which have been categorized to varying degrees in terms of complexity and ease of use. In this regard, numerous hydrodynamic models of one, two, three-dimensional and dual hydraulics have been developed in the field of flood zoning. These models are used in different regions according to the desired purpose, required data, streamflow characteristics, and performance. One-dimensional models examine flow properties only in the general direction of flow. These models, using a minimum amount of field data and a small computational volume, in most cases provide reliable results from the average characteristics of flow in cross-sections.

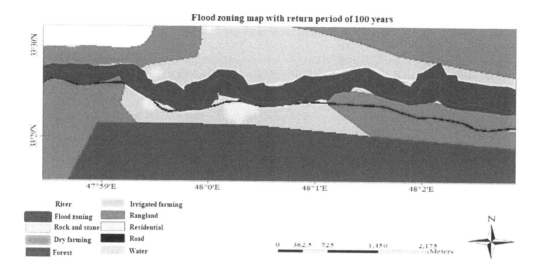

FIGURE 19.30 Flood zoning map with the return period of 100 years at Duab Veysian River.

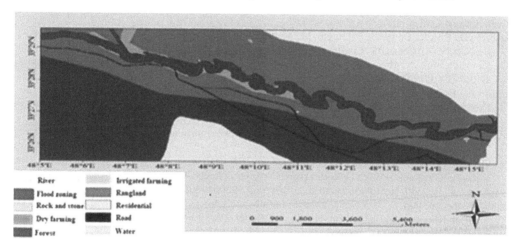

FIGURE 19.31 Flood zoning map with the return period of 2 years at Duab Veysian River.

One-dimensional models cannot simulate the structures of the secondary stream of the river. In the two-dimensional models of the horizon (mean depth), the velocity distribution in the vertical direction and vertical acceleration are ignored. Therefore, the effects of secondary spiral flows in the erosion of the bed and walls are ignored. However, they are better suited for simulating flow in spiral and flood plains (with compound cross-section). Three-dimensional models are often developed to study the current hydrodynamics and provide a complete definition of flow equations and secondary flows. The complexity and high computation of these models have made them not widely used. As a result, two-dimensional models are more widely used in studies of flood hazard zonation and flow properties than with a one-dimensional model and less complexity than 3D models. One of the most popular two-dimensional models is the HEC-RAS hydraulic model.

REFERENCES

Alaghmand, S., Bin Abdullah, R., Abustan, I., and Eslamian, S. 2012. Comparison between capabilities of HEC-RAS and MIKE11 hydraulic models in river flood risk modelling (a case study of Sungai Kayu Ara River basin, Malaysia), *Int. J. Hydrol. Sci. Technol.*, 2: 270–291.

Amini, M.A., Torkan, G., Eslamian, S., Zareian, M.J., and Adamowski, J.F. 2019. Analysis of deterministic and geostatistical interpolation techniques for mapping meteorological variables at large watershed scales, *Acta Geophys.*, 67: 191–203.

Arcement, G.J., and Schneider, V.R. 1989. *Guide for Selecting Manning's Roughness Coefficients for Natural Channels and Flood Plains*, Water Supply Paper (ed.), U. S. Geological Survey.

Bates, P.D. 2004. Remote sensing and flood inundation modelling, *Hydrol. Process.*, 18: 2593–2597, https://doi.org/10.1002/hyp.5649.

Brunner, G.W. 2010. *HEC-RAS River Analysis System: Hydraulic Reference Manual*, US Army Corps of Engineers, Institute for Water Resources, Hydrologic Engineering Center, USA.

Cowan, W.L. 1956. Estimating hydraulic roughness coefficients, *Agric. Eng.* 37: 473–475.

Degiorgis, M., Gnecco, G., Gorni, S., Roth, G., Sanguineti, M., and Taramasso, A.C. 2012. Classifiers for the detection of flood-prone areas using remote sensed elevation data, *J. Hydrol.*, 470–471: 302–315, https://doi.org/10.1016/j.jhydrol.2012.09.006.

Ding, Y., Zhang, Y., and Jia, Y. 2016. *CCHE2D-Coast: Model Description and Graphical User Interface.*

Hardy, R.J., Bates, P.D., and Anderson, M.G. 1999. The importance of spatial resolution in hydraulic models for floodplain environments, *J. Hydrol.*, 216: 124–136.

Horritt, M.S., and Bates, P.D. 2002. Evaluation of 1D and 2D numerical models for predicting river flood inundation, *J. Hydrol.*, 268: 87–99, https://doi.org/10.1016/S0022-1694(02)00121-X.

Hunter, N.M., Bates, P.D., Horritt, M.S., and Wilson, M.D. 2007. Simple spatially-distributed models for predicting flood inundation: A review, *Geomorphology*, 90: 208–225, https://doi.org/10.1016/j.geomorph.2006.10.021.

Iranian Statistic Center. https://www.amar.org.ir/

Jalayer, F., De Risi, R., De Paola, F., Giugni, M., Manfredi, G., Gasparini, P., Topa, M.E., Yonas, N., Yeshitela, K., Nebebe, A., Cavan, G., Lindley, S., Printz, A., and Renner, F. 2014. Probabilistic GIS-based method for delineation of urban flooding risk hotspots, *Nat. Hazards*, 73: 975–1001, https://doi.org/10.1007/s11069-014-1119-2.

Jaydari, A., Fathzadeh, A., Taghizadeh Mehrjardi, R., Dastorani, M., and Fatahi Ardakani, A. 2012. Comparison of the efficiency of different methods for reconstruction and prolongation of instantaneous peak flow data. *J. Range Watershed Manag.*, 64(4): 387–399.

Kirkby, M.J., and Beven, K.J. 1979. A physically based, variable contributing area model of basin hydrology, *Hydrol. Sci. J.*, 24: 43–69.

Lazaridou, P.L., Daniil, E.I., Michas, S.N., Papanicolaou, P.N., and Lazarides, L.S. 2004. Integrated environmental and hydraulic design of Xerias river, Corinthos, Greece, training works, *Water, Air, Soil Pollut. Focus*, 4: 319–330, https://doi.org/10.1023/B:WAFO.0000044808.41691.dd.

Mazidi, A., and Kooshki, S. 2015. Simulation of rainfall-runoff process and estimate of flood with HEC-HMS model in Khorramabad catchment area. *Iran. J. Geogr. Dev.*, 13(41): 1–10, doi: 10.22111/gdij.2015.2236.

McCuen, R.H. 2016. *Hydrologic Analysis and Design*, Prentice Hall.

Mohamadi, E., Montaseri, M., and Sokooti Oskoei, R. 2009. Zonation of flood dangers in urban regions, using WMS and HEC-RAS, case study: Oshnavieh, Western Azerbyjan Province, *J. Watershed Eng. Manag.*, 1(1): 61–69.

Motevalli, A., and Vafakhah, M. 2016. Flood hazard mapping using synthesis hydraulic and geomorphic properties at watershed scale, *Stoch. Environ. Res. Risk Assess.*, 30: 1889–1900, https://doi.org/10.1007/s00477-016-1305-8.

Moussa, R., and Bocquillon, C. 2009. On the use of the diffusive wave for modelling extreme flood events with overbank flow in the floodplain, *J. Hydrol.*, 374: 116–135, https://doi.org/10.1016/j.jhydrol.2009.06.006.

de Risi, R., Jalayer, F., and De Paola, F. 2015. Meso-scale hazard zoning of potentially flood prone areas q, *J. Hydrol.*, 527: 316–325, https://doi.org/10.1016/j.jhydrol.2015.04.070.

Swets, J.A. 1988. Measuring the accuracy of diagnostic systems, *Science*, 240: 1285–1293.

Thompson, J.R., Sørenson, H.R., Gavin, H., and Refsgaard, A. 2004. Application of the coupled MIKE SHE/MIKE 11 modelling system to a lowland wet grassland in southeast England, *J. Hydrol.*, 293: 151–179, https://doi.org/10.1016/j.jhydrol.2004.01.017.

Villazón, M.F., Timbe, L.M., and Willems, P. 2013. Comparative analysis of 1-D river flow models applied in a quasi 2-D approach for floodplain inundation prediction, *Maskana*, 4: 107–126.

Warner, J.C., and Brunner, G.W. 2001. *HEC-RAS River Analysis System: Applications Guide*, US Army Corps of Engineers, Institute for Water Resources, Hydrologic Engineering Center, USA.

Zeraatkar, Z., Hasanpour, F., and Tabe, M. 2015. Assessment of the estimation methods of flood peak discharge in urban catchment for controlling flood. *J. Rainw. Catch. Syst.*, 2(4): 23–32.

20 Flood Hazard, Vulnerability, and Risk Mapping in GIS
Geodata Analytical Process in Boolean, AHP, and Fuzzy Models

Ali Akbar Jamali, Mina Arianpour,
Saied Pirasteh, and Saeid Eslamian

CONTENTS

20.1 INTRODUCTION

In the past decades, the concept of spatial decision support systems (SDSS) was discussed as a field of research, development, and operations (Malczewski, 1999). Now, this concept has grown to such an extent that it is composed of many approaches and frameworks such as group spatial decision support systems (GSDSS), collaborative spatial decision-making (CSDM), spatial knowledge-based systems (SKBS), spatial expert systems (SES), intelligent spatial decision support system (ISDSS), spatial expert support systems (SESs) (Malczewski, 1999). Every decision is ranging from a highly structured to an unstructured decision (Figure 20.1), and the decision support system (DSS) is in the middle of this range.

Flooding is one of the most destructive natural hazards that cause damage to both life and property every year, and therefore the development of a flood model to determine an inundation area in watersheds is important for decision-makers. In recent years, data mining approaches (Bakhtiyarikia et al., 2012; Jun et al., 2013; Jamali et al., 2018, Dalezios and Eslamian, 2016) have been applicable in mapping and zoning studies, such as fuzzy modeling and decision-making in GIS.

Spatial decision support systems have evolved in parallel with common decision support systems. In general, decision support systems could be defined as computer-based interactive systems, which are designed in order to support a user or group of users to achieve greater impact while solving a semi-structured spatial decision-making problem.

DOI: 10.1201/9780429463938-26

FIGURE 20.1   Different structural degrees of decision-making issues.

Kazakis et al. (2015) introduced a flood hazard criteria index to asses areas on a regional scale. Accordingly, a Flood Hazard Index (FHI) has been defined and spatial analysis in a GIS environment has been applied for the estimation of the associated value.

Ntajal et al. (2017) have combined GIS, remote sensing, and indicator-based flood risk assessment techniques in mapping flood disaster risk.

A decision support system allows the user to combine his/her personal judgment with the outputs obtained from the computer and thereby provides useful information for decision-making.

Therefore, decision support systems are the systems that apply users' intellectual resources along with the computer's ability to improve the quality of their decision and in most cases, these systems are used to solve the semi-structured problems and assist the users with the use of data and models to address the semi-structured problems. The application of decision support systems using spatial data has also increased considerably. With the development of crisis management in the world and especially in Iran, awareness of the crisis sources and mechanisms could play a major role in the management of risks and loss of life and property.

Flood is one of the hazard sources and causes much damage each year in different parts of Iran. Determining the flood potential of sub-watersheds is one of the fundamental studies that would be a significant step in the context of reducing flood damage and comprehensive planning.

In research by Zaharia et al. (2017), the aim of this paper was to develop an approach that allows for the identification of flash flood and flood-prone susceptible areas based on computing and mapping of two indices: the FFPI (Flash-Flood Potential Index) and the FPI (Flooding Potential Index). These indices are obtained by integrating into a GIS environment several geographical variables, which control runoff (in the case of the FFPI) and favor flooding (in the case of the FPI).

Awareness of the flood potential of sub-watersheds can be useful for developing the various programs for crisis management, required budget allocation, water resources management, watershed management, and programs that aim to reduce erosion. Several studies with different methods have been conducted in the past few decades (Table 20.1).

Fernandez and Lutz (2010) conducted a study on urban flood hazard zoning in Tucuman province, Argentina, using GIS and multi-criteria decision analysis and, finally, the flood hazard map obtained using this method. Prawiranegara (2014) has conducted a study on the spatial multi-criteria evaluation techniques (SMCA) for basin-wide flood risk assessment as a tool in improving the spatial planning and urban resilience policy-making in the Marikina river basin, Metro Manila, Philippines. Shaw (2015) attempted to highlight the concepts of hazard, vulnerability, capacity, and risk. Additionally, various tools and techniques have been discussed for carrying out the hazard, vulnerability, and risk assessment and management

The research employed spatially explicit methods of assessment by using the spatial multi-criteria analysis (SMCA). Analytical frameworks were developed for basin-wide flood and landslide risk. The necessary steps were taken, including the geodatabase preparation, variable definition, standardization of parameters, weight assignment of indicators, and sensitivity analysis. Analysis was applied to conduct risk mapping. Identified very high flood risk areas were also validated by satellite images (Foudi et al., 2015; Geneletti, 2010; Umar and Acharya, 2016; Jamali and Ghorbani Kalkhajeh, 2019).

In this research, the watersheds based on risk and vulnerability have been clustered. This approach has a significant potential to influence and support regional decision-making.

TABLE 20.1

Comparing Some Methods in Subwatershed Prioritizing

Reference	Year	Focus	Results	Advantages
Malczewski	1999	Spatial decision support systems	Spatial decision support system	Decision
Amirahmadi et al.	2013	Zonation map of flood risk	Flood risk	AHP model and multi-criteria decision-making techniques
Sharifi Retsios	2006	Selection of proper sites	Potential sites	Determination of potential sites
Antonella et al.	2008	Site for the construction of a local park	Site selection	Use decision support system and AHP
Jamali et al.	2018	Subsurface dam study by SMCE	Identify areas suitable for subsurface dam	Decision-making techniques
Fernandez et al.	2010	Flood hazard zoning	Urban flood hazard zoning	Multi-criteria decision analysis
Jun et al.	2013	Flood vulnerability	Places vulnerable to flooding	Fuzzy, TOPSIS, and WSM
Zhaoli et al.	2012	Flood risk	Flood risk assessment map	Spatial model for assessment
Prawiranegara	2014	Flood risk assessment	Very high flood risk areas	SMCE
Foudi et al.	2015	Flood risk assessment	Integrated spatial assessment of the flood risk	Integrated spatial assessment

The objective of this research is to develop a spatial decision support system (SDSS), geodata analytical process (GAP) in Boolean, AHP, and fuzzy (BAF) to decide on the maps of flood risk assessment in a GIS environment and to prioritize them. Nevertheless, this paper contributes a multi-criteria decision-making technique by BAFs.

20.2 MATERIALS AND METHOD

20.2.1 STUDY AREA DESCRIPTION

The Omidieh watershed with an area of 34,297 hectares is located in the southeast of the Khuzestan province between east longitudes of 49°26'22" to 49°54'22" and north latitudes of 30°30'56" to 31°01'23" (Figure 20.2). The elevation varies from 20 m to 298 m from mean sea level. The average annual rainfall is about 250 mm and vegetation consists mainly of species of bushes and shrubs (Sadeghi et al., 2017).

20.2.2 DATA COLLECTION

In this study, ArcGIS software version 10.2 has been used. The spatial features including the linear features (i.e., contour lines and streams) and points (i.e., towns and villages) were extracted from topographic maps. Google Earth images (June 10, 2016) were used for investigating the recent land uses. The Landsat 8 imagery (OLI) (June 2016) and ASTER digital elevation model (DEM) with a resolution of 32 m in the GIS software have been used.

In the study multispectral images of Landsat 8 Operational Land Imager (OLI) and Landsat 7 Enhanced Thematic Mapper Plus (ETM+) sensors were utilized along with the Shuttle Radar

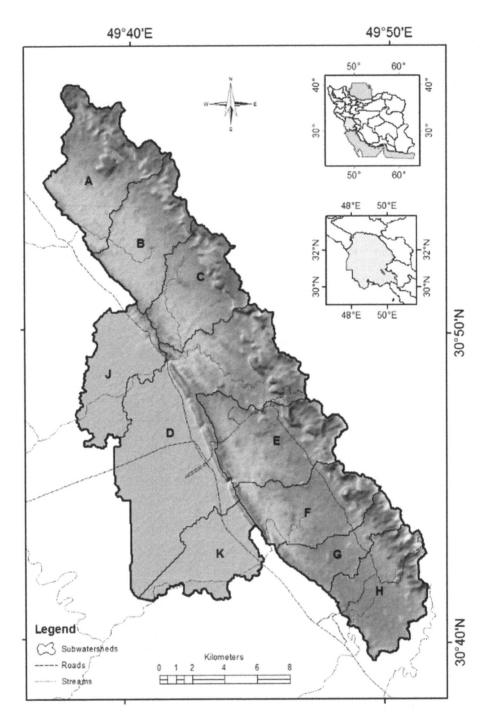

FIGURE 20.2   Location of the study area in Omidieh, Khuzestan, Iran.

Topography Mission (SRTM) digital elevation model (DEM) to derive the flood hazard and elements at risk. The linear combination of the normalized flood depth, mean turbidity, and locational
probability of flood parameters was taken to the map of the flood hazard.

In this study, erosion potential models (EPM) were used to extract the erosion map. The drainage
density map of the area is divided into several regions based on the density of streams and rivers.

In research by Franci et al. (2016), this work focuses on the exploitation of very high-resolution (VHR) satellite imagery coupled with multi-criteria analysis (MCA) to produce flood hazard maps. The MCA methodology was performed selecting five flood-conditioning factors: slope, distance to channels, drainage texture, geology, and land cover.

Drainage density is the total length of all of the streams and rivers divided by the total area of the drainage basin, and it has been considered as a determining factor in the runoff flow (Vittala et al., 2004). The study flowchart is depicted in Figure 20.3.

20.3 METHODOLOGY

The spatial multi-criteria evaluation (SMCE) is designed in tree schema in the ILWIS software version 3.3, and the choice of factors and constraints was according to what was mentioned in the introduction and background section (Figure 20.4). To apply GAP in BAF as the flood issues techniques in hydrology science also the spatial multi-criteria evaluation techniques in the GIS, the spatial data layers have been used and maps were selected for further processing such as integration and planning (Jamali and Abdolkhani, 2009; Jamali et al., 2018; He et al., 2020; He et al., 2021). Using a contour map, the slope and aspects were prepared and the NDVI map was obtained by using bands 5 and 4 of the Landsat Imagery 8 (OLI). Compositing of layers is designed by the SMCE with GIS software ILWIS 3.3. Then in order to homogenize the layers, all layers were standardized in the value range between 0 and 1, or in other words, Boolean and fuzzy were standardized using the related equations (Figure 20.5).

By using the AHP pair-wise comparison and expert choice software, criteria were able to get weight and to import them by the direct methods into the criteria's tree model. In AHP, the factors are compared in pair-wise form, and their relation is important in determining. Only, two criteria are compared at the same time. Factor weights can be determined by using the analytic hierarchy process in subgroups and two groups by the direct method (Figures 20.6 and 20.7). Normal factor weights (range from zero to one) and the sum of the factor weights must be incorporated into numerical integration. Since the constraints are removed directly, they will not be given weight. For example, the slope of less than 1% (with a value of zero) has been removed and these slopes are less prone to generate a flood hazard. Finally, by creating the composite index map (risk map) with the fuzzy value from zero to one, the priorities were diagnosed, such that every point that is closer to one has a greater potential for flooding. In research by Hu et al. (2017(, flood risk was defined as the product of hazard

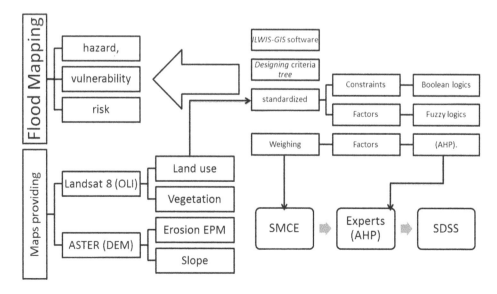

FIGURE 20.3 Flowchart of the study.

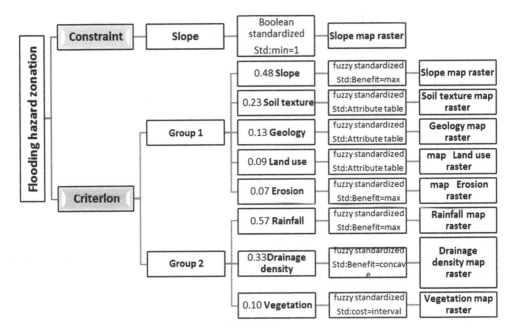

FIGURE 20.4 The tree model designed by the ILWIS software.

and vulnerability based on the disaster risk theory. A risk assessment index system was established, and the analytic hierarchy process method was used to determine the index weight.

Then, the final map is used to generate a flood hazard map. Through positioning, the situation of the vulnerable villages and cities to flooding, including three cities and 13 villages were determined. Many factors are used in determining a flood hazard (Safaripour et al., 2012). Nevertheless, this study focuses on five factors, including proximity to cities (the populations vulnerable to flooding), villages, utilities, gas and oil pipelines, and road networks. Flood damages are the best indicators for flood hazards (Safaripour et al., 2012). Due to the different effects of the mentioned factors on flood hazard, based on the experts' views and earlier research, the proper scores were given to each factor. Afterward, the rate of flood hazard was determined according to the sum of the scores (Figure 20.8 and Table 20.2).

20.4 RESULTS AND DISCUSSIONS

20.4.1 FLOOD VULNERABILITY

The flood vulnerability is a composite index map, derived from the maps of the standardized distance from cities, utilities, pipelines, roads, and villages. By integrating the layers that have values of 0 to 1, the flood vulnerability priorities were identified. Thus each point that is closer to 1 has a greater potential for flooding and each point closer to zero is less prone to flooding (Figure 20.8). Examining the distribution of residential centers, including towns, villages, and oil and gas facilities in the area show that most of the villages and three towns are located in sub-watersheds D, E, H, J, and K while sub-watersheds A, B, C, F, and G have no villages or have a few villages (Figure 20.5K and L). Moreover, D and A sub-watersheds have the most and the least populations vulnerable to flooding, respectively (Table 20.3).

20.4.2 FLOOD HAZARD

This composite index map is derived by integration of the slope, soil texture, geology, land use, erosion, rainfall, drainage density, and the NDVI layers. Layers standardized by fuzzy logic were

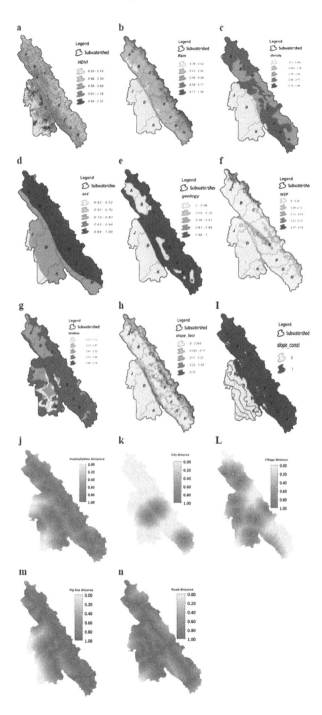

FIGURE 20.5 Standardized factors and constraints layers in tree models. (a(Std:cost=interval: fuzzy standardized vegetation map. (b) Std:Benefit=max: fuzzy standardized precipitation map. (c) Std:Benefit=concave: fuzzy standardized drainage density map. (d) Std:Attribute table: fuzzy standardized soil map. (e) Std:Attribute table: fuzzy standardized geology map. (f) Std:Benefit=max: fuzzy standardized erosion EPM map. (g) Std:Attribute table: fuzzy standardized land use map. (h) Std:Benefit=max: fuzzy standardized slope (factor) map. (i) Std:min=1: Boolean standardized slope (constraint) map. (j) Std:cost=Goal: fuzzy standardized distance from utilities map. (k) Std:cost=Goal: fuzzy standardized distance from cities map. (l(Std:cost=Goal: fuzzy standardized distance from villages map. (m(Std:cost=Goal: fuzzy standardized distance from pipelines map. (n) Std:cost=Goal: fuzzy standardized distance from roads map.

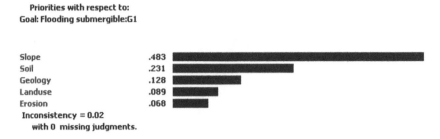

FIGURE 20.6 Graphical representation of effective factors weighting (Group 1) in the standard matrix using Expert Choice Software.

FIGURE 20.7 Graphical representation of effective factors weighting (Group 2) in the standard matrix using Expert Choice Software.

integrated, and a criteria tree model was designed with compositing these layers with the ILWIS-GIS software (Jamali et al., 2018; Parsasyrat and Jamali, 2015). The flood hazard map with the fuzzy value from zero to one is produced. It is determined that every point closer to one is more prone to flooding than those closer to zero (Figure 20.9).

The flood hazard map of the study area shows that the sub-watersheds Aghajari 2 and Meiankuh are the riskiest part of the Omidieh sub-watershed and have the highest risk of flooding and sub-basins J, K, and D have the lowest risk of flooding (Figure 20.9). According to the results obtained from Figure 20.8, the severity of the flooding was classified into five classes in the GIS environment (Table 20.4).

20.4.3 FLOOD RISK

The flood vulnerability map and flood hazard maps were integrated to generate a flood risk map. The study area in different flood risk classes is classified. It can be concluded that the risk of flooding in the Yeresie, Hur, and part of the Omidieh sub-watersheds have the lowest risk (very low risk) and Aghajari 1 and 2, Meiankuh, and part of the Omidieh sub-watersheds have the greatest risk of flooding (Figure 20.10).

The final composite index map (Figure 20.10) and spatial decision support systems (SDSS) were used as powerful tools to prioritize the flood risk in sub-watersheds. By using the developed decision support system, the necessary steps to achieve a final flood risk map in the same study areas have been built.

In addition to the flooding which is significant to prioritizing sub-watersheds, the risk of flooding in sub-watersheds is also important for decision-making and planning for the flood control measures. In this study, the factors such as proximity to urban areas, a population vulnerable to floods, rural areas, utilities, oil and gas pipelines, and road network have been considered. The result of the study in conjunction with field observations of the sub-watersheds has shown in Table 20.5. According to this table, the sub-watersheds Aghajari 1, 2, Meiankuh, and Omidieh have the highest

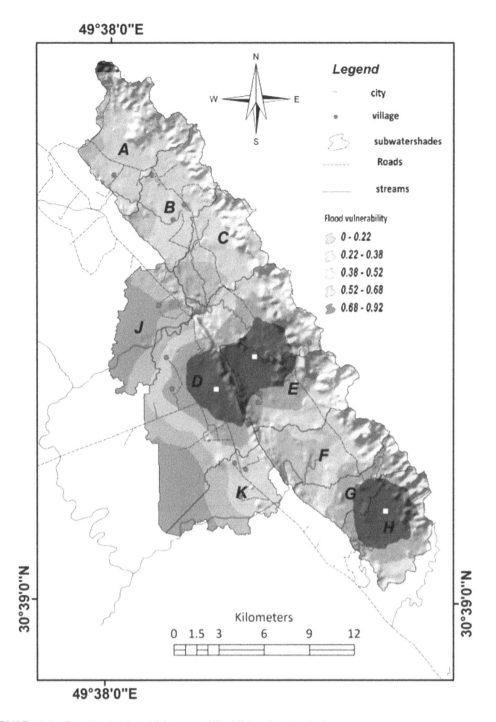

FIGURE 20.8 The flood vulnerability map of Omidieh sub-watersheds.

risk of flooding and due to the highly vulnerable population, multiple installations of oil and gas utilities, construction of industrial estates, and public infrastructure, they have to be considered as the top priorities of flood control measures. Moreover, field observations show that due to the presence of the Aghajary Formation (comprising of marl and sandstone) (Ali et al,. 2003; Hu et al. 2017), indiscriminate exploitation of mines and damages increases the flood risk.

TABLE 20.2

Range of Scores Given to the Effective Factors in Flood Hazard

Flood Vulnerability Factors	Score
Proximity to cities	46
Proximity to utilities	26
Proximity to gas and oil pipelines	16
Proximity to villages	9
Proximity to road network	4
Total scores	100

TABLE 20.3

The Areas of Floodplain Sub-Watersheds in the Study Area

Code	Sub-Watershed	Flood Plain Area (Km²)	Population
A	Kamp abas	35.983	20
B	Fay 1	29.328	150
C	Fay 2	27.337	50
D	Omidieh	93.203	50,000
E	Meiankuh	34.358	5,000
F	Chah salem	28.520	23
G	Aghajari 1	15.687	25
H	Aghajari 2	29.006	20,000
J	Yeresie	26.621	1,200
K	Hur	22.926	4,500

In addition to the resident population, in the winter pastures and grazing season (fall and winter), the nomads with their cattle move into the sub-watersheds and therefore the vulnerable population increased. In contrast, the sub-watershed such as Yeresie, Hur, and a major part of the Omidieh sub-watershed placed on the lowest class (risk is very low), which is due to the lowest concentration of facilities, infrastructures, the residential area (Table 20.5).

The results of this research verify consistency with the outcomes from the previous studies because the different factors and constraints were used in multi-criteria analysis techniques and there are no human errors. The final maps were obtained accurately because the combination of raw data has been avoided.

20.5 CONCLUSIONS

In this study, a decision-supporting system and geodata analytical process (GAP) in Boolean, AHP, and fuzzy (BAF) as the flood issues techniques in hydrology science are developed to deal with the spatial decision-making problems in the context of spatial information systems. This system evaluates and controls the expert's view on effective criteria in the occurrence of a physical phenomenon with an emphasis on the analytic hierarchy process. The system provides the ability to make paired comparisons of the results from experts, and if there has been no contradiction with the criteria, they would be able to apply their decisions. In general, facilitating the decision-making process and the simple method of weighting are the main features of the system.

This research concluded where we have an issue with the appropriate and sufficient data availability, the approach for multi-criteria decision-making could be a good solution for emergency

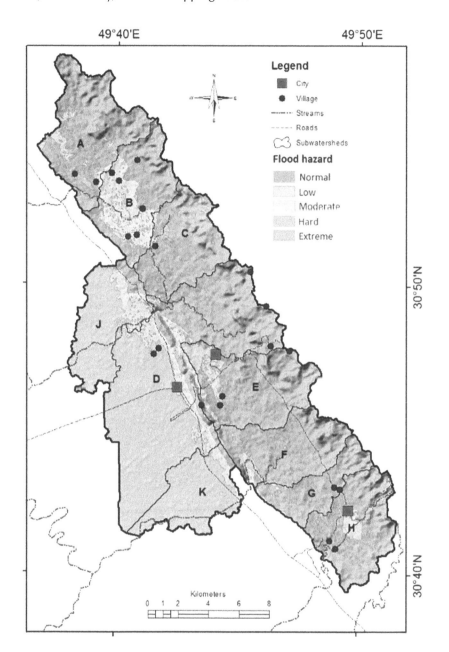

FIGURE 20.9 The flood hazard map of the Omidieh sub-watersheds.

TABLE 20.4
Index Classification of Flood Hazard Map

Flood Intensity Classification	Flood Control Priority	Variety Range of Flood Intensity (m³/s/km²)
Very low	V	0–0.111
Low	IV	0.111–0.298
Moderate	III	0.298–0.422
Nearly heavy	II	0.422–0.519
Heavy	I	0.519–0.75

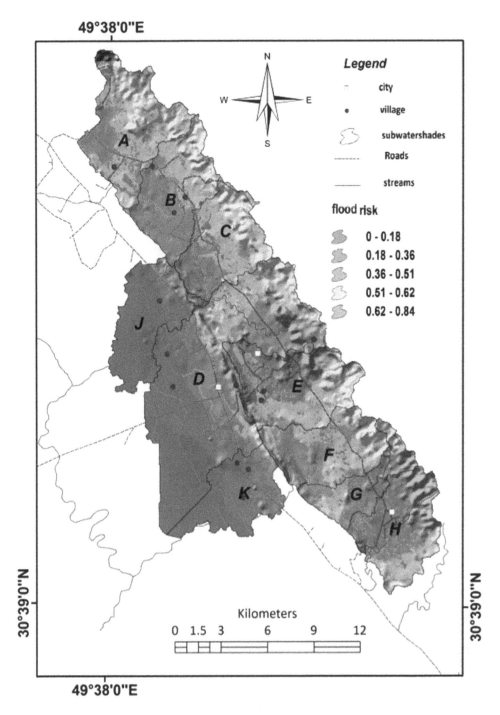

FIGURE 20.10 Map of flood risk priority of the Omidieh sub-watersheds.

responses. For example, in this study, the mentioned system was used to prepare a flood hazard map. However, this system is applicable in other deciding issues, such as environmental problems (Jamali et al., 2022), land use planning, site selection, and in general in a process in which the weighting criteria and determining priorities are the associated options. The results of the SDSS evaluation show that sub-watersheds of Aghajari 1 and 2, Meiankuh, and some parts of the Omidieh watershed

TABLE 20.5

Flood Risk Classification Based on the Flood Intensity in Study Area

Sub-Watershed	Flood Intensity	Flood Control Priority	Class of Vulnerability	Class of Hazard	
Kamp abas	Moderate	III	Moderate	Moderate	High
Fay 1	Heavy	I	Moderate	Extreme	High
Fay 2	Nearly heavy	II	Moderate	Hard	High
Omidieh	Moderate	III	Very high–very low	Moderate	Extreme–very low
Meiankuh	Nearly heavy	II	Very high	Hard	Extreme
Chah salem	Nearly heavy	II	Moderate	Hard	High
Aghajari 1	Heavy	I	Very high	Extreme	Extreme
Aghajari 2	Heavy	I	Very high	Extreme	Extreme
Yeresie	Very low	V	Very low	Normal	Very low
Hur	Low	IV	Very low	Low	Very low

cover about 12.45% of the watershed. They are the most dangerous and high-risk areas of flooding. The rest of the sub-watersheds have been prioritized based on the risk of flooding. A complex decision-making process and modern fuzzy techniques were achieved by using a spatial decision support system. The results indicate the suitability of this system to aid the multi-criteria spatial decision-making process and have quick access to sophisticated spatial goals.

REFERENCES

Ali, S.A., Rangzan, K., and Pirasteh, S. 2003. Remote sensing and GIS study of tectonics and net erosion rates in the Zagros Structural Belt, Southwestern Iran, *GISciences and Remote Sensing Journal*, 40(4): 253–262.

Antonella, Z., Sharifi, A.M., and Andrea, G.F. 2008. Application of spatial multi-criteria analysis to site selection for a local park: A case study in the Birgamo Province, Italy, *Journal of Operational Research*, 158: 1–18.

Bakhtiyarikia, M., Pirasteh, S., Pradhan, B., Mahmud, A.R., Sulaiman, W.N.A., and Moradi, A. 2012. An artificial neural network model for flood simulation using GIS: Johor River Basin, Malaysia, *Environmental Earth Sciences*, 67(1): 251–264.

Dalezios, N.R., and Eslamian, S. 2016. Regional design storm of Greece within the flood risk management framework, *International Journal of Hydrology Science and Technology*, 6(1): 82–102.

Fernandez, D.S., and Lutz, M.A. 2010. Urban flood hazard zoning in Tucumán Province, Argentina, using GIS and multicriteria decision analysis, *Engineering Geology*, 111(1–4): 90–98.

Franci, F., Bitelli, G., Mandanici, E., Hadjimitsis, D., and Agapiou, A. 2016. Satellite remote sensing and GIS-based multi-criteria analysis for flood hazard mapping, *Natural Hazards*, 83(1): 31–51.

Foudi, S., Osés-Eraso, N., and Tamayo, I. 2015. Integrated spatial flood risk assessment: The case of Zaragoza, *Land Use Policy*, 42: 278–292.

Geneletti, D. 2010. Combining stakeholder analysis and spatial multicriteria evaluation to select and rank inert landfill sites, *Waste Management*, 30: 328–337.

He, S., Wang, D., Li, Y., Peng Zhao, Lan, H., Chen, W., Jamali, A.A., and Chen, X. 2021. Social-ecological system resilience of debris flow alluvial fans in the Awang basin, China, *Journal of Environmental Management*, 286: 112230.

He, S., Wang, D., Zhao, P., Li, Y., Lan, H., Chen, W., and Jamali, A.A. 2020. A review and prospects of debris flow waste-shoal land use in typical debris flow areas, China, *Land Use Policy*, 99: 105064.

Hu, S., Cheng, X., Zhou, D., and Zhang, H. 2017. GIS-based flood risk assessment in suburban areas: A case study of the Fangshan District, Beijing. *Natural Hazards*, 87(3): 1525–1543.

Jamali, A.A., and Abdolkhani, A. 2009. Preparedness against landslide disasters with mapping of landslide potential by GIS-SMCE (Yazd-Iran), *International Journal of Geoinformatics*, 5(4): 25–31.

Jamali, A.A., and Kalkhajeh, R.G. 2019. Urban environmental and land cover change analysis using the scatter plot, kernel, and neural network methods, *Arabian Journal of Geosciences*, 12(3): 100.

Jamali, A.A., Kalkhajeh, R.G., Randhir, T.O., and He, S. 2022. Modeling relationship between land surface temperature anomaly and environmental factors using GEE and Giovanni. *Journal of Environmental Management*, 302: 113970.

Jamali, A.A., Randhir, T.O., and Nosrati, J. 2018. Site suitability analysis for subsurface dams using Boolean and fuzzy logic in arid watersheds, *Journal of Water Resources, Planning and Management*, 144(8): 0401804.

Jun, K.S., Chung, E.S., Kim, Y.G., and Kim, Y. 2013. A fuzzy multi-criteria approach to flood risk vulnerability in South Korea by considering climate change impacts, *Expert Systems with Applications*, 40(4): 1003–1013.

Kazakis, N., Kougias, I., and Patsialis, T. 2015. Assessment of flood hazard areas at a regional scale using an index-based approach and Analytical Hierarchy Process: Application in Rhodope–Evros region, Greece, *Science of the Total Environment*, 538: 555–563.

Malczewski, J. 1999. *GIS and Multicriteria Decision Analysis*, John Wiley and Sons.

Ntajal, J., Lamptey, B.L., Mahamadou, I.B., and Nyarko, B.K. 2017. Flood disaster risk mapping in the lower mono river basin in Togo, West Africa, *International Journal of Disaster Risk Reduction*, 23: 93–103.

Parsasyrat, L., and Jamali, A.A. 2015. The effects of impermeable surfaces on the flooding possibility in Zarrin-Shahr, Isfahan Municipal Watershed, *Journal of Applied Environmental and Biological Sciences*, 5(1): 28–38.

Prawiranegara, M. 2014. Spatial multi-criteria analysis (SMCA) for basin-wide flood risk assessment as a tool in improving spatial planning and urban resilience policy making: A case study of Marikina River Basin, Metro Manila – Philippines, *Procedia - Social and Behavioral Sciences*, 135: 18–24.

Sadeghi, S.H., Nouri, H., and Faramarzi, M. 2017. Assessing the spatial distribution of rainfall and the effect of altitude in Iran (Hamadan Province). *Air, Soil and Water Research*, 10: 1178622116686066.

Sharifi, M.A., and Retsios, V. 2006. *Site Selection for Waste Disposal through Spatial Multiple Criteria Decision Analysis*, Warsaw, Poland.

Safaripour, M., Monavari, M., Zare, M., Abedi, Z., and Gharagozlou, A. 2012. Flood risk assessment using GIS. Case study: Golestan Province, Iran, *Polish Journal of Environmental Studies*, 21: 1817–1824.

Shaw, R. 2015. Hazard, vulnerability and risk: The Pakistan context. In: *Disaster Risk Reduction Approaches in Pakistan*, pp. 31–52, Springer, Japan.

Umar, R., and Acharya, P. 2016. Flood hazard and risk assessment of 2014 floods in Kashmir Valley: A space-based multisensor approach, *Natural Hazards*, 84(1): 437–464.

Vittala, S.S., Govindaiah, S., and Gowda, H.H. 2004. Morphometric analysis of sub-watersheds in the Pavagada area of Tumkur district, South India using remote sensing and GIS techniques. *Journal of the Indian Society of Remote Sensing*, 32(4): 351–362.

Zaharia, L., Costache, R., Pravalie, R., and Ioana-Toroimac, G. 2017. Mapping flood and flooding potential indices: A methodological approach to identifying areas susceptible to flood and flooding risk. Case study: The Prahova catchment, Romania, *Frontiers of Earth Science*, 1–19.

Zhaoli, W., Hongliang, M., Chengguang, L., and Haijuan, S. 2012. Set pair analysis model based on GIS to evaluation for flood damage risk, *Procedia Engineering*, 28: 196–201.

Part VII

Flood Regionalization

21 Determining High-Flood-Risk Regions Using Rainfall-Runoff Modeling

Iftekhar Ahmed

CONTENTS

21.1 INTRODUCTION

The use of computer models to represent the hydrological cycle began in 1958 when the United States Army Corps of Engineers (USACE) began investigating the use of computer techniques in streamflow analysis and reservoir regulation. From this, several generations of the Stanford Watershed Model were developed. A conceptual representation of the hydrologic cycle was provided by solving physical equations to describe the watershed processes and water balance to account for watershed storage of water. The 1958 initiative by the USACE was ahead of its time because up to 2002, a developed country like the United Kingdom's approach to flood estimation was based on the statistical analysis of existing flood records or an interpretation of the relationship between rainfall and runoff based on watershed characteristics (Fleming and Frost, 2002). The accuracy of statistical methods is limited by the quality and availability of historic data. Peak flows can be widely inaccurate because of the tendency to calibrate gauge stations during low flow conditions as it is often impossible to access a site at the height of a flood. Suitable models are many and in the last 15 years, their use has seen benefits around the developed countries of the world and other countries are not too far behind in this effort. Figure 21.1 illustrates how flood data and flood estimation techniques have developed over the last 200 years and a projection of expected developments through 2050.

The merits of conceptual, parametrically parsimonious, hydrological models for investigating the dominant pathways and processes in watersheds are discussed in Refsgaard and Henriksen (2004) and Sivapalan (2003). Model parameter identification is a fundamental challenge for hydrologists

DOI: 10.1201/9780429463938-28

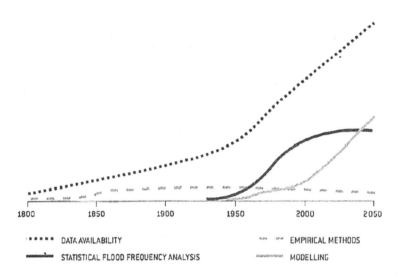

1800   1850   1900   1950   2000   2050

· · · · · DATA AVAILABILITY   EMPIRICAL METHODS

STATISTICAL FLOOD FREQUENCY ANALYSIS   MODELLING

FIGURE 21.1   Timeline of flood data and flood estimation techniques development. (Source: Fleming, G., and Frost, L. 2002. Flooding and flood estimation. In: *Flood Risk Management*, Thomas Telford, London, UK.)

(Sivapalan, 2003; Duan et al., 2006). The presence of parameter interactions in conceptual rainfall-runoff (CRR) models can make the a priori parameter prediction methods unreliable (Wagener and Wheater, 2006). Ideally, a model should be parametrically parsimonious while still capturing the dominant processes of a watershed with limited parameter interactions. Many hydrological models have been developed and used for decades for both research and operational hydrology. However, new model structures are still being developed to incorporate a new conceptual understanding of specific watershed processes and places and to facilitate the demands of new pressures on water resources, including nutrient enrichment (Futter et al., 2014). The research focus has also recently shifted to runoff initiation processes controlled by topography and geomorphology of the land-scape. The relationship between the field-mapped flow-initiation points (FIPs) and model-predicted floodplain extent for selected locations in the United States is shown in Clubb et al. (2014, 2017). In order to compare the field-mapped FIPs to the predicted floodplain extents, the flow distance between the field-mapped point and the furthest upstream point of the nearest predicted floodplain patch is measured.

Surface runoff varies by time and location, with about one-third of the precipitation that falls on land turning into runoff; the other two-thirds is evaporated, transpired, or infiltrated into the soil (Perlman, 2016). The water balance equation governs the hydrological cycle by describing the flow of water into and out of a system for a specific period of time:

$$Q = P - ET - \Delta SM - \Delta GW \tag{21.1}$$

Where,

Q is the surface runoff volume
P is the precipitation
ET is the evapotranspiration
ΔSM is the change in soil moisture
ΔGW is the change in groundwater storage

Runoff is generated by a combination of two mechanisms: saturation excess and infiltration excess (Srinivasulu and Jain, 2008; Yang et al., 2015). Saturation excess occurs when the soil becomes fully saturated with water, exceeding the water holding capacity of the soil; when the surplus rainfall can no longer be held in the soil, the water is directed to another location through overland flow (Johnson et al., 2003). Infiltration excess occurs when rainfall intensity exceeds the maximum rate that water can infiltrate into the soil, and water must flow over the land to a lower elevation (Yang et al., 2015). Excess rainwater flowing overland picks up debris and chemicals along the flow path. The debris may include sediments, organic matter, nutrients, pesticides, and other materials which impact the quality of receiving surface water (Huffman et al., 2011). Surface runoff is therefore an important area of interest for monitoring water resources, as well as solving water quality and quantity problems such as flood forecasting and ecological and biological relationships in the water environment (Kokkonen et al., 2001).

21.2 RAINFALL-RUNOFF MODEL TYPES

An overview of rainfall-runoff model types can be found in (EPA, 2017). Hydrologic model performance is mainly limited by the model structure, not by model parameters (Melsen et al., 2016). The size, complexity, and diversity of the infrastructure systems pose considerable challenges in gaining a comprehensive understanding of flood system elements and their relationships. The detail that such studies will require makes it impractical to rely solely on numerical models. With a larger number of parameters, there is a greater possibility that some parameters may become too site-specific (Viney et al., 2009). A model is a simplified representation of a real-world system (Moradkhani and Sorooshian, 2009). The best model is the one that gives results close to reality with the use of the least parameters and model complexity. Models are mainly used to predict system behavior and to understand various hydrological processes. Many models overlap within this classification of model structure (Pechlivanidis et al., 2011). The three *structural* categories of runoff models, with strengths and weaknesses for each, are displayed in Table 21.1. A detailed description of each category of models can be found in EPA (2017).

The spatial processes in runoff models provide a means of representing the watershed for modeling. They are based on input data and how runoff is generated and routed over the watershed. Variability in geology, soils, vegetation, and topography affects the relationship between rainfall and runoff within a watershed and needs to be considered in modeling (Beven, 2001). The spatial structure of watershed processes in rainfall-runoff models can be categorized as lumped, semi-distributed, and fully distributed (Table 21.2). Lumped models do not consider spatial variability

---

TABLE 21.1

Comparison of the *Basic* Structure of Rainfall-Runoff Models

	Empirical	Conceptual	Physical
Method	Non-linear relationship between inputs and outputs, black box concept	Simplified equations that represent water storage in catchment	Physical laws and equations based on real hydrologic responses
Strengths	Small number of parameters needed, can be more accurate, fast run time	Easy to calibrate, simple model structure	Incorporates spatial and temporal variability, very fine scale
Weaknesses	No connection between physical catchment, input data distortion	Does not consider spatial variability within catchment	Large number of parameters and calibration needed, site specific
Best Use	In ungauged watersheds, runoff is the only output needed	When computational time or data are limited.	Have great data availability on a small scale

Source: EPA. 2017. An overview of rainfall-runoff model types. *Technical Report*, US Environmental Protection Agency Office of Research and Development, Athens, GA.

TABLE 21.2

Comparison of the *Spatial* Structure of Rainfall-Runoff Model

	Lumped	Semi-Distributed	Distributed
Method	Spatial variability is disregarded; entire catchment is modeled as one unit	Series of lumped and distributed parameters	Spatial variability is accounted for
Inputs	All averaged data by catchment	Both averaged and specific data by sub-catchment	All specific data by cell
Strengths	Fast computational time, good at simulating average conditions	Represents important features in catchment	Physically related to hydrological processes
Weaknesses	A lot of assumptions, loss of spatial resolution, not ideal for large areas	Averages data into sub-catchment areas, loss of spatial resolution	Data intense, long computational time

Source: EPA. 2017. An overview of rainfall-runoff model types. *Technical Report*, US Environmental Protection Agency Office of Research and Development, Athens, GA.

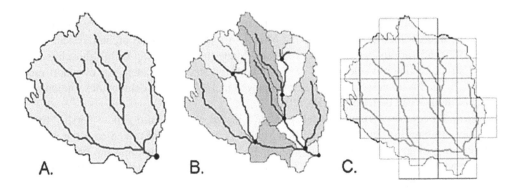

FIGURE 21.2   Visualization of the spatial structure in runoff models: **A.** Lumped model, **B.** Semi-distributed model, and **C.** Distributed model by grid cell. (Source: EPA. 2017. An overview of rainfall-runoff model types. *Technical Report.* US Environmental Protection Agency Office of Research and Development, Athens, GA.)

within the watershed; semi-distributed models reflect some spatial variability; and fully distributed models process spatial variability by grid cells (Figure 21.2).

21.3   THE UTILITY OF FLOOD RISK MAPS

Flood risk maps are created on a watershed basis. The overall objectives of a flood risk mapping study, also known as Flood Insurance Study (FIS), are to (FEMA, 2009):

- Identify areas subject to flooding from riverine sources and accurately define the flood-frequency relation at locations within those flood-prone areas.
- Depict the data and analyses results with maps, graphs, tables, and explanatory narratives in order to support flood insurance decisions and sound floodplain management.
- Document data and analyses in a digital format to the extent possible to enable the results to be readily checked, reproduced, and updated.
- Maintain (or establish) consistency and continuity within the national inventory of Flood Insurance Rate Maps (FIRMs) and FIS reports.

Riverine analyses consist of hydrologic analyses to determine discharge-frequency relations along the flooding source and hydraulic analyses to determine the extent of floodwaters (floodplain) and the elevations associated with the water-surface of each frequency studied. The base (1% annual chance or the 100-year event) flood is delineated on the FIRM as the Special Flood Hazard Area (SFHA). When determined, the 0.2%-annual-chance floodplain and/or floodway are also depicted on the maps. The analyses must be based on existing ground conditions in the watershed and floodplain. A community that conducts its own future-conditions analysis may request that FEMA reflect these results on the FIRM. Such reports are called the Letter of Map Revision (LOMR) or Conditional Letter of Map Revision (CLOMR) in the United States.

Documentation of input data must describe the methods of measurements and sources from which data were obtained or measured. Documentation of parameters used in analyses, including initial and boundary conditions, must describe the derivation of those parameters, and methods of measurements and sources from which data supporting those parameters were obtained or measured. The selection of some model parameters is self-explanatory, such as the selection of the "normal depth" method as the downstream boundary condition in a step-backwater hydraulic analysis. However, the selection of many input parameters may depend on a number of physical, topographic, and hydrologic properties of the watershed and floodplain under study. For example, the selection of Soil Conservation Service (SCS) Curve Numbers as loss parameters of a sub-watershed will depend on its geological, land use, and hydrologic properties. Therefore, documentation for the selection of Curve Numbers should provide all of the parameters used for the selection (FEMA, 2009).

21.4   CASE STUDY: THE UPPER NEW RIVER ADMP, ARIZONA

21.4.1   Project Synopsis

Flood control management and policies that formed the foundation of the Upper New River Area Drainage Master Plan (UNR-ADMP) development in Phoenix, Arizona, USA, took into account engineering, environmental, landscape, social, and economic considerations. The Upper New River watershed encompasses approximately 450 square kilometers of the New River watershed above New River Dam in north-central Maricopa County and southern Yavapai County (Figure 21.3) in the semiarid climate of Arizona. The UNR-ADMP "study area" encompassed approximately 246 square kilometers (95 square miles) of the watershed. The eastern boundaries of the study watershed are contiguous with the Skunk Creek watershed. The western boundaries of the study watershed are contiguous with the Agua Fria River watershed. The New River watershed, referred to as the study catchment in this chapter, is naturally divided at the US Geological Survey (USGS) Rock Springs gauge into an upper and lower sub-watershed with very different hydrologic characteristics. The upper sub-watershed is primarily steep and rocky with well-defined channels and areas of perennial flow. It was almost entirely undeveloped with natural vegetation and undisturbed drainage paths during the existing conditions study period of 2006–2008.

In contrast, the lower sub-watershed is characterized by moderate elevation change across basins. There was more development in the lower sub-watershed which affected the drainage paths, especially through the community of New River and along the Interstate Highway-17 (I-17) corridor. The main channel of the New River tends to be braided and distributive through portions of the lower sub-watershed. The ADMP study area includes those areas of the watershed which were subject to future development and is referred to as the "study area" in this chapter. The study area falls within the jurisdictional areas of the City of Phoenix, City of Peoria, and unincorporated Maricopa County lands. The project owner was the Flood Control District of Maricopa County (District) (Figure 21.4).

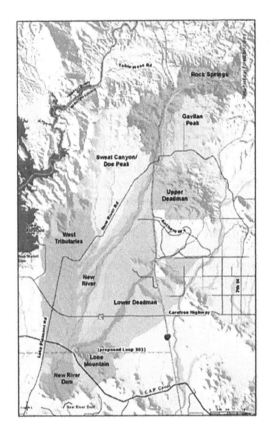
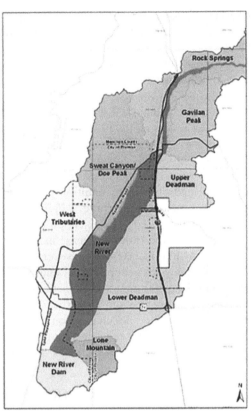

FIGURE 21.3 Upper New River ADMP planning areas with relief map and GIS layers (*figure not to scale*). (Source: Stantec. 2008d. Upper New River ADMP recommended alternative report, Report prepared for the Flood Control District of Maricopa County, Phoenix, AZ.)

21.4.2 CHALLENGES

The UNR-ADMP watershed presented the modelers with the challenge of various distributary flow features that made it essential to input the rainfall-runoff model generated hydrographs on a two-dimensional platform to delineate the flood risk areas and observe the effect of the post-FIS (post–Flood Insurance Study) in-channel sand and gravel mine pits on the watershed storage. Storage played a significant role in the shape of the hydrographs at critical locations. A hydrograph's time-to-peak is typically short if the watershed does not have notable storage capacity. The post-FIS in-channel mine pits provided significant storage areas that established control on the time-to-peak and flow area distribution at the downstream locations from the pits. It was therefore necessary to model the distribution of flow in two-dimension. This was accomplished by using the FEMA (Federal Emergency Management Agency) approved FLO-2D software package (FLO-2D, 2006) to track the time-to-peak of various hydrographs from the main channel and its tributaries. The HEC-1 (HEC, 1990) flood hydrograph package discharges were adjusted based on the FLO-2D output.

21.4.3 THE HYDROLOGIC PARAMETERS

Hydrologic models conventionally have been one-dimensional. That is, the total watershed area is considered lumped together at a point since both the input and the output are assumed to be uniformly distributed across the area. These are termed "lumped-parameter" models since the effect of all of the spatial variation is lumped into the parameters and only the vertical variation is considered

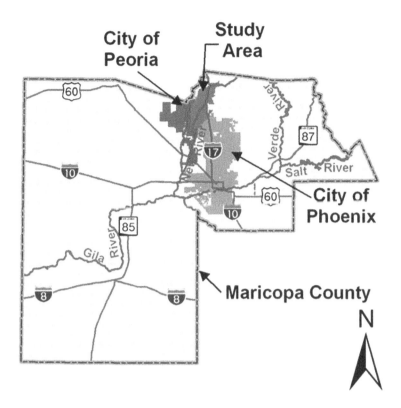

FIGURE 21.4 Project location in Maricopa County, Arizona (*figure not to scale*). (Source: Ahmed, I., and Gerlach, R.M. 2012. Distributed flow guided hydraulic modeling of a desert river system for flood control, *Proc., RiverFlow-2012: International Conference on Fluvial Hydraulics*, Sponsor: IAHR, September 5–7, San Jose, Costa Rica.)

(Beven, 2001; Bedient et al., 2001; Feldman, 1995; Kirby, 1978). The USACE's HEC-1 is a lumped-parameter watershed modeling software rooted in the kinematic wave theory (Singh, 1996) The software seeks numerical solutions based on the infiltration excess overland flow runoff mechanism (Feldman, 1995; Smith et al., 1995). Such models treat a watershed area as a sequence of hillslope segments, bounded by streamlines (Feldman, 1995; Kirby, 1978; Kibler and Woolhiser, 1970). The flow is treated as one-dimensional in the downslope direction.

Point rainfall values: A total of six storm events were modeled; the 10-, 50-, and 100-year recurrence periods were each evaluated for the 6- and 24-hour storms. In order to obtain depth-duration-frequency estimates for the watershed, values for the 2-, 6-, and 24-hour storms, as well as the 100-year, 6-, and 24-hour storms were estimated at the centroid of the watershed from isopluvials published in NOAA Atlas II (NOAA, 2003). The resulting values were input into the district's Drainage Design Management System (DDMSW, 2002) pre-processor for HEC-1 model input files creation; detailed output and discussion can be found in (Gerlach et al., 2016).

Areal rainfall reduction: Rainfall reduction for the six-hour storm was based on the depth-area curve developed for the historic storm of 1954 in Central Arizona (FCDMC, 1997). The rainfall patterns and the corresponding area reduction factors were automatically selected by the DDMSW program. The precipitation reduction factors used for the 24-hour storms are given in (FCDMC, 1997) and those values were derived from information contained in the NOAA Technical Memorandum HYDRO-40 (NOAA, 1984). The appropriate depth area reduction for all of the storms and accumulated drainage areas was simulated in HEC-1. Summary of the basin areas, reduction factors, and the aerially reduced point precipitation values used for the 24-hour HEC-1 models can be found in

(Gerlach et al., 2016). The six-hour aerial reduction factors were generated by the DDMSW program for input into the HEC-1 model.

Temporal rainfall distribution: Both 6- and 24-hour duration storms were modeled for each frequency under the existing conditions. Two storm durations were used in order to determine which storm results in a higher magnitude of discharge at the various locations in the watershed. The rainfall distributions for the six-hour duration storm were based on watershed area, and are given in (FCDMC, 1997). Each precipitation pattern is valid for a certain watershed area and is automatically coded into the model by DDMSW. The Soil Conservation Service (SCS) Type II rainfall distribution was used for the 24-hour storm (USDA, 1986). Type II is the most intense short-duration rainfall contributing to flash floods in the study area.

Rainfall loss parameters: Rainfall loss for the study area was estimated using the Green and Ampt loss rate method as implemented in HEC-1. This method combines the Green and Ampt infiltration equation with surface retention to estimate the rainfall loss after accounting for the impervious area. The Green and Ampt infiltration equation requires three parameters (Ferguson, 1994; Chow et al., 1988; Brakensiek and Rawls, 1983): XKSAT (hydraulic conductivity at natural saturation), PSIF (wetting front capillary suction head), and DTHETA (volumetric soil moisture deficit). Values of XKSAT are selected based on soil texture and can be modified to reflect the effects of vegetative cover. PSIF and DTHETA were estimated as a function of bare ground XKSAT. Values for initial abstraction (IA) were estimated based on landform and land use conditions present within the study area. Impervious area (RTIMP) was estimated to reflect naturally occurring rock outcrop and land use conditions. The initial abstraction was satisfied prior to rainfall infiltration as follows (HEC, 1990):

$$r(t) = 0 \quad \text{for} \quad P(t) \leq IA \quad T > 0 \tag{21.2}$$

$$r(t) = r_0(t) \quad \text{for} \quad P(t) > IA \quad T > 0 \tag{21.3}$$

Where,
   $P(t)$ is the cumulative precipitation over the watershed
   $r(t)$ is the rainfall intensity adjusted for surface losses
   t is the time since the start of rainfall $r_0(t)$
   IA is the initial abstraction.

The method is applied to the remaining rainfall using the following equations:

$$F(t) = \frac{PSIF * DTHETA}{\left[\dfrac{f(t)}{XKSAT} - 1 \right]} \qquad f(t) > XKSAT \tag{21.4}$$

$$f(t) = r(t) \qquad f(t) \leq XKSAT \tag{21.5}$$

where
   $F(t)$ is the cumulative infiltration
   $f(t) = dF(t)/dt$ is the infiltration rate

Soil parameters: Soil properties and texture classifications were used to estimate the bare ground XKSAT parameter of the Green and Ampt method. Values of XKSAT and RTIMP were assigned to each map unit based on interpretation of the NRCS soil surveys and were used as default values in DDMSW. The study area falls entirely within the Aguilla/Carefree soil survey zone. A summary of the soil map units found in the detailed study area and their default XKSAT and RTIMP values can be found in (Gerlach et al., 2016). The SCS Survey soil unit mapping for the detailed study area was supplied in digital format by the District. Within ArcGIS, the digital soil mapping was intersected with the delineated sub-watersheds. The resulting GIS layer provided the areas of each soil map unit including those in each sub-watershed. This information was input into the DDMSW program, which calculated a log-area weighted XKSAT value and impervious/rock outcrop percentages (RTIMP) for each sub-watershed.

Land-use parameters: Under the existing conditions, the detailed watershed was primarily undeveloped desert mountain/rangeland with only a few isolated pockets of developed land. The natural desert consisted of mountain, terrace, and floodplain areas, while the sparsely developed land areas included residential and commercial/industrial properties. Land use polygons within the study area were delineated to represent the areas of unique physical characteristics which influence rainfall loss parameters. The delineation was based on circa 2006 aerial photography supplied by the district and was confirmed during the field reconnaissance trips. The land use classifications were assigned values for surface retention or initial abstraction (IA), percent of impervious area (RTIMP) and vegetative cover based on guidelines in FCDMC (1997), and the default values available in DDMSW. Natural desert areas were divided into four classifications based on landform characteristics and vegetation density. A summary of the land-use classification identifiers and their descriptions along with the land use classifications for the detailed study area can be found in (Gerlach et al., 2016). Land use polygon mapping was intersected with sub-watershed boundaries using GIS. The resulting GIS layer provided the areas of each land-use type for each sub-watershed. This information was input into the DDMSW program, which calculated composite IA, RTIMP, XKSAT, DTHETA, PSIF, and the K_n values for each sub-watershed (Gerlach et al., 2016).

Unit hydrograph definition: Using the kinematic wave theory to generate unit hydrograph within HEC-1 can lead to conservative (higher peak discharge) runoff hydrographs with a steep rising limb and a short time to peak (Stantec, 2008a). Given the mild terrain slope in distributive flow areas with subcritical flow regimes, such a short time-to-peak can be misleading. Therefore, the local S-graphs were used in this ADMP study for direct user-defined unit hydrograph input to HEC-1. For the watershed area, rainfall excess was transformed into runoff hydrograph by applying the Phoenix Mountain and Phoenix Valley S-graphs. An S-graph is a form of unit hydrograph and is often used in performing flood studies. On an S-graph, discharge on the y-axis is expressed in percent of ultimate discharge, and time on the x-axis is expressed in percent of lag. S-graphs are usually defined by the reconstitution of recorded flood events and thus, numerous S-graphs are available (Gerlach et al., 2016; Sabol, 1987).

The K_n (mean watershed Manning's n-value) values for each land use classification were selected based on the guidance provided in (FCDMC, 1997). The Phoenix Mountain S-graph was derived from reconstitutions of flood events in the Upper New River watershed. The Phoenix Valley S-graph was derived from reconstitutions of flood events that included the New River watershed. Determination of K_n values specific to the flood events was part of the reconstitutions. The use of those values for undeveloped land use classifications present within the detailed study area provided consistency to the source data. Where the default K_n values differed from the reconstituted values, the reconstituted values were used. The K_n values are listed in (Gerlach et al., 2016).

21.4.4 THE ROUTING PARAMETERS

Reach routing parameters: Routing of sub-watershed hydrographs was performed using the Modified Puls channel routing option in HEC-1. In general, routing along the main stem of the New River was modeled using channel storage routes based on stage-discharge relations, while routing in the detailed study but outside the main stem used the normal depth option. Routing reach flow paths were delineated from the 3.048-meter contour interval county-wide topographic maps (Stantec, 2008a).

21.4.4.1 Normal Depth Routing

Cross-sectional geometry: Representative cross-sectional geometry for each normal depth channel route was developed from the project-specific 0.61-meter contour interval topographic maps. For routing reaches where the cross-sectional geometry throughout the reach was relatively uniform and could easily be characterized with only eight points (across the channel), the geometry was extracted directly from the topographic maps. For reaches where the cross-sectional geometry varied along the reach and/or was too complex to be directly described with an eight-point cross-section (i.e., multiple shallow channels inset in a broad floodplain), a representative cross-section was approximated by inspection of the topography along the entire reach, i.e., based on the channel morphology upstream and downstream of the location of interest.

Manning's n estimates: The normal depth option of the channel storage routing method requires a Manning's n-value estimate for the main channel and both the left and right overbanks. Representative Manning's n-value estimates for each routing reach were made based on photographs taken during the field reconnaissance and experience with the environment. For the routing reaches within the limits of the detailed study watercourse, the Manning's n estimates for the hydraulic analyses (using HEC-RAS) were used. Representative routing reach Manning's n-values were determined by the selection of the predominant values for all cross-sections located within each routing reach. In general, all routing reaches were characterized by a well-defined low flow channel with bed material ranging in size from sand to large cobbles. Overbank areas were typically flat with fine-grained soils and sparse vegetation.

Routing reach length and slope: The normal depth option parameters of reach length and energy slope (assumed to be the channel slope) were measured from the 0.61-meter contour interval topographic maps (Stantec, 2008a).

21.4.4.2 Stage-Storage-Discharge Routing

The Modified Puls routing method was used for channel routing along the main stem of the New River. The use of the method for channel routing requires caution because the degree of peak discharge attenuation varies depending on the river reach lengths chosen or alternatively, on the number of routing steps specified for a single reach (FLO-2D, 2006). The storage indication function is computed from given storage and outflow data:

$$\text{STRI}(i) = C * \frac{\text{STOR}(i)}{\Delta t} + \frac{\text{OUTFL}(i)}{2} \tag{21.6}$$

where,
  STRI is the storage indication in m^3/s
  STOR is the storage volume in the routing reach for a given outflow in m^3
  OUTFL is the outflow from the routing reach in m^3/s
  C is a conversion factor
  Δt is the time interval in a consistent unit
  i is the index of storage and outflow values

The storage indication at the end of each time interval is given by:

$$\text{STRI}_2 = \text{STRI}_1 + Q_{in} - Q_{out} \tag{21.7}$$

where
  Q_{in} is the average inflow
  Q_{out} is the outflow

The subscripts 1 and 2 indicate the beginning and the end of the current time interval. The outflow at the end of each time interval is interpolated from a chart of storage indication (STRI) versus outflow (OUTFL). The storage (STR) is then computed from:

$$STR = \left(STRI - \frac{Q}{2}\right) * \frac{\Delta t}{C} \tag{21.8}$$

The Modified Puls storage-discharge routing for the study watershed channels was based on results from an existing HEC-RAS river hydraulic model. The one-dimensional HEC-RAS model was run for 20 discharges ranging from 14 cms to 1,444 cms. For each discharge, water surface elevations and cumulative channel storage volumes were extracted from the model for the cross-sections that corresponded to the upstream and downstream limits of the routing reaches. The stage at each discharge was determined from the water surface elevation at the upstream cross-section of each reach, while the channel storage was taken as the difference between the upstream and downstream cumulative volume. The estimation of the number of routing steps for input to the HEC-1 models was an iterative process. The number of routing steps for each reach may vary with the storm duration being considered. The procedure used to estimate the number of routing steps and the indirect verification of HEC-1 model discharges are explained in (Gerlach et al., 2016).

21.4.5 2D FLOW DISTRIBUTION

In order to better understand the split and braided portions of the detailed study area, a FLO-2D model was developed for an area extending approximately 9.6 kilometers upstream of the Carefree Highway (Stantec, 2008b,c; Ahmed and Gerlach, 2012). Hydrographs from the HEC-1 model were used as inputs for the FLO-2D study, which generated output hydrographs for cross-sections upstream and downstream of the flow split locations based on the two-dimensional hydraulic analysis of the river. The results of the FLO-2D model were then used to modify the HEC-1 model. Specifically, the FLO-2D results indicated distinct flow splits or areas where the flow from a single concentration point traveled to two downstream concentration points. These distributed flow areas are shown in Figure 21.5 and are termed the "West Split," the "Sweat Canyon Split," and the "Deadman Diversion." Additionally, the FLO-2D model results indicated that a majority of the flow reaching the Carefree Highway Bridge passed under the west bridge as free flow.

In HEC-1, each split flow was simulated using a diversion rating curve. The rating curves were derived from hydrographs extracted from the FLO-2D model results. The FLO-2D model was run for the 100-year, 24-hour storm only. The diversion rating curves developed from the FLO-2D results were applied to all of the other storm events and the durations were modeled as part of this study under the assumption that the hydraulic characteristics for multiple inflow conditions can be represented by a single rating curve (Stantec, 2008b,c). The FLO-2D hydrographs showed slightly different diversion flow for the rising and the falling limbs of the flood event; however, HEC-1 permits only one diversion flow value for each discharge. Initial diversions were based on the rising limb of the FLO-2D hydrographs, and results were compared to the FLO-2D output. If the rising limb diversion curve did not provide a downstream hydrograph that was reasonably close to the FLO-2D hydrograph, the diversion curves were adjusted until the FLO-2D and HEC-1 outputs converged. Details on the FLO-2D model input parameters can be found in (Ahmed and Gerlach, 2012).

21.4.6 FLOOD MAPS

FLO-2D being a distributed parameter model, substantial information can be gathered from the depth coverage on backwater storage conditions in gravel/borrow pits, backwater condition upstream of the bridges under design discharge, and floodplain flow limits. The velocity plan provides

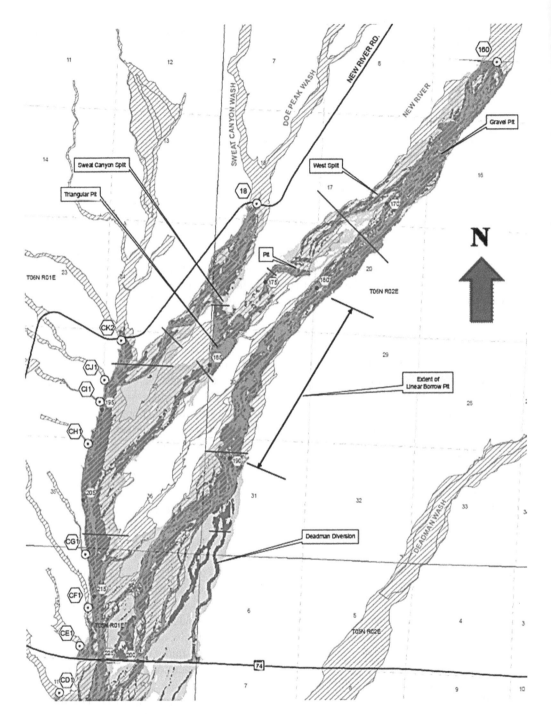

FIGURE 21.5 Flow split locations: West split, Sweat Canyon Wash split, and Deadman Wash diversion with flow concentration points shown with hexagonal IDs (*figure not to scale*). (Source: Gerlach, R.M., Ahmed, I., Beckman, N.D., and Karim, A. 2016. The utility of HEC-1 flood hydrograph package in distributed flow guided area drainage master plan discharges verification in a desert environment, *International Journal of Hydrology Science and Technology*, 6(3): 199–225.)

corresponding information on locations with potential for scouring, deposition, and possible lateral bank migration. The combined study of these two plans helps identify the locations where in-depth design calculations should be considered to maintain the existing channel morphology, and protect the riverine habitats, and any adjacent developments.

A study of the plans reveals that there are two major features that tend to control the flow dynamics within the Upper New River main stem. These are the flow split location to the West Branch, and the linear gravel pit just downstream from the split (Figures 21.6 and 21.7). Upstream of the west flow split location the flow is contained, depth is deep in most parts, and the corresponding velocity is high. At the split flow location with the West Branch, there are two distinct island features in the main stem that causes the flow to split within the main channel. This causes the high-velocity water from upstream to lose energy, and thus, some of the water simply spills over into the West Branch. At the farthest downstream point of the linear pit, there is a man-made wall, constructed perpendicular to the flow path, which constricts the flow to a 40-m opening. This constriction causes backwater condition and storage in the lower 366 m length of the linear pit behind the walls (Figures 21.6 and 21.7). The maximum flow depth in the storage area upstream of the constriction is approximately 4 m. This condition results in a delayed response in velocity distribution downstream of the pit. The flow attains a velocity of 2.4 to 3 m/s downstream of the pit.

At the bend downstream of the pit constriction, split flow directs water into two diversion channels (Figure 21.6–21.8). The west breakout flow is directed towards the SR-74 Highway East Bridge at the rate of 52 cms, and the east breakout flow is directed towards the Deadman Wash at the rate of 46 cms. These breakouts reduce the velocity farther downstream in the Upper New River main stem to 1.2 to 1.8 m/s. Both FLO-2D and HEC-RAS analyses supported that the majority of the flow is conveyed by the channel under the West SR-74 Highway Bridge (Figure 21.8). The FLO-2D estimate of the flow under the West Bridge is 510 cms whereas the flow under the East Bridge is only 226 cms. Flow depth at the West and the East Bridges are approximately 2.4 to 3m, and 1.8 to 2.4m, respectively. The corresponding velocity range is approximately 1.8 to 2.4 m/s, and 1.2 to 1.8 m/s.

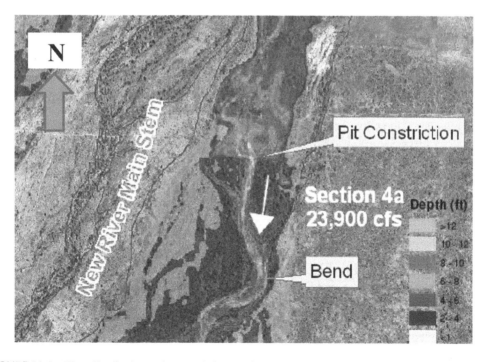

FIGURE 21.6 Flow distribution at the gravel pit constriction (*figure not to scale*). (To convert *feet* to *meters*, multiply by 0.3048. To convert *cfs* to *cms*, multiply by 0.02831685.)

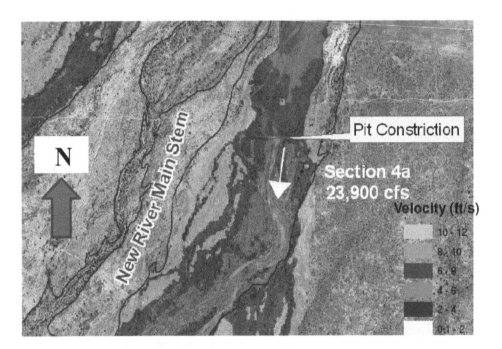

FIGURE 21.7   Flow split at the bend downstream from the pit constriction (*figure not to scale*). (To convert *feet* to *meters*, multiply by 0.3048. To convert *cfs* to *cms*, multiply by 0.02831685.)

Upstream of the West Bridge, in the main channel that conveys combined Upper New River and Sweat Canyon Wash flow, the velocity range is approximately 1.2 to 2.4 m/s. The flow depth in this reach varies spatially between 1.2 and 3 m. Downstream of the bridges, high flow depth continues for a short distance before the flow spreads over a wider portion of the main stem.

21.4.7   HIGH-FLOOD-RISK REGIONS

Four major existing condition flow breakout locations calling for flood risk mitigation were identified within the New River, downstream of the Interstate-17 (I-17) highway. The locations

of these flow splits are shown in Figure 21.9. Two of the flow splits are in the upper portion of the planning area. Flow that breaks out of New River at the West Split combines with break-out flow from the Sweat Canyon Wash. This combined break-out flow travels southwesterly between the Sweat Canyon Wash and the New River, to eventually discharge into the Sweat.

Canyon Wash, just upstream of the SR 74 West Bridge. The third flow split occurs along New River just downstream of the borrow pit. Break-out flow at this location travels parallel to the New River, crosses SR 74, and eventually enters the Deadman Wash. The fourth flow split is at SR 74. At this location flow in the New River starts to spread out and combine with the flow in the Sweat Canyon Wash. A portion of the New River flow passes SR 74 at the East Bridge. The majority of the flow combines with the Sweat Canyon Wash flow before crossing SR 74 at the West Bridge. A major element of the recommended alternative is the use of levees to eliminate the existing flow splits along the New River and the Sweat Canyon Wash. Elimination of the flow splits results in changes to the magnitude and timing of peak flow rates along the two watercourses. A summary of those changes at key locations in the planning area is provided in (Stantec, 2008d). The locations of those key flow split points are shown in Figure 21.9.

The proposed alternatives report called for levees to eliminate the flood risk from future land developments (Stantec, 2008d). A portion of the levee alignment along the New River and the Sweat Canyon Wash to the west of New River are shown in Figure 21.10. For areas with proposed

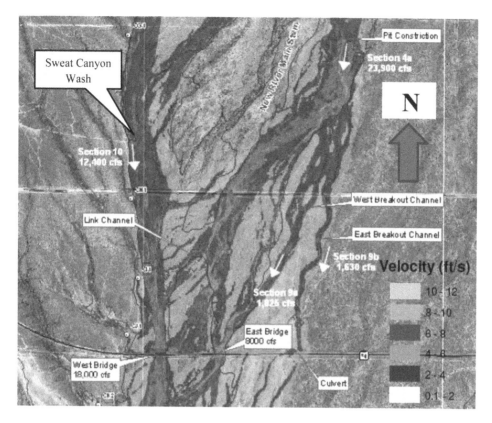

FIGURE 21.8 2D flow distribution in the New River main stem and the Sweat Canyon Wash (*figure not to scale*). (Source: Ahmed, I., and Gerlach, R.M. 2012. Distributed flow guided hydraulic modeling of a desert river system for flood control, *Proc., RiverFlow-2012: International Conference on Fluvial Hydraulics*, Sponsor: IAHR, September 5–7, San Jose, Costa Rica.)

soft-structural flood defense solution reaches, the encroachment option in HEC-RAS was used to represent the general features of the levee. Soft structural alternatives are similar to a full structural levee with the exception that the structural features are hidden or softened by landscape aesthetic treatments. For this level of analysis, it was assumed that fill slopes for the landscape aesthetic treatments would have a negligible impact on the hydraulic characteristics given the magnitude in the dimension of the proposed alternative and, therefore, not coded into the HEC-RAS model. For the recommended alternative, the levee alignments were refined to more closely fit the existing form and function of the land surface.

21.5 CONCLUSIONS

The public generally has a short memory span. Complacency can set in with time long after a notable flood event (Samuels, 2000). Hydrologists and flood control engineers are uniquely placed in society to deal with river basin management including long-term issues. Their role in resolving flooding problems needs to be portrayed to the public more forthrightly (Knight, 2006). Therefore, it is vital that the modeling results be brought to light through accurate mapping of the flood risk zones under the existing conditions before any long-term land development project is undertaken by a community. Thankfully, there are examples of communities and flood control agencies taking the long view, and adding value to society in a way that other professionals cannot.

Rainfall-runoff models linked with the hydraulic models allow us to explore the watershed response to specific rainfall scenarios, consequent excess runoff, resultant water levels, and possible

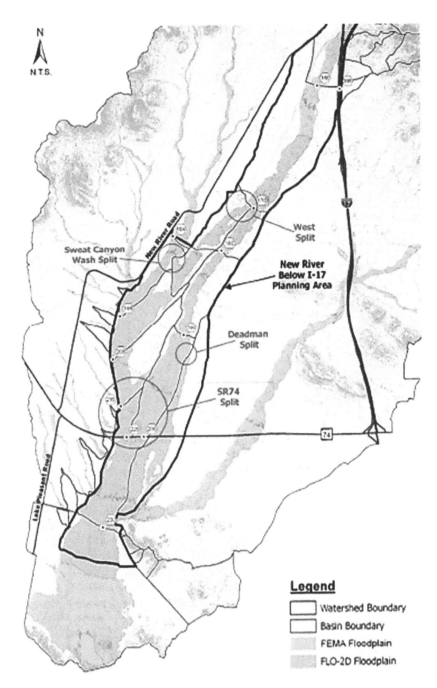

FIGURE 21.9   The major flood risk regions shown using the flow split locations (*figure not to scale*). (Source: Stantec. 2008d. Upper New River ADMP recommended alternative report, Report prepared for the Flood Control District of Maricopa County, Phoenix, AZ, USA.)

flood defense options. However, a popular misconception is that general understanding of the complex physical processes of the hydrologic cycle is at such a sufficiently advanced state that one only needs to select and join together appropriate models in such a way that a seamless whole will emerge. If only such a step would give the complete description of the spatial and temporal movement of flood waves through a watershed! There are inherent scientific and philosophical limitations

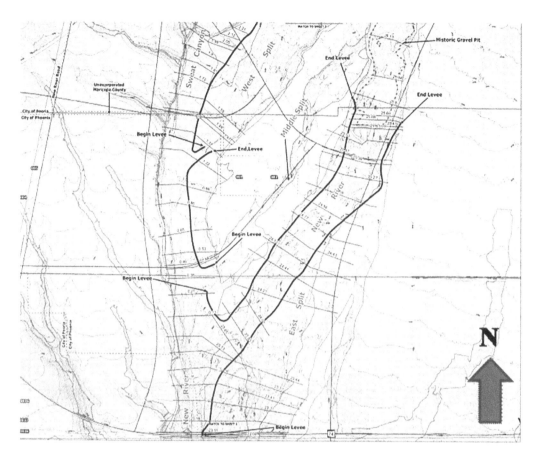

FIGURE 21.10 Proposed levee alignment alternatives to mitigate potential floods (*figure not to scale*). (Source: Stantec. 2008d. Upper New River ADMP recommended alternative report, Report prepared for the Flood Control District of Maricopa County, Phoenix, AZ, USA.)

in the modeling process (Beven, 2001; Knight, 2006). The HEC-1 flood hydrograph package models the true physical processes of most watersheds in central Arizona, but the HEC-HMS hydrologic modeling system does not. Such is the nature of rainfall-runoff modeling options available for high-flood-risk region mapping.

ACKNOWLEDGMENTS

The author acknowledges the Upper New River ADMP project's engineering and GIS team members at Stantec Consultants, and the valuable input provided by the managers of the Flood Control District of Maricopa County, Phoenix, Arizona, USA.

REFERENCES

Ahmed, I., and Gerlach, R.M. 2012. Distributed flow guided hydraulic modeling of a desert river system for flood control. *Proc., RiverFlow-2012: International Conference on Fluvial Hydraulics, Sponsor: IAHR*, 5–7 September, San Jose, Costa Rica.

Bedient, P.B., Huber, W.C., and Vieux, B.E. 2001. *Hydrology and Floodplain Analysis*, 5th edition. Pearson, New York.

Beven, K.J. 2001. *Rainfall-Runoff Modelling: The Primer*. Wiley, Chichester, UK.

Brakensiek, D.L., and Rawls, W.J. 1983. Green-ampt infiltration model parameters for hydrologic classification of soils. In: Borrelli, J., Hasfurther, V.R., and Burman, R.D. (Eds.) *Advances in Irrigation and Drainage Surviving External Pressures, Proceedings of Am. Soc. Civ. Eng. Specialty Conference*, New York, 226–233.

Chow, V.T., Maidment, D.W., and Mays, L.W. 1988. *Applied Hydrology*. McGraw-Hill, New York.

Clubb, F.J., Mudd, S.M., Milodowski, D.T., Hurst, M.D., and Slater, L.J. 2014. Objective extraction of channel heads from high-resolution topographic data. *Water Resources Research*, 50: 4283–4304.

Clubb, F.J., Mudd, S.M., Milodowski, D.T., Valters, D.A., Slater, L.J., Hurst, M.D., and Limaye, A.B. 2017. Geomorphometric delineation of floodplains and terraces from objectively defined topographic thresholds. *Earth Surface Dynamics*, 5: 369–385.

DDMSW. 2002. *User's Manual, Vesrion 2.1.0, HEC-1 Model Pre-processor Software*. Flood Control District of Maricopa County, Phoenix, AZ.

Duan, Q., Schaake, J., Andreassian, V., Franks, S., Goteti, G., Gupta, H.V., Gusev, Y.M., Habets, F., Hall, A., Hay, L., Hogue, T., Huang, M., Leavesley, G., Liang, X., Nasonova, O.N., Noihan, J., Oudin, L., Sorooshian, S. … Wood, E.F. 2006. Model Parameter Estimation Experiment (MOPEX): An overview of science strategy and major results from the second and third workshops. *Journal of Hydrology*, 320(1–2): 3–17.

EPA. 2017. *An Overview of Rainfall-Runoff Model Types*. Technical Report, U.S. Environmental Protection Agency, Office of Research and Development, Athens, GA.

FCDMC. 1997. *Drainage Design Manual. Vol. I: Hydrology, Effective Version*. Flood Control District of Maricopa County, Phoenix, AZ.

Feldman, A.D. 1995. HEC-1 flood hydrograph package. In: Singh, V.P. (Ed.), *Computer Models of Watershed Hydrology*. Water Resource Publications, Highlands Ranch, CO, 119–150.

FEMA 2009. *Guidelines and Specifications for Flood Hazard Mapping Partners, Appendix C: Guidance for Riverine Flooding Analyses and Mapping*. Federal Emergency Management Agency, Washington, DC.

Ferguson, B.K. 1994. *Stormwater Infiltration*. CRC Press, Boca Raton, FL.

Fleming, G., and Frost, L. 2002. Flooding and flood estimation. In: *Flood Risk Management*. Thomas Telford, London, UK.

FLO-2D. 2006. *User's Manual, Version 2006.01*. FLO-2D Inc., Nutrioso, AZ.

Futter, M.N., Erlandsson, M.A., Butterfield, D., Whitehead, P.G., Oni, S.K., and Wade, A.J. 2014. PERSiST: A flexible rainfall–runoff modeling toolkit for use with the INCA family of models. *Hydrology and Earth Systems Science*, 18(2): 855–873.

Gerlach, R.M., Ahmed, I., Beckman, N.D., and Karim, A. 2016. The utility of HEC-1 flood hydrograph package in distributed flow guided area drainage master plan discharges verification in a desert environment. *International Journal of Hydrology Science and Technology*, 6(3): 199–225.

HEC. 1990. *HEC-1 Flood Hydrograph Package User's Manual*. Hydrologic Engineering Center, U.S. Army Corps of Engineers, Davis, CA.

Huffman, R., Fangmeier, D., Elliot, W., Workman, S., and Schwab, G. 2011. Infiltration and runoff. In: *Soil and Water Conservation Engineering*, 6th edition. American Society of Agricultural Engineers, 81–111.

Johnson, M.S., Coon, W.F., Mehta, V.K., Steenhuis, T.S., Brooks, E.S., and Boll, J. 2003. Application of two hydrologic models with different runoff mechanisms to a hillslope dominated watershed in the northeastern US: A comparison of HSPF and SMR. *Journal of Hydrology*, 284(1–4): 57–76.

Kibler, D.F., and Woolhiser, D.A. 1970. *The Kinematic Cascade as a Hydrologic Model*. Hydrology Paper, No. 39, Colorado State University, Fort Collins, CO.

Kirby, M.J. 1978. *Hillslope Hydrology*. Wiley, Chichester, UK.

Knight, D.W. 2006. Introduction to flooding and river basin modeling. In: Knight, D.W. and Shamseldin, A.Y. (Eds.), *River Basin Modelling for Flood Risk Mitigation*. Taylor & Francis, London, UK.

Kokkonen, T., Koivusalo, H., and Karvonen, T. 2001. A semi-distributed approach to rainfall-runoff modelling—a case study in a snow affected watershed. *Environmental Modelling and Software*, 16(5): 481–493.

Melsen, L., Teuling, A., Torfs, P., Zappa, M., Mizukami, N., Clark, M., and Uijlenhoet, R. 2016. Representation of spatial and temporal variability in large-domain hydrological models: Case study for a mesoscale pre-Alpine basin. *Hydrology and Earth System Sciences*, 20(6): 2207–2226.

Moradkhani, H., and Sorooshian, S. 2009. General review of rainfall-runoff modeling: Model calibration, data assimilation, and uncertainty analysis. In: Sorooshian S., Hsu KL., Coppola E., Tomassetti B., Verdecchia M., Visconti G. (eds.) *Hydrological Modelling and the Water Cycle*. Water Science and Technology Library, Springer, Berlin, Heidelberg, Germany, Vol. 63. https://doi.org/10.1007/978-3-540-77843-1_1

NOAA. 1984. *Depth-Area Ratio in the Semi-Arid Southwest United States*. NOAA Technical Memorandum NWS HYDRO-40, National Weather Service, National Oceanic and Atmospheric Administration, Washington, DC.

NOAA. 2003. *NOAA Atlas II: Precipitation Frequency Estimates*. National Weather Service, National Oceanic and Atmospheric Administration, Washington, DC.

Pechlivanidis, I.G., Jackson, B.M., Mcintyre, N.R., and Wheater, H.S. 2011. Watershed scale hydrological modelling: A review of model types, calibration approaches and uncertainty analysis methods in the context of recent developments in technology and applications. *Global NEST Journal*, 13(3): 193–214.

Perlman, H. 2016. *The Water Cycle- USGS Water Science School*. Retrieved 5/18/2017, from https://water.usgs.gov/edu/watercycle.html.

Refsgaard, J.C., and Henriksen, H.J. 2004. Modelling guidelines—Terminology and guiding principles. *Advances in Water Resources*, 27(1): 71–82.

Sabol, G.V. 1987. *S-Graph Study*. Report prepared for the Flood Control District of Maricopa County, Phoenix, AZ.

Samuels, P.G. 2000. An overview of flood estimation and flood prevention. In: *Proceedings, International Symposium on Flood Defence*, University of Kassel, Kassel, Germany, September, 1–10.

Singh, V.P. 1996. *Kinematic Wave Modeling in Water Resources*. Wiley-Interscience, New York.

Sivapalan, M. 2003. Prediction in un-gauged basins: A grand challenge for theoretical hydrology. *Hydrological Process*, 17(15): 3163–3170.

Smith, R.E., Goodrich, D.C., Woolhiser, D.A., and Unkrich, C.L. 1995. KINEROS – A KINematic runoff and EROSion model. In: Singh V.P. (Ed.), *Computer Models of Watershed Hydrology*. Water Resource Publications, Highlands Ranch, CO, 697–732.

Srinivasulu, S., and Jain, A. 2008. Rainfall-runoff modelling: Integrating available data and modern techniques. In: R.J. Abrahart, et al. (Eds.), *Principle Hydroinformatics*, 68: 59–70. Springer, Berlin, Germany.

Stantec. 2008a. *Upper New River Area Drainage Master Plan: Hydrology Report*. Report prepared for the Flood Control District of Maricopa County, Phoenix, AZ.

Stantec. 2008b. *Upper New River ADMP Alternative Formulation Report*, Vol. I. Report prepared for the Flood Control District of Maricopa County, Phoenix, AZ.

Stantec. 2008c. *Upper New River ADMP Alternative Formulation Report*, Vol. II. Report prepared for the Flood Control District of Maricopa County, Phoenix, AZ.

Stantec. 2008d. *Upper New River ADMP Recommended Alternative Report*. Report prepared for the Flood Control District of Maricopa County, Phoenix, AZ.

USDA. 1986. *Urban Hydrology for Small Watersheds*. TR-55 (Revised), Natural Resources Conservation Service (NRCS), U.S. Department of Agriculture, Washington, DC.

Viney, N.R., Vaze, J., Chiew, F.H.S., Perraud, J., Post, D.A., and Teng, J. 2009. Comparison of multi-model and multi-donor ensembles for regionalization of runoff generation using five lumped rainfall-runoff models. In: *18th World Imacs Congress and Modsim International Congress on Modelling and Simulation: Interfacing Modelling and Simulation with Mathematical and Computational Sciences*, Cairns, Australia, July 13–17, 3428–3434.

Wagener, T., and Wheater, H.S. 2006. Parameter estimation and regionalization for continuous rainfall–runoff models including uncertainty. *Journal of Hydrology*, 320(1–2): 132–154.

Yang, W.-Y., Li, D., Sun, T., and Ni, G.-H. 2015. Saturation-excess and infiltration-excess runoff on green roofs. *Ecological Engineering*, 74: 327–336.

22 Geostatistics and Flooding
Homogeneous Regions Delineation for Multivariate Regional Frequency Analysis

Emna Gargouri-Ellouze, Rim Chérif, and Saeid Eslamian

CONTENTS

22.1 INTRODUCTION

Flood event data records (volume, maximum discharge, duration, time to peak, etc.) are generally short and are not constantly available at the target site. In the case of ungauged basins, where at-site frequency analysis is often inaccurate, regionalization is used to overcome the lack of information. Regionalization can be defined as the transfer of information from one catchment to another (Bloschl and Sivapalan, 1995). It is often adopted, in hydrology, to help transfer information from the gauged sites to others poorly or ungauged. Its support is the investigation of the connections between hydrological responses and the elements affecting them, identifying regions with similar behavior (Gargouri-Ellouze and Bargaoui, 2009).

There are diverse approaches in regionalization; the main ones are synthesized in Razavi and Coulibaly (2013) who presented their purposes, techniques, study areas, and major findings. Rao

DOI: 10.1201/9780429463938-29

and Srinivas (2008) pointed to the absence of set criteria to establish the superiority of any approach, indicating that the different approaches have their own strengths, limitations, and challenges such as the strategy to group the sites that differ from one method to another, which leads to the creation of the different sets of regions.

Amongst these approaches, regional frequency analysis (RFA) allows assessing the extreme hydrological events such as floods at sites where little or no hydrological data are available (Gargouri et al., 2016). RFA is used to estimate the flood characteristics quantiles for ungauged basins. According to Wazneh et al. (2015), the two main steps of RFA are (1) the pooling homogeneous hydrological basins and (2) the regional estimation within these pools.

The hierarchical cluster analysis is one of the RFA approaches, which agglomerate the basins using a distance showing their similarity including a set of attributes of the gauging site. Usually, the set of attributes used for the RFA approaches includes: physiographical characteristics (e.g., Ahn and Palmer, 2016; Basu and Srinivas, 2015; Razavi and Coulibaly, 2013; Wazneh et al., 2015; Burn and Goel, 2000), climatic features (e.g., Chebana and Ouarda, 2008), hydrometric data (e.g., Burn et al., 1997), topographic information (e.g. Chérif and Bargaoui, 2013) or a combination of these (e.g., Hall and Minns, 1999; Rao and Srinivas, 2006; Ali et al., 2010). Several distances are used in hierarchical clustering such as Euclidean, squared Euclidean, Manhattan, Chebyshev, cosine, Canberra, Minkowski, and Mahalanobis (Rao and Srinivas, 2008) with many algorithms: single linkage, complete linkage, average linkage, and Ward method.

Burn and Goel (2000) applied the weighted Euclidean distance to identify the regions for regional flood frequency analysis in India. They developed for identifying groups, a technique that uses a clustering algorithm as a starting point for partitioning the collection of catchments. Rao and Srinivas (2006) assessed the flood quantiles for ungauged watersheds using a fuzzy clustering algorithm (FCA) on attributes and flow records of 245 gauging stations in Indiana, USA. They found that FCA derives the homogeneous regions that are operational for flood frequency analysis. Di Prinzio et al. (2011) estimated the streamflow indices (mean annual runoff, mean annual flood, and flood quantiles in 300 ungauged Italian catchments. They used the self-organizing maps (SOM) on the available catchment descriptors and derived the variables obtained by applying PCA and canonical correlation analysis (CCA). They showed that PCA and CCA on the available catchment descriptors before applying SOM improve the effectiveness of classifications. Ssegane et al. (2012) predicted the flow in the ungauged watersheds using the K-means clustering with geographic proximity, watershed hypsometry, causal selection algorithms, principal components analysis (PCA), and stepwise regression, in three Mid-Atlantic ecoregions within the USA. They found that classification performance was the highest using causal algorithms. Chérif and Bargaoui (2013) regionalized the catchments using topographical, soil, and hydro-climatological characteristics by performing both trellis and hierarchical classification partitioning methods. They constructed the regional frequency flood index curve. Chebana et al. (2014) introduced the generalized additive model in the estimation step of RFA to take into account the nonlinearity. the homogenous regions delineation is dealt with a neighborhood approach using CCA. The regional model is applied on a dataset of 151 hydrometrical stations located in the province of Québec, Canada. Durocher et al. (2016) investigated the benefit of using the spatial copula approach in RFA. They based their methodology on the spatial copula framework to provide a more general framework for the prediction of flood quantiles at ungauged basins. The spatial copula methodology leads to a simple expression of the plugin predictive distribution, which allows the straightforward calculations of nonlinear predictors at ungauged locations in southern Quebec, Canada. Hailegeorgis and Knut (2017) performed the regional flood frequency analysis using the L-moments method and annual maximum series of mean daily streamflow observations for the flood quantiles prediction.

Furthermore, a flood's genesis involves strong nonlinearities (Ali et al., 2010, see references therein). To be correctly treated, the latter should be considered jointly in a multivariate framework. The introduction of copulas in hydrology (De Michele and Salvadori, 2003) has greatly facilitated the multivariate modeling of hydrological complexity; see, among others, Sraj et al. (2015), Aissia et al. (2014, 2012), Candela et al. (2014), Fu and Kapelan (2013), Li et al. (2012), Lian et al. (2013),

Kuchment et al. (2013), Salarpour et al. (2013), Ellouze-Gargouri and Bargaoui (2012), Golian et al. (2012), Zhang and Singh (2012), Kao and Chang (2011), Kao and Govindaraju (2007), and Balistrocchi and Bacchi (2001). Copulas theory allows the description of (1) dependence structure between the random variables without information on marginal distributions and (2) multivariate distributions with any kind of marginal distributions (Gargouri-Ellouze and Bargaoui, 2009).

This chapter presents the hierarchical cluster analysis with Ward's algorithm using Euclidean distance based on the geomorphological and physiographical basin characteristics. A step in regionalization is homogeneity testing. Here, it is performed by computing the silhouette index (Rousseeuw, 1987) which for an individual is a measure of how similar that individual is to individuals in its own cluster (cohesion) compared to individuals in other clusters (separation). This approach is applied for 22 watersheds located in Tunisian ridge.

This chapter is organized as follows. Section 22.2 gives a brief background needed to introduce the suggested approach. The methodology is detailed in section 22.3, and the application in a Tunisian watershed is presented in section 22.4. The last section presents the conclusions.

22.2 BACKGROUND

This section gives the prerequisite background material so as to introduce and apply the proposed method. Let consider a matrix X of size $n \times p$: rows are the individuals (n) and columns are the variables (p).

$$X = \begin{bmatrix} x_1^1 & \cdots & x_1^p \\ \vdots & x_i^j & \vdots \\ x_n^1 & \cdots & x_n^p \end{bmatrix}. \tag{22.1}$$

With x_i: ith row and x^j: jth column.

22.2.1 EUCLIDIAN DISTANCE

Euclidian distance $d(x_i, x_j)$ is defined as a dissimilarity measure between two random vectors x_i and x_j $(i, j = 1...n)$ in the p-dimensional space $\mathbb{R}^p$. It is expressed by:

$$d(x_i, x_j) = \sqrt{(x_i - x_j) \times (x_i - x_j)^t} \tag{22.2}$$

x^t: transpose vector

Euclidian distance satisfies the following properties:

$$d(x_i, x_j) = d(x_j, x_i) \tag{22.3}$$

$$d(x_i, x_j) > 0 \text{ if } x_i \neq x_j \tag{22.4}$$

$$d(x_i, x_j) = 0 \text{ if } x_i = x_j \tag{22.5}$$

$$d(x_i, x_j) \leq d(x_i, x_k) + d(x_k, x_j) \text{ (triangle inequality)} \tag{22.6}$$

22.2.2 HIERARCHICAL CLUSTER ANALYSIS

Hierarchical cluster analysis (HCA) is a method for finding the homogeneous clusters of individuals based on the measured characteristics. HCA seeks to build a hierarchy of clusters that can be

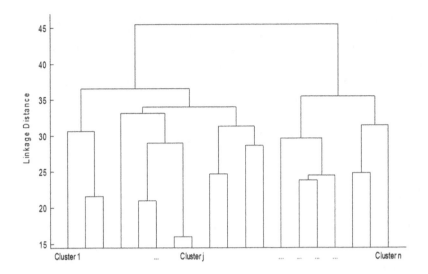

FIGURE 22.1 Dendrogram example.

agglomerative or divisive. Agglomerative algorithms merge the clusters. On the contrary, divisive hierarchical clustering algorithms split clusters. The result of clustering is presented as either the distance or the dissimilarity measure.

HCA agglomerative and divisive can be illustrated as a nested sequence or tree diagram, called a dendrogram. A dendrogram is a tree-structured graph or branching diagram (Figure 22.1) used to visualize the result of a HCA. It shows the linkage points and clusters which are connected at increasing the levels of dissimilarity. The height of the branch points indicates how similar or different they are from each other: the greater the height, the greater the difference.

22.2.3 Copulas

A copula is a multivariate joint cumulative distribution defined on the m-dimensional unit cube $[0, 1]^m$ so that every marginal distribution is uniform on the interval $[0, 1]$. Considering a vector $Y = (Y_1, Y_2 \cdots, Y_k, \dots Y_m)$ of $m \geq 2$ continuous random variables, we call the function C a copula if:

$$P(Y_1 \leq y_1, Y_2 \leq y_2 \dots, Y_k \leq y_k, \dots Y_m \leq y_m) = C(F_1(y_1), F_2(y_2) \dots F_k(y_k), \dots F_m(y_m)) \quad (22.7)$$

Such as: $F_k(y_k) = \Pr(Y_k \leq y_k)$ is the marginal distribution of Y_k. C is the cumulative distribution function of a vector $(U_1, U_2 \cdots, U_k, \dots U_m)$ of dependent uniform random variables on the interval $[0,1]$:

$$C(F_1(y_1), F_2(y_2) \dots F_k(y_k), \dots F_m(y_m)) = C(U_1, U_2 \cdots, U_k, \dots U_m) = F(y_1, y_2 \cdots, y_k, \dots y_m) \quad (22.8)$$

The associated Copula to $Y = (Y_1, Y_2 \cdots, Y_k, \dots Y_m)$ is, for all $(u_1, u_2 \cdots, u_k, \dots u_m) \in [0,1]^m$:

$$C(u_1, u_2 \cdots, u_k, \dots u_m) = F(F_1^{-1}(u_1), F_2^{-1}(u_2) \dots F_k^{-1}(u_k), \dots F_m^{-1}(u_m))$$

$$= P(U_1 \leq u_1, U_2 \leq u_2 \dots, U_k \leq u_k, \dots U_m \leq u_m) \quad (22.9)$$

Specifically, $C: [0,1]^m \to [0,1]$ is a m-dimensional copula if:

$$C(u_1, u_2 \cdots, 0, \dots u_m) = 0 \text{ and } C(1, 1 \cdots, u_k, \dots 1) = u_k; k = 1 \dots m \quad (22.10)$$

C is bounded and m-increasing.

The density function f of F is expressed by:

$$f\left(y_1, y_2 \cdots, y_k, \ldots y_m\right) = c\left(F_1\left(y_1\right), F_2\left(y_2\right) \ldots F_k\left(y_k\right), \ldots F_m\left(y_m\right)\right)$$
$$\times f_1\left(y_1\right) \times f_2\left(y_2\right) \ldots \times f_k\left(y_k\right) \ldots \times f_m\left(y_m\right) \tag{22.11}$$

f_k is a density function of Y_K and $c\left(u_1, u_2 \cdots, u_k, \ldots u_m\right)$ is copula density defined by:

$$c\left(u_1, u_2 \cdots, u_k, \ldots u_m\right) = \frac{\partial^m C\left(u_1, u_2 \cdots, u_k, \ldots u_m\right)}{\partial u_1 \partial u_2 \ldots \partial u_k \ldots \partial u_m} \tag{22.12}$$

In addition, it is worth noticing that each copula is bounded by Frechet-Hoeffding bounds so that:

$$\max\left(F_1\left(y_1\right), F_2\left(y_2\right) \ldots F_k\left(y_k\right), \ldots F_m\left(y_m\right) - 1\right) \le F\left(y_1, y_2 \cdots, y_k, \ldots y_m\right)$$
$$\le \min\left(F_1\left(y_1\right), F_2\left(y_2\right) \ldots F_k\left(y_k\right), \ldots F_m\left(y_m\right)\right) \tag{22.13}$$

There are several types of copula: non-parametric and parametric ones (one or multi-parameters). Non-parametric copula are empiric copula, independence copula, periodic copula, and Fréchet copula ("upper bound" maximum copula and "lower bound" minimum copula). For parametric copula, several families exist, such as Archimedean copula, meta-elliptical copula (in particular Gaussian copula, Cauchy copula, and student copula), exponential copula and contamination copula, survival copula, and extreme value copula (Joe BB5 and Galambos).

In hydrology, one often uses families of multivariate distributions that are extensions of univariate and suffer from several limitations and constraints, such as the marginal distributions that may belong to the same probability family. Thus to avoid these limitations, the copulas models are used. Indeed, copulas allow the description of the dependence structure between random variables without information on the marginal distributions and the description of the multivariate distributions with any kind of marginal distributions. It is worth noticing that Archimedean copula and meta-elliptical copula families remain the most used.

22.3 METHODOLOGY

The two of the main methodologies are the hydrological homogeneous region's delineation in a multidimensional space, combining geomorphological and physiographical variables using Euclidian distance, and a regional runoff-rainfall frequency curve construction within the formed regions via copula. Hierarchical classification, involving Euclidean distance, is applied to basin attributes in order to agglomerate watersheds, following these steps:

- Attributes selection: This step allows analyzing data related to several basin attributes. Attributes that mostly influence basin behavior are selected.
- Matrix construction: Matrix attributes are established from watersheds data, in which each column corresponds to a vector of attributes data. Lines symbolize the watershed identification.
- Matrix attributes standardization: Data related to each attribute are standardized by subtracting the average of attribute vector from each value and dividing the transformed values by the standard deviation. Therefore, each feature vector consists of rescaled (dimensionless) attributes.
- Cluster delineation: This step involves the hierarchical clustering method, which agglomerate watersheds according to their matrix attributes: Euclidean distance and a clustering criterion characterize this clustering.

- Selecting the optimum number of regions: The clusters formed in the previous step are interpreted visually by using a dendrogram. To determine cluster optimum number a criterion is calculated (Mardia et al., 1979).
- Testing consistency regions: The delineated regions are tested for consistency by using a statistical consistency test (silhouette index).
- Multivariate flood frequency quantiles: The aim of this step is to perform regional goodness-of-fit tests to identify and fit a suitable flood frequency copula to rainfall-runoff data of sites in a region. The fitted copula is then employed to obtain rainfall-runoff frequency curves.

22.3.1 Delineation of Hydrologically Homogeneous Basins

22.3.1.1 Clustering Algorithm

The clustering adopted approach was agglomerative hierarchical. It used Euclidian distance to assess the dissimilarity between any two clusters and Ward's algorithm (Ward, 1963) to merge clusters. Ward's algorithm is one of the clustering algorithms most used to delineate homogeneous sub-regions in RFA analysis. The algorithm uses a variance analysis approach to measure the similarity between the sites in a sub-region. The objective function of Ward's algorithm minimizes the sum of squares of errors (ESS) of the feature vectors from the centroid of their respective clusters. It starts with singleton clusters. At each step in the analysis, the union of every possible pair of clusters is considered and two clusters whose fusion results in the smallest increase in ESS are merged. Afterward, a dendrogram is held to illustrate the mergers at different levels, where the vertical axis represents the value of the ESS.

Let $\{x_i | i = 1, ...n\}$ denotes a set of n feature vectors in p-dimensional attribute space $\left(x_i = \left[x_i^1, ...x_i^p\right] \in \mathbb{R}^p\right)$, each of which characterizes one watershed. The ESS in each region r is expressed by:

$$\text{ESS}_r = \sum_{i=1}^n \left(x_i - \overline{x_r}\right)^t \left(x_i - \overline{x_r}\right) \tag{22.14}$$

22.3.1.2 Cluster's Homogeneity

After regions delineation, it is indispensable to confirm that delineated regions are hydrologically homogeneous. By performing the silhouette indices, homogeneity is verified.

Silhouette index characterizes each cluster by comparing its tightness and separation. It illustrates which feature vectors belong to their cluster, and which ones are only between the clusters. Cluster's silhouettes are plotted in a chart showing consistency within clusters and providing assessing cluster quality (Rousseeuw, 1987).

Each feature vector x_i, the corresponding silhouette index $s(i)$ is defined as:

$$s(i) = \frac{b(i) - a(i)}{\max\left[a(i), b(i)\right]} \tag{22.15}$$

Where, for a given x_i belonging to cluster A (with $Card(A) \geq 2$) and a distance $d(.,.)$,

$$a(i) = \frac{1}{Card(A) - 1} \sum_{\substack{x_j \in A \\ i \neq j}} d(x_i, x_j) \tag{22.16}$$

$$b(i) = \min_{A \neq C} \frac{1}{Card(C)} \sum_{x_k \in C} d(x_i, x_k)$$

Where $a(i)$ is the average distance from the ith feature vector to all of the other feature vectors in cluster A; $b(i)$ is the minimum average distance from the ith feature vector to all of the feature vectors in another cluster C. From this formula, it follows that $-1 \leq s(i) \leq 1$. If $s(i)$ is large thus the ith feature vector is well assigned to the cluster. On the other hand, when $s(i)$ is close to -1, the ith feature vector is misclassified.

22.3.2 Regional Curve Construction

In each homogeneous region, correlations between the runoff and rainfall variables were investigated. For the most significant and pertinent correlation for applied engineering, regional runoff-rainfall frequency curves were constructed using copulas.

22.3.2.1 Correlation Investigation

To measure the association between variables, the rank correlation coefficient Kendall's τ is used for the characterization of dependence. It allows revealing the dependence of two qualitative characteristics if sample elements can be ordered with respect to these characteristics. This coefficient, which measures nonlinear dependence, integrates the rank of observations rather than their value. This may be interesting when the values vary widely in scale also to exempt data from uncertainties. In summary, the higher Kendall's τ is, the more important the dependence is. A test of independence can be adopted for Kendall's τ, since under the null-hypothesis H_0, this statistic is close to normal distribution with zero mean and variance $2(2n+5)/[9n(n-1)]$ (n size of sample). As a result, H_0 would be rejected at an approximate level α if $|\tau| > z_{\alpha/2}\sqrt{[2(2n+5)]/[9n(n-1)]}$. For $\alpha = 5\%$, $z_{\alpha/2}$ = 1.96. Let z^* represent the quantity $z_{\alpha/2}\sqrt{[2(2n+5)]/[9n(n-1)]}$.

22.3.2.2 Copula Modeling

The Archimedean copulas with one parameter (a) are chosen (Gumbel, Frank, and Clayton). (a) directly depends on Kendall's τ. or the goodness-of-fit, the log-likelihood method was adopted. Runoff-rainfall couples were then generated which allowed the construction of the isolines which correspond to the runoff-rainfall frequency curves. The latter provide for a fixed frequency and diverse possible runoff-rainfall pairs.

22.4 APPLICATION

22.4.1 Case Study

Twenty-two basins situated in the Tunisian ridge are considered, monitored since 1992. Headwater dams control these basins. Latitudes vary from 35°N to 37°N and longitudes from 8°E to 11°E, and areas range between 1 km² and 10 km² (Figure 22.2). These basins are located in a semi-arid zone, with an annual average rainfall fluctuating between 280 mm and 500 mm. The relief is fairly high-to-high for the majority of the basins, which helps rapid runoff. These basins are slightly permeable to impermeable.

The rain gauge network consists of 22 gauges, located at each headwater dam. Observed hyetographs are divided into five minutes series. An event is selected by considering two successive rainy events that are separated by a dry period over 60 minutes. In addition, only the events for which total rainfall depths exceed 2 mm are considered. Hydrographs are reconstituted through the reservoir water budget for each rainy event.

The clustering variables used to delineate the homogeneous regions (Table 22.1) describe the geography, geomorphology, and soil occupation: latitude ($LatN$), longitude ($LongE$), area (A), perimeter (P), specific height (D_S), global slope index (Ip), Gravellus index (Ig), the percentage of forest cover (Pf), the percentage of cereal culture area (Pc), the percentage of arboriculture area (Pa), the percentage of area affected by anti-erosive practices (Aae), and pasture (Par).

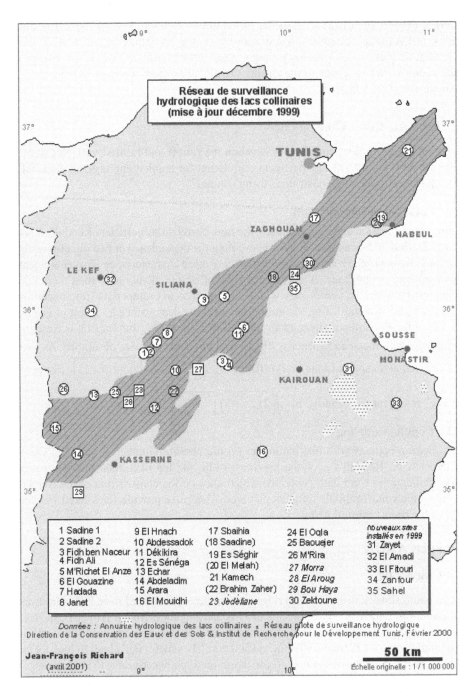

FIGURE 22.2 Localization of Headwater dams (Source: hydrological directories, DGACTA, IRD, Dams' classification is in chronological order creation).

22.4.2 Results and Discussion

22.4.2.1 Delineation

HCA based on Euclidian distance and Ward's algorithm with clustering variables (attributes), described above, was applied. Consequently, $n = 22$ feature vectors (watersheds) are obtained in $p = 12$-dimensional attribute space.

TABLE 22.1

Hydro-geomorphological characteristics

Watershed	Identification (Id)	LatN	LongE	A (km²)	P (km)	Ds (m)	Ip (m/km)	I_G	Pp (%)	Pf (%)	Pc (%)	Pa (%)	Aae (%)
Abdeladhim	14	35°13'01"	8°33'02"	6.42	11.58	194	45	1.28	9	40	50	1	5
Abdessadok	10	35°40'52"	9°14'49"	3.07	7.95	374	128	1.27	37	0	51	1	20
Arara	15	35°22'09"	8°24'25"	7.08	13.85	442	78	1.46	25	37	19	0	5
BrahimZaher	22	35°33'12"	9°14'00"	4.64	12.8	445	80	1.66	17	0	0	45	0
Dékikira	11	35°53'04"	9°40'53"	3.07	7.65	99	37	1.22	54	5	25	8	0
Echar	13	35°33'11"	8°40'45"	9.17	15.5	220	35	1.43	21	0	0	0	0
El hanech	9	36°04'01"	9°26'55"	3.95	9.55	387	104	1.35	54	0	0	2	5
El Maleh	20	36°28'01"	10°39'13"	0.85	4.13	54	36	1.25	0	0	47	11	0
El mouidhi	16	35°14'24"	9°50'42"	2.66	7.7	128	43	1.32	84	0	0	0	5
Es séghir	19	36°29'08"	10°41'05"	4.31	10.05	161	41	1.36	0	65	0	25	50
Es sénéga	12	35°29'21"	9°06'18"	3.63	8.48	265	87	1.25	39	0	40	2	20
Fidh Ali	4	35°42'04"	9°36'13"	4.12	8.6	109	38	1.19	47	0	5	0	5
Fidhbenaceur	3	35°43'26"	9°35'20"	1.69	5.75	112	55	1.24	20	0	72	1	50
Hadada	7	35°50'25"	9°07'42"	4.69	9.9	346	94	1.28	0	0	76	0	0
Janet	8	35°52'16"	9°11'35"	5.21	12.95	371	67	1.59	38	0	60	2	5
Kamech	21	36°52'18"	10°52'08"	2.45	7.25	108	40	1.3	16	0	74	7	0
MrichetAnza	5	36°05'37"	9°35'41"	1.58	5.5	140	72	1.23	48	0	32	1	0
Mrira	26	35°36'34"	8°28'37"	6.13	12.3	170	35	1.39	0	14	82	4	0
Saadine	18	36°06'55"	9°56'36"	2.72	8.28	307	93	1.4	0	36	0	0	20
Saddine 1	1	35°47'49"	9°03'58"	3.84	9.7	5	106	1.39	49	0	51	0	20
Saddine 2	2	35°47'53"	9°04'42"	6.53	16.8	436	58	1.84	22	50	28	0	5
Sbaihia	17	36°29'43"	10°12'31"	3.24	7.38	173	77	1.15	0	47	42	0	20
μ				4.14	9.71	229.36	65.86	1.36	26.36	13.36	34.27	5.00	10.68
Cv				2.06	2.98	1.68	2.37	8.27	1.13	0.63	1.20	0.47	0.72

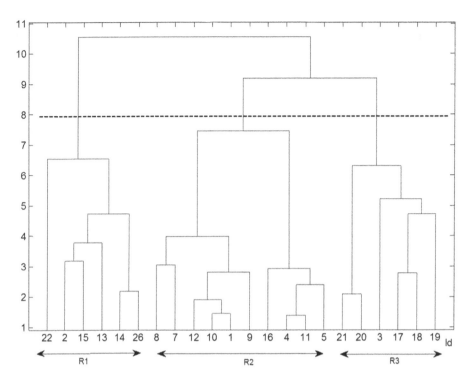

FIGURE 22.3 Dendrogram of hierarchical clustering watersheds with Euclidean distance. The dash line represents the cutoff to define regions.

The dendrogram (Figure 22.3) shows the HCA results. It presents the linkage points and clusters which are connected at increasing the levels of dissimilarity. Cluster (region) number k adopted depends on watershed number N. Mardia et al. (1979) suggested that $k \leq \sqrt{\dfrac{N}{2}}$. Consequently, Euclidian distance (Figure 22.3) led to three homogeneous regions ($k=3$).

22.4.2.2 Homogeneity Test

Silhouette index is computed on the feature vectors of basins attributes. Figure 22.4 presents the silhouette indices corresponding to basins. The values of $s(i)$ are mostly positive values. The average silhouette indices per region vary $\overline{s(i)}$ from 0.094 to 0.266. This shows that consistency within clusters and a good cluster quality (Table 22.2).

It is worth noting that the basins in a given region are not necessarily geographically contiguous. Indeed, the hydro-physiographical regions are principally delineated and not geographical ones (Figure 22.5).

22.4.2.3 Regional Curves

After the homogeneous regions delineation using hierarchical clustering with Euclidian distance, in each homogeneous region, the correlations between runoff and rainfall variables are investigated: rainfall depth (Rd), average rainfall intensity (I_{moy}), rainfall duration (D), runoff volume (V), average discharge (Q_{moy}), and maximum discharge (Q_{max}).

For each pair of rainfall-runoff variables, Table 22.3 illustrates the different values of τ, their corresponding z^* statistic, and the rejection or the acceptation at $\alpha=5\%$ confidence level. It reveals that the hypothesis of independence between (Q_{max}, I_{moy}) variables is rejected and that the correlation is mostly the highest for the three regions and it is pertinent for design engineering.

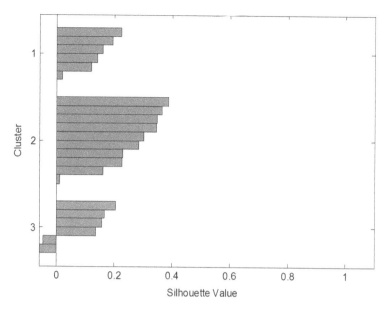

FIGURE 22.4 Silhouette chart.

TABLE 22.2
Silhouette index values

Regions	Silhouette average
1	0.143
2	0.266
3	0.094

22.4.2.4 Copula Modeling

After goodness-of-fit, a Gumbel copula was adopted for (Q_{max}, I_{moy}) for the three regions. Table 22.4 shows the copula parameter $a_i (i=1\ to\ 3)$ and its confidence interval at 95% level for each region, and Table 22.4 presents the marginal distributions.

Figure 22.6a, 22.6b, and 22.6c illustrate the joint cumulative distributions respectively for regions 1, 2, and 3. It is noticed that observed couples were reconstituted.

22.4.2.5 Curves

Figures 22.7a, 22.7b, and 22.7c display the isolines of the simulated couples (Q_{max}, I_{moy}) with regional Gumbel copula for the three regions. These isolines correspond to Q_{max}-I_{moy}-*Frequency* curves and provide for a fixed frequency, the diverse possible couples of (Q_{max}, I_{moy}). The exploitation of these curves: for a fixed intensity, a Q_{max} distribution is obtained rather than one value such as in univariate analysis.

22.5 CONCLUSIONS

The case study illustrated the methodology of homogeneous regions delineation for multivariate frequency analysis, based on the Euclidian distance and Ward's algorithm. Twenty-two basins situated in the Tunisian ridge are examined. The delineation was based on geographical, geomorphological, and physiographical watershed characteristics (latitude, longitude, area, perimeter, specific

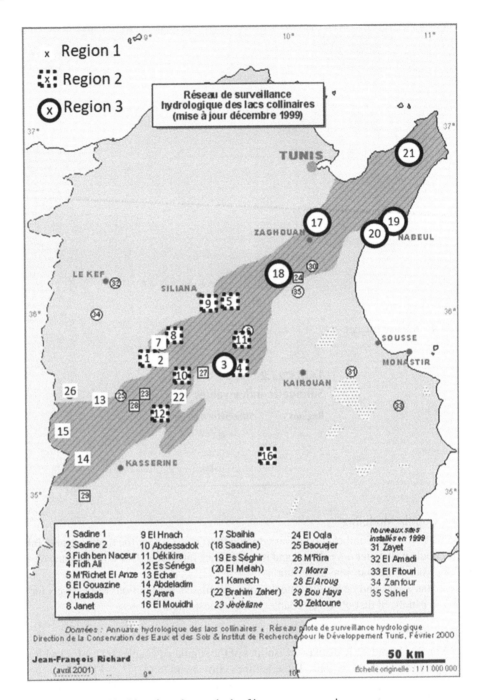

FIGURE 22.5 Geographical location of watersheds of homogeneous regions.

height, global slope index, Gravellus index, the percentage of forest cover, the percentage of cereal culture area, the percentage of arboriculture area, the percentage of area affected by anti-erosive practices, and pasture (Table 22.5).

This application was conducted for three homogeneous regions and silhouette indices confirmed the consistency of clustering. Indeed, the average values in each region were positive. The negative values would have indicated that the watersheds were misclassified.

TABLE 22.3
Independence test results H_0

	Characteristic	τ	z^* (95% confidence)	H_0
Region1	(Q_{max}, D)	-0.01220	0.09919	accepted
	(Q_{max}, Rd)	0.20650	0.09919	rejected
	(Q_{max}, I_{moy})	0.23840	0.09919	rejected
	(Q_{moy}, D)	0.01160	0.09919	accepted
	(Q_{moy}, Rd)	0.15590	0.09919	rejected
	(Q_{moy}, I_{moy})	0.20900	0.09919	rejected
	(V, D)	0.07350	0.09919	accepted
	(V, Rd)	0.21500	0.09919	rejected
	(V, I_{moy})	0.38250	0.09919	rejected
Region2	(Q_{max}, D)	-0.01150	0.05942	accepted
	(Q_{max}, Rd)	0.26880	0.05942	rejected
	(Q_{max}, I_{moy})	0.35000	0.06160	rejected
	(Q_{moy}, D)	0.00860	0.06160	accepted
	(Q_{moy}, Rd)	0.24590	0.06160	rejected
	(Q_{moy}, I_{moy})	0.34940	0.06160	rejected
	(V, D)	0.06570	0.06160	rejected
	(V, Rd)	0.20870	0.06160	rejected
	(V, I_{moy})	0.38750	0.06160	rejected
Region3	(Q_{max}, D)	0.06830	0.09297	accepted
	(Q_{max}, Rd)	0.17690	0.09297	rejected
	(Q_{max}, I_{moy})	0.27860	0.09297	rejected
	(Q_{moy}, D)	0.01110	0.09297	accepted
	(Q_{moy}, Rd)	0.16220	0.09297	rejected
	(Q_{moy}, I_{moy})	0.20370	0.09297	rejected
	(V, D)	0.12420	0.09297	rejected
	(V, Rd)	0.14760	0.09297	rejected
	(V, I_{moy})	0.32920	0.09297	rejected

τ: Kendall's tau; z^*: test statistic; H_0: null hypothesis

TABLE 22.4
Copula's parameter

Region	Parameter a_i	Confidence Interval (95%)
1	1.199	[1.058, 1.339]
2	1.352	[1.257, 1.448]
2	1.253	[1.114, 1.392]

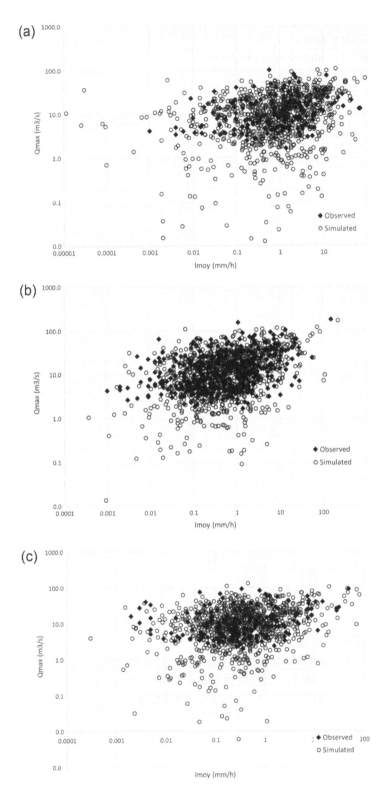

FIGURE 22.6 a) Joint cumulative distribution Region 1. b) Joint cumulative distribution Region 2. c) Joint cumulative distribution Region 3.

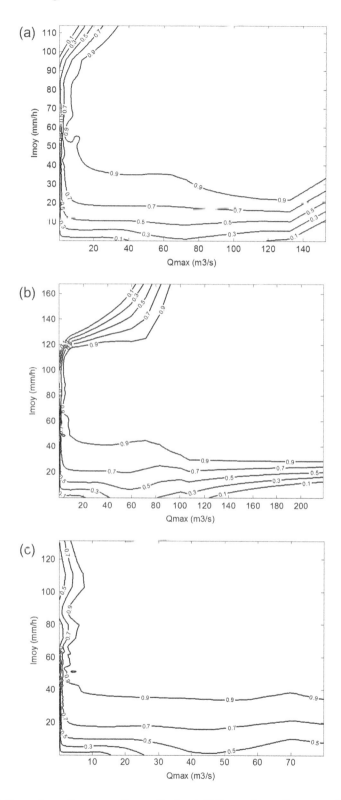

FIGURE 22.7 a) Qmax- Imoy-Frequency Curves Region 1. b) Qmax- Imoy -Frequency Curves Region 2. c) Qmax- Imoy -Frequency Curves Region 3.

TABLE 22.5
Marginal distributions

Region	Q_{max}	I_{moy}
1	Burr	Generalized Pareto
	(k = 3.7167 α = 0.65674 β = 10.33)	(k = 0.15547 σ = 10.641 μ = 2.3497)
2	LogNormal	Generalized Pareto
	σ = 1.8969 μ = −0.68799	k = 0.18475 σ = 12.905 μ = 1.7677
3	Burr	Generalized Pareto
	k = 1.0099 α = 0.98232 β = 0.3475	k=0.28651 σ = 8.9575 μ = 2.056

In each homogeneous region, the correlations between runoff and rainfall characteristics were examined. The rainfall and runoff characteristics explored were Rd, I_{max}, D, V, Q_{moy}, and Q_{max}. This analysis revealed that the correlation (Q_{max}, I_{moy}) was significant in comparison with the other pairs of variables for the three regions. Gumbel copula was adopted for (Q_{max}, I_{moy}) for the regions. The isolines were established corresponding to Q_{max}-I_{moy}-*Frequency* curves, which provide for a fixed frequency, the diverse possible couples of (Q_{max}, I_{moy}). These curves gave a Q_{max} distribution for a fixed intensity, rather than one value such us in univariate analysis.

As a final point, Euclidean distance can be applied for any *n* feature vectors in any *p*-dimensional attribute space. The climatic, hydrometric data could be added in the clustering algorithm and study the influence of such variables on delineation. By combining features, we may better explain watershed behavior.

REFERENCES

Ahn, K.-H., and Palmer, R. 2016. Regional flood frequency analysis using spatial proximity and basin characteristics: Quantile regression vs, parameter regression technique, *Journal of Hydrology*, 540: 515–526, https://doi.org/10.1016/j.jhydrol.2016.06.047.

Aissia, M.A.B., Chebana, F., Ouarda, T., Roy, L., Bruneau, P., and Barbet, M. 2014. Dependence evolution of hydrological characteristics, applied to floods in a climate change context in Quebec, *Journal of Hydrology*, 519: 148–163, https://doi.org/10.1016/j.jhydrol.2014.06.042.

Aissia, M.A.B., Chebana, F., Ouarda, T.B.M.J., Roy, L., Desrochers, G., Chartier, I., and Robichaud, É. 2012. Multivariate analysis of flood characteristics in a climate change context of the watershed of the Baskatong reservoir, Province of Québec, Canada, *Hydrological Processes*, 26(1): 130–142, https://doi.org/10.1002/hyp.8117.

Ali, G.A., Roy, A.G., Turmel, M.C., and Courchesne, F. 2010. Multivariate analysis as a tool to infer hydrologic response types and controlling variables in a humid temperate catchment, *Hydrological Processes*, 24(20): 2912–2923, https://doi.org/10.1002/hyp.7705.

Balistrocchi, M., and Bacchi, B. 2001. Modelling the statistical dependence of rainfall event variables through copula functions, *Hydrology and Earth System Sciences*, 15: 1959–1977, https://doi.org/10.5194/hess-15-1959-2011.

Basu, B., and Srinivas, V.V. 2015. A recursive multi-scaling approach to regional flood frequency analysis, *Journal of Hydrology*, 529(Part 1): 373–383, https://doi.org/10.1016/j.jhydrol.2015.07.037.

Bloschl, G., and Sivapalan, M. 1995. Scale issues in hydrological modeling, a review, *Hydrological Processes*, 9: 251–290, https://doi.org/10.1002/hyp.3360090305.

Burn, D.H., and Goel, N.K. 2000. The formation of groups for regional flood frequency analysis, *Hydrological Sciences Journal*, 45(1): 97–112, https://doi.org/10.1080/02626660009492308.

Burn, D.H., Zrinji, Z., and Kowalchuk, M. 1997. Regionalization of basins for regional flood frequency analysis, *Journal of Hydrologic Engineering*, 2: 76–82, https://doi.org/10.1061/(ASCE)1084-0699.

Candela, A., Brigandi, G., and Aronica, G.T. 2014. Estimation of synthetic flood design hydrographs using a distributed rainfall–runoff model coupled with a copula-based single storm rainfall generator, *Natural Hazards Earth System Sciences*, 14: 1819–1833, https://doi.org/10.5194/nhess-14-1819-2014.

Chebana, F., Charron, C., Ouarda, T.B.M.J., and Martel, B. 2014. Regional frequency analysis at ungauged sites with the generalized additive model, *Journal of Hydrometeorology*, 15: 2418–2428, https://doi.org/10.1175/JHM-D-14-0060.1.

Chebana, F., and Ouarda, T.B. 2008. Depth and homogeneity in regional flood frequency analysis, *Water Resources Research*, 44(11). https://doi.org/10.1029/2007WR006771.

Chérif, R., and Bargaoui, Z. 2013. Regionalization of maximum annual runoff using hierarchical and trellis methods with topographic information, *Water Resources Management*, 27(8): 2947–2963, https://doi.org/10.1007/s11269-013-0325-0.

De Michele, C., and Salvadori, G. 2003. A generalized Pareto intensity-duration model of storm rainfall exploiting 2-copulas, *Journal of Geophysical Research: Atmospheres*, 108(D2): 1984–2012, https://doi.org/10.1029/2002JD002534.

Di Prinzio, M., Castellarin, A., and Toth, E. 2011. Data-driven catchment classification: Application to the pub problem, *Hydrology and Earth System Sciences*, 15: 1921–1935, https://doi.org/10.5194/hess-15-1921-2011.

Durocher, M., Chebana, F., and Ouarda, T.B.M.J. 2016. Delineation of homogenous regions using hydrological variables predicted by projection pursuit regression, *Hydrology and Earth System Sciences*, 20, 4717–4729, https://doi.org/10.5194/hess-20-4717-2016.

Ellouze-Gargouri, E., and Bargaoui, Z. 2012. Runoff estimation for an ungauged basin using geomorphological instantaneous unit hydrograph (GIUH) and copulas, *Water Resources Management*, 26(6): 1615–1638, https://doi.org/10.1007/s11269-012-9975-6.

Fu, G., and Kapelan, Z. 2013. Flood analysis of urban drainage systems: Probabilistic dependence structure of rainfall characteristics and fuzzy model parameters, *Journal of Hydro-informatics*, 15(3): 687–699, https://doi.org/10.2166/hydro.2012.160.

Gargouri, E., Chérif, R., and Eslamian, S. 2016. Delineation of homogeneous regions for multivariate regional frequency analysis using modified Mahalanobis distance based on Kendalls tau. In: *STAHY*, 26–27 Sep., Québec, Canada.

Gargouri-Ellouze, E., and Bargaoui, Z. 2009. Investigation with Kendall plots of infiltration index–maximum rainfall intensity relationship for regionalization, *Physics and Chemistry of the Earth, Parts A/B/C*, 34(10): 642–653, https://doi.org/10.1016/j.pce.2009.02.001.

Golian, S., Saghafian, B., and Farokhnia, A. 2012. Copula-based interpretation of continuous rainfall–runoff simulations of a watershed in northern Iran, *Canadian Journal of Earth Sciences*, 49(5): 681–691, https://doi.org/10.1139/e2012-011.

Hailegeorgis, T.T., and Knut, A. 2017. Regional flood frequency analysis and prediction in ungauged basins including estimation of major uncertainties for mid-Norway, *Journal of Hydrology: Regional Studies*, 9, 104–126, ISSN 2214-5818, https://doi.org/10.1016/j.ejrh.2016.11.004.

Hall, M.J., and Minns, A.W. 1999. The classification of hydrologically homogeneous regions, *Hydrological Sciences Journal*, 44(5): 693–704, https://doi.org/10.1080/02626669909492268.

Kao, S.C., and Govindaraju, R.S. 2007. Probabilistic structure of storm surface runoff considering the dependence between average intensity and storm duration of rainfall events, *Water Resources Research*, 43(6): 1–15, W06410, https://doi.org/10.1029/2006WR005564.

Kao, S.C., and Chang, N.B. 2011. Copula-based flood frequency analysis at ungauged basin confluences: Nashville, Tennessee, *Journal of Hydrologic Engineering*, 17(7): 790–799, https://doi.org/10.1061/ASCE/HE.1943-5584.0000477.

Kuchment, L.S., and Demidov, V.N. 2013. On the application of copula theory for determination of probabilistic characteristics of spring flood, *Russian Meteorology and Hydrology*, 38(4): 263–271, https://doi.org/10.3103/S1068373913040080.

Li, T., Guo, S., Chen, L., and Guo, J. 2012. Bivariate flood frequency analysis with historical information based on copula, *Journal of Hydrologic Engineering*, 18(8): 1018–1030, https://doi.org/10.1061/ASCE/HE.1943-5584.0000684.

Lian, J.J., Xu, K., and Ma, C. 2013. Joint impact of rainfall and tidal level on flood risk in a coastal city with a complex river network: A case study of Fuzhou City, China, *Hydrology and Earth System Sciences*, 17(2): 679–689, https://doi.org/10.5194/hess-17-679-2013.

Mardia, K.V., Kent, J.T., and Bibby, J.M. 1979. *Multivariate Analysis*, Academic Press, London-New York-Toronto-Sydney-San Francisco, https://doi.org/10.1002/bimj.4710240520.

Rao, A.R., and Srinivas, V.V. 2006. Regionalization of watersheds by hybrid-cluster analysis, *Journal of Hydrology*, 318(1): 37–56, https://doi.org/10.1016/j.jhydrol.2005.06.003.

Rao, A.R., and Srinivas, V.V. 2008. *Regionalization of Watersheds: An Approach Based on Cluster Analysis*, Edition Springer, Science and Business Media, 245, ISBN 978-1-4020-6852-2.

Razavi, T., and Coulibaly, P. 2013. Classification of Ontario basins based on physical attributes and streamflow series, *Journal of Hydrology*, 493: 81–89, https://doi.org/10.1016/j.jhydrol.2013.04.013.

Rousseeuw, P.J. 1987. Silhouettes: A graphical aid to the interpretation and validation of cluster analysis, *Journal of Computational and Applied Mathematics*, https://doi.org/10.1016/0377-0427(87)90125-7.

Salarpour, M., Yusop, Z., Yusof, F., Shahid, S., and Jajarmizadeh, M. 2013. Flood frequency analysis based on t-copula for Johor river, Malaysia, *Journal of Applied Sciences*, 13(7): 1021–1028, https://doi.org/10.3923/jas.2013.1021.1028.

Sraj, M., Bezak, N., and Brilly, M. 2015. Bivariate flood frequency analysis using the copula function: A case study of the Litija station on the Sava River, *Hydrological Processes*, 29(2): 225–238, https://doi.org/10.1002/hyp.10145.

Ssegane, H., Tollner, E.W., Mohamoud, Y.M., Rasmussen, T.C., and Dowd, J.F. 2012. Advances in variable selection methods I: Causal selection methods versus stepwise regression and principal component analysis on data of known and unknown functional relationships, *Journal of Hydrology*, 438–439, 16–25, ISSN 0022-1694, https://doi.org/10.1016/j.jhydrol.2012.01.008.

Ward, J.H., Jr. 1963. Hierarchical grouping to optimize an objective function, *Journal of the American Statistical Association*, 58(301): 236–244, https://doi.org/10.1080/01621459.1963.10500845.

Wazneh, H., Chebana, F., and Ouarda, T.B.M.J. 2015. Delineation of homogeneous regions for regional frequency analysis using statistical depth function, *Journal of Hydrology*, 521: 232–244, https://doi.org/10.1016/j.jhydrol.2014.11.068.

Zhang, L., and Singh, V.P. 2012. Bivariate rainfall and runoff analysis using entropy and copula theories, *Entropy*, 14(9): 1784–1812, https://doi.org/10.3390/e14091784.

23 Application of Index Flood Approach in Regional Flood Estimation Using L-Moments in Both Fixed Region and Region of Influence Framework

Ayesha S. Rahman, Zaved K Khan,
Fazlul Karim, and Saeid Eslamian

CONTENTS

23.1 INTRODUCTION

Flooding is a natural disaster that causes significant financial and economic damage. For the purpose of planning and designing different types of hydraulic structures, flood plain zoning, and economic analysis of flood protection projects, it is essential to estimate the magnitude and frequency of design floods. Flood frequency analysis (FFA) is a common method for design flood estimation where flood quantiles are estimated at a gauged site for a given annual exceedance probability (AEP) using the annual maximum (AM) flood data or the peak over threshold (POT) flood data. If the period of record for the gauging sites is longer, the flood quantiles can be reliably estimated from the gauged data using FFA. However, a problem arises when the gauging sites have a very short period of data or no data at all (ungauged sites). Regional flood frequency estimation (RFFE) is usually applied in such cases with the goal of transferring information from gauged sites to the ungauged target site within a homogeneous region to estimate design flood for ungauged catchments.

RFFE is generally carried out in three steps: (1) identification of a hydrologically homogeneous region; (2) development of a suitable regional flood estimation model for the identified homogeneous region to estimate flood quantile and; and (3) validation of the developed RFFE model. A homogeneous group can be defined as a group of sites with similar standardized flood frequency curves within a certain margin of sampling variability [1]. Index flood method (IFM) is one of the commonly adopted RFFE techniques, which assumes that, for a homogenous region, the same probability distribution can be adopted for all the sites in the region via a site-specific scaling factor (Q_M) and a common regional growth factor. Q_M, also known as the index flood, depends on the physiographic and climatic characteristics of a given site [2]. The IFM developed in [2] uses the product of at-site mean flood, which serves as the Q_M and a dimensionless growth factor, which is constant for a homogeneous region.

Index frequency approach is a popular technique and is widely adopted in both flood and rainfall estimation [3, 4]. This method has been applied to many countries [2]. In IFM, typically, the mean of the at-site AM floods is considered to be the index flood, Q_M. The median AM flood can also be used as Q_M instead of the mean [5]. The IFM is not recommended by the Australian Rainfall and Runoff (ARR) [6], the national guide. The reason behind this is that it is difficult to form homogeneous regions in Australia [7–10]. It should also be noted that in the case of IFM, the coefficient of variation (C_v) of the AM flood series can vary approximately inversely with catchment area, and in the case of larger and humid catchments, the flood frequency curves are generally found to be flatter [5, 11, 12].

The IFM received renewed attention during the late 1980s after the introduction of L moments where the regional average L-coefficient of variation (L-CV) and L-coefficient of skewness (L-SK) were used to fit a generalized extreme value (GEV) or an alternative three-parameter distribution [13]. This approach is applicable if AM flood record lengths are relatively small and the region is homogeneous [14]. If the at-site record length is not long enough to estimate the shape parameter (L-SK) accurately but sufficient to define the at-site L-CV, adoption of a regional GEV shape parameter based on the regional average value is more suitable than the at-site estimate [13, 15, 16]. The sample size largely defines the adequacy of using either the at-site or regional estimator. An improvement can also be made by combining the at-site and the regional estimators based on the accuracy of each estimator. In Bulletin 17b [17], this approach was proposed where a regional estimate of the shape parameter of the log Pearson type 3 (LP3) distribution is combined with the at-site estimator [15, 18].

The performance of IFM is expected to be dependent on the degree of homogeneity of the proposed region. Numerous researchers have adopted homogeneity tests in RFFE [2, 13, 14, 19–29]. For example, the method proposed in [2] is based on the sampling distribution of the standardized 10% AEP flood (Q_{10}), assuming an Extreme Value 1 (EV1) distribution; whereas [27, 30] presented a test based on the sampling distribution of C_v of the AM flood data. However, the problem with these distribution-specific tests is that, when the hypothesis of homogeneity is rejected, the group of sites can still be homogeneous with a different parent distribution [14]. A problem can also arise due to the adoption of conventional moments to estimate the parameters of the hypothesized distribution as these moments are subject to higher sampling variability due to the squaring and cubing of the flood observations. Also, in the case of the maximum likelihood method, a smaller size sample may under- or overestimate the parameters. Application of L-moments-based homogeneity analysis can be more effective in these cases as L-moments are analogous to conventional moments with measures of location (mean), scale (standard deviation), and shape (skewness and kurtosis) and do not involve squaring or cubing the observations. L-moments are linearly transformed probability-weighted moments (PWM) that can estimate the parameters of a distribution; this feature of the L-moments makes them more robust and less sensitive to outliers.

Numerous studies have been carried out using L-moments based homogeneity analysis [1, 3, 4, 31–40]. L-moments can derive reliable estimates of characteristics and parameters of a distribution for a wide range of hydrological data and it is superior to the conventional moments due to its unique linear aspects [16, 28, 41]. For example, in Turkey, an investigation by [42] using L-moments-based heterogeneity measure resulted in the identification of three homogeneous subregions. In India, the application of L-moments-based heterogeneity measure on the 12 gauged sites resulted in identifying the hypothesized region to be homogeneous [43]. Application of this method on 51 streamflow gauging sites in Sicily resulted in gross heterogeneity for Sicily Island [44]. An investigation to identify homogeneous regions in the case of a large region in northern-central Italy with 36 sites by applying L-moments-based heterogeneity measure resulted in the area being grossly heterogeneous [45]. L-moments based homogeneity test for 18 sites in the Zayandehrood catchment showed that these 18 catchments form a homogeneous region [46]. Attempts to form homogeneous regions in south-east Australia based on [14] method by [1, 32, 47] were not successful to establish any homogeneous regions. From an Australian context, it has been found that homogeneity cannot be achieved generally in the case of Australian states/regions [7–10].

Although IFM was not recommended in Australia due to difficulties in forming homogeneous regions, it can provide improved quantile estimation [15]. Hence, the motivation of this research is whether an approximate IFM can be developed in Australia even though a homogeneous region cannot be established in Australia. Hence, in this chapter, the focus is to develop an approximate IFM using the 10% AEP flood (Q_{10}) as the index variable. This index flood is estimated by employing a multiple linear regression technique that uses the catchment and climatic characteristics data as predictor variables. For this study, a total of 88 stations and eight catchment and climatic characteristics data are used to develop and test the approximate IFM for New South Wales (NSW), Australia. The developed method need to be tested for other Australian states to test its general applicability.

23.2   STUDY AREA AND DATA

For this study, AM flow data from 88 catchments from New South Wales (NSW) are selected. The use of AM series is preferable to peaks over threshold (POT) as AM data series are easy to extract and the data points more easily satisfy the independence criterion. At smaller return periods, the AM and POT provide different quantile estimates, but at higher return periods, they produce similar quantile estimates. This study uses AM series; however, POT can be used in future studies. The locations and some important information on these catchments are presented in Figure 23.1 and Table 23.1, respectively. These sites are mostly unregulated and not affected by major land-use changes. All the selected catchments are small to medium in size and the catchment area for the selected sites varies from 8 to 1,010 km^2 with a mean of 352 km^2 and median of 260 km^2. Record length of AM flood data for these selected sites ranges from 25 to 82 years, with a mean of 41.5 years and median of 37 years. For infilling the gaps in AM flow series, a linear regression model is used by utilizing the AM flow data from nearby stations to estimate the missing data point. About 3% of the AM data points were filled by regression. The physiography of the selected catchments varies from mountainous to coastal plain areas with the mean annual rainfall ranging from 625 to 1,955 mm/y (mean: 1,000 mm/y and median: 910 mm/y).

23.3   METHODS

23.3.1   L-MOMENTS

In order to estimate flood quantiles from the available at-site AM flood data, in the beginning, it is necessary to select an appropriate probability distribution. The conventional "method of moments" to estimate the parameters of a probability distribution can put unnecessary weights on the observed

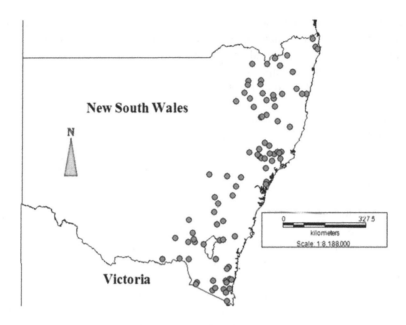

FIGURE 23.1   Location of the selected 88 catchments in NSW, Australia.

data by squaring and cubing it. The method based on the probability-weighted moments (PWM) yields less biased estimates as it linearly transforms the data to estimate the moments of the data [48]. Although this method does not put undue weights on the observed data, the method lacks the capacity of interpretation of different parameters of a distribution. In this circumstance, the L-moments-based parameter estimation technique can be more successful to yield less biased estimates of the moments of the data [49]. L-moments are linear combinations of PWMs and can directly derive the scale and location parameters of a probability distribution. The procedure to derive the L-moments is briefly given below:

Probability-weighted moments (PWMs), introduced in [48] are defined as:

$$\beta_r = E\left[x\{F(x)^r \right] \tag{23.1}$$

which can be rewritten as:

$$\beta_r = \int_0^1 x(F) F^r dF \tag{23.2}$$

where $F = F(x)$ is the cumulative distribution function (CDF) of x, $x(F)$ is the inverse CDF of x evaluated at the probability F, and $r = 0, 1, 2, ..., s$ is a non-negative integer. When $r = 0$, β_0 is equal to the mean of the distribution $\mu = E[x]$. The rth L-moment λ_r is related to the rth PWM through:

$$\lambda_{r+1} = \sum_{k=0}^{r} \beta_k (-1)^{r-k} \binom{r}{k}\binom{r+k}{k} \tag{23.3}$$

Therefore, the first four L-moments can be written using PWMs as follows:

$$\lambda_1 = \beta_0 \tag{23.4}$$

$$\lambda_2 = 2\beta_1 - \beta_0 \tag{23.5}$$

TABLE 23.1
Important Information on the Selected Stations

Station ID	Latitude	Longitude	Area (km²)	Record Length (yrs)	Period of Record
201001	−28.36	153.29	213	54	1958–2011
203002	−28.64	153.41	62	35	1977–2011
203012	−28.71	153.50	39	34	1978–2011
204025	−30.26	153.03	135	42	1970–2011
204026	−30.25	152.85	80	29	1956–1985
204030	−30.26	152.01	200	34	1978–2011
204036	−28.93	152.22	236	59	1953–2011
204037	−30.09	152.63	62	40	1972–2011
204056	−29.49	152.45	104	36	1976–2011
206009	−31.19	151.83	261	57	1955–2011
206025	−30.68	151.71	594	39	1973–2011
206026	−30.42	151.66	8	37	1975–2011
207006	−31.39	152.33	363	36	1976–2011
208001	−32.03	151.47	20	57	1955–2011
209001	−32.24	151.82	203	34	1946–1979
209002	−32.25	151.98	156	36	1976–2011
209003	−32.48	151.95	974	43	1969–2011
209018	−32.28	151.90	300	32	1980–2011
210011	−32.32	151.69	194	80	1932–2011
210014	−32.15	151.05	395	52	1960–2011
210017	−31.94	151.28	103	71	1941–2011
210022	−32.31	151.51	205	71	1941–2011
210040	−32.27	150.64	676	56	1956–2011
210042	−32.40	151.05	170	30	1967–1996
210044	−32.45	151.15	466	55	1957–2011
210068	−32.80	151.33	25	41	1965–2005
210076	−32.34	150.98	13	37	1969–2005
210079	−32.55	151.59	956	37	1975–2011
210080	−32.47	151.28	80	35	1977–2011
211009	−33.27	151.36	236	39	1973–2011
211013	−33.35	151.34	83	35	1977–2011
212008	−33.43	150.08	199	60	1952–2011
212018	−33.12	150.28	1010	40	1972–2011
212040	−34.61	149.54	96	32	1980–2011
215004	−35.15	150.03	166	82	1930–2011
218005	−36.20	149.76	900	47	1965–2011
218007	−36.26	149.69	122	37	1975–2011
219003	−36.67	149.65	316	68	1944–2011
219017	−36.60	149.81	152	45	1967–2011
219022	−36.73	149.68	202	40	1972–2011
219025	−36.62	149.88	717	35	1977–2011
220001	−36.96	149.56	272	26	1955–1980
220003	−36.94	149.82	105	45	1967–2011
220004	−37.07	149.66	745	41	1971–2011
221002	−37.37	149.71	479	40	1972–2011
222004	−37.00	149.09	604	70	1942–2011

(Continued)

TABLE 23.1 (CONTINUED)
Important Information on the Selected Stations

Station ID	Latitude	Longitude	Area (km²)	Record Length (yrs)	Period of Record
222009	−36.92	149.21	559	43	1952–1994
222015	−36.73	148.43	187	27	1976–2002
222016	−36.79	148.40	155	35	1976–2010
222017	−36.66	149.11	313	33	1979–2011
401009	−35.93	148.10	220	62	1950–2011
401013	−35.90	147.69	378	39	1973–2011
401015	−35.92	146.98	316	37	1975–2011
410038	−35.02	148.25	411	43	1969–2011
410048	−35.20	147.51	530	48	1939–1986
410057	−35.33	148.35	673	54	1958–2011
410061	−35.33	148.07	155	64	1948–2011
410076	−35.92	149.24	212	37	1975–2011
410088	−35.42	148.73	427	44	1968–2011
410112	−34.58	148.09	14	36	1976–2011
410114	−35.24	148.31	23	35	1977–2011
411001	−35.28	149.39	16	25	1960–1984
411003	−35.26	149.54	65	33	1979–2011
412050	−34.31	149.17	740	34	1970–2003
412063	−34.74	149.29	570	39	1961–1999
412081	−33.80	149.19	145	33	1969–2001
412083	−34.02	149.33	321	33	1969–2001
416003	−29.03	151.72	570	33	1979–2011
416008	−29.22	151.38	866	40	1972–2011
416016	−29.79	151.13	726	40	1972–2011
416020	−29.23	150.76	402	33	1979–2011
416023	−29.29	151.92	505	33	1979–2011
418005	−29.92	151.11	259	40	1972–2011
418014	−30.47	151.36	855	37	1971–2007
418017	−29.80	150.58	842	33	1979–2011
418021	−30.23	151.19	311	34	1978–2011
418025	−29.94	150.57	156	32	1980–2011
418027	−30.21	150.43	220	40	1972–2011
418034	−30.30	151.64	14	29	1976–2004
419010	−30.97	151.35	829	32	1980–2011
419016	−31.06	151.13	907	34	1978–2011
419029	−30.71	150.83	389	33	1979–2011
419051	−30.50	150.08	454	35	1977–2011
419053	−30.42	150.65	791	37	1975–2011
419054	−31.04	151.17	391	37	1975–2011
421026	−33.08	149.69	883	38	1974–2011
421036	−33.75	149.94	112	25	1956–1980
421050	−33.03	148.95	365	37	1975–2011

$$\lambda_3 = 6\beta_2 - 6\beta_1 + \beta_0 \tag{23.6}$$

$$\lambda_4 = 20\beta_3 - 30\beta_2 + 12\beta_1 - \beta_0 \tag{23.7}$$

The L-moment ratios defined in [49] using the L-moments are as follows:

$$L-C_v = \tau = \lambda_2 \big/ \lambda_1 \tag{23.8}$$

$$L\text{-skew} = \tau_3 = \lambda_3 \big/ \lambda_2 \tag{23.9}$$

$$L\text{-kurt} = \tau_4 = \lambda_4 \big/ \lambda_2 \tag{23.10}$$

λ_1 is considered as the mean of the distribution because when $r=0$, β_0 is equal to the mean of the distribution as mentioned above. λ_2 is the scale parameter and τ, τ_3 and τ_4 are the measures of coefficients of variation (L-CV), skewness (L-SK), and kurtosis (L-KURT), respectively.

However, the fundamental problem in FFA is that the available data is just a sample of the parent distribution; hence the estimated parameters do not belong to the parent distribution rather to a sample. For this purpose, considering an ordered sample of data $x_1 \leq x_2 \leq \ldots \leq x_n$, we can estimate the sample L moments as below:

$$l_{r+1} = \sum_{k=0}^{r} p^*_{r,k} b_k \tag{23.11}$$

Where

$$b_r = n^{-1} \sum_{j=1}^{n} \frac{(j-1)(j-2)\ldots(j-r)}{(n-1)(n-2)\ldots(n-r)} \tag{23.12}$$

From the above equation, l_r can be considered as unbiased as λ_r; however, the estimators $t_r = l_r/l_2$ for τ_r and l_2/l_1 for τ are consistent but not unbiased. These statistical estimators can be used successfully to derive the suitable distribution that can be fitted to the available sample. For regionalization of the L-moments, if there are N sites in a sample data and i is one of the sites, the sample L-moment ratio for site i can be denoted as $t^{(i)}$, $t_3^{(i)}$, $t_4^{(i)}$. Group average L-moment ratios can be computed in the following manner by weighting against their individual period of record.

$$\bar{t} = \sum_{i=1}^{N} n_i t^{(i)} \Big/ \sum_{i=1}^{N} n_i \tag{23.13}$$

23.3.2 DATA SCREENING

Before performing any statistical analysis, it is of utmost importance to ensure the quality of the data. For any analysis, the collected flood data should be an accurate reflection of the estimated value; thereby, they can be derived from the same frequency distribution. However, the measured data may be subjected to two types of errors; (1) incorrect recording of the data, this may result in outliers in the data and cause misinterpretation of results; (2) the circumstance of collecting the data may change with time (e.g., location change of streamflow gauging site), which alters the frequency distribution of the sample. A preliminary investigation should be carried out during data collection to ensure the quality of the data; however, the collected data may still have some errors.

In RFFE, data from several sites are utilized, and three types of checking should be carried out to ensure the quality of the data [13]. Individual records should be checked for gross errors. An outlier test and trend analysis should be carried out to ensure the reliability of the data. There are many established statistical techniques to carry out these investigations [50–53]. In the case of the application of L-moments in RFFE, a discordancy measure is developed using L-moment ratios to identify discordant sites that may contain data errors. This technique detects a site to be discordant based on the availability of incorrect data, outliers, trends, and shifting of the mean of the sample from the study data set.

It is possible to look for discordant sites using L-moments diagrams. Graphical representations of the L-moments are defined as L-moment ratio diagrams. To check for the discordant sites, the L-moment ratios (L-CV, L-SK, and L-KURT) are plotted in a three-dimensional space, which creates a cloud of the selected sites. Any site far from the center of the cloud is considered to be discordant. "Far" is measured in terms of allowance of the correlation between the L-moments. Mathematically, let i be a site from the proposed region that has N number of sites. For site i, the vector containing the L-moment ratios t, t_3, and t_4 are represented as:

$$u_i = \begin{bmatrix} t^{(i)} & t_3^{(i)} & t_4^{(i)} \end{bmatrix}$$ (23.14)

The unweighted group average $\bar{u}$ is calculated as follows:

$$\bar{u} = \frac{\sum_{i=1}^{N} u_i}{N}$$ (23.15)

The sample covariance matrix can be found using the following:

$$S = \frac{\sum_{i=1}^{N} (u_i - \bar{u})(u_i - \bar{u})^T}{N-1}$$ (23.16)

The discordancy measure for site i is:

$$D_i = \frac{1}{3}(u_i - \bar{u})^T S^{-1}(u_i - \bar{u})$$ (23.17)

D_i is considered to be a standard measure to identify whether a site is discordant in RFFA. To identify a site to be discordant, the criteria $D_i \geq 3$ is suggested in [14]. However, they recommend that the data for sites with the largest D_i values should still be examined for possible data error.

23.3.3 Heterogeneity Measure

For a region to be homogeneous, the population L-moments should be the same for all the sites in the region. However, when samples are pooled from the at-site data for quantile estimation, their sample L-moments will differ from the population. An obvious question is whether the between-sites dispersion of the sample L-moments is too great for the group of sites under consideration to be homogeneous. To examine the homogeneity of a group of sites, a heterogeneity measure is suggested in [14]. The method can be mathematically described as, let site i have record length n_i and sample L-moment ratios are $t^{(i)}$, $t_3^{(i)}$, and $t_4^{(i)}$. If the regional average L-CV, L-SK, and L-KURT weighted by the sites' record length are represented as t^R, t_3^R, and t_4^R where:

$$t_R = \frac{\sum_{i=1}^{N} n_i t^{(i)}}{\sum_{i=1}^{N} n_i} \tag{23.18}$$

The weighted standard deviation can be found as follows:

$$V = \left\{ \frac{\sum_{i=1}^{N} n_i (t^{(i)} - t^R)}{\sum_{i=1}^{N} n_i} \right\}^{1/2} \tag{23.19}$$

A kappa distribution is fitted to the regional average L-moments and a large number of simulations are run to generate N_{sim} of realizations of a region with N sites each having the same kappa distribution as their frequency distribution. These simulated regions are homogeneous and have no cross-correlation or serial correlation; they have the same record length as their real counterparts. For each simulated region, V is calculated. Heterogeneity measure or the H statistics can be computed based on the V calculated for each simulated region by using the following equation:

$$H = \frac{(V - \mu_v)}{\sigma_v} \tag{23.20}$$

Where μ_v and σ_v are the mean and standard deviation of N_{sim} values of V respectively. A region is declared as heterogeneous if H has a large value [54]. Values of H less than 1 can be labeled as "acceptably homogeneous," values lying between +1 to +2 can mean "possibly heterogeneous" and $H \geq 2$ indicates a "definitely heterogeneous" region [13, 14].

23.3.4 INDEX FLOOD METHOD

The IFM is one of the most popular techniques to derive regional flood quantiles [2]. This method is applicable under the assumption that the flood data at different sites in a homogeneous region is identical through a site-specific scaling factor (index flood). Let, Q_M be the "index flood" and X_T be the dimensionless growth factor for a return period of T. If Q_T is the flood quantile at a site:

$$Q_T = X_T Q_M \tag{23.21}$$

Q_M can be derived from the mean of the at-site AM flow [13] or the median of the at-site AM flow [55]. In the case of ungauged sites, where very little or no data is available, Q_M is derived by performing a multiple regression analysis between the mean annual flood flow and the catchment and climatic characteristics of the region. X_T can be computed once an adequate probability distribution is fitted to the available at-site data. If a distribution can fit all the gauged flood data Q_{ij} in a homogeneous region with N sites, where $i = 1, 2, 3, ..., N$, $j = 1, 2, 3, ..., L_i$, and L_i is the record length at site i. The growth factor of the region can be computed by:

$$X_{ij} = \frac{Q_{ij}}{Q_{im}} \tag{23.22}$$

Here Q_{im} is the observed mean or median record length at site i.

Five different probability distributions are assessed in finding homogeneous regions using the L-moments-based method: Pearson type 3 (P3), generalized extreme value distribution (GEV), generalized Pareto (GPA), generalized normal (GN), and generalized logistic (GLO).

23.3.5 Formation of Region (Fixed Region and Region of Influence Approach)

Application of RFFE requires the formation of regions and often the required region is formed based on state/political boundaries. In [6], the fixed region (FR) approach was used to form regions to apply RFFE to different states in Australia. Each state has its own method to estimate flood quantiles due to the FR approach. However, this causes a major problem, as different methods can provide quite different flood estimates at state boundaries. Hence, the goal should be to develop a single flood estimation technique, which is not restricted by the state boundaries. In this case, regionalization based on the region of influence (ROI) method can be more effective.

Formation of regions without fixed boundaries was firstly carried out in [56, 57]. Based on their work, the ROI approach was introduced [19, 20, 46, 58]. According to this method, each site of interest (i.e., catchment where flood quantiles are to be estimated) forms its own region. These defined regions may overlap and gauged sites can be part of more than one ROI for different sites of interest. The ROI is formed for the site of interest using the group of sites in close proximity. A weighted Euclidean distance in an M-dimensional space may be used to measure this proximity. The distance metric can be defined by:

$$D_{i,j} = \left[\sum_{m=1}^{M} W_m \left(X_i^m - X_j^m \right)^2 \right]^{1/2} \tag{23.23}$$

here, $D_{i,j}$ is the weighted Euclidean distance between site i and j, M is the number of attributes included in the distance measure, and the X terms denote standardized values for attribute m at site i and site j, and W_m is a weight applied to attribute m reflecting the relative importance of the attribute. Standardization of attributes is performed to remove units; therefore, bias due to scaling of the attributes can be avoided.

23.3.6 Evaluation Statistics

To assess the performance of the developed RFFE model a leave-one-out (LOO) validation technique is undertaken in this chapter. Based on LOO, a site is left out in each step while constructing the model; this site is treated as an ungauged site. The following performance statistics are calculated from the FR and ROI analyses. For predicted flood quantile (Q_{pred}) and observed flood quantile (Q_{obs}), the following error statistics are adopted: relative error (RE), median absolute relative error (RE$_r$), Q_{pred}/Q_{obs} ratio, median Q_{pred}/Q_{obs} ratio, mean square error (MSE), root mean square error (RMSE), bias (BIAS), relative bias (RBIAS), relative root mean square error (RRMSE), root mean square normalised error (RMSNE), and Nash-Sutcliffe efficiency (NSE):

$$RE = \frac{Q_{pred} - Q_{obs}}{Q_{obs}} \times 100 \tag{23.24}$$

$$RE_r = median \left[abs \left(RE \right) \right] \tag{23.25}$$

$$MSE = mean \left[\left(Q_{pred} - Q_{obs} \right)^2 \right] \tag{23.26}$$

$$RMSE = \sqrt{MSE} \tag{23.27}$$

$$Bias = mean \left(Q_{pred} - Q_{obs} \right) \tag{23.28}$$

$$\text{RBias} = \left[\text{mean}\left(\frac{Q_{\text{pred}} - Q_{\text{obs}}}{Q_{\text{obs}}} \right) \right] \times 100 \tag{23.29}$$

$$\text{RRMSE} = \frac{\sqrt{\text{mean}\left[\left(Q_{\text{pred}} - Q_{\text{obs}} \right)^2 \right]}}{\text{mean}\left(Q_{\text{obs}} \right)} \tag{23.30}$$

$$\text{RMSNE} = \sqrt{\text{mean}\left[\left(\frac{Q_{\text{pred}} - Q_{\text{obs}}}{Q_{\text{obs}}} \right)^2 \right]} \tag{23.31}$$

$$\text{NSE} = 1 - \frac{\text{mean}\left[\left(Q_{\text{pred}} - Q_{\text{obs}} \right)^2 \right]}{\text{mean}\left[\left\{ \text{mean}\left(Q_{\text{obs}} \right) - Q_{\text{obs}} \right\}^2 \right]} \tag{23.32}$$

Q_{obs}, the observed flood quantile at site i, is obtained from at-site flood frequency analysis based on LP3 distribution carried out using FLIKE [59].

23.4 RESULTS

23.4.1 L-MOMENTS COMPUTATION

To investigate the performance of the selected techniques for this chapter, the L-moments and L-moment ratios for each selected site are computed. Table 23.2 represents the mean, median, and range of the L-moments ratios for each site. The range of L-CV is found to be from 0.258 to 0.729; whereas the ranges of L-SK and L-KURT, are 0.083 to 0.695 and −0.001 to 0.554, respectively. The mean of L-CV, L-SK, and L-KURT are found to be 0.541 and 0.386 and 0.198, respectively; whereas, the median of L-CV, L-SK, and L-KURT are found to be 0.553 and 0.397 and 0.195, respectively.

23.4.2 INVESTIGATION FOR HOMOGENEOUS REGION

Based on the computed L-moments and L-moment ratios the discordancy measure is carried out. The discordant site investigation is done in three steps. Firstly, the whole data set of 88 sites is considered to be one group, all the discordant sites that are found after the application of discordancy measure are removed together at the same time, and homogeneity of the region is checked by applying the heterogeneity measure. In the second step, each discordant site found from the discordancy

TABLE 23.2

Mean, Median, and Range of the L-Moment Ratios of the Selected Sites

	L-CV	L-SK	L-KURT
Mean	0.541	0.386	0.198
Median	0.553	0.397	0.195
Min	0.258	0.083	−0.001
Max	0.729	0.695	0.554

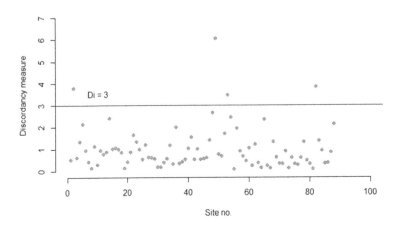

FIGURE 23.2   Values of discordancy measure (D_i) for the 88 sites.

measure (considering 88 sites as one group) is removed one at a time and the heterogeneity measure is carried out to investigate the homogeneity of the region. In the last step, the data set is divided into two groups based on their drainage area (a line separating the neighboring drainage catchments), and discordancy and heterogeneity measures are applied to both groups.

Figure 23.2 shows the discordancy measures D_i for each of the 88 sites computed by equation 17. Amongst these 88 sites, four sites (203002, 222016, 401015, and 419029) are found to have D_i values greater than 3. The values of D_i for sites 203002, 401015, and 419029 are 3.78, 3.48, and 3.86, respectively; the value of D_i for site 222016 is found to be the highest, i.e., 6.05. Although according to the suggested criteria, these sites can be considered discordant, further investigations are carried out to be certain about their discordancy status.

All the AM flows for these selected sites are examined for gross data errors as the first step of the investigation in the data collection stage and no data error is found. Then discordancy measure is applied to further check for any abnormality in the data. The sites that are found to have D_i greater than 3, are removed as they can largely affect the H statistic. The H statistics, i.e., H_1, H_2, and H_3 are computed for the 88 selected sites. For the group consisting of 88 sites H_1, H_2, and H_3 are found to be 13.44, 10.06, and 5.96, respectively. Removal of all the four sites that have D_i greater than 3, results in H values being equal to 11.74, 8.73, and 4.74, respectively. For a homogeneous group, H values have to be less than 1 [49]. This means that, as there are no significant reductions on the H statistics, this group of 84 sites (after removal of four discordant sites) are still highly heterogeneous. Furthermore, after removing the four discordant sites, two more sites are found to have D_i greater than 3. These two new sites are 208001 and 222015 having D_i values of 3.37 and 3.69, respectively. Next, the two new discordant sites are removed from the group of 84 sites and D_i and H statistics are computed for the new group of 82 sites. H values for the new group of 82 sites are found to be 12.35, 8.91, and 5.01, respectively. This step also renders one more discordant site (204026) having D_i of 3.5. The new discordant site having D_i of 3.5, is removed from the 82 sites forming a new group of 81 sites and again the D_i and H statistics are computed for the group. This new computation does not result in any more discordant sites. However, H_1, H_2, and H_3 for this latest group of 81 sites do not show any significant reduction in their H values, which are 11.1, 8.31, and 4.83, respectively. It seems that the H statistics are increasing rather than decreasing after the removal of the discordant sites, which is a bit unusual.

In this step of homogeneous region investigation, each discordant site is removed one at a time from the 88 sites, e.g., site 203002 is first removed and the H statistics are computed for the remaining 87 sites. Afterward, station 203002 is put back into the group and the next site is removed and the H statistics are again computed for the 87 sites as before, and so on. Figure 23.3 plots the value of the discordancy measure for the 87 sites. There are four plots (a), (b), (c), and (d) in Figure 23.3;

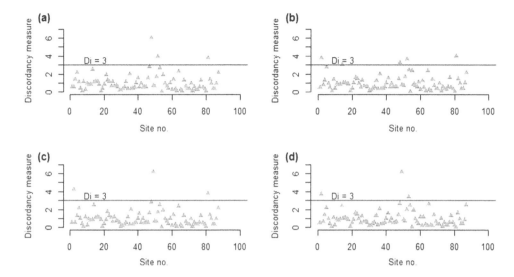

FIGURE 23.3 Values of discordancy measure (D_i) for the 87 sites; (a) after removing site 203002, (b) after removing site 222016, (c) after removing site 401015, and (d) after removing site 419029.

each of them is derived by removing one discordant site at a time. In each of these plots, it is visible that no matter which discordant site is removed from the group there still are a few discordant sites appearing in the group.

Table 23.3 shows the results from the investigation carried out to determine the effect of individual discordant sites on H values of the 87 sites. When the first discordant site 203002 is removed from the 88 sites, the other three discordant sites 222016, 401015, and 419029 still remain discordant, having D_i values of 6.05, 39.5, and 3.81, respectively, and the H values show no significant reduction, which indicates that this group of sites is highly heterogeneous. The removal of 222016 ($D_i = 6.05$) provides four more discordant sites and shows no significant effect on the H values. Removal of the other two sites 401015 and 419029 does not give any better results as can be seen in the table. Results after removing the two discordant sites, one at a time from a group of 84 sites (the first four discordant sites are removed from the 88 sites, thus rendering a new group with 84 sites) are recorded in Table 23.4. It is visible from the table that removals of these discordant sites yield no significant effect on the reduction of H values.

In the third step of this investigation, all 88 sites are divided into two groups based on their drainage divisions. There are two drainage divisions here, drainage divisions 2 and 4. Group 1 consists of sites that are in drainage division 2 and there are a total of 50 sites in this group; whereas, in group 2 there are a total of 38 sites belonging to drainage division 4. Both D_i and H statistics are examined for these two groups. Application of discordancy measure to both of the groups show two discordant sites from each group (i.e., sites 203002 and 222016 from group 1, and sites 401015 and

TABLE 23.3

Sites Removed One at a Time from the Initial Group with 88 Sites

Removed Station No.	H_1	H_2	H_3
203002	13.43	9.71	5.57
222016	12.16	9.53	5.72
401015	13.32	9.54	5.67
419029	13.31	9.54	5.52

TABLE 23.4

Sites Removed One at a Time from the Group with 84 Sites

Removed Station No.	H_1	H_2	H_3
208001	11.22	8.28	4.82
222015	12.22	8.82	4.92

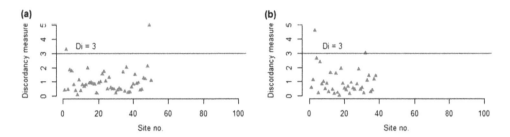

FIGURE 23.4 Values of discordancy measure (a) showing two discordant sites in group 1 (50 sites); and (b) showing two discordant sites in group 2 (38 sites).

TABLE 23.5

_H_ Values for Group 1 Sites Obtained from Drainage Division 2

No. of Sites	Sites Removed	Removed Site Names	H_1	H_2	H_3
50	2	203002, 222016	12.81	9.62	6.13
48	1	222015	11.56	9.29	5.83
47	1	208001	11.35	9.09	5.88
46	1	204026	10.66	9.20	6.46
45	1	204056	10.05	8.36	5.85
44	None	N/A	9.96	8.28	5.89

419029 from group 2). The values of discordancy measure for groups 1 and 2 are plotted in Figure 23.4 (a) and (b), respectively.

The _H_ values for group 1 are found to be 12.81, 9.62, and 6.13, respectively and in the case of group 2, they are found to be 5.84, 3.89, and 1.64, respectively. Division of groups according to their drainage divisions yields no notable reduction in _H_ values. Moreover, an additional application of the discordancy measure after removing the previously found discordant sites reveals no remarkable changes. Tables 23.5 and 23.6 record the _H_ values for groups 1 and 2 respectively after the removal of the discordant sites.

In summary from the discordancy measure, it may be stated that although there are a few sites found as discordant from the case of 88 sites treated as a single group and two groups based on their drainage divisions, their removal of discordant sites does not show any significant effect on the H statistics. In addition, there are no gross data errors discovered for these sites and no other discordant site is found, whose removal can result in a significant reduction in H values. Thus, it can be concluded from these results that acceptable homogeneous regions cannot be established in NSW. In the subsequent analyses, all the 88 sites are included as a single region.

TABLE 23.6

H Values for Group 2 Sites Obtained from Drainage Division 4

No. of Sites	Sites Removed	Removed Site Names	H_1	H_2	H_3
38	2	401015, 419029	5.84	3.89	1.64
36	1	410038	5.55	2.92	0.61
35	1	411001	5.00	2.63	0.59
34	None	N/A	4.64	2.18	0.33

23.4.3 DEVELOPMENT AND APPLICATION OF APPROXIMATE INDEX FLOOD METHOD

Since no homogeneous region is identified, the ideal IFM is not applicable to this data set. Hence, the applicability of an approximate IFM is developed for this data set. This section describes the development and application of this approximate IFM. For this technique, the 10% AEP flood (Q_{10}) is selected as the "index flood" instead of the mean or median of AM flows. For an ungauged catchment, the design flood can be estimated by using the approximate IFM by using the following equation:

$$Q_T = X_T Q_{10} \tag{23.33}$$

where Q_T is the AEP flood of 1 in T years for the ungauged catchment, X_T is the regional growth function.

23.4.3.1 Applicability of Index Flood Method

Although it is a widely used method, the IFM can be criticized on the basis of the inverse relationship between the coefficient of variation of the annual flood series (C_v) and the catchment area (A, km²). For a larger catchment, the flood frequency curve can become very flat due to this relationship, which can produce outliers in design flood estimation in that homogenous region.

Therefore, it is seemingly essential to confirm that the study catchments have no inverse relationship between A and C_v. For this investigation, two plots of C_v versus A and C_v versus log(A) are prepared for all the 88 catchments (Figure 23.5[a] and [b]). Figure 23.5(a) plots all the C_v values for all 88 stations against A; whereas Figure 23.5(b) plots the C_v values versus the logarithm (base 10) of A. No evidence of any relationship between the C_v and A is found from these plots. Therefore, it

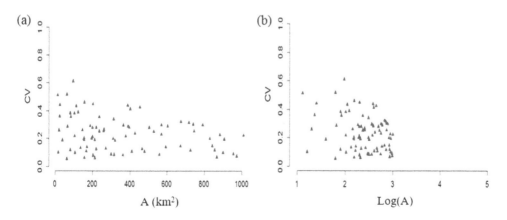

FIGURE 23.5 Relationship between (a) the coefficient of variation (C_V) of the annual maximum flood series and A; (b) the coefficient of variation (C_V) and log (A) for all the 88 catchments.

is possible to apply the approximate IFM to the catchments up to 1,010 km² without violation of its underlying assumption.

23.4.3.2 Development of Approximate Index Flood Method

23.4.3.2.1 Development of Prediction Equation for Q_{10}

In this approximate IFM, Q_{10} is selected as the index variable as mentioned earlier. The prediction equation is developed for Q_{10} using quantile regression technique (QRT) based on the eight selected catchment characteristics, i.e., catchment area (A), rainfall intensity (I_{62}), shape factor (SF), stream density ($sden$), mean annual rainfall (MAR), mean annual evapotranspiration (MAE), slope ($S1085$), and forest ($forest$). To develop the prediction equations, all the response and predictor variables are natural log-transformed to achieve near-normal distribution. Two procedures are taken into account to select the appropriate predictor variables in order to generate the Q_{10} prediction equations. First, a stepwise procedure based on their level of significance is applied to select the best set of predictor variables. This procedure results in the following sets of equations for Q_{10}, which are presented in general form as below (equations 34 to 41). A second set of regression models are also developed to examine the significance of the predictor variable "forest" on the predicted value of Q_{10}. This results in the set of equations (equations 42 to 47) presented below.

Model set 1:

$$\ln Q_{10} = \beta_0 + \beta_1\left(\ln\left(A\right)\right) + \beta_2\left(\ln\left(I_{62}\right)\right) + \beta_3\left(\ln\left(SF\right)\right) + \beta_4\left(\ln\left(sden\right)\right)$$
$$+ \beta_5\left(\ln\left(forest\right)\right) + \beta_6\left(\ln\left(MAR\right)\right) + \beta_7\left(\ln\left(MAE\right)\right) + \beta_8\left(\ln\left(S1085\right)\right) \tag{23.34}$$

$$\ln Q_{10} = \beta_0 + \beta_1\left(\ln\left(A\right)\right) + \beta_2\left(\ln\left(I_{62}\right)\right) + \beta_3\left(\ln\left(SF\right)\right) + \beta_4\left(\ln\left(sden\right)\right)$$
$$+ \beta_5\left(\ln\left(MAR\right)\right) + \beta_6\left(\ln\left(MAE\right)\right) + \beta_7\left(\ln\left(S1085\right)\right) \tag{23.35}$$

$$\ln Q_{10} = \beta_0 + \beta_1\left(\ln\left(A\right)\right) + \beta_2\left(\ln\left(I_{62}\right)\right) + \beta_3\left(\ln\left(SF\right)\right) + \beta_4\left(\ln\left(sden\right)\right)$$
$$+ \beta_5\left(\ln\left(MAR\right)\right) + \beta_6\left(\ln\left(MAE\right)\right) \tag{23.36}$$

$$\ln Q_{10} = \beta_0 + \beta_1\left(\ln\left(A\right)\right) + \beta_2\left(\ln\left(I_{62}\right)\right) + \beta_3\left(\ln\left(SF\right)\right) + \beta_4\left(\ln\left(sden\right)\right) + \beta_5\left(\ln\left(MAE\right)\right) \tag{23.37}$$

$$\ln Q_{10} = \beta_0 + \beta_1\left(\ln\left(A\right)\right) + \beta_2\left(\ln\left(I_{62}\right)\right) + \beta_3\left(\ln\left(SF\right)\right) + \beta_4\left(\ln\left(sden\right)\right) \tag{23.38}$$

$$\ln Q_{10} = \beta_0 + \beta_1\left(\ln\left(A\right)\right) + \beta_2\left(\ln\left(I_{62}\right)\right) + \beta_3\left(\ln(SF)\right) \tag{23.39}$$

$$\ln Q_{10} = \beta_0 + \beta_1\left(\ln\left(A\right)\right) + \beta_2\left(\ln\left(I_{62}\right)\right) \tag{23.40}$$

$$\ln Q_{10} = \beta_0 + \beta_1\left(\ln\left(A\right)\right) \tag{23.41}$$

Model set 2:

$$\ln Q_{10} = \beta_0 + \beta_1\left(\ln\left(A\right)\right) + \beta_2\left(\ln\left(I_{62}\right)\right) + \beta_3\left(\ln\left(SF\right)\right) + \beta_4\left(\ln\left(sden\right)\right)$$
$$+ \beta_5\left(\ln\left(forest\right)\right) + \beta_6\left(\ln\left(MAR\right)\right) + \beta_7\left(\ln\left(MAE\right)\right) + \beta_8\left(\ln\left(S1085\right)\right) \tag{23.42}$$

$$\ln Q_{10} = \beta_0 + \beta_1 \left(\ln \left(A \right) \right) + \beta_2 \left(\ln \left(I_{62} \right) \right) + \beta_3 \left(\ln \left(SF \right) \right) + \beta_4 \left(\ln \left(sden \right) \right)$$
$$+ \beta_5 \left(\ln \left(MAR \right) \right) + \beta_6 \left(\ln \left(MAE \right) \right) + \beta_7 \left(\ln \left(forest \right) \right) \tag{23.43}$$

$$\ln Q_{10} = \beta_0 + \beta_1 \left(\ln \left(A \right) \right) + \beta_2 \left(\ln \left(I_{62} \right) \right) + \beta_3 \left(\ln \left(SF \right) \right) + \beta_4 \left(\ln \left(sden \right) \right)$$
$$+ \beta_5 \left(\ln \left(MAE \right) \right) + \beta_6 \left(\ln \left(forest \right) \right) \tag{23.44}$$

$$\ln Q_{10} = \beta_0 + \beta_1 \left(\ln \left(A \right) \right) + \beta_2 \left(\ln \left(I_{62} \right) \right) + \beta_3 \left(\ln \left(SF \right) \right) + \beta_4 \left(\ln \left(sden \right) \right) + \beta_5 \left(\ln \left(forest \right) \right) \tag{23.45}$$

$$\ln Q_{10} = \beta_0 + \beta_1 \left(\ln \left(A \right) \right) + \beta_2 \left(\ln \left(I_{62} \right) \right) + \beta_3 \left(\ln \left(SF \right) \right) + \beta_4 \left(\ln \left(forest \right) \right) \tag{23.46}$$

$$\ln Q_{10} = \beta_0 + \beta_1 \left(\ln \left(A \right) \right) + \beta_2 \left(\ln \left(I_{62} \right) \right) + \beta_3 \left(\ln \left(sden \right) \right) + \beta_4 \left(\ln \left(forest \right) \right) \tag{23.47}$$

The significance levels for each of the selected predictor variables in stepwise regression are presented in Table 23.7. The second column in Table 23.7 shows the number of variables selected for each model and the third column shows the names of each predictor variable. The next columns (column 4 to column 12) show the significance levels (p-values) for each selected predictor variable including the intercept of the regression. The first model uses all the selected predictor variables to generate the required 10% AEP flood. For each of the models, all the variables except A are removed one by one based on their p-values leaving the last model having only A as the predictor variable. A p-value of maximum 0.10 is selected for a variable to be considered as a significant predictor for the regression equation. From the first model (model 1-a) it is visible that the variable *forest* has the maximum p-value of 0.863 which means this variable shows the least likelihood of being a meaningful addition to the regression equation. The variables, A and I_{62} have the lowest p-values for all the models; making these two variables to be the most important predictor variables. Looking at the p-values of SF and *sden*, it can be seen that, in case of all the eight models, their p-values remain less than 0.10. These two variables may yield significance in the quantile regression procedure. The addition of the other three variables, i.e., *MAE*, *MAR*, and *S1085*, may not improve the model performance as their p-values are as high as 0.238, 0.369, and 0.719, respectively, and as low as 0.137, 0.363, and 0.648, respectively. These p-values are far greater than 0.10, making the addition of these variables, i.e., *forest*, *S1085*, *MAR*, and *MAE* to the regression models, pointless.

Despite the large p-value of the variable *forest*, another set of regression equations are developed to check the effect of this variable on the overall regression. For this stage of regression, the variable *forest* is left untouched and the other predictor variables are removed one by one depending on their p-values. Table 23.8 has the same number of columns as Table 23.7; however, there are six regression models rather than eight in Table 23.8. Again, the first model (model 1-b) consists of all the predictor variables included in the regression generating the same p-values for each of the variables as shown in Table 23.7. For model 2-b, although *forest* has a p-value of 0.863, it is left in the model and the variable having a second maximum p-value of 0.719 (*S1085*) is removed from the model. This process is carried out for *MAR* and *MAE* in models 3 and 4, respectively keeping *forest* intact in the modeling despite having large p-values. For models 5 and 6, there are four predictor variables in each model keeping variables A, I_{62}, and *forest* in common in each case and changing variables SF and *sden*. The p-values for *forest* in the case of models 5 and 6 have been found to be as high as 0.251 and 0.718, respectively. This proves that the addition of the variable *forest* does not have much value to overall regression analysis and it also provides further confirmation on the variables A, I_{62}, SF, and *sden* being valuable additions to this modeling as previously seen from Table 23.7.

TABLE 23.7

Significance Level of Each of the Selected Predictor variables

Model No.	No. of Variables in the Model	Variables Used in the Model	Intercept	A	I_{62}	SF	sden	MAE	MAR	S1085	Forest
									Level of Significance of the Selected Variables (p-Values)		
1-a	8	$A, I_{62}, SF, sden, MAE, MAR, S1085, forest$	0.836	0	0	0.014	0.025	0.238	0.363	0.719	0.863
2-a	7	$A, I_{62}, SF, sden, MAR, MAE, S1085$	0.847	0	0	0.014	0.014	0.198	0.365	0.648	–
3-a	6	$A, I_{62}, SF, sden, MAE, MAR$	0.969	0	0	0.015	0.012	0.137	0.369	–	–
4-a	5	$A, I_{62}, SF, sden, MAE$	0.739	0	0	0.019	0.013	0.155	–	–	–
5-a	4	$A, I_{62}, SF, sden$	0	0	0	0.013	0.025	–	–	–	–
6-a	3	A, I_{62}, SF	0	0	0	0.017	–	–	–	–	–
7	2	A, I_{62}	0	0	0	–	–	–	–	–	–
8	1	A	0	0	–	–	–	–	–	–	–

Note: Blue indicates statistically significant variables.

TABLE 23.8

Significance Level of Each of the Selected Predictor Variables (Keeping "Forest" Variable in All the Models)

Model No.	No. of Variables in the Model	Variables Used in the Model	Intercept	A	I_{62}	forest	SF	sden	MAE	MAR	S1085
						Level of Significance of the Selected Variables (p-Values)					
1-b	8	A, I_{62}, SF, sden, MAE, MAR, S1085, forest	0.836	0	0	0.863	0.014	0.025	0.238	0.363	0.719
2-b	7	A, I_{62}, SF, sden, MAR, MAE, forest	0.912	0	0	0.742	0.015	0.025	0.209	0.342	–
3-b	6	A, I_{62}, SF, sden, MAE, forest	0.740	0	0	0.957	0.020	0.022	0.196	–	–
4-b	5	A, I_{62}, SF, sden, forest	0	0	0	0.564	0.012	0.046	–	–	–
5-b	4	A, I_{62}, SF, forest	0	0	0	0.251	0.014	–	–	–	–
6-b	3	A, I_{62}, sden, forest	0	0	0	0.718	–	0.053	–	–	–

TABLE 23.9

Coefficient of Determination (R^2), Adjusted Coefficient of Determination (adj-R^2), and Residual Standard Error for the Models Described in Table 1.7

Model No.	No. of Variables in the Model	R^2	Adj-R^2	Residual Standard Error
1-a	8	0.77	0.75	0.24
2-a	7	0.77	0.76	0.24
3-a	6	0.77	0.76	0.24
4-a	5	0.77	0.76	0.24
5-a	4	0.77	0.75	0.24
6-a	3	0.75	0.74	0.25
7	2	0.73	0.73	0.25
8	1	0.48	0.47	0.35

The results shown in Tables 23.7 and 23.8 confirm the significance of the variables A, I_{62}, SF, and *sden* in the regression analysis. Table 23.9 presents the coefficient of determination (R^2), adjusted coefficient of determination (*adj-R^2*), and residual standard error for each of the models shown in Table 23.8. From Table 23.9, it is visible that models 1-a to 5-a have the same R^2 values of 0.77 and the *adj-R^2* and residual standard error values are also not much different. For models 6-a and 7, the values of R^2 and *adj-R^2* are decreased and residual standard error is increased; however, for model 8, values of R^2 and *adj-R^2* are decreased to 0.48 and 0.47, respectively. Also, the residual standard error is increased to 0.35. This proves that, although variable A has the maximum significance in the regression, the use of this variable as the only predictor variable does not deliver a good prediction equation for Q_{10}. For models 1-a to 5-a the R^2 and *adj-R^2* values do not change very much; however, the values should be decreasing with the removal of each predictor variable as has happened for models 6-a to 8. This is not the case for model 5-a. Considering the results presented in Tables 23.7, 23.8, and 23.9, model 5-a seems to have the most significant predictor variables (A, I_{62}, SF, and *sden*). In addition, this model seems to qualify as the best model amongst all the models tested having included the models presented in Table 23.9 with a good R^2, *adj-R^2*, and residual standard error values.

Furthermore, the standardized residual vs. the fitted value plots and normal probability plots for the standardized residuals are also examined. Figure 23.6 shows the standardized residual vs. the fitted value plots for models 1-a to 8. All the standardized residual values are almost equally distributed around the zero line and lie between the range of ±2 and also no specific pattern (heteroscedasticity) can be identified; however, in case of model 5-a, all standardized residuals lie between ±2 having fewer outliers. Figure 23.7 shows the normality plots for the standardized residual values for models 1-a to 8. It can be seen that in model 5-a (Table 23.6) where all the predictors have p-values $\leq$0.10, the standardized residuals closely follow a straight line, which indicates that the assumption of normality and the homogeneity of variance of the standardized residuals have largely been satisfied. It is clear from Figures 23.6 and 23.7 that the regression model with the four variables (A, I_{62}, SF, and *sden*) is the best model to predict the Q_{10} for the approximate IFM.

23.4.3.2.2 Generation of Regional Growth Factors

Generation of growth factor (X_T) for each of the quantiles are derived by a weighted average method (weighted by the record length) as shown below:

$$X_T = \frac{\Sigma\left[\left(Q_T\middle/Q_{10}\right)\times\text{record length}\right]}{\Sigma\,\text{record length}}$$

(23.48)

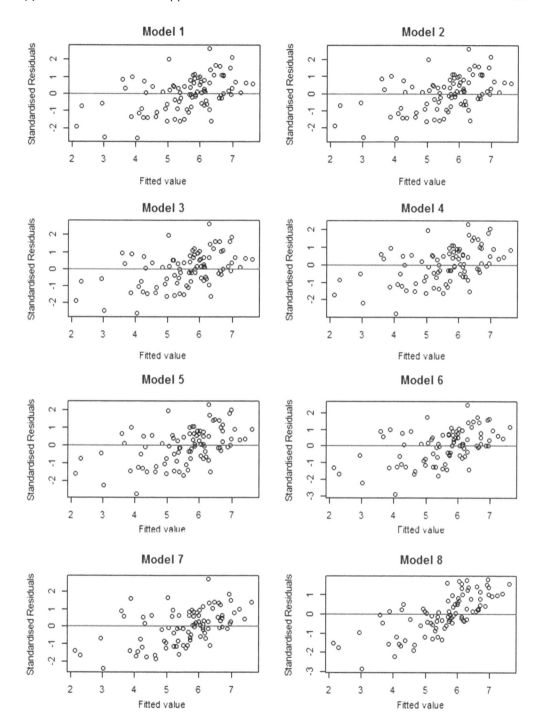

FIGURE 23.6 Plot of the standardised residual vs. fitted value.

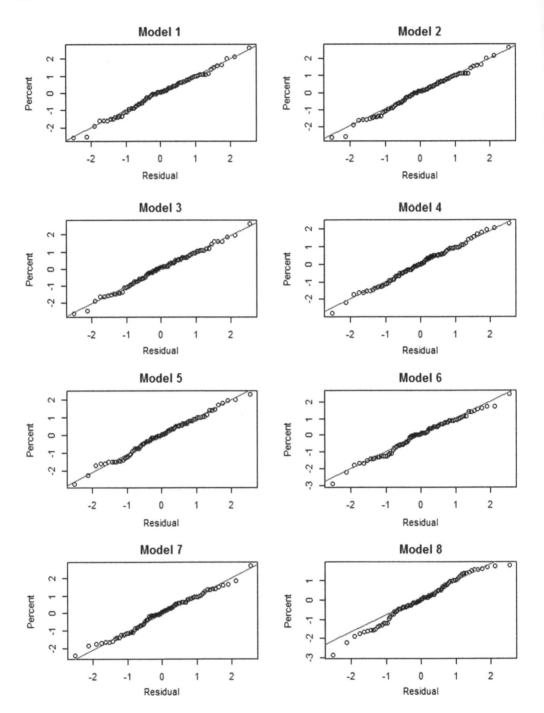

FIGURE 23.7 Normal probability plots for standardised residuals.

The minimum, maximum, mean, and median values for record length and the selected AEPs (50%, 20%, 10%, 5%, 2%, and 1%) with their growth factor (X_T) are presented in Table 23.10. The quantile values in Table 23.10 are the observed values that are derived from LP3 distribution using the FLIKE software; the observed Q_{10} is used as the dependent variable in the case of the regression analysis. The growth factors achieved from the above equation are then multiplied by the Q_{10} achieved from the regression analysis to generate the predicted quantiles.

TABLE 23.10

Details of the AM Flow Record Length and Quantile Values with the Growth Factors

	Record Length	50%	20%	10%	5%	2%	1%
Min	25	1.25	5.17	8.48	11.65	15.79	18.77
Max	82	602.21	1214.13	2028.36	2984.83	4306.82	5300.24
Mean	41.44	100.11	266.09	415.52	582.50	833.35	1051.45
Median	37	62.44	184.76	319.58	490.56	682.39	802.65
X_T	–	0.26	0.639	1	1.43	2.10	2.71

23.4.3.2.3 Comparison among Fixed Region and Region of Influence Approaches for the Approximate Index Flood Method

In this section of this chapter, the comparison between FR and ROI is presented. In the fixed region approach, all the 88 catchments are considered to be forming one region, however, one catchment is left out as the test catchment and the procedure is repeated 88 times to implement the leave-one-out (LOO) validation. Hence, at each iteration, the model data set contains 87 catchments. In the ROI approach, an optimum region is formed for each of the 88 catchments by starting with ten stations in the first proposed region and then consecutively adding five stations at each iteration step until the number of stations reached 30. After that, at each iteration, ten stations are consecutively added until they reach 80 stations in one region.

Tables 23.11 and 23.12 compare the Q_{pred}/Q_{obs} ratio and the RE (%) values for all the AEPs (50% to 1%) for both the FR and ROI, respectively. The first 18 rows of Tables 23.11 and 23.12 show the mean, median and standard deviation (Std_Dev) of the selected AEPs for both the FR and ROI; the last three rows show the overall mean, median, and Std_Dev for the six quantiles. The lowest values in the case of both Q_{pred}/Q_{obs} ratio and RE (%) are presented with blue in both the table. One can see that for each quantile a different model (either FR or ROI) performs better. If the median is to be chosen to be the selection criteria for the best model, it is evident that from Table 23.11, FR performs the best in case of 50% and 5% AEPs (0.98 and 0.99, respectively) and also in the overall median, with Q_{pred}/Q_{obs} ratio being 1.00. However, from Table 23.12, KNN15 performs better with the lowest RE (%) 30.48%, 48.57%, and 30.50% in the case of 50%, 20%, and 10%, respectively but in the case of the overall median, KNN50 has the lowest RE (%) (0.6%). It should be mentioned here that all the mean, median, and Std_Dev presented in Table 23.12 are calculated as absolute RE.

Table 23.13 summarizes the evaluation statistics from LOO, i.e., MSE, RMSE, BIAS, RBIAS, RRMSE, RMSNE, NSE, RE_r(%), and median Q_{pred}/Q_{obs} ratios based on the observed and predicted values for the 5% AEP flood using the equations 24 to 32 provided in section 23.1.3.7. The reason behind choosing 5% AEP flood is that it is the most commonly used flood quantile in flood frequency analysis. A small value of MSE is preferable; however, from Table 23.13, it is seen that MSE values for a fixed region and all the other regions from ROI are quite high. The lowest value (92,562.70) for MSE is found in the case of KNN25. The MSE values for KNN40, KNN50, KNN60, and KNN70 range from 92,562 to 96,619. For a model to be acceptable, RMSE should be relatively small. In the case of the selected models, RMSE values lie in the range of 304.24 to 703.09; which is quite high. The lowest RMSE (304.24) is found for KNN25. Looking into bias, one can see that the lowest positive BIAS (0.1) is found in the case of KNN20 and a small negative BIAS (–3.79) is found in the case of KNN30. KNN70 is showing the highest negative BIAS of –55.05, which indicates an overall underestimation by the model and KNN10 shows the highest positive BIAS of 74.20, which indicates an overall overestimation by the model. In terms of RBIAS, the lowest value is found in the case of KNN70 (19.24%). For the rest of the models, the RBIAS values range from 21.72% to

TABLE 23.11

Mean, Median, and Standard Deviation of Q_{pred}/Q_{obs} Ratio Values for FR and ROI

AEP		FR	KNN10	KNN15	KNN20	KNN25	KNN30	KNN40	KNN50	KNN60	KNN70	KNN80
50%	Mean_abs	1.33	1.34	1.37	1.35	1.31	1.34	1.33	1.32	1.34	1.32	1.32
	Median_abs	0.98	1.08	1.14	1.12	1.11	1.24	1.10	1.06	1.09	1.04	0.94
	Std Dev_abs	0.84	1.08	1.01	0.93	0.90	0.84	0.87	0.88	0.92	0.93	0.87
20%	Mean_abs	1.20	1.23	1.23	1.21	1.18	1.20	1.20	1.20	1.19	1.17	1.19
	Median_abs	0.93	1.04	1.01	0.97	0.94	0.96	0.97	0.96	0.94	0.87	0.91
	Std Dev_abs	0.72	0.96	0.81	0.83	0.80	0.73	0.76	0.79	0.78	0.75	0.74
10%	Mean_abs	1.19	1.24	1.23	1.22	1.18	1.21	1.20	1.20	1.19	1.17	1.19
	Median_abs	0.98	1.01	1.00	0.98	0.91	0.94	0.98	0.98	0.94	0.95	0.96
	Std Dev_abs	0.79	0.97	0.84	0.91	0.88	0.81	0.83	0.88	0.85	0.79	0.79
5%	Mean_abs	1.22	1.26	1.26	1.25	1.21	1.25	1.23	1.24	1.22	1.19	1.22
	Median_abs	0.99	1.01	1.07	1.01	0.97	0.98	0.99	0.98	0.96	0.95	0.98
	Std Dev_abs	0.89	1.01	0.91	1.01	0.99	0.92	0.94	1.01	0.95	0.87	0.89
2%	Mean_abs	1.29	1.33	1.33	1.31	1.29	1.33	1.31	1.31	1.30	1.26	1.29
	Median_abs	1.02	1.07	1.17	1.05	1.07	1.06	1.01	1.02	1.01	0.98	1.00
	Std Dev_abs	1.05	1.12	1.05	1.18	1.17	1.09	1.12	1.20	1.13	1.03	1.05
1%	Mean_abs	1.36	1.41	1.40	1.39	1.37	1.42	1.39	1.39	1.38	1.33	1.36
	Median_abs	1.04	1.15	1.20	1.12	1.17	1.13	1.08	1.08	1.10	1.01	1.06
	Std Dev_abs	1.20	1.25	1.20	1.32	1.33	1.24	1.28	1.36	1.28	1.16	1.20
	Overall mean	1.26	1.30	1.30	1.29	1.26	1.29	1.28	1.28	1.27	1.24	1.26
	Overall median	1.00	1.06	1.08	1.05	1.03	1.05	1.01	1.01	0.99	0.97	0.99
	Overall Std Dev	0.93	1.07	0.98	1.04	1.02	0.96	0.98	1.04	1.00	0.93	0.93

TABLE 23.12

Mean, Median, and Standard Deviation of RE (%) for FR and ROI

AEP		FR	KNN10	KNN15	KNN20	KNN25	KNN30	KNN40	KNN50	KNN60	KNN70	KNN80
50%	Mean_abs	63.47	65.78	61.14	62.04	60.65	63.55	65.63	64.62	67.01	66.20	64.25
	Median_abs	44.01	37.13	30.50	41.44	40.62	44.31	45.49	45.85	46.16	44.91	43.40
	Std Dev_abs	64.16	91.85	88.30	77.12	73.49	64.10	65.94	67.13	71.21	73.06	66.18
20%	Mean_abs	53.05	55.06	48.57	50.50	48.94	49.97	53.58	52.85	53.87	53.41	53.57
	Median_abs	36.35	31.87	29.34	29.90	31.57	34.80	34.45	33.07	34.38	34.95	34.42
	Std Dev_abs	52.37	82.04	69.15	69.29	65.95	57.29	56.91	61.64	59.45	54.71	53.57
10%	Mean_abs	52.81	56.97	50.63	51.87	50.64	51.67	53.16	52.46	52.77	52.07	53.07
	Median_abs	35.97	31.52	30.48	31.72	30.77	31.48	35.29	32.82	31.54	32.68	36.86
	Std Dev_abs	61.13	81.87	70.50	77.16	74.04	66.08	66.97	73.50	58.65	60.99	61.88
5%	Mean_abs	55.35	60.62	55.12	56.11	54.59	55.81	55.79	55.45	54.77	53.33	55.38
	Median_abs	35.44	36.97	36.60	34.10	36.11	32.59	33.17	31.01	30.69	31.04	34.23
	Std Dev_abs	72.53	84.88	76.30	87.27	85.07	77.36	79.58	87.18	80.89	71.51	73.06
2%	Mean_abs	62.10	68.29	63.09	64.23	62.68	64.37	62.88	63.10	51.40	59.51	61.90
	Median_abs	36.57	42.51	40.49	45.16	42.63	44.47	35.91	35.93	33.42	35.47	37.16
	Std Dev_abs	89.35	94.75	89.97	103.30	102.44	94.27	97.98	106.44	98.90	87.24	89.76
1%	Mean_abs	69.22	76.96	71.44	72.73	71.58	74.01	70.43	70.89	59.85	66.21	69.13
	Median_abs	43.13	45.26	47.12	46.98	48.37	46.66	44.01	44.16	42.27	41.54	41.02
	Std Dev_abs	103.75	107.11	104.27	117.16	117.42	108.21	113.61	122.34	113.45	101.11	103.94
	Overall mean	58.15	63.95	58.33	59.58	58.18	59.90	60.25	27.73	59.94	58.46	59.55
	Overall median	37.06	37.80	35.53	37.34	37.91	38.79	37.97	0.60	36.42	37.03	38.23
	Overall Std Dev	72.62	90.73	83.97	90.01	88.19	79.99	82.44	103.58	34.04	76.25	76.55

TABLE 23.13

Statistical Evaluations of the Fixed Region Approach and ROI (in Case of 5% AEP Flood) Approaches

	FR	KNN10	KNN15	KNN20	KNN25	KNN30	KNN40	KNN50	KNN60	KNN70	KNN80
MSE	117984.19	494337.79	412034.01	114600.42	92562.70	124705.17	94795.92	95551.08	95347.81	96618.81	113601.55
RMSE	343.49	703.09	641.90	338.53	304.24	353.14	307.89	309.11	308.78	310.84	337.05
BIAS	−40.86	74.20	66.05	0.10	−24.64	−3.79	−33.53	−37.80	−49.62	−55.05	−45.64
RBIAS	21.93	26.46	25.79	24.50	21.49	24.76	23.25	23.64	22.34	19.24	21.72
RRMSE	0.07	0.13	0.11	0.00	0.04	0.01	0.06	0.06	0.09	0.09	0.08
RRMSNE	0.91	1.04	0.94	1.03	1.01	0.95	0.97	1.03	0.97	0.89	0.91
NSE	0.55	−0.89	−0.58	0.56	0.65	0.52	0.64	0.63	0.64	0.63	0.57
median_RE_AbS	35.44	36.97	36.60	34.10	36.10	32.59	33.17	31.01	30.68	31.03	34.23
median_Q_{pred}/Q_{abs}	0.99	1.01	1.07	1.01	0.97	0.98	0.99	0.98	0.96	0.94	0.98

26.46%. In the case of RRMSE, except KNN10 and KNN15, the other models are performing quite well with RRMSE being less than 10%.

A statistical model can have an NSE value in the range of $-\infty$ to 1; where values closer to 1 indicate a good model. Most of the models (except KNN10 and KNN15) presented in Table 23.12 have acquired an NSE value close to 0.60 which indicates that these models have the potential to derive a reliable flood estimate. Examination of the RE_r (%) and median Q_{pred}/Q_{obs} ratios for each model provides a clear view of KNN50, KNN60, and KNN70 having the lowest RE_r (%) (31.01%, 30.68%, and 31.03%, respectively). However, KNN50 has a median Q_{pred}/Q_{obs} ratio of 0.98 whereas KNN60 and KNN70 have median Q_{pred}/Q_{obs} ratios of 0.96 and 0.94, respectively which indicates underestimation of flood quantiles. KNN20 also shows a median Q_{pred}/Q_{obs} ratio of 1.01. The overall result portrays KNN20 and KNN50 to be the better performing candidates for the approximate IFM. Therefore, the next section further investigates the adequacy of all these models for approximate IFM.

To find the best model, Figures 23.8 and 23.9 are produced. Figures 23.8 and 23.9 show the boxplots of RE (%) and the Q_{pred}/Q_{obs} ratios of predicted to observed values for 5% AEP flood in case of both FR and ROI. In the case of Figure 23.8 the expected RE (%) value is chosen to be 0, indicating an unbiased model. For Figure 23.9, it is set at 1, as it indicates an unbiased model. The box contains the middle 50% of the data; whereas the bars contain the first and last 25% of the data; anything outside of the box and bars is considered an outlier. Small box size indicates a less spread which also means less variability in error. The middle line in the box indicates the median value of the data. In Figure 23.8, all the boxes are small in size indicating that the RE (%) values for all models show small variability. The top and bottom bars are in between the values of -100 and 200. There are a few outliers visible on the top side of the plot; however, the plot has a fixed margin of -300 to $+300$. Therefore, there may be possible positive outliers indicating overestimation by the models for some sites. The smallest box size is visible in the case of KNN60; however, the median line is slightly away from the zero line. KNN50 is showing a smaller box size than KNN20; however, the median lines for both the models (KNN20 and KNN50) pass through the zero line.

For Figure 23.9 the boundary is set from 0 to 3. In the case of Figure 23.9, the boxes are quite tall with KNN60 having the smallest box size. However, the median line is away from the expected line. All the boxes show a disproportional size of the top and bottom part of the box from the median line. This indicates that the Q_{pred}/Q_{obs} ratios may be skewed. Again, KNN50 shows a smaller box

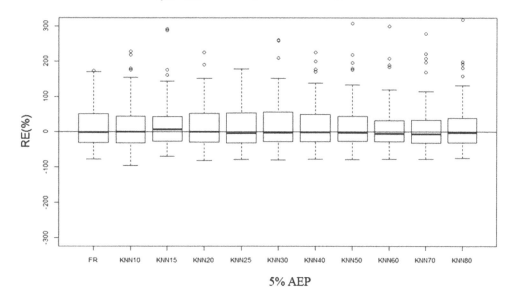

FIGURE 23.8 Boxplot of the RE values (%) for 5% AEP for both FR and ROI.

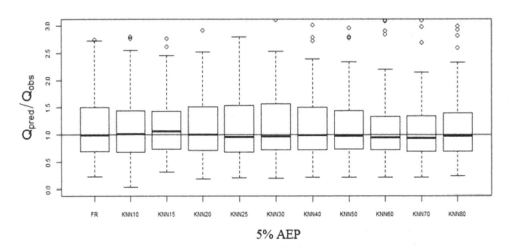

FIGURE 23.9 Boxplot of Q_{pred}/Q_{obs} ratios of predicted to observed for floods 5% AEP for both the FR and ROI.

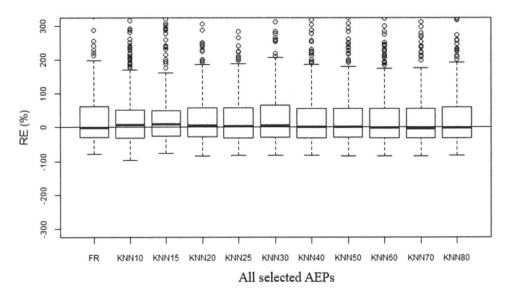

FIGURE 23.10 Boxplots of the RE values (%) for the six selected AEPs (50%, 20%, 10%, 5%, 2%, and 1%) for both FR and ROI.

than KNN20 with a median line going through the expected line. KNN10, KNN40, and KNN80 models have their median line touching the expected line; however, their error bars are widely spaced with possible outliers. There are no negative outliers found in both cases of boxplots for RE (%) and Q_{pred}/Q_{obs} ratios. However; there can be possible under- and overestimation for some stations as the bars are on the negative side of the plot in the case of RE (%) and there are a number of positive outliers visible in the figures with more outside the boundary of the figures.

Figures 23.10 and 23.11 present the boxplots of RE (%) and Q_{pred}/Q_{obs} ratios for all the selected AEPs (50% to 1%) for both FR and ROI. It is visible that, although the boxes are small in the case of RE (%), the boxes for Q_{pred}/Q_{obs} ratios are bigger and disproportionate. This further confirms that the error in prediction is skewed. It is visible from both the figures that although KNN20, KNN40, KNN50, and KNN60 have small box sizes, KNN50 has smaller error bars, and the median RE (%) and median Q_{pred}/Q_{obs} ratios both cross through expected values (0 and 1 respectively). In the case

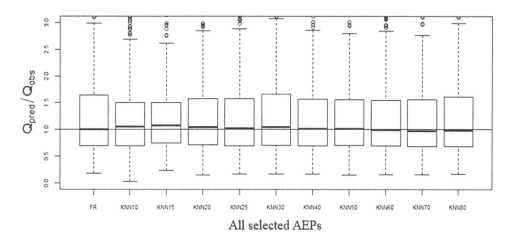

Q_{pred}/Q_{obs}

All selected AEPs

FIGURE 23.11 Boxplots of Q_{pred}/Q_{obs} ratios of the predicted to observed ones for the six selected AEPs (50%, 20%, 10%, 5%, 2%, and 1%) both FR and ROI.

of Figure 23.10, the median line for KNN20 does not touch the expected line. From all these figures it is evident that KNN10 and KNN15 are the worst-performing models, which supports the previous findings mentioned in the earlier part of this section.

Tables 23.14 and 23.15 showcases the number of stations performing "fair," "poor," and "good" in the case of KNN20 and KNN50 based on their median RE (%) and median Q_{pred}/Q_{obs} ratios. The range of "fair," "poor," and "good" has been arbitrarily decided based on the following values: RE (%) from −30% to +30% is "good," from −60% to +60% is "fair," and beyond ±60% is "poor"; Q_{pred}/Q_{obs} ratio from 0.80 to 1.3 is "good," from 0.60 to 2 "fair," and less than 0.6 to greater than 2 is "poor." It can be seen from both the tables that, for 50% and 20% AEPs KNN20 display a greater number of sites performing "good" (36 and 30 stations in case of median RE (%) and 26 and 38 stations in case of Q_{pred}/Q_{obs} ratio). However, for 10% to 1% AEPs KNN50 display a higher number of "good" performing sites ranging from 28 to 33 in the case of median RE (%) and 33 to 36 in the case of Q_{pred}/Q_{obs} ratio. Therefore, it is clear from the above discussion that, the best performing model is the model with 50 closest sites forming a region with four predictor variables.

23.4.4 COMPARISON WITH ARR RFFE MODEL

In this section, a comparison of absolute RE (%) values between the ARR RFFE model [54] and a newly developed approximate IFM KNN50 model here is presented. A Bayesian generalized least

TABLE 23.14
Grouping of Stations on the Basis of RE Values (%)

	KNN20			KNN50		
	Number of stations			Number of stations		
Quantiles	Fair	Poor	Good	Fair	Poor	Good
50%	25	27	36	44	28	16
20%	35	23	30	41	24	26
10%	38	22	28	32	23	33
5%	42	23	23	32	24	32
2%	38	25	25	32	26	30
1%	36	33	19	31	26	28

TABLE 23.15
Grouping of Stations on the Basis of Q_{pred}/Q_{obs} Ratio Values

	KNN20			KNN50		
	Number of Stations			Number of Stations		
Quantiles	Fair	Poor	Good	Fair	Poor	Good
50%	13	49	26	34	31	23
20%	32	18	38	31	27	30
10%	38	19	31	25	27	36
5%	16	41	31	26	27	35
2%	29	29	30	25	30	33
1%	28	34	26	21	33	34

TABLE 23.16
Comparison of Absolute RE (%) Values between the ARR RFFE Model and Newly Developed Approximate IFM KNN50 Model

AEPs	ARR RFFE Model Absolute RE (%)	Approximate IFM KNN50 Model Absolute RE (%)
50%	63.07	45.85
20%	57.25	33.06
10%	57.48	32.82
5%	58.85	31.01
2%	60.39	35.93
1%	64.06	44.16

square parameter regression technique is used to develop the RFFE model to estimate regional flood quantile for the ARR RFFE model in the case of Australian flood data [54]. The absolute RE (%) values for both the ARR RFFE model and the approximate IFM KNN50 model are compared in Table 23.16. It is evident from Table 23.16 that the RE (%) values for the ARR RFFE model (ranging from 56% to 64%) are greater than the approximate IFM KNN50 model (ranging from 30% to 46%). One possible reason for these differences in RE values by the two methods is that the ARR RFFE model is developed with the data from 558 stations from NSW, Victoria, and Queensland; whereas, the approximated IFM KNN50 is developed for 88 sites in NSW. However, it is promising to see that the RE (%) values from the approximate IFM KNN50 model are comparable to the RE (%) values of the ARR RFFE model. From this research, it can be said that approximate IFM can be a viable model to estimate regional flood quantiles in the case of Australia even though the homogeneity criteria are not fulfilled. Further research is required to ensure the reliability of approximate IFM in regional flood estimation in Australia.

23.5   CONCLUSIONS

A total of 88 stations from NSW, Australia, has been used in this chapter to develop an approximate index flood method (IFM). Homogeneous region formation in the case of these data set is unsuccessful, which is the most important criterion to be satisfied for application of the IFM. An approximate IFM is developed for the selected sites ignoring the homogeneity issue. For this approximate IFM, the 10% AEP flood is selected as the index flood and the prediction equation for the 10% AEP flood is developed using the regression approach, where catchment characteristics are selected as predictor variables. Out of the eight candidate characteristics, four are selected in the prediction equation, i.e., catchment area, rainfall intensity, shape factor of the catchment, and stream density based on their significance level in the regression analysis. The developed equation is then tested in both FR and ROI frameworks. To assess the performances of the developed prediction equation, a LOO validation procedure is adopted. Several statistical evaluations are carried out in order to find the best-performing model from both FR and ROI models. Based on the results presented in this chapter, it can be said that a total of 50 stations forming one optimum region generates better prediction for flood quantiles using the approximate IFM. The approximate IFM outperforms the ARR RFFE 2016 model. This demonstrates that strict adherence to the "homogenerous region" assumption may not be needed in IFM. This investigation is carried out using 88 sites from NSW and eight selected catchment characteristics. It is recommended to add more stations and catchment characteristics for future investigation.

REFERENCES

1. Rahman, A., Weinmann, P.E., and Mein R.G. 1999b. At-site flood frequency analysis: LP3-product moment, GEV-L moment and GEV-LH moment procedures compared. In: Proceedings of 2nd International Conference on Water Resources and Environmental Research, I.E. Aust., 6–8 July, 2, 715–720.
2. Dalrymple, T. 1960. *Flood Frequency Analysis*, Water Supply Paper 1543-A, U.S. Geological Survey, Reston, VA, USA.
3. Di Baldassarre, G., Castellarin, A., and Brath A. 2006. Relationships between statistics of rainfall extremes and mean annual precipitation: An application for design-storm estimation in northern Italy, *Hydrology and Earth System Sciences*, 10: 589–601.
4. Madsen, H., Mikkelsen, P.S., Rosbjerg, D., and Harremoes, P. 2002. Regional estimation of rainfall intensity duration curves using generalised least squares regression of partial duration series statistics, *Water Resources Research*, 38(11): 1–11.
5. Riggs, H.C., 1973. *Regional Analyses of Streamflow Techniques. Techniques of Water Resources Investigations of the U.S. Geol. Surv., Book 4, Chapter B3*, U.S. Geol. Surv., Washington, DC, USA.
6. Institution of Engineers Australia (I.E. Aust.) 1987. *Australian Rainfall and Runoff: A Guide to Flood Estimation*, edited by D.H. Pilgrim, Vol. 1, I.E. Australia, Canberra, Australia.
7. Haddad, K. 2008. *Design Flood Estimation in Ungauged Catchments Using a Quantile Regression Technique: Ordinary and Generalised Least Squares Methods Compared for Victoria*, Masters (Honors) thesis, School of Engineering, The University of Western Sydney, New South Wales, Australia.
8. Haddad, K., and Rahman, A. 2012. Regional flood frequency analysis in eastern Australia: Bayesian GLS regression-based methods within fixed region and ROI framework: Quantile regression vs. parameter regression technique, *Journal of Hydrology*, 430–431: 142–161.
9. Ishak, E., Haddad, K., Zaman, M., and Rahman, A. 2011. Scaling property of regional floods in New South Wales Australia, *Natural Hazards*, 58: 1155–1167.
10. Rahman, A. 1997. *Flood Estimation for Ungauged Catchments: A Regional Approach Using Flood and Catchment Characteristics*, PhD thesis, Department of Civil Engineering, Monash University, Australia.
11. Benson, M.A. 1962. *Evolution of Methods for Evaluating the Occurrence of Floods*, U.S. Geological Survey, Water Supply Paper, 1580-A, 30.
12. Smith, J.A. 1992. Representation of basin scale in flood peak distributions, *Water Resources Research*, 28(11): 2993–2999.
13. Hosking, J.R.M., and Wallis, J.R. 1997. *Regional Frequency Analysis: An Approach Based on L-Moments*, Cambridge University Press, New York, UK.
14. Hosking, J.R.M., and Wallis, J.R. 1993. Some statistics useful in regional frequency analysis, *Water Resources Research*, 29(2): 271–281.
15. Fill, H.D., and Stedinger, J.R. 1998. Using regional regression within IF procedures and an empirical Bayesian estimator, *Journal of Hydrology*, 210: 128–145.
16. Stedinger, J.R., Vogel, R.M., and Foufoula-Georgiou, E. 1993. Frequency analysis of extreme events. In *Handbook of Hydrology*, McGraw Hill Book Company, New York, USA, pp. 18.1–18.66.

17. Interagency Advisory Committee on Water Data (IAWCD). 1982. *Guidelines for Determining Flood Flow Frequency: Bulletin17-B (Revised and Corrected).* Hydrological Sub-commission, Washington, DC, USA, March 1982, 28.

18. Griffis, V.W., and Stedinger, J.R. 2004. LP3 Flood quantile estimators using at-site and regional information. In: Sehlke, G., Hayes, D.F. Stevens, D.K. (eds) Critical Transitions in Water and Environmental Resources Management, Proceedings of World Water and Environmental Resources Congress, Salt Lake City, Utah, June 27–July 1, ASCE, Reston, VA, USA.

19. Burn, D.H. 1990a. An appraisal of the "region of influence" approach to flood frequency analysis, *Hydrological Sciences Journal*, 35(2): 149–165.

20. Burn, D.H. 1990b. Evaluation of regional flood frequency analysis with a region of influence approach, *Water Resources Research*, 26(10): 2257–2265.

21. Cavadias, G.S. 1990. The canonical correlation approach to regional flood estimation. In: Regionalization in Hydrology, Proceedings of the Ljubljana Symposium, April 1990, Ljubljana, Slovenia.

22. Chowdhury, J.U., Stedinger, J.R., and Lu, L.H. 1991. Goodness of fit tests for regional flood distributions, *Water Resources Research*, 27(7): 1765–1776.

23. Fill, D.H., and Stedinger J.R. 1995b. L moment and PPCC goodness-of-fit tests for the Gumbel distribution and effect of autocorrelation, *Water Resources Research*, 31(1): 225–229.

24. Fill, D.H., and Stedinger J.R. 1995a. Homogeneity tests based upon Gumbel distribution and a critical appraisal of Darymple's test, *Journal of Hydrology*, 166: 81–105.

25. Lu, L.H., and Stedinger, J.R. 1992. Sampling variance of normalized GEV/PWM quantile estimators and a regional homogeneity test, *Journal of Hydrology*, 138(1–2): 223–245.

26. Ouarda, T.B., Girard, C., Cavadias, G.S., and Bobée, B. 2001. Regional flood frequency estimation with canonical correlation analysis, *Journal of Hydrology*, 254(1–4): 157–173.

27. Wiltshire, S.E. 1986a. Identification of homogeneous regions for flood frequency analysis, *Journal of Hydrology*, 84(3–4): 287–302.

28. Ulrych, T.J., Velis, D.R., Woodbury, A.D., and Sacchi, M.D. 2000. L-moments and C-moments, *Stochastic Environmental Research and Risk Assessment*, 14(1): 50–56.

29. Zrinji, Z., and Burn, D.H. 1996. Regional flood frequency with hierarchical region of influence, *Journal of Water Resources and Planning Management*, 122(4): 245–252.

30. Wiltshire, S.E. 1986b. Regional flood frequency analysis I: Homogeneity statistics, *Hydrological Sciences Journal*, 31(3): 321–333.

31. Alila, Y.P., Adamowski, K., and Pilon, J. 1992. Regional homogeneity testing of low-flows using L moments. In: Proceedings of 12th Conference on Probability and Statistics in the Atmospheric Sciences, 5th International Meeting on Statistical Climatology, 22–26 June 1992, Toronto, ON, Canada.

32. Bates, B.C., Rahman, A., Mein, R.G., and Weinmann, P.E. 1998. Climatic and physical factors that influence the homogeneity of regional floods in south-eastern Australia, *Water Resources Research*, 34(12): 3369–3382.

33. Castellarin, A., Vogel, R.M., and Matalas, N.C. 2007. Multivariate probabilistic regional envelopes of extreme floods, *Journal of Hydrology*, 336: 376–390.

34. Chebana, F., and Ouarda, T.B.M.J. 2008. Depth and homogeneity in regional flood frequency analysis, *Water Resources Research*, 44(11): W11422.

35. Gaume, E., Gaál, L., Viglione, A., Szolgay, J., Kohnová, S., and Blöschl., G. 2010. Bayesian MCMC approach to regional flood frequency analyses involving extraordinary flood events at ungauged sites, *Journal of Hydrology*, 394(1–2): 101–117.

36. Guttman, N.B. 1993. The use of L-moments in the determination of regional precipitation climates, *Journal of Climatology*, 6: 2309–2325.

37. Kjeldsen, T.R., and Rosbjerg, D. 2002. Comparison of regional index flood estimation procedures Based on the extreme value type I distribution, *Stochastic Environmental Resources Risk A*, 16: 358–373.

38. Pearson, C.P., 1991. New Zealand regional flood frequency analysis using L moments, *Journal of Hydrology*, 30(2): 53–64.

39. Thomas, Jr., W.O., and Olsen, S.A. 1992. Regional analysis of minimum streamflow. In: Proceedings of 12th Conference on Probability and Statistics in the Atmospheric Sciences, 5th International Meeting on Statistical Climatology, Toronto, ON, Canada, 22–26 June, pp. 261–266.

40. Zrinji, Z., and Burn, D.H. 1994. Flood frequency analysis for ungauged sites using a region of influence approach, *Journal of Hydrology*, 153(1–4): 1–21.

41. Vogel, R.M., and Fennessey, N.M. 1993. L moment diagrams should replace product moment diagrams, *Water Resources Research*, 29(6): 1745–1752.

42. Saf, B. 2009. Regional flood frequency analysis using L-moments for the West Mediterranean Region of Turkey, *Water Resources Management*, 23(3): 531–551.

43. Parida, B.P., Kachroo, R.K., and Shrestha D.B. 1998. Regional flood frequency analysis of Mahi-Sabarmati basin (subzone 3-a) using index flood procedure with L-moments, *Water Resources Management*, 12: 1–12.

44. Noto, L.V., and La Loggia, G. 2009. Use of L-moments approach for regional flood frequency analysis in Sicily, Italy, *Water Resources Management*, 23(11): 2207–2229.

45. Castellarin, A., Burn, D.H., and Brath, A. 2001. Assessing the effectiveness of hydrological similarity measures for regional flood frequency analysis, *Journal of Hydrology*, 241(3–4): 270–285.

46. Eslamian, S.S. 1995. *Regional Flood Frequency Analysis Using a New Region of Influence Approach*, Ph.D. Thesis, University of New South Wales, School of Civil Engineering, Department of Water Engineering, Sydney, NSW, Australia, 1995, Supervised by: Professor David H. Pilgrim, 380.

47. Micevski, T., Hackelbusch, A., Haddad, K., Kuczera, G., and Rahman, A. 2015. Regionalisation of the parameters of the log-Pearson 3 distribution: a case study for New South Wales, Australia, Hydrological Processes, 29(2): 250–260.

48. Greenwood, J.A., Landwehr, J.M, Matalas, N.C., and Wallis, J.R. 1979. Probability weighted moments: Definition and relation to parameters of several distributions expressible in inverse form, *Water Resources Research*, 15: 1049–1054.

49. Hosking, J.R.M. 1990. L moments: analysis and estimation of distributions using linear combinations of order statistics, *Journal of Royal Statistical Society Series B*, 52(1): 105–124.

50. Cohn, T.A., England, J.F., Berenbroc, C.E., Mason, R.R., Stedinger, J.R., and Lamontagne, J.R. A generalized Grubbs-Beck test statistic for detecting multiple potentially influential outliers in flood series, *Water Resources Research*, 49: 5047–5058.

51. Grubbs, F.E. (1969). Procedures for detecting outlying observations in samples, *Technometrics*, 11(1): 1–21.

52. Kendall, M.G. 1970. *Rank Correlation Methods*, 4th edition, Griffen, London, 202.

53. Mann, H.B. 1945. Nonparametric tests against trend, *Econometrica*, 13(3): 245–259.

54. Rahman, A., Haddad, K., Haque, M., Kuczera, G., and Weinmann, P.E. 2015. *Australian Rainfall and Runoff: Project 5: Regional Flood Methods: Stage 3 Report* (No. P5/S3, p. 025). Technical Report, Australia.

55. Ribeiro-Corréa, J., Cavadias, G.S., Clément, B., and Rousselle, J. 1995. Identification of hydrological neighbourhoods using canonical correlation analysis, *Journal of Hydrology*, 173(1–4): 71–89.

56. Acreman, M.C. 1987. *Regional Flood Frequency Analysis in the UK: Recent Research-New Ideas*, Institute of Hydrology, Wallingford, UK.

57. Acreman, M.C., and Wiltshire, S.E. 1987. Identification of regions for regional flood frequency analysis, *EOS*, 68(44): 1262 (Abstract).

58. Eslamian, S.S., and Hosseinipour, E.Z. 2010, A modified region of influence approach for flood regionalization. In: 2010 World Water and Environmental Resources Congress, Providence, Rhode Island, USA.

59. Kuczera, G. 1999a. Comprehensive at-site flood frequency analysis using Monte Carlo Bayesian inference, *Water Resources Research*, 35(5): 1551–1557.

24 NASA Global Near-Real-Time and Research Precipitation Products for Flood Monitoring, Modeling, Assessment, and Research

Zhong Liu, Dana Ostrenga, Andrey Savtchenko,
William Teng, Bruce Vollmer, Jennifer Wei, and D. Meyer

CONTENTS

24.1 INTRODUCTION

Floods are among the most powerful and deadliest natural disasters in the world and can cause severe infrastructure damage and deaths, particularly in developing countries due to the inadequate flood-resistant infrastructure and early warning systems. For example, the deadliest floods in China were often associated with two major rivers, the Yellow River and the Yangtze River, particularly in the last century (e.g., the floods in 1931 [Courtney, 2018]). Even in developed countries, local floods are frequently reported and can be deadly as well, such as the flash flood on June 11, 2010, in the

Albert Pike campgrounds of Arkansas, USA, where at least 16 people were killed when torrential rains caused the water in a nearby river to surge over its normal level.

Floods are closely associated with heavy precipitation and snowmelt. Heavy rains often occur in the regions of lower and middle latitudes and are associated with mesoscale and synoptic weather systems (e.g., squall lines, tropical cyclones, monsoon lows, and stationary lows). By contrast, floods from snowmelt are associated with heavy snowfall accumulation during winter and abnormal warming in springtime that causes snow on the ground to melt more rapidly than normal and flood the watershed and its downstream surrounding areas. Nonetheless, precipitation data are essential for flood monitoring, modeling, assessment, and research. From the examples above, it is seen that the sizes of watersheds can vary a lot, ranging from small creeks to large river systems across a country. Furthermore, floods can happen at a very short time scale (e.g., flash floods). All these flood characteristics require a high spatiotemporal resolution, accurate and timely precipitation datasets available in order to accurately develop models and products for flood monitoring and prediction as well as flood-related research and applications.

It is well known that precipitation is notoriously difficult to measure and predict. Traditional methods (e.g., gauges, radars) of precipitation measurement can be costly and difficult to deploy, particularly in remote and mountainous regions, resulting in sparse observations available for flood operation, research, and applications in those regions. In the past several decades, techniques to estimate satellite-based precipitation have been developed (Kloub et al., 2010). In particular, with a constellation of satellites, the spatiotemporal resolution of satellite-based precipitation products and abilities to detect the light rain and snow with new sensor technology have been greatly improved (e.g., Huffman et al., 2007; Hou et al., 2014), compared to traditional methods of using gauges and radars. As a result, satellite-based precipitation products are widely used in flood research and applications.

It is even more challenging for numerical weather prediction models to accurately predict precipitation in space and time as well as in both short- and long-terms. Many obstacles still exist for improving numerical prediction of precipitation in the model, such as the accurate representation of complex cloud microphysics and precipitation processes which are still not well-observed, understood, and heavily rely on parameterization to represent these processes in the numerical model. Improvements will continue to depend on new observations in those areas to better understand and represent cloud microphysics and precipitation processes in the model. In short, new and improved observational theories and techniques will continue to play a key role to advance the model capability for precipitation prediction.

Currently, there are still issues about data quality in both land- and satellite-based precipitation estimates. For multi-satellite, multi-sensor products, the matter can be further complicated. For example, different sensors with different data quality are used in multi-sensor and multi-satellite products and sensor calibration is a necessary step to ensure systematic differences are minimized in the final products. Biases are common in satellite-based products and evaluation with gauge data is often the first step before conducting flood research and application activities. Nonetheless, improvements (e.g., bias correction) in precipitation estimates can provide more accurate measurement of precipitation as well as further reduce uncertainty and biases for flood models, thereby helping model development and operations.

In this chapter, NASA global satellite-based near-real-time and research precipitation products at the NASA Goddard Earth Sciences Data and Information Services Center (GES DISC), including the Integrated Multi-satellitE Retrievals for GPM (IMERG) product suite that consists of global, half-hourly, near-real-time, and research precipitation products on a uniform 0.1 deg.×0.1 deg. grid (Huffman et al., 2017), are introduced. The GES DISC is home to the data archive for two major NASA satellite precipitation measurement missions (Tropical Rainfall Measuring Mission [TRMM] and Global Precipitation Measurement [GPM]) as well as other satellite missions and projects. Other global and regional precipitation and their ancillary products at the GES DISC are also listed.

To facilitate data access, the GES DISC has developed user-friendly data services together with a user support system, including an online visualization and analysis tool, Giovanni, for rapid product evaluation and exploration without downloading data and software. We describe these services along with examples. As mentioned earlier, bias and systematic differences are common in satellite-based products, which is important to understand and quantify in flood research and applications. An example is presented to use Giovanni for conducting a preliminary investigation on systematic differences between the IMERG near-real-time and research products. This book chapter is organized as follows: section 24.2 contains an overview of global and regional precipitation products at GES DISC; section 24.3 describes the data services for accessing and evaluating precipitation products; examples are in section 24.4 and followed by a brief summary in section 24.5.

24.2 OVERVIEW OF NASA GLOBAL AND REGIONAL PRECIPITATION PRODUCTS AT GES DISC

The GES DISC is one of 12 NASA discipline-based Distributed Active Archive Centers (DAACs) in the USA, managed by the NASA Earth Observing System Data and Information System (EOSDIS [NASA, 2018a]). The GES DISC archives global satellite-based precipitation products, including two of NASA's Earth-observing satellite missions, TRMM and GPM, as well as other reanalysis and assimilation projects. In addition, the GES DISC also archives the global and regional satellite-based interdisciplinary data products from solar irradiance, atmospheric composition and dynamics, global modeling, etc. Currently, over 2,700 unique data products are available at the GES DISC for public distribution. Timely available precipitation products are crucial for flood operation and modeling. Our emphasis here is on near-real-time precipitation products from TRMM and GPM missions since only these two missions provide near-real-time precipitation products at GES DISC.

24.2.1 TRMM AND GPM MISSIONS

The NASA-JAXA's (the Japan Aerospace Exploration Agency) TRMM and GPM missions are two of NASA's Earth-observing satellite missions that provide global precipitation measurement and more. Launched on November 27, 1997 in Japan, TRMM (Simpson et al., 1988; Kummerow et al., 1998, 2000; Liu et al., 2012) was designed to: (1) advance understanding of the global energy and water cycles by providing distributions of rainfall and latent heating over the global tropics; (2) understand the mechanisms through which changes in tropical rainfall influence global circulation and to improve ability to model these processes in order to predict global circulations and rainfall variability at monthly and longer timescales; (3) provide the rain and latent heating distributions to improve the initialization of models ranging from 24-hour forecasts to short-range climate variations; (4) help to understand, diagnose, and predict the onset and development of the El Niño, Southern Oscillation, and the propagation of the 30–60-day oscillations in the tropics; (5) help to understand the effect that rainfall has on the ocean thermohaline circulations and the structure of the upper ocean; (5) allow cross calibration between TRMM and other sensors with life expectancies beyond that of TRMM itself; (6) evaluate the diurnal variability of tropical rainfall globally; and (7) evaluate a space-based system for rainfall measurement.

Instruments carried by TRMM (Table 24.1) were used to retrieve precipitation information: (a) the first space-borne Precipitation Radar (PR) that operated at Ku band (13.8 GHz) and provided three-dimensional rain distribution over land and ocean surfaces; (b) the TRMM Microwave Imager (TMI); (c) the Visible and Infrared Scanner (VIRS); and (d) the Lightning Imaging Sensor (LIS). Combined with measurements from other satellites, TRMM played an important role in providing the baseline measurements for satellite inter-calibration and more. Furthermore, its multi-satellite and multi-sensor merged global precipitation product suite, the TRMM Multi-Satellite Precipitation Analysis (TMPA), has been widely used in many disciplines and applications. The TMPA was among

TABLE 24.1
TRMM Precipitation-Related Instruments

Instrument Name	Band Frequencies/Wavelengths	Spatial Resolution		Swath Width	
		Pre-Boost	Post-Boost	Pre-Boost	Post-Boost
Visible and infrared scanner (VIRS)	5 channels (0.63, 1.6, 3.75, 10.8, and 12 μm)	2.2 km	2.4 km	720 km	833 km
TRMM microwave imager (TMI)	5 frequencies (10.7, 19.4, 21.3, 37, and 85.5 GHz)	4.4 km at 85.5 GHz	5.1 km at 85.5 GHz	760 km	878 km
Precipitation radar (PR)	13.8 GHz	4.3 km. Vertical: 250 m	5 km. Vertical: 250 m	215 km	247 km
Lightning imaging sensor (LIS)	0.7774 μm	3.7 km	4.3 km	580 km	668 km

Source: Kummerow et al., 1998, 2000.

the highly cited product suite in the *Journal of Hydrometeorology* of the American Meteorological Society. The TRMM satellite mission ended on June 16, 2015, when the satellite reentered the Earth's atmosphere; however, the TRMM data collection continues to be an important asset for research and application activities.

Built on the success of TRMM, GPM was launched on February 27, 1994, to further advance global precipitation measurements in particular in light rain and snow for improving our understanding of precipitation distribution and processes especially in the Polar Regions (Hou et al., 2014; Kirschbaum et al., 2017; Skofronick-Jackson et al., 2017; Liu et al., 2017). Table 24.2 lists instruments onboard the GPM satellite. New instruments (Table 24.2) are (a) a dual-frequency PR (DPR) that contains a new Ka-band frequency (35 GHz) for measuring frozen precipitation and light rain and (b) new high-frequency channels in the microwave instrument to enhance the capability for measuring precipitation intensity and type through all cloud layers.

The GES DISC is home to the permanent archive of TRMM and GPM data. The following sections provide further details with emphasis on datasets that are most frequently used in flood research and applications.

TABLE 24.2
GPM Instruments

Instrument Name	Band Frequencies/Wavelengths	Spatial Resolution	Swath Width
GPM microwave imager (GMI)	13 frequencies (10–183 GHz)	5.1 km at 85.5 GHz	885 km
Dual-frequency precipitation radar (DPR)	13.8 GHz (KuPR), 33.5 GHz (KaPR)	5 km. Vertical: 250 m (KuPR); 250 m/500 m (KaPR)	245 km (KuPR), 120 km (KaPR)

Source: Hou et al., 2014.

24.2.2 NEAR-REAL-TIME, MULTI-SATELLITE, AND MULTI-SENSOR MERGED GLOBAL PRECIPITATION PRODUCTS

Data gaps commonly exist from observations of a single low Earth orbit (LEO) satellite due to long and infrequent revisit times. Infrared (IR) sensors onboard geostationary satellites, on the other hand, can provide abundant observations; however, the quality of precipitation estimates from the IR sensors is considered to be less than those from active and passive microwave measurements. In order to fill in data gaps, techniques or retrieval algorithms for different sensors have been developed over the years to utilize strengths of multi-satellites and multi-sensors (i.e., microwave and geostationary infrared sensors) and improve limited spatiotemporal coverage from a single LEO satellite (Adler et al., 2003; Huffman et al., 2007, 2009, 2010, 2012, 2013; Joyce et al., 2004; Mahrooghy et al., 2012, Hong et al., 2007, Sorooshian et al., 2000; Behrangi et al., 2009; Aonashi et al., 2009). These near-global precipitation products are available uniformly in space and time and over a long time period (i.e., from 1998 onward for TMPA), with very few missing data and thereby are suitable for hydrometeorological research and applications.

The TMPA products in Table 24.3 (Huffman 2017; Huffman et al., 2007, 2010, 2012, 2013), developed by the Mesoscale Atmospheric Processes Laboratory at NASA Goddard Space Flight Center, provide the precipitation estimates at 3-hourly and monthly temporal resolutions on a 0.25 deg. × 0.25 deg. grid available from January 1998 to the present. The TMPA consists of two products: near-real-time (3B42RT, spatial coverage: 60° N–S, latency: ~8 hours) and research-grade (3B42, spatial coverage: 50° N–S). The former is less accurate, but provides quick precipitation estimates suitable for near-real-time monitoring and modeling activities (e.g., Wu et al., 2012, 2014). The latter, available approximately two months after observation, is processed with additional input data and calibrated with gauge data, different sensor calibration, and additional post-processing in the algorithm. The resulting products are more accurate and suitable for research (Huffman et al., 2007, 2010). Table 24.3 lists all parameters in TMPA products. Over the years, the TMPA products

---

TABLE 24.3
Summary of Variables in TMPA Products

3-Hourly Near-Real-Time Product (3B42RT)
> Calibrated precipitation
> Calibrated precipitation error
> Satellite source identifier
> Uncalibrated precipitation

3-Hourly Research Product (3B42)
> Multi-satellite precipitation
> Multi-satellite precipitation error
> Satellite observation time
> PMW precipitation
> IR precipitation
> Satellite source identifier

Monthly Research Product (3B43)
> Satellite-gauge precipitation
> Satellite-gauge precipitation error
> Gauge relative weighting

Source: Huffman and Bolvin, 2014.
Note: Daily products are also available and provided by GES DISC as value-added products.

TABLE 24.4

Summary of Variables in IMERG Products

Half-Hourly Products (IMERG Early, Late, and Final)

Calibrated multi-satellite precipitation

Uncalibrated multi-satellite precipitation

Calibrated multi-satellite precipitation error

PMW precipitation

PMW source identifier

PMW source time

IR precipitation

IR KF weight

Probability of liquid-phase precipitation

Quality index for calibrated multi-satellite precipitation

Monthly Research Product (IMERG Final)

Satellite-gauge precipitation

Satellite-gauge precipitation error

Gauge relative weighting

Probability of liquid-phase precipitation

Quality index for calibrated multi-satellite precipitation

Source: Huffman et al., 2017.

Note: Daily products are also available and provided by GES DISC as value-added products.

have been widely used in various hydrological research and applications (e.g. Wu et al., 2012; Bitew et al., 2012; Gourley et al., 2011; Su et al., 2011; Gianotti et al., 2012; Tekeli, 2017; Engel et al., 2017; Tan and Duan, 2017).

During the GPM era, the Integrated Multi-satellitE Retrievals for GPM (IMERG) product suite (Huffman et al., 2017) has several improvements compared to the TMPA products: (1) the grid spatial resolution has been increased from 0.25 degrees to 0.1 degrees; (2) the temporal resolution is improved from three-hourly to half-hourly; (3) the spatial coverage has been expanded from 50° N–S to 60° N–S (the full global coverage is planned in future releases); (4) light rain detection has been improved with advanced instruments and algorithms; (5) frozen precipitation is available for the first time; and (6) more parameters have been added to help diagnose data quality issues. The IMERG suite (Table 24.4) contains three output products, "Early Run" (lag time: ~4 hours), "Late Run" (lag time: ~12 hours), and "Final Run" (lag time: ~3.5 months) along with additional new input and intermediate files. The first two contain half-hourly products and the last both half-hourly and monthly. To facilitate data access, the GES DISC has developed daily products for research and applications. All parameters are listed in Table 24.4. Compared to those in TMPA (Table 24.3), it is seen that more ancillary parameters are included in IMERG (Table 24.4) to help diagnose data quality issues such as identifying the source of errors, which is very important for multi-sensor, multi-satellite products. A detailed comparison between TMPA and IMERG is available (Huffman, 2017). The retro-processing of the IMERG product suite back to the TRMM era has been finished to allow the users to develop baseline products for anomaly detection and other applications.

The data latencies (from satellite acquisition) for near-real-time TMPA and IMERG are approximately eight and four hours, separately. Flash floods usually occur within six hours after a heavy rainfall event, depending on rainfall intensity, duration, and size of the watershed. Apparently, the latency of near-real-time TMPA has exceeded this time period for flash floods and may not be suitable for smaller-scale flash flood operation, compared to ground radars that can provide observations much faster when available. Improvements on data latency are still needed for satellite-based multi-satellite and multi-sensor precipitation products. On the other hand, the latency of near-real-time IMERG is improved and significantly shorter than TMPA and could be used in some flash flood applications. For other large-scale floods, both near-real-time TMPA and IMERG are suitable (e.g., Wu et al., 2014).

As mentioned earlier, the TRMM ended on June 16, 2015. As a result, the TRMM Microwave Imager (TMI) and precipitation radar (PR) that were used to intercalibrate passive microwave precipitation estimates in TMPA are no longer available (replaced by a climatological calibration instead). Product transition from TRMM to GPM has begun since the launch of the GPM satellite in February 2014. The retro-processing of the IMERG suite back to the TRMM era is completed and satisfactory (Huffman, 2017). The detailed transition plan from TMPA to IMERG is available in Huffman (2017).

24.2.3   RESEARCH-GRADE MULTI-SATELLITE AND MULTI-SENSOR MERGED GLOBAL PRECIPITATION PRODUCTS

Non-operational research and development, in general, do not require data with short latency. Since biases are common in satellite-based products, research-grade precipitation products are developed with gauge data for bias correction. In both TMPA and IMERG, the Global Precipitation Climatology Centre (GPCC) monthly dataset is used. Tables 24.3 and 24.4 list research-grade precipitation products from TMPA and IMERG. Research-grade precipitation products can be evaluated with gauge data obtained from investigators in their area of interest, which is often the first step before using satellite-based products. Research-grade products can be used in hydrological model development as a forcing dataset and methods can be developed to correct systematic differences and biases in near-real-time products.

24.2.4   OTHER PRECIPITATION PRODUCTS

Other global and regional precipitation products are available at the GES DISC, including (a) global and regional land data assimilation products; (b) Modern-Era Retrospective Analysis for Research and Applications (MERRA) products; and (c) precipitation-related ancillary products. Precipitation products from the Global Precipitation Climatology Project (GPCP; Adler et al., 2003) Version 3.1 are also archived at the GES DISC (Huffman et al., 2020).

Global and regional land data assimilation products consist of quality-controlled, and spatially and temporally consistent, land-surface model (LSM) datasets from the best available observations and model output to support the modeling activities (NASA, 2018b). Advanced land surface modeling and data assimilation techniques are used to ingest and process satellite- and ground-based observational data products.

The precipitation field in the forcing group of the North American Land Data Assimilation System (NLDAS-2) is derived from a temporal disaggregation (from the daily analysis to hourly intervals) of a gauge-only NOAA CPC (Climate Prediction Center) analysis of daily precipitation (Higgins et al., 2000) with an orographic adjustment based on the widely-applied PRISM climatology (Daly et al., 1994; NASA, 2018b). On the other hand, more precipitation observation products (NASA, 2018b) are used in the Global Land Data Assimilation System (GLDAS): (a) satellite-based observed precipitation products from the Naval Research Laboratory, the NASA Goddard TMPA near-real-time algorithm (3B42RT), and PERSIANN; and (b) the merged satellite and gauge algorithm (CMAP).

The GES DISC archives precipitation products from GLDAS, NLDAS, NCA-LDAS, and FLDAS. Spatiotemporal resolutions of these products vary and none of these precipitation products is near-real-time. Latencies vary with products.

The Modern-Era Retrospective Analysis for Research and Applications, Version 2 (MERRA-2), developed at the NASA Global Modeling and Assimilation Office (GMAO) at the NASA Goddard Space Flight Center (Gelaro et al., 2017; Suarez and Bacmeister, 2015), provides global precipitation and other datasets beginning in 1980 and runs a few weeks behind real-time (Gelaro et al., 2017). MERRA-2 focuses on historical analyses of the hydrological cycle on a broad range of weather and climate time scales and places the NASA Earth Observing System (EOS) suite of observations in a climate context (Gelaro et al., 2017). MERRA-2 includes many atmospheric processes including an interactive analysis of aerosols that feed back into the circulation, NASA's observations of stratospheric ozone and temperature (when available), and steps towards representing cryogenic processes (Gelaro et al., 2017). Data in MERRA-2 ranges from 1980 to the present.

There are two types of precipitation parameters in MERRA-2: (a) precipitation from the atmospheric model and (b) observation-corrected precipitation (Reichle and Liu 2014; Bosilovich et al., 2015). Observational data are introduced in the latter parameter due to considerable errors that propagate into land surface hydrological fields and beyond (Reichle et al., 2011). A general evaluation of MERRA-2 precipitation estimates includes precipitation climatology, interannual variability, diurnal cycle, Madden-Julian Oscillation (MJO) events, global water cycle, and US summertime variability. Major findings can be found in Bosilovich et al. (2015). A special collection of research papers (Gelaro et al., 2017), Modern-Era Retrospective Analysis for Research and Applications, Version 2 (MERRA-2), has been published by the American Meteorological Society with an overview article of MERRA-2 and articles about MERRA-2 assessment in different disciplines and regions around the world (AMS, 2018).

In addition to global and regional precipitation products, the GES DISC also archives many interdisciplinary products for flood-related research and applications. As one of 12 NASA DAACs, the focus areas at GES DISC are atmospheric composition, water and energy cycles, climate variability, as well as carbon cycle and ecosystem. Major NASA missions and projects are TRMM, GPM, MERRA, NLDAS, MODIS-Aqua, MODIS-Terra, etc. Users can search and download these datasets at GES DISC with a newly released web interface (NASA, 2018c). NASA DAACs are organized based on disciplines. Users who want more datasets that are not available at GES DISC can search the Earthdata portal (NASA, 2018d) that has been developed by the EOSDIS (NASA, 2018a). The Earthdata portal (NASA, 2018d) provides access to datasets archived at all 12 NASA DAACs.

24.3   PRECIPITATION DATA SERVICES

Data services play an important role in facilitating data access and utilization in order to maximize the use of satellite data products, which is particularly true for precipitation datasets that are often used in many interdisciplinary research and applications by a very diverse group of users, compared to other parameters in Earth science. For non-expert users, especially novices (Liu and Acker, 2017), significant investment in time and resources is needed due to obstacles users may encounter such as heterogeneous data formats, complex data structures, large-volume data storage, special programming requirements, diverse analytical software options, and other factors. Point-and-click online tools such as Giovanni are ideal for them to explore and evaluate before committing valuable time and resources. For experienced users, essential data services include dataset/parameter search and discovery, documentation, dataset citation, data subsetting (e.g., spatial, parameter), format conversion, web services, user services, and more. Advanced data services or analysis-ready services can provide even more capabilities to deliver customized data or results to users who need fewer or no post-processing after they download data/results. For example, Giovanni can provide accumulated rainfall for a watershed or country.

Over the years, various disciplinary-based data services have been developed at NASA's EOSDIS DAACs to improve NASA's data discovery and access. First, an EOSDIS web search interface (NASA, 2018d) or Earthdata has been developed and anyone with a web browser can access NASA

FIGURE 24.1 Web portal of Earthdata, showing the search results for "precipitation."

data products at 12 DAACs through this interface. Figure 24.1 is a screenshot of the EOSDIS web search interface, showing a text search box where precipitation is entered, and below is a list of related search results. User registration is required for downloading data from all EOSDIS data services including those from the GES DISC. Users can also visit each individual DAAC and use their web interfaces to access discipline-oriented data products and services. Furthermore, special discipline-oriented data tools (NASA, 2018e) have been developed at DAACs and they are organized in the following categories: search and order, data handling, subsetting and filtering, geolocation, reprojection, and mapping, and data visualization and analysis (NASA, 2018e). EOSDIS DAACs also support web services and various web protocols for machine-to-machine data access and applications such as OPeNDAP (Open Source Project for a Network Data Access Protocol [OPeNDAP, 2018), WMS (Web Map Service), GDS (GrADS Data Server [GDS, 2018]), THREDDS, https, etc. To address data and science-related issues and inquiries from users, DAACs provide the user services including frequently asked questions (FAQs), data recipes, user forums, email or phone inquiries, etc. Due to page limitations, it is difficult to describe all of the data services at the DAACs in detail. Since our focus is on floods, precipitation data services at the GES DISC are presented next.

24.3.1 POINT-AND-CLICK ONLINE TOOL: NASA GES DISC GIOVANNI

Giovanni stands for the Geospatial Interactive Online Visualization and Analysis Infrastructure (Giovanni [NASA, 2018f]), developed by the GES DISC to assist a wide range of users around the world with data access and evaluation, as well as with scientific exploration and discovery (Liu and Acker, 2017; Acker and Leptoukh, 2007). User surveys (e.g., Kearns 2017) and experience from user support services at the GES DISC show that non-expert users and those who occasionally use satellite-based products prefer point-and-click data tools in order to obtain graphic and data assessment results. As mentioned earlier, dataset assessment activities may not be straightforward and can be time-consuming and costly. Point-and-click tools provide fast and easy access to satellite-based data products for all users without the need for coding and downloading data and software. Furthermore, they can be further developed for in-depth data analysis and visualization. As of this writing, there are 82 measurement groups from eight disciplines and over 1,900 variables are available in Giovanni and more are being added. Since Giovanni was released to the public, over 1,300 peer-reviewed papers across various Earth science disciplines and other areas were published with assistance from Giovanni (Liu and Acker, 2017).

It can be a challenge to locate a variable of interest from over 1,900. For example, precipitation estimates can come from the different missions and projects with different spatiotemporal resolutions, availability, and more. Giovanni provides several simple ways to make searching variables easy by offering both keyword and faceted search capabilities in its web interface (Figure 24.2). A search for "NRT" returns the GPM-related near-real-time variables (Figure 24.2).

Commonly used analytical and plotting capabilities (Liu and Acker, 2017) are available in Giovanni. Giovanni provides mapping options including time-averaging, animation, accumulation (precipitation), time-averaged overlay of two datasets, and user-defined climatology. For time series, options include area-averaged, differences, seasonal, and Hovmöller diagrams. For data comparison, Giovanni has built-in processing code for data sets that require measurement unit conversion and re-gridding. Commonly used comparison functions include map and time-series differences, as well as correlation maps and X–Y scatterplots (area-averaged or time-averaged).

Giovanni contains a collection of visualization features (Liu and Acker, 2017), including interactive map area adjustment; animation; interactive scatter plots; adjustments of data range; change of color palette; contouring; and scaling (linear or log). The on-the-fly area adjustment feature allows an interactive and detailed examination of a result map without re-plotting data. Interactive scatterplots allow the identification of the geolocation of a point of interest in a scatterplot. Adjustments of any of these plots provide custom options to users.

Giovanni supports the vector shapefiles for countries, states in the United States, and major watersheds around the world. Available functions for shapefiles are time-averaged and accumulated maps, area-averaged time series, and histograms. Land-sea masks are available.

All data files involved in Giovanni processing are listed and can be downloaded on the lineage page. Available image formats are PNG, GEOTIFF, and KMZ (Keyhole Markup Language). All input and output data are available in NetCDF, which can be handled by many off-the-shelf software

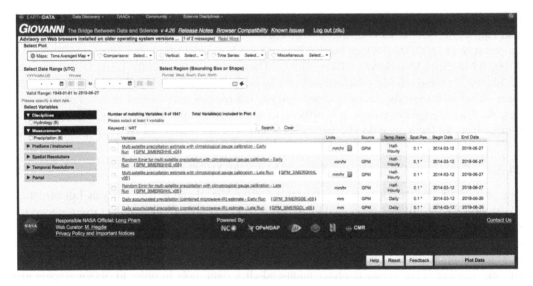

FIGURE 24.2 The Giovanni web portal, showing search results of near-real-time precipitation variables for "NRT."

packages. Furthermore, URLs generated by Giovanni can be bookmarked for reference, documentation, or sharing with other colleagues.

24.3.2 Data Rod Service

Providing long time series data with high temporal resolution (e.g., hourly precipitation over 20 years) to the hydrology community can be a challenge (Teng et al., 2016). In hydrology, earth surface features are expressed as discrete spatial objects such as watersheds, river reaches, and point observation sites; and time-varying data are contained in time series associated with these spatial objects (Teng et al., 2016). Long-time histories of data may be associated with a single point or feature in space. Most remote sensing precipitation products are expressed as continuous spatial fields, with data sequenced in time from one data file to the next. Hydrology tends to be narrow in space and deep in time (Teng et al., 2016), which poses a challenge during the GPM era. For example, to generate a one-year time series, one needs to pull all the 0.1 deg., half-hourly IMERG product, which can be time-consuming and not suitable for online data services due to the large volume of data.

The concept of data rods (Teng et al., 2016; Gallaher and Grant, 2012; Rui et al., 2012, 2013) can be applied to this challenge. Teng et al. (2016) proposed two general solutions: (1) retrieve multiple time series for short time periods and stitch the multiple time series into desired single long time series and (2) reprocess (parameter and spatial subsetting) and archive data as one-time cost approach. The resultant time series files would be geospatially searchable and could be optimally accessed and retrieved by any user at any time (Teng et al., 2016). One drawback for the data rod approach is that there are a lot of files to be generated and maintained. The data rod concept has been implemented in CUAHSI-HIS (Consortium of Universities for the Advancement of Hydrologic Science, Inc. – Hydrologic Information System) and other hydrologic community tools (Rui et al., 2013) where three-hourly TMPA time series data can be accessed.

24.3.3 Subsetting and Format Conversion Services

For advanced users who can handle more complex data processing, it is necessary to directly work on data products in their native formats. All datasets at GES DSIC can be searched and downloaded from a newly designed web interface (NASA, 2018c). Compared to previous data ordering systems at GES DISC, significant efforts have been carried out to make sure all of the dataset-related services and information are available and accessible on the dataset landing page that serves as a one-stop shop.

The new GES DISC web interface (NASA, 2018c) provides a search capability for data collections, data documentation, product alerts, FAQs, glossary, how-to recipes, and more. For example, a search for "imerg" returns all of the IMERG related products. Each product has its own landing page that contains a product summary, dataset citation, documentation, and data access. A product summary basically lists basic information about the product such as version, format, spatial and temporal resolutions, etc. Dataset citation provides necessary information for citing the dataset in publications. The documentation section provides related technical documents in case one needs to know details about the product. Data access lists different tools and methods to access data.

Subsetting and format conversion services are essential. Users normally do not need a full set of parameters in a dataset and parameter subsetting helps reduce data volume for download. Because the spatial coverage of NASA data products is normally global, the spatial subsetting capability is also helpful for data reduction. Some users are not familiar with the Hierarchical Data Format (HDF) that is used in many NASA products and providing format conversion (to NetCDF, ASCII, etc.) will help these users to get their work done with ease.

The GES DISC has developed data subsetting services for popular precipitation datasets. This feature has been included in the GES DISC search and downloads. When "Subset/Get Data" icon in the dataset landing page is clicked on, a list of options is available for product subsetting, format conversion, and more.

24.3.4 OTHER WEB DATA SERVICES AND PROTOCOLS

Advanced users also prefer to work with popular community-based web services and protocols. Remote data access to web services and protocols can be embedded into processing software for research and operations, supporting machine-to-machine activities. Most data products at the GES DISC are also accessible via other web services and protocols (Table 24.5) including https (the data archive), OPeNDAP, WMS, GDS, etc. The https method provides direct access to product archives. OPeNDAP, WMS, and GDS (NASA, 2018g) provide remote access to the individual variables within datasets in a form usable by many tools and software packages such as IDV, McIDAS-V, Panoply, Ferret, GrADS, etc.

24.4 EXAMPLES

Since this book chapter is about satellite-based precipitation data for floods, examples in this section focus on using Giovanni to obtain the maps, plots, and data of the near-real-time IMERG Early product. The first two examples describe steps of generating maps and plots as well as downloading data from Giovanni. The last example shows steps to intercompare IMERG Early and Late products with Giovanni.

24.4.1 OBTAIN MAP OF NEAR-REAL-TIME IMERG EARLY PRECIPITATION ESTIMATES

With ~4-hour latency, the half-hourly IMERG Early dataset on a 0.1 deg. × 0.1 deg. grid can be used in flash flood operation and modeling activities. Giovanni can generate maps of IMERG Early precipitation estimates, including the time-averaged and time-accumulated maps for a rectangular area (user selected bounding box) or a shape (watersheds, US states, countries, etc.). The procedure in Giovanni is as follows,

1) Search "IMERG Early half-hourly" in the keyword search box;
2) Select "Multi-satellite precipitation estimate with climatological gauge calibration – Early Run" from the variable list;
3) Select area/country/state (US only)/watershed;

TABLE 24.5

Other Precipitation Data Web Services at GES DISC

Service	Description
GrADS Data Server	Stable, secure data server that provides subsetting and analysis services across the internet. The core of GDS is OPeNDAP (also known as DODS), a software framework used for data networking that makes local data accessible to remote locations.
OPeNDAP	The Open Source Project for a Network Data Access Protocol (OPeNDAP) provides remote access to individual variables within data sets in a form usable by many tools, such as IDV, McIDAS-V, Panoply, Ferret, and GrADS.
OGC Web Map Service	The Open Geospatial Consortium (OGC) Web Map Service (WMS) provides map depictions over the network via a standard protocol, enabling clients to build customized maps based on data coming from a variety of distributed sources.

Source: NASA, 2018g.

4) Select "Time Averaged Map" from Maps;
5) Select a time period for averaging;
6) Click on the "Plot Data" button.

Figure 24.3 is the rainfall-triggered flood event that is associated with the landfall of Hurricane Harvey in Houston, USA during August 25–28, 2017. Figure 24.3 contains the time-averaged rainfall map (Figure 24.3a) and the time-accumulated rainfall map (Figure 24.3b), respectively. It is seen that a large amount of rainfall caused the flood event in Houston. The data for downloads are available on the output page (Figure 24.3c). User registration is required for downloading all data and Giovanni provides the information and steps for facilitating registration.

24.4.2 TIME SERIES ANALYSIS OF PRECIPITATION IN A WATERSHED

Time series information is essential for flood research. With Giovanni, one can obtain time series plots and data for a single point, a rectangular area, watersheds, states in the United States, and countries. Meanwhile, users can use data rods to obtain time-series data in ASCII as mentioned earlier. The procedure of using Giovanni is,

1) Search "IMERG Early half-hourly" in the keyword search box;
2) For a single point: type in the geolocation information (latitude, longitude) in this format: west, south, east, north. For an area: select area/country/state (US only)/watershed;
3) Select Area-Averaged from "Time Series";
4) Select a time period for the time series;
5) Click on the "Plot Data" button.

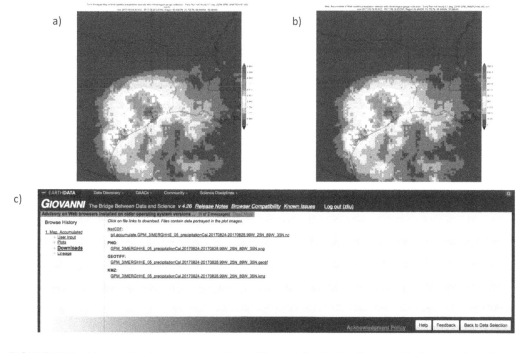

FIGURE 24.3 Maps and webpage generated from Giovanni for the flooding event in Houston due to heavy rainfall from Hurricane Harvey during August 24–28, 2017: (a) time-averaged rainfall map in mm/hr; (b) time-accumulated rainfall map in mm; and (c) webpage for downloading data.

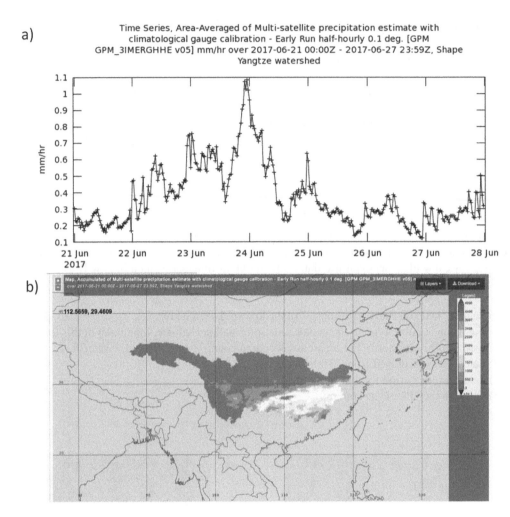

FIGURE 24.4 Time series plot (a) for the precipitation events in the Yangtze River watershed during June 21–27, 2017, and the accumulated rainfall (b) in the same watershed.

For example, Figure 24.4a is a time series plot of IMERG Early averaged over the Yangtze River watershed during June 21–27, 2017 when heavy floods were reported, showing several rainfall events during the period in the watershed. Figure 24.4b is the accumulated rainfall map of the watershed and it is seen that a large amount of rainfall is found in the southeastern part of the watershed. The time-series data, available on the output page, can be downloaded in CSV (Microsoft Excel) for further analysis.

24.4.3 Intercomparison of IMERG Early and Final Products

Due to input product availability, calibration, and more, differences exist in IMERG Early and Final products. With Giovanni, one can investigate their differences and several analysis functions are available such as map difference, scatter plots, etc. The procedure is,

1) Search "IMERG" from the keyword box;
2) Select both IMERG Early and Late from the variable list;
3) Select a region;
4) Select a time period;

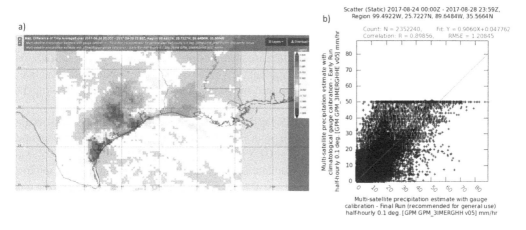

FIGURE 24.5 Intercomparison between IMERG Early and Final for the flooding event in Houston during August 24–28, 2017: (a) time-averaged difference (Final minus Early) and (b) their scatterplot.

5) Select a function for comparison (e.g., Difference of Time Averaged from Maps);
6) Click on "Plot Data."

Figure 24.5 is a difference map between the IMERG Early and Final products for the flood event in Houston in section 24.4.1. Their differences in Figure 24.5 can be summarized as: (a) differences are not evenly distributed (Figure 24.5a); (b) larger differences are found over land than over ocean, especially in high rain rate areas (Figure 24.5a); (c) positive differences (IMERG Final > Early) in high rain rates dominate over land (Figure 24.5a); (d) differences are clustered and not randomly distributed (Figure 24.5a); (e) the highest rain rates in the IMERG Early are limited below 50 mm/hr during this period (Figure 24.5b).

24.5 CONCLUSIONS

In this chapter, we introduce a list of global and regional precipitation products archived and distributed at the GES DISC, with emphasis on the near-real-time half-hourly 0.1-deg. IMERG precipitation products from the NASA fundamental GPM mission. The IMERG suite contains three output products, "Early Run" (lag time: ~4 hours), "Late Run" (lag time: ~12 hours), and "Final Run" (lag time: ~3.5 months) along with additional new input and intermediate files. The retro-processing of the IMERG product suite back to the TRMM era has been finished.

Global and regional land data assimilation system data products include optimal fields of land surface states and fluxes. They are generated by ingesting satellite- and ground-based observational data products and using advanced land surface modeling and data assimilation techniques. Both forcing data and model results support water resources applications, numerical weather prediction studies, numerous water and energy cycle investigations, and also serve as a foundation for interpreting satellite and ground-based observations. Spatial and temporal resolutions vary with products.

The MERRA-2 has been developed at the NASA GMAO at the NASA Goddard Space Flight Center. MERRA-2 provides the global data beginning in 1980 and runs a few weeks behind real-time. Alongside the meteorological data assimilation using a modern satellite database, MERRA-2 includes an interactive analysis of aerosols that feed back into the circulation, uses NASA's observations of stratospheric ozone and temperature (when available), and takes steps towards representing cryogenic processes. The MERRA project focuses on historical analyses of the hydrological cycle on a broad range of weather and climate time scales and places the NASA EOS suite of observations in a climate context. Observation-corrected precipitation data are available in MERRA-2.

The GES DISC also provides the data products from other science disciplines: solar irradiance, atmospheric composition and dynamics, global modeling, etc. Currently, there are over 2,700 unique data products archived at the GES DISC. More NASA Earth Observing products including NASA's EOS and other past NASA satellite mission data are available and distributed by the EOSDIS, with major facilities at the 12 DAACs located throughout the United States. Science disciplines at twelve DAACs include atmosphere, cryosphere, human dimensions, land, ocean, and calibrated radiance and solar radiance, etc.

The data services are available at the GES DISC to facilitate precipitation data access, evaluation, visualization, and analysis. Giovanni is a point-and-click web-based tool. One can access popular precipitation data and the other datasets at GES DISC without downloading data and software, making it very suitable for first-timers to evaluate and explore global and regional precipitations from different satellite missions and projects. A new GES DISC web-based search and download interface has been introduced. Users can search over 2,700 datasets archived at the GES DISC as well as their documents, FAQs, how-to recipes, etc. A dataset landing page offers a one-stop place for data access, dataset citation, technical documents, and more. Data subsetting and format conversion are essential for conducting research and applications. A user-friendly subsetting tool has been described. Advanced users can use different open-access methods (OPeNDAP, GDS, etc.) to remotely access precipitation data at GES DISC.

Several examples are given to demonstrate the basic data access with Giovanni such as the time-averaged maps and time series plots. A dataset comparison between the IMERG Early and Final has been presented. Although the analysis is pretty preliminary, users can quickly learn the basic characteristics of the different precipitation products in their areas of interest.

ACKNOWLEDGMENTS

We recognize the team effort of all past and current members at the GES DISC for their contributions to the development of data services and tools such as Giovanni. We extend our thanks to data set algorithm developers and many users for their feedback and suggestions. The GES DISC is funded by NASA's Science Mission Directorate (SMD).

REFERENCES

Acker, J.G., and Leptoukh, G. 2007. Online analysis enhances use of NASA earth science data, *Eos. Trans. Amer. Geophys. Union*, 88(2): 14–17.
Adler, R.F., Huffman, G.J., Chang, A., Ferraro, R., Xie, P., Janowiak, J., Rudolf, B., Schneider, U., Curtis, S., Bolvin, D., Gruber, A., Susskind, J., Arkin, P., and Nelkin, E. 2003. The version 2 global precipitation climatology project (GPCP) monthly precipitation analysis (1979-present), *J. Hydrometeor.*, 4: 1147–1167.
AMS. 2018. A special collection for MERRA-2, available online, https://journals.ametsoc.org/topic/merra-2, last accessed: June 30, 2018.
Aonashi, K., Awaka, J., Hirose, M., Kozu, T., Kubota, T., Liu, G., Shige, S., Kida, S., Seto, S., Takahashi, N., and Takayabu, Y.N. 2009. GSMaP passive, microwave precipitation retrieval algorithm: Algorithm description and validation, *J. Meteor. Soc. Japan*, 87A: 119–136.
Behrangi, A., Hsu, K.-L., Imam, B., Sorooshian, S., Huffman, G.J., and Kuligowski, R.J. 2009. PERSIANN-MSA: A precipitation estimation method from satellite-based multispectral analysis, *J. Hydrometeor*, 10: 1414–1429, http://doi.org/10.1175/2009JHM1139.1.
Bitew, M., Gebremichael, M., Ghebremichael, L.T., and Bayissa, Y.A. 2012. Evaluation of high-resolution satellite rainfall products through streamflow simulation in a hydrological modeling of a small mountainous watershed in Ethiopia, *J. Hydrometeor*, 13: 338–350, http://doi.org/10.1175/2011JHM1292.1.
Bosilovich, M., Akella, S., Coy, L., Cullather, R., Draper, C., Gelaro, R., Kovach, R., Liu, Q., Molod, A., Norris, P., Wargan, K., Chao, W., Reichle, R., Takacs, L., Vikhliaev, Y., Bloom, S., Collow, A., Firth, S., Labow, G., Partyka, G., Pawson, S., Reale, O., Schubert, S., and Suarez, M. 2015. *Technical Report Series on Global Modeling and Data Assimilation*, Vol. 43, R. Koster (Ed.), Available online: http://gmao.gsfc.nasa.gov/pubs/tm/docs/Bosilovich803.pdf

Courtney, C. 2018. *The Nature of Disaster in China: The 1931 Yangzi River Flood*, Cambridge University Press, UK, ISBN 978-1-108-41777-8.

Daly, C., Neilson, R.P., and Phillips, D.L. 1994. A statistical-topographic model for mapping climatological precipitation over mountainous terrain, *J. Appl. Meteor.*, 33: 140–158, http://doi.org/10.1175/1520-04 50(1994)033<0140:ASTMFM>2.0.CO;2.

Engel, T., Fink, A.H., Knippertz, P., Pante, G., and Bliefernicht, J. 2017. Extreme precipitation in the West African cities of Dakar and Ouagadougou-atmospheric dynamics and implications for flood risk assessments, *J. Hydrometeor.*, 18: 2937–2957.

Gallaher, D., and Grant, G. 2012. Data rods: High speed, time-series analysis of massive cryospheric data sets using pure object databases. In: *Geoscience and Remote Sensing Symposium (IGRASS)*, 22–27 July 2012, Available: http://ieeexplore.ieee.org/xpl/articleDetails.jsp?reload=true&arnumber=6352413 &contentType=Conference+Publications.

GDS. 2018. The GrADS data server (GDS), available online, http://cola.gmu.edu/grads/gds.php, last accessed, June 30, 2018.

Gelaro, R., and Coauthors. 2017. The modern-era retrospective analysis for research and applications, version 2 (MERRA-2), *J. Climate*, 30: 5419–5454, https://doi.org/10.1175/JCLI-D-16-0758.1.

Gianotti, R.L., Zhang, D., and Eltahir, E.A.B. 2012. Assessment of the regional climate model version 3 over the maritime continent using different cumulus parameterization and land surface schemes, *J. Climate*, 25: 638–656, http://dx.doi.org/10.1175/JCLI-D-11-00025.1.

Gourley, J.J., Hong, Y., Flamig, Z.L., Wang, J., Vergara, H., and Anagnostou, E.N. 2011. Hydrologic evaluation of rainfall estimates from Radar, Satellite, Gauge, and Combinations on Ft. Cobb Basin, Oklahoma, *J. Hydrometeor.*, 12: 973–988, http://doi.org/10.1175/2011JHM1287.1.

Higgins, R.W., Shi, W., Yarosh, E., and Joyce, R. 2000. *Improved United States Precipitation Quality Control System and Analysis*. NCEP/Climate Prediction Center Atlas No. 7, USA.

Hong, Y., Gochis, D., Cheng, J., Hsu, K.-L., and Sorooshian, S. 2007. Evaluation of PERSIANN-CCS rainfall measurement using the NAME event rain gauge network, *J. Hydrometeor.*, 8: 469–482, http://doi.org /10.1175/JHM574.1.

Hou, A.Y., Kakar, R.K., Neeck, S., Azarbarzin, A., Kummerow, C.D., Kojima, M., Oki, R., Nakamura, K., and Iguchi, T. 2014. The global precipitation measurement mission, *Bull. Amer. Meteor. Soc.*, 95: 701–722, http://doi.org/10.1175/BAMS-D-13-00164.1.

Huffman, G.J. 2017. The transition in multi-satellite products from TRMM to GPM (TMPA to IMERG), Available at, https://pmm.nasa.gov/sites/default/files/document_files/TMPA-to-IMERG_transition _170810.pdf. Assessed on June 19, 2018.

Huffman, G.J., Adler, R.F., Bolvin, D.T., and Gu, G. 2009. Improving the global precipitation record: GPCP version 2.1, *Geophys. Res. Lett.*, 36: L17808, http://doi.org/10.1029/2009GL040000.

Huffman, G.J., Adler, R.F., Bolvin, D.T., Gu, G., Nelkin, E.J., Bowman, K.P., Hong, Y., Stocker, E.F., and Wolff, D.B. 2007. The TRMM multi-satellite precipitation analysis: Quasi-global, multi-year, combined-sensor precipitation estimates at fine scale, *J. Hydrometeor.*, 8(1): 38–55.

Huffman, G.J., Adler, R.F., Bolvin, D.T., and Nelkin, E.J. 2010. The TRMM multi - satellite precipitation analysis (*TAMPA*). Chapter 1 in *Satellite Rainfall Applications for Surface Hydrology*, F. Hossain and M. Gebremichael (Eds.), Springer Verlag, ISBN: 978-90-481-2914-0, 3–22.

Huffman, G.J., Behrangi, A., Bolvin, D.T., and Nelkin, E.J. 2020. *GPCP Version 3.1 Satellite-Gauge (SG) Combined Precipitation Data Set*, edited by Huffman, G.J., A. Behrangi, D.T. Bolvin, E.J. Nelkin, NASA GES DISC, Greenbelt, Maryland, USA, Accessed: June 24, 2021, http://doi.org/10.5067/ DBVUO4KQHXTK.

Huffman, G.J., and Bolvin, D.T. 2012. Real-time TRMM multi-satellite precipitation analysis data set documentation, Available online: ftp://trmmopen.gsfc.nasa.gov/pub/merged/V7Documents/3B4XRT_doc _V7.pdf. Accessed on 8 June 2014.

Huffman, G.J., and Bolvin, D.T. 2013. TRMM and other data precipitation data set documentation, Available online: ftp://meso-a.gsfc.nasa.gov/pub/trmmdocs/3B42_3B43_doc.pdf. Accessed on September 6, 2017.

Huffman, G.J., and Bolvin, D.T. 2014. TRMM and other data precipitation data set documentation, Available at: ftp://meso-a.gsfc.nasa.gov/pub/trmmdocs/3B42_3B43_doc.pdf. Accessed on September 6, 2017.

Huffman, G.J., Bolvin, D., Braithwaite, D., Hsu, K., Joyce, R., Kidd, C., Nelkin, E., Sorooshian, S., Tan, J., and Xie, P. 2017. IMERG algorithm theoretical basis document (ATBD), Available online: https://pmm.nasa .gov/sites/default/files/document_files/IMERG_ATBD_V4.6.pdf, Last accessed: September 6, 2017.

Joyce, R.J., Janowiak, J.E., Arkin, P.A., and Xie, P. 2004. CMORPH: A method that produces global precipitation estimates from passive microwave and infrared data at high spatial and temporal resolution, *J. Hydrometer.*, 5: 487–503.

Kearns, E. 2017. Improving access to open data through NOAA's big data project, Available online: https://bigdatawg.nist.gov/Day2_08_NIST_Big_Data-Kearns.pdf, Last Accessed, September 6, 2017.

Kirschbaum, D., Huffman, G., Skofronick-Jackson, G., Braun, S., Stocker, E., Garrett, K., Jones, E., Adler, R., Wu, H., McNally, A., and Zaitchik, B. 2017. NASA's remotely-sensed precipitation: A reservoir for applications users, *Bull. Amer. Meteor. Soc.*, 98: 1169–1184, http://doi.org/10.1175/BAMS-D-15-00296.1.

Kloub, N., Matouq, M., Krishan, M., Eslamian, S.S., and Abdelhadi, M. 2010. Monitoring of water resources degradation at al-azraq oasis, jordan using remote sensing and GIS techniques, *Int. J. Glob. Warm.*, 2(1): 1–16.

Kummerow, C., Barnes, W., Kozu, T., Shiue, J., and Simpson, J. 1998. The tropical rainfall measuring mission (TRMM) sensor package, *J. Atmos. Oceanic Technol.*, 15: 809–817, https://doi.org/10.1175/1520-0426(1998)015<0809:TTRMMT>2.0.CO;2.

Kummerow, C., Simpson, J., Thiele, O., Barnes, W., Chang, A.T.C., Stocker, E., Adler, R.F., Hou, A., Kakar, R., Wentz, F., et al. (2000). The status of the tropical rainfall measuring mission (TRMM) after two years in Orbit, *J. Appl. Meteorol.*, 39: 1965–1982, Bibcode:2000JApMe.39.1965K, http://doi.org/10.1175/1520-0450(2001)040<1965:TSOTTR>2.0.CO;2.

Liu, Z., and Acker, J. 2017. Giovanni: The bridge between data and science, *Eos*, 98: https://doi.org/10.1029/2017EO079299, Published on 24 August 2017.

Liu, Z., Ostrenga, D., Teng, W., and Kempler, S. 2012. Tropical rainfall measuring mission (TRMM) precipitation data and services for research and applications, *Bull. Amer. Meteorol. Soc.*, http://doi.org/10.1175/BAMS-D-11-00152.1.

Liu, Z., Ostrenga, D., Vollmer, B., et al. 2017. Global precipitation measurement mission products and services at the NASA GES DISC, *Bull. Amer. Meteorol. Soc.*, 98(3): 437–444, https://doi.org/10.1175/bams-d-16-0023.1.

Mahrooghy, M., Anantharaj, V.G., Younan, N.H., Aanstoos, J., and Hsu, K.-L. 2012. On an enhanced PERSIANN-CCS algorithm for precipitation estimation, *J. Atmos. Oceanic Technol.*, 29: 922–932, http://doi.org/10.1175/JTECH-D-11-00146.1.

NASA. 2018a. *NASA Earth Science Data and Information System Project*, Available online, https://earthdata.nasa.gov/about/esdis-project, Last Accessed, June 30, 2018.

NASA. 2018b. *Land Data Assimilation Systems*, Available online, https://ldas.gsfc.nasa.gov/gldas/, Last Accessed, June 30, 2018.

NASA. 2018c. *The GES DISC Web Portal*, Available online, https://disc.gsfc.nasa.gov/, Last Accessed, June 30, 2018.

NASA. 2018d. *Earthdata Search*, Available online, https://search.earthdata.nasa.gov/search, Last Accessed, June 30, 2018.

NASA. 2018e. EOSDIS Data Tools, Available online, https://earthdata.nasa.gov/earth-observation-data/tools, Last Accessed, June 30, 2018.

NASA. 2018f. *NASA GES DISC Giovanni*, Available online, https://giovanni.gsfc.nasa.gov/, Last Accessed, June 30, 2018.

NASA. 2018g. *Web Services at GES DIC*, Available online, https://disc.gsfc.nasa.gov/information/tools?title=OPeNDAP%20and%20GDS, Last Accessed, June 30, 2018.

OPeNDAP. 2018. *OPeNDAP – Advanced Software for Remote Data Retrieval*, Available online, https://www.opendap.org/, Last Accessed, June 30, 2018.

Prat, O.P., and Nelson, B.R. 2013. Precipitation contribution of tropical cyclones in the Southeastern United States from 1998 to 2009 using TRMM satellite data, *J. Climate*, 26: 1047–1062, http://doi.org/10.1175/JCLI-D-11-00736.1.

Reichle, R.H., and Liu, Q. 2014. *Observation-Corrected Precipitation Estimates in GEOS-5*. NASA/TM–2014-104606, Vol. 35.

Rienecker, M.M., Suarez, M.J., Gelaro, R., Todling, R., Bacmeister, J., Liu, E., Bosilovich, M.G., Schubert, S.D., Takacs, L., Kim, G.-K., Bloom, S., Chen, J., Collins, D., Conaty, A., da Silva, A., et al. 2011. MERRA: NASA's modern-era retrospective analysis for research and applications, *J. Climate*, 24: 3624–3648, http://doi.org/10.1175/JCLI-D-11-00015.1.

Rui, H., Strub, R., Teng, W.L., Vollmer, B., Mocko, D.M., Maidment, D.R., and Whiteaker, T.L. 2013. *Enhancing Access to and Use of NASA Earth Sciences Data via CUAHSI-HIS (Hydrologic Information System) and Other Hydrologic Community Tools*. In: *AGU Fall Meeting*, San Francisco, CA, USA, Dec. 9–13, 2013.

Rui, H., Teng, B., Strub, R., and Vollmer, B. 2012. *Data Reorganization for Optimal Time Series Data Access, Analysis, and Visualization*. In: *AGU Fall Meeting*, San Francisco, CA, USA, Dec. 3–7, 2012.

Simpson, J., Adler, R.F., and North, G.R. 1988. A proposed tropical rainfall measuring mission (TRMM) satellite, *Bull. Amer. Meteor. Soc.*, 69: 278–295, http://doi.org/10.1175/1520-0477(1988)069<0278:APTRMM>2.0.CO;2.

Skofronick-Jackson, G., Petersen, W.A., Berg, W., Kidd, C., Stocker, E.F., Kirschbaum, D.B., Kakar, R., Braun, S.A., Huffman, G.J., Iguchi, T., Kirstetter, P.E., Kummerow, C., Meneghini, R., Oki, R., Olson, W.S., Takayabu, Y.N., Furukawa, K., and Wilheit, T. 2017. The global precipitation measurement (GPM) mission for science and society, *Bull. Amer. Meteor. Soc.*, 98: 1657–1672, http://doi.org/10.1175/BAMS-D-15-00306.1.

Sorooshian, S., Hsu, K.-L., Gao, X., Gupta, H.V., Imam, B., and Braithwaite, D. 2000. Evaluation of PERSIANN system satellite–based estimates of tropical rainfall, *Bull. Amer. Meteor. Soc.*, 81: 2035–2046, https://doi.org/10.1175/1520-0477(2000)081<2035:EOPSSE>2.3.CO;2.

Su, F., Gao, H., Huffman, G.J., and Lettenmaier, D.P. 2011. Potential utility of the real-time TMPA-RT precipitation estimates in streamflow prediction, *J. Hydrometeor.*, 12: 444–455, http://doi.org/10.1175/2010JHM1353.122.

Suarez, M., and Bacmeister, J. 2015. Development of the GEOS-5 atmospheric general circulation model: Evolution from MERRA to MERRA2, *Geosci. Model Dev.*, 8: 1339–1356, http://doi.org/10.5194/gmd-8-1339-2015.

Tan, M.L., and Duan, Z. 2017. Assessment of GPM and TRMM precipitation products over Singapore, *Remote Sens.*, 9: 720.

Tekeli, A.E. 2017. Exploring jeddah floods by tropical rainfall measuring mission analysis, *Water*, 9: 612.

Teng, W., Rui, H., Strub, R., and Vollmer, B. 2016. Optimal reorganization of NASA earth science data for enhanced accessibility and usability for the hydrology community, *Journal of American Water Resources Association (JAWRA)*, 1–11, http://doi.org/10.1111/1752-1688.12405.

Wu H., Adler, R.F., Hong, Y., Tian, Y., and Policelli, F. 2012. Evaluation of global flood detection using satellite-based rainfall and a hydrologic model, *J. Hydrometeor.*, 13: 1268–1284.

Wu, H., Adler, R.F., Tian, Y., Huffman, G.J., Li, H., and Wang, J. 2014. Real-time global flood estimation using satellite-based precipitation and a coupled land surface and routing model, *Water Resour. Res.*, 50: 2693–2717, http://doi.org/10.1002/2013WR014710.

Part VIII

Flood Soft Computing

25 Real-Time Flood Hydrograph Predictions Using the Rating Curve and Soft Computing Methods

Gokmen Tayfur

CONTENTS

25.1 INTRODUCTION

Flood routing is a means to obtain flow depths, velocities, volumes, and discharges at a river section. When a flood wave enters a river through an upstream section, by the flood routing method(s), one can trace the movement of flood wave along a channel length, and thereby he/she can calculate flood hydrograph at any downstream section of the river. This information is needed for designing flood control structures, such as levees, and also for channel improvements, navigation, and assessing flood effects.

There are basically two flood routing methods: (1) hydraulic and (2) hydrologic. Hydraulic methods are based on numerical solutions of St.Venant equations of continuity and momentum. They can handle the lateral flow contributions. Hydrologic methods are, on the other hand, based solely on the conservation of mass principle. Both methods require substantial field data, such as cross-sectional surveying, roughness, flow depth, and velocity measurements that are costly and time-consuming. When the lateral flow comes into the picture, the hydrologic model needs to be modified to handle such a case which becomes often problematic.

With the first applications of ANN and GA in hydrology in the late 1990s, researchers have taken advantage of these methods for the purpose of flood routing in natural channels, such that these methods have overcome some of the major problems of the existing ones.

DOI: 10.1201/9780429463938-33

25.2 FLOOD ROUTING METHODS

25.2.1 Hydraulic Flood Routing

It is based on the numerical solutions of the St.Venant equations (Chaudhry, 1993):

$$\frac{\partial A}{\partial t} + \frac{\partial Q}{\partial x} = q_l \tag{25.1}$$

$$\frac{\partial Q}{\partial t} + \frac{\partial (Qu)}{\partial x} + gA\frac{\partial h}{\partial x} = gA\left(S_o - S_f\right) \tag{25.2}$$

where A is the cross-sectional area (L^2), Q is the flow rate (L^3T^{-1}), q_l is the unit lateral flow (L^2T^{-1}), g is the gravitational acceleration (LT^{-2}), u is the flow velocity (LT^{-1}), S_o is the channel bed slope (LL^{-1}), and S_f is the friction slope (energy gradient) (LL^{-1}). There are simplified versions of these equations such as the kinematic wave and the diffusion wave. The simplifications are done in the momentum equation (25.2). In the case of the kinematic wave approximation (KWA), all the inertia and the depth gradient terms are neglected, i.e., only the slope terms (i.e., $S_o - S_f = 0$) are preserved. In the diffusion wave approximation (DWA), in addition to the slopes terms, the depth gradient term $\left(\text{i.e., } \frac{\partial h}{\partial x} = S_o - S_f\right)$ is also conserved. Friction slope term can be related to flow discharge by using either the Manning $\left(Q = \frac{1}{n}AR^{2/3}S_f^{1/2}\right.$, where n is the Mannings roughness coefficent$\left.\right)$ or Chezy $\left(Q = C_z AR^{1/2}S_f^{1/2}\right.$, where C_z is the roughness coefficent$\left.\right)$ relations (Henderson, 1966). In both roughness relations, R is the hydraulic radius (L), which is the ratio of the flow cross-sectional area to the wetted perimeter of the channel. KWA is generally employed when flow occurs in steep channels, i.e., $S_o \geq 2\%$ (Tayfur et al., 1993). DWA can produce results as good as the full St.Venant equations on mild slopes (Tayfur et al., 1993).

St.Venant equations are highly nonlinear and therefore they can be solved numerically by employing either the finite difference, finite volume, or finite element method. The numerical solutions might have convergence and stability problems (Chaudhry, 1993). Specifically, for flood wave routing, Lax, Lax-Wendroff, and MacCormack schemes were developed in the literature to handle the numerical solutions of the St. Venant equations (Chaudhry, 1993). The solutions of these equations can give, at any time at any section of a river; the flow depth, flow velocity, and flow rate provided that roughness, slope, and lateral flow information is provided accurately. It can allow the variability in roughness, slope, and lateral flow along a channel length. It is quite comprehensive and physics-based and therefore when possible they should be preferred.

25.2.2 Hydrologic Flood Routing

It is based on the conservation of mass, i.e., "mass in minus mass out is equal to change in mass per unit time." This, without lateral flow contribution, can be expressed as follows (Henderson, 1966):

$$Q_{in}(t) - Q_{out}(t) = \frac{dS(t)}{dt} \tag{25.3}$$

where $Q_{in}(t)$ is the inflow discharge at time t (L^3T^{-1}), $Q_{out}(t)$ is the outflow discharge at time t (L^3T^{-1}), and S is the storage (L^3). When equation (25.3) is averaged over a Δt time interval, the equation becomes (Henderson, 1966);

$$\frac{Q_{in}(t+\Delta t) + Q_{in}(t)}{2} - \frac{Q_{out}(t+\Delta t) + Q_{out}(t)}{2} = \frac{S(t+\Delta t) - S(t)}{\Delta t} \tag{25.4}$$

When storage is computed as a function of only outflow using the critical flow assumption, the routing method is called "the storage flood routing method" and it is employed for flood routing in reservoirs. For hydrologic flood routing in rivers, the Muskingum method is employed. It is based on equation (25.4) and relates the storage as a function of inflow and outflow rates, as follows (Henderson, 1966):

$$S(t) = K\left[xQ_{in}(t) + (1-x)Q_{out}(t)\right]$$
(25.5)

where K and x are the parameters. K is considered as a wave travel time thus it has a unit of time and x is called the storage coefficient and it is generally taken as 0.2, although there are several methods in the literature to accurately compute x for a given river reach (Henderson, 1966).

Gill (1978) advocated an alternative approach for accounting nonlinearity in the channel routing process by modifying the storage equation of the classical Muskingum method. He replaced the linear storage equation of the classical Muskingum method with the nonlinear storage equation of the following form:

$$S(t) = K\left[xQ_{in}(t) + (1-x)Q_{out}(t)\right]^m$$
(25.6)

where m is the exponent of the weighted discharge. In addition to the two parameters (K and x) employed in the classical Muskingum method, Gill's storage equation employs a third parameter in the form of nonlinear exponent m. This method is called "the nonlinear Muskingum (NLM) method." When $m = 1$, the storage equation of the NLM method reduces to the storage equation of the classical Muskingum method.

Perumal and Price (2013) advanced the NLM by varying the parameters (K, x, and m) at every routing time interval based on the hydrodynamic principles. They called the method "the Variable Parameter McCarthy-Muskingum (VPMM) method." Perumal et al. (2013) evaluated and compared the channel routing performance of a physically based variable parameter McCarty-Muskingum (VPMM) method proposed by Perumal and Price (2013), and the NLM method (Mohan, 1997). They concluded that the use of the VPMM method, which is capable of accounting for the nonlinear behavior of flood wave movement process without involving the model calibration process, is more reliable for field applications than the conceptually based NLM method which performs relatively poorly in calibration as well as in verification modes of the routing process.

25.2.3 Rating Curve Method

Moramarco et al. (2005) developed the rating curve method (RCM) which can handle the lateral flow contributions. They related discharge at the downstream station to measured flow variables at the upstream station as follows:

$$Q_d(t) = \alpha \frac{A_d(t)}{A_u(t - T_L)} Q_u(t - T_L) + \beta$$
(25.7)

where Q_u is the upstream discharge(L^3T^{-1}), Q_d is the downstream discharge (L^3T^{-1}), A_d and A_u are the effective downstream and upstream cross-sectional flow areas (L^2) obtained from the observed stages, respectively; T_L is the wave travel time depending on the wave celerity, c; and α and β are the model parameters that can be estimated as (Moramarco and Singh, 2001):

$$Q_d(t_b) = \alpha \frac{A_d(t_b)}{A_u(t_b - T_L)} Q_u(t_b - T_L) + \beta$$
(25.8)

$$Q_d(t_p) = \alpha \frac{A_d(t_p)}{A_u(t_p - T_L)} Q_u(t_p - T_L) + \beta \tag{25.9}$$

where $Q_d(t_b)$ is the base flow rate at the downstream section (L^3T^{-1}), $Q_d(t_p)$ is the peak discharge at the downstream section (L^3T^{-1}), t_p and t_b are the times when the peak stage and baseflow occurs at the downstream section, respectively. In particular, t_b is assumed to be the time just before the start of the rising limb of the hydrograph.

Baseflow rate $Q_d(t_b)$ can be computed from the velocity measurements during low flows. The peak discharge $Q_d(t_p)$ is surmised as the contribution of two main elements: (1) the upstream discharge delayed for the wave travel time T_L, $Q_u(t_p - T_L)$ with its attenuation, Q^*, due to flood routing along the reach of length L; and (2) the lateral inflows, q_pL, during the time interval $(t_p - T_L, t_p)$ (Moramarco et al., 2005):

$$Q_d(t_p) = \left(Q_u(t_p - T_L) - Q^*\right) + q_p L \tag{25.10}$$

In equation (25.10), T_L is implicitly assumed as the time to match the rising limb and the peak region of the upstream and downstream dimensionless hydrographs. The flood attenuation (Q^*) is computed from the Price formula (Raudkivi, 1979). The lateral inflow contribution, q_pL, is obtained from the solution of the characteristic form of the continuity equation (Moramarco et al., 2005). q_p is estimated by assuming that along the characteristic corresponding to the downstream peak stage, the following relationship holds (Moramarco and Singh, 2000):

$$\frac{A_d(t_p) - A_u(t_p - T_L)}{T_L} = q_p \tag{25.11}$$

Once $Q_d(t_b)$ and $Q_d(t_p)$ are known, parameters α and β are obtained from the solution of equations 25.8 and 25.9.

The hydraulic and hydrologic methods briefly described above require substantial data such as the measurements of flow depth, flow velocity, and topographical surveying that are time-consuming and costly. When lateral contribution becomes significant, which is the general case in large river basins, the procedures become more complicated and the problem of parameter estimation emerges. For example, Franchini et al. (1999) developed a methodology based on a variable parameter Muskingum-Cunge model with a specific parameterization scheme. The application of this model is, however, complex and requires the estimation of nine parameters. The RCM model, for each event that occurs in the same river reach, has to determine the wave travel time and the model parameters. In other words, the model uses different values of model parameters and wave travel time for each event at each river reach. With the advent of the soft computing methods applications in hydrology in the last two decades, the researchers have tried to overcome the difficulties of the hydrologic and hydraulic flood routing methods, briefly summarized above, by utilizing especially the genetic algorithm (GA) and artificial neural network (ANN).

25.3 SOFT COMPUTING METHODS IN FLOOD ROUTING

25.3.1 GENETIC ALGORITHM (GA)

The genetic algorithm (GA) was first introduced by Holland (1975) who was inspired by the biological processes of natural selection (i.e., Darwinian evolutionary view), and the genetic operations (i.e., Mendelian genetics). The first engineering application of GA by Goldberg (1983) triggered the use of it in different disciplines, including hydraulic, hydrology, and water resources engineering.

25.3.1.1 GA Basics

GA has four basic units of *bit, gene, chromosome*, and *gene pool*. The *bit* is the basic building block which is represented by a digit of 1 or 0. The combination of bits forms a *gene* representing a model parameter. The attachment of genes forms a *chromosome* standing for a possible solution. The *gene pool* is formed by many individual chromosomes (Tayfur, 2012).

The GA algorithm has five basic operations of *generation of the initial gene pool, calculation of fitness for each chromosome, selection of chromosomes, operation of cross-over*, and *mutation* (Tayfur, 2012). Uniform distribution or a normal distribution can be employed to randomly generate initial chromosomes for the gene pool (Sen, 2004). The fitness of each chromosome can be evaluated by first substituting each chromosome into the objective function to find their values $f(C_i)$ and then obtaining their fitness by equation 25.12 (Sen, 2004) as follows:

$$F\left(C_i\right) = \frac{f(C_i)}{\sum_{i=1}^{N} f\left(C_i\right)} \quad (25.12)$$

where N is the number of chromosomes in the gene pool, C_i is the chromosome i, $f(C_i)$ is the value of the objective function for chromosome i, and $F(C_i)$ is the fitness value for chromosome i.

The selection of chromosomes after the evaluation of their fitness values can be performed randomly. There are methods available for the selection process such as roulette wheel and ranking (Sen, 2004). After the selection process, pairs (parent chromosomes) are first formed and they are then subjected to the cross-over operation by interchanging the genes. The last operation in a single iteration is the mutation by which bits are reversed (i.e., *1 to 0* or *0 to 1*).

By these operations, it intends to search the solution space thoroughly. Figure 25.1 shows an example for crossover and mutation operations. As seen, the first two chromosomes (parent chromosome I and parent chromosome II) are subjected to the cross-over by the single cut from the third digit on the left, yielding new chromosomes (off-spring I and off-spring II) at the bottom. The values of 186 and 107 become 122 and 171, respectively after the crossover operation. By the mutation operation, the fourth digits from the left of the off-springs are reversed. Thus, the final version of the off-springs becomes 106 and 187, respectively at the bottom. By the basic GA operations, a large

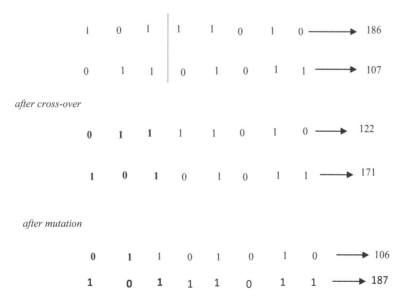

FIGURE 25.1 Example for crossover and mutation operations.

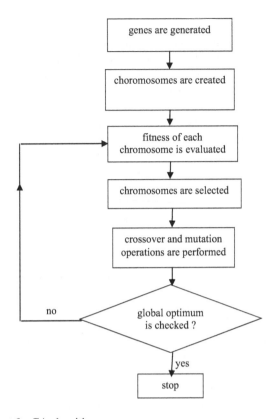

FIGURE 25.2   Flow chart for GA algorithm.

portion of solution space is searched to reach the global optimum. Figure 25.2 shows the flowchart on how the GA algorithm works. One can find more details on GA elsewhere (Goldberg, 1999; Sen, 2004; Tayfur, 2012).

25.3.1.2   GA-RCM Model for Real-Time Flood Hydrograph Prediction

The model employs the physics-based formulation (equation 25.7) of the RCM and finds its parameters using basic processes of genetic algorithms. Unlike the RCM, the GA-RCM model, for each river reach, employs an average wave travel time and finds a single set of average values for the model parameters. Tayfur et al. (2009) employed the rating curve method (RCM) (equation 25.7) as a basis to apply GA for hydrograph predictions. The GA simply finds the optimal values of the coefficients (α and β) of the RCM by minimizing the MAE function.

$$MAE = \frac{1}{N} \sum_{i=1}^{N} |Q_m - Q_p| \tag{25.13}$$

where N is the number of observations, Q_m is the measured flow discharge (L³T⁻¹), and Q_p is the predicted flow discharge (L³T⁻¹), which is $Q_d(t)$ computed by equation (25.7). They called their model GA-RCM. They tested their model on three equipped river reaches of the Upper Tiber River in Central Italy (Figure 25.3).

They considered severe storm events that occurred in the three different reaches for GA-RCM model calibration and testing. For the sake of brevity, only events that occurred in one of the river reaches, namely the reach in between Santa Lucia and Ponte Felcino, is presented herein. The main properties of the selected flood events are summarized in Table 25.1. The wave travel time is about

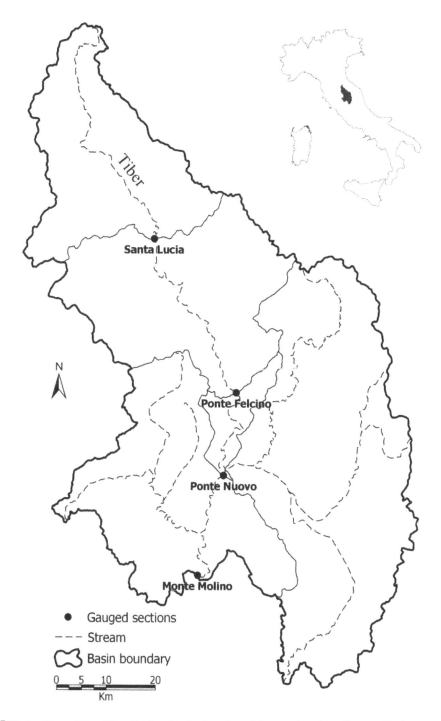

FIGURE 25.3   Upper Tiber River Basin with the location of the gauging sites. (Source: Tayfur et al., 2009.)

four hours and α and β parameter values used by the RCM model for each event are also shown in Table 25.1.

In the GA modeling, Tayfur et al. (2009) employed 100 chromosomes in the initial gene pool, 75% cross-over rate, 5% mutation rate, and 10,000 iterations. The range for α was constrained in [–5, 5] while β was in [–50, 50] for each iteration. Four events marked by * in Table 25.1 were used

TABLE 25.1

Main Characteristics of Flood Events Observed at Santa Lucia and Ponte Felcino Stations

	Santa Lucia Station			Ponte Felcino Station				
Event	Q_b (m³s⁻¹)	Q_p (m³s⁻¹)	V (10⁶ m³)	Q_b (m³s⁻¹)	Q_p (m³s⁻¹)	V (10⁶ m³)	α	β
December 1990	9.0	419	48	5.0	404	57	1.2	−9.0
January 1994	36	108	8.0	51	241	18	1.7	−35
May 1995*	4.0	71.0	10	9.0	139	19	1.2	−2.9
January 1997*	18	120	24	36	225	52	1.3	−3.2
June 1997*	5.0	346	28	11	450	49	1.0	0.3
January 2003	24	40.0	13	49	113	38	1.7	−24
February 2004*	22	92.0	12	55	278	44	1.5	−13

for calibrating the model parameters by GA for the reach and their optimal values were obtained as $\alpha = 1.2$ and $\beta = -6.0$. Figure 25.4 presents simulations measured at the Felcino station by the GA-RCM model. As seen, the model shows good performance in capturing the trend, time to peak, and the peak rates of the hydrographs.

The percentage error in peak discharge and the error in time to peak for each event simulated by the GA-RCM and standard RCM models are summarized in Table 25.2. Note that, in the case of the peak rate, a negative error value indicates under-estimation whereas a positive value indicates overestimation. In the case of time to peak, the negative error value indicates an early rise in reaching the peak rate, while the positive value indicates delay. According to Table 25.2, the GA-RCM and RCM models make about 10% over-prediction error of the peak rate of event December 1990. The GA-RCM model predicts the peak rates of the other two events with less than 4% error, while RCM produces, on average, about 14%. Especially, the peak rate of January 1994 is almost exactly predicted by the GA-RCM model, while RCM over-predicts it with about 24% error. The time to peak for each event is exactly predicted by the GA-RCM model, while RCM has, on average, a 40 min delay (Table 25.2).

The GA-RCM model successfully simulates hydrographs at each river reach having different wave travel times and lateral inflows. It can closely capture trends, time to peaks, and peak rates. It outperforms the standard RCM, although using substantially less data. Tayfur et al. (2009) also performed sensitivity analysis by using low peak hydrographs in the calibration stage and predicting the high peak hydrographs. In a similar fashion, they employed shorter wave travel time events in the calibration stage and predicted longer wave travel time events. In all those analyses, they concluded that the GA-RCM model has a good extrapolation capability.

Tayfur and Moramorca (2008) took advantage of the GA optimization method by further proposing the following equations:

$$Q_d(t) = \alpha \frac{h_d^\beta(t)}{h_u^\gamma (t - T_L)} + \eta \tag{25.14}$$

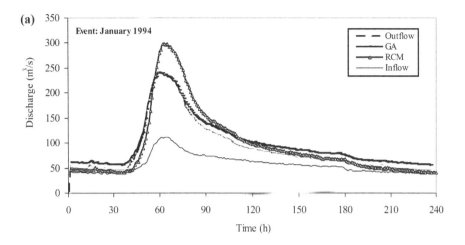

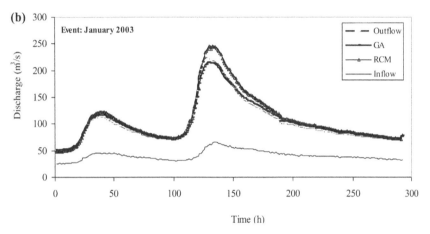

FIGURE 25.4 GA-RCM and RCM model simulations at Ponte Felcino station on (a) January 1994 and (c) January 2003. (Source: Tayfur et al., 2009.)

$$Q_d(t) = \alpha_1 h_u^{\beta_1}\left(t - T_L\right) + \alpha_2 h_d^{\beta_2}\left(t\right) + \eta \qquad (25.15)$$

where h_d is the flow depth at the downstream station, h_u is the flow depth at the upstream station, and α, β, γ, η, α_1, α_2, β_1, and β_2 are the model parameters, whose optimal values are found by the GA model. Equation (25.14) is based on the formulation of equation (25.7) and is called "GA-Stage I" herein. Here, only flow stage information is required. As opposed to the RCM, it does not require

TABLE 25.2

Percentage Errors in Peak Discharge (E_{Qp}) and Time to Peak (E_{Tp}) for the Events in Figure 25.4 Observed at Ponte Felcino Station

Event	E_{Qp} (%)		E_{Tp} (%)	
	GA-RCM	RCM	GA_RCM	RCM
January 1994	−0.6	23.5	0.0	8.3
January 2003	−3.8	9.70	0.0	2.3

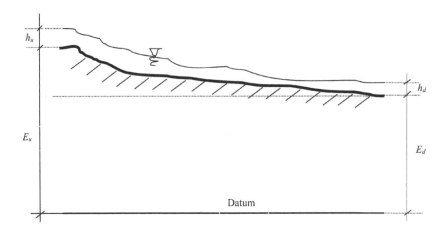

FIGURE 25.5 Schematic representation of water surface elevation from a reference datum (in equations 25.16 and 25.17): $E = (E_u - E_d) + h_u$. (Source: Tayfur and Moramarco, 2008.)

velocity, discharge, and cross-sectional information. From a data reduction point of view, this is a major advantage. Equation (25.15) is based on the kinematic wave approximation and is called "GA-Stage II" herein.

Tayfur and Moramarco (2008) also proposed a formulation that predicts flow rate at a downstream station using only water surface elevation data. The elevation consists of the elevation of a station from a reference datum plus flow depth as schematically presented in Figure 25.5. Considering the reference datum at the downstream station enables one to need measurement of only flow depth at the downstream station. They proposed the formulations as follows:

$$Q_d(t) = \alpha \frac{h_d^\beta(t)}{E^\gamma(t - T_L)} + \eta \tag{25.16}$$

$$Q_d(t) = \alpha_1 E^{\beta_1}(t - T_L) + \alpha_2 h_d^{\beta_2}(t) + \eta \tag{26.17}$$

where E is the elevation, as shown in Figure 25.5. Models, expressed by equation (25.16) and equation (25.17), are called herein "GA-Elevation I" and "GA-Elevation II" models, respectively.

These proposed four models were applied to predict real-time flood hydrographs in Tiber River reaches, shown in Figure 25.3. For each river reach, four events were employed to calibrate the model parameters. The calibrated parameter values were used in the related equations to make predictions of new events. For the sake of brevity, Figure 25.6 shows a prediction of single events at each river reach.

These results imply that one can construct a model in the form of equation (25.14) and then find the optimal values of the parameters of the equation using the GA model to perform hydrograph simulations. Such a model would require easily measurable flow stage data at upstream and downstream stations only. In a similar fashion, we can construct a model in the form of equation (25.16) and obtain the optimal values of the parameters of the equation by the GA to do hydrograph simulations. Such a model is even better since it would require only the elevation data of both stations. In a way, the elevation data can be easily obtained from satellite maps. This has an important implication for ungauged sites (Ghasemizade et al., 2011), where flow rates can be easily predicted.

It is realized that GAs are strong optimization methods, which can perform quite satisfactorily with substantially less but easily measurable data. This, in turn, has a positive economic implication. That means by this method, we can save time, labor, and money and do, at the same time, very efficient work.

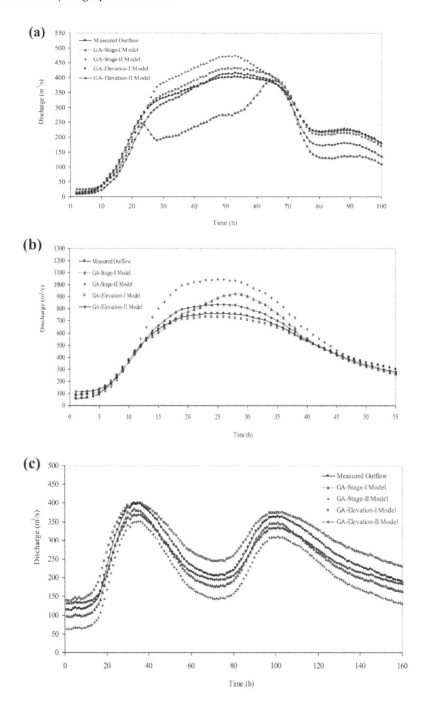

FIGURE 25.6 Hydrograph predictions (a) at the Lucia-Felcino River reach in January 1994, (b) at the Lucia-Nuovo River reach in February 1999, and (c) at the Lucia-Molino River reach in January 2001. (Source: Tayfur and Moramarco, 2008.)

25.3.2 Artificial Neural Network (ANN)

Artificial neural network (ANN) is inspired by the information processing of the brain. There are basically two aspects of how ANN is deduced from the nervous system of the brain. The first aspect is the physical deduction where the brain nervous system is overly simplified and resembled in an artificial network consisting of layers of neurons that process information. The second aspect is the learning ability of a brain through experience and experiments. This learning ability of a human being and his/her biological nervous system have inspired researchers to form a system made up of artificial layers containing neurons that process information and produce a system output. The whole system is called the artificial neural network. The developments in the training algorithms have made ANN be employed in many disciplines from finance to engineering to solve classification, prediction, forecasting, and optimization problems. These methods are very powerful in mapping inputs to outputs. They are very powerful to capture trends among several variables. Hence, they are commonly employed in solving highly nonlinear engineering problems, including the ones in the water resources engineering field.

ANN is attractive for discharge prediction and flood forecasting because they can accommodate the nonlinearity of the watershed runoff process and uncertainty in parameter estimation, have the capability to extract the relationship between input and output of the process without explicitly considering the physics of the process, find relationships between different input samples, and generalize a relationship from small subsets of data while remaining robust in the presence of noisy or missing input.

25.3.2.1 ANN Basics

In hydrologic applications, a three-layer-feedforward type of artificial neural network is commonly considered (Figure 25.7). In a feedforward network, the input quantities are fed into input layer neurons, which, in turn, pass them on to the hidden layer neurons after multiplication by a weight. A hidden layer neuron adds up the weighted input received from each input neuron, associates it with a bias, and then passes the result on through a non-linear transfer function. The output neuron does the same operation as does a hidden neuron.

The back-propagation algorithm finds the optimal weights by minimizing a predetermined error function (E) of the following form (Tayfur, 2012):

$$E = \sum_P \sum_p \left(y_i - t_i\right)^2 \qquad (25.18)$$

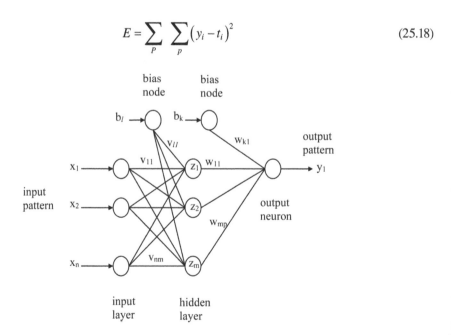

FIGURE 25.7 Representation of three-layer feed-forward ANN.

where y_i = component of a network output vector Y; t_i = component of a target output vector T; p = number of output neurons; and P = number of training patterns.

In the backpropagation algorithm, optimal weights would generate an output vector $Y = (y_1, y_2, \ldots, y_p)$ as close as possible to the target values of the output vector $T = (t_1, t_2, \ldots, t_p)$ with a selected accuracy level. The backpropagation algorithm employs the gradient-descent method, along with the chain rule of differentiation, to modify the network weights as (Tayfur, 2012):

$$v_{ij}^{new} = v_{ij}^{old} - \delta \frac{\partial E}{\partial v_{ij}} \qquad (25.19)$$

where v_{ij} = weight between ith neuron in a layer to the jth neuron in the neighbor layer; v_{ij}^{new} is the new (updated) value of the connection weight at current iteration, v_{ij}^{old} is the old (previous) value of the connection weight at previous iteration, and δ = learning rate.

The network learns by adjusting the biases and weights that link its neurons. Before training begins, a network's weights and biases are set equal to small random values. Also, due to the nature of the activation function (sigmoid function or tangent hyperbolic function) used in the back-propagation algorithm, all of the external input and output values before passing them into a neural network are accordingly standardized. Without standardization (compaction), large values input into an ANN would require extremely small weighting factors to be applied and this could cause a number of problems (Dawson and Wilby, 1998). The details of the compaction methods and ANN are available in the literature (Tayfur, 2012).

25.3.2.2 ANN for Real-Time Flood Hydrograph Prediction

Tayfur et al (2007) applied ANN to predict event-based flood hydrographs measured at different gauging stations in Upper Tiber Basin in central Italy (see Figure 25.3). Several accurate flow measurements were available which allowed the estimation of the rating curve for each section (Moramarco et al., 2005). Seven severe storm events were available and four events (June 1997, May 1995, January 1997, February 2004) were chosen by Tayfur et al. (2007) for training the ANN. They used the remaining three events for testing the model. The main properties of the selected flood events are summarized in Table 25.3. It is seen that the lateral inflow contribution was significant in some of the events.

TABLE 25.3
Main Characteristics of Observed Flood Events at Stations on the Tiber River

	Santa Lucia Station			Ponte Felcino Station			
Event	Q_b $(m^3 s^{-1})$	Q_p $(m^3 s^{-1})$	V $(10^6 m^3)$	Q_b $(m^3 s^{-1})$	Q_p $(m^3 s^{-1})$	V $(10^6 m^3)$	T_L (h)
December 1990	8.0	418	50	10	404	60	2.0
January 1994	36	108	19	51	241	35	3.0
May 1995*	4.0	71	10	9	139	19	4.0
January 1997*	18	120	24	36	225	52	3.5
June 1997*	5.0	346	28	11	450	49	5.0
January 2003	24	58	14	50	218	41	3.5
February 2004*	22	91	7	55	276	27	3.5

Notes: Q_b = base flow; Q_p = peak discharge; V = direct runoff volume; T_L = travel time.
* Used for ANN model training.

The river reach between Santa Lucia and Ponte Felcino gauging stations (Figure 25.3) was considered by Tayfur et al. (2007) for testing the models for flood prediction. The ANN model used flow stage data at Santa Lucia station (upstream station) and the flow stage data at Ponte Felcino station (downstream station) to predict the flow discharge at Ponte Felcino station. The travel time between the two stations is about four hours. As pointed out earlier, since the flow stage is an easily measurable variable, engineers usually tend to relate flow rate to flow stage as is the case in the rating curve method.

Tayfur et al. (2007) trained the ANN with a learning rate of 0.01 and 2000 iterations. The network had two neurons in the input layer, five neurons in the inner layer, and one neuron in the output layer. The number of neurons in the hidden layer was decided by the commonly employed trial and error procedure. For this purpose, the mean error (ME) and mean relative error (MRE) were used as error measures. Accordingly, the number of iterations that provided the minimum ME and MRE values were the stopping criteria for terminating the iterations. For example, for this particular problem of flood hydrograph prediction application, the values of the error measures started with $ME = 181.7$ m^3/s and $MRE = 63.4$ at the first iteration and rapidly decreased to 31.1 m^3/s and 32.4, respectively, after the 100th iteration. The errors then gradually decreased to and stabilized at $ME = 10.4$ m^3/s and $MRE = 16.5$ after 2,000 iterations.

Tayfur et al. (2007), then, employed the trained ANN and the linear RCM to predict hydrographs of the three testing events (December 1990, January 1994, January 2003) measured at the Ponte Felcino station (Table 25.3). Figure 25.8 shows the predicted hydrographs. It is seen that ANN satisfactorily predicted the hydrographs in terms of the overall trend, time to peak, and peak discharges. Overall, it yielded better results than did RCM which, in general, overestimated the discharge.

For the two-peak hydrograph of January 2003, shown in Figure 25.8c, ANN under-predicted the lower peak but closely captured the higher peak, whereas RCM better predicted the lower peak but over-predicted the higher peak. The percentage error in peak discharge and time to peak was computed for each event and is given in Table 25.4. Note that a negative error value indicates underestimation, whereas a positive value indicates overestimation. ANN predicted the peak discharge of each event with less than a 5% error, while RCM had more than 10% error. For the January 2003 event, RCM over-predicted the peak discharge with about 24% error, while ANN had a 5% error. The time to peak was exactly predicted by ANN, while RCM had about a 4% error.

Tayfur et al. (2007) carried out a sensitivity analysis by considering the flow stage at the upstream station as the only input variable. The results revealed the poor performance of the network, especially for a river reach receiving significant lateral flow. Also, they investigated the extrapolation capability of the network by considering low peak hydrographs in the input vector. The results revealed that ANN is not a good extrapolator. They cannot predict high peak hydrographs when they are trained with low peak ones.

ANN, by using only flow stage data that is easily measurable, can make good predictions of real flood hydrographs. This is an advantage over the RCM model which requires measurements of cross-sections and flow velocities, in addition to the flow stage. Also, the RCM employs different travel times and tries to find values of α and β for each event, even for the same river reach.

25.4 CONCLUSIONS

Flood routing has a long history of more than a century. The original methods are based on the conservation of mass and applied to flood routing in reservoirs and in river reaches. With the advent of computers, the St. Venant equations were solved in their simplest form. As the computers were developed, so as the numerical solutions of the St. Venant equations, leading to very comprehensive numerical flood routing methods. For these models to be very effective, substantial data and parameter estimation are required, which is often problematic and costly, apart from the numerical convergence and instability problems. Engineers tend to use simple but problem solver models. For

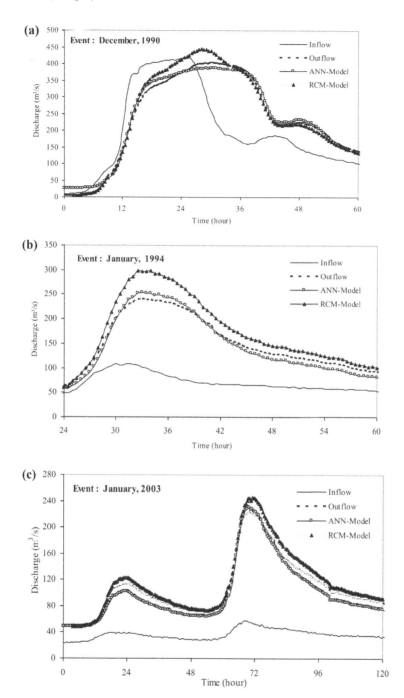

FIGURE 25.8 ANN and RCM model simulations of flood hydrographs measured at Ponte Felcino gauging station in (a) December 1990; (b) January 1994; and (c) January 2003. (Source: Tayfur et al., 2007.)

that reason, with the advent of soft computing methods such as ANN and GA, they were eager to employ them in flood routing.

ANN is a black box model optimization algorithm. Given a pair of input-output data, it is trained by finding optimal values for the connection weights between neurons located in neighboring layers. It hence cannot reveal any information or yield any mathematical relation between input variables

TABLE 25.4

Percentage Errors in Peak Discharge, E_{Qp}, and Time to Peak, E_{Tp}

Event	E_{Qp} (%)		E_{Tp} (%)	
	ANN	RCM	ANN	RCM
Dec 1990	−4	10	0	−6.7
Jan 1994	3	9	0	2.2
Jan 2003	5	24	0	3.0
Average	4.2	14.3	0	4.0

and the output variable. They are strong interpolators but not extrapolators. They need to be trained with the flood hydrograph events for each river reach separately. Despite all these shortcomings, they are powerful to yield answers in a short period of time, which may be crucial for authorities for handling flood situations.

GA is an optimization algorithm, which maximizes or minimizes an objective function under some constraints. It often finds optimal values of some parameters of mathematical equations. Therefore, the relation between input and output variables is given mathematically. Hence, we can say that GA is not a black-box model, as ANN. Since there is mathematical relation, it has both interpolation and extrapolation capabilities. One more advantage of GA is that one can propose a new equation, provided that it physically makes sense, and find its optimal values of the coefficients and exponents.

Soft computing methods (such as GA, ANN) can make good simulations of real-time flood hydrographs using substantially less data, such as the easily measurable flow stage. In that sense, they are quite advantages. The use of soft computing methods, together with physics-based models, would be the way forward for solving problems in the water resources engineering field, including flood routing. Also, it would be beneficial to compare the performances of these methods against those of other soft computing methods, such as the SVM (support vector machine) (Karahan et al., 2014), the HS (harmony search) (Perumal et al., 2017), the AC (ant colony) (Tayfur, 2017), the DE (differential evolution) and the PS (particle swarm) (Perumal et al., 2017).

ACKNOWLEDGMENTS

The author thanks the Department of Environment, Planning, and Infrastructure of Umbria Region for providing Tiber River data for the already published papers and the co-authors of the papers cited in this book chapter.

REFERENCES

Chaudhry, M.H. 1993. *Open-Channel Flow*, Prentice Hall, New Jersey, USA.

Dawson, W.C., and Wilby, R. 1998. An artificial neural network approach to rainfall-runoff modeling, *Hydrological Sciences J.*, 43(1): 47–66.

Franchini, M., Lamberti, P., and Di Giammarco, P. 1999. Rating curve estimation using local stages, upstream discharge data and a simplified hydraulic model, *Hydrology Earth Syst. Sci.*, 3(4): 541–548.

Ghasemizade, M., Mohammadi, K., and Eslamian, S.S. 2011. Estimation of design flood hydrograph for an ungauged watershed, *J. Flood Engineering*, 2(1): 27–36.

Gill, M.A. 1978. Flood routing by the Muskingum method, *J. Hydrol.*, 36: 353–363.

Goldberg, D.E. 1983. *Computer-Aided Gas Pipeline Operation Using Genetic Algorithms and Rule Learning.* Ph.D. Thesis, University of Michigan, Ann Arbor, MI, USA.

Goldberg, D.E. 1999. *Genetic Algorithms*, Addison-Wesley, USA.

Henderson, F.M. 1966. *Open Channel Flow*, MacMillan, New York.

Holland, J.H. 1975. *Adaptation in Natural and Artificial Systems*, University of Michigan Press, Michigan, USA.

Karahan, H., Iplikci, S., Yasar, M., and Gurarslan, G. 2014. River flow estimation from upstream flow records using support vector machines, *Journal of Applied Mathematics*, 2014: 7, http://dx.doi.org/10.1155/2014 /714213.

Mohan, S. 1997. Parameter estimation of nonlinear Muskingum models using genetic algorithm, *J. Hydraul. Engineering*, 123(2): 137–142.

Moramarco, T., and Singh, V.P. 2000. A practical method for analysis of river waves and for kinematic wave routing in natural channel networks, *Hydrological Processes*, 14: 51–62.

Moramarco, T., and Singh, V.P. 2001. Simple method for relating local stage and remote discharge, *J. Hydrol. Engrg., ASCE*, 6(1): 78–81.

Moramarco, M., Barbetta, S., Melone, F., and Singh, V.P. 2005. Relating local stage and remote discharge with significant lateral inflow, *J. Hydrol. Engrg., ASCE*, 10(1): 58–69.

Perumal, M., Tayfur, G., Rao, C.M., and Gurarslan, G. 2017. Evaluation of a physically based quasi-linear and a conceptually based nonlinear Muskingum methods, *Journal of Hydrology*, 546: 437–449.

Perumal, M., and Price, R.K. 2013. A fully volume conservative variable parameter McCarthy-Muskingum method: Theory and verification, *J. Hydrol.*, 502: 89–102.

Perumal, M., Naren, A., and Ch.Madhusudana, R. 2013. Appraisal of two forms of non-linear Muskingum flood routing methods. In: *6th International Perspective on Water Resources and the Environment*. January 7–9, Izmir, Turkey, Proceeding paper number: IPWE2013-304pdf.

Raudkivi, A.J. 1979. *Hydrology: An Advanced Introduction to Hydrological Processes and Modeling*, Pergamon, New York, UK.

Sen, Z. 2004. *Genetic Algorithm and Optimization Methods*. Su Vakfı Yayınları, Istanbul, Turkey. (in Turkish).

Tayfur, G. 2017. Modern optimization methods in water resources planning, engineering and management, *Water Resources Management*, 31(10): 3205–3233.

Tayfur, G. 2012. *Soft Computing in Water Resources Engineering: Artificial Neural Networks, Fuzzy Logic, and Genetic Algorithm*, WIT Press, Southampton, U.K.

Tayfur, G., Barbetta, S., and Moramarco, T. 2009. Genetic algorithm-based discharge estimation at sites receiving lateral inflows, *J. Hydrologic Engineering*, 14(5): 463–474.

Tayfur, G., and Moramarco, T. 2008. Predicting hourly-based flow discharge hydrographs from level data using genetic algorithms, *J. Hydrology*, 352(1–2): 77–93.

Tayfur, G., Moramarco, T., and Singh, V.P. 2007. Predicting and forecasting flow discharge at sites receiving significant lateral inflow, *Hydrol. Processes*, 21: 1848–1859.

Tayfur, G., Kavvas, M.L., Govindaraju, R.S., and Storm, D.E. 1993. Applicability of St.Venant equations for 2-Dimensional overland flows over rough infiltrating surfaces, *J. Hydraulic Engineering*, 119(1): 51–63.

26 Application of Integral Transforms in Flood Studies

Vahid Nourani, Mehran Dadashzadeh, and Saeid Eslamian

CONTENTS

26.1 INTRODUCTION

Mathematics has played and continues to play a critical role in expanding fields of water sciences and hydrology, so much so that mathematics is referred to as the language of hydrology. The functional aspect of mathematics is extensively applied to solve hydrological problems. Of these functional and significant tools that can be mentioned is integral transforms. Integral transforms have been successfully used for almost two centuries in solving many problems in applied mathematics and engineering science. The importance of the integral transforms is that they transform a difficult mathematical problem into a relatively easy problem, which can easily be solved. In addition, they provide robust operational methods for solving initial value problems and initial-boundary value problems for linear differential and integral equations. The integral transforms are also powerful and thus common tools for analytical and numerical investigations in a wide range of applications in the field of hydrology, such as climatic and hydrological time series analysis, rainfall-runoff modeling, streamflow modeling, groundwater modeling, etc.

DOI: 10.1201/9780429463938-34

26.1.1 INTEGRAL TRANSFORMS

The origin of the integral transforms, including the Laplace and Fourier transforms, can be traced back to the distinguished work of P. S. Laplace on probability theory in the 1780s and to the remarkable treatise of Joseph Fourier on the analytical theory of heat published in 1822. Laplace's classic book on *La Théorie Analytique des Probabilités* included some basic results of the Laplace transform which was one of the oldest and most commonly used integral transforms available in the mathematical literature. On the other hand, in his treatise, Fourier stated a remarkable result that is universally known as the Fourier integral theorem. In an attempt to extend his new ideas to functions defined on an infinite interval, he discovered the Fourier transform and the inverse Fourier transform.

There are many other integral transforms, including the Mellin transform, the Hankel transform, the Hilbert transform, and the wavelet transform, which are widely used to solve a broad range of problems in mathematics, science, and engineering. Although Mellin (1854–1933) presented an elaborate discussion of his transform and its inversion formula, it was G. Bernhard Riemann (1826–1866) who first recognized the Mellin transform and its inversion formula. Hermann Hankel (1839–1873) introduced the Hankel transform with the Bessel function as its kernel. Although the Hilbert transform was named after one of the greatest mathematicians of the 20th century, David Hilbert (1862–1943), this transform and its properties are basically studied by G. H. Hardy (1877–1947) and E. C. Titchmarsh (1899–1963). The original idea of wavelet transform belongs to Fourier with his theories of frequency analysis. In 1980, Grossman and Morlet provided a way of thinking for wavelets based on physical intuition. Then, Stephane Mallat gave wavelets an additional jump-start through his work in digital signal processing in 1985. Ingrid Daubechies used Mallat's work to construct a set of wavelet orthonormal basis functions that are perhaps the most elegant and have become the cornerstone of wavelet applications today.

The integral transform of a function $f(x)$ defined in $a \le x \le b$ is denoted by $\mathcal{I}\{f(x)\} = F(k)$, and defined by (Debnath and Bhatta, 2014):

$$\mathcal{I}\{f(x)\} = F(k) = \int_a^b K(x,k)f(x)dx, \tag{26.1}$$

where $K(x, k)$, given function of two variables x and k, is called the kernel of the transform. The operator $\mathcal{I}$ is usually called an integral transform operator or simply an integral transformation. The transform function $F(k)$ is often referred to as the image of the given object function $f(x)$, and k is called the transform variable.

As well, the integral transform of a function of several variables is defined by (Debnath and Bhatta, 2014):

$$\mathcal{I}\{f(x)\} = F(\kappa) = \int_S K(x,\kappa)f(x)dx, \tag{26.2}$$

where $x = (x_1, x_2, \ldots, x_n)$, $\kappa = (k_1, k_2, \ldots, k_n)$, and $S \subset \mathbb{R}^n$.

Evidently, there are a number of important integral transforms that are defined by choosing different kernels K(x, k) and different values for a and b involved in (26.2). In section 26.2, some of the most notable integral transforms are described.

26.2 BASIC DEFINITION AND PROPRETIES

26.2.1 LAPLACE TRANSFORM

Although the Laplace transform has been discovered by Pierre-Simon Laplace in the 19th century, it was the British electrical engineer Oliver Heaviside (1850–1925) who made the Laplace transform very popular by using it to solve ordinary differential equations of electrical circuits and systems, and then to develop modern operational calculus. However, Heaviside's treatment was not very systematic and lacked rigor which was later on attended to and recapitulated by Bromwich and Carson.

The Laplace Transform can be interpreted as a transformation from time domain where inputs and outputs are functions of time to the frequency domain where inputs and outputs are functions of complex angular frequency. It provides an alternative functional description that often simplifies the process of analyzing the behavior of the system, or in synthesizing a new system based on a set of specifications. Many problems of physical interest are described by ordinary or partial differential equations with appropriate initial or boundary conditions. These problems are usually formulated as initial value problems, boundary value problems, or initial-boundary value problems that seem to be mathematically more rigorous and physically realistic in applied and engineering sciences. The Laplace transform method is particularly useful for finding solutions to these problems. It can also be used effectively for evaluating certain definite integrals. The Laplace transform can be exploited as an effective tool for analyzing the basic characteristics of a linear system governed by the differential equation in response to initial data and/or to an external disturbance.

The Laplace transform of $f(t)$ is formally defined by (Debnath and Bhatta, 2014):

$$\mathcal{L}\left\{f\left(t\right)\right\} = \bar{f}\left(s\right) = \int_{0}^{\infty} e^{-st} f\left(t\right) dt, \ Res > 0, \tag{26.3}$$

where e^{-st} is the kernel of the transform and s is the transform variable which is a complex number.

The inverse Laplace transform is also given by the following complex integral (Debnath and Bhatta, 2014):

$$\mathcal{L}^{-1}\left\{\bar{f}\left(s\right)\right\} = f\left(t\right) = \frac{1}{2\pi i} \int_{c-i\infty}^{c+i\infty} e^{st} \bar{f}\left(s\right) ds, \ c > 0. \tag{26.4}$$

26.2.2 FOURIER TRANSFORM

The Fourier transform originated from the Fourier integral theorem that was stated in Fourier's treatise entitled *The Analytical Theory of Heat*, and its deep significance has subsequently been recognized by mathematicians and physicists. Fourier transforms is one of the most remarkable discoveries in the mathematical sciences and has widespread applications in mathematics, physics, and engineering. Many linear boundary values and initial value problems in applied mathematics, mathematical physics, and engineering science can be effectively solved by the use of the Fourier transform. This transform is also very useful for solving differential or integral equations.

In summary, the Fourier transform maps a function (or signal) of time t to a function of frequency ω. It generates a function (or signal) of a continuous variable whose value represents the frequency content of the original signal. Many applications, including the analysis of stationary signals and real-time signal processing, make effective use of the Fourier transform in time and frequency domains.

The Fourier transform of $f(x)$ is denoted by $\mathcal{F}\left\{f\left(x\right)\right\} = F\left(k\right), k \in \mathbb{R}$, and defined by the integral (Debnath and Bhatta, 2014):

$$\mathcal{F}\left\{f\left(x\right)\right\} = F\left(k\right) = \frac{1}{\sqrt{2\pi}} \int_{-\infty}^{\infty} e^{-ikx} f\left(x\right) dx, \tag{26.5}$$

and the inverse Fourier transform, denoted by $\mathcal{F}^{-1}\left\{F\left(k\right)\right\} = f\left(x\right)$, is defined by (Debnath and Bhatta, 2014):

$$\mathcal{F}^{-1}\left\{F\left(k\right)\right\} = f\left(x\right) = \frac{1}{\sqrt{2\pi}} \int_{-\infty}^{\infty} e^{ikx} F\left(k\right) dk, \tag{26.6}$$

The remarkable success of the Fourier transform analysis is due to the fact that, under certain conditions, the signal $f(t)$ can be reconstructed by the Fourier inversion formula. Thus, the Fourier transform theory has been very useful for analyzing harmonic signals or signals for which there is no need for local information.

Fourier transform is also frequently used in uncertainty analysis of a model that involves exponentiation of stochastic variables. That is particularly useful when stochastic variables are independent and linearly related. Examples of its applications can be found in probabilistic cash flow analysis and probabilistic modeling of pollutant decay.

26.2.3 Hankel Transform

Hermann Hankel (1839–1873), a German mathematician, is remembered for his numerous contributions to mathematical analysis including the Hankel transformation, which occurs in the study of functions that depend only on the distance from the origin. This transform can easily be derived from the two-dimensional Fourier transform when circular symmetry is assumed. The Hankel transform involving Bessel functions as the kernel arises naturally in axisymmetric problems and solving boundary value problems formulated in cylindrical polar coordinates (Singh et al., 2010a). A large number of axisymmetric problems in cylindrical polar coordinates are solved with the aid of the Hankel transform. They are extremely useful in solving a variety of partial differential equations in cylindrical polar coordinates. Hankel transforms also can be used to solve a class of linear time-varying differential equations and make the solution much easier than by classical methods, especially in finding the particular solution. They are widely used in several fields like elasticity (Kulkarni and Deshmukh, 2008), optics (Magni et al., 1992), fluid mechanics (Eldabe et al., 2004), seismology (Patella, 1980), astronomy, and image processing (Cavanagh and Cook, 1979; Hansen, 1985; Higgins and Munson, 1988). The Hankel transforms become very useful in the analysis of wave fields where it is used in the mathematical handling of radiation, diffraction, and field projection.

Hankel transform is also advantageous in the solution of boundary value problems. Since, in the use of Hankel transform, only one input boundary condition is required while in the Laplace transform technique, an additional boundary condition is required to solve the second-order partial differential equation (Singh et al., 2010b).

$\widetilde{f_n}(\kappa)$ is called the Hankel transform of $f(r)$ and is defined formally by (Debnath and Bhatta, 2014):

$$\mathcal{H}_n\{f(r)\} = \widetilde{f_n}(\kappa) = \int_0^\infty rJ_n(\kappa r)f(r)dr, \qquad (26.7)$$

and the inverse Hankel transform is defined by (Debnath and Bhatta, 2014):

$$\mathcal{H}_n^{-1}\left[\widetilde{f_n}(\kappa)\right] = f(r) = \int_0^\infty \kappa J_n(\kappa r)\widetilde{f_n}(\kappa)d\kappa, \qquad (26.8)$$

where κ is Hankel transform variable and $J_n(\kappa r)$ is the Bessel function of the first kind of order n.

26.2.4 Mellin Transform

Historically, Riemann (1876) first recognized the Mellin transform in his famous memoir on prime numbers. Its explicit formulation was given by Cahen (1894). Almost simultaneously, Mellin (1896, 1902) gave an elaborate discussion of the Mellin transform and its inversion formula. The Mellin transform and its inverse can be derived from the complex Fourier transform and its inverse. Also, this transform is closely related to the bilateral Laplace transform and the theory of Dirichlet series.

The Mellin transform of $f(x)$ and the inverse Mellin transform are defined as (Debnath and Bhatta, 2014):

$$\mathcal{M}\{f(x)\} = \tilde{f}(p) = \int_0^\infty x^{p-1} f(x)\, dx, \tag{26.9}$$

$$\mathcal{M}^{-1}\{\tilde{f}(p)\} = f(x) = \frac{1}{2\pi i} \int_{c-i\infty}^{c+i\infty} x^{-p} \tilde{f}(p)\, dp, \tag{26.10}$$

where $f(x)$ is a real-valued function defined on $(0,\infty)$ and the Mellin transform variable p is a complex number.

Similar to other integral transform methods, the Mellin transform has been known to be of great importance in the solution of initial and boundary value problems for partial differential and integral equations (Butzer and Jansche, 1997).

The Mellin transform is also potentially useful in uncertainty analyses of hydrologic and hydraulic problems. Many equations used in engineering designs in hydrology and hydraulics are empirically derived involving parameters subject to uncertainty. The Mellin transform is especially attractive and simple to use when the dependent random variable is related to several independent random variables in a multiplicative manner. Under such circumstances, exact values of the moments of any order can be derived with simple algebraic manipulations. In fact, many equations in hydrologic and hydraulic computations are of this nature, to which the Mellin transform is applicable. The Mellin transform yields an expression of uncertainty in model output that is amenable for analytical sensitivity analysis. Results of sensitivity analysis provide useful information in directing future data-collection efforts in an attempt to reduce uncertainty in model output (Tung, 1990; DeChant and Moradkhani, 2014).

26.2.5 Hilbert Transform

In his 1912 famous paper on integral equations, David Hilbert (1862–1943) introduced an integral transformation, which is now known as the Hilbert transform. Although it was named after Hilbert, the Hilbert transform and its basic properties were developed mainly by G. H. Hardy (1924) and simultaneously by E. C. Titchmarsh during 1925–1930.

Hilbert transform arises in many problems in applied mathematics, mathematical physics, and engineering science. It plays an important role in fluid mechanics, aerodynamics, and electronics. The Hilbert transform is important in signal processing, where it derives the analytic representation of a real-valued signal $f(t)$ (Klingspor, 2015).

The Hilbert transform is also useful in calculating instantaneous attributes of a time series, especially the envelope amplitude and instantaneous frequency. These properties can be applied to identify dynamic characteristics of a linear as well as a nonlinear system (Luo et al., 2009). In this regard, Simon and Tomlinson (1984) provided a tool for a nonlinear detection method from the measured frequency response function based on the fact that the frequency-response function of a linear system is invariant under a Hilbert transform.

If $f(t)$ is defined on the real line $-\infty < t < \infty$, its Hilbert transform, denoted by $\hat{f}_H(x)$, is defined by (Debnath and Bhatta, 2014):

$$H\{f(t)\} = \hat{f}_H(x) = \frac{1}{\pi} \oint_{-\infty}^\infty \frac{f(t)}{t-x}\, dt, \tag{26.11}$$

where x is real and the integral is treated as a Cauchy principal value, that is,

$$\oint_{-\infty}^{\infty} \frac{f(t)\,dt}{t-x} = \lim_{\varepsilon \to 0} \left[\int_{-\infty}^{x-\varepsilon} + \int_{x+\varepsilon}^{\infty} \right] \frac{f(t)\,dt}{t-x}. \tag{26.12}$$

The inverse Hilbert transform is given by (Debnath and Bhatta, 2014):

$$f(t) = H^{-1}\left\{ \hat{f}_H(x) \right\} = -H\left\{ \hat{f}_H(x) \right\} = -\frac{1}{\pi} \oint_{-\infty}^{\infty} \frac{\hat{f}_H(x)\,dx}{t-x}. \tag{26.13}$$

26.2.6 HILBERT-HUANG TRANSFORM

The Hilbert-Huang Transform (HHT) was proposed by Huang et al. (1996, 1998). The HHT method is specially developed for analyzing nonlinear and nonstationary data. The method consists of two parts: (1) the empirical mode decomposition (EMD), and (2) the Hilbert spectral analysis. The key part of the method is the first step, the EMD, with which any complicated data set can be decomposed into a finite and often small number of intrinsic mode functions (IMF). An IMF is defined as any function having the same number of zero-crossing and extrema, and also having symmetric envelopes defined by the local maxima, and minima respectively. The IMF also thus admits well-behaved Hilbert transforms. This decomposition method is adaptive, and, therefore, highly efficient. Since the decomposition is based on the local characteristic time scale of the data, it is applicable to non-linear and non-stationary processes. With the Hilbert transform, the IMF yield instantaneous frequencies as functions of time that give sharp identifications of the imbedded structures. The final presentation of the results is an energy–frequency–time distribution, which is designated as the Hilbert Spectrum. The EMD is also useful as a filter to extract the variability of different scales. In contrast to other common transforms like the Fourier transform, the HHT is more like an algorithm (an empirical approach) that can be applied to a data set, rather than a theoretical tool. However, comparisons with Wavelet and Fourier analyses show the HHT method offers much better temporal and frequency resolutions.

In recent years, the HHT is applied in several scientific and engineering disciplines, for example, in biomedical science (Balocchi et al., 2004; Liang et al., 2005; Ponomarenko et al., 2005; Pachori, 2008), in the financial market (Huang et al., 2003), in testing structures (Quek et al., 2003; Peng et al., 2005; Douka and Hadjileontiadis, 2005), in structural health monitoring (Pines and Salvino, 2002; Yang et al., 2004), in seismic studies (Huang et al., 2001; Zhang et al., 2004;), in atmospheric and geophysical sciences (Pan et al., 2003; Gloersen and Huang, 2003), in coastal engineering applications (Veltcheva, 2002; Hwang et al., 2002), in meteorological and atmospheric studies (Salisbury and Wimbush, 2002; Pan et al., 2003), etc.

26.2.7 WAVELET TRANSFORM

The concept of "wavelets" or "ondelettes" started to appear in the literature only in the early 1980s. This new concept can be viewed as a synthesis of various ideas which originated from different disciplines including mathematics, physics, and engineering. In 1982 Jean Morlet, a French geophysical engineer, first introduced the idea of wavelet transform as a new mathematical tool for seismic signal analysis. It was Alex Grossmann, a French theoretical physicist, who quickly recognized the importance of the Morlet wavelet transform which is something similar to coherent states formalism in quantum mechanics, and developed an exact inversion formula for the wavelet transform. In 1984 the joint venture of Morlet and Grossmann led to a detailed mathematical study of the continuous wavelet transforms and their various applications. It has become clear from their work that, analogous to the Fourier expansions, the wavelet theory has provided a new method for decomposing a function or a signal (Morlet et al., 1982; Grossmann and Morlet, 1984; Goupillaud et al., 1984; Grossmann and Morlet, 1985a; Grossmann et al., 1985b).

In 1985 Yves Meyer, a French mathematician, gave a mathematical foundation for wavelet theory. Inspired by the work of Meyer, Ingrid Daubechies (1988) made a new remarkable contribution to wavelet theory by constructing families of compactly supported orthonormal wavelets with some degree of smoothness. Her 1988 paper had a tremendous positive impact on the study of wavelets and their diverse applications. This work significantly explained the connection between the continuous wavelets on $\mathbb{R}$ and the discrete wavelets on $\mathbb{Z}$.

The continuous wavelet transform (CWT) is similar to the Fourier transform in the sense that it is based on a single function ψ and that this function is scaled. But unlike the Fourier transform, we also shift the function, thus generating a two-parameter family of functions $\psi_{a,b}$. It is convenient to define $\psi_{a,b}$ as follows:

$$\psi_{a,b}(x) = |a|^{-\frac{1}{2}} \psi\left(\frac{x-b}{a}\right). \tag{26.14}$$

Then the continuous wavelet transform is defined by (Debnath and Bhatta, 2014):

$$\left(W_\psi f\right)(a,b) = \int_{-\infty}^{\infty} f(t)\overline{\psi_{a,b}(t)}dt = |a|^{-\frac{1}{2}} \int_{-\infty}^{\infty} f(t)\overline{\psi\left(\frac{t-b}{a}\right)}dt. \tag{26.15}$$

The function ψ is often called the mother wavelet or, the analyzing wavelet. The parameter b can be interpreted as the time translation and a is a scaling parameter that measures the degree of compression. The parameters a and b vary continuously over $\mathbb{R}(a \neq 0)$.

Furthermore, the orthogonal (discrete) wavelets are employed because this method associates the wavelets to orthonormal bases of $L^2(\mathbb{R})$. In this case, the wavelet transform is performed only on a discrete grid of the parameters of dilation and translation, i.e., a and b take only integral values. Within this framework, an arbitrary signal $x(t)$ of finite energy can be written using an orthonormal wavelet basis (Debnath and Bhatta, 2014):

$$x(t) = \sum_m \sum_n d_n^m \psi_n^m(t), \tag{26.16}$$

where the coefficients of the expansion are given by (Debnath and Bhatta, 2014):

$$d_n^m = \int_{-\infty}^{\infty} x(t)\psi_n^m(t)dt. \tag{26.17}$$

The orthonormal basis functions are all dilations and translations of a function referred to as the analyzing wavelet $\psi(t)$, and they can be expressed in the form (Debnath and Bhatta, 2014):

$$\psi_n^m(t) = 2^{m/2}\psi\left(2^m t - n\right), \tag{26.18}$$

with m and n denoting the dilation and translation indices, respectively. The contribution of the signal at a particular wavelet level m is given by:

$$d_m(t) = \sum_n d_n^m \psi_n^m(t), \tag{26.19}$$

which provides information on the time behavior of the signal within different scale bands. Additionally, it provides knowledge of their contribution to the total signal energy (Olkkonen, 2011).

The discrete wavelet transform (DWT) has a window size that varies frequency scale. This is advantageous for the analysis of signals containing both discontinuities and smooth components. Short, high-frequency basis functions are needed for the discontinuities, while at the same time, long, low-frequency ones are needed for the smooth components.

The wavelet transform provides a time scale representation of the signal and has recently been used for addressing problems in signal processing (especially non-stationary signals), image processing, data and image compression, pattern recognition, transient detection, noise/trend reduction, seismology, turbulence, computer vision, digital communication, approximation theory, quantum optics, biomedical engineering, sampling theory, matrix theory, operator theory, differential equations solving, numerical analysis, statistics and multiscale segmentation of well logs, etc.

26.3   NEW FINDINGS FACILITATED BY INTEGRAL TRANSFORMS IN HYDROLOGY AND FLOOD STUDIES

26.3.1   LAPLACE TRANSFORM APPLICATIONS

In hydrology, one of the major works using Laplace transforms was performed by Diskin (1967, 1968), who made a detailed study of the transfer function (the Laplace transform of the instantaneous unit hydrograph [IUH]) and a special transfer function in relation to a number of mathematical models of the IUH. He also investigated the dependence of the special transfer function on different storms and different catchments. Moreover, Johnson (1970) and Delleur and Rao (1971) developed instantaneous unit hydrographs using Laplace transform. Wang and Chen (1996) presented a linear spatially distributed model for a surface rainfall-runoff system using Laplace transform. The system approach and the model developed have certain advantages over Diskin's approaches and models and are capable of predicting runoff from non-uniform rainfall and different geographic conditions over a whole watershed.

Laplace transforms are also frequently used in solving groundwater flow and well hydraulics problems. In the 1950s, Hantush introduced the method of Laplace transform into the field of groundwater flow. He (1955) employed the Laplace transform to find the transient drawdown distribution in a leaky aquifer drained by a flowing well. Since then, the Laplace transform has been used in the analytical solution of almost all transient groundwater problems. Verma (1969) obtained an analytical expression for the moisture content distribution, in a problem of one-dimensional vertical groundwater recharge by using the Laplace transform method. Chen and Stone (1993) offered a new method for determining the correct asymptotic calculation of the Laplace inverse in analytical solutions of groundwater problems. Mathias and Zimmerman (2003) applied the Laplace transform to the problem of water influx into a slab-like matrix block in a dual-porosity medium. In doing so, they cleared up a long-standing discrepancy between the fracture/matrix transfer coefficient that has been calculated by Laplace methods. Several other studies are conducted using Laplace transform for problems in the well hydraulic literature (Zhan and Zlotnik, 2002; Neuman and Li, 2004).

Laplace transform has been commonly used to obtain analytical solutions for solute transport in uniform flow. This is because of being simpler than other methods and the analytical solutions using the Laplace transform being more reliable in verifying the numerical solutions in terms of accuracy and stability. There is a considerable body of literature in this case. Rasmuson (1981) developed a solution of the two-dimensional differential equation of dispersion from a disk source, coupled with a differential equation of diffusion and sorption in particles. The solution was obtained by the use of the Laplace and the Hankel transforms. Analytical solutions for convective-dispersive transport in confined aquifers subject to various initial and boundary conditions were presented by Lindstrom and Boersma (1989). Flury et al. (1998) derived an analytical solution for solute transport with depth-dependent transformation or sorption coefficient. Singh et al. (2010c) formulated a solute transport model with time-dependent source concentration in a homogeneous finite aquifer and obtained an analytical solution to predict contaminant concentration along and against transient groundwater flow in a homogeneous finite aquifer. Kumar et al. (2012) obtained analytical solutions for temporally dependent solute dispersion for uniform and increasing input source in a semi-infinite one-dimensional longitudinal domain using the Laplace transform technique.

26.3.2 Fourier Transform Applications

Fourier transform encompasses a broad scope of applications in water sciences and the studies of hydrological phenomena. Some of these applications are outlined in the following. Fleming et al. (2002) provided an overview of the Fourier transform and spectral analysis and presented examples of how these methods may be applied to practical hydrologic problems: determination of the frequency content of a time series; inference of the physical mechanisms responsible for this frequency content; and evaluation of the performance of a process-based simulation model used for water resource management. Bayazit et al. (2001) presented a nonparametric streamflow simulation model that is based on Fourier analysis. They applied the model to the annual flow series at Homa station on the Manavgat River in Turkey. Periodic (seasonal or monthly) flow series is also simulated. Kendal and Hyndman (2007) extracted quantitative information from precipitation, stream discharge, and groundwater head data from watersheds in northern-lower Michigan using Fourier transform methods. They also demonstrate how unit hydrographs can be efficiently and non-parametrically derived using the Fourier transform. Anderson et al. (2007) developed a parsimonious method of parameter fitting for high-resolution periodic autoregressive moving average (PARMA) time series models, using discrete Fourier transforms. Fourier-PARMA models can be fruitfully applied to generate high-resolution synthetic flows that faithfully reproduce a weekly series. Tularam and Mahbub (2010) used the fast Fourier fransform (FFT) function to analyze the variations in rainfall and temperature activity over the last 50–60 years in eastern Queensland, Australia. Faye et al. (2014) examined the fine-scale spatiotemporal heterogeneity in air, crop canopy, and soil temperatures of agricultural landscapes in the Ecuadorian Andes and compared them to predictions of global interpolated climatic grids. They measured temperature time-series in air, canopy, and soil for 108 localities at three altitudes and analyzed using Fourier transform. Mukhopadhyay et al. (2013) applied Fourier transform to study the wind speed data of the East Midnapore district of West Bengal. They compared wind speed time series of winter with summer using Fourier analysis. Wavelet transform is also used for a clear distinction between these two seasonal data. Elmaghraby et al. (2016) performed a 20-year investigation of the evolution of regional climate by applying the fast Fourier transform signal decomposition technique to the discrete meteorological data. They applied the technique to the pressure, temperature, humidity, wind direction, and solar irradiance signals.

26.3.3 Hankel Transform Applications

The Hankel transform is a powerful tool for solving reservoir and groundwater flow problems in the cylindrical coordinate system. Hantush (1960) was the first to apply Hankel transform for solving groundwater flow problems. He used a combination of the Laplace and Hankel transforms to find the equation of drawdown and the rate of induced leakage. Russell and Prats (1962) applied the combination of the finite Hankel and the Laplace transforms to find a solution for the problem of flow of fluid in layered reservoirs with crossflow. Hantush (1967) used successive finite Hankel and Laplace transforms to solve the problem of flow in a system of two aquifers separated by a semi-pervious layer.

As mentioned in section 26.2.1.3, another area of the Hankel transform application is in the solution of boundary value problems. Ilias et al. (2008) presented a new analytical solution of a linearized form of the Boussinesq equation to describe the water table fluctuation in unconfined aquifers overlying a semi-impervious layer in response to transient recharge. They used the finite Hankel transform to solve the boundary value problem. Xie et al. (2010) obtained closed-form solutions for discontinuous boundary-condition problems of water flow due to a circular source, which was located on the upper surface of a confined aquifer, with the aid of the Hankel transform technique. Singh et al. (2010b) derived an analytical solution, using the Hankel transform, for the two-dimensional space-time distribution of contaminant concentration in unsteady flow velocity in a homogeneous finite aquifer, where the aquifer is contaminated with a time-dependent point source at the far end.

Furthermore, the Hankel transform is useful in shallow water acoustic fields which relates it to Green's function. Frisk and Lynch (1984) presented a technique for acoustically characterizing shallow water waveguides using Hankel transform. Wengrovitz (1986) developed a hybrid method, referred to as the Hilbert-Hankel transform, for accurate shallow-water synthetic data generation. The Hilbert-Hankel transform is a unilateral version of the Hankel transform and its application to this problem is based on the outgoing nature of the acoustic field.

26.3.4 MELLIN TRANSFORM APPLICATIONS

The Mellin transform has seen limited application in hydrology and water resources studies. The Mellin transform is particularly an attractive uncertainty analysis approach that was first introduced to the field of hydrology and hydraulics by Tung (1990). He used the Mellin transform in uncertainty analyses of hydraulic computations in channel flood routing and the design of a storm sewer system. Also, Chen and Tung (2002) applied the Mellin transform to determine total hydrologic and hydraulic uncertainties in reliability analysis of local scour around bridge piers.

On the other hand, the close relationship between Fourier and Mellin transforms suggests that the theorems given for the Fourier transform theory can be rewritten and applied to Mellin transforms. In this regard, the Mellin transform can be expressed in the solution of boundary value problems. Analytical solutions for steady-state drawdown due to a single well pumping at a constant rate from a homogeneous aquifer derived by Chan et al. (1978) using Mellin transform under a variety of boundary conditions. Craster (1997) solved several free boundary conditions encountered in groundwater flow problems using Mellin transform.

26.3.5 HILBERT TRANSFORM APPLICATIONS

The Hilbert transform has been widely used to investigate oscillatory signals that resemble a noisy periodic oscillation, because it allows instantaneous phase and frequency to be estimated, which in turn reveals interesting properties of the underlying process that generates the signal. For instance, the Hilbert transform is a quite powerful way to use in wind speed prediction. Zhu and Yang (2002) applied the discrete Hilbert transform to estimate the wind speed with the sample data sequence obtained in Hong Kong in June 1989. Alpay et al. (2006) exploited the discrete Hilbert transform to characterize the wind sample data in Turkey, in March 2001 and 2002. They also tried to estimate the hourly wind data using a daily sequence by the Hilbert transform.

The Hilbert transform has been also used to examine the local properties of ocean waves by Huang et al (1992). Mercier et al. (2008) applied the Hilbert transform to the physics of internal waves in two-dimensional fluids. Using this demodulation technique, internal waves propagating in different directions can be discriminated.

Zappalà et al. (2016) tried to explore the potential of the Hilbert transform to analyze an atmospheric dataset consisting of noisy oscillatory signals with a typically annual periodicity. They applied Hilbert instantaneous frequency analysis to daily surface air temperature time series recorded over a regular grid of locations covering the Earth's surface. Their results demonstrated that Hilbert analysis is a promising tool for the study of other climatological variables.

Despite the capabilities of the Hilbert transform, the application of this transform by itself is associated with restrictions in the field of hydrology. For example, it is not usable for general random data and its applications have been limited to narrow-band data in the past (Long et al., 1993).

Especially problems arise when the signal is in fact a superposition of oscillating components with different time scales – a common situation in the real world – oscillations (Lam et al., 2005). In order to overcome the limitations of the method, Huang and coworkers combined the Hilbert spectral analysis with empirical mode decomposition. A number of Hilbert-Huang transform applications in hydrology are given below.

26.3.6 HILBERT-HUANG TRANSFORM APPLICATIONS

In the past few years, applications of the Hilbert-Huang transform (HHT) have been notably expanded in scientific research and engineering. Since the HHT has a high potential in analyzing nonlinear and nonstationary data, it is seen to be an excellent tool to investigate the characteristics of environmental and hydrologic phenomena. In this regard, a brief survey is provided on discoveries facilitated by HHT with putting much emphasis on the analysis of climatic and hydrological (rainfall, streamflow, wind, and temperature) time series and study of water wave problems.

The nonstationarity and periodicity in the rainfall data make the HHT applicable and efficient in this field. Baines (2005) applied HHT to analyze the time series of several of the key variables involved in the temporal behavior of the changes that relate to southwest Western Australia (SWWA) rainfall variations. He found that the long-term SWWA rainfall variations were directly related to the African monsoon. Molla et al. (2006a) used empirical mode decomposition (EMD) and Hilbert spectrum to analyze the properties of nonlinear and nonstationary daily rainfall time-series data. Their study suggested that the recent global warming along with decadal climate variability contributes to more frequent, long-lasting drought and flood. Sinclair and Pegram (2005) introduced the concept of two-dimensional EMD to the hydrometeorological literature as a tool for the analysis of spatio-temporal rainfall data. The reason behind this is that in a single dimension, EMD analysis produces a set of intrinsic mode functions (IMFs) that are very nearly orthogonal. McMahon et al. (2008) applied the approach to generate six-monthly rainfall totals for six rainfall stations located near Canberra, Australia. They compared the results with those obtained using a traditional autoregressive lag-one [AR(1)] and found that the new EMD stochastic model performed satisfactorily. Srikanthan et al. (2011) applied the ensemble empirical mode decomposition (EEMD), which was proposed by Wu and Huang (2009), to monthly rainfall and temperature data from 44 Australian rainfall and ten climate stations respectively. They obtained dominant cycles in the climate indices in order to determine any relationship between the rainfall and climate indices. Of course, they found that there may be some relationship between the two. Roushangar and Alizadeh (2018) proposed an EEMD-based multiscale entropy (EME) approach. The proposed model was used to analyze and gauge the variability of the annual precipitation series and spatially classify rain gauges in Iran. They decomposed historical annual precipitation data during 1960–2010 from 31 rain gauges using EEMD.

Another area that takes advantage of the power of HHT is the study of streamflow data. Chiew et al. (2005) applied EMD analysis to the annual streamflow time series from the 20 catchments (on the original time series and the bootstrap replicates) to identify oscillations in the data and test their statistical significance. Huang et al. (2009) applied the EMD method to analyze two long time series of daily river flow data, 32 years recorded in the Seine River, and 25 years recorded in the Wimereux River, France. Cao and Lin (2010) used the HHT to analyze the annual average streamflow near the Three Gorges Dam on the Yangtze River in China. Xuehua et al. (2011) analyzed the runoff time series of the Yellow River and Fenhe River based on the HHT. Di et al. (2014) simulated and analyzed the multi-scale characteristics, variation periods, and trends of the annual streamflow series in the Haihe River Basin (HRB), China, using the Hilbert-Huang transform.

The use of the HHT has the potential to provide more information about wind behavior. Pan et al. (2003) used HHT to interpret satellite scatterometer wind data over the northwestern Pacific and compared the results to vector empirical orthogonal function (VEOF) results. Xu and Chen (2004) applied the HHT approach to the nonstationary wind data recorded in the field during a nearby typhoon and compared the resulting wind characteristics with those obtained by the traditional approach based on a stationary wind model. Rao and Hsu (2008) applied the HHT to analyze both hourly and daily wind speed data measured at four National Weather Service stations in the state of Indiana in a 14-year period (1988–2002) and compared the results to other conventional methods. They revealed that the HHT provides a better picture of the time and frequency behavior. Vincent et al. (2011) investigated climatological patterns in wind speed fluctuations over the North Sea, and the HHT is applied to create conditional spectra which demonstrate patterns in the occurrence of severe wind variability. Hsieh and Dai (2012) used the HHT method to analyze the wind characteristics of offshore islands in Taiwan. They analyzed inter-monthly and inter-annual variation in the 47 years of daily wind speed data. Jing-Jing and Fei (2014) applied the HHT to analyze non-stationary wind fluctuations in mountain terrain. They employed the method to obtain conditionally averaged spectra, and to investigate climatological patterns between wind fluctuations and several potential explanatory variables including time of year, time of day, wind direction, and pressure tendency in mountain terrain compared with offshore sites.

With consideration of the nonlinear and nonstationary properties of the temperature time series, the HHT can be a very efficient method to reveal new features of these data. Molla et al. (2006b) applied EMD to daily temperature time series data in order to explore the properties of the multi-year air temperature. They focused on the relation between temperature variability and global warming. Rao and Hsu (2008) applied HHD to long-term monthly temperature time series in Europe and the state of Indiana, USA. They detected several periodicities in the temperature series which bring out the possibility of predicting the time series with the fundamental frequencies and the residual. Ma et al. (2015) used the HHT to analyze the oscillating characteristics of the monthly mean temperature time series from 1959 to 2012 in the Fengxian district of Shanghai. Their results showed that the EMD method has a good adaptivity for both nonlinear and linear cases and the relationship among the time, the frequency, and the energy was obtained by the HHT. Kbaier Ben Ismail et al. (2016) used the HHT for the spectral analysis of a high-frequency sampled time series in nearshore waters of the Réunion Island, located in the Indian Ocean. They focused particularly on temperature records, sea-level fluctuations, and current data sets and compared the obtained results with the continuous wavelet transform approach.

Since all water waves are nonlinear, HHT has been also used in the study of water wave problems. It was first introduced by Huang et al. (1999) in this field. Afterward, Hwang et al. (2003) applied the HHT to calculate the spectrum of nonlinear and nonstationary ocean wave data and compared the results with those obtained by Fourier-based techniques. Their study indicated that the HHT spectral level is considerably higher than the Fourier-based spectra in the lower frequency region. These differences in the wave spectral properties affect many engineering applications such as the frequency response of marine structures. Dätig and Schlurmann (2004) applied the HHT to evaluate computed nonlinear irregular water waves based on Stokes perturbation expansion approach and measurements on fully nonlinear irregular water waves recorded in a laboratory wave flume. Ortega and Smith (2009) used the method for the spectral analysis of a North Sea storm that took place in 1997 and lasted over five days. Veltcheva and Soares (2016) studied the wave groups by the HHT method. For this, full-scale wave records containing abnormal waves were used.

To catch more insights into the subject, an example for the Hilbert-Huang transform application is presented herein. The HHT method is applied to the field of sediment transport for the time scale analysis (Kuai and Tsai, 2012). The biweekly hydraulic and suspended load concentration data is studied in the Rio Grande at Cochiti between March 10, 1954, and July 10, 1956, reported in Nordin and Beverage (1965). They used this dataset to indicate the existence of multiple time scales in sediment transport in alluvial rivers and variation of time scales with respect to floods.

The original time-series data of the suspended sediment concentration and flow rate are shown in Figure 26.1. It is not easy to identify the time scales directly from the recorded data. By implementing the EMD package, the IMFs and HSA of sediment data can be obtained as shown in Figures 26.2 and 26.3. It can be found from the decompositions in Figure 26.2 that the first IMF is

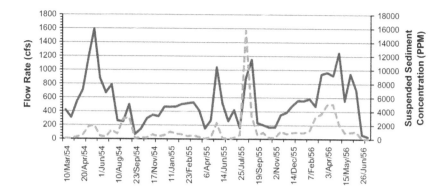

FIGURE 26.1 Original time series for flow rate and suspended sediment concentration (dashed line). (Source: Kuai and Tsai, 2012.)

representative of highly fluctuating values (monthly) and the rest may represent trimonthly (IMF 2), quarterly (IMF 3), semiyearly (IMF 4), and yearly (IMF 5) variation trends. The last monotonic component is the residual that shows the predominant trend in the time series data after the removal of noise and seasonality. It can be concluded from the residual function that the predominant trend is increasing for suspended sediment concentrations in this observation.

Hilbert spectral analysis (HSA) can provide a clear and sharp picture of time–frequency–energy representation as shown in Figure 26.3. Each point represents the energy level over time (x-axis) and over the frequency (y-axis). The darker colored points at the low-frequency range are typical of

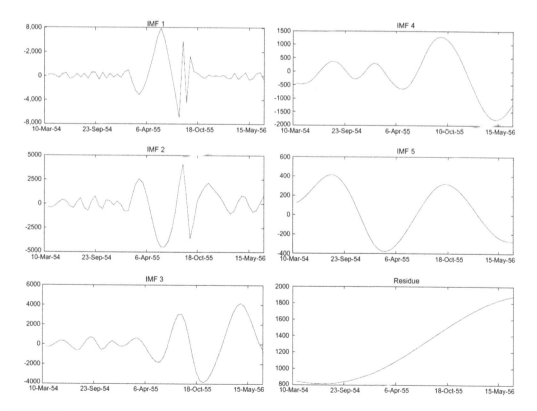

FIGURE 26.2 Decomposition of time series concentration data into IMFs and residual function. (Source: Kuai and Tsai, 2012.)

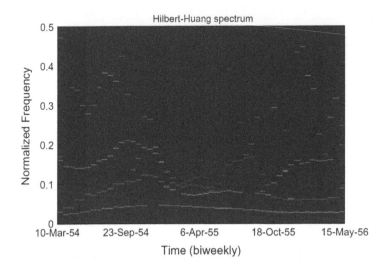

FIGURE 26.3   Hilbert spectrum of the sediment concentration. (Source: Kuai and Tsai, 2012.)

higher energy levels. The collection of the most energy points represents the dominant frequency or time scale. The scattering points usually at the high-frequency range represent different individual time scales. It can be identified from Figure 26.3 that the dominant time scale is (1/dominant relative frequency)(measuring time interval) = (1/0.1)(14 days) =140 days (semiyearly variation). Note that the dominant relative frequency is the y-axis value where most energy points can be located. It is noted that the dominant time scale is not constant but varies with time.

The difference of time scales of the flow rate and suspended sediment concentration is also difficult to be retrieved from Figure 26.1. We can identify from the Hilbert spectrum in Figure 26.4 that the dominant time scale of flow rate is about 0.05 cycles biweekly (about 280 days) (yearly variation). It can be found that the dominant time scale of the flow rate changes with time. The undulations of the dominant time scales (for flow and also for sediment) correspond to the variation of flood waves in Figure 26.1. For example, there is a large flood between March 23, 1954, and August 24, 1954, as shown in Figure 26.1. There is a high increase of dominant frequencies around September 23, 1954, in HSAs of both sediment and flow. This manifests the strong impact of flow

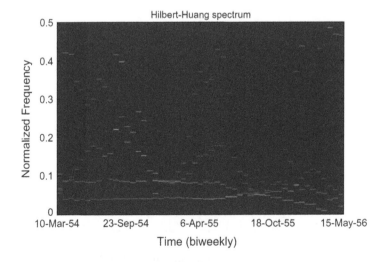

FIGURE 26.4   Hilbert spectrum of the flow rate. (Source: Kuai and Tsai, 2012.)

properties on the time scale of sediment transport. Both flow and sediment experienced a big drop in the dominant time scale after the passage of a flood.

26.3.7 Wavelet Transform Applications

In recent years, the application of wavelet transform has been increased since its inception in the early 1980s. Wavelet transform analysis has become a popular analysis tool due to its ability to elucidate simultaneously both spectral and temporal information within the signal. It is applicable in extracting nontrivial and potentially useful information, or knowledge, from the large data sets available in experimental sciences (Smith et al., 1998; Nason and Sachs, 1999). In the field of earth sciences, Grossmann and Morlet (1984) introduced the wavelet transform application, which worked especially on geophysical seismic signals. A comprehensive literature survey of wavelet in geosciences was then presented by Foufoula-Georgiou and Kumar (1994) and Labat (2005).

The ability of the wavelet transform to decompose nonstationary signals into sub-signals at different temporal scales is helpful in better interpreting hydrological processes. There are plenty of issues with respect to the use of wavelet analysis within the areas of water resources engineering and hydrology. In a review of the applications of the wavelet transform in hydrologic time series modeling, Sang (2013) categorized the information drawn from this approach in terms of six aspects: (1) characterization and understanding of hydrologic series' multi-temporal scales; (2) identification of hydrologic series' deterministic components such as trend and periods; (3) estimating the temporal evolution and quantifying the complexity degrees of hydrologic systems; (4) data de-noising in hydrologic series; (5) wavelet cross-correlation analysis of hydrologic series; (6) and wavelet-based hydrologic simulation and forecasting. A number of other general reviews of wavelet applications in hydrology (Schaefli et al., 2007; Dadu and Deka, 2016; Potočki et al., 2017).

The widespread application of wavelet analysis in hydrology began in the early 1990s. Kumar and Foufoula-Georgiou (1993) analyzed rainfall data for scaling characteristics without an a priori assumed model. They studied the behavior of rainfall fluctuations obtained at several scales, via orthogonal wavelet transform of the data, to infer the precise nature of scaling exhibited by spatial rainfall. They (1994) also provided an overview of the nature and scope of wavelet transforms and presented their applications to geophysical phenomena.

Furthermore, Labat et al. (2000) used wavelet transform to study the nonstationarity of Karstic watersheds which appear as highly nonlinear and nonstationary systems. They applied the wavelet methods to rainfall rates and runoffs measured at different sampling rates, from daily to half-hourly sampling rates. Massel (2001) used wavelet analysis for the processing of ocean surface wave records. He tried to demonstrate its capability to give a full time-frequency representation of the wave signals. Gaucherel (2002) exploited the potential of continuous wavelet transform on flow curves continuously recorded over a period of 23 years at the major hydrological stations of French Guyana watersheds in South America to detect new periodicities or time annual features. Kantelhardt et al. (2003) studied the multifractal temporal scaling properties of river discharge and precipitation records and compared the results for the multiracial detrended fluctuation analysis (MF-DFA) method with the results for the wavelet analysis. Chang and Wang (2003) used the discrete wavelet transform to detect acoustic signals underwater. Massei et al. (2006) highlighted the usefulness of wavelet analysis for studying time series in karst hydrosystems by applying wavelet analyses to time series of rainfall (the input signal) and water level, specific conductance, and turbidity (the output signals) measured at a karst spring system. Adamowski (2008) developed a new method of stand-alone short-term spring snowmelt river flood forecasting based on wavelet and cross-wavelet analysis which was shown to be useful for one- and two-day lead-time forecasting. Özger et al. (2010) employed a large set of monthly precipitation data from 43 stations throughout Texas and used wavelet transform to obtain significant low-frequency patterns of precipitation. Liu et al. (2011) presented a new approach based on the cross wavelet transform technique to assess and quantify timing error in hydrologic predictions. They used the method to assess timing error in streamflow predictions for 11 headwater basins from the US state of Texas. Avdakovic et al. (2011) applied a wavelet transform to analyze the wind speed data in the context of insight into the characteristics of the wind and the selection of suitable locations that could be the subject of a wind farm construction. Tang et al. (2012) adopted wavelet methods to analyze the inter-annual, annual, and periodic variation characteristics of the streamflows in the upper Manasi River Basin for the last 50 years (1957–2006). Nalley et al. (2012) co-utilized the discrete wavelet transform technique and the Mann–Kendall trend tests to analyze and detect trends in monthly, seasonally-based, and annual mean flow and total precipitation data from eight flow stations and seven meteorological stations in southern Ontario and Quebec during 1954–2008. They (2013) also applied the techniques to detect trends in the mean surface air temperature in the desired area for the period of 1967–2006. Onderka et al. (2013) introduced the continuous wavelet transform, cross wavelet spectra, and wavelet phase difference as a tool for the extraction of diurnal temperature signals from time series of streambed temperatures and to use the time lag and amplitude ratios calculated from the wavelet analysis to calculate seepage velocities. Liu et al. (2014) used the wavelet method to analyze the period characteristics of tidal range, river discharge, and salinity and the impact of tidal range and river discharge on the salinity for explaining the nonlinear characteristics of saltwater intrusion in the Modaomen Waterway, a major outlet of the Pearl River Estuary, China. Araghi et al. (2015) applied the discrete wavelet transform, the Mann–Kendall trend test, and the sequential Mann–Kendall test to temperature series at different time scales in order to detect the long-term trends (1956–2010) in synoptic-scale surface temperatures in Iran. Joshi et al. (2016) used discrete wavelet transform in conjunction with the Mann-Kendall test to analyze trends and dominant periodicities associated with the drought variables for 30 rainfall subdivisions in India over 141 years (1871–2012). Ouyang et al. (2017) employed the wavelet analysis technique to identify the temporal trend of air temperature and its impact upon forest stream flows in the Lower Mississippi River Alluvial Valley. Marcolini et al. (2017) performed a wavelet analysis of the mean seasonal snow depth and of the snow cover duration at four elevation classes in the Adige catchment (Italy) to identify changes occurring in the snow depth and snow cover duration signals at various temporal scales. Subsequently, they applied the wavelet coherence analysis in order to investigate the relationship between the variations of the mean seasonal snow depth and that of climate indices. Guo et al. (2019) analyzed the joint distribution of trivariate extreme climate events, which is of great significance for prediction and early warning of abrupt dry–wet transitions and for water resource management, based on daily meteorological data in Northeast China. They adopted the continuous wavelet transform to calculate periodicities of individual extreme climate indices.

In recent years, many methods have been combined with wavelet transform for hydrologic series analysis, such as the combinations of artificial neural networks with wavelet theory. Present studies and practical applications have demonstrated the better performances of hybrid methods compared with a single method, and it is due to the complementary effects among various methods (Sang, 2013). A hybrid ANN-wavelet model which uses multi-scale signals as input data may present more probable forecasting rather than a single pattern input. In the field of hydrology, Wang and Ding (2003) firstly applied a wavelet-network model to forecast shallow groundwater levels and daily discharge. Afterward, Anctil and Tape (2004) used the neuro-wavelet hybrid system to one-day-ahead streamflow forecasting. Partal and Cigizoglu (2008) used a neuro-wavelet technique for forecasting river daily suspended sediment load. Nourani et al. (2009a) linked wavelet analysis to a nonlinear inter-extrapolator ANN for monthly precipitation prediction. Also, Nourani et al. (2009b) employed coupled wavelets and ANN to model the rainfall-runoff process. In combination with other methods, Nourani et al. (2011) investigated the rainfall-runoff process using two hybrid wavelet-AI models (i.e., WANN[1] and WANFIS[2]). Kisi and Cimen (2011) investigated the accuracy of discrete wavelet transform and support vector machine conjunction models in monthly streamflow forecasting. They (2012) also used the coupled model for the prediction of daily precipitation. The linkage of wavelet analysis to genetic programming (GP) in constructing a hybrid model to detect seasonality patterns

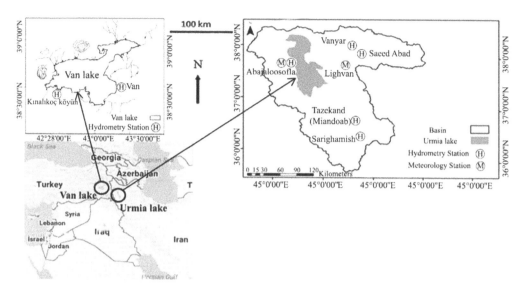

FIGURE 26.5   Locations of Lake Urmia and Lake Van and their basins.

in rainfall-runoff was investigated by Nourani et al. (2012). Several other representative contributions about the application of hybrid models in hydrology can be followed in Kisi and Shiri (2012), Nourani and Parhizkar (2013), Yu and Liu (2015) Ravansalar et al. (2017), Zare and Koch (2018), Shiri (2018), and Jeihouni et al. (2019).

Also, one of the wavelet transform applications is extensively investigated herein. In this study, wavelet transform coherence (WTC) is implemented to examine the impacts of hydroclimatological variables on water level fluctuations in two large saline lakes in the Middle East with a similar geographical location, namely, Urmia Lake in northwest Iran and Van Lake in northeast Turkey (Figure 26.5) (Nourani et al., 2019). The aim of this study is to investigate the complexity of the Urmia Lake water level time series which could lead to decrease fluctuations of time series. Hence, the strength and relationships between five hydroclimatological variables, including rainfall, runoff, temperature, and relative humidity, as well as evaporation and water level fluctuations in the lakes, were determined and discussed in terms of high common power region, phase relationships, and local multi-scale correlations.

Water level fluctuations in the lakes for the period 1960–2014 and the rainfall and runoff time series of the stations are shown in Figures 26.6 and 26.7, respectively. The original time series were also standardized before applying wavelet transform to have zero mean and unit variance. The results are presented and discussed in the following.

Figures 26.8 and 26.9 illustrate the results of WTC analysis between rainfall/runoff time series and water level fluctuations of Urmia Lake, respectively. Regarding the coherence results given for the stations Saeed Abad, Tazekand (Miandoab), Abajaloosofla, and Sarighamish, Figures 26.8a, 26.8c, and 26.8d show that the common periodicities between rainfall and water level signals have an 8–16-month frequency band, which indicates the most coherency with Urmia Lake water level fluctuations. An anti-phase correlation between rainfall and water level signals was observed and one time series leads to another by a quarter period length in the 8–16-month frequency band which means rainfall time series variations affect water level fluctuation with a two-to-four-month time lag. As shown in Figure 26.8(a), the rainfall effect has been stretched on the lake water level from the beginning of the study period (excluding 1971) on the 8–16-month frequency band. This indicates the absence of small fluctuations and the effect of rainfall in larger periods during this period; while the impact of rainfall data has been changed to a smaller period from 1984 which indicates the existence of small fluctuations and, consequently, increases the effect of rainfall parameter.

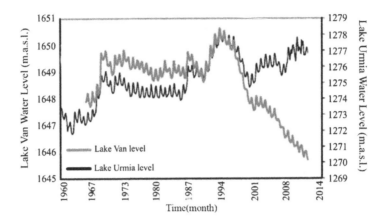

FIGURE 26.6 Urmia and Van lakes water level fluctuations time series. (Source: Nourani et al., 2019.)

Considering the coherence graph between Sarighamish station rainfall and the lake water level (Figure 26.8c), this parameter has almost constant behavior and high coherency from the beginning to the end of the investigated period. Also, it is clear that wavelet coherence between rainfall and water level fluctuations in Abajaloosofla station (Figure 26.8d) has almost a constant trend and there is no considerable change in terms of the frequency band, arrows' direction, and overall coherence graph. Therefore, it can be said that the impact of rainfall is almost constant even though it gradually increased during the studied time period while the fluctuations of lake water level decreased. Thus, it can be proven that the role of this factor on Urmia Lake water level decreasing is not so important. Figure 26.9 also illustrates that runoff signals affect the water level fluctuations in the 8–16-month frequency band with high coherency in all the stations. The correlations between runoff and water level fluctuations time series in the 8–16-month frequency band are mainly anti-phase, so that runoff leads to water level fluctuations time series by a quarter period length and effects on water level fluctuations time series by a two-to-four-month time lag. Figure 26.9b illustrates that the coherence between runoff and water level fluctuations has increased from 8–14-month frequency band in 1974 to 8–17-month frequency band in 1981; also from 1984, runoff effect on the lake water level period was stretched to 88–132-month frequency band which shows the period increasing and phase lag impact on water level fluctuations. As shown in Figure 26.9e, the runoff is effective on water level fluctuations from the beginning up to 2001; however, since 1981, its impact has larger frequency bands and long-term periods. This represents the relevance of runoff effect on decreasing water level and so the fluctuations which increase the phase lag are confirmatory on this output.

According to wavelet coherence graphs of rainfall and lake water level time series for the Zarine basin (Figure 26.8c), it can be seen that the rainfall parameter has a high coherency value from the beginning to the end of the investigated time intervals. In addition to rainfall data, as can be seen in WTC graphs between runoff and water level time series in Zarine river (Figure 26.9d), the effect of this variable from the beginning of the investigated time in an 8–16-month period is gradually reduced and even has become negligible since 1981 whereas the impact of runoff has been changed to a period of 46–84 months since 1977 to 1997. In other words, in this time interval, the runoff has affected the water level fluctuations in larger periods, which causes a decrease in fluctuations and, consequently, a decreasing water level in the lake. Also, phase lag in this period increases compared to the previous periods. Therefore, since the Zarine River originates from the Chelcheshme Mountains in Sagez, it can be concluded that runoff parameters and human factors have less impact than natural factors in the Kurdistan region on the lake water level fluctuations.

Similar to the above-mentioned results, the wavelet coherence graphs between rainfall/runoff time series and water level fluctuations of Van Lake are shown in Figure 26.10. Rainfall time series in the 8–16-month frequency band during 1946–1955 and 1988–1992, 4–16-month frequency band

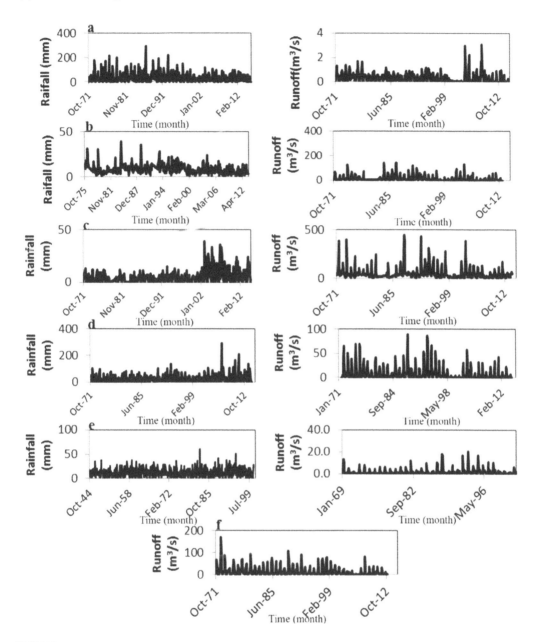

FIGURE 26.7   Rainfall/runoff time series in (a) Saeed Abad, (b) Tazekand (Miandoab), (c) Sarighamish, (d) Abajaloosofla, (e) Van stations, and (f) Vanyar runoff time series. (Source: Nourani et al., 2019.)

during 1958–1969 and 1973–1986, and four-to-eight-month frequency band during 1976–1980 and 1988–1993 with 0.8–0.9 coherency value showed the most effects on the lake water level fluctuations (Figure 26.10a). Also, coherency values between 0.8 and 0.9 have been obtained for the periods 1969–1958 and 1986–1973 at a 4–16-month frequency band. Arrows in Figure 26.10a with approximately 180° angle represent an anti-phase relationship between rainfall hydrological parameters and Van Lake water level fluctuations. Considering the coherence graph between runoff and Van Lake water level (Figure 26.10b), the 8–17-month frequency band during 1971–2002 with 0.9–1 coherency value reflects the impact of the fluctuations in most recent years. Also, in the four-to-eight-month frequency band between the periods 1970–1976, 1992–1995, 1998–2000, and

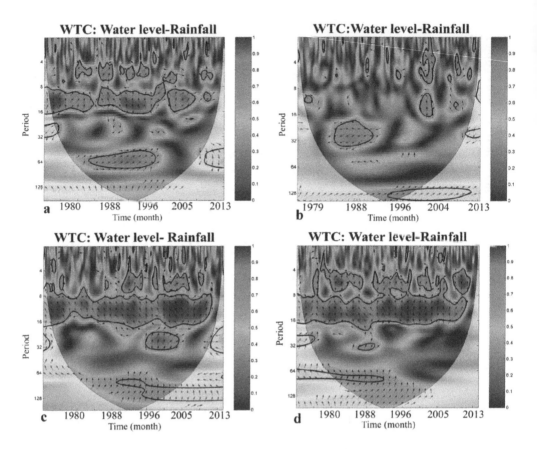

FIGURE 26.8 Wavelet coherence between Urmia Lake water level fluctuation and rainfall hydroclimatological parameters for (a) Saeed Abad, (b) Simine, (c) Zarine, and (d) Nazloo rivers. The thick black contour designates the 95% confidence level using red noise as background spectrum, and the cone of influence where edge effects affect interpretation is shown as a lighter shade.

2001–2002, runoff and water level fluctuations have shown 0.8–0.9 wavelet coherency value. In the 24–40-month frequency band during 1985–1993, 40–60 months during 1974–1976 with 0.8–0.9 wavelet coherency, and 7–14 months during 1969–1970 with 0.7–0.8 wavelet coherency, runoff is effective on Van Lake water level fluctuations and out of these bands, there is no common power between the runoff and the water level (Figure 26.10b). The phase lag between runoff and water level fluctuations is shown by the direction arrows within a 5% significance level. As can be seen in Figure 26.10b, the same direction of the arrows between the two time series reflects that runoff time series impact on Van Lake water level fluctuations without a significant time lag (delay). The obtained results for Van Lake demonstrate that rainfall influence on water level fluctuation has the same behavior as Urmia Lake. The effective periods of rainfall on Van Lake fluctuations gradually become smaller; also according to phase arrows, the phase lag in these periods is gradually decreasing (Figure 26.10a). Regarding the coherence between runoff and Van Lake water level fluctuations, runoff is almost constant during the study period. The results indicate that the 8–16-month period is extended throughout the entire interval. Even in 1989, the frequency band has become smaller. The direction of phase arrows confirms the absence of phase lag in runoff change and its impact on Van Lake water levels, while Urmia Lake in early 1981 has experienced a larger frequency band. In addition, the direction of phase arrows at significant levels indicates the impact of runoff phase lag on Urmia Lake water level fluctuations. General comparison between the water level time series of Urmia and Van Lakes shows stable behavior in the same period. As a representative of climatic

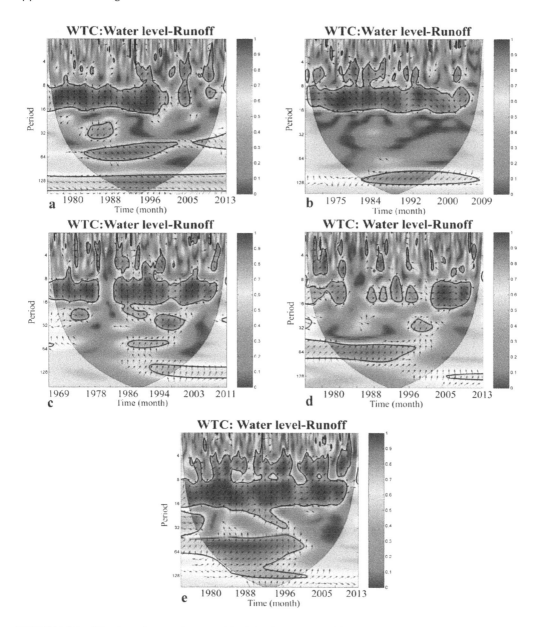

FIGURE 26.9 Wavelet coherence between Urmia Lake water level fluctuation and runoff hydroclimatological parameters for (a) Saeed Abad, (b) Vanyar, (c) Simine, (d) Zarine, and (e) Nazloo rivers. The thick black contour designates the 95% confidence level using red noise as background spectrum, and the cone of influence where edge effects affect interpretation is shown as a lighter shade. (Source: Nourani et al., 2019.)

factors, rainfall results show that this variable has the same behavior on both the lakes' levels. Therefore, it can be concluded that rainfall has less impact on the water level fluctuations of Urmia Lake. On the other hand, runoff, as a representative factor of human influence on the lake level, has an almost constant trend until the end of the study period.

According to WTC graphs between the water level of Lake Van and runoff (Figure 26.10b), the fluctuations have gradually become smaller since 1989 and the Van Lake water level has risen, thus lack of phase lag in the runoff parameter changes confirms consistency with Van Lake water level fluctuations. The runoff parameter in stations located around Urmia Lake has experienced greater periods in the same interval since early 1981 leading to decreasing water levels. Overall, it is clear

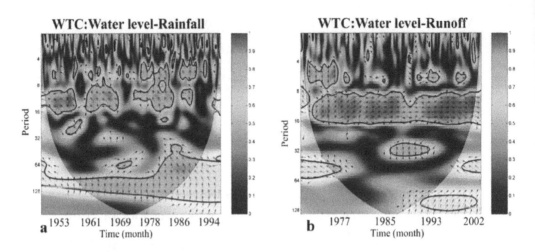

FIGURE 26.10 Wavelet coherence between Van Lake water level fluctuation and (a) rainfall and (b) runoff hydrological parameters. The thick black contour designates the 95% confidence level using red noise as background spectrum, and the cone of influence where edge effects affect interpretation is shown as a lighter shade. (Source: Nourani et al., 2019.)

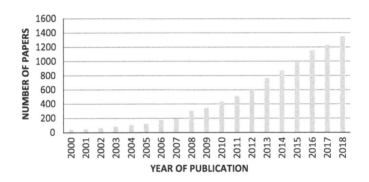

FIGURE 26.11 Number of published papers in regard to integral transforms applications in hydrology (indexed in Scopus) according to year of publication.

that the impact of rainfall data on the two lakes' fluctuations shows the same behavior but the runoff parameter has almost the dominant role in decreasing water level fluctuations.

26.4 DISCUSSIONS

In the last few decades, the advances in hydrological modeling and simulation achieved through integral transforms have led to an increase in related research and resulting publication numbers (Figure 26.11). In the previous section, a number of these studies have been presented in several fields of hydrology for each integral transform, with some examples listed in Table 26.1. The functionality of each transform in any considered field has also been investigated. Despite the capabilities of traditional frequency analysis techniques based on Fourier transforms, they tend to spread the energy of the signal into several frequencies, which sometimes leads to misinterpretations of the characteristics of the data. In particular, if the signal is nonstationary or nonlinear, the spectrum obtained from Fourier-based methods blurs in the low frequency and results in some spurious oscillations. With respect to the nonstationary nature of most hydrological phenomena, Hilbert-Huang and wavelet transforms are great of importance due to the abilities of decomposing nonstationary signals into sub-signals at different temporal scales and helping to better interpret them.

TABLE 26.1

Examples of Some Integral Transforms Applications in Hydrological Process Modeling

Hydrologic Process	Laplace	Fourier	Hankel	Mellin	Hilbert	HHT	Wavelet
Rainfall-runoff modeling	Diskin (1968);Wang and Chen (1996)	—	—	—	—	Xuehua et al. (2011)	Labat et al. (2000);Nourani et al. (2009b)
Streamflow modeling	Bayazit et al. (2001)	—	—	—	—	Chiew et al. (2005);Di et al. (2014)	Anctil and Tape (2004);Tang et al. (2012)
Groundwater flow problems	Hantush and Jacob (1955);Chen and Stone (1993)	Kendal and Hyndman (2007)	Hantush (1960);Russell and Prats (1962)	Chan et al. (1978);Craster (1997)	—	—	Wang and Ding (2003)
Solute transport problems	Flury et al. (1998);Singh et al. (2010c)	—	—	—	—	—	—
Temperature data analysis	—	Faye et al. (2014); Gurudeo and Mahbub (2010)	—	—	Zappalà et al. (2016)	Molla et al. (2006b);Ma et al. (2015)	Nalley et al. (2013);Araghi et al. (2015)
Wind data analysis	—	Mukhopadhyay et al. (2013);Elmaghraby et al. (2016)	—	—	Zhu and Yang (2002);Alpay et al. (2006)	Xu and Chen (2004);Jing-Jing and Fei (2014)	Avdakovic et al. (2011)
Precipitation data analysis	—	Kendall and Hyndman (2007)	—	—	—	Baines (2005);Srikanthan et al (2011)	Kantelhardt et al. (2003);Özger et al. (2010)
Ocean waves study	—	—	—	—	Huang et al (1992);Mercier et al. (2008)	Hwang et al. (2003);Veltcheva and Soares (2016)	Massel (2001)
Underwater acoustic signal detection	—	—	Frisk and Lynch (1984);Wengrovitz (1986)	—	—	Wang et al. (2006)	Chang and Wang (2003)
Hydrologic uncertainty analysis	—	—	—	Tung (1990);Chen and Tung (2002)	—	—	—

The HHT is based on the local characteristic time scale of the data. With the concept of instantaneous frequency, HHT provides a better representation in the frequency domain than the conventional Fourier spectra, which have fixed frequency bins; hence the HHT provides a finer resolution in time-frequency representation. It provides not only the local, adaptive and efficient information directly from the data without any linearity or stationary assumptions or restrictions, but also a more physically meaningful interpretation of the underlying dynamic processes. Therefore it can be applied to nonlinear time-variant signals. These strengths make it a powerful tool to investigate the characteristics of environmental and hydrologic time series. However, it is purely empirical and lacks a theoretical basis.

Wavelet transform is another analysis technique suitable to nonstationary time series which has been already become a widely used tool. One of the main advantages of WT is that it may decompose a signal directly according to the frequency and represents it in the frequency domain distribution state in the time domain. As for the wavelet transformation, both time and frequency information of the signal are retained. It is thus a more powerful transformation for time-frequency analysis and its better performance compared with traditional methods has been adequately demonstrated.

26.5 CONCLUSIONS

Integral transforms are one of the most-used mathematical tools in various fields of hydrology. They have been widely applied for hydrological process modeling and time series analysis. In this chapter, it has also been tried to present a brief introduction to integral transforms (such as Laplace, Fourier, Hankel, Mellin, Hilbert, and wavelet) and investigate several dozen successful applications of these transforms for hydrological process modeling (e.g., streamflow, rainfall-runoff, groundwater, solute transport, temperature, wind). Meanwhile, Hilbert-Huang and wavelet transforms have been taken into consideration due to their ability to better interpret climatic and hydrological time series that are often nonstationary. Hereupon their applications are extensively studied and some applied examples are presented.

On the other hand, since each transform has its own deficiencies and application restrictions, exact results of a hydrologic series analysis may not be gained using any transform alone. To dominate this problem, using hybrid methods is an effective approach. For example, many methods have been combined with wavelet transform for hydrologic series analysis, such as the combinations of information entropy theory, artificial neural network, genetic programming, and support vector machine with wavelet theory. Studies and practical applications have also indicated the better performances of hybrid methods compared to a single method. Certainly, with further development of hybrid methods, the overall performance of models will be enhanced in the future.

NOTES

1. Hybrid wavelet-ANN model.
2. Hybrid wavelet-ANFIS model.

REFERENCES

Adamowski, J.F. 2008. Development of a short-term river flood forecasting method for snowmelt driven floods based on wavelet and cross-wavelet analysis, *Journal of Hydrology*, 353(3–4): 247–266.

Alpay, S., Bilir, L., Ozdemir, S., and Ozerdem, B. 2006. Wind speed time series characterization by Hilbert transform, *International Journal of Energy Research*, 30: 359–364.

Anctil, F., and Tape, D.G. 2004. An exploration of artificial neural network rainfall– runoff forecasting combined with wavelet decomposition, *Journal of Environmental Engineering and Science*, 3: 121–128.

Anderson, P.L., Tesfaye, Y.G., and Meerschaert, M.M. 2007. Fourier-PARMA models and their application to river flows, *Journal of Hydrologic Engineering*, 12: 462–472.

Araghi, A., Mousavi Baygi, M., Adamowski, J., Malard, J., Nalley, D., and Hasheminia, S.M. 2015. Using wavelet transforms to estimate surface temperature trends and dominant periodicities in Iran based on gridded reanalysis data, *Atmospheric Research*, 155: 52–72.

Avdakovic, S., Lukac, A., Nuhanovic, A., and Music, M. 2011. Wind speed data analysis using wavelet transform, *World Academy of Science, Engineering and Technology*, 51: 829–833.

Baines, P.G. 2005. Long-term variations in winter rainfall of Southwest Australia and the African monsoon, *Australian Meteorological Magazine*, 54: 91–102.

Balocchi, R., Menicucci, D., Santarcangelo, E., Sebastiani, L., Gemignani, A., Ghelarducci, B., and Varanini, M. 2004. Deriving the respiratory sinus arrhythmia from the heartbeat time series using empirical mode decomposition, *Chaos, Solitons and Fractals*, 20: 171–177.

Bayazit, M., Önöz, B., and Aksoy, H. 2001. Nonparametric streamflow simulation by wavelet or fourier analysis, *Hydrological Sciences Journal*, 46(4): 623–634.

Butzer, P., and Jansche, S. 1997. A direct approach to the mellin transform, *Journal of Fourier Analysis and Applications*, 3: 325–376.

Cahen, E. 1894. Sur la fonction $\zeta(s)$ de Riemann et sur des functions analogues, *Annales Scientifiques de l'École Normale Supérieure*, 11: 75–164.

Cao, L.Q., and Lin, Z.S. 2010. Application of the new technology HHT to the forecasting of annual average streamflow near the three gorges dam on the Yangtze River in China, *IAHS Publ. 350*, 2011(1): 597–602.

Cavanagh, E.C., and Cook, B.D. 1979. Numerical evaluation of Hankel Transform via Gaussian–Laguerre polynomial expressions, *IEEE Trans. Acoust. Speech Signal Process. (ASSP)*, 27: 361–366.

Chan, Y.K., Mullineux, N., Reed, J.R., and Wells, G.G. 1978. Analytic solutions for drawdowns in wedge-shaped artesian aquifers, *Journal of Hydrology*, 36(3–4): 233–246.

Chang, S.H., and Wang, F.T. 2003. The application of the robust discrete wavelet transform to underwater sound, *International Journal of Electronics*, 90: 361–371.

Chen, C.S., and Stone, W.D. 1993. Asymptotic calculation of Laplace inverse in analytical solutions of groundwater problems, *Water Resources Research*, 29(1): 207–209.

Chen, X., and Tung, Y.K. 2002.Time-dependent reliability analysis of local scour around bridge piers. In: *Conference: Advances in Hydraulics and Water Engineering-13th IAHR-APD Congress*, 266–271.

Chiew, F.H.S., Peel, M.C., Amirthanathan, G.E., and Pegram, G.G.S. 2005. Identification of oscillations in historical global streamflow data using empirical mode decomposition. In: *Regional Hydrological Impacts of Climatic Change – Hydroclimatic Variability, Proceedings of Symposium S6 Held during the Seventh IAHS Scientific Assembly at Foz do Iguacu*, IAHS Publication, Brazil, Vol. 296, 53–62.

Craster, R.V. 1997. The solution of a class of free boundary problems, *Proceedings: Mathematical, Physical and Engineering Sciences*, 453(1958): 607–630.

Dadu, K.S., and Deka, P. 2016. Applications of wavelet transform technique in hydrology - A brief review. In: Sarma, A., Singh, V., Kartha, S., and Bhattacharjya, R. (Eds.), *Urban Hydrology, Watershed Management and Socio-Economic Aspects*, 241–253.

Dätig, M., and Schlurmann, T. 2004. Performance and limitations of the Hilbert–Huang transformation (HHT) with an application to irregular water waves, *Ocean Engineering*, 31(14–15): 1783–1834.

Daubechies, I. 1988. Orthogonal bases of compactly supported wavelets, *Communications on Pure and Applied Mathematics*, 41: 909–996.

Debnath, L., and Bhatta, D. 2014. *Integral Transforms and Their Applications* (3rd ed.), Chapman and Hall/CRC, New York, p. 818.

DeChant, C.M., and Moradkhani, H. 2014. Hydrologic prediction and uncertainty quantification. In: Eslamian, S. (ed.), *Handbook of Engineering Hydrology: Modeling, Climate Change, and Variability*, CRC Press, Taylor and Francis Group, Boca Raton, FL, 389–416.

Delleur, J.W., and Rao, R.A. 1971. Linear system analysis in hydrology-the transform approach, the kernel function and effect of noise. In: *Proceedings of the first bilateral U.S.–Japan Seminar on Hydrology*, Water Resources Publications, Fort Collins, CO, USA, 116–142.

Di, C., Yang, X., Zhang, X., He, J., and Mei, Y. 2014. Multi-scale analysis of streamflow using the hilbert-huang transform, *International Journal of Numerical Methods for Heat and Fluid Flow*, 24(6): 1363–1377.

Diskin, M.H. 1967. A laplace transform proof of the theories of moments for the instantaneous unit hydrograph, *Water Resources Research*, 3: 385–388.

Diskin, M.H. 1968. Transfer functions for the analysis of rainfall-runoff relations. In: *International Association of Hydrological Sciences Publication, No. 85*.

Douka, E., and Hadjileontiadis, LJ. 2005. Time-frequency analysis of the free vibration response of a beam with a breathing crack, *NDT and E International*, 38: 3–10.

Eldabe, N.T., El-Shahed, M., and Shawkey, M. 2004. An extension of the finite Hankel transform, *Appl. Math. Comput.*, 151: 713–717.

ELmaghraby, E.K., Abu Khadra, S.A., and Eissa, H.S. 2016. Using the fast fourier transform technique for climate time series decomposition, *Arab Journal of Nuclear Sciences and Applications*, 49(3): 78–85.

Faye, E., Herrera, M., Bellomo, L., Silvain, J.F., and Dangles, O. 2014. Strong discrepancies between local temperature mapping and interpolated climatic grids in tropical mountainous agricultural landscapes, *PLoS ONE*, 9(8): e105541.

Fleming, S.W., Lavenue, A.M., Aly, A.H., and Adams, A. 2002. Practical applications of spectral analysis to hydrologic time series, *Hydrological Processes*, 16: 565–574.

Flury, M., Wu, Q.J., Wu, L., and Xu, L. 1998. Analytical solution for solute transport with depth dependent transformation or sorption coefficient, *Water Resources Research*, 3411: 2931–2937.

Foufoula-Georgiou, E., and Kumar, P. 1994. *Wavelets in Geophysics*, Academic Press, San Diego, California, USA, Vol. 4.

Frisk, G.V., and Lynch, J.F. 1984. Shallow water waveguide characterization using the Hankel transform, *Journal of the Acoustical Society of America*, 76(1): 205–216.

Gaucherel, C. 2002. Use of wavelet transform for temporal characterization of remote watersheds, *Journal of Hydrology*, 269(3–4): 101–121.

Gloersen, P., and Huang, N. 2003. Comparison of interannual intrinsic modes in Hemispheric sea ice covers and other geophysical parameters, *IEEE Transactions on Geoscience and Remote Sensing*, 41(5): 1062–1074.

Goupillaud, P., Grossmann, A., and Morlet, J. 1984. Cycle-octave and related transforms in seismic signal analysis, *Geoexploration*, 23(1): 85–102.

Grossmann, A., and Morlet, J. 1984. Decomposition of Hardy function into square integrable wavelets of constant shape, *Journal of Mathematical Analysis*, 5: 723–736.

Grossmann, A., and Morlet, J. 1985a. Decomposition of functions into wavelets of constant shape and related transforms. In: *Mathematics and Physics, Lectures on Recent Results*, World Scientific, Singapore.

Grossmann, A., Morlet, J., and Paul, T. 1985b. Transforms associated to square integrable group representations, *Journal of Mathematical Physics*, 26: 2473–2479.

Guo, E., Zhang, J., Wang, Y., Quan, L., Zhang, R., Zhang, F., and Zhou, M. 2019. Spatiotemporal variations of extreme climate events in Northeast China during 1960–2014, *Ecological Indicators*, 96(1): 669–683.

Hansen, E.V. 1985. Fast Hankel transform algorithms, *IEEE Transactions on Acoustics, Speech, and Signal Processing (ASSP)*, 33: 666–671.

Hantush, M.S. 1960. Modification of the theory of leaky aquifers, *Journal of Geophysical Research*, 64: 1043–1052.

Hantush, M.S. 1967. Flow to wells in aquifers separated by a semipervious layer, *Journal of Geophysics Research*, 72(6): 1709–1720.

Hantush, M.S., and Jacob, C.E. 1955. Non-steady radial flow in an infinite leaky aquifer, *American Geophysical Union Transcripts*, 36: 95–100.

Hardy, G. 1924. On Hilbert transforms, *Messenger of Mathematics*, 54: 20–27, 81–88.

Higgins, W.E., and Munson, D.C. 1988. A Hankel transform approach to tomographic image reconstruction, *IEEE Transactions on Medical Imaging*, 7: 59–72.

Hsieh, C.H., and Dai, C.F. 2012. The analysis of offshore islands wind characteristics in Taiwan by Hilbert–Huang transform, *Journal of Wind Engineering and Industrial Aerodynamics*, 107–108: 160–168.

Huang, N.E., Chern, C.C., Huang, K., Salvino, L.W., Long, S.R., and Fan, K.L. 2001. A new spectral analysis of station TCU129, Chi-Chi, Taiwan, 21 September 1999, *Bulletin of the Seismological of America*, 91(5): 1310–1338.

Huang, N.E., Long, S.R., and Shen, Z. 1996. The mechanism for frequency downshift in nonlinear wave evolution, *Advances in Applied Mechanics*, 32: 59–111.

Huang, N.E., Long, S.R., Tung, C.C., Donelan, M.A., Yuan, Y., and Lai, R.J. 1992. The local properties of ocean surface waves by the phase-time method, *Geophysical Research Letters*, 19(7): 685–688.

Huang, N.E., Shen, Z., and Long, S.R. 1999. A new view of nonlinear water waves—The Hilbert spectrum, *Annual Review of Fluid Mechanics*, 31: 417–457.

Huang, N.E., Shen, Z., Long, S.R., Wu, M.L.C., Shih, H.H., Zheng, Q., Yen, N.C., Tung, C.C., and Liu, H.H. 1998. The empirical mode decomposition and the Hilbert spectrum for nonlinear and non-stationary time series analysis, *Proceedings of the Royal Society of London. Series A: Mathematical, Physical and Engineering Sciences*, 454: 903–995.

Huang, N.E., Wu, M., Qu, W., Long, S.R., and Shen, S.S.P. 2003. Applications of Hilbert-Huang transform to non-stationary financial time series analysis, *Applied Stochastic Models in Business and Industry*, 19: 245–268.

Huang, Y., Schmitt, F.G. Lu, Z and Liu, Y. 2009. Analysis of daily river flow fluctuations using Empirical Mode Decomposition and arbitrary order Hilbert spectral analysis, *Journal of Hydrology*, 373(1): 103–111.

Hwang, P.A., Huang, N.E., and Wang, D.W. 2003. A note on analyzing nonlinear and nonstationary ocean wave data, *Applied Ocean Research*, 25(4): 187–193.

Hwang, P.A., Kaihatu, J.M., and Wang, D.W. 2002. A comparison of the energy flux computation of shoaling waves using Hilbert and wavelet spectral analysis techniques. In: *7th Int. Workshop on Wave Hindcasting and Forecasting*, Vol. 10, 21–25.

Ilias, T.S., Thomas, Z.S., and Andreas, P.C. 2008. Water table fluctuation in aquifers overlying a semi-impervious layer due to transient recharge from a circular basin, *Journal of Hydrology*, 348(1–2): 215–223.

Jeihouni, E., Eslamian, S., Mohammadi, M., and Zareian, M.J. 2019. Simulation of groundwater level fluctuations in response to main climate parameters using a wavelet–ANN hybrid technique for the Shabestar Plain, Iran, *Environmental Earth Science*, 78: 293.

Jing-Jing, X., and Fei, H. 2014. Analysis of nonstationary wind fluctuations using the hilbert-huang transform, *Atmospheric and Oceanic Science Letters*, 7(5): 428–433.

Johnson, P. 1970. Calculation of the instantaneous unit hydrograph using Laplace transforms, *Journal of Hydrology (New Zealand)*, 9(2): 307–322.

Joshi, N., Gupta, D., Suryavanshi, S., Adamowski, J., and Madramootoo, C.A. 2016. Analysis of trends and dominant periodicities in drought variables in India: A wavelet transform based approach, *Atmospheric Research*, 182: 200–220.

Kantelhardt, J.W., Rybski, D., Zschiegner, S.A., Braun, P., Koscielny-Bunde, E., Livina, V., Havlin, Sh., and Bunde, A. 2003. Multifractality of river runoff and precipitation: Comparison of fluctuation analysis and wavelet methods, *Physica A: Statistical Mechanics and Its Applications*, 330(1–2): 240–245.

Kbaier Ben Ismail, D., Lazure, P., and Puillat, I. 2016. Application of Hilbert-Huang decomposition to temperature and currents data in the Réunion island. In: *OCEANS 2016 MTS/IEEE Monterey*, Monterey, CA, USA, 1–9.

Kendal, A.D., and Hyndman, D.W. 2007. Examining watershed processes using spectral analysis methods including the scaled-widowed fourier transform. In: *AGU Monograph, Data Integration in Subsurface Hydrology*, Vol. 171, 183–200.

Kisi, O., and Cimen, M. 2011. A wavelet-support vector machine conjunction model for monthly stream-flow forecasting, *Journal of Hydrology*, 399: 132–140.

Kisi, O., and Shiri, J. 2012. Wavelet and neuro-fuzzy conjunction model for predicting water table depth fluctuations, *Hydrology Research*, 43(3): 286–300.

Klingspor, M. 2015. *Hilbert Transform : Mathematical Theory and Applications to Signal Processing*. M.S. thesis, Linköping University, Linköping, Sweden.

Kuai, K.Z., and Tsai, C.W. 2012. Identification of varying time scales in sediment transport using the Hilbert–Huang Transform method, *Journal of Hydrology*, 420–421: 245–254.

Kulkarni, V.S., and Deshmukh, K.C. 2008. An inverse quasi-static steady-state in a thick circular plate, *Journal of The Franklin Institute*, 345(1): 29–38.

Kumar, A., Jaiswal, D.K., and Yadav, R.R. 2012. Analytical solutions of one-dimensional temporally dependent advection-diffusion equation along longitudinal semi-infinite homogeneous porous domain for uniform flow, *IOSR Journal of Mathematics (IOSRJM)*, 2(1): 1–11, ISSN: 2278-5728.

Kumar, P., and Foufoula-Georgiou, E. 1993. A multi-component decomposition of spatial rainfall fields. 1. Segregation of large- and small-scale features using wavelet transforms, *Water Resources Research*, 29(8): 2515–2532.

Labat, D. 2005. Recent advances in wavelet analyses: Part 1. A review of concepts, *Journal of Hydrology*, 314: 275–288.

Labat, D., Ababou, R., and Mangin, A. 2000. Rainfall–runoff relations for karstic springs. Part II: Continuous wavelet and discrete orthogonal multiresolution analyses, *Journal of Hydrology*, 238(3–4): 149–178.

Lam, W.S., Ray, W., Guzdar, P.N., and Roy, R. 2005. Measurement of Hurst exponents for semiconductor laser phase dynamics, *Physical Review Letters*, 94(1): 010602.

Liang, H., Bressler, S.L., Desimone, R., and Fries, P. 2005. Empirical mode decomposition: A method for analyzing neural data, *Neurocomputing*, 65–66: 801–807.

Lindstrom, F.T., and Boersma, L. 1989. Analytical solutions for convective dispersive transport in confined aquifers with different initial and boundary conditions, *Water Resources Research*, 25(2): 241–256.

Liu, B., Yan, S., Chen, X., Lian, Y., and Yanbo, X. 2014. Wavelet analysis of the dynamic characteristics of saltwater intrusion-A case study in the Pearl River Estuary of China, *Ocean & Coastal Management*, 95: 81–92.

Liu, Y., Brown, J., Demargne, J., and Seo, D.-J. 2011. A wavelet-based approach to assessing timing errors in hydrologic predictions, *Journal of Hydrology*, 397(3–4): 210–224.

Long, S.R., Huang, N.E., Tung, C.C., Wu, M.L., Lin, R.Q., et al 1993. The Hilbert techniques: An alternate approach for non-steady time series analysis, *IEEE, Geoscience and Remote Sensing Letters*, 3: 6–11.

Luo, H., Fang, X., and Ertas, B. 2009. Hilbert transform and its engineering applications, *AIAA Journal*, 47(4): 923–932.

Ma, H., Qiu, X., Luo, J., Gu, P., and Liu, Y. 2015. Analysis of temperature time series based on Hilbert-Huang Transform, *Journal of Hydrodynamics, Ser. B*, 27(4): 587–592.

Magni, V., Cerullo, G., and De Silvestri, S. 1992. High-accuracy fast Hankel transform for optical beam propagation, *Journal of the Optical Society of America*, 12: 2031–2033.

Marcolini, G., Bellin, A., Disse, M., and Chiogna, G. 2017. Variability in snow depth time series in the Adige catchment, *Journal of Hydrology: Regional Studies*, 13: 240–254.

Massei, N., Dupont, J.P., Mahler, B.J., Laignel, B., Fournier, M., Valdes, D., and Ogier, S. 2006. Investigating transport properties and turbidity dynamics of a karst aquifer using correlation, spectral, and wavelet analyses, *Journal of Hydrology*, 329(1–2): 244–257.

Massel, S.R. 2001. Wavelet analysis for processing of ocean surface wave records, *Ocean Engineering*, 28(8): 957–987.

Mathias, S.A., and Zimmerman, R.W. 2003. Laplace transform inversion for late-time behavior of groundwater flow problems, *Water Resources Research*, 39: 1283.

Mcmahon, T.A., Kiem, A.S., Peel, M.C., Jordan, P.W., and Pegram, G.G.S. 2008. A new approach to stochastically generating six-monthly rainfall sequences based on empirical mode decomposition, *Journal of Hydrometeorology*, 9(6): 1377–1389.

Mellin, H. 1896. Über die fundamentale Wichtgkeit des Satzes von Cauchy fur die Theorien der Gamma-und der hypergeometrischen funktionen, *Acta Societas Scientiarum Fennicae*, 21: 1–115.

Mellin, H. 1902. Über den Zusammenhang zwischen den linearen Differential-und Differezengleichugen, *Acta Mathematica*, 25: 139–164.

Mercier, M.J., Garnier, N., and Dauxois, T. 2008. Reflection and diffraction of internal waves analyzed with the Hilbert transform, *Physics of Fluids*, 20(8): 086601.

Molla, M.K.I., Rahman, M.S., Sumi, A., and Banik, P. 2006a. Empirical mode decomposition analysis of climate changes with special reference to rainfall data, *Discrete Dynamics in Nature and Society*, 2006: 1–17, Article ID 45348.

Molla, M.K.I., Sumi, A., and Rahman, M.S. 2006b. Analysis of temperature change under global warming impact using empirical mode decomposition, *International Journal of Information Technology*, 3(2): 131–139.

Morlet, J., Arens, G., Fourgeau, I., and Giard, D. 1982. Wave propagation and sampling theory, *Geophysics*, 47: 203–236.

Mukhopadhyay, S., Dash, D., Mitra, A., and Bhattacharya, P. 2013. *A Comparative Study between Seasonal Wind Speed by Fourier and Wavelet Analysis*. India institute of Science Research Kolkata, Mohanpur Campus, Nadia, India.

Nalley, D., Adamowski, J., and Khalil, B. 2012. Using discrete wavelet transforms to analyze trends in streamflow and precipitation in Quebec and Ontario (1954–2008), *Journal of Hydrology*, 475: 204–228.

Nalley, D., Adamowski, J., Khalil, B., and Ozga-Zielinski, B. 2013. Trend detection in surface air temperature in Ontario and Quebec, Canada during 1967–2006 using the discrete wavelet transform, *Atmospheric Research*, 132–133: 375–398.

Nason, G.P., and Sachs, R. 1999. Wavelets in time series analysis, *Philosophical Transactions of the Royal Society*, 357: 2511–2526.

Neuman, S.P., and Li, Y.H. 2004. Flow to a well in a five-layer system with application to the oxnard basin, *Ground Water*, 45(6): 672–682.

Nordin, C.F., and Beverage, J.P., 1965. *Sediment Transport in the Rio Grande, New Mexico-USGS Professional Paper 462-F*, US Government Printing Office, Washington, DC.

Nourani, V., Alami, M.T., and Aminfar, M.H. 2009a. A combined neural-wavelet model for prediction of Ligvanchai watershed precipitation, *Engineering Applications of Artificial Intelligence*, 22(3): 466–472.

Nourani, V., Ghasemzade, M., Danande Mehr, A., and Sharghi, E. 2019. Investigating the effect of hydroclimatological variables on Urmia Lake water level using wavelet coherence measure. *Journal of Water and Climate Change*, 10(1): 13–29.

Nourani, V., Kisi, O., and Komasi, M. 2011. Two hybrid artificial intelligence approaches for modeling rainfall–runoff process, *Journal of Hydrology*, 402(1–2): 41–59.

Nourani, V., Komasi, M., and Alami, M. 2012. Hybrid Wavelet-genetic programming approach to optimize ANN modeling of rainfall–runoff process, *Journal of Hydrologic Engineering*, 16(6): 724–741.

Nourani, V., Komasi, M., and Mano, A. 2009b. A multivariate ANN-wavelet approach for rainfall–runoff modeling, *Water Resources Management*, 23(14): 2877–2894.

Nourani, V., and Parhizkar, M. 2013. Conjunction of SOM-based feature extraction method and hybrid wavelet–ANN approach for rainfall–runoff modeling, *Journal of Hydroinformatics*, 15(3): 829–848.

Olkkonen, J. 2011. *Discrete Wavelet Transforms - Theory and Applications*, Janeza Trdine 9, Rijeka, Croatia.

Onderka, M., Banzhaf, S., Scheytt, T., and Krein, A. 2013. Seepage velocities derived from thermal records using wavelet analysis, *Journal of Hydrology*, 479: 64–74.

Ortega, J., and Smith, G.H. 2009. Hilbert–Huang transform analysis of storm waves, *Applied Ocean Research*, 31(3): 212–219.

Ouyang, Y., Parajuli, P.B., Li, Y., Leininger, T.D., and Feng, G. 2017. Identify temporal trend of air temperature and its impact on forest stream flow in Lower Mississippi River Alluvial Valley using wavelet analysis, *Journal of Environmental Management*, 198: 21–31, ISSN 0301-4797.

Özger, M., Mishra, A.K., and Singh, V.P. 2010. Scaling characteristics of precipitation data in conjunction with wavelet analysis, *Journal of Hydrology*, 395(3–4): 279–288.

Pachori, R.B. 2008. Discrimination between ictal and seizure-free EEG signals using empirical mode decomposition, *Research Letters in Signal Processing*, 2008: 1–5, Article ID 293056.

Pan, J., Yan, X., Zheng, Q., Liu, W.T., and Klemas, V.V. 2003. Interpretation of scatterometer ocean surface wind vector EOFs over the Northwestern Pacific, *Remote Sensing of Environment*, 84(1): 53–68.

Partal, T., and Cigizoglu, H.K. 2008. Estimation and forecasting of daily suspended sediment data using wavelet-neural networks, *Journal of Hydrology*, 358: 317–331.

Patella, D. 1980. Gravity interpretation using the Hankel transform, *Geophysical Prospecting*, 28: 744–749.

Peng, Z.K., Tse, P.W., and Chu, F.L. 2005. A comparison study of improved Hilbert-Huang transform and wavelet transform: Application to fault diagnosis for rolling bearing, *Mechanical Systems and Signal Processing*, 19(5): 974–988.

Pines, D.J., and Salvino, L.W. 2002. Health monitoring of one-dimensional structures using empirical mode decomposition and the Hilbert-Huang transform, *Smart Structures and Materials 2002: Smart Structures and Integrated Systems*, 4701: 127–143.

Ponomarenko, V.I., Prokhorov, M.D., Bespyatov, A.B., Bodrov, M.B., and Gridnev, V.I. 2005. Deriving main rhythms of the human cardiovascular system from the heartbeat time series and detecting their synchronization, *Chaos. Solitons and Fractals*, 23: 1429–1428.

Potočki, K., Gilja, G., and Kunštek, D. 2017. An overview of the applications of wavelet transform for discharge and suspended sediment analysis, *Technical Gazette*, 24: 1561–1569.

Quek, S.T., Tua, P.S., and Wang, Q. 2003. Detecting anomalies in beams and plate based on Hilbert-Huang transform of real signals, *Smart Materials and Structures*, 12: 447–460.

Rao, A.R., and Hsu, E.C. 2008. *Hilbert-Huang Transform Analysis of Hydrological and Environmental Time Series*, Springer, Dordrecht, the Netherlands, 10.1007/978-1-4020-6454-8.

Rasmuson, A. 1981. Diffusion and sorption in particles and two dimensional dispersion in porous medium, *Water Resources Research*, 17(2): 321–328.

Ravansalar, M., Rajaee, T., and Kisi, O. 2017. Wavelet-linear genetic programming: A new approach for modeling monthly streamflow, *Journal of Hydrology*, 549, 461–475.

Riemann, B. 1876. Über die Anzahl der Primzahlen unter eine gegebenen Grosse, *Gesammelte Mathematische Werke*, 136–144.

Roushangar, K., and Alizadeh, F. 2018. Entropy-based analysis and regionalization of annual precipitation variation in Iran during 1960–2010 using ensemble empirical mode decomposition, *Journal of Hydroinformatics*, 20(2): 468–485.

Russell, D.G., and Prats, M. 1962. Performance of layered reservoirs with crossflow--Single-compressible-fluid case, *Society of Petroleum Engineers Journal*, 2(1): 53–67.

Salisbury, J.I., and Wimbush, M. 2002. Using modern time series analysis techniques to predict ENSO events from the SOI time series, *Nonlinear Processes in Geophysics*, 9(3): 341–345.

Sang, Y.F. 2013. A review on the applications of wavelet transform in hydrology time series analysis, *Atmospheric Research*, 122: 8–15.

Schaefli, B., Maraun, D., and Holschneider, M. 2007. What drives high flow events in the Swiss Alps? Recent developments in wavelet spectral analysis and their application to hydrology, *Advances in Water Resources*, 30: 2511–2525.

Shiri, J. 2018. Improving the performance of the mass transfer-based reference evapotranspiration estimation approaches through a coupled wavelet-random forest methodology, *Journal of Hydrology*, 561: 737–750.

Simon, M., and Tomlinson, G.R. 1984. Use of the Hilbert transform in modal analysis of linear and nonlinear structures, *Journal of Sound and Vibration*, 96(4): 421–436.

Sinclair, S., and Pegram, G.G.S. 2005. Empirical mode decomposition in 2-D space and time: A tool for space-time rainfall analysis and nowcasting, *Hydrology and Earth System Sciences*, 9: 127–137.

Singh, M.K., Singh, P., and Singh, V.P. 2010b. Analytical solution for two dimensional solute transport in finite aquifer with time dependent source concentration, *Journal of Engineering Mechanics-ASCE*, 136(10): 1309–1315.

Singh, M.K., Singh, P., and Singh, V.P. 2010c. Analytical solution for solute Transport along and against time dependent source concentration in homogeneous finite aquifers. *Advances in Theoretical and Applied Mechanics*, 3: 99–119.

Singh, V., Pandey, R., and Singh, S. 2010a. A stable algorithm for Hankel transforms using hybrid of Block-pulse and Legendre polynomials, *Computer Physics Communications*, 181: 1–10.

Smith, L.C., Turcotte, D.L., and Isacks, B.L. 1998. Stream flow characterization and feature detection using a discrete wavelet transform, *Hydrological Processes*, 12(2): 233–249.

Srikanthan, R., Peel, M.C., McMahon, T.A., and Karoly, D.J. 2011. Ensemble empirical mode decomposition of Australian monthly rainfall and temperature data. In: *MODSIM2011, 19th International Congress on Modelling and Simulation*, December 2011, edited by Chan, F., Marinova, D., and Anderssen, R.S., Modelling and Simulation Society of Australia and New Zealand, 3643–3649.

Tang, X.L., Li, J.F., Lv, X., and Long, H.L. 2012. Analysis of the characteristics of runoff in manasi river basin in the past 50 years, *Procedia Environmental Sciences*, 13: 1354–1362.

Tularam, A., and Mahbub, I. 2010. Time series analysis of rainfall and temperature interactions in coastal catchments, *Journal of Mathematics and Statistics*, 6(3): 372–380.

Tung, Y.K. 1990. Mellin transform applied to uncertainty analysis in hydrology/hydraulics, *Journal of Hydraulic Engineering-ASCE*, 116: 659–674.

Veltcheva, A., and Soares, C.G. 2016. Analysis of wave groups by wave envelope-phase and the Hilbert Huang transform methods, *Applied Ocean Research*, 60: 176–184.

Veltcheva, A.D. 2002. Wave and group transformation by a Hilbert Spectrum, *Coastal Engineering Journal*, 44(4): 283–300.

Verma, A.P. 1969. The Laplace transform solution of a one dimensional groundwater recharge by spreading, *Annals of Geophysics*, 22(1): 25–31.

Vincent, C.L., Pinson, P., and Giebela, G. 2011. Wind fluctuations over the North Sea, *International Journal of Climatology*, 31: 1584–1595.

Wang, F.T., Chang, S.H., and Lee J.C.Y. 2006. Signal detection in underwater sound using the empirical mode decomposition, *IEICE Transactions on Fundamentals of Electronics, Communications and Computer Sciences*, E89-A: 2415–2421.

Wang, G.T., and Chen, S. 1996. A linear spatially distributed model for a surface rainfall-runoff system, *Journal of Hydrology*, 185(1–4): 183–198.

Wang, W., and Ding, S. 2003. Wavelet network model and its application to the predication of hydrology, *Nature and Science*, 1(1): 67–71.

Wengrovitz, M.S. 1986. *The Hilbert-Hankel Transform and Its Application to Shallow Water Ocean Acoustics*, Massachusetts Institute of Technology, Research Laboratory of Electronics, Technical Report No. 513.

Wu, Z., and Huang, N.E. 2009. Ensemble empirical mode decomposition: A noise-assisted data analysis method, *Advances in Adaptive Data Analysis*, 1(1): 1–41.

Xie, K., Wang, Y., Wang, K., and Cai, X. 2010. Application of Hankel transforms to boundary value problems of water flow due to a circular source, *Applied Mathematics and Computation*, 216(5): 1469–1477.

Xu, Y.L., and Chen, J. 2004. Characterizing nonstationary wind speed using empirical mode decomposition, *Journal of Structural Engineering*, 130: 912–920.

Xuehua, Z., Qiang, H., and Jingping, Z. 2011. Periodic analysis of runoff based on Hilbert-Huang Transform. In: *2011 International Symposium on Water Resource and Environmental Protection*, Xi'an, 633–637.

Yang, J.N., Lei, Y., Lin, S., and Huang, N. 2004. Hilbert-Huang based approach for structural damage detection, *Journal of Engineering Mechanics*, 130(1): 85–95.

Yu, H.L., and Lin, Y.C. 2015. Analysis of space–time non-stationary patterns of rainfall–groundwater interactions by integrating empirical orthogonal function and cross wavelet transform methods, *Journal of Hydrology*, 525: 585–597.

Zappalà, D., Barreiro, M., and Masoller, C. 2016. Global atmospheric dynamics investigated by using Hilbert frequency analysis, *Entropy*, 18: 408. 10.3390/e18110408.

Zare, M., and Koch, M. 2018. Groundwater level fluctuations simulation and prediction by ANFIS- and hybrid Wavelet-ANFIS/Fuzzy C-Means (FCM) clustering models: Application to the Miandarband plain, *Journal of Hydro-environment Research*, 18: 63–76.

Zhan, H., and Zlotnik, V.A. 2002. Groundwater flow to a horizontal or slanted well in an unconfined aquifer, *Water Resources Research*, 38(7): 13.1–13.11.

Zhang, R.R., VanDemark, L., Liang, J., and Hu, Y. 2004. On estimating site damping with soil non-linear from earthquake recordings, *International Journal of Non-linear Mechanics*, 39: 1502–1517.

Zhu, Z., and Yang, H. 2002. Discrete Hilbert transformation and its application to estimate the wind speed in Hong Kong, *Journal of Wind Engineering and Industrial Aerodynamics*, 90(1): 9–18.

27 Multi-Criteria Decision Analysis for Flood Risk Assessment

Daniel Jato-Espino, Mariana Madruga de Brito, and Saeid Eslamian

CONTENTS

27.1 INTRODUCTION

Floods were the natural disaster leading to the highest number of fatalities during the 20th century, amounting to 6.8 million deaths (Doocy et al., 2013). The devastating effects of floods are especially remarkable in Asia, where almost half of these events took place (Jonkman, 2005). The occurrence and intensity of these phenomena are being favored by the action of urban sprawl, resulting in an increase in flood damage over the last years (Elmer et al., 2012). In addition, climate change is another potential catalyzer for the frequency of flooding in areas where mean precipitation and wet extremes are expected to rise (Guhathakurta et al., 2011).

In order to mitigate the negative impacts of floods, the Sendai framework for disaster risk reduction recommends that the design and implementation of risk management strategies should be based on a comprehensive understanding of risk in all its dimensions, including the hazard characteristics, vulnerability, coping capacity, and exposure of persons and assets (UNISDR, 2015a). The assessment of risk, when carried out holistically, can provide the floodplain managers with better tools to make informed decisions for flood mitigation at various levels. It can assist decision-makers to elaborate land use planning policies and to identify areas where preventive and corrective measures are needed, and, if so, which option is most suitable. Additionally, it can help to raise public awareness by providing an understandable visualization of the flooding risks (Dalezios and Eslamian, 2016).

The variables favoring flood hazard relate to conditioning aspects, such as the orography and permeability of the terrain, as well as to trigger factors, mainly represented by precipitation. Another relevant parameter boosting the probability of occurrence of floods is the proximity to watercourses, whose rise can threaten the areas located in their vicinity. In the end, the integrated consideration of all of these elements makes urban areas particularly prone to floods, especially if they are close to the coast (Neumann et al., 2015).

The consequences stemming from the occurrence of floods are usually expressed as costs, which can be grouped into four categories depending on the assessment methods they require (Meyer et al.,

DOI: 10.1201/9780429463938-35

2013): direct, business interruption, indirect, and intangible. One way or another, all these categories refer to impacts that increase the vulnerability of people, goods, services, and the environment, either in the form of physical damages or through interruptions, disruptions, and depreciations.

The dual nature of floods, whereby they must be considered in terms of both hazard and vulnerability, provoked a conceptual evolution from a traditional approach only focused on protection to a more comprehensive framework focused on risk management (Schanze, 2006). Risk management must, in turn, be founded on risk assessment, which is the process that enables identifying hazards and how they affect the vulnerability of people and goods to their occurrence. In this context, risk assessment encompasses both the determination of hotspots in what concerns the susceptibility to floods (hazard) and the quantification of the human and material consequences stemming from these events (vulnerability). Therefore, the definition of risk proposed for this chapter considers it as the product between hazard and vulnerability. This is equivalent to expressing risk as the combination of probability and consequences or susceptibility and impact, which are two of the most widely used approaches in the literature (Bell and Glade, 2004).

Given the spatial condition of the variables involved in flood hazard and vulnerability, the development of risk assessment methodologies is usually supported by the use of Geographic Information Systems (GIS) (McMaster et al., 1997). GIS enables importing and geoprocessing the data required for mapping hazard and vulnerability variables. The aggregation of these individual layers to produce integrated risk maps can be assisted using multi-criteria decision analysis (MCDA) methods, since some of the factors to combine might be in conflict (Carver, 1991). One of the strengths of MCDA is that it provides a suitable platform to involve relevant stakeholders and gain insight into their priorities in terms of flood hazards and vulnerability (Pelling, 2007). Furthermore, it makes the criteria evaluation process more explicit and rational, by making subjective judgments visible in a transparent and fair way (San Cristóbal Mateo, 2012).

Under these premises, the aim of this chapter is to provide an integrated and generic approach to producing flood risk maps through the combination of GIS and MCDA. After establishing the current state of the art in terms of flood risk assessment through a literature review, the different steps forming the proposed framework are presented sequentially, highlighting their potential replicability due to the use of open-access data and participatory methods. The main limitation associated with this approach lies precisely in the worldwide availability of the data proposed, whose resolution might be too coarse in case of conducting detailed or small-scale studies.

27.2 LITERATURE REVIEW

In order to highlight the increasing relevance that flood risk assessment has gained over the years, this section provides an overview of the most relevant scientific outputs produced during the last two decades in this field of research. Table 27.1 summarizes the main features of these investigations, which are presented in descending order according to their current number of citations in the Scopus database.

In addition to the number of citations achieved by each contribution so far, Table 27.1 includes the name, year, and title of the scientific works addressed, as well as the country where the flood-related studies were conducted. The three remaining fields forming Table 27.1 focus on the most relevant aspects of the literature review from a conceptual point of view, since they indicate the way in which flood risk was approached, the MCDA methods used, and whether the research item was participatory or not.

Regarding the geographic distribution of the studies addressed, almost half of them took place in Asia (48.39%), followed by Europe (35.48%), America (9.68%), Oceania (3.23%), and Africa (3.23%). The predominance of Asia in this sense is consistent with the data reported by the United Nations between 1995 and 2015 (UNISDR, 2015b), which highlighted the sensitivity of this continent to weather-related disasters due to the concentration of population in the surroundings of river basins and floodplains.

TABLE 27.1

Overview of the Main Existing Research Items Related to Flood Risk Assessment through Multi-Criteria Decision Analysis (MCDA)

Reference	Country	Cites	Title	Approach	MCDA Methods	Participatory
(Meyer et al., 2009)	Germany	133	A multicriteria approach for flood risk mapping exemplified at the Mulde river, Germany	Vulnerability assessment; mitigation measures	Disjunctive approach; MAUT	No
(Raaijmakers et al., 2008)	Spain	95	Flood risk perceptions and spatial multi-criteria analysis: An exploratory research for hazard mitigation	Vulnerability assessment	SAW	Yes
(Kienberger et al., 2009)	Austria	75	Spatial vulnerability units – Expert-based spatial modelling of socio-economic vulnerability in the Salzach catchment, Austria	Vulnerability assessment	Delphi; SAW	Yes
(Levy, 2005)	China	70	Multiple criteria decision making and decision support systems for flood risk management	Mitigation measures	ANP	No
(Wang et al., 2011)	China	69	A GIS-based spatial multi-criteria approach for flood risk assessment in the Dongting Lake region, Hunan, Central China	Risk assessment	FAHP	No
(Kubal et al., 2009)	Germany	58	Integrated urban flood risk assessment – Adapting a multicriteria approach to a city	Vulnerability assessment	SAW	No
(Kenyon, 2007)	Scotland	51	Evaluating flood risk management options in Scotland: A participant-led multi-criteria approach	Mitigation measures	Rank sum; rank order centroid; SAW	Yes
(Levy et al., 2007)	Japan	42	Multi-criteria decision support systems for flood hazard mitigation and emergency response in urban watersheds	Mitigation measures	ANP	Yes
(Lee et al., 2013)	South Korea	34	Integrated multi-criteria flood vulnerability approach using fuzzy TOPSIS and Delphi technique	Risk assessment	Delphi; FTOPSIS	No
(Kandilioti and Makropoulos, 2012)	Athens	30	Preliminary flood risk assessment: The case of Athens	Risk assessment	AHP; SAW; OWA	Yes
(Scolobig et al., 2008)	Italy	26	Integrating multiple perspectives in social multicriteria evaluation of flood-mitigation alternatives: The case of Malborghetto-Valbruna	Vulnerability assessment; mitigation measures	NAIADE	Yes

(Continued)

TABLE 27.1 (CONTINUED)
Overview of the Main Existing Research Items Related to Flood Risk Assessment through Multi-Criteria Decision Analysis (MCDA)

Authors (Year)	Country	Cites	Title	Approach	MCDA Methods	Participatory
(Sharifi et al., 2002)	Bolivia	22	Application of GIS and multicriteria evaluation in locating sustainable boundary between the Tunari national park and Cochabamba city (Bolivia)	Vulnerability assessment; mitigation measures	SAW	Yes
(Haque et al., 2012)	Bangladesh	19	Participatory integrated assessment of flood protection measures for climate adaptation in Dhaka	Mitigation measures	SAW	Yes
(Chen et al., 2015)	Australia	16	A spatial assessment framework for evaluating flood risk under extreme climates	Hazard assessment	AHP	No
(Solín, 2012)	Slovakia	14	Spatial variability in the flood vulnerability of urban areas in the headwater basins of Slovakia	Vulnerability assessment	MADM	No
(Sowmya et al., 2015)	India	13	Urban flood vulnerability zoning of Cochin City, southwest coast of India, using remote sensing and GIS	Risk assessment	SAW	No
(Malekian and Azarnivand, 2016)	Iran	12	Application of integrated Shannon's entropy and VIKOR techniques in prioritization of flood risk in the Shemshak watershed, Iran	Hazard assessment	Entropy; VIKOR	Yes
(Yang et al., 2011)	China	12	Spatial multicriteria decision analysis of flood risks in aging-dam management in China: A framework and case study	Vulnerability assessment	SAW	Yes
(Xiao et al., 2017)	China	8	Integrated flood hazard assessment based on spatial ordered weighted averaging method considering spatial heterogeneity of risk preference	Hazard assessment	FAHP; OWA	No
(Ghanbarpour et al., 2013)	Iran	8	A comparative evaluation of flood mitigation alternatives using GIS-based river hydraulics modelling and multicriteria decision analysis	Mitigation measures	TOPSIS	No

(Continued)

TABLE 27.1 (CONTINUED)

Overview of the Main Existing Research Items Related to Flood Risk Assessment through Multi-Criteria Decision Analysis (MCDA)

Authors (Year)	Country	Cites	Title	Approach	MCDA Methods	Participatory
(Fernandez et al., 2016)	Portugal	6	Social vulnerability assessment of flood risk using GIS-based multicriteria decision analysis: A case study of Vila Nova de Gaia	Vulnerability assessment	AHP; OWA; SAW	No
(Seekao and Pharino, 2016)	Thailand	3	Assessment of the flood vulnerability of shrimp farms using a multicriteria evaluation and GIS: A case study in the Bangpakong Sub-Basin, Thailand	Hazard assessment	AHP; SAW	No
(Tang et al., 2018)	China	1	Incorporating probabilistic approach into local multi-criteria decision analysis for flood susceptibility assessment	Hazard assessment	AHP; SAW	No
(Hazarika et al., 2018)	India	1	Assessing and mapping flood hazard, vulnerability and risk in the Upper Brahmaputra River valley using stakeholders' knowledge and multicriteria evaluation (MCE)	Risk assessment	SAW	Yes
(Panhalkar and Jarag, 2017)	India	1	Flood risk assessment of Panchganga River (Kolhapur district, Maharashtra) using GIS-based multicriteria decision technique	Hazard assessment	AHP	No
(de Brito et al., 2018)	Brazil	1	Participatory flood vulnerability assessment: A multi-criteria approach	Vulnerability assessment	AHP; ANP	Yes
(Loos and Rogers, 2016)	US	1	Understanding stakeholder preferences for flood adaptation alternatives with natural capital implications	Vulnerability assessment; mitigation measures	MAUT	Yes
(Luu and von Meding, 2018)	Vietnam	0	A flood risk assessment of Quang Nam, Vietnam using spatial multicriteria decision analysis	Risk assessment	AHP	Yes
(Patrikaki et al., 2018)	Greece	0	Assessing flood hazard at river basin scale with an index-based approach: The case of Mouriki, Greece	Hazard assessment	AHP; SAW	No

(Continued)

TABLE 27.1 (CONTINUED)

Overview of the Main Existing Research Items Related to Flood Risk Assessment through Multi-Criteria Decision Analysis (MCDA)

Authors (Year)	Country	Cites	Title	Approach	MCDA Methods	Participatory
(Zeleňáková et al., 2018)	Slovakia	0	Flood vulnerability assessment of Bodva cross-border river basin	Hazard assessment	AHP; SAW	No
(Mallouk et al., 2016)	Morocco	0	A multicriteria approach with GIS for assessing vulnerability to flood risk in urban area (case of Casablanca city, Morocco)	Vulnerability assessment	AHP	No

In addition to the strict components of risk, either as a whole or isolated (hazard and vulnerability), investigations concerning the evaluation of mitigation measures were also considered, since their inclusion can have attenuating effects on the occurrence and impact of floods. With this in view, the review yielded rather balanced results, whereby vulnerability emerged as the most addressed aspect (34.29%), but not very far from mitigation measures (25.71%), hazard (22.86%), and risk (17.14%). On the one hand, these results prove the relevance of the social dimension of floods, highlighting the importance of identifying critical areas in terms of exposure to these events. On the other hand, the fact that risk assessment was the approach taken less frequently suggests the need for developing accessible and replicable frameworks for evaluating flood risk integrally.

The predominance of the simple additive weighting method (SAW) and the analytic hierarchy process (AHP), which were present in 70% of the investigations reviewed, indicate a clear trend towards the application of simple and widely used MCDA methods (Jato-Espino et al., 2014). SAW is the easiest technique to aggregate the different factors involved in the assessment of either hazard or vulnerability, especially in a context in which these variables must be processed with the support of GIS. Similarly to SAW, AHP is a straightforward, flexible, and easily understandable method (Cinelli et al., 2014). Thanks to these characteristics, it can be adapted to different problems without requiring previous knowledge from the analyst.

The application of the AHP method is usually linked to the use of participatory approaches, which are often based on establishing priorities in relation to flood risk according to the opinions provided by a group of stakeholders. The integration of participatory methods and the MCDA tools may facilitate the achievement of consensus, which is essential for finding solutions that reconcile conflicting interests and can be accepted by the majority (de Brito and Evers, 2016; Malczewski and Rinner, 2015; Simão et al., 2009). Despite this importance, more than half of the studies consulted (54.84%) disregarded this aspect, suggesting that there is still room for increasing the involvement of participants in the design of flood risk management strategies.

27.3 INTEGRATED FLOOD RISK ASSESSMENT

The main elements forming the framework conceived to assess flood risk are illustrated in Figure 27.1. On the one hand, the processing and combination of a series of morphologic and hydrogeological factors are proposed to determine flood hazard. On the other hand, the vulnerability of people, goods, and natural areas to floods is also examined, in order to evaluate their socioeconomic and environmental consequences. To get insight into the real dimension of floods, the use of participatory methods involving multiple stakeholders is suggested to both identify the hotspots in terms of flood hazard and prioritize the vulnerable elements requiring protection. Hence, the coupled and inclusive consideration of hazard and vulnerability enables determining flood risk integrally.

FIGURE 27.1 Outline of the generic framework proposed to assess flood risk.

For the sake of maximizing the replicability of the proposed approach, the characterization of all the variables involved in the calculation of flood risk is addressed through open data available at a worldwide scale. From a technical point of view, the only requirement for implementing this methodology is to use (GIS and MCDA methods to manage these data.

27.3.1 DEFINITION OF FACTORS INVOLVED IN FLOOD RISK USING OPEN DATA

The development of flood risk maps roughly consists of the processing and aggregation of a series of factors or criteria with the support of GIS and participatory MCDA methods. The characterization of these factors requires the acquisition and further processing of a series of spatial data. To boost the replicability of the proposed framework, the factors suggested for assessing flood risk in Table 27.2 meet two main requirements in terms of the data from which they stem: worldwide scale availability and open accessibility. Hence, the underlying aim of this approach is to enable its application

TABLE 27.2
List of Datasets and Factors Suggested to Assess Flood Risk

Category	Data	Source	Factor(s)	Units
Hazard	Digital Elevation Model (DEM)	(LP DAAC, 2014)	Elevation, slope, flow accumulation	m, °, no. of cells
	Lithology	(Hartmann and Moosdorf, 2012)	Soil permeability	Score
	Land cover	(Jun et al., 2014)	Curve number	Score
	Groundwater level	(Fan et al., 2013)	Water table depth	m
	Precipitation	(Hijmans et al., 2005)	Precipitation	mm
	Water bodies	(Geofabrik Download Server, 2018)	Proximity to water bodies	m
Vulnerability	Population density	(SEDAC, 2017)	Population density	km²
	Protected areas	(IUCN, 2016)	Protected areas density	km²
	Buildings	(Geofabrik Download Server, 2018)	Building density	km²
	Infrastructures	(Geofabrik Download Server, 2018)	Infrastructure density	km²

all around the globe; however, the list of factors proposed in Table 27.2 can also be produced using different or complementary local or regional data with finer resolutions.

Consistent with the two main components involved in flood risk, the factors needed for its evaluation can also be divided into hazard and vulnerability. Hence, one of the cornerstones in the determination of flood hazards is the elevation of the terrain in the study area. This data can be obtained from the United States Geological Survey (USGS) Earth Explorer, which made available the Advanced Spaceborne Thermal Emission and Reflection Radiometer (ASTER) Global Digital Elevation Model (GDEM), as a result of the collaboration between the US National Aeronautics and Space Administration (NASA) and Japan's Ministry of Economy, Trade, and Industry (METI) (LP DAAC, 2014). The ASTER GDEM is provided in raster format (GeoTIFF) with a pixel size of 30 m. The relationship between this factor and flooding is inversely proportional, since low elevation areas are prone to receive large amounts of runoff.

Two other hazard-related criteria stem from the DEM. On the one hand, the slope of the terrain describes its steepness according to changes in elevation. Again, this variable is inversely related to the occurrence of floods, because the presence of flat slopes favors the concentration of water. On the other hand, flow accumulation represents the contributing area flowing to the same location with a descendent slope. Hence, this factor is directly associated with flood hazard, such that the higher the value of flow accumulation, the more likely water to stagnate.

Soil permeability is a factor indicating the ability of subsurface layers to transmit water infiltrated from the ground. Hence, this variable has a negative effect on floods, in that high values of permeability facilitate water percolation and, therefore, reduce flood probability. The permeability of the underlying soil can be determined according to its characteristics and composition, which are available at the Global Lithological Map (GLiM) produced by the Institute for Biogeochemistry and Marine Chemistry of the University of Hamburg, with an average resolution of 1:3,750,000 (Hartmann and Moosdorf, 2012).

The next factor relates to the threshold runoff of the surface, i.e., the amount of excess rainfall accumulated over the ground after a storm event. The quantification of this variable can be approached using the curve number (CN) (Garen and Moore, 2005), which is an empirical parameter developed by the US Department of Agriculture (USDA) Natural Resources Conservation Service (USDA, 2018). CN ranges from 30 to 100, such that high values of CN indicate high runoff potential. Thus, this factor has a direct correspondence to flood hazards. CN stems from the combination of the land cover type and hydrologic soil group (HSG) of the study area. The latter can be determined from a lithological map as described above, whilst the former is addressable from the data included in the GlobeLand30 initiative, an open-access map of Earth's land cover with a resolution of 30 m donated by China to the United Nations (Jun et al., 2014).

Groundwater is the water contained in the voids and fractures of the soil beneath the Earth's surface. This variable can contribute to flooding when groundwater rises above its common level and reaches the surface. Therefore, the shallower the groundwater level, the more likely the occurrence of floods. A study conducted by Fan et al. (2013) used measurements of water table depth from 1,603,781 sites, either provided by governments or published in the literature, to produce a regionalized 1 km grid dataset in NetCDF format that can be used to characterize this factor.

The next factor symbolizes the precipitation patterns in the study area. Precipitation is a crucial variable in determining the amount of water that the terrain has to deal with, such that high rainfall rates hinder the capacity of filtration of the ground and contribute to provoking floods. The data required to compute this factor can be obtained from version 1.4 of WorldClim (Hijmans et al., 2005), which provides global climate maps with a cell size of 1 km. These data are available both under current (stationarity) and future (non-stationarity) conditions, enabling the projection of variations in flood hazard due to the impacts of climate change according to the different Global Circulation Models (GCM).

The last variable contributing to flood hazard concerns the proximity of the study area to watercourses. Rainfall during a continued period can cause the overflow of water bodies and, by extension, the inundation of their surroundings. In this case, the probability of flooding increases as the distance to watercourses is reduced, especially if their volume is high. This information is available via the OpenStreetMap project, which includes polygonal and vector layers indicating the location of water bodies (Geofabrik Download Server, 2018).

The first factor related to flood vulnerability is population density, which provides an indicator of hotspots in terms of the concentration of people. Hence, this aspect accounts for the vulnerability of crowded areas, such that the higher the population density, the greater the impacts caused by floods. The data needed for the creation of this factor is supplied by the Socioeconomic Data and Applications Center (SEDAC) via a global map containing the Gridded Population of the World (GWP) (SEDAC, 2017). This map provides the estimates of population density with a 1 km resolution for several years based on national censuses, population registers, and United Nations counts.

The next aspect to consider for the assessment of vulnerability encompasses the environmental dimension of flood management, represented by the terrestrial and marine protected areas that might be subject to these phenomena. This factor can also be represented based on its density, such that the higher the presence of protected areas, the greater the impacts of floods on the environment. The World Database on Protected Areas (WDPA) (IUCN, 2016), jointly prepared by the United Nations Environment Programme (UNEP) and the International Union for Conservation of Nature (IUCN), includes the data required to model this factor.

Finally, the last two vulnerability-related variables focus on the main assets that might be damaged during the occurrence of floods: buildings and infrastructures. Affections to these elements may limit accessibility and cause traffic disruptions, hindering the transit of people and vehicles across dense areas in terms of buildings and infrastructures and increasing their vulnerability to flooding events. Again, the data involved in the processing of these factors can be downloaded from OpenStreetMap (Geofabrik Download Server, 2018), which provides two layers symbolizing the spatial arrangement of these facilities.

27.3.2 Participatory Assessment of Flood Hazard and Vulnerability

The previous reviews showed that the assessment of flood hazards and vulnerability using MCDA is seldom conducted in a systematic way (de Brito and Evers, 2016). The reasoning for the model assumptions, such as the selection of the input criteria, standardization of the data to a common scale, and definition of criteria weighs, is typically unstated and these decisions are restricted to researchers conducting the study (Beccari, 2016; Müller et al., 2011; Rufat et al., 2015; Tate, 2012). Even when stakeholders are involved, their participation is fragmented and constrained to information dissemination and consultation at specific stages (de Brito and Evers, 2016; Evers et al., 2018). Consequently, vulnerability and hazard MCDA models are commonly perceived as black boxes by end-users, which limits the use and implementation of their results.

To overcome these problems, participatory approaches for flood hazard and vulnerability assessment can be used to go beyond the limited perspective of a single expert by acknowledging the multiple standpoints and explicitly showing the rationale for model decisions. The key generic steps of the approach are illustrated in Figure 27.2, which shows that expert stakeholders should collaborate throughout the entire process. This allows the building of trust among participants, facilitates information sharing, and improves the model transparency, thus enhancing the results' acceptance. A detailed description of the proposed methodology and its application in two case studies is provided by de Brito et al. (2017, 2018) and Katz et al. (2017).

The first step comprehends the identification of relevant expert stakeholders that have an in-depth knowledge of flood vulnerability and hazard assessment. For this purpose, the snowball sampling technique can be used (Wright and Stein, 2005). The basic idea is that initially sampled experts (i.e., starting seeds) indicate other specialists in the field, which in turn lead to the other prospective participants until the desired sample size is reached. Alternatively, iterative stakeholder analysis involving focus groups and brainstorming exercises can be conducted to ensure that all of

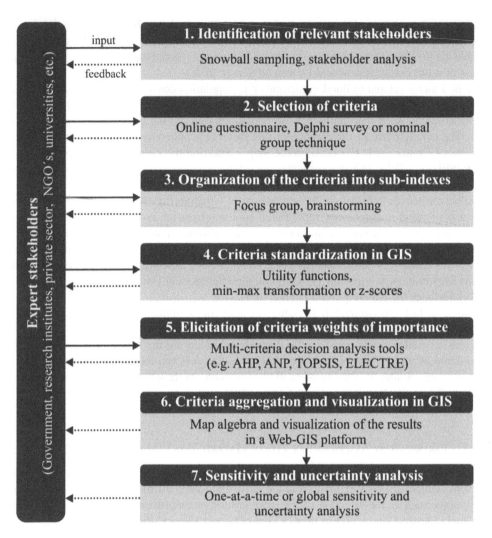

FIGURE 27.2 Methodological framework for flood vulnerability and risk assessment. Key steps and suggested methods are shown in boxes.

the relevant actors have been included. The use of the previously mentioned tools can be complemented with social network analysis to investigate social structures and identify key experts that (1) have the unique positions in the network, hence, occupying non-redundant communication roles in the network; (2) come from the different stakeholder categories, thus allowing capturing contrasting opinions; and (3) are relatively well-connected to the others and tend to break across different segments of the network (Prell et al., 2008).

Step 2 involves the identification of a set of criteria or factors that are going to be incorporated into the model. The selection of the evaluation criteria is a crucial step in the development of any indicator, as the inclusion or exclusion of relevant criteria can have a dramatic impact on the model results. Hence, these should be preferentially independent, complete, concise, and operationally meaningful. The nominal group technique (NGT) can be applied to obtain the consensus among experts on the set of criteria (Harvey and Holmes, 2012). Alternatively, to avoid group effects (e.g., group-thinking or anticipatory consensus) obscuring individual preferences, anonymous questionnaires or the Delphi survey can be used.

The third step consists of organizing the selected criteria into sub-indexes (e.g., social, economic, and environmental dimensions). The organization scheme, i.e., hierarchical or network, depends on the MCDA technique considered. For this purpose, the brainstorming sections can be conducted to have an unstructured discussion of the problem. In this setting, the participants propose solutions and the group actively debates what the best course of action is.

Step 4 comprises the standardization of the spatial data into a common scale. There is a number of methods for standardizing raw data to the comparable units, including value functions, min-max transformation, and z-score. Since vulnerability and hazard criteria usually do not have linear behavior, the use of value functions is recommended. The value function is a mathematical representation of human judgment (Malczewski and Rinner, 2015). It relates possible decision outcomes (criterion or attribute values) to a scale that reflects the decision maker's preferences. The type and shape of the function can be defined individually, i.e., one value function per criterion defined by each participant, or consensus regarding the function type can be achieved based on focus group discussion. Nevertheless, due to the complexity of the task at hand, these meetings need to be restricted to a small number of participants.

In step 5, the importance of the criteria for the vulnerability analysis needs to be assessed. This is a critical phase, given that even small changes in weights may have a significant impact on the model results, leading to inaccurate outcomes (Feizizadeh and Blaschke, 2014). In the field of flood hazard and vulnerability assessment, tThe selection of the MCDA method to be used depends on the time and resources available and decision makers' objectives. Regardless of the MCDA method used, the weights can be elicited using either online or in-person questionnaires. Alternatively, since assigning weights requires a significant mental effort for most stakeholders, serious games (i.e., games including a non-entertaining purpose) in combination with MCDA could be used instead (Aubert et al., 2018). Voinov et al. (2016) argued that serious games are promising tools for participatory modeling due to (1) the stakeholders' engagement through intrinsic game motivational features, (2) the potential for interactive visualization, and (3) the ability to create social learning.

Step 6 comprehends the aggregation of the criterion maps and decision maker's preferences (criterion weights) in a GIS environment using a combination rule. In general, the combination rules can be compensatory or non-compensatory, where the former takes into account the trade-offs between criteria, while the latter ignores the value of trade-offs. The compensatory methods allow trade-off of a low value on one criterion against a high value on another (Malczewski and Rinner, 2015). It is recommendable to display the aggregated results in a Web-GIS platform, where participants can compare their results with the maps from other actors.

The final step consists of a post-analysis study to check for the model inaccuracies. Uncertainty analysis (UA) investigates how uncertainty in model inputs translates into uncertainty in model outputs (Tate, 2012). Similarly, sensitivity analysis (SA) investigates how the results vary when the criteria are changed. This helps identify crucial variables in the model and allows disagreements between individuals to be examined to see if they make a difference in the final results. At the end of the process, the outcomes of the MCDA analysis should be made available to all interested parties through the reports and other channels of communication.

Although the participatory MCDA phases are presented here as a logical sequence of steps, it should be emphasized that, in reality, the development of hazard and vulnerability indices process may be far from sequential and continuous. In practice, the whole process is iterative, possibly having internal conflicts that require an ongoing review of the index structure to ensure that the results will be accepted by the majority of the participants.

27.3.3   Geoprocessing and Aggregation of Factors to Assess Flood Risk

The determination of flood risk is based on the geoprocessing and aggregation of the factors listed in Table 27.2. This must be carried out with the support of GIS, which provides the tools needed for the management of the datasets used to produce the factors. Although the functionalities available in different GIS might vary from each other, there are several open-source programs that include all the capabilities required to undertake the tasks for modeling flood hazards and vulnerability (Jato-Espino, 2016).

One common consideration for all the data involved in flood risk assessment is the clipping of the original vector or raster layers to the boundaries of the study area, in order to delimit further calculations to the workspace. The aggregation of different layers also requires their projection to the same reference system, as well as their resampling to the minimum cell size of the original data, in order to boost the accuracy of the results to achieve. Finally, those datasets originally available in vector format must be transformed into raster to enable their eventual combination with the remaining layers.

The DEM-related factors depend on calculations concerning the relative elevation of the cells of a raster map with respect to their adjacent pixels. This course of action enables determining the slope of the terrain, either in degrees or as a percentage, and a flow direction map according to the eight direction model proposed by Jenson and Domingue (1988), which assigns a value to each cell in the neighborhood of the processing cell based on the changes in elevation. The further processing of the flow direction map through hydrology tools serves to aggregate the number of cells flowing to each cell in the workspace (Tarboton et al., 1991), yielding the flow accumulation map of the study area.

Soil permeability can be calculated from the description of the rocks in the study area, which enables classifying them in different levels according to their properties. To this end, first is the creation of a new field in the attribute table of the layer corresponding to the lithologic map, such that values of permeability or scores are allocated to each group depending on their characteristics. The number of groups into which divided the lithology of the workspace should be preferably four, in order to meet the HSG classification and, therefore, facilitate the processing of the CN of the terrain surface. The joint selection by attributes of the HSG and the land cover types in the study area leads to the production of the CN map sought (Jato-Espino et al., 2016b).

Water table depth is one of the most complex factors to manage, since it is provided in NetCDF format as a grid of points. Hence, the first step in the processing of this variable is the extraction and arrangement of the original raw data in tabular format, in order to use a GIS readable format. Once imported, the grid of points must be interpolated to generate a continuous surface of the values of water table depth. This can be carried out using both deterministic and geostatistical techniques, such that the goodness-of-fit of the resulting maps is measured by comparison between the interpolated and observed values (Jato-Espino et al., 2016a).

Unlike water table depth, precipitation and population density are already available as a continuous raster map in the data source suggested in Table 27.2, such that their processing only requires generic clipping and resampling tasks. Instead, the calculations of proximity and density associated with the remaining factors are based on the application of specific spatial tools with which to create the raster layers from the polygons and polylines defining water bodies, protected areas, buildings, and infrastructures.

Once all of the factors have been processed and converted into raster layers, they must be normalized to enable their joint aggregation. Normalization is the step whereby the ratings r_{ij} in each cell i of the workspace in relation to a factor j, measured in the units indicated in Table 27.2, are adjusted to a common scale by applying different transformations including value functions, min-max transformation, or z-score, as described before.

The aggregation of the normalized ratings n_{ij} can be undertaken using different MCDA techniques, but always taking into account the weights w_j of the factors. The use of one MCDA method or another, including distance-based, outranking, scoring or utility/value approaches, may involve different equations and calculations; however, they all are strongly dependent on the determination of the weights, which is carried out independently. These weights represent the relative importance of the factors in the computation of flood hazards and vulnerability. Consequently, the values of w_j must also be divided into hazard $\left(w_{H_j}\right)$ and vulnerability $\left(w_{V_j}\right)$, such that their coupled consideration leads to determining flood risk.

The weights of the vulnerability-related factors can be determined straightforwardly through participatory tools, either by direct allocation or using MCDA methods, from the opinions collected

TABLE 27.3

Weighting Scenarios Proposed to Fit the Observed Ranking of Flood-Prone Areas

Scenario	w_{H_1}	w_{H_2}	w_{H_3}	w_{H_4}	w_{H_5}	w_{H_6}	w_{H_7}	w_{H_8}
1	0.125	0.125	0.125	0.125	0.125	0.125	0.125	0.125
2	0.650	0.050	0.050	0.050	0.050	0.050	0.050	0.050
3	0.050	0.650	0.050	0.050	0.050	0.050	0.050	0.050
4	0.050	0.050	0.650	0.050	0.050	0.050	0.050	0.050
5	0.050	0.050	0.050	0.650	0.050	0.050	0.050	0.050
6	0.050	0.050	0.050	0.050	0.650	0.050	0.050	0.050
7	0.050	0.050	0.050	0.050	0.050	0.650	0.050	0.050
8	0.050	0.050	0.050	0.050	0.050	0.050	0.650	0.050
9	0.050	0.050	0.050	0.050	0.050	0.050	0.050	0.650
10	0.200	0.200	0.200	0.080	0.080	0.080	0.080	0.080
11	0.080	0.080	0.080	0.260	0.260	0.080	0.080	0.080
12	0.080	0.080	0.080	0.080	0.080	0.200	0.200	0.200

from a panel of stakeholders, as described before. Instead, the information about flood hazards that can be obtained through public engagement may consist of an ordinal ranking of flood-prone areas, based on the experience gathered from historical events in the study area. Hence, the goal with respect to flood hazards is to maximize the fit between the observed ranking of sensitive areas and the values determined via MCDA and GIS.

To this end, first is the definition of a series of weighting scenarios to provide both balanced and biased situations. Table 27.3 shows a potential list of weights to use, which apart from eight scenarios in which the importance of each factor clearly predominates over the others, contains three additional combinations where the factors are grouped by type (morphology, permeability, and hydrology) and prioritized accordingly.

The next step in the calculation of flood hazard is the modeling of the association between the ranking of prone areas obtained through participatory methods and the summary statistics computed at such areas by aggregating the factors according to the weights in Table 27.3. The definition of the sensitive areas can be accomplished by establishing a buffer of 250 m around the streets where the floods are frequent (Jato-Espino et al., 2018; van Hove et al., 2015). Thus, the summary statistics refer to the mean, minimum, maximum, and sum values enclosed by these buffer areas.

Since observed data about flood-prone areas is in ordinal format, its relationship to the summary statistics can be modeled using Spearman's correlation coefficient (ρ), which measures the strength of the monotonic association between two variables. Hence, the goodness-of-fit of the hazard maps obtained for each scenario is represented by a value between −1 and 1, which indicates whether its correlation with the ranking of flood-prone areas is perfectly negative or positive.

Then, the identification of the factors proving to be statistically significant for explaining flood hazard can be undertaken with the application multiple regression analysis (MRA), which enables modeling the relationship between the list of values of ρ associated with the combinations of weights proposed in Table 27.3. In addition to linear terms, first-order interactions should also be included to model potential combined effects, since some of the hazard factors suggested in Table 27.2 are related to each other. The results obtained from this analysis must be validated through the verification of the assumptions of normality, homoscedasticity, multicollinearity, and independence of residuals (Osborne and Waters, 2003).

The final step to take for producing a validated flood hazard map might be addressed through two different approaches. The simplest option consists of determining the optimal weights for the factors based on their relative contribution to the MRA model, such that the optimal weight $\bar{w}_{H_j}$

of a factor j is computed as its contribution as a linear term plus half the sum of its contributions in the interaction terms in which it is included. Another approach might involve the application of optimization methods to solve the problem formulated in equation 27.1, which seeks to maximize Spearman's correlation coefficient, while complying with the restrictions associated with the values of ρ and w_{H_j}. Due to the inclusion of interaction terms, the resolution of this optimization problem requires the use of nonlinear methods, such as the generalized reduced gradient (GRG) (Abadie and Carpentier, 1969) or evolutionary algorithms (Elbeltagi et al., 2005).

$$
\begin{aligned}
&\text{Maximise} \quad \rho \\
&\text{subject to:} \quad -1 \le \rho \le 1 \\
&\qquad\qquad 0 \le w_{H_j} \le 1,\ \forall\ f_j \text{ in the MRA model}
\end{aligned}
\tag{27.1}
$$

In consequence, the integrated flood risk assessment sought is provided by the multiplication of the validated hazard map, based on the calculation of these optimal weights, by that of vulnerability produced by aggregating the factors corresponding to this aspect (Table 27.2) according to their weights, which are obtained through participatory methods from the priorities of relevant stakeholders in the study area. The resulting flood risk map is highlighted by its foundations on open data globally available, which are processed with the support of the MCDA methods incorporated into GIS.

27.4 CONCLUSIONS

This chapter presents an integrated framework for flood risk assessment founded on the coupling of multi-criteria decision analysis (MCDA) with geographic information systems (GIS). Both tools are supported with the inclusion of participatory methods to help the processing of a list of hazard and vulnerability-related factors built from open data sources and involved in the probability of occurrence and potential impacts of flooding phenomena.

The use of open datasets contributes to boosting the replicability of the proposed approach, since the sources explicitly suggested are available at a worldwide scale and have enough resolution to produce satisfactory results. However, the flexibility of this framework enables either the addition of new hazard and vulnerability factors or the replacement of some of them by others with higher precision, depending on the quality of the data available at regional or local scales.

The combination of MCDA with participatory tools for flood hazard and vulnerability assessment can lead to an increased, shared understanding of the problem by avoiding the limited perspective of a single expert. The methodology described here can enhance the credibility and deployments of hazard and vulnerability indicators, as stakeholders' opinions, expert judgment, and local knowledge are taken into consideration throughout the entire process. Furthermore, its transdisciplinary nature might support the social learning processes and develop the capacity through awareness-raising.

In summary, the content included in this chapter is intended to provide a complete guide on how to assess flood risk integrally without requiring either restricted data or proprietary software. Instead, data are suggested to be either acquired from global open sources or generated through participatory methods and then processed using MCDA techniques incorporated into free GIS, resulting in a generic resource that can help improve flood management all around the globe in an easy, accessible, and inclusive manner.

REFERENCES

Abadie, J., and Carpentier, J. 1969. Generalization of the Wolfe reduced gradient method to the case of nonlinear constraints. In: R. Fletcher (Ed.) *Optimization*, Academic Press, New York, USA, 37–47.

Abdullah, L., and Adawiyah, C.W.R. 2014. Simple additive weighting methods of multi criteria decision making and applications: A decade review, *Int. J. Inf. Process. Manag.*, 5: 39–49.

Aubert, A.H., Bauer, R., and Lienert, J. 2018. A review of water-related serious games to specify use in environmental multi-criteria decision analysis, *Environ. Model. Softw.*, 105: 64–78, https://doi.org/10.1016/j.envsoft.2018.03.023.

Beccari, B. 2016. A comparative analysis of disaster risk, vulnerability and resilience composite indicators, *PLoS Curr. Disasters*, 14. https://doi.org/10.1371/currents.dis.453df025e34b682e9737f95070f9b970.

Behzadian, M., Otaghsara, S.K., Yazdani, M., and Ignatius, J. 2012. A state-of the-art survey of TOPSIS applications, *Expert Syst. Appl.*, 39: 13051–13069, https://doi.org/10.1016/j.eswa.2012.05.056.

Bell, R., and Glade, T. 2004. Quantitative risk analysis for landslides - Examples from Bíldudalur, NW-Iceland, *Nat. Hazards Earth Syst. Sci.*, 4: 117–131.

Carver, S.J. 1991. Integrating multi-criteria evaluation with geographical information systems, *Int. J. Geogr. Inf. Syst.*, 5: 321–339, https://doi.org/10.1080/02693799108927858.

Chen, Y., Liu, R., Barrett, D., Gao, L., Zhou, M., Renzullo, L., and Emelyanova, I. 2015. A spatial assessment framework for evaluating flood risk under extreme climates, *Sci. Total Environ.* 538: 512–523, https://doi.org/10.1016/j.scitotenv.2015.08.094.

Cinelli, M., Coles, S.R., and Kirwan, K. 2014. Analysis of the potentials of multi criteria decision analysis methods to conduct sustainability assessment, *Ecol. Indic.*, 46: 138–148, https://doi.org/10.1016/j.ecolind.2014.06.011.

Dalezios, N.R., and Eslamian, S. 2016. Regional design storm of Greece within the flood risk management framework, *Int. J. Hydrology Science and Technology*, 6(1): 82–102.

de Brito, M.M., and Evers, M. 2016. Multi-criteria decision-making for flood risk management: A survey of the current state of the art, *Nat. Hazards Earth Syst. Sci.*, 16: 1019–1033, https://doi.org/10.5194/nhess-16-1019-2016.

de Brito, M.M., Evers, M., and Almoradie, A.D.S. 2018. Participatory flood vulnerability assessment: A multi-criteria approach, *Hydrol. Earth Syst. Sci.*, 22: 373–390, https://doi.org/10.5194/hess-22-373-2018.

de Brito, M.M., Evers, M., and Höllermann, B. 2017. Prioritization of flood vulnerability, coping capacity and exposure indicators through the Delphi technique: A case study in Taquari-Antas basin, Brazil, *Int. J. Disaster Risk Reduct.*, 24: 119–128, https://doi.org/10.1016/j.ijdrr.2017.05.027.

Doocy, S., Daniels, A., Murray, S., and Kirsch, T.D. 2013. The human impact of floods: A historical review of events 1980–2009 and systematic literature review, *PLoS Curr.* https://doi.org/10.1371/currents.dis.f4deb457904936b07c09daa98ee8171a.

Elbeltagi, E., Hegazy, T., and Grierson, D. 2005. Comparison among five evolutionary-based optimization algorithms, *Adv. Eng. Informatics*, 19: 43–53, https://doi.org/10.1016/j.aei.2005.01.004.

Elmer, F., Hoymann, J., Düthmann, D., Vorogushyn, S., and Kreibich, H. 2012. Drivers of flood risk change in residential areas, *Nat. Hazards Earth Syst. Sci.*, 12: 1641–1657, https://doi.org/10.5194/nhess-12-1641-2012.

Evers, M., Almoradie, A., and de Brito, M.M. 2018. Enhancing flood resilience through collaborative modelling and multi-criteria decision analysis (MCDA). In: *Urban Disaster Resilience and Security*, 221–236, https://doi.org/10.1007/978-3-319-68606-6_14.

Fan, Y., Li, H., and Miguez-Macho, G. 2013. Global patterns of groundwater table depth, *Science*, 339: 940–943, https://doi.org/10.1126/science.1229881.

Feizizadeh, B., and Blaschke, T. 2014. An uncertainty and sensitivity analysis approach for GIS-based multicriteria landslide susceptibility mapping, *Int. J. Geogr. Inf. Sci.*, 28: 610–638, https://doi.org/10.1080/13658816.2013.869821.

Fernandez, P., Mourato, S., and Moreira, M. 2016. Social vulnerability assessment of flood risk using GIS-based multicriteria decision analysis. A case study of Vila Nova de Gaia, *Geomatics, Nat. Hazards Risk*, 7: 1367–1389, https://doi.org/10.1080/19475705.2015.1052021.

Garen, D.C., and Moore, D.S. 2005. Curve number hydrology in water quality modeling: Uses, abuses, and future directions, *J. Am. Water Resour. Assoc.*, 41: 377–388, https://doi.org/10.1111/j.1752-1688.2005.tb03742.x.

Geofabrik Download Server. 2018. OpenStreetMap Data Extracts [WWW Document]. URL http://download.geofabrik.de/index.html. Aaccessed 7.5.18.

Ghanbarpour, M.R., Salimi, S., and Hipel, K.W. 2013. A comparative evaluation of flood mitigation alternatives using GIS-based river hydraulics modelling and multicriteria decision analysis, *J. Flood Risk Manag.*, 6: 319–331, https://doi.org/10.1111/jfr3.12017.

Guhathakurta, P., Sreejith, O.P., and Menon, P.A. 2011. Impact of climate change on extreme rainfall events and flood risk in India, *J. Earth Syst. Sci.*, 120: 359–373, https://doi.org/10.1007/s12040-011-0082-5.

Haque, A.N., Grafakos, S., and Huijsman, M. 2012. Participatory integrated assessment of flood protection measures for climate adaptation in Dhaka, *Environ. Urban.*, 24: 197–213, https://doi.org/10.1177/0956247811433538.

Hartmann, J., and Moosdorf, N. 2012. The new global lithological map database GLiM: A representation of rock properties at the Earth surface, *Geochemistry, Geophys. Geosystems*, 13, https://doi.org/10.1029/2012GC004370.

Harvey, N., and Holmes, C.A. 2012. Nominal group technique: An effective method for obtaining group consensus, *Int. J. Nurs. Pract.*, 18: 188–194, https://doi.org/10.1111/j.1440-172X.2012.02017.x.

Hazarika, N., Barman, D., Das, A.K., Sarma, A.K., and Borah, S.B. 2018. Assessing and mapping flood hazard, vulnerability and risk in the Upper Brahmaputra River valley using stakeholders' knowledge and multicriteria evaluation (MCE), *J. Flood Risk Manag.*, 11: S700–S716, https://doi.org/10.1111/jfr3.12237.

Hijmans, R.J., Cameron, S.E., Parra, J.L., Jones, P.G., and Jarvis, A. 2005. Very high resolution interpolated climate surfaces for global land areas, *Int. J. Climatol.*, 25: 1965–1978, https://doi.org/10.1002/joc.1276.

IUCN. 2016. World Database on Protected Areas [WWW Document]. URL https://www.iucn.org/theme/protected-areas/our-work/world-database-protected-areas. Accessed 7.5.18.

Jato-Espino, D. 2016. *Hydrological Modelling of Urban Catchments under Climate Change for the Design of a Spatial Decision Support System to Mitigate Flooding Using Pervious Pavements Meeting the pRinciples of Sustainability*. Universidad de Cantabria, Spain.

Jato-Espino, D., Castillo-Lopez, E., Rodriguez-Hernandez, J., and Ballester-Muñoz, F. 2018. Air quality modelling in Catalonia from a combination of solar radiation, surface reflectance and elevation, *Sci. Total Environ.*, 624: 189–200, https://doi.org/10.1016/j.scitotenv.2017.12.139.

Jato-Espino, D., Castillo-Lopez, E., Rodriguez-Hernandez, J., and Canteras-Jordana, J.C. 2014. A review of application of multi-criteria decision making methods in construction, *Autom. Constr.*, 45: 151–162, https://doi.org/10.1016/j.autcon.2014.05.013.

Jato-Espino, D., Sillanpää, N., Charlesworth, S.M., and Andrés-Doménech, I. 2016a. Coupling GIS with stormwater modelling for the location prioritization and hydrological simulation of permeable pavements in urban catchments, *Water (Switzerland)*, 8, https://doi.org/10.3390/w8100451.

Jato-Espino, D., Sillanpää, N., Charlesworth, S.M., and Rodriguez-Hernandez, J. 2016b. A simulation-optimization methodology to model urban catchments under non-stationary extreme rainfall events, *Environ. Model. Softw.*, https://doi.org/10.1016/j.envsoft.2017.05.008.

Jenson, S.K., and Domingue, J.O. 1988. Extracting topographic structure from digital elevation data for geographic information system analysis, *Photogramm. Eng. Remote Sensing*, 54: 1593–1600.

Jonkman, S.N. 2005. Global perspectives on loss of human life caused by floods, *Nat. Hazards*, 34: 151–175, https://doi.org/10.1007/s11069-004-8891-3.

Jun, C., Ban, Y., and Li, S. 2014. Open access to earth land-cover map, *Nature*, 514: 434–434, https://doi.org/10.1038/514434c.

Kandilioti, G., and Makropoulos, C. 2012. Preliminary flood risk assessment: The case of Athens, *Nat. Hazards*, 61: 441–468, https://doi.org/10.1007/s11069-011-9930-5.

Katz, E.C., Niedzwiedz, J., and Steyer, L. 2017. Visualisierung sozialer vulnerabilität Kölns: Eine ArcGIS-gestützte untersuchung, *Zeitschrift für studentische wasserbezogene Forsch.*, I: 15–22.

Kenyon, W. 2007. Evaluating flood risk management options in Scotland: A participant-led multi-criteria approach, *Ecol. Econ.*, 64: 70–81, https://doi.org/10.1016/j.ecolecon.2007.06.011.

Kienberger, S., Lang, S., and Zeil, P. 2009. Spatial vulnerability units – Expert-based spatial modelling of socio-economic vulnerability in the Salzach catchment, Austria, *Nat. Hazards Earth Syst. Sci.*, 9: 767–778, https://doi.org/10.5194/nhess-9-767-2009.

Kubal, C., Haase, D., Meyer, V., and Scheuer, S. 2009. Integrated urban flood risk assessment - Adapting a multicriteria approach to a city, *Nat. Hazards Earth Syst. Sci.*, 9: 1881–1895, https://doi.org/10.5194/nhess-9-1881-2009.

Lee, G., Jun, K.S., and Chung, E.S. 2013. Integrated multi-criteria flood vulnerability approach using fuzzy TOPSIS and Delphi technique, *Nat. Hazards Earth Syst. Sci.*, 13: 1293–1312, https://doi.org/10.5194/nhess-13-1293-2013.

Levy, J.K. 2005. Multiple criteria decision making and decision support systems for flood risk management, *Stoch. Environ. Res. Risk Assess.*, 19: 438–447, https://doi.org/10.1007/s00477-005-0009-2.

Levy, J.K., Hartmann, J., Li, K.W., An, Y., and Asgary, A. 2007. Multi-criteria decision support systems for flood hazard mitigation and emergency response in urban watersheds, *J. Am. Water Resour. Assoc.*, 43: 346–358, https://doi.org/10.1111/j.1752-1688.2007.00027.x.

Loos, J.R., and Rogers, S.H. 2016. Understanding stakeholder preferences for flood adaptation alternatives with natural capital implications, *Ecol. Soc.* 21: art32, https://doi.org/10.5751/ES-08680-210332.

LP DAAC. 2014. ASTGTM: ASTER Global Digital Elevation Model V002 [WWW Document]. URL https://lpdaac.usgs.gov/node/1079. Accessed 7.4.18.

Luu, C., and von Meding, J. 2018. A flood risk assessment of quang nam, vietnam using spatial multicriteria decision analysis, *Water*, 10: 461, https://doi.org/10.3390/w10040461.

Malczewski, J., and Rinner, C. 2015. *Multicriteria Decision Analysis in Geographic Information Science*. 1st ed., Springer-Verlag, Berlin Heidelberg, Heidelberg (Germany), https://doi.org/10.1007/978-3-540-74757-4.

Malekian, A., and Azarnivand, A. 2016. Application of integrated shannon's entropy and VIKOR techniques in prioritization of flood risk in the Shemshak Watershed, Iran, *Water Resour. Manag.*, 30: 409–425, https://doi.org/10.1007/s11269-015-1169-6.

Mallouk, A., Lechgar, H., Malaainine, M.E., and Rhinane, H. 2016. A multicriteria approach with GIS for assessing vulnerability to flood risk in urban area (Case of Casablanca City, Morocco), Springer Verlag, 257–266, https://doi.org/10.1007/978-3-319-30301-7_27.

McMaster, R.B., Leitner, H., and Sheppard, E. 1997. GIS-based environmental equity and risk assessment: Methodological problems and prospects, *Cartogr. Geogr. Inf. Sci.*, 24: 172–189, https://doi.org/10.1559/152304097782476933.

Meyer, V., Becker, N., Markantonis, V., Schwarze, R., van den Bergh, J.C.J.M., Bouwer, L.M., Bubeck, P., Ciavola, P., Genovese, E., Green, C., Hallegatte, S., Kreibich, H., Lequeux, Q., Logar, I., Papyrakis, E., Pfurtscheller, C., Poussin, J., Przyluski, V., Thieken, A.H., and Viavattene, C. 2013. Review article: Assessing the costs of natural hazards – State of the art and knowledge gaps, *Nat. Hazards Earth Syst. Sci.*, 13: 1351–1373, https://doi.org/10.5194/nhess-13-1351-2013.

Meyer, V., Scheuer, S., and Haase, D. 2009. A multicriteria approach for flood risk mapping exemplified at the Mulde river, Germany, *Nat. Hazards*, 48: 17–39, https://doi.org/10.1007/s11069-008-9244-4.

Müller, A., Reiter, J., and Weiland, U. 2011. Assessment of urban vulnerability towards floods using an indicator-based approach-a case study for Santiago de Chile, *Nat. Hazards Earth Syst. Sci.*, 11: 2107–2123, https://doi.org/10.5194/nhess-11-2107-2011.

Neumann, B., Vafeidis, A.T., Zimmermann, J., and Nicholls, R.J. 2015. Future coastal population growth and exposure to sea-level rise and coastal flooding - A global assessment, *PLoS One*, 10: e0118571, https://doi.org/10.1371/journal.pone.0118571.

Osborne, J.W., and Waters, E. 2003. Four assumptions of multiple regression that researchers should always test, *Pract. Assessment, Res. and Eval.*, 8: 2.

Panhalkar, S.S., and Jarag, A.P. 2017. Flood risk assessment of panchganga river (Kolhapur District, Maharashtra) using GIS-based multicriteria decision technique, *Curr. Sci.*, 112: 785, https://doi.org/10.18520/cs/v112/i04/785-793.

Patrikaki, O., Kazakis, N., Kougias, I., Patsialis, T., Theodossiou, N., and Voudouris, K. 2018. Assessing flood hazard at river basin scale with an index-based approach: The case of Mouriki, Greece, *Geosciences*, 8: 50, https://doi.org/10.3390/geosciences8020050.

Pelling, M. 2007. Learning from others: The scope and challenges for participatory disaster risk assessment, *Disasters*, 31: 373–385, https://doi.org/10.1111/j.1467-7717.2007.01014.x.

Prell, C., Hubacek, K., Quinn, C., and Reed, M. 2008 "Who's in the network?' When stakeholders influence data analysis, *Syst. Pract. Action Res.*, 21: 443–458, https://doi.org/10.1007/s11213-008-9105-9.

Raaijmakers, R., Krywkow, J., and van der Veen, A. 2008. Flood risk perceptions and spatial multi-criteria analysis: An exploratory research for hazard mitigation, *Nat. Hazards*, 46: 307–322, https://doi.org/10.1007/s11069-007-9189-z.

Rufat, S., Tate, E., Burton, C.G., and Maroof, A.S. 2015. Social vulnerability to floods: Review of case studies and implications for measurement, *Int. J. Disaster Risk Reduct.*, 14: 470–486, https://doi.org/10.1016/j.ijdrr.2015.09.013.

Saaty, T.L. 1980. *The Analytic Hierarchy Process*, McGraw-Hill, New York, USA.

San Cristóbal Mateo, J.R. 2012. *Multi Criteria Analysis in the Renewable Energy Industry, Green Energy and Technology*, Springer, London, https://doi.org/10.1007/978-1-4471-2346-0.

Schanze, J. 2006. Flood risk management - A basic framework. In Schanze, J., Zeman, E., Marsalek, J. (Eds.), *Flood Risk Management: Hazards, Vulnerability and Mitigation Measures*, Springer, Dordrecht, Netherlands, 1–20.

Scolobig, A., Broto, V.C., and Zabala, A. 2008. Integrating multiple perspectives in social multicriteria evaluation of flood-mitigation alternatives: The Case of Malborghetto-Valbruna, *Environ. Plan. C Gov. Policy*, 26: 1143–1161, https://doi.org/10.1068/c0765s.

SEDAC. 2017. Gridded Population of the World (GPW), v4 [WWW Document]. URL http://sedac.ciesin.columbia.edu/data/collection/gpw-v4/sets/browse. Accessed 7.5.18.

Seekao, C., and Pharino, C. 2016. Assessment of the flood vulnerability of shrimp farms using a multicriteria evaluation and GIS: A case study in the Bangpakong Sub-Basin, Thailand, *Environ. Earth Sci.*, 75: 308, https://doi.org/10.1007/s12665-015-5154-4.

Sharifi, M.A., van den Toorn, W., Rico, A., and Emmanuel, M. 2002. Application of GIS and multicriteria evaluation in locating sustainable boundary between the tunari National Park and Cochabamba City (Bolivia), *J. Multi-Criteria Decis. Anal.*, 11: 151–164, https://doi.org/10.1002/mcda.323.

Simão, A., Densham, P.J., and (Muki) Haklay, M. 2009. Web-based GIS for collaborative planning and public participation: An application to the strategic planning of wind farm sites, *J. Environ. Manage.*, 90: 2027–2040, https://doi.org/10.1016/j.jenvman.2007.08.032.

Solín, Ľ. 2012. Spatial variability in the flood vulnerability of urban areas in the headwater basins of Slovakia, *J. Flood Risk Manag.*, 5: 303–320, https://doi.org/10.1111/j.1753-318X.2012.01153.x.

Sowmya, K., John, C.M., and Shrivasthava, N.K. 2015. Urban flood vulnerability zoning of Cochin City, southwest coast of India, using remote sensing and GIS, *Nat. Hazards*, 75: 1271–1286, https://doi.org/10.1007/s11069-014-1372-4.

Tang, Z., Yi, S., Wang, C., and Xiao, Y. 2018. Incorporating probabilistic approach into local multi-criteria decision analysis for flood susceptibility assessment, *Stoch. Environ. Res. Risk Assess.*, 32: 701–714, https://doi.org/10.1007/s00477-017-1431-y.

Tarboton, D.G., Bras, R.L., and Rodriguez-Iturbe, I. 1991. On the extraction of channel networks from digital elevation data, *Hydrol. Process.*, 5: 81–100, https://doi.org/10.1002/hyp.3360050107.

Tate, E. 2012. Social vulnerability indices: A comparative assessment using uncertainty and sensitivity analysis, *Nat. Hazards*, 63: 325–347, https://doi.org/10.1007/s11069-012-0152-2.

UNISDR. 2015a. *Sendai Framework for Disaster Risk Reduction 2015–2013*. Geneva, Switzerland.

UNISDR. 2015b. *The Human Cost of Weather-Related Disasters 1995–2015*. Brussels (Belgium) and Geneva (Switzerland).

USDA. 2018. Natural Resources Conservation Service [WWW Document]. URL https://www.nrcs.usda.gov/wps/portal/nrcs/site/national/home/. Accessed 7.5.18.

van Hove, L.W.A., Jacobs, C.M.J., Heusinkveld, B.G., Elbers, J.A., Van Driel, B.L., and Holtslag, A.A.M. 2015. Temporal and spatial variability of urban heat island and thermal comfort within the Rotterdam agglomeration, *Build. Environ.*, 83: 91–103, https://doi.org/10.1016/j.buildenv.2014.08.029.

Voinov, A., Kolagani, N., McCall, M.K., Glynn, P.D., Kragt, M.E., Ostermann, F.O., Pierce, S.A., and Ramu, P. 2016. Modelling with stakeholders - Next generation, *Environ. Model. Softw.*, 77: 196–220, https://doi.org/10.1016/j.envsoft.2015.11.016.

Wang, Y., Li, Z., Tang, Z., and Zeng, G. 2011. A GIS-based spatial multi-criteria approach for flood risk assessment in the Dongting Lake Region, Hunan, Central China, *Water Resour. Manag.*, 25: 3465–3484, https://doi.org/10.1007/s11269-011-9866-2.

Wright, R., and Stein, M. 2005. Snowball sampling. In: *Encyclopedia of Social Measurement*, Elsevier, 495–500, https://doi.org/10.1016/B0-12-369398-5/00087-6.

Xiao, Y., Yi, S., and Tang, Z. 2017. Integrated flood hazard assessment based on spatial ordered weighted averaging method considering spatial heterogeneity of risk preference, *Sci. Total Environ.*, 599–600, 1034–1046, https://doi.org/10.1016/j.scitotenv.2017.04.218.

Yang, M., Qian, X., Zhang, Y., Sheng, J., Shen, D., and Ge, Y. 2011. Spatial multicriteria decision analysis of flood risks in aging-dam management in China: A framework and case study, *Int. J. Environ. Res. Public Health*, 8: 1368–1387, https://doi.org/10.3390/ijerph8051368.

Zeleňáková, M., Dobos, E., Kováčová, L., Vágo, J., Abu-Hashim, M., Fijko, R., and Purcz, P. 2018. Flood vulnerability assessment of Bodva cross-border river basin, *Acta Montan. Slovaca*, 23: 53–61.

Index

Note: Locators in *italics* represent figures and **bold** indicate tables in the text.

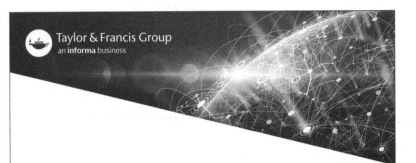